Key Methods in Geography

SAGE was founded in 1965 by Sara Miller McCune to support the dissemination of usable knowledge by publishing innovative and high-quality research and teaching content. Today, we publish over 900 journals, including those of more than 400 learned societies, more than 800 new books per year, and a growing range of library products including archives, data, case studies, reports, and video. SAGE remains majority-owned by our founder, and after Sara's lifetime will become owned by a charitable trust that secures our continued independence.

Los Angeles | London | New Delhi | Singapore | Washington DC | Melbourne

THIRD EDITION

Edited by Nicholas Clifford, Meghan Cope, Thomas Gillespie & Shaun French

Key Methods in Geography

Los Angeles I London I New Delhi
Singapore I Washington DC I Melbourne

Los Angeles | London | New Delhi
Singapore | Washington DC | Melbourne

SAGE Publications Ltd
1 Oliver's Yard
55 City Road
London EC1Y 1SP

SAGE Publications Inc.
2455 Teller Road
Thousand Oaks, California 91320

SAGE Publications India Pvt Ltd
B 1/I 1 Mohan Cooperative Industrial Area
Mathura Road
New Delhi 110 044

SAGE Publications Asia-Pacific Pte Ltd
3 Church Street
#10-04 Samsung Hub
Singapore 049483

Editor: Robert Rojek
Editorial assistant : Matthew Oldfield
Production editor: Katherine Haw
Copyeditor: Catja Pafort
Indexer: Adam Pozner
Marketing manager: Lucia Sweet
Cover design: Wendy Scott
Typeset by: C&M Digitals (P) Ltd, Chennai, India
Printed and bound in Great Britain by Bell and
    Bain Ltd, Glasgow

Third edition first published 2016
First edition published 2003, reprinted in 2005, 2006, 2007, 2008 & 2009
Second edition published 2010, reprinted 2012, 2013, 2014 & 2015

Library of Congress Control Number: 2015955789

British Library Cataloguing in Publication data

A catalogue record for this book is available from the British Library

ISBN 978-1-4462-9858-9
ISBN 978-1-4462-9860-2 (pbk)

At SAGE we take sustainability seriously. Most of our products are printed in the UK using FSC papers and boards.
When we print overseas we ensure sustainable papers are used as measured by the PREPS grading system.
We undertake an annual audit to monitor our sustainability.

'This excellent book remains the most comprehensive and accessible introduction to research methods in Geography today. The new edition will be essential reading for any student undertaking independent research as part of their degree.'

Noel Castree, Professor of Geography, The University of Manchester

'A new generation of scholars explain how who to get started with geographical research. Geography is an academic subject that rapidly evolves. What was in vogue one year can soon be passé. However, *Key Methods in Geography* provides a mixture that both preserves the best of the past and introduces what is both more new and most promising. Unnecessary verbiage is out – clarity is in. Practical, accessible careful and interesting, this greatly updated and revised volume brings the subject up-to-date and explains in bite sized chunks the hows and whys of modern day geographical study that begins to brings together physical and human approaches again in a new synthesis.'

Danny Dorling, Professor of Geography, University of Oxford

'This collection is a valuable resource for geography students and researchers at all levels, in human geography, physical geography, and GIScience. It stands apart from many methods texts by speaking to the complete research process – from conceptualizing and designing geographical research, to collecting and analyzing many different forms of evidence, to representing results and finding in diverse ways. The volume is noteworthy in the breadth of methods it includes, spanning classic approaches that have been foundational to geographical scholarship and some of the discipline's newest and most innovative methodological practices.'

Sarah Elwood, Professor of Geography, University of Washington

Contents

List of Figures

List of Tables

Notes on Contributors

Ben Anderson is Reader in human geography at Durham University, where he has worked since completing a PhD at Sheffield University in 2004. His research explores the importance of affect to cultural and political life, initially through work on spaces of boredom and hope. A 2014 monograph – *Encountering Affect: Capacities, Apparatuses, Conditions* (Ashgate) – summarises this work. Over the past five years, his research has shifted to explore the implications of affect-based research for how we understand the geographies of power. Supported by a Philip Leverhulme Prize, his current work focuses on the emergence of novel ways of governing emergencies and their implications for power and authority. Occasionally he tweets at @BenAndersonGeog.

Stewart Barr is Professor of Geography at the University of Exeter. He graduated from the University of Exeter's Geography Department in 1998 and continued his studies at Exeter undertaking a PhD exploring household waste practices. Building on this research, he worked for two years in the Department as a post-doctoral researcher on an ESRC-funded project entitled 'Environmental Action in and Around the Home'. He became a Lecturer in Geography in 2003, Senior Lecturer in 2008, Associate Professor in 2012 and has been Professor of Geography since 2015. Within the Department Stewart undertakes research in the Environment and Sustainability and Spatial Responsibilities Research Groups and teaches modules at all levels of undergraduate study.

Myrna M. Breitbart is Professor of Geography and Urban Studies and Director of the Community Engagement and Collaborative Learning Network at Hampshire College where she has taught since 1977. Her teaching and research interests as well as publications focus on the broad themes of participatory planning and social action (with a special interest in young people), community economic development, and the role of the built environment in social change. She recently published the book, *Creative Economies in Post-Industrial Cities: Manufacturing a (different) Scene* (Ashgate, 2013) with a special focus on smaller cities. Professor Breitbart has a strong commitment to community-based learning and participatory action research. She works closely with a number of community development and housing organizations as well as urban youth and community art organizations in Western Massachusetts.

Joanna Bullard is Professor of Aeolian Geomorphology at Loughborough University. She completed her undergraduate degree at Edinburgh University and her PhD at the University of Sheffield, specializing in the relationships between sand dune geomorphology, vegetation and climate. Fieldwork has taken her over glaciers, through tropical forests, along various coastlines, and into sandy and rocky deserts around the world. Her current research focuses primarily on aeolian dust emissions in cold and hot deserts. She is currently the Chair of the British Society for Geomorphology and the President of the International Society for Aeolian Research. She is also an Associate Dean for Teaching at Loughborough University and a Senior Fellow of the Higher Education Academy.

Nick Clifford is Professor of Physical Geography at King's College, London, and from July 2016 will be Dean of the School of Social, Political and Geographical Sciences at Loughborough University. He received his BA and PhD from the University of Cambridge. His principal research interests are in fluvial geomorphology, and the history of ideas and methods in Geography. He is a member of the editorial board of River Research and Applications, an editorial collective member of the Geographical Association's flagship journal, *Geography*, and is Managing Editor of *Progress in Physical Geography*. He has produced diverse publications, such as *Turbulence: Perspectives on Flow and Sediment Transport* (with J.R. French and J. Hardisty; John Wiley, 1993), and *Incredible Earth* (DK Books, 1996). He teaches courses in river form and processes, river restoration, and the history and philosophy of Geography.

Meghan Cope is a Professor in the Department of Geography at the University of Vermont. Her interests lie in the areas of youth geographies, urban social problems, and qualitative research methods. She has worked on projects such as participatory research on children's conceptualizations of urban space in inner-city Buffalo, teens' independent mobility in Vermont, and a new project involving critical historical geography called 'Mapping American Childhoods'. She has published in a wide range of human geography journals and is co-editor with Sarah Elwood of *Qualitative GIS: A Mixed-Methods Approach* (Sage, 2009).

Ruth Craggs is Lecturer in Cultural and Historical Geography at King's College, London. She completed her undergraduate, Master's and PhD study at the University of Nottingham, focusing on engagements with the 'modern' Commonwealth in Britain in the post-war period. Since then her work has explored cultures of decolonisation and post-colonial geopolitics; her current research explores the impact of decolonisation on Britain's post-war urban reconstruction. She is on the editorial boards of *The London Journal* and *The Round Table: The Commonwealth Journal of International Affairs*.

Mike Crang is Professor of Geography at Durham University. He has worked across issues of landscape, tourism, temporality, memory and waste, as well as work on new media. His first work on new media was the edited collection *Virtual Geographies* (with Phil Crang and Jon May), since when he has worked on the role of ICTs in urban economies, governance and social life. His focus has been the imbrications of digital media into the logistics of everyday life in the city.

Eric Delmelle is Associate Professor in the Department of Geography and Earth Sciences, University of North Carolina. His teaching and research interests include GIS Algorithms, Spatial Analysis and Modelling, Spatial Optimization, Geovisualization and Geostatistics. He is the co-author of *Spatial Analysis in Health Geography* (with P. Kanaroglou and A. Páez, Ashgate, 2015).

Marcus A. Doel is Professor of Human Geography, and the Deputy Pro Vice Chancellor for Research and Innovation, at Swansea University. He studied Geography as an undergraduate at the University of Bristol, and pursued his PhD in Human Geography there. He is a Fellow of the Royal Geographical Society with the Institute of British Geographers as well as being a member of the Association of American Geographers. In addition to poststructuralist spatial theory, his research interests include graphic novels and film, modernity and postmodernity, and consumer culture and risk society. His most recent (edited) books are entitled *Moving Pictures/Stopping Places: Hotels and Motels on Film* (Lexington Books, 2009) and *Jean Baudrillard: Fatal Theories* (Routledge, 2009). He is currently researching the geographies of violence.

Nick A. Drake is Professor of Geography at King's College London. His research interests include remote sensing, geomorphological analysis and geochemical analysis of landforms and processes of landform change in semi-arid and arid environments. During the last few years Nick's interests have focused on applying his expertise to arid lands and, in particular, the Sahara Desert. He has published 65 single and joint authored journal articles, 22 book chapters, 31 published conference proceedings and was co-editor of *Spatial Modelling of the Terrestrial Environment* (Wiley, 2004).

Richard Field is Associate Professor of Biogeography in the School of Geography at the University of Nottingham. He gained an MA from the University of Oxford, an MSc from the University of Durham, and a PhD from Imperial College London. His main research interests are in biodiversity patterns at scales from global to local (particularly plants), macroecology, island biogeography and conservation ecology. He has been on ten tropical and subtropical scientific expeditions and is a Trustee of the Operation Wallacea Trust ('conservation through business partnerships'). He is Deputy Editor-in-Chief of *Global Ecology* and *Biogeography: A Journal of Macroecology* and Deputy Editor-in-Chief of *Frontiers of Biogeography*. He is a member of five learned societies, including being a Fellow of the Royal Geographical Society, a Fellow of the Higher Education Academy and Secretary of the International Biogeography Society. He teaches courses in biogeography, quantitative methods and environmental management.

Shaun French is Associate Professor in Economic Geography at the University of Nottingham. He gained his PhD from the University of Bristol and was subsequently seconded to the Bank of England to work on a study of access to finance for small and medium-sized enterprises in deprived areas. He has longstanding research interests in the geographies of financial risk technologies, financial subjects and processes of financial exclusion. Recent foci include the long-term insurance sector (life assurance, health insurance and pensions); socially responsible investment; and financial centres. His work engages with wider theories of industrial clustering and agglomeration, with professions, professionals and business knowledge communities. He is especially interested in e-commerce and the increasing significance of software and information communications technologies for business, including the ways in which business practice becomes embedded in technology and the ways in which hardware and software themselves produce new economic practices and relationships.

Bradley L. Garrett is Lecturer in Human Geography at the University of Southampton. His research interests revolve around heritage, place, urbanity, ruins and waste, ethnography, spatial politics, subversion and creative (mostly audio/visual) methods. Dr Garrett also holds a Visiting Research Associate position in the School of Geography and the Environment at the University of Oxford. He is the author of *Explore Everything: Place Hacking the City* (Verso, 2013) *Subterranean London: Cracking the Capital* (Prestel Publishing, 2014), *Global Undergrounds: Exploring Cities Within* (Reaktion Books, 2016) and *London Rising: Illicit Photos from the City's Heights* (Prestel Publishing, 2016).

Thomas W. Gillespie is Professor of Geography at the UCLA. His interests focus on using geographic information systems (GIS) and remote sensing data for predicting patterns of species richness and rarity for plants and birds at a regional spatial scale.

Peter Glaves is Principal Lecturer in Ecology and Environmental Management at the University of Northumbria, Newcastle. His main research interest is environmental management and planning, especially the link between ecology and economy and the practical application of ecosystem services, environmental assessment and environmental auditing to achieve more sustainable solutions.

Mark Graham is an Associate Professor and Senior Research Fellow at the Oxford Internet Institute in the University of Oxford. He has published articles in major geography, communications, and urban studies journals, and his work has been covered by the media in dozens of countries. In 2014, he was awarded a European Research Council Starting Grant to lead a team to study 'knowledge economies' in Sub-Saharan Africa over five years. This will entail looking at the geographies of information production, low-end (virtual labour and microwork) knowledge work, and high-end (innovation hubs and bespoke information services) knowledge work across Africa.

Peter L. Guth teaches geology, GIS, and research methods in the Oceanography Department at the US Naval Academy. His research interests focus on geomorphometry, measuring the landscape from digital topography, military terrain analysis, and moving terrain analysis to ever smaller devices – the personal computer starting in the 1980s, and now tablets and cell phones.

Siti Mazidah Haji Mohamad is a Geography lecturer in the Faculty of Arts and Social Sciences (FASS) at Universiti Brunei Darussalam. She graduated from Durham University with a PhD in Human Geography. Her PhD thesis titled *Rooted Muslim Cosmopolitanism: An Ethnographic Study of Malay Malaysian Students' Cultivation and Performance of Cosmopolitanism on Facebook and Offline* analyses the potential of Facebook, a social network site, as well as offline social interactions and experiences in cultivating cosmopolitan sensibilities and the performance of cosmopolitanism in both online and offline spaces. She is currently working on a number of research topics within the online contexts: privacy in the context of new media use and engagements; mapping the religious landscape; and performance of religiosity in the online space.

Iain Hay is Matthew Flinders Distinguished Professor of Geography at Flinders University, South Australia. He is a recent past President of the Institute of Australian

Geographers and is currently Vice-President of the International Geographical Union. His research focuses on geographies of domination and oppression. He is the author or editor of ten books including *Qualitative Research Methods in Human Geography* (4th edn, Oxford, 2016) and *Handbook on Wealth and the Super-Rich* (Elgar, 2016) and has had editorial roles with journals including *Applied Geography, Ethics, Place and Environment* and *Social and Cultural Geography*. In 2006 Iain received the Prime Minister's Award for Australian University Teacher of the Year, and in 2014 was admitted as a Fellow of the Academy of Social Sciences (UK).

Mick Healey was Professor of Geography at the University of Gloucestershire until 2010. He is now a Higher Education Consultant and Researcher. He holds several positions including Emeritus Professor, University of Gloucestershire; Visiting Professor, University College London; The Humboldt Distinguished Scholar in Research-Based Learning, McMaster University; Adjunct Professor, Macquarie University; and International Teaching Fellow, University College Cork. He is on the International Editorial Advisory Panel for the *Journal of Geography in Higher Education*. In 2000 he was awarded one of the first National Teaching Fellowships in the UK. In 2004 the Council of the Royal Geographical Society conferred on him the Taylor and Francis Award for 'contributions to the promotion of learning and teaching in higher education'. He was also one of the first people to be awarded a Principal Fellowship of the Higher Education Academy.

Ruth L. Healey is a Senior Lecturer in Human Geography at the University of Chester. She was awarded her PhD from the University of Sheffield in 2009. She is a social geographer with research interests in refugees and asylum seekers in the UK and in teaching and learning. Dr Healey is also a member of the editorial board of the *Journal of Geography in Higher Education* and a Senior Fellow of the Higher Education Academy.

Benjamin W. Heumann is Director of the Center for Geographic Information Science and a member of the Institute for Great Lakes Research at Central Michigan University. His research focuses on mapping and quantifying biodiversity in the Laurentian Great Lakes region using vegetation surveys, GIS remote sensing, with statistical and simulation modelling.

Jennifer Hill is an Associate Professor in Teaching and Learning of Geography at the University of the West of England, Bristol. She completed her undergraduate degree at the University of Oxford and her PhD at Swansea University, researching the effects of forest fragmentation on tree species diversity in the tropics. She has undertaken fieldwork in the tropical forests of Ghana, Peru and Australia, the hot deserts of Tunisia, and the glacial forelands of Norway. Dr Hill's pedagogic research focuses on participatory pedagogies: enhancing student empowerment, notably via their active integration in the scholarship of teaching and learning. She is a member of the editorial boards for *Journal of Geography in Higher Education* and *Geography*.

Hilda Kurtz is a Professor of Geography at the University of Georgia. Her research examines the geographic dimensions of political practice on the part of marginalized social actors in relation to environmental justice, food justice and food sovereignty, and racialization in US urban contexts.

Stuart N. Lane is Professor of Geomorphology and Director of the Institute of Earth Surface Dynamics at the Université de Lausanne, Switzerland. He completed his undergraduate degree at the University of Cambridge, and his PhD at Cambridge and City University, London. He has wide-ranging interests in geographical methodology and history. He is a fluvial geomorphologist and hydrologist, with major research projects in the process–form relationships in sand- and gravel-bedded rivers, the measurement and numerical simulation of river flows using complex three-dimensional computer codes, mountain geomorphology, and flood risk and diffuse pollution modelling. He received the EGU Bagnold medal in 2011 and the Royal Geographical Society Victoria Medal in 2012.

Alan Latham is Lecturer in Geography at University College London. He received a BA from Massey University, and a PhD from the University of Bristol. He is an urban geographer, with interests in sociality and urban life, globalization and the cultural economy of cities. He has recently co-edited *Key Concepts in Urban Geography* (Sage, 2009).

Eric Laurier is Reader in Geography and Interaction at the University of Edinburgh. His research projects have been on cafés and their place in civic life; cars and how we inhabit them; the workplace skills required for video-editing; wayfinding with paper and digital maps; and the maintenance and transformation of personal relationships. Members' methods have been an abiding interest – the shared methods of everyday life and workplaces in parallel with the more arcane methods of the arts, sciences and social sciences.

Robyn Longhurst is Deputy Vice-Chancellor Academic and Professor of Geography at the University of Waikato, New Zealand. Her areas of teaching and research include 'the body', feminist geography, the politics of knowledge production and qualitative methodologies. She is author of *Maternities: Gender, Bodies and Space* (Routledge, 2008) and *Bodies: Exploring Fluid Boundaries* (Routledge, 2001), and co-author of *Space, Place and Sex: Geographies of Sexualities* (Rowman & Littlefield, 2010) and *Pleasure Zones: Bodies, Cities, Spaces* (Syracuse University Press, 2001).

George P. Malanson is Coleman-Miller Professor of Geographical and Sustainability Sciences at the University of Iowa, and Program Director of the Population and Community Ecology Cluster at the US National Science Foundation. His current research focus is on the response of alpine tundra and alpine treeline to climate change. His research integrates fieldwork, including quantitative vegetation sampling, computer simulations, and statistical analyses.

Alan Marshall is Lecturer in Human Geography at the University of St Andrews. His research is concerned with understanding and describing social and spatial inequalities in health. It explores how individual characteristics and contextual factors (from national to neighbourhood) interact to influence health across the life course and particularly in later life.

Sara L. McLafferty is Professor of Geography and GIScience at the University of Illinois at Urbana–Champaign. Her research explores the use of spatial analysis methods and GIS in analysing health and social issues in cities and women's access to social services and employment opportunities.

Scott A. Mensing is a biogeographer and palaeoecologist at the University of Nevada, Reno. He has extensive experience reconstructing Quaternary environments in the Great Basin and California. His primary research tools are pollen and charcoal analysis and he maintains the department palynology laboratory. He also has experience with tree ring analysis and woodrat middens. He enjoys field work and is always anxious to explore new corners in the intermountain west.

James D. A. Millington is Lecturer in Physical and Quantitative Geography at King's College London. He is a broadly trained geographer and landscape ecologist with expertise in developing bespoke modelling tools to investigate spatial ecological and socio-economic processes and their interaction. His work has focused on vegetation succession-disturbance dynamics and human decision-making in multifunctional forest and agricultural landscapes of North America and Europe, with an emphasis on natural hazards such as wildfire. He also has interests in the different epistemological roles models and modelling can play in furthering geographical understanding. James is a Fellow of the Royal Geographical Society and a long-standing member of the International Association of Landscape Ecology.

Chris Perkins is Reader in Geography in the School of Environment Education and Development, University of Manchester, and from 2007–2015 was Chair of the International Cartographic Association Maps and Society Commission. He is the author of seven books and numerous academic papers and is currently researching the social lives of mapping as part of the ERC-funded Charting the Digital Project.

Ate Poorthuis is an Assistant Professor in the Humanities, Arts and Social Sciences at Singapore University of Technology and Design. His research is focused on the possibilities and limitations of the analysis and visualization of big data to better understand how our cities work. He is the technical lead on The DOLLY Project, a repository of billions of geolocated social media, that strives to address the difficulties of using big data within the social sciences.

Shelly A. Rayback is Associate Professor of Geography at the University of Vermont. She is interested in the response of trees and shrubs to climate and other environmental changes across varying temporal and spatial scales. She uses dendrochronological and stable isotope analysis techniques on long time series derived from arctic and alpine dwarf shrubs, and from eastern North American tree species, to understand the influence of climate on plant growth and reproduction, and to reconstruct climate.

Liz Roberts is a cultural geographer currently working at the University of the West of England on an RCUK project on digital storytelling and water scarcity (www.dry project.co.uk). She undertook her PhD at the University of Exeter working on an AHRC-funded project titled 'Spectral Geographies of the Visual'. Liz's ongoing work has a focus on visual images and visual methods, the interaction of narrative and the visual, non-representational theory, cultural theory, and philosophical and political engagements with the visual within geography.

Taylor Shelton is a Postdoctoral Fellow in the Center for Urban Innovation at the Georgia Institute of Technology. He is a broadly-trained human geographer with interests in the myriad ways that data are reshaping how we understand and intervene in urban processes. This research has been especially focused on developing new conceptual and methodological approaches to mapping social media data. Taylor received BA and MA degrees in Geography from the University of Kentucky and his PhD from the Graduate School of Geography at Clark University.

Fiona M. Smith is a Lecturer in Human Geography at the University of Dundee, UK. Her research currently focuses on contemporary political and cultural geographies in Germany, and on geographies of volunteering in the UK. She is also interested in language and cross-cultural research. She has published widely on these issues and is co-author of *Geographies of New Femininities* (with Nina Laurie, Claire Dwyer and Sarah Holloway; Longman, 1999).

Thomas Smith is Lecturer in Physical and Environmental Geography at King's College London. Tom is interested in the role of wildfires in the Earth system. His research spans field, laboratory and computer simulation modelling applications, and focuses on measuring greenhouse and reactive gas emissions from biomass burning in tropical and temperate ecosystems, including the savannas of northern Australia, UK heather moorlands, and the tropical peatlands of Southeast Asia. His work makes use of open-path and solar occultation Fourier Transform Infrared spectroscopy for the field and laboratory study of gases.

Laura N. Stahle is a PhD student in Paleoecology at Montana State University. Her research focuses on the relationship between human and climatic drivers of Holocene environmental change in Australia.

Monica Stephens is an Assistant Professor of GIScience in the Department of Geography at the University at Buffalo (SUNY Buffalo) in New York. She obtained her doctoral degree from the University of Arizona in 2012, and worked as a visiting scholar at the University of Kentucky and an Assistant Professor at Humboldt State University in California. Her research integrates methodologies in Geographic Information Science (GIS) with Social Network Analysis (SNA) and Big Data. She harnesses and critiques these methodologies with data from social media and user-generated content to trace inequalities across gender, race and economic status.

Liz Taylor is Senior Lecturer in Education at the University of Cambridge. Her research interests focus on young people's representations and experiences of place and space, geography education and understanding of place and space in children's literature.

Naomi Tyrrell is a Lecturer in Human Geography at Plymouth University. Her teaching and research interests are in the broad field of population geography, with a focus on family migration processes and children's geographies. Her current research projects are focused on the experiences of settled migrant children in the UK, researcher mobility and family life in Europe, the impacts of child migration on later-life migration, and the mobilities of military families.

Nigel Walford is Professor of Applied GIS and Director of Research and Enterprise in the School of Natural and Built Environments at Kingston University. His main research and teaching interests focus on geographical and spatial data, especially in relation to geodemographics, population ageing, counterurbanisation, agriculture and environmental planning. Over some 30 years he has accrued an extensive publication record of journal papers, books and book chapters, computer-based teaching materials, conference proceedings and conference papers. His latest book, *Practical Statistics for Geographers and Earth Scientists* (Wiley-Blackwell), was published in January 2011.

Helen Walkington is Principal Lecturer in Geography at Oxford Brookes University. She completed her undergraduate degree at Durham University and an MSc in Soils and PhD in Geography Education at the University of Reading. She has specialised in working with sediments and soils in archaeological contexts. She has also worked on university, national and international initiatives to enhance the student learning experience through research-based learning. Dr Walkington is Editor in Chief of *GEOverse*, a national undergraduate research journal, an editorial board member of the *Journal of Geography in Higher Education*, and co-chair of the International Network for Learning and Teaching in Higher Education.

Cathy Whitlock is Professor of Earth Sciences and Director of the Montana Institute on Ecosystems at Montana State University. Her research focuses on long-term environmental and climate change, and she has published over 150 scientific papers describing the vegetation, fire and climate history of the western US and other temperate regions. Whitlock is also actively involved in efforts to use paleoscience to inform natural resource and management decision making.

Matthew W. Wilson is Associate Professor of Geography at the University of Kentucky. He is interested in the relationship between information technologies and the urban. His research agenda bridges GIScience and critical geography, with current interests in the proliferation of locative media for consumer handheld devices. He is currently co-authoring *Understanding Spatial Media* (with Rob Kitchin and Tracey P. Lauriault; Sage, 2016).

Martin J. Wooster is Professor of Earth Observation Science at King's College, London, and a Divisional Director of the NERC National Centre for Earth Observation (NCEO). His research interests include satellite Earth Observation, remote sensing and infrared spectroscopy, and many of his research projects have used these techniques to better quantify and understand global landscape fires and the contribution these make to the chemical composition of Earth's atmosphere. His research group are currently responsible for two operational satellite-based fire monitoring products generated from European satellites. Prior to academia, he worked for the UK Department of International Development (DFID) on the application of satellite remote sensing to environmental monitoring in developing countries, and he has been a recipient of the Royal Geographical Society Cuthbert Peak Award, the London Development Agency-NERC Environmental Science Knowledge Transfer Award, and the Daiwa-Adrian Award for Excellence in UK-Japan Science.

Matthew Zook is a Professor in the Geography department at the University of Kentucky where he also serves as the Director of GIS Initiative and heads the The DOLLY Project, a repository of billions of geolocated social media. His research interest centres on the spatiality of technology and innovation, particularly the ways in which it interacts with the organization of the economy. Other work focuses on the interaction of user-generated data with code, space and place in the construction of everyday, lived geographies.

Preface

The third edition of this volume marks a significant revision and extension to its highly successful predecessors. Two new Editors have joined the editorial team, and have helped revise, update and add to the substantive chapters. Some new highlights include working with virtual communities; the inclusion of considerations of affect and emotion; video and audio technologies; case studies; biogeography and landscape ecology; and a fresh look at enduring themes such as remote sensing and GIS. The authorship is even more international and diverse than before, and I trust the appeal of the volume will be to an even wider student audience and geographical community.

Just as importantly, the printed volume is accompanied by an entirely new online resource (see **https://study.sagepub.com/keymethods3e**) providing additional case studies, data and tools to assist in various forms of geographical analysis. The third edition thus reflects a more contemporary subject matter, and a more contemporary way of approaching geographical enquiry through mixed methods and multiple resources.

I thank my co-editors and the team at SAGE for all of their hard work over the past two years, and I wish readers well in using and encountering the information and perspectives which the new volume and its companion website contain.

N J C
King's College, London, February 2016.

About the Companion Website

Specially developed for the third edition, the *Key Methods in Geography* companion website can be found at https://study.sagepub.com/keymethods3e.

Visit the site for:

- **VIDEOS** – including introductions to the new edition and research methods in Geography.

- **JOURNAL ARTICLES** – FREE access to a brilliant collection of Progress Reports from the *Progress in Human Geography* and *Progress in Physical Geography* journals. These have been selected by the chapter authors for their relevance and quality.

- **FURTHER READING** – the best print and online resources relating to each chapter.

- **EXERCISES** – a range of activities to test your understanding of different methods.
- **SUPPLEMENTARY MATERIALS** – everything from slideshows to datasets.

As you read the book, you'll find a short guide to the relevant companion website content located at the end of each chapter.

SECTION ONE

Planning a Research Project: Getting Started and Putting Your Research into Context

1 Getting Started in Geographical Research: How This Book Can Help

Nick Clifford, Meghan Cope, Tom Gillespie, Shaun French and Gill Valentine

SYNOPSIS

Geography is a very diverse subject that includes studies of human behaviour, the built and the imagined world, the physical environment extending to biological, ecological and geophysical systems, and increasingly, to the intersections between these. It is also a discipline that embraces a very diverse range of philosophical approaches to knowledge (from positivism to post-structuralism) and from the representation of social and physical relationships to those which focus much more on experience and the practices which embody this. As such, geographers employ *quantitative methods* (statistics and mathematical modelling) and *qualitative methods* (a set of techniques such as interviewing, participant observation and visual imagery that are used to explore subjective meanings, values and emotions) or a combination of the two. These methods can be used in both *extensive research designs* (where the emphasis is on pattern and regularity in large 'representative' data sets, which is assumed to represent the outcome of some underlying (causal) regularity or process) and *intensive research designs* (where the emphasis is on describing a single case study, or small number of case studies, with the maximum amount of detail). Geographical research may also be undertaken individually or in teams, and adopt a multi-, inter- or transdiciplinary research methodology. Yet, despite this diversity, all geographers, whatever their philosophical or methodological approach, must make common decisions and go through common processes when they are embarking on their research. This means doing preparatory work (a literature review, thinking about health and safety and research ethics); thinking through the practicalities of data collection (whether to do original fieldwork or rely on secondary sources; whether to use quantitative or qualitative methods or a combination of both); planning how to manage and analyse the data generated from these techniques; and thinking about how to present/write up the findings of the research. This chapter aims to guide you through these choices if you are doing research for a project or dissertation. In doing so, it explains the structure and content of this book and points you in the direction of which chapters to turn to for advice on different forms of research techniques and analysis.

This chapter is organized into the following sections:

- Introduction: The nature of geographical research
- Quantitative and qualitative approaches to geography
- Designing a geographical research project
- The philosophy of research and importance of research design
- Conclusion: How this book can help you get started

1.1 Introduction: The Nature of Geographical Research

This book aims to help you prepare for, design and carry out geographical research, and to analyse and present your findings. Geographers have given attention to an enormous range of subject matters. Most aspects of the world, whether physically or environmentally determined, or politically, economically or culturally constructed and experienced, have been considered as suitable for geographical research. Moreover, the range of geographical enquiry continues to increase. Traditionally, geographers considered the contemporary human and physical world together with their historical configurations, thus extending geographies to the past as well as to the present. Now, in both physical and human geography, the range is even greater (see, for example, Walford and Haggett, 1995; Gregory, 2000; Thrift, 2002; Gregory, 2009). Physical geographers have access to new techniques of absolute and relative environmental dating and mathematical modelling, and physical geography is increasingly conducted under the umbrella of 'Earth System Science' which stresses interconnections between bio-physical atmospheric and earth science, and also includes human activity as a driver of, or response to, earth and environmental change (Pitman, 2005). In human geography, technological advances in areas such as GIS allow more flexible and more creative analysis of data, facilitating 'virtual geographies' which exist only in 'hyper-space'. Both physical and human geography are, in common with other academic subjects, in the midst of a revolution or 'fourth paradigm' in research marked by the appearance of Big Data approaches where the large amounts of data routinely captured in an increasingly networked and digital world can be used in combination with analytical and visualization technologies to reinterpret more enduring themes of people, place and process (Graham and Shelton, 2013; Kitchin, 2014; Wyly, 2014). This Big Data revolution follows what have been seen as three other major shifts or paradigms in research techniques and research environment: experimental, theoretical and computational. The challenges and opportunities presented by ubiquitous digital data are, in turn, related to the phenomenon of globalization and to some fundamental reconceptualizations of more traditional questions relating to nature, culture and society, and to questions of the scale of enquiry (Clifford, 2009). 'The Anthropocene' (Crutzen and Stoermer, 2000) is now a commonplace shorthand to encapsulate an epoch of unprecedented human influence on the environment. In a period of unprecedented technical, economic and environmental change, it is crucial to remember that Geography is by necessity a subject of essential hybridities and of new discourses of enquiry (Whatmore, 2002; Lorimer, 2012; Castree, 2014; Johnston and Moorhouse, 2014). As such, the subject is probing areas traditionally within the domains of psychology and cultural anthropology: there are now, for example, imagined and mystic geographies, whose foundations, or connections with the 'real' world, are almost entirely interpretational, rather than empirical, and where feeling and emotion (the 'affective') are essential avenues of research (Thien, 2005; Pile, 2010). Physical geographers, too, are exploring the role of narrative, discourse and ethnographies in describing and explaining otherwise natural processes and landforms (Tadaki et al., 2012; Brierley et al., 2013; Wilcock et al., 2013); and more radically, in linking to questions of resource allocation and equity in emerging critical physical geographies (Lave et al., 2014). Nevertheless, older, but enduring questions of scale and place remain (see, for example, Richards and Clifford, 2008). All of these new areas of geographical exploration bring challenges of interpretation, as methods of research associated with them may be radically different (even fundamentally irreconcilable with one another), or so new that they have yet to be formalized into transferable schemes to inform other research programmes. As such, an open, methodological pluralism is increasingly being championed (DeLyser and Sui, 2014).

Until the 1980s, geography was cast largely as either a physical (environmental or geological) science, a social science, or some combination of the two. This implied a commonality of objective, if not entirely of method: there was a shared commitment to the goal of 'general' explanation. More recently, however, some would dispute the use of the term 'science' in any of its forms in certain areas of subject (for an excellent introduction to debates surrounding science – its meaning, construction and application – see Chalmers, 2013). Instead, the 'cultural' turn in human geography (which in part reflects the growing influence of feminist and post-structural approaches) has brought a new emphasis on meanings, representation, emotions and so on that is more readily associated with the arts. Non-representational theory has challenged traditional approaches insofar as the purpose and focus of research are less on representing and explaining the outcomes or products of social interactions, and more on experiences and practices (Thrift, 2007). (These issues are discussed in more detail in texts such as Cloke et al., 2013; Cresswell, 2014, and Nayak and Jeffrey, 2011, and in various chapters in Clifford et al., 2009.)

Given the breadth of geographical enquiry, it is not surprising that the subject is similarly broad with respect to the methods it employs, and the philosophical and ethical stances it adopts. This book reflects the diversity of contemporary geography, both in the number, and the range of chapters which it contains. In this chapter, we want to briefly introduce you to different approaches to research methods and design, and to offer some guidance on how you might develop your own research design for a geographical project using this book.

1.2 Quantitative and Qualitative Approaches to Geography

The chapters in the book loosely deal with two forms of data collection/analysis: **quantitative** and **qualitative** methods and techniques. Quantitative methods involve the use of physical (science) concepts and reasoning, mathematical modelling and statistical techniques to understand geographical phenomena. These form the basis of most research in physical geography. They first began to be adopted by human geographers in the 1950s, but it was in the 1960s – a period dubbed the 'quantitative revolution' – that their application became both more widespread and more sophisticated in Anglo-American geography. It was at this time that, influenced by the 'scientific' approaches to human behaviour that were being adopted by social sciences such as economics and psychology, some human geographers began to be concerned with scientific rigour in their own research. In particular, they began to use quantitative methods to develop hypotheses and to explain, predict and model human spatial behaviour and decision making (Johnston, 2003; Barnes, 2014). (Collectively, the adoption of 'objective', quantified means of collecting data, hypothesis testing and generalizing explanations is known as positivism.) Much of this work was applied to planning and locational decision making (see foundational early works by Haggett, 1965; Abler et al., 1971).

In the 1970s, however, some geographers began to criticize positivist approaches to geography, particularly the application of 'objective' scientific methods that conceptualized people as rational actors (Cloke et al., 2013). Rather, the geographers adopting a humanistic approach argued that human behaviour is, in fact, subjective, complex, messy, irrational and contradictory. As such, humanistic geographers began to draw on methods that would allow them to explore the meanings, emotions, intentions and values that make up our taken-for-granted lifeworlds (Ley, 1974; Seamon, 1979). These included methods such as in-depth interviews, participant observation and focus groups. At the same time, Marxist geographers criticized the apolitical nature of positivist approaches, accusing those who adopted them for failing to recognize the way that scientific methods, and the spatial laws and models they produced, might reproduce capitalism (Harvey, 1973). More recently, feminist and post-structuralist approaches to geography have criticized the 'grand theories' of positivism and Marxism, and their failure to recognize people's multiple subjectivities. Instead, the emphasis is on refining qualitative methods to allow the voice of informants to be heard in ways which are non-exploitative or oppressive (WGSG, 1997; Moss, 2001) and to focus on the politics of knowledge production, particularly in terms of the positionality of the researcher and the way 'other' people and places are represented (see articles in two special issues (1/2) of *Environment and Planning: D*, 1992; Moss, 2001).

Humanistic inquiry did not just prompt interest in people's own account of their experiences, but also in how these experiences are represented in texts, literature, art, fiction and so on (Pocock, 1981; Daniels, 1985). Again, such visual methodologies have also been informed and developed by the emergence of post-structuralist approaches to geography which have further stimulated human geographers' concerns with issues of representation.

Despite the evolving nature of geographical thought and practice, both quantitative and qualitative approaches remain important within the discipline of Geography. While taken at face value they appear to be incompatible ways of 'doing' research, it is important not to see these two approaches as binary opposites. Subjective concerns often inform the development and use of quantitative methods. Likewise, it is also possible to work with qualitative material in quite scientific ways. Whatever methods are adopted, some degree of philosophical reflection is required to make sense of the research process. Equally, the two approaches are often combined in research designs in a process known as mixing methods (see below).

1.3 Designing a Geographical Research Project

Faced with a bewildering array of possibilities, both in what to study and in how to approach this study, it may seem that geographical research is difficult to do well. However, the very range of geographical enquiry is also a source of excitement and encouragement. The key is to harness this variety, rather than to be overwhelmed by it. Essentially, geographical research requires perhaps more thought than any of the other human or physical academic disciplines. Whether this thought is exercised with the assistance of some formal scheme of how to structure the research programme, or whether it is exercised self-critically, or reflexively in a much less formal sense, is less important than the awareness of the opportunities, limitations and context of the research question chosen, the appropriateness of the research methods selected, the range of techniques used to gather, sort and display information, and ultimately, the manner and intent with which the research findings are presented. For student projects, these questions are as much determined by practical considerations, such as the time available for the project, or the funding to undertake the research. These limitations should be built into the project at an early stage, so that the likely quality of the outcomes can be judged in advance. None of the constraints should be used after the research is completed to justify a partial answer or unnecessarily restricted project.

The 'scientific' view

Conventionally, geographical research programmes have been presented as a sequence of steps, or procedures (Haring and Lounsbury, 1992 – see below). These steps were based upon the premise that geography was an essentially scientific activity, that is, a subject identifying research questions, testing hypotheses regarding possible causal relationships, and presenting the results with some sort of more general (normative) statement or context. The aim of separating tasks was to enable time (and money) to be budgeted effectively between each, and to encourage a structuring of the thought processes underpinning the research.

The steps identified in this form of 'scientific geographic research' (Haring and Lounsbury, 1992) are as follows:

- *Formulation of the research problem* – which means asking a question in a precise, testable manner, and which requires consideration of the place and time-scale of the work.

- *Definition of hypotheses* – the generation of one or more assumptions which are used as the basis of investigation, and which are subsequently tested by the research.

- *Determination of the type of data to be collected* – how much, in what manner is sampling or measurement to be done.

- *Collection of data* – either primary from the field or archive, or secondary, from the analysis of published materials.

- *Analysis and processing of the data* – selecting appropriate quantitative and presentational techniques.

- *Stating conclusions* – nowadays, this might also include the presentation of findings verbally or in publication.

Today, there is more recognition that these tasks are not truly independent, and that an element of reflexivity might usefully be incorporated in this process. In some areas of the subject – particularly human geography – the entire notion of a formalized procedure or sequence would be considered unnecessary, and the notion of normative, problem-solving science would, at best, be considered applicable to a restricted range of subjects and methodology. Rather, as outlined above, many human geographers now reject or are sceptical of scientific approaches to human behaviour, preferring to adopt a more subjective approach to their research. Nevertheless, having said this, most qualitative research also involves many of the same steps outlined in the mechanical or scientific formulation above – albeit not conceptualized in quite the same way.

For example, qualitative researchers also need to think about what research questions to ask, what data need to be collected and how this material should be analysed and presented. In other words, all research in Geography – whatever its philosophical stance – involves thinking about the relationships between methods, techniques, analysis and interpretation. This important role is filled by **research design**.

The importance of research design

In its broadest sense, research design results from a series of decisions we make as researchers. These decisions flow from our knowledge of the academic literature (see Chapter 4), the research questions we want to ask, what is safe and ethical, especially in contexts which might be less familiar to ourselves (see Chapters 2, 3 and 6) our conceptual framework, and our knowledge of the advantage and disadvantages of different techniques (see Section 2 on Human Geography, Section 3 on Physical Geography, and Section 4 on some common aspects of geographical analysis). The research design should be an explicit part of the research: it should show that you have thought about how, what, where, when and why!

There are at least six key points you need to keep in mind to formulate a convincing research design:

(1) Think about what research questions to ask

On the basis of your own thinking about the topic, the relevant theoretical and empirical literatures (see Chapter 4), and consulting secondary material (see especially Chapter 7, 16 and 30) – and if possible having discussed it with other students and your tutor – you need to move towards framing your specific research questions. For a human geographer, these might include questions about what discourses you can identify, what patterns of behaviour/activity you can determine, what events, beliefs, and attitudes are shaping people's actions, who is affected by the issue under consideration and in what ways, and how social relations are played out, and so on. For a physical geographer, these might include questions concerning the rate of operation and location of a certain geomorphological process, the morphology of a selected set of landforms, or the abundance and diversity of particular plant or animal species in a given area (many of the chapters in this volume provide examples of research problems).

It is important to have a strong focus to your research questions rather than adopting a scatter-gun approach asking a diverse range of unconnected questions. This also means bearing in mind the time and resource constraints on your research (see below). At the same time as you develop a set of core aims it is also important to remain flexible, and to remember that unanticipated themes can emerge during the course of fieldwork which redefine the relevance of different research questions, likewise, access or other practical problems can prevent some research aims being fulfilled and lead to a shift in the focus of the work. As such, you should be aware that your research questions may evolve during the course of your project.

(2) Think about the most appropriate method(s) to employ

There is no set recipe for this: different methods have particular strengths and collect different forms of empirical material. The most appropriate method(s) for your research will therefore depend on the questions you want to ask and the sort of information you want to generate. Chapters 7–29 outline the advantages and disadvantages of core methods used by human and physical geographers. While many projects in human and physical geography involve going out into the field – for example to interview or observe people, or to take samples or measurements – it is also possible to do your research without leaving your computer, living room or the library. For example, research can be based on visual imagery such as films and television programmes (Chapters 15 and 38); the virtual worlds of the Internet (Chapters 16 and 17); secondary sources including contemporary and historical/archival material (Chapter 7); or GIS and remote sensing (Chapters 18, 25, 27 and 37). Some human geographers are also experimenting with conducting research using diaries (Chapter 10) and extending research into the domains of emotion and deeper analysis of texts (Chapters 12 and 14).

In the process of research design it is important not to view each of these methods as an either/or choice. Rather, it is possible (and often desirable) to mix methods. This process of drawing on different sources or perspectives is known as *triangulation*. The term comes from surveying, where it describes using different bearings to give the correct position. Thus, researchers can use multiple methods or different sources of information to try and maximize an understanding of a research question. These might be both qualitative and quantitative (see, for example, Sporton, 1999). Different techniques should each contribute something unique to the project (perhaps addressing a different research question or collecting a new type of data) rather than merely being repetitive of each other.

(3) Think about what data you will produce and how to manage these

An intrinsic element of your choice of method should not only involve reflecting on the technique itself, but also how you intend to analyse and interpret the data that you will produce. Chapters 8 and 9 demonstrate how to bring a rigorous analysis to bear on interview transcripts/diary material, while Chapters 19 and 30–37 discuss some of the issues you need to think about when deciding which statistical techniques to apply to quantitative data. Chapters 20–29 concentrate on methods most likely to be important in Physical Geography, and Chapter 29 looks at putting some of these into practice in environmental audit and appraisal. While qualitative techniques (Chapters 11–15 and 36 and 38) emphasize quality, depth, richness and understanding, instead of the statistical representativeness and scientific rigour which are associated with quantitative techniques, this does not mean that they can be used without any thought. Rather, they should be approached in as rigorous a way as quantitative techniques.

(4) Think about the practicalities of doing fieldwork

The nitty-gritty practicalities of who, what, when, where, and for how long inevitably shape the choices we can make about our aims, methods, sample size, and the amount of data we have the time to analyse and manage (see Chapter 19 on sampling and Chapter 8 on handling large amounts of qualitative data using computer software). Increasingly, the kind of work which is permissible is constrained by changing attitudes and legislative requirements with respect to safety and risk, which ultimately define the range and scale of what you can achieve (see, for example, Chapter 2 on the health and safety and risk assessment in fieldwork). It is important to bear in mind that the research which is written up by academics in journals and books is often conducted over several years and is commonly funded by substantial grants. Thus, the scale this sort of research is conducted on is very different to that at which student research projects must be pitched. It is not possible to replicate or fully develop in a three-month student dissertation or project all of the objectives of a two-year piece of academic research that you may have uncovered in your literature review! Rather, it is often best to begin by identifying the limitations of your proposed study and recognizing what you will and will not be able to say at the end of it. Remember that doing qualitative or quantitative work in human geography, just like fieldwork in physical geography, requires a lot of concentration and mental energy. It is both stressful and tiring, so there is a limit to how much you can achieve in the field in any one day. Other practicalities such as the availability of field equipment, tape recorders, cameras, transcribers or access to transport can also define the parameters of your project, and there are important considerations when working overseas and in different cultures (Chapter 6) as well as when using details from case studies to make wider generalisations (Chapter 33).

Drawing up a time-management chart or work schedule at the research-design stage can be an effective way of working out how much you can achieve in your study, and later on can also serve as a useful indicator if you are slipping behind. While planning ahead (and in doing so, drawing on the experience of your tutor and other researchers) is crucial to developing an effective research design, it is also important to remember that you should always remain flexible.

(5) Think about the ethical issues you need to consider

An awareness of the ethical issues which are embedded in your proposed research questions and possible methodologies must underpin your final decisions about the research design. The most common ethical dilemmas in human geography focus around: participation, consent, confidentiality/safeguarding personal information, and giving something back (see Alderson, 1995; Valentine, 1999). In physical geography, ethical issues involve not only questions of consent (for example to access field sites on private land) but also the potential impacts of the research techniques on the environment (e.g. pollution). Thus, while ethical issues may seem routine or moral questions rather than anything which is intrinsic to the design of a research project, in practice they actually underpin what we do. They can shape what questions we can ask, where we make observations, who we talk to, and where, when, and in what order. These choices in turn may have consequences for what sort of material we collect, how it can be analysed and used, and what we do with it when the project is at an end. As such ethics are not a politically correct add-on but should always be at the heart of any research design (see Chapter 3 for an overview of ethical issues, Chapters 12 and 13 for the specific ethics of participatory research, Chapter 6 about the specific ethical issues involved working in different cultural contexts and Chapter 16 regarding ethical issues in Internet-mediated research).

(6) Think about the form in which your research is to be presented

The scale and scope of your research design will partially be shaped by your motivations for doing the research and what you intend to use the findings for. If you are presenting your findings in a dissertation it will very different to a piece of work than if it is to be presented as a report or in a verbal seminar. Chapter 5 outlines presentational strategies and styles, and these are illustrated in detail in the various chapters of Section 4.

1.4 The Philosophy of Research and Importance of Research Design

Intensive and Extensive Research

The most basic, formal, distinction in research design is that between **extensive** and **intensive** approaches (also known as 'cross-sectional' and 'longitudinal'). The important aspects of these contrasting approaches have been explored in some detail for the social sciences by Sayer (2010). This book is an ideal and thought-provoking introduction to the ways in which research seeks to make sense of a complex world. Sayer reviews theories of causation and explanation in which 'events' (what we observe) are thought to reflect the operation of 'mechanisms' which, in turn, are determined by basic, underlying 'structures' in the world. The way in which explanations are obtained reflects differing degrees of 'concrete' and 'abstract' research – that is, how much our generalizations rely on observation, and how much they rely on interpretation of the ways in which events, mechanisms and structures are related. In the physical sciences, more attention has been given to the broader topic of scientific explanation of which research design is a part, but recently the implications of extensive and intensive research designs and the varying philosophies of research have begun to be explored within physical geography as well (e.g. Richards, 1996).

Both extensive and intensive designs are concerned with the relationship between individual observations drawn from measurement programmes or case studies, and the ability to generalize on the basis of these observations. The detailed distinctions are illustrated in Table 1.1 but the essential differences are as follows:

- In an **extensive** research design, the emphasis is on pattern and regularity in data, which are assumed to represent the outcome of some underlying (causal) regularity or process. Usually, large numbers of observations are taken from many case studies so as to ensure a 'representative' dataset, and this type of design is sometimes referred to as the 'large-n' type of study.

- In an **intensive** research design, the emphasis is on describing a single, or small number of case studies with the maximum amount of detail. This approach is therefore sometimes known as the 'small-n' type of study. In anthropology, the term 'thick description' has been used (Geertz, 2000). In an intensive design, by thoroughly appreciating the operation of one physical or social system, or by immersion in one culture or social group, elements of a more fundamental, causal nature are sought. 'Explanation' is therefore concerned with disclosing the links between events, mechanisms and structures. General explanations are derived from identification of the structures underlying observation, and from the possible transferring of the linkages discovered from detailed 'instantiations'.

Table 1.1   The essential differences between extensive and intensive research designs

Notes	Intensive	Extensive
Research question	How? What? Why? In a certain case or example	How representative is a feature, pattern, or attribute of a population?
Type of explanation	Causes are elucidated through in-depth examination and interpretation	Representative generalizations are produced from repeated studies or large samples
Typical methods of research	Case study. Ethnography. Qualitative analysis	Questionnaires. Large-scale surveys. Statistical analysis.
Limitations	Relationships discovered will not be 'representative' or an average/ generalization	Explanation is a generalization – it is difficult to relate to the individual observation Generalization is specific to the group/ population in question
Philosophy	Method and explanation rely on discovering the connection between events, mechanisms and causal properties	Explanation based upon formal relations of similarity and identification of taxonomic groups

Source: Based on Sayer (2010, Figure 13, p. 163–4)

Importantly, both approaches may be undertaken in quantitative or qualitative fashions – there is no necessary distinction in the *techniques* used. The two approaches are, however, separated to some extent in their philosophical underpinnings, and, more obviously, in the practical, logistical requirements they impose.

Philosophically, the extensive approach relies on the idea that a data pattern necessarily reflects an underlying cause or process, which is obscured only by measurement error, or 'noise'. However, in the 'real' world, it is rare that one cause would lead directly, or simply, to another 'effect' – the chain of causation is more obscure, and 'noise' may be an essential part of the 'causation', reflecting the presence of some other (unknown or uncontrolled) effect which merely mimics the apparent pattern. There is the related problem of being unable to explain individual occurrences on the basis of 'average' behaviour of entire groups – the so-called ecological fallacy. In an intensive research design, there is a deeper appreciation of the 'layers' which separate observations from an underlying (causal) reality. As such (and at the risk of considerable oversimplification) extensive approaches have often been linked to positivist methodology and philosophy, and intensive approaches to realist methodologies and philosophies.

Practically, the different types of research design have clearly different requirements in both data type and amount, and with respect to cost and time. The extensive design lends itself to situations where large amounts of data are already published, or where large amounts of data can be generated from secondary sources. In many student projects, the need for many observations across comparative or contrasting field sites may be too daunting or logistically impossible if an attempt is made to mount an extensive research design-based upon primary data sources. An exception to this is in laboratory-type studies, where a series of experiments may quickly build up a dataset representative of a wider range of conditions. The intensive design is perhaps more common, but care is needed to 'tease out' those aspects of the study which might disclose basic, causal processes.

Multi-, Inter- and Transdisciplinary Research

Geography, along with most other subjects investigating the intersection of the human and natural worlds, is increasingly faced with problems which are both more complex, and those which are less clearly defined and where outcomes are less easy to predict. Climate change provides perhaps the best example, but issues of water, energy and food security – now often seen as 'nexus studies' – or urbanization, development and sustainability are similar examples. It is likely that your own research may be connected broadly with such issues, even if only to provide a context for the project you are doing, or if you are taking your place in a larger research team or study. Faced with problems of increasing scale and reduced certainty, and with a need to maintain an applied or policy focus, research is increasingly categorized in terms of its degree of 'integration' between disciplines, perspectives or methodologies, and, as well, by the degree to which the research problems, issues and research designs lie in the hands of the researcher themselves, or are co-designed (that is, the extent to which they are 'participatory' of academics and non-academics, such as stakeholders – see Figure 1.1).

Explained in more detail, the degree of integration describes firstly the degree to which methods, concepts, techniques, data and other knowledge are integrated across different disciplines; and secondly the extent to which there is integration across scales and sectors from local to global, from nations and cultures and across science and society (Mauser et al., 2013). Participation in the design, delivery and interpretation of the research process includes: co-design of a research agenda, to identify questions, methods and techniques which are of relevance and which are manageable and useful; co-production of knowledge in which investigators and stakeholders are in a dialogue to ensure that expertise, methods and results remain relevant to an original agenda; and finally, co-dissemination of results, where specialist, formal results are presented to, translated for, and shared by the widest possible audience (Mauser et al., 2013: 427–8).

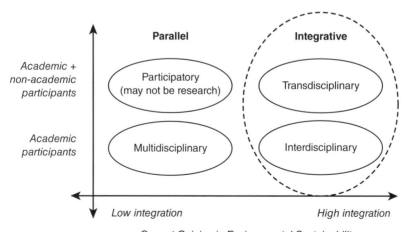

Current Opinion in Environmental Sustainability

Figure 1.1   Multi-, inter- and transdisciplinary research characteristics defined in terms of integration and participation. From Tress et al. (2004, Figure 2) as reproduced in Mauser et al. (2013)

Overlaying this generic classification of research are the terms multi-, inter- and transdisciplinarity. The following short descriptions are based largely on the article by Stock and Burton (2011). Multi-disciplinary research brings together work and workers from various academic disciplines, each of which has their own or disciplinary perspective on a shared issue. Research is coordinated, in that results and perspectives are brought together in the end reporting of the results and conclusions, but it is not truly integrated. Interdisciplinary research is explicitly integrated – pursuit of a shared goal forces those engaged in research to generate new knowledge. Instead of simply contributing a set of multiple knowledges or approaches to a problem, the problem is shared and each knowledge set is seen from a different perspective. Interdisciplinary research is, therefore, both more collaborative/participatory and more integrated – there is a jointly identified problem, methodology and analysis. Transdisciplinarity moves beyond this – it involves multiple academic perspectives and participants and, crucially, non-academics. The aim is to generate (social) scientific work and decision-making capabilities, focusing on the problem rather than the discipline involved, and bridging across between experts, lay, science and non-science. Transdisciplinarity is the clearest attempt to address complex issues, with combination of knowledge types and plurality of methods, and a combination of participants in the research. Lang et al. in part summarize the work of other authors to highlight three essential aspects of transdisciplinary study:

> (a) focusing on societally relevant problems; (b) enabling mutual learning processes among researchers from different disciplines (from within academia and from other research institutions), as well as actors from outside academia; and (c) aiming at creating knowledge that is solution-oriented, socially robust … and transferable to both the scientific and societal practice. With regards to the latter, it is important to consider that transdisciplinarity can serve different functions, including capacity building and legitimization… (2014: 27)

Key characteristics of the three kinds of research are presented in Figure 1.2.

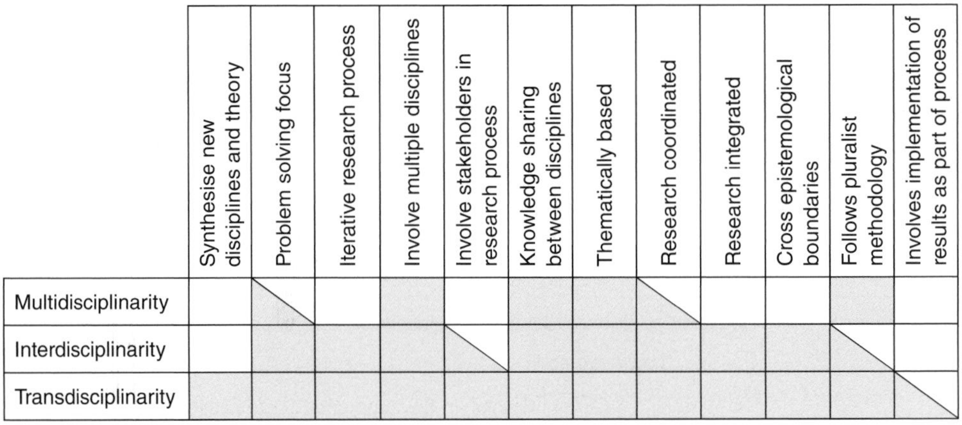

	Synthesise new disciplines and theory	Problem solving focus	Iterative research process	Involve multiple disciplines	Involve stakeholders in research process	Knowledge sharing between disciplines	Thematically based	Research coordinated	Research integrated	Cross epistemological boundaries	Follows pluralist methodology	Involves implementation of results as part of process
Multidisciplinarity												
Interdisciplinarity												
Transdisciplinarity												

Figure 1.2 Characteristics of multi-, inter- and transdisciplinary research. Stock and Burton (2011, Figure 1)

1.5 Conclusion: How This Book Can Help You Get Started

Geographical research is complicated, because geographical phenomena are many and varied, and because they may transcend multiple scales in space and time. Further, over the last few decades, geographers have adopted a diverse range of philosophic stances, methods and research designs in their efforts to understand and interpret the human and physical worlds. In order to make sense of this variety in a research context, a considerable amount of thought must go into geographical research at all of its stages. Although the prospect of embarking on your own research might seem daunting, this book has been put together to help make it easier for you. The first six chapters all offer advice and guidance about how to plan, prepare and contextualize your research project. This chapter has explained the process of research design; Chapter 2 explains the practical and logistical issues you need to plan for in terms of your own health and safety in the field; Chapter 3 raises some of the ethical issues you might need to consider in your research design; Chapter 4 describes how to do a literature search to help define your topic and research questions; ; and Chapter 5 outlines some of the ways you might eventually present your work. Chapter 6 considers these issues within the specific situations of working in different cultural contexts and doing participatory research.

Other sections of the book focus on generating and working with data, first in Human Geography (Chapters 7–19), and then Physical Geography (Chapters 20–29); the final section offers more detail on techniques to analyse, visualize and interpret geographical data across the widest subject range. You will obviously not want to use all of the chapters, but by reading across them you will get a feel for the different ways that you might approach the same topic, and the advantages and disadvantages of different methods. When you have developed your own research design the chapter(s) appropriate to your chosen method(s) will give you practical advice about how to go about your research.

As with this chapter, each chapter in the book contains a synopsis at the beginning which briefly defines the content of the chapter and outlines the way it is structured. At the end of each chapter is a summary of key points and an annotated list of key further readings. Several of the chapters also contain boxed tips or useful exercises to develop your understanding of the topic in question. For each chapter, there is a WWW resource of material taken from the journals *Progress in Human Geography* and *Progress in Physical Geography*. These provide an essential follow-up on the key topics you will encounter and would wish to explore further. For some chapters, there is additional material online as detailed after the conclusions.

Doing your own research can be one of the most rewarding aspects of a degree. It is your chance to explore something that really interests or motivates you and to contribute to geographical knowledge, so enjoy it. And good luck!

SUMMARY

- Geographers are faced with a vast array of potential subject matters and techniques for data collection, visualization and analysis.
- Research design is crucial to link data collection, methods and techniques together, and to produce convincing, meaningful results.

- The basic choice in research design is between extensive and intensive designs. These have very different implications in terms of data collection, analysis and interpretation, although both quantitative and qualitative methods may be used in either. You may also consider the wider context of your research and research methodology, and consider whether your research represents part of a multi-, inter- or transdisciplinary approach.
- Whichever form of research design is adopted, the ethical dimension of research must be considered.
- Frequently, practical issues of land access, time or financial constraints, or field safety will, together with ethical issues, determine the scope and kind of research adopted. While these are constraints, prior consideration of their likely effects will minimize the loss of intellectual integrity and merit in the project.

Further Reading

- The different philosophies underlying human geography research are outlined in numerous volumes including Aitkin and Valentine (2015), Cresswell (2014) and Johnston and Siddaway (2015). Trudgill and Roy (2014) attempt a similar review of methods and meanings in physical geography. Various chapters in Clifford et al. (2009) explain the tensions between geography as a physical science, social science or arts subject and show how geographers' understandings of key concepts have evolved as approaches to geographical thought have developed.

- Sayer (2010) is a thorough and extensive treatment of 'methodology' and its connections to the way in which we make sense of the world through observations, experiments, surveys and experiences. It argues that our philosophy of the way in which the world is structured (how things come to be as they are seen) must inform our choice of research design, and our choice of techniques for generalizing on the basis of the information which we collect from putting the research design into practice. The book depends on a particular ('realist') approach, but is an excellent starting point from which to relate philosophy to the practicalities of doing research, and to the strengths and weakness of particular kinds of research methods.

- Chalmers (2013) is a student-centred volume which covers an enormous range of material dealing with the various approaches to the philosophy of science and the status of scientific knowledge. It examines the nature of scientific explanation (the relations between observations, experiments and generalization), and gives both a historical and contemporary overview of contrasting viewpoints and approaches.

- The edited book by Hay (2010) provides a detailed guide to selecting, operationalizing and presenting the results of qualitative methodologies in geographical research. It has chapters dealing with research design, ethics, differing methodologies, and the use of Internet and computing-related resources. Moss (2001) is a collection of essays exploring feminist geography in practice. These essays share a particular concern with the ethics and politics of knowledge production, notably in terms of the positionality of the researcher and the way 'other' people and places are represented.

- The journal *Anthropocene Review* spans material from earth, environmental and social science and the humanities, while the journals *Progress in Human Geography* and *Progress in Physical Geography* provide substantive research articles, summaries of up-to-date developments in progress reports, reviews of research resources, and retrospectives on classic works in Geography.

Note: Full details of the above can be found in the references list below.

References

Abler, R.F., Adams, J.S. and Gould, P.R. (1971) *Spatial Organization: The Geographer's View of the World.* Englewood Cliffs, NJ: Prentice-Hall.

Aitkin, S. C. and Valentine, G. (eds) (2015) *Approaches to Human Geography: Philosophies, Theories, People and Practices* (2nd edition). London: Sage.

Alderson, P. (1995) *Listening to Children: Children, Ethics and Social Research*. Ilford: Barnardos.

Barnes, T. (2014) 'What's old is new, and new is old: History and geography's quantitative revolutions', *Dialogues in Human Geography* 4 (1): 50–3.

Brierley, G., Fryirs, K., Cullum, C., Tadaki, M., Huang, H. and Blue, B. (2013) 'Reading the landscape: Integrating the theory and practice of geomorphology to develop place-based understandings of river systems', *Progress in Physical Geography* 37, 5: 601–21.

Castree N. (2014) 'The Anthropocene and Geography I: The Back Story', *Geography Compass* 8 (7): 436–49. (See also follow-up companion articles II and III)

Chalmers, A. F. (2013) *What Is This Thing Called Science?* (4th edition). Cambridge, MA: Hackett.

Clifford, N. J. (2009) 'Globalisation: a physical geography perspective', *Progress in Physical Geography* 33: 5–16.

Clifford, N.J., Holloway, S.L., Rice, S.P. and Valentine, G. (2009) *Key Concepts in Geography* (2nd edition). London: Sage.

Cloke, P., Crang, P. and Goodwin (2013) *Introducing Human Geographies* (3rd edition). London: Routledge.

Creswell, J. W. (2014) *Research Design. Qualitative, Quantitative, and Mixed Methods Approaches*. 4th edition. Thousand Oaks, CA: Sage.

Crutzen, P. J., and Stoermer, E. F. (2000) 'The "Anthropocene"', *Global Change Newsletter* 41: 17–18.

Daniels, S. (1985) 'Arguments for a humanistic geography', in R. J. Johnston (ed.) *The Future of Geography*. London: Methuen. pp. 143–58.

DeLyser, D. and Sui, D. (2014) 'Crossing the qualitative-quantitative chasm III: Enduring methods, open geography, participatory research and the fourth paradigm', *Progress in Human Geography* 38: 294–307.

Geertz, C. (2000) *The Interpretation of Cultures* (2nd edition). New York: Basic Books.

Graham, M and Shelton, T. (2013) 'Geography and the future of big data, big data and the future of geography', *Dialogues in Human Geography* 3 (3): 255–61.

Gregory, D.J. (2009) 'Geography', in D.J. Gregory, R.J. Johnson, G. Pratt, M.J. Watts and S. Whatmore (eds) *The Dictionary of Human Geography* (5th edition). Chichester: Wiley-Blackwell. pp. 287–95.

Gregory, K.J. (2000) *The Changing Nature of Physical Geography*. London: Arnold.

Haggett, P. (1965) *Locational Analysis in Human Geography*. London: Edward Arnold.

Haring, L.L. and Lounsbury, J.F. (1992) *Introduction to Scientific Geographical Research* (4th edition). Dubuque, WC: Brown.

Harvey, D. (1973) *Social Justice and the City*. London: Edward Arnold.

Hay, I. (ed.) (2010) *Qualitative Research methods in Human Geography* (3rd edition). Oxford: Oxford University Press.

Johnston, R.J. (2003) 'Geography and the social science tradition', in S.L. Holloway, S. Rice and G. Valentine (eds) *Key Concepts in Geography*. London: Sage. pp. 51–72.

Johnson, E. and Morehouse, H. (with other forum authors) (2014) 'After the Anthropocene: Politics and geographic inquiry for a new epoch', *Progress in Human Geography* 38: 439–56.

Johnston, R.J. and Siddaway, J. (1997) *Geography and Geographers: Anglo-North American Human Geography since 1945*. London: Routledge.

Johnston, R. and Siddaway, J. (2015) *Geography and Geographers: Anglo-American human geography since 1945* (7th edition). Abingdon: Taylor & Francis.

Kitchin, R. (2014) *The Data Revolution: Big Data, Open Data, Data Infrastructures & Their Consequences*. Thousand Oaks, CA: Sage.

Lang, D. J., Wiek, A., Bergmann, M., Stauffacher, M., Martens, P., Moll, P. Swilling, M. and Thomas, C. J. (2012) 'Transdisciplinary research in sustainability science: practice, principles, and challenges.', *Sustainability Science* 7 (Supplement 1): 25–43

Lave, R., Wilson, M.W., Barron, E.S., Biermann, C., Carey, M.A., Duvall, C.S., Johnson, L., Lane, K.M., McClintock, N., Munroe, D., Pain, R., Proctor, J., Rhoads, B.L., Robertson, M.M., Rossi, J., Sayre, N.F., Simon, G., Tadaki, M. and van Dyke, C. (2014) 'Intervention: Critical physical geography', *The Canadian Geographer* 58 (1): 1–10.

Ley, D. (1974) *The Black Inner City as Frontier Outpost: Image and Behaviour of a Philadelphia Neighbourhood*. Monograph Series 7. Washington, DC: Association of American Geographers.

Lorimer, J. (2012) 'Multinatural geographies for the Anthropocene', *Progress in Human Geography* 36 (5): 593–612.

Mauser, W., Klepper, G., Rice, M., Schmalzbauer, B., Hackmann, H., Leemans, R. and Moore, H. (2013) 'Transdisciplinary global change research: The co-creation of knowledge for sustainability', *Current Opinion in Environmental Sustainability* 5: 420–31.

Moss, P. (ed.) (2001) *Feminist Geography in Practice*. Oxford: Blackwell.

Nayak, A. and Jeffrey, A. (2011) G*eographical Thought: An Introduction to Ideas in Human Geography*. London: Routledge.

Pile, S. (2010) 'Emotions and affect in recent human geography', *Transactions of the Institute of British Geographers* 35 (1): 5–20.

Pitman, A.J. (2005) 'On the role of geography in earth system science', *Geoforum* 36: 137–48. (See also subsequent discussions.)

Pocock, D.C.D. (ed.) (1981) *Humanistic Geography and Literature: Essays on the Experience of Place*. London: Croom Helm.

Rhoads, B.L. and Thorn, C.E. (eds) (1997) *The Scientific Nature of Geomorphology*. Chichester: John Wiley and Sons.

Richards, K. (1996) 'Samples and cases: Generalisation and explanation in geomorphology', in B.L. Rhoads and C.E. Thorn (eds) *The Scientific Nature of Geomorphology*. Chichester: John Wiley and Sons. pp. 171–90.

Richards, K. S. and Clifford, N. J. (2008) 'Science, systems and geomorphologies: why *LESS* may be more', *Earth Surface Processes and Landforms* 33 (9): 1323–40.

Sayer, A. (2010) *Method in Social Science: A Realist Approach* (revised 2nd edition). London: Routledge.

Seamon, D. (1979) *A Geography of a Lifeworld*. London: Croom Helm.

Sporton, D. (1999) 'Mixing methods of fertility research', *The Professional Geographer*, 51: 68–76.

Stock, P. and Burton, R. J. F. (2011) 'Defining terms for integrated (multi-inter-trans-disciplinary) sustainability research', *Sustainability* 3(8): 1090–113.

Tadaki, M., Salmond, J., Le Heron, R. and Brierley, G. (2012) 'Nature, culture, and the work of physical geography', *Transactions of the Institute of British Geographers* 37 (4): 547–62.

Thien, D. (2005) 'After or beyond feeling? A consideration of affect and emotion in *geography*', *Area* 37 (4): 450–4.

Thrift, N. (2002) 'The future of Geography', *Geoforum*, 33: 291–8.

Thrift, N. (2007) *Non-representational Theory: Space, Politics, Affect*. London: Routledge.

Tress, B., Tress, G. and Fry, G. (2004) 'Clarifying integrative research concepts in landscape ecology', *Landscape Ecology* 20: 479–93.

Trudgill, S. J. and Roy, A. R. (eds) (2003) *Contemporary Meanings in Physical Geography: From What to Why?* London: Arnold.

Trudgill, S. and Roy, A. (eds) (2014) *Contemporary Meanings in Physical Geography: From What to Why?* Abingdon: Routledge.

Valentine, G. (1999) 'Being seen and heard? The ethical complexities of working with children and young people at home and at school', *Ethics, Place and Environment*, 2: 141–55.

Walford, R. and Haggett, P. (1995) 'Geography and geographical education: some speculations for the twenty-first century', *Geography*, 80: 3–13.

Whatmore, S. (2002) *Hybrid Geographies: Natures, Cultures, Spaces*. London: Sage.

Wilcock, D., Brierley, G. and Howitt, R. (2013) 'Ethnogeomorphology', *Progress in Physical Geography* 37 (5): 573–600.

WGSG (Women and Geography Study Group) (1997) *Feminist Geographies: Explorations in Diversity and Difference*. London: Longman.

Wyly, E. (2014) 'The new quantitative revolution', *Dialogues in Human Geography* 4 (1): 26–38.

ON THE COMPANION WEBSITE...

Visit **https://study.sagepub.com/keymethods3e** for links to the best resources relating to this chapter.

2 Health, Safety and Risk in the Field

Joanna Bullard

SYNOPSIS

Health and safety in the field concerns the practical steps that geographers can take to lessen the chances of an incident or accident causing harm to themselves and others during fieldwork. This chapter focuses on minimizing risks, reducing hazards and taking responsibility for health and safety both in group and independent fieldwork.

This chapter is organized into the following sections:

- Introduction
- Managing health and safety
- Know your field environment
- Know your limitations
- Know your equipment
- What if something goes wrong?

2.1 Introduction

Fieldwork is often one of the most rewarding aspects of being a geography student. Many residential trips and one-day outings are remembered by students long after graduation. Fieldwork can, however, also be time consuming, frustrating, difficult and potentially dangerous. This chapter offers some guidance about minimizing the risks and hazards involved in fieldwork. Following this guidance should reduce the dangers inherent in your chosen field activity and consequently should also make your time in the field more rewarding, more enjoyable and possibly a little easier.

As a geography student you may undertake fieldwork in a variety of contexts including supervised and unsupervised, group and individual field activities, residential and non-residential field studies, local and overseas environments. These different contexts present a range of health and safety considerations and the roles and responsibilities of both students and staff may change accordingly.

This chapter discusses the assessment and management of risks associated with undertaking fieldwork. It covers evaluating and minimizing risk in a range of common field environments and emphasizes the importance of planning ahead. In particular, the chapter focuses on your responsibility as an individual and as a member of a group undertaking fieldwork to ensure that appropriate health and safety precautions are observed.

In addition to reading this chapter, note that every higher education institution (and frequently school or department within this) has its own health and safety policy, now most commonly available in its website provision. You may have been required to sign a declaration stating that you have read and understood these guidelines. One of your responsibilities as a student is to ensure that you follow any institutional or departmental guidelines when you are out in the field, and often this is a compulsory part of planning research projects such as dissertations.

2.2 Managing Health and Safety

In most countries, national legislation provides the legal framework for health and safety management. For example, universities in the UK abide by the Health and Safety at Work Act 1974, while in Australia, the Workplace Health and Safety Act 1995 fulfils a similar role. These Acts place a duty upon employers to take steps to ensure, as far as possible, the health and safety of their employees and any other people affected by their activities including, in the case of universities, all students and members of the public. Additional or complementary legislation and guidelines may also exist at the organizational, state or national scale. In the UK, for higher education this includes the USHA-UCEA (2011) 'Guidance on Health and Safety in Fieldwork' which has a very broad definition of fieldwork as 'any work carried out by staff or students for the purposes of teaching, research or other activities while representing the institution off-site' (p.7). There is also a British Standard standard concerning fieldwork, expeditions, and adventurous activities outside the United Kingdom that your university may follow (BS 8848:2014 – see Further Reading). These laws and guidelines typically require that 'risk assessments' are undertaken to identify what should be done in order to manage safety. Assessments of risk are usually focused around 'risk' and 'hazard'. Put simply, hazards result from working in potentially dangerous environments and refer to environmental conditions, agents or substances that can cause harm. Risk is the chance that someone might be harmed by the hazard. During fieldwork hazards and risks can change rapidly, for example, as a result of changing weather conditions or political actions, and should be continually reassessed.

Many education institutions adopt a five-step approach to risk assessment:

1. *Identify the hazards*: During the fieldwork, what could cause harm – for example, slippery ground, high altitude, weather conditions, civil or political unrest?

2. *Identify who might be harmed and how*: This includes all fieldworkers and members of the public.

3. *Evaluate and minimize the risks*: Once suitable precautions have been taken, for example wearing appropriate protective clothing, how likely is it that someone will be harmed?

4. *Record the findings*: Write down the identified hazards and precautions to be taken.

5. *Review the assessment periodically*: Situations change and so do the potential hazards and risks, so risk assessment is not an isolated task but an ongoing evaluation to be updated and revised as often as necessary.

A risk assessment should be completed for fieldwork conducted as part of your degree. The extent of your involvement in actually assessing the risks will vary according to the way in which the fieldwork is organized, but you have a responsibility to follow any precautions or safety measures laid down in the risk assessment.

Supervised and unsupervised group fieldwork

Group fieldwork will usually be undertaken as part of a course or module designed by a member of academic staff at your university. During supervised fieldwork, one member of staff will take overall responsibility for field activities and will complete the risk assessment. The leader of the fieldwork will be responsible for ensuring that appropriate health and safety matters have been taken into consideration and are complied with by all accompanying members of staff and students (Kent et al., 1997). You are not, however, absolved of responsibility. Be aware that as an individual you have a responsibility for your own safety and also for the safety of others who may be affected by your actions, including staff, other students and members of the public. You should follow any instructions given by staff members. If you witness any accidents or foresee any unexpected problems, these should be reported to the fieldwork leader as soon as possible. Failure to uphold your responsibilities may result in disciplinary measures and, in extreme cases, in a criminal prosecution. It is worth noting that your activities prior to fieldwork, for example having a few alcoholic drinks the evening before, can adversely affect your performance and reliability in the field. Staff may refuse to let you participate in fieldwork if you are considered a danger to yourself or others.

On arrival at university, you may have completed a confidential medical questionnaire. You may be asked to complete a more detailed questionnaire prior to undertaking any fieldwork, especially if it will be conducted overseas. These questionnaires should be used to give details of any diseases, conditions, disabilities or susceptibilities that you may have, such as diabetes, severe allergies, impaired hearing, vertigo. If you are involved in an incident or accident whilst on fieldwork this information may be vital in ensuring that you receive prompt and appropriate medical treatment. If you did not complete a questionnaire or equivalent, and do have medical needs that could compromise the health and safety of yourself or others, you should let staff know well before fieldwork commences so that they can take appropriate action where necessary.

During a field course you may undertake unsupervised group fieldwork. This commonly involves students working in groups of two or more. A member of staff may visit you during the field activities; for example, they may move from group to group along the course of a river or arrange to meet you at a particular location within a city. Alternatively you may be left entirely to your own devices and asked to book in or out of a field centre or to conduct work on campus or in your university city. In these situations you have a responsibility to the group of students with whom you are working. Keep together as a group, discuss any difficulties that you might have and pay attention to anyone who is feeling unwell or is having difficulty keeping up. You should follow any guidelines issued; for example, take a first aid kit, have contact numbers for the university department or individual members of staff, and know where you should be working and what time you are expected to check back in.

Individual, unsupervised fieldwork

You may be involved in individual, unsupervised fieldwork at any time during your university career, but this is most likely to be the case when you undertake fieldwork as part of a dissertation or other independent research projects. Many universities now require every student to complete an individual risk assessment for their intended project. This can take many forms but will usually be organized around the five-step plan outlined above. Detailed discussion and examples of this procedure are provided by Higgitt and Bullard (1999), templates are now frequently to be found on institutions' websites and most departments will have a nominated Safety Officer who can give advice.

The most important health and safety concerns associated with individual, unsupervised fieldwork arise from lone working which should be avoided wherever possible. If you are conducting an individual project or dissertation a good strategy is to persuade a responsible friend or sibling to act as your field assistant. You might pair up with a friend from university and agree to help them survey a beach if they, in turn, will accompany you delivering questionnaires. If lone working is unavoidable or you are working with just one other person it is vital that you develop a strategy for dealing with emergencies should they arise. At the very least you should leave with a responsible adult written details of where you are going – for example, the map co-ordinates of your field site or the name and address of the person you intend to interview – what you intend to do and what time you expect to return. You should also write down what you expect the person to do in the event that you do not return on time. If your plans change you must let this person know. In risk assessment, lone working always counts as a hazardous activity. Where the site or conditions are themselves hazardous, then lone working is liable to be dangerous.

Whether you are working in a group or as an individual there are some basic steps that you can always take to minimize any health or safety risks. First know the telephone numbers that can be used to contact the emergency services if necessary. In the UK, 999 or 112 will allow you to contact the fire, police and ambulance services, the coastguard and mine, mountain, cave and fell rescue services. Similar services can be contacted in the USA and Canada by dialling 911, in Australia 000, and in New Zealand 111. If you are working overseas, find out local emergency numbers and the contact details of your embassy, high commission or consulate before you start working and keep them written down in a safe place (e.g. on a card inside your first aid kit). Other emergency communication methods include the international Alpine Distress signal – six long whistle blasts or torch flashes or shouts for help in succession repeated at one-minute intervals – and the Morse code signal 'SOS' signalled by three short whistle blasts/torch flashes, three long blasts/flashes, three short blasts/flashes, pause, then repeat as necessary.

Second, have a basic kit that you take into the field with you every day. Never start any fieldwork in any environment without having with you a pen/pencil and paper and any personal medication. If you are working in a rural or remote environment you will also need a first aid kit, mobile phone, whistle, compass, map, additional clothing (hat, gloves, socks, spare sweater), food and drink as a minimum. If you are working in a built-up or urban environment your field kit may include a first aid kit, mobile phone, street map and personal alarm.

Third, make sure that you always know exactly where you are – this is relatively easy in a town centre but can be more problematic in an unfamiliar large city or a remote area with few landmarks. If the worst happens and you need help from the emergency services this will be a vital piece of information that you can provide. Borrowing or buying a GPS (Global Positioning System) unit (see below) is a good idea.

Fourth, consider whether insurance and other forms of cover for travel accidents and breakdowns are necessary. Frequently, cover can be provided for support and assistance before emergency services need to be involved (breakdown assistance, for example, and advice and support for foreign travel, travel incidents and health). Universities may provide this, or you may be expected to make your own provision. There may also be health requirements to meet before travel can be undertaken.

When you carry out the risk assessment for your field project or set off on a field trip with other students there are a number of other things that you need to take into account to ensure the health and safety of everyone involved. These fall broadly into the categories of knowing your field environment, knowing your own limitations and knowing your equipment.

2.3 Know Your Field Environment

There are many different locations in which you might find yourself doing fieldwork, from a suburban street to a kibbutz, from a quarry to the top of a mountain. No matter where you intend to go, there will always be some information that you can acquire before you start, that will provide guidance as to how to act, dress and behave. Gathering this information together is important, because in many cases you will need to complete a risk assessment for your field project *before* you reach the site. If the field site is nearby or very familiar to you then you can draw on your existing experiences when completing the risk assessment, but if the site is not somewhere you have been before or is overseas, you will need some help in anticipating the risks and hazards involved.

A key control on many field activities is the weather or climate of the area in which you are working. As part of your ongoing risk assessment you should consult a daily weather forecast throughout your fieldwork (see Further Reading). Specific hazards associated with weather and climate include hypothermia, frostbite, sunburn, dehydration, heat stroke and heat exhaustion. In most situations being aware of the predicted weather for the local area and having the correct clothing and equipment can reduce the risks associated with these hazards. Specialist information for mountain and coastal environments is available in many countries. When working at the coast make sure you know where the nearest coastguard station is and check local tide timetables.

For any fieldwork, but particularly those occasions when you will come into close or regular contact with people, such as when conducting interviews, questionnaires, or participant observation, you are advised to do some research on the local area and to make yourself familiar with any customs, political issues and religious beliefs that may affect the ways in which you conduct your research and how people react to your project. The aim is to minimize the risk of causing offence, which may lead to personal attack/injury or abuse. Take advice from local contacts about how to respect local customs and about any unrest in the area and dress appropriately for the environment and culture within which you are conducting your research. You should also consider whether your individual characteristics (e.g. gender, race) make you particularly vulnerable (Ross, 2015). If possible avoid areas that are known to be 'unpleasant' and try not to enter unfamiliar neighbourhoods alone. A good street map is invaluable, plan your route, and walk purposefully and with confidence.

When taking part in a supervised field trip, the trip leader should check local conditions and is likely to brief you about them. If you are told tide times or other specific pieces of information, such as neighbourhoods to avoid, make sure you write them down. Prior to the field trip you may be given a list of clothing and equipment to bring with you that is suitable for the expected conditions. Depending on the environment, this could include sturdy walking boots, waterproof jacket, gloves, sunglasses, sunhat, water bottle, long sleeves, and so on. In some institutions you will not be allowed to take part in the field activities if you do not have appropriate clothing with you – this is not a minor issue, it is one with potentially profound health and safety implications.

Many government websites provide detailed information for nationals travelling overseas. This can include information on countries or regions to avoid, visa requirements, local customs, pre-travel inoculations and what to do should you be the victim of crime, fall ill or get into trouble of any sort. Make sure you are aware of any political, military or civil unrest in the country – local newspapers and Internet sites can be an invaluable source of information. If you are working alone or as a small independent group you should inform your nearest embassy, high commission or consulate of your presence in the country when you arrive to enable them to keep you informed of any hazardous situations. Nash (2000) provides a discussion of things to consider when planning independent overseas fieldwork and includes a list of essential information sources. Another useful reference both whilst planning and during an overseas field visit will be a guidebook to that country or area. Most popular guidebooks, such as those in the Rough Guide or Lonely Planet series, include sections on health risks, safety precautions and appropriate dress codes.

2.4 Know Your Limitations

Many people underestimate the time needed to complete fieldwork tasks. The majority of incidents and accidents occur when people are tired. Make sure that the aims of your fieldwork are realistic and that you have allowed enough time to achieve them. Weather conditions can dramatically affect your efficiency in the field – it is hard to survey for six hours in the pouring rain but much easier (and more fun) if the sun is shining, although you should take note of risks from sun and wind burn, and take appropriate precautions. A useful approach is to make a list of the minimum amount of work that you need do to achieve your basic goals and rank these tasks in order of priority. Then make a list of what would be useful or desirable if you have the time. Work through the list in order, achieve the minimum and then add to it if possible. Do not compromise your safety by trying to work longer hours in unsuitable conditions. If you are doing supervised fieldwork and are uncomfortable about undertaking certain field activities, tell the leader and ask whether or not there is an alternative. For example, if you are scared of heights, and are very nervous about walking along a narrow path with a steep drop, find out if there is an alternative route, or whether someone more experienced with the environment can walk with you.

Although you should try to work within your limitations, fieldwork is about facing challenges and doing things you might not normally do, for example interviewing complete strangers, working in a challenging physical environment. There are many things that you can do before you head off into the field that will extend your skills and abilities and so reduce the risks involved. Some organizations offer courses and workshops and produce publications that may be useful (see Further Reading). If you intend to work in sparsely populated areas, you should consider taking a first aid train-ing course. These are run on a regular basis by bodies such as the St John Ambulance and St Andrew's Ambulance Association and may be available at your college or uni-versity as an evening class. Advice on expedition planning and organization is available from a number of sources including the Expedition Advisory Centre (UK) and The Explorers Club (USA) (see Further Reading). More specialist courses, such as winter mountain skills and sea navigation, are run by specific organizations and societies.

2.5 Know Your Equipment

Whatever equipment you intend to use for your fieldwork, whether a tape recorder, a soil auger or something electronic, make sure that you know how to use it and how to carry it. Some items of equipment are very heavy and need two people to move them safely. Even a simple surveying kit comprising staff, level and tripod can be awkward for one person to manage. Take advice from academic and support staff in your institu-tion about the safest ways to collect samples from a river, to dig a deep soil pit, to conduct a house-to-house survey. If your equipment requires batteries and/or exchange-able media such as cassettes, take spare ones; if you are delivering questionnaires house-to-house make sure you have more copies than you think you need and a good street map. These types of precautions will prevent delays in your field programme and reduce the chance of you being tempted to work at unsuitable times, for example deliv-ering questionnaires after dark or working on a beach with a rapidly rising tide.

Mobile devices such as smartphones and tablets are increasingly being used for field data collection (e.g. Glass, 2015; Medzini et al., 2015) and may also be useful for reducing risk by providing means to get emergency assistance if required. You should, however, check that your phone can be used throughout your field area as coverage can be patchy especially in rural areas or where the topography of the landscape blocks the signal (such as valleys). In addition, technological failure and (more commonly) lack of power can reduce your mobile phone to a useless piece of plastic or metal in any location. The same is true for personal alarms. Carrying a mobile phone or personal alarm is no substitute for a well-thought out system for informing other people of your whereabouts (your lone working strategy).

If you are working in a rural or remote land area or at sea you may decide to take a hand-held GPS with you. This type of device can provide accurate latitude/longitude co-ordinates and can be used for mapping and/or navigation when referenced to national mapping conventions. If you are using a GPS for navigation you must be aware of its limitations. Most hand-held GPSs are only accurate to a few metres latitude/longitude and are notoriously unreliable with regard to altitude readings.

In addition they run on batteries, which may expire at an inconvenient moment and, like any technological device, can malfunction or break. If you do take a GPS with you into the field, you should still ensure that you have a navigational compass and a map with you and that you know how to use them. If working overseas, bear in mind that most compasses are calibrated for specific parts of the world (e.g. one of three compass, or balancing, zones – Magnetic North, Magnetic Equator and Magnetic South) and if used outside the zone for which it is designed, your compass will be inaccurate.

2.6 What if Something Goes Wrong?

If you take the precautions outlined in this chapter and carry out a full and careful assessment of risks before undertaking fieldwork the chances of any problems should be minimized. However, risks can never be eliminated altogether. Vehicle accidents and breakdowns happen, people trip over, the weather changes. If you do find yourself in an unfortunate situation try to keep calm, assess your options and get help if necessary. If you have insurance cover, consult this, depending on the type and severity of the incident or circumstance. Keeping calm is always easier said than done, but it is important that you try not to panic and not to induce panic in other people. Take some time to think about the situation, use your common sense and never place yourself or others in danger. If an incident or accident does happen that jeopardizes health and safety there are two key things that you, or members of your party, need to do. First, understand what has happened and, second, assess the severity of the situation.

When things start to go wrong try to understand what has happened as fully and quickly as possible. If you are lost on a hillside do you know where you took a wrong turning? If so, can you retrace your steps and get back on the right path? If your vehicle has broken down, do you know why, for example a flat tyre? Can you mend it safely? If someone is injured, do you know how the accident occurred? Was there a rockfall? Did they slide down a steep slope? This will help you to understand not only what their injuries might be, but also whether or not the danger is still present. For example an unstable cliff face may collapse as a series of multiple rockfalls. Look out for any dangers to yourself or to the casualty and never put yourself at risk. If necessary, can you treat the casualty using your first aid kit?

If you cannot resolve your predicament then you need to assess the situation as accurately as possible before you seek help. If you are dealing with a casualty or casualties that you cannot treat, try to determine what their injuries are. The emergency services will need to know how many casualties there are, whether they are conscious or unconscious and whether or not they are breathing and/or have a pulse. If the person has a known condition, such as asthma or a heart condition, and if you know how the incident occurred, for example immersion in water or a traffic collision, inform the emergency services of this.

You also need to know where the incident has occurred. Take a note of your grid reference and landmarks if you are in a remote area, note any road numbers or junction details and have an idea of how far away from the nearest habitation you are. If you have to leave the scene to get help be sure to write this information down before you set off. Emergency services can trace your telephone call to any call box or motorway telephone but if you have to leave the site or are using a mobile phone to make the call you should be extra vigilant about establishing your exact location. If you are lost in hilly terrain on foot and cannot retrace your steps, stop and consider the safest route off the hill or mountain. Use your compass to set a bearing in that direction. Heading towards buildings, a water course or a road is often a good option. If your vehicle has broken down in a remote area the best option is usually to stay with the vehicle until help arrives, especially in extreme weather conditions. Finally, if hazardous conditions such as poor visibility, unstable slopes, flooding, high tides or electricity have contributed to the accident or may affect the response from emergency services, then inform them of this when you call.

When the crisis is over, try to recall and write down what went wrong and why, including times and dates, if possible. You should give this information to the safety officer in your department as it may be useful for establishing procedures to prevent similar future incidences.

SUMMARY

- Fieldwork can be a risky business, but considering the variety of locations in which geographers undertake fieldwork and the amount of time they spend in the field, few incidents and accidents actually take place.
- Many of the precautions that are taken to protect health and safety in the field are common sense and you would take the same precautions whilst shopping at the weekend or going walking in the hills with friends – remember to apply this common sense when undertaking fieldwork in all situations (whether as part of a supervised group or as an unsupervised individual).
- Risk-assessment techniques can be used to anticipate likely fieldwork hazards. Identify any precautions that could minimize the risks and ensure that these are fully implemented.
- Occasionally there may be special procedures associated with particular equipment or environments that you need to learn and apply – seek advice from tutors and/or specialist advisory services.
- You have a responsibility to initiate and/or follow appropriate health and safety guidelines and to take all reasonable precautions to ensure the health and safety of yourself and others in the field.
- By following the above advice, you may not only learn new skills and take on new responsibilities, but, more importantly, you may also prevent any serious incidents or accidents occurring.
- Always consult health and safety information for your university, and specific information for your department within this. This should now be easily accessible from your institution's internal website.

Further Reading

There is not much written specifically for students about health and safety in the field but two useful sources are:

- Higgitt and Bullard (1999) which details why and how risk assessments are undertaken for undergraduate dissertations. Using two geography case studies, one human and one physical, the paper illustrates the types of hazards and risks that need to be considered.
- Nash (2000) which is the first part of a guide to doing independent overseas fieldwork. It discusses where to go, what to do when you get there, and also some of the health, safety and insurance issues that you should consider before setting off.

Note: Full details of the above can be found in the references list below.

Many government websites present comprehensive information about travelling overseas, including country-by-country guides to safety and security, local travel, entry requirements and health concerns. Some also include more general sections within the websites, such as travellers' tips, how to get your mobile phone to work overseas and what to do if it all goes wrong. These sites are updated regularly, particularly with regard to political disturbances and natural disasters. The sites for UK, US and Australian citizens are listed below. Other nationals should consult the travel section of their home-government website.

Foreign and Commonwealth Office website (UK): https://www.gov.uk/foreign-travel-advice

Center for Disease Control and Prevention (health advice) (USA): http://wwwnc.cdc.gov/travel/

United States Department of State (travel advice): http://travel.state.gov/

Australian Department for Foreign Affairs and Trade: http://www.smartraveller.gov.au/

Up-to-date local weather forecasts are usually available in newspapers, by telephone (check local papers for the number) or from websites, for example:

The Meteorological Office (UK): http://www.metoffice.gov.uk/

National Weather Service (USA): http://www.weather.gov

Bureau of Meteorology (Australia): http://www.bom.gov.au/

If you are planning an expedition and need advice or want to develop particular skills before you leave useful contacts include:

The Royal Geographical Society (with the Institute of British Geographers) (http://www.rgs.org/OurWork/Fieldwork+and+Expeditions/Fieldwork+Expeditions.htm) which runs workshops and seminars on topics such as four-wheel drive training, people-oriented research techniques, risk assessment and crisis management, and produces publications offering advice on logistics and safety in a range of environments from tropical forests to deserts;

The Explorers Club (USA) (http://www.explorers.org) which can provide expedition planning assistance and has a lecture series which occasionally features sessions on field techniques;

The International Mountaineering and Climbing Federation (http://www.theuiaa.org) and British Mountaineering Council (http://www.thebmc.co.uk) which provide advice on all aspects of high altitude travel and safety.

Fieldwork opportunities arise from a variety of sources including not only educational institutions but also through working with charities or taking part in expeditions. The British Standards Institution first published 'BS8848 Specification for the provision of visits, fieldwork, expeditions, and adventurous activities, outside the United Kingdom' in 2007 and revised it in 2014. This document sets out the requirements to be met by those organizing adventurous trips in order to comply with good practice. BS8848 (which takes its name from the metric height of Everest) is aimed primarily at 'providers', but the 2014 revisions specifically highlight the responsibilities of participants to commit to 'actively engage in:

- Taking reasonable care of themselves and others, including actions required of them arising from risk assessment;

- Following instructions from the leadership team;

- Bringing concerns about their own health, safety and well-being and those of others to the attention of the leadership team or supervisors;

- Complying with the code of conduct [as set out by the fieldtrip organizer].' (p. 13)

If you are organizing an expedition or fieldwork, there is a useful checklist of things to consider included in the appendices to BS8848, which is available from libraries or from the British Standards Institute (http://www.bsigroup.com/en-GB/).

References

BS 8848:2014 (2014) *Specification of the provision of visits, fieldwork, expeditions and adventurous activities outside the United Kingdom.* British Standards Institution, London.

Glass, M.R. (2015) 'Enhancing field research methods with mobile survey technology', *Journal of Geography in Higher Education*, 39 (2): 288–98.

Higgitt, D. and Bullard, J.E. (1999) 'Assessing fieldwork risk for undergraduate projects', *Journal of Geography in Higher Education*, 23: 441–9.

Kent, M., Gilbertson, D.D. and Hint, C.O. (1997) 'Fieldwork in geography teaching: A critical review of the literature and approaches', *Journal of Geography in Higher Education*, 21 (3): 313–32.

Medzini, A., Meishor-Tal, H. and Sneh, Y. (2015) 'Use of mobile technologies as support tools for geography field trips', *International Research in Geographical and Environmental Education*, 24 (1): 13–23.

Nash, D.J. (2000) 'Doing independent overseas fieldwork 1: Practicalities and pitfalls', *Journal of Geography in Higher Education*, 24: 139–49.

Ross, K. (2015) '"No Sir, she was not a fool in the field": Gendered risks and sexual violence in immersed cross-cultural fieldwork', *The Professional Geographer*, 67 (2): 180–6.

USHA-UCEA (2011) 'Guidance on Health and Safety in Fieldwork including Offsite Visits and Travel in the UK and Overseas.' Universities Safety and Health Association (USHA) in association with the Universities and Colleges Employers Association (UCEA). Available from http://www.ucea.ac.uk/en/publications/index.cfm/guidance-on-health-and-safety-in-fieldwork (accessed 8 December 2015).

ON THE COMPANION WEBSITE...

Visit **https://study.sagepub.com/keymethods3e** for links to the best resources relating to this chapter.

3 On Being Ethical in Geographical Research

Iain Hay

SYNOPSIS

Ethical research in geography is characterized by practitioners who behave with integrity and who act in ways that are just, beneficent and respectful. Ethical geographers are sensitive to the diversity of moral communities within which they work and are ultimately responsible for the moral significance of their deeds. This chapter explains the importance of behaving ethically, provides some key advice on the conduct of ethical research and provides some examples of ethical dilemmas.

This chapter is organized into the following sections:

- Introduction
- Why behave ethically?
- Principles of ethical behaviour and common ethical issues
- Truth or consequences? Teleological and deontological approaches to dealing with ethical dilemmas in your research
- Conclusion

3.1 Introduction

To behave ethically in geographical research requires that you and I act in accordance with notions of right and wrong – that we conduct ourselves morally (Mitchell and Draper, 1982).[1] Ethical research is carried out by thoughtful, informed and reflexive geographers who act honourably because it is the 'right' thing to do, not just because someone is making them do it (see, for example, Cloke, 2002; Dowling, 2016).[2]

This chapter seeks to heighten your awareness of the reasons for, and principles underpinning, ethical research as well as to provide some guidance on ways of dealing with those ethical dilemmas you might encounter in your work. Although I am conscious of the need to avoid ethical prescription, leaving you the opportunity to exercise and act on your own 'moral imagination' (Hay, 1998b), it is important to set out a range of specific ethical matters that colleagues and communities will commonly expect you to consider when preparing and conducting research. As a geographer, you must consider carefully the ethical significance of your actions in those contexts within which they have meaning and be prepared to take responsibility for your actions. In some instances (e.g. cross-cultural research), ethically reflexive practice includes acknowledging and working with (negotiating) different groups' ethical expectations in ways that yield satisfactory approaches for all parties involved (see Howitt and Stevens, 2016; Johnson and Madge, 2016; and Chapter 6 of this volume for more).

The first part of the chapter introduces a few of the reasons why geographers need to behave ethically. I follow this with a discussion of some fundamental principles underpinning ethical behaviour and offer a few points of guidance that might assist you in your practice as an ethically reflexive geographer. As I note, however, no matter how well you – or even an ethics committee – might try to anticipate the issues that might arise in your work, you are still likely to encounter difficult dilemmas. In recognition of this, the chapter sets out a strategy that might help you resolve those dilemmas you confront. I conclude with some suggestions on ways in which you might continue to develop as an ethical geographer. A number of real ethical cases are included in the chapter to illustrate points made and to provide some material for you to discuss.

In starting, there are two cautions I would like to offer. First, decisions about ethical practice are made in specific contexts. While all people and places deserve to be treated with integrity, justice and respect (see Smith, 2000a), ethical behaviour requires sensitivity to the expectations of people from diverse moral communities and acknowledgement of the webs of physical and social relationships within which the work is conducted (see Chapter 6). For these reasons, unbending rules for ethical practice cannot be prescribed, particularly in work involving people (Hay, 1998a; Hay, 1998b; Hay and Israel, 2005; Israel and Hay, 2006; Israel, 2015). And as you will see from the real quandaries encountered by geographers that are discussed as case studies in this chapter, the 'correct' resolution of most ethical dilemmas cannot be dictated. Second, simply because your peers, colleagues and institution (e.g. fellow students, professional geographers, university ethics committee) say that some behaviour or practice is ethical does not necessarily mean it is always so. That determination is ultimately up to you, your conscience and the people with whom you are working. In the end, you must take responsibility for the decisions you make.

3.2 Why Behave Ethically?

Aside from any moral arguments that as human beings we should always act ethically, there are important practical arguments for geographical researchers to behave ethically. These fall into three main categories.

First, ethical behaviour protects the rights of individuals, communities and environments involved in, or affected by, our research. As social and physical scientists interested in helping to 'make the world a better place', we should avoid (or at least minimize) doing harm.

Second, and perhaps a little more self-interestedly, ethical behaviour helps assure a favourable climate for the continued conduct of scientific inquiry. For example, Walsh (1992: 86) noted in an early discussion of ethical issues in Pacific Island research that incautious practice and a lack of cultural awareness led to community denial of research privileges. More recently, Indigenous peoples and communities around the globe are responding to problematic research behaviours by developing their own research authorization and conduct protocols (Howitt and Stevens, 2016: 68). By behaving ethically, we maintain public trust. From that position of trust we may be able to continue research and to do so without causing suspicion or fear amongst those people who are our hosts (see Case Study 3.1). Moreover, our ethical conduct supports trust within research communities. Not only is it important that we feel sure we can depend on the integrity of colleagues' work but also trustworthy work helps ensure the continuing support of agencies upon whom we depend to fund our research.

Case Study 3.1   Waving, not drowning

While he was conducting surveys of flood hazard perception in the USA, Bob Kates found that in rare cases his questions raised anxieties and fears in the people to whom he was speaking. Even the actions of the research team in measuring street elevations to calculate flood risk created rumours that led some people to believe that their homes were to be taken for highway expansion (Kates, 1994: 2).

For discussion
Is it Kates' responsibility to quash the rumours and suspicions his research engenders? Justify your answer.

Third, growing public demands for accountability and the sentiment that institutions such as universities must protect themselves legally from the unethical or immoral actions of a student or employee mean there is greater emphasis on acting ethically than ever before (see, for example, Israel, 2015: 45–78).

Clearly, then, there are compelling moral and practical reasons for conducting research ethically. How are these brought to life?

3.3 Principles of Ethical Behaviour and Common Ethical Issues

Around the world a growing number of organizations supporting research have established committees to scrutinize research proposals to ensure that work is conducted ethically (Israel, 2015). Because these committees regularly consider the possible ethical implications of research, their workings and the principles behind their operation might provide you with a useful starting point to ensure that your own research is ethical.

In general, the central guiding value for these committees is integrity, meaning a commitment to the honest conduct of research and to the communication of results. However, committees usually emphasize the desirability of three fundamental principles,[3] as set out in Box 3.1. These principles lead to a set of core questions (right-hand column) to guide your personal consideration of ethical matters. You will find it helpful always to consider these simple questions when reflecting on your work. They provide the beginnings of ethical practice. For instance, if you can sustain the argument that your work is just, is doing good and you are demonstrating respect for others, you are probably well on the way to conducting an ethical piece of work. If, by comparison, you believe that your work is just, yet you are not showing respect for others or you are doing harm, you may have an ethical dilemma on your hands. I set out a strategy for dealing with such situations later in this chapter.

Box 3.1   Principles of ethical behaviour

Justice: this gives emphasis to the distribution of benefits and burdens.

Is this just?

Beneficence/non-maleficence: respectively, these mean 'doing good' and 'avoiding harm'.
Our work should maximize benefits and minimize physical, emotional, economic and environmental harms and discomfort.

Am I doing harm?

Am I doing good?

Respect: individuals should be regarded as autonomous agents and anyone of diminished autonomy (e.g. intellectually disabled) should be particularly protected. It is important to have consideration for the welfare, beliefs, rights, heritage and customs of people involved in research. Of course, respect should also extend to consideration of any discomfort, trauma or transformation affecting organisms or environments involved in the research.

Am I showing respect?

The principles and questions set out in Box 3.1 are a good general framework but if you are just beginning to engage in research, some more specific advice on how to determine whether your research might be understood to be ethical will be helpful (see Case Study 3.2).

Typically, ethics committees give detailed consideration to five major issues when evaluating research proposals. These issues are set out in Box 3.2. I have attached to each of these a set of 'prompts' for moral contemplation. As you can see, the 'prompts' offer no specific direction about the actions you should or should not take in any or all situations. To do so would be virtually impossible given the enormous variability of social and geographic research (see Bosk and de Vries, 2004: 260), our (global) woven togetherness (see for instance Popke, 2007), and the associated need for flexible research practices. It would also deny the need to negotiate specific issues with participants.[4]

When you are thinking about the issues in Box 3.2, consider them in terms of the value of integrity and the principles set out earlier – justice, beneficence and respect for others. For example, let us think about the prompt 'time for consideration of the study before consent is provided'. A human geographer might be expected to allow people more time to consider their involvement in a complex, long-term observational study than for a two-minute interview about their grocery shopping behaviour in a retail mall. In another example concerning harm minimization, a physical geographer might think about, and act on, the potential harm caused to nesting birds as a result of fieldwork involving weed clearance. Indeed, there could be no need to conduct the work if similar studies have been conducted previously.[5]

Case Study 3.2   'They did that last week'

University student Ali bin Ahmed bin Saleh Al-Fulani carefully prepares a questionnaire survey for distribution to two groups of 16-year-old students in 'home groups' at two local high schools. The survey is central to the comparative work Ali is conducting as part of his thesis. In compliance with government regulations, Ali secures permission from the students' parents to conduct the survey. He also gets permission for his work from the university's Ethics Committee. The Ethics Committee requires him to include a covering letter to students which states that their participation in the study is voluntary and that no one is obliged to answer any of the questions asked. A few weeks before he intends to administer the questionnaire survey, Ali leaves near-final drafts of it with the students' teachers for comment. The draft copy of the questionnaire does not include the covering letter. It is Ali's intention to revise the questionnaire in the light of each teacher's comments and then return to the schools to administer the questionnaire during 'home group' meeting times. About a week after he leaves the survey forms with the teachers Ali calls them to find out if they have had an opportunity to comment on the questionnaire. The first teacher has just returned the questionnaire – with no amendments – by post. However, Ali finds that the second teacher had already made multiple copies of the forms and had administered the questionnaire to her student 'home group'. She asks Ali to come along to collect the completed forms. Ali scuttles off to the school immediately. He finds that the questionnaires had been completed fully by every student present in the home group. Only one student from the class of 30 had been absent so the response rate was 97 per cent – an extraordinarily high rate. Ali feels he cannot ask the teacher to readminister the survey because she has already indicated several times that she is tired of his requests for assistance and access to the class.

For discussion

It would appear from the circumstances and from the very high response rate that students were not free to refuse to participate in the study. Is it ethical for Ali to use results that have been acquired without free and informed consent? Would your view change if the survey had dealt with some sensitive issue such as sexual assault or if the results had been acquired, without consent, through the use of physical force?

Box 3.2   Prompts for contemplation and action

Before and during your research, have you considered the following?

Consent

- Amount of information provided to participants on matters such as purposes, methods, other participants, sponsors, demands, risks, time involved, discomforts, inconveniences and potential consequences
- Accessibility and comprehensibility to prospective participants of information upon which consent decisions are made
- Time for consideration of the study before consent is provided
- Caution in research requiring deceit[6]
- Caution in obtaining consent from people in dependent relationships
- Recording of consent
- Informed consent issues for others working on the project

Confidentiality

- Disclosing identity of participants in the course of research and in the release of results
- Consent and the collection of private or confidential information
- Relationships between relevant privacy laws and any assurances made by researchers of confidentiality or privacy
- Data storage during and after research (for example, field notes, completed surveys and recorded interviews)

Harm

- Potential physical, psychological, cultural, social, financial, legal and environmentally harmful effects of the study or its results
- Extent to which similar studies have been performed previously
- Issues of harm for dependent populations
- Relationship between the risks involved and the potential advantage of the work
- Opportunities for participants to withdraw from the research after it has commenced
- Competence of researchers and appropriateness of facilities
- Representations of results

Cultural awareness

- Personality, rights, wishes, beliefs and ethical views of the individual subjects and communities of which they are a part

Dissemination of results and feedback to participants

- Availability and comprehensibility of results to participants
- Potential (mis)interpretations of the results
- Potential (mis)uses of results
- 'Ownership' of results
- Sponsorship
- Debriefing
- Authorship[7]

3.4 Truth or Consequences? Teleological and Deontological Approaches to Dealing with Ethical Dilemmas in Your Research

As scholars such as Price (2012) and Ritterbusch (2012) make clear, consideration of human subjects' protection cannot be confined to your conversations with ethics committees. No matter how well prepared you are, no matter how thoroughly you have prepared your research project, and no matter how properly you behave, it is likely that in your geographical research you will have to deal with a variety of unanticipated ethical dilemmas (see, for example, Case Studies 3.3 and 3.4).

Case Study 3.3   You are being uncooperative

Catriona McDonald has recently completed an undergraduate research project that involved sending out a confidential questionnaire to members of a non-government welfare organization called ANZAC Helpers. Members of this organization are mainly World War II veterans aged over 70 years, and a good deal of their volunteer work requires them to drive around familiar parts of a large metropolitan area. Catriona works part-time for the organization and feels that her employment tenure is somewhat precarious. The research project has been completed and assessed formally by Catriona's professor and she is planning to present a modified report to ANZAC Helpers. In the work car-park one afternoon, Catriona meets Mr Montgomery Smythe, one of the organization's members and a man known to Catriona to be something of a trouble-maker. After some small talk, Smythe asks what the study results are. Catriona outlines some of the findings, such as the percentage of the membership who are having trouble performing their voluntary duties for the organization due to old age and ill-health. Smythe then asks the student to tell him the names of those members who are having difficulties with their duties. It is possible that he could use the information to help encourage the implementation of strategies to help those members experiencing difficulties. It is also possible that he could campaign to have the same members redirected to less demanding volunteer roles that many of them are likely to find less fulfilling.

For discussion
1. Should Catriona give Montgomery the information he wants?
2. Given that some of the members of ANZAC Helpers might actually be putting their lives at risk by driving around the city, should Catriona disclose the names of those people she has discovered to be experiencing sight and hearing problems, for example, to the organization?

Case Study 3.4   The power of maps

Dr Tina Kong has recently commenced work as a post-doctoral fellow with a research organization applying GIS (geographical information systems) to illustrate and resolve significant social problems. This is a position in which full-time research is possible. It is also a position that can allow someone to begin to forge the beginnings of a noteworthy academic career for themselves. However, much of that promise depends on producing good results and having them published in reputable journals. Tina decides to conduct work on environmental carcinogens (cancer-producing substances) in a major metropolitan area. She spends about two months of her two-year fellowship conducting some background research to assess the need for, and utility of, the work. After this early research, Tina resolves to use GIS to produce maps which will illustrate clearly those areas in which high levels of carcinogenic materials are likely to be found. At a meeting of interested parties to discuss the proposed research, one of the participants makes the observation that, if broadcast, the results of the study may cause considerable public alarm. For example, there may be widespread individual and institutional concern about public health and welfare; property values in areas with high levels of carcinogenic material may be adversely affected; past and present producers of carcinogenic pollutants may be exposed to liability suits; and local government authorities might react poorly to claims that there are toxic materials in their areas. Tina is cautioned by senior members of the research organization to proceed cautiously, if at all.

For discussion
Should Tina proceed with the project? Justify your answer. Should Tina 'cut her losses' and move into a less controversial area that might be supported by her colleagues?

How will you deal with such situations? To answer this fully, we will have to make a short excursion into the work of philosophers. But, first, I should point out that to resolve any dilemma you encounter it is likely that you will have to 'violate' one of the three ethical principles – justice, beneficence or respect for others – set out earlier in this chapter. That is why it is a dilemma! But you will probably feel more confident about the difficult decision you will have to make if it is well considered and informed by a basic appreciation of two key normative approaches to behaviour.

Box 3.3   Steps to resolving an ethical dilemma

How do you decide what to do if you are presented with an ethical dilemma? There are two major approaches that you might draw from. One focuses on the practical consequences of your actions (*teleological* or consequentialist approach) and might be summed up brutally in the phrase 'no harm, no foul'. In contrast, the *deontological* approach would lead you to ask whether an action is, in itself, right. For example, does an action uphold a promise or demonstrate loyalty? The essence of deontological approaches can be said to be captured by the phrase 'let justice be done though the heavens fall' (e.g. Quinton, 1988: 216). These two positions serve as useful starting points for a strategy for coping with ethical dilemmas.

1 What are the options?

List the full range of alternative courses of action available to you.

2 Consider the consequences

Think carefully about the range of positive and negative consequences associated with each of the different paths of action before you:

- Who/what will be helped by what you do?
- Who/what will be hurt?
- What kinds of benefits and harms are involved and what are their relative values? Some things (e.g. healthy bodies and beaches) are more valuable than others (e.g. new cars). Some harms (e.g. a violation of trust) are more significant than others (e.g. lying in a public meeting to protect a seal colony).
- What are the short- and long-term implications?

Now, on the basis of your answers to these questions, which of your options produces the best combination of benefits maximization and harm minimization?

3 Analyse the options in terms of moral principles

You now have to consider each of your options from a completely different perspective. Disregard the consequences, concentrating instead on the actions and looking for that option which seems problematic. How do the options measure up against such moral principles as honesty, fairness, equality and recognition of social and environmental vulnerability? In the case you are considering, is there a way to see one principle as more important than the others?

(Continued)

(Continued)

4   Make your decision and act with commitment

Now, bring together both parts of your analysis and make *your* informed decision. Act on your decision and assume responsibility for it. Be prepared to justify *your* choice of action. No one else is responsible for this action but you.

5   Evaluate the system and your own actions

Think about the circumstances which led to the dilemma with the intention of identifying and removing the conditions that allowed it to arise. Don't forget to think about your own behaviours and any part they may have played.

Source: Adapted from Stark-Adamec and Pettifor (1995); Israel and Hay (2012)

Many philosophers suggest that two categories, teleological and deontological, exhaust the possible range of theories of right action (Davis, 1993). In summary, the former sees acts judged as ethical or otherwise on the basis of the consequences of those acts. The latter suggests, perhaps surprisingly, that what is 'right' is not necessarily good.

In the terms of teleology – also known as consequentialism – an action is morally right if it produces more good than evil (Israel, 2015). It might therefore be appropriate to violate and make public the habitat of a rare animal which you were shown in confidence if doing so prevents the construction of a bridge through its habitat, because disclosure yields a greater benefit than it 'costs'.

Deontological approaches reject this emphasis on consequences, suggesting that the balance of good over evil is insufficient to determine whether some behaviour is ethical. Instead, certain acts are seen as good in themselves and must be viewed as morally correct because, for example, they keep a promise, show gratitude or demonstrate loyalty to an unconditional command (Kimmel, 1988). It is possible, therefore, for something to be ethically correct even if it does not promote the greatest balance of good over evil. To illustrate the point: if we return to the example of the researcher made aware of the location of a rare animal's habitat, a deontological view might require that researcher to maintain the trust with which he or she had been privileged, even if non-disclosure meant that construction of the bridge would destroy that habitat.

Thus, we have two philosophical approaches that can point to potentially contradictory ethical ways of responding to a particular situation. This is not as debilitating as you might think. As Box 3.3 sets out, you can draw from these approaches to ensure that your response to an ethical dilemma is at least well considered, informed and defensible.

You can really give this scheme a workout by considering the (in)famous work of Laud Humphreys, set out as Case Study 3.5.

Case Study 3.5   The 'watch queen' in the 'tea room'

In 1966–7, and as part of a study of homosexual behaviours in public spaces, Laud Humphreys acted as voyeuristic 'lookout' or 'watch queen' in public toilets ('tea rooms'). As a 'watch queen' he observed homosexual acts, sometimes warning men engaged in those acts of the presence of intruders. In the course of his observations Humphreys recorded the car licence plate numbers of the men who visited the 'tea room'. He subsequently learnt their names and addresses by presenting himself as a market researcher and requesting information from 'friendly policemen' (Humphreys, 1970: 38). One year later, Humphreys had changed his appearance, dress and car and got a job as a member of a public health survey team. In that capacity he interviewed the homosexual men he had observed, pretending that they had been selected randomly for the health study. This latter deception was necessary to avoid the problems associated with the fact that most of the sampled population were married and secretive about their homosexual activity (Humphreys, 1970: 41). After the study, Humphreys destroyed the names and addresses of the men he had interviewed in order to protect their anonymity. His study was subsequently published as a major work on human sexual behaviour (Humphreys, 1970).[8]

For discussion

1. It might be said that Humphreys' research in 'tea rooms' was a form of participant observation, a type of research which is often most successful when 'subjects' do not know they are being observed. Was it unethical for Humphreys to observe men engaged in homosexual acts in the 'tea room'? Does the fact that the behaviour was occurring in a public place make a difference to your argument? Why?
2. Was it ethical for Humphreys to seek and use name and address information – details that appear commonly in telephone books – from police officers who should not have released those details for non-official reasons? Would it have been acceptable if he had been able to acquire that same information without deceit?
3. Upon completion of the research should Humphreys have advised those men who had been observed and interviewed that they had been used for the study? Why? Discuss the significance of your answer. Should only some research results be 'returned' to participants and not others? What criteria might one employ to make that determination? Why are those criteria important/more important than others?
4. Should Humphreys have destroyed the name and address information he used? How do we know he was not making the whole story up? How can someone else replicate or corroborate his findings without that information?
5. Humphreys' work offered a major social scientific insight into male homosexual behaviours. It might be argued that his book *Tearoom Trade* contributed to growing public understanding of one group in the broader community. Moreover, no apparent harm was done to those people whose behaviour was observed. Do you think, then, that the ends may have justified the means?

3.5 Conclusion

Being an ethical geographer is important. It helps to protect those people, places and organisms affected by our research and helps to ensure that we are able to continue to conduct socially and environmentally valuable work. The steps set out in this chapter to help you prepare for research and to deal with the ethical problems you may encounter should go some way to helping achieve these ends. However, your development as an ethically responsible geographer cannot stop with this chapter. It is important that you continue to heighten your awareness of ethical issues and develop your ability to act thoughtfully when confronted with dilemmas like those set out in this chapter. To that end, I shall conclude with some thoughts on ways in which you can continue to become a more ethical geographer.

Good luck!

SUMMARY

What can you do to become a more ethical geographer?

- *Make sure your 'moral imagination' is active and engaged.* There will always be ethical issues in your research. Make ethics as normal a part of your research project discussions as how the stream gauge or questionnaire is working. Discuss ethical issues and possibilities with your colleagues. Read journals like *Ethics, Policy and Environment*. Learn to recognize ethical issues in context. Think about the potential moral significance of your own actions and those of other people. Remember that we live in a vast network of moral relationships (Appiah, 2007). The meanings of particular behaviours and moral positions may sometimes be given or understood far from the places they might be expected (Smith, 1998) and interpreted from different ethical standpoints. Look for hidden value biases, moral logic and conflicting moral obligations. Make yourself aware of (local) ethical practices (Mehlinger, 1986).
- *Develop your philosophical and analytical skills.* What is 'right' or 'good'? On what bases are those decisions made? Be prepared to think hard about difficult questions. For example, how can you evaluate prescriptive moral statements such as 'endangered species should (or should not) be protected' or 'research should (or should not) be conducted with the consent of all participants'?
- *Heighten your sense of moral obligation and personal responsibility.* 'Why should I be moral?' or 'Why should I think about ethics?' Embrace ethical thought and action as an element of your professional and social identity as a geographer. Come to terms with the idea that you need to act morally because it is the 'right' thing to do, not just because someone is making you do it (Mehlinger, 1986).
- *Expect – but do not passively accept – disagreement and ambiguity.* Ethical problems are almost inevitably associated with disagreements and ambiguities. However, do not let that expectation of ambiguity and disagreement provide the justification for abandoning debate and critical thought. Learn to seek out the core of differences to see if disagreement might be reduced. Be committed enough to follow through on your own decisions.

Notes

1   Although most of the principles discussed in this chapter can be applied to environmental research ethics, the chapter focuses most heavily on research involving humans. Readers especially interested in environmental research ethics are strongly advised to consult ASTEC (1998) which remains one of the few published statements on environmental research ethics.

2   Nevertheless, geographers and geography students – as well as other social scientists – increasingly have to take account of the formal codes of ethical practice drawn up by departments, universities and research funding councils (see Israel and Hay, 2006). When planning your own research you should always consult your department's or school's own ethical policy and guidelines.

3   These principles place a strong emphasis on individual autonomy. It is important to note that in some societies and situations (e.g. work with some indigenous groups and children) individual autonomy may be limited and influenced by related groups and other individuals who have authority over that individual (NHMRC et al., 2007). *It is imperative, therefore, to consider the specific local contexts within which rights are understood.*

4   Ethical dimensions of a research project should be negotiated between the researcher and the participant(s) to ensure that the work satisfies the moral and practical needs and expectations of those individuals and communities involved. However, the negotiations will be influenced by participants' different social and geographical positions and different levels of power between them. In some situations a researcher will have more power than the informant (e.g. research work involving children) while in others (e.g. an interview with the CEO of a large organization whose comments are critical to your study) informants may have more power. (For examples of these important issues and ways of dealing with them, see Chouinard, 2000; Cloke et al., 2000; England, 2002; Harvey, 2010; Kezar, 2003; Dowling, 2016; Smith, 2006).

5   For detailed discussions of some of the ethical responsibilities of researchers involved in environmental fieldwork, see ASTEC (1998: esp. pp. 21–4) and Smith (2002).

6   Wilton (2000) and Routledge (2002) provide fascinating and thought provoking reflections on the ethical dilemmas of geographical work involving deception.

7   Kearns et al. (1996) offer a concise and very helpful discussion of the ethics of authorship/ownership of knowledge in geography. In other fields the International Committee of Medical Journal Authors (2015) has formulated a comprehensive statement on writing and editing for biomedical journals. Used with caution, this offers some very helpful suggestions for practice in geography.

8   In a postscript to his book, Humphreys (1970) convincingly addresses issues of misrepresentation, confidentiality, consequentiality and situation ethics associated with his research. His vocation as a religious minister might add, for some, strength to his arguments. For other discussion of this example, Diener and Crandall (1978) is helpful.

Further Reading

A very good place to start is Iain Hay's (2013) comprehensive review of geography and ethics. This offers an overview of practical and philosophical engagements between the two areas, discussing useful reference resources, relevant journals, as well as topics such as social and spatial justice, geographies of care, postcolonial and cosmopolitan ethics, and professional and research ethics. It is especially useful if you have an interest in exploring ethical issues broader than those covered in this chapter.

- Dowling (2016) provides a helpful introduction to some central ethical issues in geographical research (e.g. harm, consent) and makes a case for critical reflexivity (i.e. ongoing, self-conscious scrutiny of oneself as a researcher and of the research process).

- Israel and Hay (2006) and Israel (2015) elaborate fully and clearly on some of the ideas set out in this chapter and Hay and Foley (1998) offer many other examples of real ethical cases as well as a carefully thought-out strategy for learning-and-teaching about ethical conduct in geography.

- Richie Howitt's (2005) paper on ethics and cross-cultural engagement in work with Australian Indigenous people is very helpful. Another challenging but deeply engaging chapter examining issues in cross-cultural research ethics and relationships is that by Howitt and Stevens (2016) – who have vast experience working with Indigenous peoples in Australia and Chomolungma (Mt Everest) respectively.

- Mitchell and Draper (1982) is a classic reference for geographers interested in issues of relevance and research ethics. Though dated, this volume is well worth a read.

- Scheyvens and Leslie (2000) is a helpful article exploring ethical dimensions of power, gender and representation in overseas fieldwork. It is illustrated by several examples drawn from the authors' fieldwork practice.

- Daniel A. Griffith's (2008) paper on ethical considerations in geographical research provides a number of interesting examples of academic misconduct in geography and more broadly. These are used as a foundation for a discussion on the need for ethical education as part of (post)graduate education.

- Smith (2000b) explores the interface between geography, ethics and morality. At its core is Smith's longstanding concern with the practice of morality.

Note: Full details of the above can be found in the references list below.

References

Appiah, K.A. (2007) *Cosmopolitanism: Ethics in a World of Strangers*. London: Penguin.

ASTEC (Australian Science, Technology and Engineering Council) (1998) *Environmental Research Ethics*. Canberra: ASTEC.

Bosk, C.L. and de Vries, R.G. (2004) 'Bureaucracies of mass deception: Institutional Review Boards and the ethics of ethnographic research', *Annals of the American Association of Political and Social Science*, 595: 249–63.

Chouinard, V. (2000) 'Getting ethical: For inclusive and engaged geographies of disability', *Ethics, Place and Environment*, 3: 70–80.

Cloke, P. (2002) 'Deliver us from evil? Prospects for living ethically and acting politically in human geography', *Progress in Human Geography*, 26: 587–604.

Cloke, P., Cooke, P., Cursons, J., Milbourne, P. and Widdowfield, R. (2000) 'Ethics, reflexivity and research: Encounters with homeless people', *Ethics, Place and Environment*, 3: 133–54.

Davis, N.A. (1993) 'Contemporary deontology', in P. Singer (ed.) *A Companion to Ethics*. Oxford: Blackwell. pp. 205–18.

Diener, E. and Crandall, R. (1978*) Ethics and Values in Social and Behavioural Research*. Chicago: University of Chicago Press.

Dowling, R. (2016) 'Power, subjectivity and ethics in qualitative research', in I. Hay (ed.) *Qualitative Research Methods in Human Geography* (4th edition). Toronto: Oxford University Press. pp. 29–44.

England, K. (2002) 'Interviewing elites: Cautionary tales about researching women managers in Canada's banking industry', in P. Moss (ed.) *Feminist Geography in Practice: Research and Methods*. Oxford: Blackwell. pp. 200–13.

Griffith, D.A. (2008) 'Ethical considerations in geographic research: What especially graduate students need to know', *Ethics, Place and Environment*, 11: 237–52.

Harvey, W.S. (2010) 'Methodological approaches for interviewing elites', *Geography Compass*, 4: 193–205.

Hay, I. (1998a) 'From code to conduct: Professional ethics in New Zealand geography', *New Zealand Geographer*, 54: 21–27.

Hay, I. (1998b) 'Making moral imaginations: Research ethics, pedagogy and professional human geography', *Ethics, Place and Environment*, 1: 55–76.

Hay, I. (2013) 'Geography and ethics', in B. Warf (ed.) *Oxford Bibliographies in Geography*. New York: Oxford University Press. Available from http://www.oxfordbibliographies.com/view/document/obo-9780199874002/obo-9780199874002-0093.xml (accessed 5 November 2015).

Hay, I. and Foley, P. (1998) 'Ethics, geography and responsible citizenship', *Journal of Geography in Higher Education*, 22: 169–183.

Hay, I. and Israel, M. (2005) 'A case for ethics (not conformity)', in G.A. Goodwin and M.D. Schwartz (eds) *Professing Humanist Sociology* (5th edition). Washington, DC: American Sociological Association. pp. 26–31.

Howitt, R. (2005) 'The importance of process in Social Impact Assessment: ethics, methods and process for cross-cultural engagement', *Ethics, Place and Environment*, 8: 209–21.

Howitt, R. and Stevens, S. (2016) 'Cross-cultural research: Ethics, methods and relationships', in I. Hay (ed.) *Qualitative Research Methods in Human Geography,* 4th edition. Toronto: Oxford University Press. pp. 45–75.

Humphreys, L. (1970) *Tearoom Trade: A Study of Homosexual Encounters in Public Places*. London: Duckworth.

International Committee of Medical Journal Authors (2015) *Uniform Requirements for Manuscripts Submitted to Biomedical Journals: Writing and Editing for Biomedical Journals*: available at: www.icmje.org (accessed 5 November 2015).

Israel, M. (2015) *Research Ethics and Integrity for Social Scientists* (2nd edition). London: Sage.

Israel, M. and Hay, I. (2006) *Research Ethics for Social Scientists: Between Ethical Conduct and Regulatory Compliance*. London: Sage.

Israel, M. and Hay, I. (2012) 'Research ethics in criminology', in D. Gadd, S. Karsted and S. Messner (eds) *Sage Handbook of Criminological Research Methods*. London: Sage. pp. 500–14.

Johnson, J.T. and Madge, C. (2016) 'Empowering methodologies: Feminist and Indigenous approaches', in I. Hay (ed.) *Qualitative Research Methods in Human Geography* (4th edition). Toronto: Oxford University Press. pp. 76–94.

Kates, B. (1994) 'President's column', *Association of American Geographers' Newsletter*, 29: 1–2.

Kearns, R., Arnold, G., Laituri, M. and Le Heron, R. (1996) 'Exploring the politics of geographical authorship', *Area*, 28: 414–20.

Kezar, A. (2003) 'Transformational elite interviews: Principles and problems', *Qualitative Inquiry*, 9: 395–415.

Kimmel, A.J. (1988) *Ethics and Values in Applied Social Research*. London: Sage.

Mehlinger, H. (1986) 'The nature of moral education in the contemporary world', in M.J. Frazer and A. Kornhauser (eds) *Ethics and Responsibility in Science Education*. Oxford: ICSU Press. pp. 17–30.

Metzel, D. (2000) 'Research with the mentally incompetent: The dilemma of informed consent', *Ethics, Place and Environment*, 3: 87–90.

Mitchell, B. and Draper, D. (1982) *Relevance and Ethics in Geography*. London: Longman.

NHMRC (Australian National Health and Medical Research Council), Australian Research Council (ARC) and Australian Vice-Chancellors Committee (AVCC) (2007) National Statement on Ethical Conduct in Human Research (2007) – updated May 2015: available at www.nhmrc.gov.au/publications/synopses/e72syn.htm (accessed 5 November 2015).

Popke, J. (2007) 'Geography and ethics: Spaces of cosmopolitan responsibility', *Progress in Human Geography*, 31: 509–518.

Price, P.L. (2012) 'Introduction: Protecting human subjects across the geographic research process', *The Professional Geographer*, 64: 1–6.

Proctor, J. and Smith, D.M. (eds) (1999) *Geography and Ethics: Journeys in a Moral Terrain*. London: Routledge.

Quinton, A. (1988) 'Deontology', in A. Bullock, O. Stallybrass and S. Trombley (eds) *The Fontana Dictionary of Modern Thought,* 2nd edition. Glasgow: Fontana. p. 216.

Ritterbusch, A. (2012) 'Bridging guidelines and practice: towards a grounded care ethics in youth participatory action research', *The Professional Geographer*, 64: 16–14.

Routledge, P. (2002) 'Travelling east as Walter Kurtz: Identity, performance and collaboration in Goa, India', *Environment and Planning D: Society and Space*, 20: 477–98.

Scheyvens, R. and Leslie, H. (2000) 'Gender, ethics and empowerment: Dilemmas of development fieldwork', *Women's Studies International Forum*, 23: 119–30.

Smith, D.M. (1998) 'How far should we care? On the spatial scope of beneficence', *Progress in Human Geography*, 22: 15–38.

Smith, D.M. (2000a) 'Moral progress in human geography: Transcending the place of good fortune', *Progress in Human Geography*, 24: 1–18.

Smith, D.M. (2000b) *Moral Geographies: Ethics in a World of* Difference. Edinburgh: Edinburgh University Press.

Smith, K.E. (2006) 'Problematising power relations in "elite" interviews', *Geoforum*, 37: 643–653.

Smith, M. (ed.) (2002) *Environmental Responsibilities for Expeditions: A Guide to Good Practice* (2nd edition) available from http://www.rgs.org/NR/rdonlyres/2AA7AFB6-2C92-4987-8098-C5AE94091DFA/0/YETEnvResp.pdf (accessed 5 November 2015).

Stark-Adamec, C. and Pettifor, J. (1995) *Ethical Decision Making for Practising Social Scientists: Putting Values into Practice*. Ottawa: Social Science Federation of Canada.

Walsh, A.C. (1992) 'Ethical matters in Pacific Island research', *New Zealand Geographer*, 48: 86.

Wilton, R.D. (2000) '"Sometimes it's OK to be a spy": Ethics and politics in geographies of disability', *Ethics, Place and Environment*, 3: 91–7.

ON THE COMPANION WEBSITE...

Visit **https://study.sagepub.com/keymethods3e** for links to the best resources relating to this chapter.

4 How to Conduct a Literature Search

Mick Healey and Ruth L. Healey

SYNOPSIS

Identifying the most relevant, up-to-date and reliable references is a critical stage in the preparation of a whole range of assessments at university including essays, reports and dissertations, but it is a stage which is often undertaken unsystematically and in a hurry. This chapter is designed to help you improve the quality of your literature search. There is a growing interest in higher education in students undertaking research and inquiry projects and co-inquiring with academics not just in their final year but throughout their undergraduate studies (Healey and Jenkins, 2009; Healey et al., 2013, 2014a, 2014b). Undertaking a thorough literature search is a key element in undertaking a research or inquiry project.

This chapter is organized into the following sections:

- The purpose of searching the literature
- Making a start
- A framework for your search
- Managing your search
- Search tools
- Evaluating the literature

4.1 The Purpose of Searching the Literature

The purpose of this chapter is to support you in developing and using your literature search skills over a range of media, including paper and the web, not just books and journals. It is aimed primarily at undergraduate geography students needing to search the literature for research projects, dissertations and essays in human and physical geography. However, the search methods and principles are applicable to most subjects and, if you are a postgraduate geography student, you should also find it a useful refresher to help you get started. Many sources are available worldwide, though details of accessing arrangements may vary. Country-specific sources are illustrated with selected examples of those available in the UK, North America and Australia.

Information literacy is a critical skill that all students need to work on during their degrees. As Badke notes:

Academia is all about a profound discontent, about a quest to discover more, about a burning desire to solve society's problems and make a better world. Research is at the heart of this academic yearning, and our students need to be able to do it well, way beyond the uneven vagaries of a Google search. (2013: 67)

Unfortunately the perception that students, through the use of technology, are masters at information search and retrieval is a myth (Badke, 2015). For example, the ERIAL project which undertook detailed surveys of 161 students in higher education, 75 academics, and 48 librarians in the Chicago area concluded that:

Almost without exception, students exhibited a lack of understanding of search logic, how to build a search to narrow/expand results, how to use subject headings, and how various search engines (including Google) organize and display results. (Asher et al., 2010; cited by Badke, 2015)

Exercise 4.1   Why read?

Make a list of the reasons why you should read for a research project. Compare your list with those in Box 4.1. Most also apply if you are preparing an essay.

Reading the literature is an important element of academic research. It is a requirement for essays and projects as well as dissertations that you relate your ideas to the wider literature on the topic. Reading around the subject will also help you broaden and refine your ideas, see examples of different writing styles and generally improve your understanding of the discipline. When undertaking a dissertation or thesis, reading will help you identify gaps, find case studies in other areas which you may replicate and then compare with your findings, and learn more about particular research methods and their application in practice (Box 4.1). Effective reading may, of course, take many different forms depending on your purpose – from skimming, through browsing, to in-depth textual analysis – and will rarely involve just reading from the beginning to the end (Kneale, 2011).

Box 4.1   Ten reasons for reading for research

1 It will give you ideas.
2 You need to understand what other researchers have done in your area.
3 To broaden your perspectives and set your work in context.
4 Direct personal experience can never be enough.
5 To legitimate your arguments.
6 It may cause you to change your mind.

(Continued)

(Continued)

  7 Writers (and you will be one) need readers.
  8 So that you can effectively criticize what others have done.
  9 To learn more about research methods and their application in practice.
10 In order to spot areas which have not been researched.

Source: Based on Blaxter et al. (2010: 100)

4.2 Making a Start

Your literature search strategy will vary with your purpose. Sometimes you may want to search for something specific, for example a case study to illustrate an argument. In other situations a more general search may be required; for example, you might wish to identify 15 articles which have been written on a particular topic for an essay. Your search strategy may also vary with the level at which you are in the higher education system and your motivation. Identifying half a dozen up-to-date books on a topic may be appropriate at the beginning stage.

Exercise 4.2   Starting your search

You have been set a research project on a topic you know little about (e.g. organic farming). Before reading any further, write down the first three things you would do to find out what has already been written on the topic.

When we have used Exercise 4.2, the most common responses were to look in the subject section of the library catalogue, use a search engine on the Internet and ask a lecturer. These are all sensible strategies, though the usefulness of most search engines is exaggerated due to the lack of regulation and quality control on the sites found. However, apart from possibly asking the lecturer/professor who set the assignment for one or two key references to get you going, usually the first things you should do are to identify and define the key terms in the assignment and construct a list of terms to use in your literature search. Only then is it appropriate to turn to the search tools, such as library catalogues, reference books, indexes, databases and websites, and seek help from a librarian.

    Making a start is usually the most difficult stage of undertaking a research project or assignment. The issues involved in identifying your own research topic were discussed in Chapter 1. When you have a provisional idea about your topic and the research methods you may use, or when a research project or assignment is given to you, take a little while to plan your literature search. Defining the key terms in the topic or assignment is a good starting point. Dictionaries of human and physical geography are essential references for all geography students (Thomas and Goudie, 2000;

Gregory et al., 2009). The indexes of appropriate textbooks will also help. These references will further help you identify search terms, as will a thesaurus, a good English dictionary and a high-quality encyclopedia. For example, the Oxford Reference Online (www.oxfordreference.com) includes subject dictionaries and an encyclopedia as well as English dictionaries. The Geobase subject classification is another source (see the section below on abstracts and reviews). Remember to allow for American English spellings of words as well as standard English spellings.

In identifying search terms, group them into three categories: broader, related and narrower. The first will be useful in searching for books, which may contain useful sections on your topic, while the second and third will be particularly helpful in identifying journal articles and websites and using indexes to books. Box 4.2 illustrates how to make a start with searching the literature for a research project on organic farming.

Box 4.2   Defining key terms and identifying search terms: an example

Topic: Social and economic impacts of organic farming in the UK.

Definition:

> Organic farming: 'this system uses fewer purchased inputs compared with conventional farming, especially agri-chemicals and fertilizers, and consequently produces less food per hectare of farmland …, but is compensated by higher output prices' (Atkins and Bowler, 2001: 68–9).

Search terms:

Broader	Related	Narrower
Agricultural geography	Organic farming	Certified organic growers
Farm extensification	Organic agriculture	Organic organizations
Farm diversification	Organic production	Soil Association
Alternative farm systems	Organic growers	Organic food retailers
Food, geography of	Organic food	Organic food markets
Sustainable agriculture	Organic movement	Organic food shops

Note: 'Organic farming' is not listed in *The Dictionary of Human Geography*, although discussion of 'agricultural geography' and 'food, geography of' provides a useful context. The index of Atkins and Bowler (2001), which is on the 'Food and the Environment' course reading list, identifies four mentions of 'organic farming'. These lead to a useful introduction to the topic and to several recent references, and are a source for some of the above search terms.

Tip

If you are collecting data as part of your research project remember to carry out a literature search on research methodologies and techniques as well as on your main topic. Many of the later chapters in this book will help you with this.

4.3 A Framework for Your Search

A summary of how to search the literature is given in Figure 4.1 – which also provides a framework for the structure of this chapter. Where you start depends on the purpose of your search. For example, if you are looking to see whether the government has any policy documents on your topic you might begin by searching the Gov.UK website (www.gov.uk). If you want to check news stories you might go to one of the newspaper database sites given in the other literature sources section, while if you are searching for journal articles to start you on an assignment you might look at a citation index, such as Web of Science. Figure 4.1 might suggest that undertaking a literature search and writing an essay or a literature review is a linear process. The reality is much messier. There is frequent interaction between the different stages. As you begin to identify and scan the key references, your knowledge and understanding of the topic will increase, which will lead you to identify particular subtopics that you wish to investigate in more detail and equally important those topics you choose to ignore.

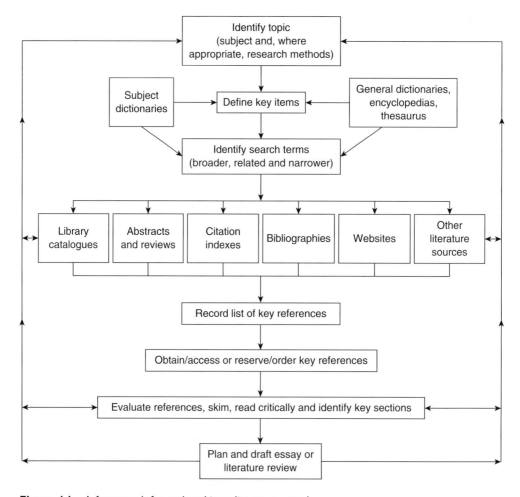

Figure 4.1 A framework for undertaking a literature search

A further search, using new key terms, may then be appropriate. Iteration is a key element of the search process. You should not give up after entering the first obvious key word in a search engine. Further thoughts are also likely to arise as you begin to draft your essay or literature review, which may call for additional searches.

> **Tip**
>
> Identify a few key references, skim read them, then revise your search criteria in the light of your new understanding.

> **Tip**
>
> Ensure that you use references appropriate for degree-level study. Your lecturer/professor will not be impressed if you use textbooks, dictionaries and magazines aimed primarily at school/college-level study, or if you cite Wikipedia (though the references Wikipedia articles are based on may be a useful start). Only a small proportion of websites are likely to be appropriate (see later). Many of the most appropriate references will be academic journal articles. Remember also to check the library catalogue to see that you have the latest edition of a textbook in the library (e.g. at the time of writing *The Dictionary of Human Geography* is in its fifth edition). Remember that earlier editions are usually the ones that your peers will leave on the open library shelves. A copy of the most recent edition of a popular textbook may be in the short loan collection.

> **Tip**
>
> A useful place to start building your list of references is the reading lists your lecturers/professors provide for their courses. In most institutions these are put on their Intranet or virtual learning environment.

4.4 Managing Your Search

The search process, as just indicated, is one that you will keep coming back to at various stages in your research. It is therefore sensible to keep a search diary, which includes the sources searched, the key words used and brief notes on the relevant references they reveal. This is, perhaps, best done using a word-processing package, which will enable you to list your key words, cut and paste the results from your online searches and keep track of which search engines and sources you have used. Alternatively, packages such as EndNote or Reference Manager may be useful in keeping track of your references. However, some programs may require further training as they can be complicated to use. For further advice on working online, Dolowitz et al. (2008) provide much useful information on search strategies using the Internet, emphasizing the need to develop a search strategy. They suggest that you should begin by asking yourself three main questions: 'What am I looking for?' 'What are the most relevant search terms?' and 'What tool will be the most useful for helping me find it?' (Dolowitz et al., 2008: 52).

> **Tip**
>
> Have a memory stick available when you are searching. Many of the databases enable you to save your searches direct to a memory stick or a PC. Some will also allow you to email them to yourself. Chapter 5 in Ridley (2008) provides useful advice on keeping records and organizing information to help you with your searches.

> **Tip**
>
> When using key words you can easily miss articles, as authors may have chosen different key words than you to describe their work. Look at the key words authors have chosen in the articles you find. Making a list of these key words will help to increase your search terms and avoid missing other useful references (Ridley, 2008).

The amount of time to spend on the search process depends on the purpose of your search. For example, are you seeking 10 key references for an essay, or 50 or more references for a dissertation? Generally, the broader the topic and the more that has been written on it, the longer the search tends to take. This is because much of the effort is spent in trying to identify the key references from what may be a list of several hundred marginal or irrelevant ones.

> **Tip**
>
> Aim to identify two to three times as many relevant references as you think you will need for your assignment/project. Many may not meet your exact needs when you obtain them and/or may not be accessible in the time you have available. If you are finding what appear to be too many relevant references, focus on the most recent ones and the references most frequently cited, and consider narrowing the search by, for example, focusing on a subtopic or restricting the geographical coverage. If you are finding too few relevant references, try some new search terms and consider broadening the topic or the geographical area. Also ask a librarian.

To avoid any possibility of plagiarism (that is, the unacknowledged use of the work of others), be sure to take down the full bibliographic details of the references you find, including, where relevant, the author(s)' name(s) and initials, year of publication (not print date), title of book, edition (if not the first), publisher and town/city of publisher (not printer), title of article/chapter, journal title, volume number and page numbers, and names of editors for edited books. Be sure to put all direct quotations in quotation marks and give the source, including the page number(s) (Mills, 1994). The same applies to material taken from websites. However, avoid direct quoting too much, and cutting and pasting information from websites; instead summarize and paraphrase material. Take particular care in citing websites, giving wherever possible the author/ organization responsible for the site, the date the page/site cited was last updated, the title of the page/site and the date you accessed the site, as well as the URL.

> **Tip**
>
> Inconsistent referencing and missing or erroneous information are some of the most common comments made on student writing. Most geographers use the Harvard style of referencing, but departments vary on the format in which they like references to be cited. Guidance on how to provide correct references, including websites, is available from Kneale (2011). It is best to acquire the habit of using one way of formatting references and to apply this consistently whenever you note a reference (down to the last comma, full stop and capital letter!).

> **Tip**
>
> When taking notes, remember to put any sentences you copy (or paste from a website) in quotation marks and note down the page number(s) so that if you decide to use the author's direct words later you can acknowledge this properly.

4.5 Search Tools

Various features are used to search databases and the web. As these vary, it is important to check their 'help' facilities. Most allow you to use exact phrases. Simply place the phrase in double quotes (" ") (e.g. "organic farming"). Most search engines support the use of Boolean operators. The basic ones are represented by the words AND, OR and NOT (e.g. "organic farming" AND "UK"; "organic farming" OR "organic food"; "organic farming" NOT "North America"). The use of wild cards (*) may also be available. For example, "farm*" will find records containing the word 'farm', including farm, farms, farmer, farmers and farming. It will also identify farmhouse, farmstead and farmyard. Bell (2014), Flowerdew and Martin (2005), and Dolowitz et al. (2008) provide further information on search tool techniques.

Library catalogues

In most cases the first place to search for relevant books is your university's library catalogue. You will usually need to use your broader list of search terms. Unfortunately the classification systems used in most libraries put geography books in several different sections of the library. But do not restrict yourself to books with 'geography' in the title. The integrative nature of the subject means that many books written, for example, for sociologists, economists, planners, earth scientists, hydrologists and ecologists, may be just as relevant. Once you have found the classification numbers of relevant books check the catalogue for other books with the same number and browse the relevant shelves. Looking at other books on the shelves near to the ones you are looking for often reveals other relevant references. Use your list of related and narrower search terms to explore the book indexes. Do not forget also to check where short loan and oversize books and pamphlets are shelved. Older books may be in store. Furthermore, a search for a topic on a website such as Amazon (www.amazon.co.uk) provides a list of a large number of the books published within that topic area (Ridley, 2008).

Tip

One of the quickest ways to generate a list of references is to find the latest book or article covering your topic and look at its reference list.

For a wider search try the combined catalogues of around 90 of the largest UK and Irish research libraries, which are available online through COPAC (copac.jisc. ac.uk). The British Library Catalogue provides a national collection (www.bl.uk). The equivalent in the USA is the Library of Congress, which may be searched along with the catalogues of many other libraries in the USA and other countries via the Z39.50 Gateway (www.loc.gov). To check book details try WorldCat (www.world cat.org). This is the world's largest network of library content and services with a huge database of over 2 billion references held in more than 72,000 libraries worldwide.

When you are away from the university it may be worth seeing whether your local university or public library has the book you require in stock. The Library Catalogue of the Royal Geographical Society (www.rgs.org/OurWork/Collections/Collections. htm) holds more than 2 million items tracing 500 years of geographical research and receives over 800 periodical publications. You can use the library free if your department has an Educational Corporate Membership, or otherwise there is a small fee charged per visit.

Abstracts and reviews

Evaluating the relevance of a book or journal article simply from its title is difficult. Generally each journal will articulate the purposes of the journal and its intended audience in each issue published, providing you with some idea of the relevance of the type of articles in that journal (Ridley, 2008). Abstracts give a clearer idea of the contents of articles. One of the most useful set of abstracts for geographers is Geobase (www.elsevier.com/solutions/engineering-village/content/geobase), which provides international coverage of the literature (particularly journal articles) on all aspects of geography, geology, ecology and international development. The database provides coverage of over 2000 journals and contains over 2.8 million records from 1980 onwards. Book reviews which appear in the journals abstracted are also included. These are useful for evaluating the significance of books and finding out what has recently been published. Environment Complete, which some libraries are using instead of Geobase, is available through EBSCOhost (www.ebscohost.com/ academic/environment-complete). It offers deep coverage in applicable areas of agriculture, ecosystems, energy, renewable energy sources, natural resources, geography, pollution and waste management, social impacts, urban planning, and more. This contains more than 3.4 million records going back to the 1940s. You should also check recent issues of review journals, particularly *Progress in Human Geography* and *Progress in Physical Geography*, for appropriate articles and updates on the literature in particular subfields.

Citation indexes

Probably the most useful tool that you will find for searching the literature is the ISI Web of Science. So it is well worth investing some time in exploring how to use it effectively. The ISI Web of Science includes proceedings of international conferences, symposia, seminars, colloquia, workshops and conventions. ISI Web of Science consists of seven online databases; the most relevant for geographers are the social sciences, the arts and humanities, and science. It is the prime source for finding articles published in high impact refereed journals. As well as providing data on the number of times articles published in a wide range of journals are cited by authors of other articles, they also provide a valuable source for identifying journal abstracts and reviews. They can thus be used for generating lists of articles with abstracts on particular topics, as well as identifying influential articles. You can also use the indexes to identify related records that share at least one cited reference with the retrieved item using the 'Citation Map' option (once you have clicked on a relevant reference). The ISI Web of Science contains sophisticated ways of restricting searches including by subject area, date of publication and country. However, not all geography or related journals are included in the ISI database. Another useful citation index is SCOPUS (www.elsevier.com/solutions/scopus).

Tip

Some articles are cited frequently because they are heavily criticized, but they have nevertheless contributed to the debate. In the world of citation analyses the only real sin is largely to be ignored, which is the fate of most published papers. However, a few papers are ahead of their time and are not 'discovered' until several years after they have been published.

Tip

Once you have identified key authors who are writing on your topic it is worth checking abstracts and citation indexes to see what else these authors have written, some of which may be on related topics. Beware when searching that the way authors' first names are cited may differ between publications. For example, the first author of this chapter appears variously as: Michael; Michael J; Mick; M; and M J.

Citation analyses are used to rank the impact that journals have on intellectual debate. They thus provide a crude guide as to which journals to browse through in the library and a possible basis for choosing between which of two, otherwise apparently equally relevant articles, to read first (Table 4.1). Lists of journal ranking may be obtained from ISI Journal Citation Reports (scientific.thomsonreuters.com/products/jcr). One of their limitations is that many of the key articles used by geographers are not published in mainline geography journals. Google Scholar (scholar.google.co.uk), the freely available web search engine, also provides citations both for individual publications and authors.

Table 4.1　Geography journals by impact factor, 2014 data

Rank	Journal title
1	*Global Environmental Change*
2	*Progress in Human Geography*
3	*Transactions of the Institute of British Geographers*
4	*Landscape and Urban Planning*
5	*Economic Geography*
6	*Political Geography*
7	*Journal of Transport Geography*
8	*Applied Geography*
9	*Journal of Economic Geography*
10	*Annals of the Association of American Geographers*
11	*Antipode*
12	*Regional Studies*
13	*Geographical Journal*
14	*Cultural Geographies*
15	*Population, Space and Place*

Note: Based on ranking journals by their 'impact factor' (a measure of the frequency with which the 'average article' in a journal has been cited in a particular year)

Source: Journal Citations Report, Social Sciences Edition (2014), courtesy of Taylor & Francis

Bibliographies

A range of specialized bibliographies is available. The most useful are annotated. Some are in printed form. An increasing number are available on the web at no charge and without registration (see Table 4.2). Others may be available if your university has taken out a licence.

Table 4.2　Examples of web-based Geography bibliographies

Bibliography	Comments
Australian Heritage Bibliography www.informit.org/index-product-details/AHB	Full access is provided through the Informit server at Melbourne which provides greater flexibility in search formulation and output
Bibliography of Aeolian Research www.lbk.ars.usda.gov/wewc/biblio/bar.htm	Coverage from 1930; updated monthly
Development Studies – BLDS Bibliographic Database blds.ids.ac.uk	Web-searchable version of the library catalogue and journal articles database of the British Library for Development Studies at the Institute of Development Studies, UK
Gender in Geography jgieseking.org/current-projects/gender-geography-bibliography	Alphabetical list contributed by members of the discussion list for Feminism in Geography
GIS Bibliography http://gis.library.esri.com	ESRI Virtual Campus Library provides a searchable database of over 152,000 references
International Bibliography of the Social Sciences (IBSS) www.proquest.com/libraries/academic/databases/ibss-set-c.html	This database focuses on four core social science disciplines – anthropology, economics, politics and sociology

Note: all URLs accessed 6 November 2015

> **Tip**
>
> As you generate your list of references check whether your library holds the books, and if so whether they are on loan. If they are on loan put in a reservation request. In the case of journals, check whether the library takes them or has access to them electronically. Also check whether your library has a subscription to the journals identified in your search for accessing the full text of articles online. If so, obtain passwords and check how to access them and how you can do this off-site. If the library does not hold the book or journal, consider ordering the reference on inter-library loan (ILL). Make sure that the journal article is relevant by reading the abstract first, if one is available. To check the relevance of the book it is worth doing a Google Scholar (scholar.google.co.uk) or Amazon (www.amazon.co.uk) search for the text as you may be able to read a few pages of the introduction or a sample chapter before ordering the book. An increasing number of books are available as eBooks. Journal articles can usually arrive within 24 hours electronically or between a week and ten days for a hard copy. Recalling books, or ordering them through ILL, often takes longer. Check whether your library entitles you to a certain number of free inter-library loans or, if not, what the charge is per loan.

Websites

An increasing amount of useful information is being placed on the web. However, identifying this from the huge amount of irrelevant and low-quality information is a time-consuming task. The general search engines, such as Google (www.google.co.uk) and Yahoo (www.yahoo.com), and meta search engines, such as Dogpile (www.dogpile.com) and Ixquick (www.ixquick.com), which search other search engines' databases, are indispensable when searching for specific information, such as the URL address of an institution's website. One of the problems with search engines is that 'even the largest of search engines have only indexed a small proportion of the total information available on the internet' (Dolowitz et al., 2008: 62). The 'deep web' consists of databases, non-textual files and content available on sites protected by passwords. Items such as phone directories, dictionary definitions, job postings and news are all part of the deep web. To access these may involve a two-stage process: first searching for database sites (e.g. UK newspapers) and then going to the database itself and searching for the information you want. More advanced search engines, such as Google, incorporate access to some parts of the deep web.

Google Scholar (scholar.google.co.uk) is a freely available specialized web search database which focuses upon scholarly literature and therefore the quality of the content of results should be higher, although caution is still advised. As you will see in Box 4.3, Google Scholar can produce an overwhelming number of sources which need to be carefully filtered to find the specific material you need.

With most of these websites the most relevant resources are on the first few pages. So even if your search reveals hundreds or thousands of potential references, limiting yourself to the first few pages of the search is likely to direct you to some of the most useful items. This is where Internet gateways or portals can be useful because they provide links to sites on particular subjects which have been evaluated for their quality. For example, ELDIS provides a gateway to development policy, practice and research (www.eldis.org).

Tip

Check out the advanced search facilities on search engines and databases. These allow you to focus your endeavours in a variety of ways, for example by a combination of key words, geographical area and date of publication, and make your searches much more efficient.

Box 4.3   Searching the literature: an example

Assignment: a 2,000-word essay.

Topic: Social and economic impacts of organic farming in the UK.

Library catalogues: 11 records on "agricultural geography" and 39 records on "sustainable agriculture" were listed in the University of Gloucestershire library catalogue; 11 were found on "organic farming". A search of *COPAC* found 424 books with "organic farming" in the title.

Abstracts and reviews: *Environment Complete*: a search for "organic farming" resulted in 3,856 references, but the majority were about scientific aspects. A more specific search of "organic farming AND social and economic impact AND United Kingdom OR UK OR Britain OR England OR Wales OR Scotland or Northern Ireland" since 2005 reduced it to 7 articles. The term "organic farming" was used in 25 articles in *Progress in Human Geography,* 9 since 2005.

Citation indexes: *ISI Web of Science*: a search for "organic AND farming" focusing specifically on the Social Science Citation Index since 2005 produced 743 references.

Bibliographies: No specific organic farming bibliographies were listed in *Google Scholar*, but several useful references were found from the reference lists of books and articles identified by the above search tools and the economic geography course reading list.

Websites: A search of *Google Scholar* on "organic farming" gave 85,000 hits. However if "social and economic impacts" and "UK" are added this reduced the number of references since 2005 to 214.

Summary: Given the nature of the assignment (2,000 words, 30 per cent of marks for course) and the number of references identified in early searches, it would be sensible to focus the search (e.g. to references published since 2005). It would be best to start with the most frequently cited and most up-to-date references and websites and those that appear to be the most comprehensive. Apart from sources such as ISI Web of Science, which have sophisticated ways of focusing a search, much time will be spent on weeding out non-relevant references which deal with topics such as methods of farming or environmental impacts. Expect to find further references and undertake more specific searches as you become more familiar with the topic. Make a shortlist of references (two to three times as many as you think you are likely to need) to show your lecturer to ask his or her advice on identifying key ones and any major omissions.

Tip

When doing searches using the Internet it may be useful to be more specific with your search terms and select particular fields to search (e.g. search only academic journals or peer-reviewed titles in order to focus your search). QUT (2014) provides a useful tutorial for negotiating the vocabulary to work through when considering your exact search terms.

Other literature sources

For many topics newspapers can be a useful source of information, especially for up-to-date case studies. Proquest Newspapers provides access to library subscribers to 1,500 newspapers; over 250 have same day availability (www.proquest. com/libraries/schools/news-newspapers/newsstand.html). Lexis®Library, although primarily a legal database, also provides access to national and local UK newspapers (www.lexisnexis.com/uk/legal/). Many individual newspapers provide free online searchable databases. World Newspapers provides a searchable directory of online world newspapers, magazines and news sites in English (www.world-news papers.com).

If you are undertaking your own thesis it is important to check whether anyone else has written a thesis on a similar topic by looking at ProQuest Dissertations and Theses — UK & Ireland (www.proquest.com/products-services/pqdt_uk_ireland. html). A wealth of information is available from central and local government via Gov.UK (www.gov.uk) and a DigitalGOV search in the USA (search.digitalgov.gov). A selection of UK official statistics is available via the Office for National Statistics (www.ons.gov.uk). Europa.eu offers access to European Union information (europa. eu/index_en.htm).

The use of videos and podcasts, for example, from YouTube (www.youtube.com) and iTunes (itunes.apple.com/us/genre/podcasts/id26?mt=2), is becoming increasingly common in higher education. These can provide useful insights into geographical topic areas through a different, and sometimes more accessible, medium.

> **Tip**
>
> Continue to refine your search as you progress using the cycle illustrated in Figure 4.1 in order to compile a list of useful references. Start with more general sources in the early stages of your search, moving towards more specific ones later on (Dolowitz et al., 2008).

4.6 Evaluating the Literature

Do not be put off undertaking a systematic literature search, such as is illustrated in Box 4.3, because you feel you will not have time to read all the references you find. Indeed, you will not have time to read them all. The purpose of the literature search is to identify the most appropriate references for the task in hand (Cornell University Library, 2015). Table 4.3 provides guidance on how to manage your readings. Websites in particular need to be evaluated critically for their origin, purpose, authority and credibility.

If you follow the advice above you should have reduced the list of references several fold before you have even opened a book or journal or read a newspaper article or website, for example by focusing on the most frequently cited and up-to-date references. The titles and abstracts will also help you to judge those references likely to be most relevant.

Table 4.3 Reducing your list of references to manageable proportions

Criterion	Possible (Score 4 points)	More doubtful (Score 2 points)	Probably forget it (Score 0 points)
Relevance to my topic – judged by title and/or abstract (double the score for this criterion)	High	Moderate	Tangential
Up-to-date	Last 5 years	6–15 years old	Over 15 years old
Authority – the author or paper is cited in the references I have already read	Extensively cited	Recent paper not yet had time to be cited extensively	Older paper cited infrequently or not at all
Respectability and reliability of source publication	Published in major geographical publication or that of sister subject or something very close to my topic	Publication is not in geography or an allied field	Informal publication or unreliable Internet source
Nature of publication	Peer-reviewed academic journal or monograph	Textbook or conference proceedings	Popular magazine
Originality	Primary source of information – the authors generated this information using reliable and recognized methods	The authors take their information from clearly identified and reliable secondary sources	The authors assert facts and produce information without providing appropriate supporting evidence
Accessible	Instant – by download or short walk to library	Obtainable with effort – reserve, interlibrary loan	Unobtainable

Source: Modified from an idea by Martin Haigh (personal communication 29 January 2002)

Tip

Avoid listing all the references you have found simply to try to impress your lecturer/professor. You must use some relevant idea or material from each one to justify its inclusion. Ensure that you include a reference to each in the text, according to your department's citation style (e.g. Harvard).

Exercise 4.3

Select four references from different sources on a topic that you are preparing, such as a website, a textbook, a journal article and a newspaper report, and use the criteria in Table 4.3 to evaluate their relevance, provenance and source reliability.

Exercise 4.4

Take a book or article relevant to your topic. You have five minutes to extract the key points it contains.

Researchers rarely read books from cover to cover and they read relatively few articles in their entirety. Like you, they do not have the time. They are practised at evaluating references in a few minutes by skimming the title, abstracts, executive summaries, publisher's blurbs, contents pages, indexes, introductions, conclusions and subheadings. This enables them to select the references that deserve more attention. Even then they will usually identify key sections by, for example, reading the first and last paragraphs of sections and the first and last sentences of paragraphs. This is not to suggest that all you need is a superficial knowledge of the literature; rather, that you should read selectively and critically to ensure that you obtain both a broad understanding of the topic and an in-depth knowledge of those parts of the literature that are particularly significant. If you are not familiar with the processes involved in critical and strategic reading, have a look at the relevant chapters in Blaxter et al. (2010) and Kneale (2011).

Exercise 4.5

In your next essay or research project try applying the framework (Figure 4.1) for searching the literature outlined in this chapter. Good hunting!

SUMMARY

The aim of this chapter has been to identify effective and efficient ways to systematically search and evaluate the literature:

- The first stage is to define the key terms for your topic and to identify a range of search terms.
- You should then systematically search a range of sources, including library catalogues, abstracts and reviews, citation indexes, bibliographies and websites, being careful to keep a search diary.
- Having made a record of the references you have found, you should evaluate each of them for such things as relevance, respectability, originality and accessibility, whether they are up to date or an authority.
- Although searching the literature needs to be systematic it is also iterative, and as your knowledge and understanding of your topic and of the number and quality of the references you are identifying increase, you will inevitably need to make modifications to your search and repeat and refine many of the stages several times.
- For a quick guide, look at the exercises, boxes, tables and tips in this chapter and the framework for undertaking a literature search shown in Figure 4.1.

Further Reading

Few books focus only on searching the literature, though Hart (2001) is an exception; most are guides to study skills or how to research, which put the literature search process in a broader context.

- Bell (2014) presents a chapter on finding and searching for literature. This includes a useful introduction to searches using the Internet and ways to help limit or broaden your criteria. It is part of a guide to doing research.

- Blaxter et al. (2010) provide an excellent user-friendly guide on how to research; they include chapters on reading for research and writing up.

- Cornell University (2014) provide a useful guide to searching and evaluating information sources critically.

- Dolowitz et al. (2008) present a thorough guide to searching for information on the Internet including advice on how to approach searching and different techniques to use within your search strategy.

- Flowerdew and Martin (2005) provide a guide for human geographers doing research projects; they include a chapter by Flowerdew on finding previous work and a chapter by Clark illustrating the benefits and issues of using secondary data sources which also provides further information on how to use search engines to their best advantage.

- Hart (2001) provides a comprehensive guide for doing a literature search in the social sciences.

- Ridley (2008) presents a step-by-step guide to writing a literature review. This book includes a chapter on doing a literature search and a chapter focusing on information management.

Note: Full details of the above can be found in the references list below.

References

Asher, A., Duke, L., and Green, D. (2010) 'The ERIAL project: Ethnographic research in Illinois academic libraries', *Academic Commons*. www.academiccommons.org/2014/09/09/the-erial-project-ethnographic-research-in-illinois-academic-libraries/ (accessed 6 November 2015).

Atkins, P. and Bowler, I. (2001) *Food in Society: Economy, Culture and Geography*. London: Arnold.

Badke, W. (2013) 'The path of least resistance', *Online Searcher* 37(1), 65–7.

Badke, W. (2015) 'Students as researchers: The faculty role'. http://williambadke.com/StudentsasResearchers.pdf (accessed 6 November 2015)

Bell, J. (2014) 'Literature searching', in *Doing Your Research Project: A Guide for First-Time Researchers in Education and Social Science* (6th edition). Buckingham: Open University Press. pp. 87–103.

Blaxter, L., Hughes, C. and Tight, M. (2010) *How to Research* (4th edition). Buckingham: Open University Press.

Cornell University Library (2015) *Library Research at Cornell: The Seven Steps*. guides.library.cornell.edu/sevensteps (accessed 6 November 2015).

Dolowitz, D., Buckler, S. and Sweeney, F. (2008) *Researching Online*. Basingstoke: Palgrave Macmillan.

Flowerdew, R. and Martin, D. (eds) (2005) *Methods in Human Geography: A Guide for Students Doing Research Projects* (2nd edition). Harlow: Longman. pp. 48–56.

Gregory, D., Johnston, R.J., Pratt, G., Watts, M. and Whatmore, S. (2009) *The Dictionary of Human Geography* (5th edition). Oxford: Blackwell.

Hart, C. (2001) *Doing a Literature Search: A Comprehensive Guide for the Social Sciences*. London: Sage.

Healey, M. and Jenkins, A. (2009) *Developing Undergraduate Research and Inquiry*. York: HE Academy. Available from www.heacademy.ac.uk/node/3146 (accessed 3 October 2015).

Healey, M., Lannin, L., Stibbe, A. and Derounian, J. (2013) *Developing and Enhancing Undergraduate Final Year Projects and Dissertations*. York: HE Academy.

Healey, M., Flint, A. and Harrington, K. (2014a) *Engagement through Partnership: Students as Partners in Learning and Teaching in Higher Education*. York: HE Academy. www.heacademy.ac.uk/engagement-through-partnership-students-partners-learning-and-teaching-higher-education (accessed 6 November 2015).

Healey, M., Jenkins, A. and Lea, J. (2014b) *Developing Research-Based Curricula in College-Based Higher Education*. York: HE Academy. www.heacademy.ac.uk/resources/detail/heinfe/Developing_research-based_curricula_in_CBHE (accessed 5 November 2015).

Kneale, P.E. (2011) *Study Skills for Geography, Earth and Environmental Science Students: A Practical Guide* (3rd edition). London: Arnold.

Mills, C. (1994) 'Acknowledging sources in written assignments', *Journal of Geography in Higher Education*, 18: 263–68.

QUT (Queensland University of Technology) (2014) *Study Smart: Research and Study Skills Tutorial.* studysmart.library. qut.edu.au/ (accessed 6 November 2015).

Ridley, D. (2008) *The Literature Review: Step-by-Step Guide for Students.* London: Sage.

Thomas, D. and Goudie, A. (2000) *A Dictionary of Physical Geography* (3rd edition). Oxford: Blackwell.

ON THE COMPANION WEBSITE…

Visit **https://study.sagepub.com/keymethods3e** for links to the best resources relating to this chapter.

5 Effective Research Communication

Jennifer Hill and Helen Walkington

SYNOPSIS

This chapter begins by outlining the nature of research, identifying the key steps that both comprise the research process and help you to plan the presentation of your research. It highlights the advantages that can be gained as a geography student if you complete the research process right through to communication of your findings. The chapter makes explicit the principles of effective research communication in a variety of oral, visual and written formats, including checklists that you might use to help you prepare for and feel confident in presenting your research in specific settings. The increasing number of venues in which you can make your research public are identified, moving beyond your department and institution to a variety of public audiences. Delivering verbal presentations, defending posters, writing papers for publication and authoring web pages and blogs are examples of the diverse ways in which you can communicate your research. When dissemination is aimed at external multi-disciplinary audiences, you will develop a broad range of intellectual, organizational and inter-personal skills that can help you to gain relevant employment after graduation.

This chapter is organized into the following sections:

- What is research and why engage in research as a geography student?
- What are the elements of the research process?
- Who are the possible audiences for student research?
- Communicating your research effectively to different audiences
- What are the outcomes of effective research communication?
- Conclusion

5.1 What is Research and Why Engage in Research as a Geography Student?

As a student of geography you are inherently a researcher. Research may be defined simplistically as finding the answer to a question that you don't yet know the answer to. Adopting this basic definition, you undertake research in order to produce a written answer to a particular essay question set by your tutor. But you go beyond this when preparing independent research projects (dissertations), to establish your own research question(s) and to collect primary and/or secondary data in order to answer these question(s) (see Section 5.2). Both essays and dissertations require you to organize and present material in a well-structured and accessible way. You must develop an argument, providing evidence to critically examine the argument and to come to reasoned conclusions. As the Oxford English Dictionary states, research is: 'Systematic investigation or inquiry aimed at contributing to knowledge of a theory, topic, etc., by careful consideration, observation, or study of a subject'. Answering a question set by your tutor with recourse to literature advances your knowledge, whilst answering your own questions with reference to primary data or via original application of secondary data moves towards the ultimate prize of research: the creation of new knowledge.

There are two very different motivations for undertaking research as a student geographer (Plotnik and Kouyoumdjian, 2011). Extrinsic motivations are forces external to you that encourage or reward you to undertake a task. The greatest extrinsic motivating force in university is assessment. You might carry out your research to the best of your abilities in order to receive the highest grade you can from your tutor. Intrinsic motivations, by contrast, are your internal self-driven desires to want to perform well, to know that your research is accurate and reliable and might inform others. With intrinsic motivation, you undertake research because you find it personally rewarding to develop your own understanding and that of others. Intrinsic motivation encourages you to undertake true research, moving beyond scholarship (see Table 5.1). True research is really only achieved if you communicate the knowledge and understanding you have acquired to a larger audience in order to subject it to public scrutiny. This is reiterated by the Boyer Commission, which states:

> Every university graduate should understand that no idea is fully formed until it can be communicated, and that the organisation required for writing and speaking is part of the thought process that enables one to understand material fully. Dissemination of results is an essential and integral part of the research process. (1998: 24)

In contrast to the experience of academic staff, who habitually communicate their research findings through conferences and journal articles, the student experience of the research process (or research cycle) often remains incomplete. Research written up for a dissertation or capstone project, for example, is usually submitted late in the final year and receives feedback only from the supervisor and markers. As such, the research produced is rarely disseminated. This is something that Walkington (2008) has referred to as a 'gap' in the research cycle, leaving the research process unfinished and disconnected from the skills associated with effective communication. Ideally, you will learn more by consciously considering and putting into practice the skills necessary to effectively communicate your research – in other words you should complete the research process through to the dissemination phase.

Recent changes in higher education have resulted in a continuum of opportunities for you as geography students to conduct and disseminate your research. There is an increasingly competitive higher education environment around the world, driven by the desire of governments to produce employable graduates who possess the skills and knowledge to play pivotal roles in national economies (Li et al., 2007; Hennemann and Liefner, 2010; Arrowsmith et al., 2011; Castree, 2011; Whalley et al., 2011; Erickson, 2012). As a result, moving beyond research assessed within the curriculum by your tutor and/or your peers, you can increasingly expand the exposure of your research

Table 5.1 Relationship between research and other forms of investigation (Modified from Ashwin and Trigwell, 2004: 122)

Level of investigation	Purpose of investigation	Verification of investigative process	Result of investigation
Scholarly	To inform oneself	Verified by self	Personal knowledge
Scholarship	To inform a group within a shared context	Verified by those within the same context	Local knowledge
Research	To inform a wider audience	Verified by those outside of that context	Public knowledge

findings to communicate your work to external audiences (Spronken-Smith et al., 2013) (see Section 5.3). With this in mind you need to need be able to identify and action the principles of effective research communication in a variety of oral, visual and written formats and this chapter will provide you with the information you need to work through this process (Section 5.4). Whilst internal dissemination of research builds your confidence, deepens your understanding of the discipline and develops certain inter-personal skills, external presentation of research develops additional skills only acquired by entering these more professional learning spaces (Section 5.5).

5.2 What are the Elements of the Research Process?

Research is essentially a problem-solving activity that can be presented in an ideal-ised model as a sequence of steps (Figure 5.1). The starting point in the process is to define the research problem, to establish the overall aim or intent of your research (Brause, 2000). You should essentially spell out what you are researching, clarifying the rationale for undertaking your study. Once you have defined your broad area of interest you will be able to consider the theoretical and conceptual context to your research problem. You achieve this by reading and critically evaluating the literature associated with your topic area, elucidating what is known and not known about the topic. Your literature review should make sense of the literature in terms of your research aim. If the literature review is well structured and appropriately critical then, following its completion, you will be able to define and justify hypotheses or assumptions that you will use to guide your investigations.

In order to answer your research questions or test your hypotheses you will need to design a sampling framework and select appropriate methods to collect empirical evidence. Data can be collected from primary sources accessed in the field, or from secondary published sources. These data can be collected and analysed using quantitative techniques: finding variables for concepts, measuring them and applying statistical techniques to understand geographical phenomena; and/or qualitative techniques: interpreting the subjective experience of individuals via methods such as ethnography, participant observation, in-depth interviews, focus groups, and visual and documentary analyses (Grix, 2010). Some researchers purposefully adopt a mixed methods approach, combining qualitative and quantitative approaches.

It is important that you can justify and defend your choice of research method. When utilising quantitative methods, often collecting extensive empirical data and

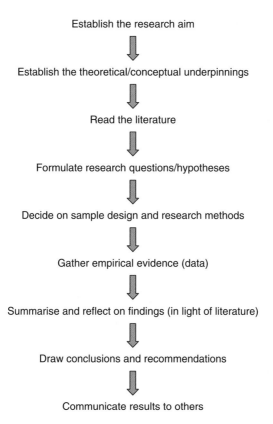

Establish the research aim

Establish the theoretical/conceptual underpinnings

Read the literature

Formulate research questions/hypotheses

Decide on sample design and research methods

Gather empirical evidence (data)

Summarise and reflect on findings (in light of literature)

Draw conclusions and recommendations

Communicate results to others

Figure 5.1   The research process expressed in a simplified linear format

generalizing results from a sample to a population, you must consider issues of sampling validity (strictness of sampling relative to the question set), rigour (precision of your overall sampling frame), representativeness (capturing variability in the population as closely as possible), error (inaccuracies inherent in counting any variable) and significance (confidence that the samples represent the population) (Creswell, 2014). Qualitative methods are often used to offer a case study approach and gather in-depth understanding of human behaviour and the reasons that govern such behaviour (Creswell, 2014). As such, the results of qualitative research may be descriptive rather than predictive, but the methods must still be applied with careful reasoning, just as with quantitative techniques. The most appropriate method for you to adopt for your research will depend on the questions you want to ask.

After data collection is complete, you must analyse your data, using appropriate descriptive or inferential techniques, in order to tease out and present your key findings. You will then interpret your data in the light of your research questions or hypotheses, referring back to the literature. Directed interpretation allows you to draw conclusions from your research, progressing beyond a summary of key findings to present a considered synthesis. You might also highlight limitations of the research, summarize its implications and make recommendations for future research. At this point of the process, you are ready to complete the research cycle by communicating your findings to others.

In reality, the research process is more usefully, and indeed more truthfully, thought of as a way of communicating research rather than as a way of carrying it out (Philips and Pugh, 2010). The process you pass through 'on the ground' is usually non-linear and iterative – you re-visit a number of the steps depicted in Figure 5.1 over the course of your research. As such, you do not communicate the reflexivity of the research process in your dissertation, or in any verbal presentation, poster or published article. Instead, you present a methodical and clear account of your main route through the process. Equally, the dissemination phase does not have to come at the end of your research, when you are communicating a finished product. Your research can be communicated 'in progress' through certain outlets, to gain feedback from others and to refine particular steps of the process as you pass through them.

5.3 Who are the Possible Audiences for Student Research?

There has been growing encouragement in higher education for students to become producers of research (Neary and Winn, 2009) and partners in research and scholarship (Little, 2011; Healey et al., 2014). This trend has broadened the scope for students to engage with and to communicate their research to external audiences. Spronken-Smith et al. (2013) have created a framework for the dissemination of student research (Figure 5.2), which allows you to consider and 'map' the level of exposure that your research achieves as you progress through your geography degree. You might start by sharing your research in courses and modules, through to departmental research conferences, institutional research poster events and exhibitions, and on to national multi-disciplinary events hosted, for example, through the British Conference of Undergraduate Research (BCUR) or the North American National Conferences on Undergraduate Research (NCUR), and perhaps even through international journals and meetings within geography. Your research in geography, even at undergraduate level, can therefore be disseminated to a variety of audiences in a range of physical settings within the university or beyond (see Table 5.2).

External communication in a multi-disciplinary context is generally perceived as requiring higher order skills in order to deliver a more polished public product when

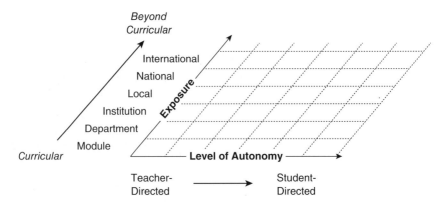

Figure 5.2   A framework for student research dissemination (Spronken-Smith et al., 2013)

compared with presentation to tutors and peers within university (Willison and O'Regan, 2008). This is partly related to the language we use as geographers. When talking to people within our discipline we can expect a level of understanding of geographical terminology, whereas if we are speaking to physicists, medics, artists and psychologists, for example, we will need to use lay language to describe our research. Many of the questions that geographers engage with are shared with people from other disciplines, so communicating clearly is essential because people will have a range of perspectives that they can bring to geographical problems. Clear communication without recourse to jargon is an important skill to develop as global problems will increasingly require large multi-disciplinary teams. Employers appreciate that geographers, with their ability to

Table 5.2 Opportunities for undergraduate research dissemination within and beyond university (all URLs accessed 9 November 2015)

Research dissemination inside your institution	Research dissemination beyond your institution
Module	*Local*
− Essay read by peers and tutors − Verbal presentations and posters in class − Research podcasts/videos − Research report/project read by peers and tutors − Student-led field research presentations − Presentations to clients within consultancy based modules	− Employer presentations/exhibitions (client presentations for consultancy projects) − General public presentations/exhibitions − Writing a book review for a public blog
Department/Faculty	*National*
− Departmental or faculty undergraduate research conferences (posters, verbal papers, podcasts) − Research posters displayed within department corridors and teaching spaces − Research showcases on department or faculty web pages − Student-led research blogs and e-magazines (e.g. blog posts by geography students at Leicester: http://studentblogs.le.ac.uk/geography and https://environmentalgeographies.wordpress.com)	− Undergraduate research conferences (e.g. British Conference of Undergraduate Research http://www.bcur.org and North American National Conferences on Undergraduate Research http://www.cur.org/conferences_and_events/student_events/ncur) − National exhibitions (e.g. Posters in Parliament http://www.bcur.org/about/posters-in-parliment and Posters on the Hill http://www.cur.org/conferences_and_events/student_events/posters_on_the_hill) − Writing for a national undergraduate research journal in geography (e.g. *GEOverse* in the UK http://geoverse.brookes.ac.uk, *Geoview* in Australia https://www.iag.org.au/publications/geoview)
Institution	*International*
− Institutional multi-disciplinary research showcases including conferences (posters, verbal papers, podcasts) and journals (e.g. *The Plymouth Student Scientist* http://bcur.org/journals/index.php/TPSS and *Diffusion* http://www.uclan.ac.uk/research/explore/projects/diffusion_the_uclan_journal_of_undergraduate_research.php	− Writing for an international multi-disciplinary undergraduate research journal (e.g. *Reinvention* http://www2.warwick.ac.uk/fac/cross_fac/iatl/ejournal) − Co-authoring a paper with an academic staff member for an academic or professional journal − Co-presenting a verbal paper at a national/international geography conference (e.g. Royal Geographical Society, Association of American Geographers) − Producing web based podcasts − Authoring Wikipedia pages − Authoring Blogs (e.g. writing for The People's Journal http://www.thepj.org)

understand both scientific and social science methodologies and discourses, are well placed to work on and communicate team-based research.

Internal audiences for student research

Small scale exposure of student research can be achieved within institutions by writing essays, reports or dissertations, presenting posters, or delivering verbal presentations to tutors and peers in class or even in the field (Marvell, 2008; Marvell et al., 2013). Posters or verbal papers can also be presented at extra-curricular departmental or institutional conferences. The audience for this research communication includes your peers, tutors and other staff and students in the university. Increasing the exposure of your research (Walkington, 2014; 2015), to include podcasts or videos of research results, makes it available to a wider group of staff and students and offers a lasting legacy. These audiences may be students who come onto a university course in geography some years after you. You might even write a paper for your institutional journal if one exists (Walkington and Jenkins, 2008; Walkington, 2012; Walkington et al., 2013).

External audiences for student research

Formats designed to maximize exposure of student research beyond institutions include: blogs, video logs and podcasts; client presentations and consultancy reports; papers or chapters within journals and books; exhibitions, displays and shows; web pages, public wikis and Wikipedia pages; and poster and paper presentations at online (virtual) or face-to-face conferences (e.g. departmental/national/international). Two online journals dedicated to the publication of undergraduate research papers in geography are *GEOverse* (in the UK) and *Geoview* (in Australia) (see Table 5.2). Postgraduate researchers should target standard academic journals. Going public in these ways can be done individually or collaboratively with other students or with academic staff. Co-authoring an article with an academic staff member in a published journal is a good way to start. If you are writing up your research as an article you will want to make sure that the research findings have been fully analysed and the research is completed, but for some less formal formats such as blogs, an ongoing update on your research findings as you analyse them might be acceptable, so long as you are not still collecting data (which could be biased by your blog posts).

Robertson and Walkington (2009) co-authored an article based on an undergraduate dissertation in geography about recycling and waste minimisation behaviour. There were significant unexpected implications for local authorities, so it was important to share the research results with an audience beyond academia. Co-authoring with academic supervisors who are familiar with the journal reviewing process can provide support with writing for a wider audience, although it often involves an investment in time beyond the university course for both people. Other examples of the co-production of geographical research include co-presenting at an academic conference such as the Royal Geographical Society annual conference (Hill et al., 2013). It is essential that ownership of material is clearly agreed when you are co-authoring and co-presenting with peers and academics. Acknowledgement of different levels of input is usually done through the order of authorship. Acknowledgements sections can also be added to papers and presentations to thank those who are not authors but who have helped, for example, with the drawing of figures, field assistance and commenting on drafts.

Opportunities for public dissemination of student research beyond universities are increasing due to the possibilities afforded by information technology. Promoting work through web pages and blogs, or podcasts and vodcasts on YouTube, is a good way to share your research findings with a global audience. When communicating your research online, the requirements for a scholarly style over and above a more popular style will need to be carefully tailored to the outlet and audience. It is important to remember that the audience for web-based resources is international, potentially non-academic and could include future generations. It is consequently a very powerful way to share your research and contribute to knowledge, but it comes with a significant responsibility in terms of accurate reporting and making realistic claims.

Finally, it is important to understand that research is a commodity. The copyright of research is sometimes signed away in the publication process. There is a growing commitment to the open sharing of research; creative commons licenses allow you to share your work under different standardised licencing conditions. It is important to understand that you must have the agreement of people whose material you use whether this is their data (via ethics procedures), or figures and maps that they have produced (copyright agreements). Just as with your university work, you must be careful to avoid plagiarism and to make sure that your own work is protected and attributed to yourself in turn.

5.4 Communicating Your Research Effectively to Different Audiences

This section highlights the principles of effective research communication that are applicable to different oral and written/visual formats delivered in a variety of contexts. We aim to make you conscious of how you might connect appropriately to different audiences, 'pitching' your work to their likely knowledge and experience. Within each context, considering the elements of the research process (Figure 5.1) will help you to structure your work logically and coherently and you will see these steps referred to repeatedly in the sub-sections that follow.

Principles of effective oral presentation

An effective oral presentation communicates a clear message in a logical manner. It considers the needs of the audience in order to capture their interest and develop their understanding. As you prepare to verbally communicate your research you should consider prior rehearsal and the context, content and delivery of your final presentation.

Rehearsal

A good way to prepare an effective presentation is to practise it on your own to check the timing and flow and then rehearse it in front of your friends to test the accessibility of your argument and gain constructive help. Rehearsing in front of peers or tutors can help you improve the content and delivery of your presentation. Rehearsing will help you to feel confident and relaxed as you stand up to present (Kneale, 2011). You can also brainstorm and practise answers to the questions that you are likely to be asked at the close of your talk.

Context

You will need to establish whether the audience members are likely to be disciplinary specialists or range across a number of disciplines. You can then match the style, language and technical content of your talk to their anticipated interests and levels of knowledge (Kneale, 2011). You should consider the layout of the room in which you are presenting and the technology available to you. Arrive early enough to check that all equipment is functioning correctly (Cryer, 2006). Run your slides through in advance to ensure they display properly with the equipment and software available to you and check that Internet links are working. Excite the interest of the audience by making connections to their experiences where you can. Explain technical terms that may not be widely understood but avoid oversimplifying and using colloquial language.

Content

Your presentation must be organized carefully as many internal and external presentations are only 15 minutes long, including time for questions from the audience. It is rare that you will have the time to say all that you want to about your research and, as such, it is just as important to consider what to omit from your talk as it is to decide what to include (Young, 2003). The most difficult part of any oral presentation is the introduction, when you must capture the attention of audience members, establish your credibility and make the audience receptive to hearing what you have to say. Begin with an opening slide that presents a clear research title and states your name. For external audiences you should also include your affiliation and email address. It is a good idea to follow this slide with an outline of your talk, setting out its content to provide a handrail through the information. Try to 'hook' in the audience early in your talk by asking a compelling question that clarifies the purpose of your research and sets this question in context personally (why is it of interest to you and why should it be to others?). You should also set the question in perspective theoretically and/or conceptually with reference to appropriate literature, and geographically by establishing the key characteristics of the 'field' location(s). Outline and justify briefly the methods you adopted and communicate your key results clearly and succinctly. Interpret your key findings (referring back to your selected literature) and draw to a clear conclusion about your research, perhaps indicating its main implications for the academic field or for practical application. Finally, thank any sponsors and finish by asking if there are questions from the audience. You may also want to display a final slide of references included throughout the presentation. Overall, keep in mind that you are trying to develop a clear structure in order to deliver an authoritative argument.

Delivery

A presentation is essentially a visual performance (Kneale, 2011), so your style of delivery is an important aspect of effective communication. Consider the appearance of both your slides and yourself as you prepare to deliver your presentation. The text and visuals on your slides should be readable from the back of the room (24 point font or larger is acceptable and images should be simple). You should not crowd slides with text but use them to communicate key issues and to highlight relevant visuals. Use a consistent background colour, font and graphic style throughout your presentation. Leave slides projected long enough for the audience to take in their content, take the time to explain complex issues and graphics, and do not be afraid to emphasize key points. Remember to lead your visuals rather than let them lead you. You can incorporate technology into your presentation, such as slide animation and embedded audio and video materials, but do not use technology simply as a gimmick (Cryer, 2006). It needs to clarify your message for the audience.

Speak with suitable pacing and varied intonation, loud enough to be heard by everyone. At the beginning of your talk you can ask if the audience can hear you and during your presentation you can check that sound levels for audio-visual clips are satisfactory. Do not allow your nerves to compel you to read from a prepared script. Instead, learn your material well enough to speak confidently and freely, perhaps referring to one or two pre-prepared cue cards. Speaking unscripted will enable you to maintain eye contact with the audience and to help them feel involved. It will also allow you to gauge whether audience members look confused, bored or are nodding attentively and taking notes (Kneale, 2011). It is a good idea to shift your focus around the room to involve as many people as possible in your talk. You can also acknowledge your audience by making verbal contact with them. During your presentation you can use inclusive language to draw your audience into your talk by noting, for example, that 'this graph shows *us*' an important fact. Feel free to move away from a lectern to projected slides in order to point out key aspects of text, graphs or tables. When you do this, ensure that your gestures are controlled so that you do not appear nervous. Remember that your presentation should express your personality and communicate your 'story' with an enthusiasm that captures the interest of the audience. You must be confident that the audience wants to listen and that you have something interesting to tell them.

Try to ignore late arrivals to the room and do not go back over material you have already covered for their benefit. As you progress through your presentation, watch the clock and keep to time. If you find you are running out of time then you need to have the confidence to cut out an aspect of your talk or to skim quickly over a slide indicating only the fundamental issues (Young, 2003). Do not just stop mid-way through a slide if you are cut off by a tutor or session convenor. Ensure you reach your concluding slide, clearly coming to the end of your argument.

Do not worry if you make a mistake during your presentation. Calmly start the sentence or explanation again with a smile. If you recognise there is an error with text on a slide just point this out and indicate what it should say. It is always best to be honest if you notice a mistake in your written or verbal communication. Even research experts are fallible when they present and they do make mistakes. They do not know everything and they are often presenting their research to find out where the weaknesses in their arguments lie. Your presentation will be no different. If you do not know the answer to an audience question, be honest and perhaps invite responses from the audience (Cryer, 2006). Never try to 'make up' an answer to appear clever as you will inevitably be found out. Offering the question back to the room is a good way to draw everyone into debate to try and solve the problem.

Finally, you should aim to look and act like a professional by wearing smart clothes and adopting a confident but relaxed posture (Young, 2003).

It is important to study the guidelines directing your presentation and to relate your work to them. These guidelines indicate what is expected of your work. When you present your research internally for assessment your tutor should provide you with assessment criteria at the outset. You should be aware of the main criteria as, if you use them as extrinsic motivators to think critically about your work, they will enable you to plan and deliver a better presentation. Table 5.3 offers a generic checklist of criteria based on presentation content and delivery. Usually, a selection of these criteria will be used to compose a full grading scheme against which your final presentation will be assessed.

Table 5.3   Checklist for oral presentations based on common assessment criteria

Self-assessment checklist for oral presentations

Presentation content
Clear indication of aim/intent
Organisation of material
Relevance of content
Breadth and depth of knowledge
Clarity and depth of argument
Level and currency of supporting evidence
Effectively drawn conclusions
Quality of referencing
Directed appropriately to audience

Presentation delivery
Pacing, audibility and intonation of speaker
Fluency of delivery and appropriate use of language
Enthusiasm and confidence of speaker, including eye contact
Appearance and demeanour of speaker
Effective use of visual aids (clarity of slides/handouts)
Audience engagement
Well managed timing
Quality of question handling
Integration of team members (where appropriate)

Principles of effective poster presentation

Posters are commonly used in academic environments to present research in a concise format and to promote discussion. If you are presenting your poster within your institution for assessment you will often be asked to stand by it so that you can answer questions posed by the tutor. Similarly, if you are presenting your poster at an external conference, you will most likely be allocated a specific time to stand by it so that you can summarize it to delegates and answer their questions. The dialogue that emerges from poster presentation is generally more informal and less threatening, but more penetrating than the question and answer sessions at the close of oral presentations. With poster presentation, it is possible to have detailed personal discussions with people who are genuinely interested in your research and whose knowledge may range from expert to novice in relation to your subject matter. The dialogue that emerges from 'defending' your poster to a diversity of audience members, particularly non-specialists, offers you an important method of enhancing your graduate competencies (see Section 5.2 of this chapter).

Posters are a highly visual medium, usually presented in large format (A0 or A1) and designed to be eye-catching but also informative. They allow you to practise both your research and creative skills (Vujakovic, 1995). Many of the criteria for effective oral presentations apply equally to effective poster presentation (Table 5.4). The fundamental challenge remains of how to reduce the complexity of your research into a small number of words that convey your key messages in a self-contained format that does not necessarily require further explanation. Posters usually contain only around 500 words of text, although some conference organisers specify a higher limit (Hay and Thomas, 1999). Within this limit your aim is to make it easy for a person unfamiliar with your research to understand it and to want to find out more about it. Decisions on content will depend on whether the audience is likely to be comprised of disciplinary specialists, who may require detailed information, or non-specialists, who may be looking for an accessible summary. As with oral presentations, set aside plenty of time for preparation and review the guidelines for poster production.

After you have distilled out your key research messages, you will need to consider the poster title, writing style, the type and size of font for your text, overall poster layout and colour scheme, and selection of supporting photographs, figures and tables. The title at the top of your poster will play a large part in attracting the attention of viewers. It should be worded appropriately for your audience such that they understand what your poster is about solely from its title. Poster conferences often have a catalogue of authors and titles and people will search for research

Table 5.4 Checklist for poster presentations based on common assessment criteria

Self-assessment checklist for poster presentations
Poster content
Clarity of the title
Clarity of aim/intent
Logical and coherent organisation/layout
Relevance of content
Focus and depth of argument
Level and currency of supporting evidence (including references)
Clarity of conclusions
Grammar, spelling and use of language
Quality of presentation – text and images
Visual appeal
Directed appropriately to audience
Poster delivery
Quality of verbal summary/defence
Ability to adapt language and terminology to the viewers
Appearance and demeanour of presenter
Ability to handle questions at a range of levels
Ability to 'add value' to the poster through discussion of ideas
Integration of team members (where appropriate)

based on titles that interest them. Your title should therefore be a clear description of your research without being too long. For the benefit of an external audience it is useful to state your name and institutional affiliation under the title. In your main text, write clearly and accurately, adopting an objective third person style (e.g. 'The research found that'), keep sentences short and simple and proof-read your work to ensure it is free from errors. Use scientific language to minimise ambiguity but define subject-specific terms on first mention if the audience is multi-disciplinary. It is usual for font type to be consistent across the poster, whereas font size tends to vary according to its function (Hay and Thomas, 1999). For example, the title should be the largest and boldest text on your poster, followed by primary headings and sub-headings, on to the main body of text. Finally, references can be in the smallest font size. Your poster is a scholarly document so the text still needs to be referenced using a standardised style. It is recommended that your main text is large enough to be read at a distance of about 1.5 metres away from your poster (Vujakovic, 1995). Good fonts to use on posters are Arial or Helvetica as they are simple and lack fine decoration (University of Leicester, 2009). Whichever font you choose, it is a good idea to use italics, bold and underlined text sparingly. You should also consider line spacing and text alignment. Apart from poster titles, a line spacing of 1.5 is ideal. You can then decide whether left or full justification suits your blocks of text.

The layout of your poster should be comprised of coherent blocks of text that are readily distinguishable and logically ordered so that the viewer's eye naturally follows the flow of information in your display. Scientific posters usually comprise the six primary components of title, introduction, methods, results, conclusions and references (Hay and Thomas, 1999). This structure can be followed effectively when presenting either physical or human geography research. These components should be balanced with images and white space to create visual stability (Figure 5.3).

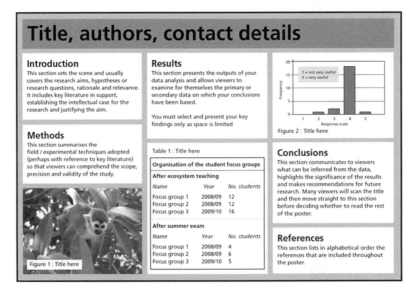

Figure 5.3   Example of a typical research poster layout

You should aim to limit yourself to two or three colours on your poster (University of Leicester, 2009), using black or dark blue for your smallest text, contrasted sufficiently with the background. Avoid textured backgrounds that make the text difficult to read and, if you use a photographic background, adjust the transparency so it does not dominate. It is a good idea to blend your colours with those in your images. Using shades of the same colour is visually appealing, as is the use of complementary colours. The images you include should have a high resolution so that when enlarged and printed they are not pixelated in appearance, and figures and tables should be simple and clear. Remember that any images you include in your poster should be relevant to your argument, labelled and, if they are not your own, should be appropriately referenced. You should have permission to use them or they should have been obtained under an open commons licence.

You will usually present your poster in a busy room, with many other posters and presenters (Figure 5.4). Your poster will need to be eye-catching to draw in viewers from an initial glance and you must be prepared to discuss your research amidst the distractions of noise and a constant flow of people moving past you. It is a good idea to practise a 30-second 'elevator pitch' about your research to convey your key messages and to draw audience members into your poster. Consider the types of questions that may be asked about your research by those who come to look at your poster and rehearse you answers. Remember the audience for your poster can ask you about any aspect of the underlying research. The sequence in which you discuss your material might differ each time, so practise with peers to become accustomed to answering unexpected questions.

Figure 5.4 Posters in Parliament (Photograph courtesy of BCUR)

Principles of writing for publication in academic journals

In contrast to a conference presentation that may only last a few minutes, writing for academic journals contributes to knowledge in a lasting way. Journal articles have a relatively long legacy and will be accessed by people well into the future, so it is important to understand that publication in this format may require a lengthy review process and several attempts before your work is accepted.

When writing for journals the most common mistake that student authors make is to submit essays and dissertations in the format of a piece of university work. A minority of student journals accept work in this style but they are institution-specific and aim to facilitate publication of their own students' work (Walkington et al., 2013). There are only a small number of universities and colleges that host journals for students and, even here, most require reformatting in a journal style. Rewriting is essential if you want your work to be taken seriously by an external peer reviewed academic journal. You need to invest time to get it right, but this is the best way to publish your work successfully beyond your own institution.

Most academic journals have teams of reviewers who provide anonymous reviews of your work against the journal's criteria. This is the peer review process. These reviewers do not see each other's comments, adding to the rigour and impartiality of reviewing. The journal editor makes a final decision and provides feedback to you on your submission, frequently with conditions that you need to fulfil in terms of additional information, analysis, explanation or reference to further literature. Submitting work after you have received feedback from within your university or college will increase your chances of publication as you can revise your work in the light of comments. In rare cases your article will be accepted as it is, in others you will be given time to make minor or major changes. Unfortunately, most journals receive many more articles than they can publish, so some articles will be rejected. Coping with rejection is part of academic life and does not necessarily mean an article will never be published. You might try to get your work published in a different journal, but it is essential that you only submit your article for consideration to one journal at a time. This can mean waiting for a significant period of time while your paper is being reviewed.

There are generally two types of research article accepted by journals: full papers, perhaps with limits of 5,000–8,000 words; and rapid communication-style articles, sometimes called 'work in progress' (more typically up to 2,000 words).

Full papers

Full journal articles are most likely to arise from a significant research endeavour such as a final year dissertation, independent study or project. In this situation you are likely to be writing alone, rather than in a group. The research you report should be formatted in the style of the journal and author guidelines are provided to help with this. Additionally, reading published articles for the journal that you are targeting is a good way to establish if you have the right kind of material. Author guidelines are specific to each journal and are rigidly enforced. Read them carefully. Your submission will be rejected if you do not conform to the word limits, referencing style, formatting for figures and tables, and the aims and scope of the journal. The research process, outlined in Section 5.2, highlights the main elements that your article will cover, beginning with an introduction to the study and perhaps the study location, the methods you have chosen, results obtained and then a discussion of your findings before you conclude in relation to the impact your study has more broadly on the field of geography. Box 5.1 outlines in more depth the principles of writing a clearly communicated geography journal article.

Box 5.1   Principles of a clearly communicated geography journal article

A journal article is not the same as an essay, thesis or project report. You need to be committed to rewriting your work in a new format. The first step is to read articles from your target journal to become familiar with the layout and style. All journals have a style guide dictating, for example, word limits, figure requirements and referencing conventions.

Academic journals publish work that addresses a gap in the existing research literature or a contribution to knowledge. *GEOverse* is a journal that publishes the very best of original undergraduate research and scholarship in physical and human geography. As a potential author, you must be very clear about how your article contributes to the literature, why it is of interest to readers and how it differs from research published previously.

Titles, key words and abstracts are crucial for making your research work discoverable through online searches. Use your title to convey concisely what your research is about. The abstract, usually about 200 words, should present a clear summary of your paper – what you did, why you did it, how you did it, what your results were and where your research is going from here. Key words are like search terms, they need to be informative and clear.

Here is an example of a title, abstract and key words from an article published in *GEOverse*:

Locavoracious: What are the impacts and feasibility of satisfying food demand with local production?
   Katie O'Sullivan, McGill University, Montreal, Canada, May 2012.
   Growing concern over global food security and environmental sustainability has popularized the local food movement within many industrialized countries. While context-specific, the impacts and feasibility of localizing diverse food systems are subject to similar parameters and analytical approaches, which have not been evaluated holistically. I conduct a systematic review of peer-reviewed literature assessing the impacts and feasibility of local food systems. I demonstrate that knowledge of environmental impacts is well-developed in comparison to economic and social elements, while feasibility analysis over-simplifies networks of food production and consumption as geographical alignment of supply and demand. As policymakers strive to improve food security at multiple scales, there is inadequate consideration of external market forces and infrastructure that are crucial in determining local food distribution and availability.

Key Words: Local, Food systems, Systematic review, Impacts, Feasibility (O'Sullivan, 2012).
   The title uses a colon to capture the imagination, but also to say clearly what the article is about. The key words include methods (systematic review) as well as content. The addition of 'policy' as a key word would increase the hit rate of the article in searches.
   You need to convey complex ideas in a clear and concise manner in a journal article. Day and Peters (1994) suggest that a research article should address the following prompts:

- What are your research questions and why are they important?
- What did you do and why is your method valid?

(Continued)

(Continued)

- What do your results mean and what has your analysis shown? What impact might this have on geography as a discipline?
- Have your discoveries opened up further questions for research?

Your article should demonstrate awareness of current research and acknowledge supporting and opposing views. If you need to use the figures, tables or maps of others you should gain permission and state that you have done this in your article (e.g. in a figure caption). If a photograph is your own, you should include something to indicate this e.g. 'from the author's own collection.' Reviewers will then understand it is yours and readers can contact you if they wish to use it. You must also demonstrate that your research is ethically sound. You must have permission from participants whose data you use in your research, even if their words are anonymised. Finally, proofread your paper carefully and make sure you follow the journal's referencing and other conventions.

Short articles

Rapid communication-style articles are not accepted by every journal so if your results are significant and it is important to share them quickly then search for journals that cater for this kind of short publication style. Alternatively, you may want to share your findings through social media (see below).

Writing for publication in academic journals hinges ultimately on having a set of novel research findings that you wish to communicate and that other people will want to read about. Not every piece of independent research that you carry out at university will be suitable, so think carefully about the big picture and what has already been published in your topic area before considering submission to a journal. If you are unsure, write to the journal editor to pitch your idea and ask their advice on whether your research is something the journal would be interested in. Journal editors will know what papers have proved popular with their readers, so they are a valuable source of advice prior to submitting an article.

Effective communication in informal arenas online

Advances in technology have allowed many groups of people to become authors, particularly of online content, and this has opened up issues of authority, liability and privacy (Flanagin and Metzger, 2008). Geography students can take on a role in research and research communication through social media, blogs, e-magazines, adding to photo-sharing sites, etc. While some blog sites will be discipline-specific, others may be generic and allow you to practise writing for a non-disciplinary audience. New blogs are being created all the time and you could make your own or write for an existing one. The People's Journal is an open publishing platform premised on citizen journalism, which aims to provide a source of news and opinion. The journal has editorial guidelines and publishes research-based stories, so you are able to disseminate research findings (either your own research or a collation of published research) communicated as a story.

Involving the wider public in your research can be fruitful if you want feedback on your interpretations and blogging is a good way to test out ideas and receive feedback. Tweeting your results is also a way to create followers who are interested in your research area, people who may eventually become collaborators or participants in future research. Box 5.2 outlines some of the media you might use and considers the style you can adopt when writing for online audiences.

Box 5.2   Using social media for effective research communication

Increasing the reach and exposure of your research through engaging in blog posts, tweets and other forms of social media is a good complement to more traditional forms of research dissemination. Social media can help with your research at all stages of the process, from the initial planning through to publicising the results, from exploring ideas for collaborative research projects to engaging participants. The following are just a few examples of the ways in which you might want to engage with social media for research purposes:

- Follow a conference remotely via Twitter
- Promote your research on YouTube
- Use a wiki for collaborative writing with others
- Use Facebook to involve the public in your research
- Receive literature alerts using RSS feeds
- Post your articles on LinkedIn, Academia.edu or your own blog or web page

Writing for digital media is different from writing essays or formal journal articles. Your writing needs to be up-to-date, contain lots of links, be informal and engaging and very inclusive (people from all over the world will be able to access your work, from a range of backgrounds and ages, so be very careful about the words you use). Some guidelines for sharing your research through social media are:

1   Make sure your writing serves a clear purpose, know who your audience is and what you want to achieve;
2   Bite-sized is best, so keep your writing concise, adopt a consistent style and use the house style if you contribute to established blogs;
3   Be user-friendly for your readers. Adopt their language, avoid jargon, use terms that will help people to find your work;
4   Use headings, short paragraphs and break up the text. Reading online is harder than in print. Readers can easily click away from your site if they are distracted so keep them on track by getting straight to the point;
5   Provide evidence for your claims and never make up statements to hook readers;
6   Ensure there are no mistakes as this will provide confidence in your readers that your writing is authoritative.

You can also include video, audio and still images to create interactive experiences and focus reader attention. As with other forms of writing, you must obtain any necessary permissions (e.g. if you are using music as a background to video or still images you must show that you have gained permission).

The impact indicators of web media are different from those of journals. With formal journals the impact factor is a significant metric to consider, whereas with online publishing the number of times that articles are shared is seen as a means of calculating impact. If you are not wishing to publish your research through a journal article and writing is not your favoured medium, then presenting your research direct to camera for a short video might suit you. There is an international competition for PhD (doctoral) students called the 'Three Minute Thesis'™ developed by the University of Queensland and YouTube houses a wealth of examples of winning presentations. Vodcasts provide an excellent model of how to communicate your research findings to a wide audience, beyond geographers, effectively and clearly. You may want to have a go at this and add your own short video clip on YouTube as a link on your CV or LinkedIn page. Done well this is also a great way of promoting your presentation skills to prospective employers.

5.5 What are the Outcomes of Effective Research Communication?

Research communication within your institution

Communication skills are an important component in the education of geography students because the many jobs that geography graduates find themselves in often require the use of such skills. Graduate students from Oxford Brookes University, for example, have noted:

> 'All the fundamental skills I have needed for my career I learnt as an undergraduate researcher. Communication is a huge part of my working life. I have to do a substantial amount of presentations. I had solid foundations through the numerous presentations I had to do at University, including the Geography undergraduate research conference.'

> 'In my opinion presenting with personality, conviction and presence can do more for your career than anything else ... My geography degree focussed on getting this right. Now I'm doing it week-in, week-out in my job.'

Employers across the globe highlight verbal, written and visual communication skills as of central importance to their decision-making when recruiting graduates (Solem et al., 2008; Arrowsmith et al., 2011; Whalley et al., 2011). Collectively, it is the sum of geographical knowledge, technical competencies and personal attributes, assembled in appropriate ways, which will define your employment capability. As a result, you are currently provided with opportunities across all levels of your degree programme to practise, and have assessed, your skills in oral, visual and written presentation of geographical research material.

You should embrace the opportunities to engage in research and research communication that are offered throughout your student journey. These should progress from research-led experiences (learning about current research in the discipline) to research-based experiences (undertaking research and inquiry) inside and outside curricula (Healey, 2005). Undertaking research and dissemination that is extrinsically motivated and assessed gains you marks and develops the skills summarized in Tables 5.3 and 5.4. Undertaking research and dissemination within your institution but beyond curricula and grades, intrinsically motivated, allows you to build a portfolio of graduate skills that include confidence, understanding of disciplinary content, oral communication skills, visual creativity, information literacy, critical thinking, reflective judgement, self-evaluation, and offering and accepting critique.

Research communication beyond your institution

It is worthwhile disseminating your research as a student beyond your university and in diverse arenas because 'post-degree activities will be in spaces that are quite different from the college lecture theatre, Virtual Learning Environment and formal examination settings' (Whalley et al., 2011: 389). Preparing and delivering your verbal presentation, defending your poster to audiences external to your institution, writing a paper for publication, or communicating through social media, especially when these contexts embrace a multi-disciplinary audience, develops a broad range of intellectual, organizational and inter-personal skills.

Students who have delivered oral and poster presentations at the British Conference of Undergraduate Research (BCUR) have made their research public, completing the research cycle (Walkington, 2008). The participating students embraced this opportunity enthusiastically, perceiving it to afford their work meaning in academia and perhaps in wider society. They gained in confidence and sense of achievement, feeling validated as professionals due to their acceptance in what they perceived to be an authentic context: a genuinely interested and diverse undergraduate community offering peer critique. Not surprisingly, the participants identified effective communication as a key skill developed through conference presentation. They referred to the importance of preparation, practice and repurposing of their work (see Box 5.3) (Hill and Walkington, 2016). Additionally, students who presented posters that had to be defended highlighted the importance of dialogue in developing their communication and wider skills. As the students shared alternative interpretations of their research they re-evaluated it, recognising that it assumed a different appearance from diverse viewpoints. Such transactional communication made explicit the tacit understanding of these students by offering them the chance to clarify and develop their ideas interactively. Overall, the students who presented their research at BCUR evidenced progress towards self-authorship (Baxter Magolda, 2004), consciously balancing the contextual nature of their disciplinary knowledge with intra-personally grounded goals, beliefs and values. The 'public' presentation of their research motivated them in ways that other 'assessment' had not. BCUR thereby offered the students an opportunity to begin to construct their professional identities during their studies, potentially helping them to navigate into their working and wider social lives.

Box 5.3   Student experiences of research communication
at a national undergraduate research conference

We have examined the experiences of geography, earth and environmental science students who have presented their research at a national undergraduate research conference (the British Conference of Undergraduate Research or BCUR). The students were very aware of honing their communication skills, including specific aspects of oral delivery such as pacing, fluency and audience engagement. They thought critically about their research content, re-purposing it for a multi-disciplinary external context. Notably, they prioritised material that would convey their core messages in a comprehensible manner:

> 'After doing a dissertation, condensing 16,000 words onto a side of A1 was quite challenging, but also I think it really makes you own your research, because you can't abbreviate that much without really having a firm understanding of what you're talking about.'

Students planned the structure of their presentations carefully, wanting to develop an accessible narrative for the many disciplines represented at the conference. They referred to the spatial layout of posters and the temporal sequencing of slides. They were particularly sensitive to language, translating technical terms:

> 'In the geography department, because we're all doing the same discipline, you can talk in the same way and they understand, but it's quite difficult to speak to say a medical student … because they don't understand the specific terminologies.'

Participating in BCUR legitimised the students as having undertaken research that mattered – to their audiences, to their disciplines and to the field of science:

> 'Coming here and presenting is a lot better than … writing your dissertation and it being put in a cupboard for how many years. Making it public is worthwhile … you feel like oh well it's being heard by other people, not just you and your supervisor.'

The majority of students self-regulated their preparations. They rehearsed in front of peers, sought feedback, and subsequently improved their poster and paper presentations. This process continued for many to benchmarking themselves against peers during the conference:

> 'I learnt things by going to the other presentations … you see people's mistakes so you think well if I am going to do a presentation I'm not going to do that.'

The students who presented posters learnt to negotiate and verbally organize their thoughts in real time, engaging in 'deeper' critical thinking:

> 'You can be asked a question and it can put you off kilter and then you have to think on your feet which is quite a good experience. There's a difference between that and being in a classroom because it's more of a controlled environment.'

The students welcomed the opportunity to present their work unconstrained from formal grading, being judged instead by peers who were 'genuinely interested' in their work:

'It's quite good that it's not assessed because that does take some of the pressure off, it gives you that real feel … there was no-one constantly writing down what you've been doing and to give you a mark on it. It lets you focus on what you're doing … am I ticking the right boxes to engage the audience rather than am I ticking the right boxes to get a first class grade?'

BCUR enables an authentic form of research communication as presentations are not *delivered to* others who already know the answers and are simply assessing to award a grade; they are *for the benefit of* others who do not know the answers and who are genuinely interested in learning from the presenters.

To conclude, students recognised the employability skills that they were practising and improving through presenting their research at BCUR. They made links from academic skills, knowledge and values to the world beyond their campuses:

'I wanted to be able to put on my CV that I have presented at a conference … presenting is such a vital part of any job now.'

For those students who have published articles in the journal *GEOverse*, there have been many positive impacts. In terms of gaining places on Master's courses and doctoral research places, several students have gone on to develop their academic career and to publish further work. Students benefited from the feedback that they were given by the journal reviewers:

'It was my first experience of publishing and I got lots of useful but critical feedback. Because it was online and anonymous, the reviewers were free to say exactly what they liked. It was a longer and more complicated process than I thought it would be at the outset but I was really proud when it was accepted.'

Successfully published student authors from *GEOverse* described the experience as an iterative learning process. They trusted the written advice of others, receiving detailed comments that helped them to improve their articles. The publication itself gave them recognition as a researcher. However, student authors disliked the impersonal nature of the feedback (via email) and the lack of face-to-face dialogue with reviewers (Walkington, 2008; 2012).

Section 5.1 described intrinsic and extrinsic motivations for communicating research findings. Research (Walkington, 2012; Hill and Walkington, 2016) has suggested that students who involve themselves in research dissemination beyond their own module/course, and particularly beyond their institution, change their priorities away from focusing on marks for pieces of assessed work, to developing a voice for their research activities. Having an impact in geography or informing society at large is empowering and students who are initially successful with research dissemination tend to begin a journey of continuing impact. These kinds of goals are both intrinsic and extrinsic. For some students recognition is a key factor, whereas for others the subject is a source of passion. In both instances, however, the sense of purpose students have to share their research findings can lead naturally into further study or into employment where these skills are valued.

5.6 Conclusion

To master the communication of your research you need to practise and take feedback on board. Grasp the opportunities available to you as a geography student to undertake research, to consider staff and peer critique, and to disseminate your findings. Operating alongside the formal curriculum, disseminating research externally can provide a means for you to see how your research has value beyond the modules and assignments that form your degree programme. Passing through the entire research cycle essentially offers you a signature learning experience (Spronken-Smith, 2013), developing a suite of graduate attributes (Barrie, 2004) and promoting self-authorship (Baxter Magolda, 2004). The skills learnt in 'going public' with your research will help you to move beyond the confines of your discipline to apply yourself productively to whatever you encounter in the dynamic, uncertain and insecure world beyond education (Barnett, 2004).

SUMMARY

- As a student of geography you are inherently a researcher. If you seize the opportunities that currently exist in higher education to communicate your research findings to a range of audiences you will develop your disciplinary knowledge and generic skills and abilities.
- Delivering effective oral and poster presentations requires practice, however confident you are. Know your audience, rehearse, and tailor your language, content and delivery appropriately.
- Develop your writing style for journal articles and online destinations according to your audience – be punchy and concise for web-based media, and scholarly and in-depth for journal articles.
- Think creatively about audiences for your research and have a go at creating an audience through sharing your work via social media.
- Do not consider research communication as a one-way process. Use communication as a means to gain further feedback on your work and to continue to develop your research for the future.

Further Reading

Two HEA commissioned resources by Waller and Schultz (2013; 2015), examining how to succeed at university in Geography and related disciplines, will support you with independent research.

Hay (2012) is an up-to-date textbook offering a practical guide for geography and environmental science students on how to communicate clearly and effectively in an academic setting. Now in its fourth edition, the text covers a wealth of written, graphic and spoken forms of communication.

The University of Leicester Student Learning Development website offers a suite of open access resources that support the delivery of effective oral and poster presentations: http://www2.le.ac.uk/offices/ld/resources/presentations. From this web address you can access user-friendly printable study guides and online tutorials. Equally, a detailed document from the University of Oxford that describes the process of creating an academic poster can be found at:

https://weblearn.ox.ac.uk/access/content/group/e05e05d2-f4ce-4a24-a008-031832bd1509/LearningRes_Open/Course_Book_Ppt_TIUD_Conference_Posters10.pdf

The third edition (2015) of 'Publishing and getting read: a guide for researchers in Geography' is now accessible online at: http://www.rgs.org/OurWork/Research+and+Higher+Education/Journals+books+and+guides/Publishing+and+getting+read.htm. This free guide, published by the RGS-IBG, provides clear practical advice about how to publish research in a wide range of forms, and how to maximize the reach of your research.

The British Conference of Undergraduate Research website offers useful material and further electronic links concerning how to publish student research in geography and other disciplines: http://www.bcur.org/.

The Reinvention website offers advice on writing your first journal article: http://www2.warwick.ac.uk/fac/cross_fac/iatl/reinvention/contributors/toptips/.

Note: Full details of the above can be found in the references list below.

References

Arrowsmith, C., Bagoly-Simó, P., Finchum, A., Oda, K. and Pawson, E. (2011) 'Student employability and its implications for geography curricula and learning practices', *Journal of Geography in Higher Education*, 35: 365–77.

Ashwin, P. and Trigwell, K. (2004) 'Investigating staff and educational development', in D. Baume and P. Kahn (eds) *Enhancing Staff and Educational Development*. London: Kogan Page. pp 117–31.

Barnett, R. (2004) 'Learning for an unknown future', *Higher Education Research and Development*, 23: 247–60.

Barrie, S.C. (2004) 'A research-based approach to generic graduate attributes policy', *Higher Education Research & Development*, 23: 261–75.

Baxter Magolda, M.B. (2004) 'Preface', in M.B. Baxter Magolda and P.M. King. (eds.) *Learning Partnerships: Theory and Models of Practice to Educate for Self-authorship*. Sterling, VA: Stylus Publishing. pp. xvii–xxvi.

Boyer Commission on Educating Undergraduates in the Research University (1998) *Reinventing Undergraduate Education: A Blueprint for America's Research Universities*. Stony Brook: State University of New York at Stony Brook.

Brause, R.S. (2000) *Writing Your Doctoral Dissertation: Invisible Rules for Success*. Abingdon: Routledge Palmer.

Castree, N. (2011). 'The future of geography in English universities', *The Geographical Journal*, 177, 294–9.

Cresswell, J.W. (2014) *Research Design: Qualitative, Quantitative and Mixed Methods Approaches* (4th edition). London: Sage.

Cryer, P. (2006) 'Giving presentations on your work', in *The Research Student's Guide to Success* (3rd edition). Buckingham: Open University Press. pp. 177–86.

Day, A. and Peters, J. (1994) 'Quality indicators in academic publishing', *Library Review*, 43: 4–72.

Erickson, R.A. (2012) 'Geography and the changing landscape of higher education', *Journal of Geography in Higher Education*, 36: 9–24.

Flanagin, A.J. and Metzger, M.J. (2008) 'The credibility of volunteered geographic information', *GeoJournal*, 72: 137–48.

Grix, J. (2010) *The Foundations of Research* (2nd edition). Basingstoke: Palgrave Macmillan.

Hay, I. (2012) *Communicating in Geography and the Environmental Sciences* (4th edition). Oxford: Oxford University Press.

Hay, I. and Thomas, S.M. (1999) 'Making sense with posters in biological science education', *Journal of Biological Education*, 33: 209–14.

Healey, M. (2005) 'Linking research and teaching exploring disciplinary spaces and the role of inquiry-based learning', in R. Barnett (ed.) *Reshaping the University: New Relationships between Research, Scholarship and Teaching*. Maidenhead: McGraw-Hill/Open University Press. pp. 30–42.

Healey, M., Flint, A. and Harrington, K. (2014) *Developing Students as Partners in Learning and Teaching in Higher Education*. York: Higher Education Academy.

Hennemann, S. and Liefner, I. (2010) 'Employability of German geography graduates: The mismatch between knowledge acquired and competencies required', *Journal of Geography in Higher Education*, 34: 215–30.

Hill, J. and Walkington, H. (2016) 'Developing graduate attributes through participation in undergraduate research conferences', *Journal of Geography in Higher Education* DOI: 10.1080/03098265.2016.1140128.

Hill, J., Blackler, V., Chellew, R., Ha, L. and Lendrum, S. (2013) 'From researched to researcher: Student experiences of becoming co-producers and co-disseminators of knowledge', *Planet*, 27: 35–41.

Kneale, P. (2011) *Study Skills for Geography, Earth and Environmental Science Students* (3rd edition). London: Hodder Education.

Li, X., Kong, Y. and Peng, B. (2007) 'Development of geography in higher education in China since 1980', *Journal of Geography in Higher Education*, 31: 19–37.

Little, S. (ed.) (2011) *Staff-Student Partnerships in Higher Education*. London: Continuum.

Marvell, A. (2008) 'Student-led presentations in situ: The challenges to presenting on the edge of a volcano', *Journal of Geography in Higher Education*, 32: 321–35.

Marvell, A., Simm, D., Schaaf, R. and Harper, R. (2013*) Journal of Geography in Higher Education*, 37: 547–66.

Neary, M. and Winn, J. (2009) 'The student as producer: Reinventing the student experience in higher education', in L. Bell, H. Stevenson and M. Neary (eds) *The Future of Higher Education: Policy, Pedagogy and the Student Experience*. London: Continuum. pp. 192–210.

O'Sullivan, K. (2012) 'Locavoracious: What are the impacts and feasibility of satisfying food demand with local production?' Geoverse. http://geoverse.brookes.ac.uk/article_resources/osullivanK/imagesKOS/Locavoracious-pdf.pdf (accessed 9 November 2015).

Philips, E.M. and Pugh, D.S. (2010) *How to get a PhD: A Handbook for Students and their Supervisors* (5th edition). Buckingham: Open University Press.

Plotnik, R. and Kouyoumdjian. H. (2011) *Introduction to Psychology*. (10th edition). Belmont, CA: Wadsworth.

Robertson, S. and Walkington, H. (2009) 'Recycling and waste minimisation behaviours of the transient student population in Oxford: Results of an online survey', *Local Environment: The International Journal of Justice and Sustainability* 14: 285–96.

Solem, M., Cheung, I. and Schlemper, B. (2008) 'Skills in professional geography: An assessment of workforce needs and expectations', *The Professional Geographer*, 60: 1–18.

Spronken-Smith, R. (2013) 'Toward securing a future for geography graduates', *Journal of Geography in Higher Education*, 37: 315–26.

Spronken-Smith, R., Brodeur, J.J., Kajaks, T., Luck, M., Myatt, P., Verburgh, A., Walkington, H. and Wuetherick, B. (2013) 'Completing the research cycle: A framework for promoting dissemination of undergraduate research and inquiry', *Teaching and Learning Inquiry*, 1: 105–18.

University of Leicester (2009) 'Poster presentations'. http://www2.le.ac.uk/offices/ld/resources/presentations (accessed 21 December 2015).

Vujakovic, P. (1995) 'Making posters', *Journal of Geography in Higher Education*, 19: 251–6.

Walkington, H. (2008) 'Geoverse: Piloting a national journal of undergraduate research in Geography', *Planet*, 20: 41–46. https://www.heacademy.ac.uk/sites/default/files/plan.1.20i.pdf (accessed 6 November 2015). doi: 10.11120/plan.2008.00200041

Walkington, H. (2012) 'Developing dialogic learning space: The case of online undergraduate research journals', *Journal of Geography in Higher Education*, 36: 547–62.

Walkington, H. (2014) *Get Published!* Keynote lecture to the British Conference of Undergraduate Research, University of Nottingham, UK.

Walkington, H. (2015) *Students as Researchers: Supporting Undergraduate Research in the Disciplines in Higher Education*. York: Higher Education Academy.

Walkington, H. and Jenkins, A. (2008) 'Embedding undergraduate research publication in the student learning experience: Ten suggested strategies', *Brookes eJournal of Learning and Teaching*, 2. http://bejlt.brookes.ac.uk/paper/embedding_undergraduate_research_publication_in_the_student_learning_experi-2/ (accessed 9 November 2015).

Walkington, H., Edwards-Jones, A. and Gresty, K. (2013) 'Strategies for widening students' engagement with undergraduate research journals', *Council on Undergraduate Research Quarterly*, 34: 24–30.

Waller, R. and Schultz, D.M. (2013) *How to Succeed at University in GEES Disciplines: Using Online Data for Independent Research*. York: HEA [online]. https://www.heacademy.ac.uk/sites/default/files/resources/gees_9_transitions_resource_wallerandschultz.pdf (accessed 9 November 2015).

Waller, R. and Schultz, D.M. (2015) *How to Succeed at University in GEES Disciplines: Enhancing Students' Information Literacy Skills*. York: HEA [online]. https://www.heacademy.ac.uk/sites/default/files/resources/how_to_succeed_in_gees_0.pdf (accessed 9 November 2015).

Whalley, W.B., Saunders, A., Lewis, R.A., Buenemann, M. and Sutton, P.C. (2011) *Journal of Geography in Higher Education*, 35: 379–83.

Willison, J. and O'Regan, K. (2008) *The Researcher Skill Development Framework*. Available at http://www.adelaide.edu.au/rsd/framework/rsd7 (accessed 9 November 2015).

Young, C. (2003) 'Making a presentation', in A. Rogers and H.A. Viles (eds) *The Student's Companion to Geography* (2nd edition). Oxford: Blackwell Publishing. pp. 185–9.

ON THE COMPANION WEBSITE...

Visit **https://study.sagepub.com/keymethods3e** for author videos, chapter exercises, resources and links, plus **free** access to the following recommended articles:

1. **Winter, C. (2013) 'Geography and education III: Update on the development of school geography in England under the Coalition Government', *Progress in Human Geography*, 37: 3.**

It's important to understand how the discipline is created in the school system, particularly for those students who may wish to go into teaching.

2. **Bridge, G. (2014) 'Resource geographies II: The resource-state nexus', *Progress in Human Geography*, 38: 1.**

This article is essential for understanding the relationship between the state and energy resource.

3. **Kent, M. (2007) 'Biogeography and landscape ecology', *Progress in Physical Geography*, 31: 3.**

This report clarifies the importance of the landscape scale in interpreting ecological patterns, processes, measurements and monitoring, highlighting the implications of such scale-related understanding for biodiversity conservation.

6 Working in Different Cultures and Different Languages

Fiona M. Smith

SYNOPSIS

Cross-cultural research is the term used to describe researching 'other' cultures, perhaps using other languages. This may involve working in distant places but can also include working with 'other' communities closer to home. It requires sensitivity to cultural similarities and differences, unequal power relations, fieldwork ethics, the practicalities and politics of language use, the position of the researcher and other participants, including interpreters, consideration of collaborative or participatory research, and care in writing up the research. All these issues are explored in this chapter.

This chapter is organized into the following sections:

- Fieldwork in 'different' cultures: Understanding difference and sameness
- Working with different languages
- Working with difference: Power relations, positionality and beyond
- (Re)presenting 'other' cultures
- Conclusion

6.1 Fieldwork in 'Different' Cultures: Understanding Difference and Sameness

The first encounters many Geography undergraduates have with fieldwork in different cultural contexts are on field courses, during dissertation fieldwork or study abroad programmes, or participating in an expedition (Nash, 2000a and 2000b; Smith, 2006). At the University of Dundee, for example, we take students to southeast Spain. It is somewhere 'different' from Scotland for students to experience, providing 'an intense learning curve of different environments and cultures', as one student put it in their feedback. This 'classic' approach to fieldwork shapes many geography undergraduate careers, while other forms such as study abroad or working on projects with local students and informants can also provide 'intensive' and 'emotional' experiences (Glass, 2014). However, encounters with 'different' cultures need not always involve travel to distant places. Nairn et al. (2000) explain how New Zealand students participated in a field trip which involved meeting migrants to New Zealand from different cultures, thereby seeing the country from a 'different' perspective, while Hammersley et al. (2014)

reflect on the cross-cultural challenges and 'unexpected benefits' of Australian students undertaking fieldwork with indigenous tourist operators in the country.

To the popular imagination, travel to 'other' places and cultures defines 'Geography' as a subject, particularly as it is presented in publications such as the *National Geographic* or in television programmes about grand adventures.[1] These ideas contribute to the romance of fieldwork (Nairn, 1996). Fieldwork, particularly if abroad or in a difficult setting, is often seen as 'character building' (Sparke, 1996: 212), a rite of passage to becoming a 'proper' geographer (Rose, 1992). As attractive as these images are, however, geographers have a responsibility to think critically about how we undertake fieldwork in different cultures, whether distant from home or just down the road (Smith, 2006). Nash (2000a: 146) argues 'all geographers undertaking fieldwork overseas [or in more local 'different' cultures] need to be sensitive to local attitudes and customs', in a manner that 'respects the cultural as well as the physical environments you encounter'. Nash suggests a code of ethics applicable to such fieldwork (Box 6.1) which highlights the need to think about the 'difference' or 'similarity' of other cultures, to consider uneven and unequal power relations, and to move away from 'ethnocentric' approaches to fieldwork (see also Chapter 3).[2] In an effort to address such issues, many geography programmes and researchers have developed more 'reflective', collaborative or community-based forms of fieldwork (Benson and Nagar, 2006; Hammersley et al., 2014; Hawthorne et al., 2014), but all recognize that these remain challenging. Cross-cultural fieldwork requires careful and on-going negotiation of 'multiple axes of difference, inequality and geopolitics' (Sultana, 2007: 374).

Box 6.1   A code of ethics for tourists which is equally applicable to fieldwork in different cultures

- Travel in a spirit of humility and with a genuine desire to learn more about the people of your host country. Be sensitively aware of the feelings of other people, thus preventing what might be offensive behaviour on your part. This applies very much to photography.
- Cultivate the habit of listening and observing, rather than merely hearing and seeing.
- Realize that often the people in the country you visit have time concepts and thought patterns different to your own. This does not make them inferior, only different.
- Instead of looking for the 'beach paradise', discover the enrichment of seeing a different way of life, through other eyes.
- Acquaint yourself with local customs. What is courteous in one country may be quite the reverse in another – people will be happy to help you.
- Instead of the western practice of 'knowing all the answers', cultivate the habit of asking questions.
- Remember that you are only one of thousands of visitors to this country and do not expect special privileges. If you really want your experience to be a 'home away from home', it is foolish to waste money on travelling.
- When you are shopping, remember that the 'bargain' you obtained was possible only because of the low wages paid to the maker.
- Do not make promises to people in your host country unless you can carry them through.
- Spend time reflecting on your daily experience in an attempt to deepen your understanding. It has been said that 'what enriches you may rob and violate others'.

Source: Nash (2000a), published originally in O'Grady (1975).

In the heading for this section the word 'different' is deliberately placed in inverted commas to raise the question of how we think of cultures as 'other' or 'different'. Cultures and their relations to each other are understood in a variety of ways (see Hall, 1995; Skelton and Allen, 1999; McEwan, 2008 for useful overviews). 'Other' cultures might sometimes be regarded as unchanging, even 'traditional' or 'primitive', with change caused by outside forces (usually from the West) (Cloke, 2014). Such approaches emphasize the 'exotic appeal' of other places or their difference and danger. This was particularly evident in the representations of other cultures produced by Europeans during the period of colonialism and imperial rule which emphasized their strangeness and exotic qualities, the apparent danger posed by 'savage' peoples, and the 'need' for the 'civilizing' and 'modernizing' influence of European colonial cultures. This mix of approaches is often summarized under the term 'orientalism', from the work of Edward Said (1978; see also Driver, 2014; Phillips, 2014).[3] Contemporary accounts where westerners assume their own experiences are the model to which other cultures aspire or see other cultures as a backdrop for their own enrichment can also be 'orientalist'. Think about how western travellers often focus on the excitement and adventure provided by other countries with little understanding of the cultures of the countries themselves. Such approaches produce 'self-centred' or 'ethnocentric' geographies, where one's own culture is set as the measure for all others (Cloke, 2014).

Other approaches regard cultures as more dynamic and interconnected (McEwan, 2008; Potter et al., 2008). Some argue globalization erases differences between cultures in the technologically and culturally interconnected 'global village', or highlight the emergence of one homogenized, global consumer culture built around brands such as McDonald's, Coca-Cola or Apple. These approaches see 'other' cultures as 'just like us', and the rapid expansion of globalized communication has helped to reinforce such ideas. However, this rather overlooks the diverse experiences of cultural, social, economic and technological globalization in which geographers are fundamentally interested. One event may be watched simultaneously on television or streamed online by many people across the world, but its significance for each of them may vary, as will their ability to influence responses to the event. Taking diversity into account provides a third set of ideas about culture where, instead of thinking of other cultures as either 'strange' and 'unchanging', or as 'just the same as us', we seek to understand other cultures in and of themselves while also understanding how local places and cultures are connected to national and global processes in uneven and unequal ways (Massey and Jess, 1995). For example, cities are important centres in the economic geographies of globalization, but they are unevenly connected to these processes – New York is a hub of global finance, Cairo is less central but still connected to it – and these roles may also change over time. Furthermore, residents in these cities and beyond vary in their ability to connect to these processes, depending on their position in the labour market, access to education, class, gender, ethnicity and so on. Researchers should consider how research participants may be situated in these uneven social relations so that the research does not exacerbate or perpetuate inequalities or stereotypes. And the researcher should be aware of how their research is often made possible by their own relatively privileged position in these wider processes (Laurie et al., 1999; Nagar et al., 2003).

A further approach to culture considers how the encounters, relocations and flows of globalization may stretch out social relations across space to produce cultural forms which are 'hybrid' or 'syncretic', reflecting the multiple forms of belonging in and across different places particularly associated with experiences of transnational communities, migration and diaspora populations (Collins, 2009).[4] Claire Dwyer (2014: 669) argues that it is often in fields such as music, fashion, writing, food and the media that diaspora cultures and transnational connections are found and that the theories of diaspora and transnationalism 'offer new ways of thinking about the connections between global-local geographies and in so doing enable us to rethink the relations between society and space beyond a national frame'. Examples of diaspora cultural analysis include Paul Gilroy's (1993) analysis of the emergence of different cultural traditions, particularly around music, in what he calls 'the black Atlantic' diaspora, or in the UK context Bhangra music, characterized by fusions of Punjabi folk music, hip-hop, soul and house, and later developing in post-Bhangra and 'Asian Kool' musical forms in diverse South Asian-British youth cultures (Dudrah, 2002). Another example is the existence of hyphenated identities such as 'Black-British', 'Afro-American' or 'Japanese-Canadian' (Hall, 1995). These suggest 'different' cultures and transnational communities are found in close proximity to each other, perhaps most obviously in the diverse ethnic geographies of major cities (Banglatown/Brick Lane in London: Dwyer, 2014; Chinatown in San Francisco: McEwan, 2008), but also increasingly in suburban spaces, as drawn out by Dwyer's discussion of places of worship in London and Vancouver, and studies of the ways in which migration and diaspora might revitalize, discontinue or transform religious practices (Levitt, 2007). However, these diasporic geographies need to be seen not just as exoticized spaces of the 'other', but as spaces of encounter and contestation between diverse peoples, and where versions of hybrid cultures are often commodified and promoted for economic gain. There is also a danger of focusing on what appear to be 'obviously different' cultures. In the North American or European context, for example, a diversity of 'white' cultures should also be analysed, rather than seeing them as the 'norm' and 'other' cultures as those that are novel or 'different'. Dwyer suggests diasporic approaches are both 'hopeful' and 'critical':

> [They are]: hopeful that diasporic lives, cultures and concepts can help to forge relations between society and space in less bounded, exclusionary ways; but critical of the suffering caused to diasporic populations in their experiences of displacement and marginalization. (2014: 675)

Central to these types of ideas is the view that rather than cultures being slow-to-change, fixed sets of beliefs, values and behaviours, with a permanent connection to places, as Stuart Hall (1995: 187) argues, culture is 'a meeting point where different influences, traditions and forces intersect', formed by 'the juxtaposition and co-presence of different cultural forces and discourses and their effects' and consisting of 'changing cultural practices and meanings'. This suggests our 'own' cultures are as caught up in change as other cultures, and that connections between cultures and places can be highly dynamic. However, sometimes people or states re-emphasize fixed, homogeneous and bounded cultures in the face of such fluidity, often claiming landscapes or territories. Nationalist movements might be one such form (Grundy-Warr and Sidaway, 2008).

Using more fluid conceptualizations of culture means we must question what ideas about similarity and difference we bring to our cross-cultural studies. Nevertheless, it is important not to overplay the changing and dynamic nature of cultures or to over-emphasize the 'hopeful' rather than the 'critical' in our approaches. Instead we need to pay attention to how cultures are embedded in, and part of, on-going global inequalities. Likewise it is important to explore how different people's experiences of culture 'will be affected by the multiple aspects of their identity – race, gender, age, sexuality, class, caste position, religion, geography and so forth – and [are] likely to alter in various circumstances' (Skelton and Allen, 1999: 4). The remainder of the chapter looks at some of the practical ways geographers grapple with the challenges of cross-cultural research, starting with work in different languages.

6.2 Working With Different Languages

Translation (dealing with written texts) and interpretation (dealing with the spoken word) are at once skilled activities involving long processes of professional training and common everyday practices for people around the globe in business, education, tourism, among migrants and diaspora communities, in diplomacy and politics. Indeed some writers suggest that working between languages and cultural references increasingly characterizes the very nature of our lives, whether in the 'translation nation' of bilingual Hispanic communities in the USA (Tobar, 2005), the multilingual, multinational European Union (Eco, 1993), or the lives of Somali young people in Sheffield, UK (Valentine et al., 2008). Many people around the world speak English, either as their first language or as a second or third language, and what Ives (2010) calls 'global English' is increasingly a central element of contemporary globalization. As a result, students who speak English are often in a fortunate position but this can make native English-speakers lazy in learning other languages. On the other hand, many students and researchers around the world are already working in and across multiple languages in the context of global higher education. Whatever the situation, there are many contexts where cross-cultural fieldwork will be hampered if you do not make some effort to communicate in the relevant language. While this is not without its challenges (Tremlett, 2009), it is possible to make a big difference to fieldwork with some practical strategies and (although this will vary between projects) by paying attention to some key questions about the use of language in research and fieldwork (Box 6.2). Here we address issues about questionnaires, working with interpreters in interviews, and working as a bilingual researcher.

Box 6.2   Key questions for working in different languages

Data collection and fieldwork

- What languages are being used? Do you speak any of the languages yourself? How did you learn them – as a native speaker, at school or university, from more informal study? How fluent are you with reading, writing or speaking?
- Are you using written translation or working with interpreters?
- Have you considered the translation strategy for your project? In particular, consider what the roles of interpreters or translators are in relation to data collection and data analysis/interpretation.
- How might this affect the research and data collected?
- Does language use itself become part of the focus of the research?

Data analysis

- Are you assuming the data will mean the same in the other language as in your own? Do you have some way to check the meaning of the data?
- If you are analysing qualitative materials, are you analysing the original speech or the translation of this material? Does this affect how you analyse the data?
- Do meanings get 'lost' as you translate, or can you consider how translation might help to highlight particular aspects of cultural meaning and significance?

Dissemination and publication

- Do you mention issues of translation in your report, with ambiguities or difficulties explained?
- Will the language of the report affect who can access it?

Ethics and political issues

- Does your use of particular languages mean some social groups are excluded or others privileged in the research?
- Have you been careful when translating sensitive topics and concepts between different languages and cultural contexts?
- Would paying attention to language issues, such as providing translated versions of the report/findings, help to make the material more accessible to those who participated in the research or others who might be interested?

Consider how some of the questions in Box 6.2 might be dealt with in a survey using questionnaires. One common strategy is to aim to have the questionnaire translated in such a way that there is 'concept equivalence' between the original questionnaire and the translated one (something even more important if the results of questionnaires administered in different languages are to be compared directly). A key technique for this is to translate first into the target language and then to have someone else 'back translate' into the original language, allowing ambiguities and mistakes to be observed and ironed out. While this may seem relatively straightforward, it is important to check if the terms being translated are meaningful in the target language. For example, Thickett et al. (2013) examine the variation in terms around alcohol consumption and alcoholism between Estonia, Poland, Hungary and the UK and point out how such varied cultural understandings make cross-national surveys on the topic difficult to develop, particularly when trying to insure that terms used are relevant to respondents in different countries and can be analysed as meaning the same in different contexts. David Simon (2006) suggests that being sensitive to different calendars, units of measurement and local frames of reference or cultural practices might help make questionnaires meaningful to research participants in cross-cultural research. Back translation also helps to weed out ambiguous, insensitive or offensive terminology.

This may seem rather daunting, but sometimes relatively simple strategies can be very useful, particularly in student projects. For example, students investigating the local labour market of a tourist resort as part of the University of Dundee field course produced a bilingual questionnaire before going to Spain, getting the English translated into Spanish by a language teacher they knew. The questionnaire explained in both languages who they were and what they were doing and then gave dual language versions of each question. Using this they were able to gather data on the employment experiences not only of expatriate English speakers and of Spaniards (and other nationalities) with high-level English skills, but also of people with limited English who found it difficult to access better paid jobs in a resort dominated by British tourists. The questionnaire also stimulated further conversations, with friends or relatives of those completing it who spoke more English joining in to translate parts of the discussions between the students and the person completing the questionnaire. Making this effort allowed the students to understand how language abilities were one of the factors structuring differential access to the labour market in international tourist resorts such as this (see Marneros and Gibbs, 2015, on this issue in Cyprus, for example), as well as to gather data about labour-market processes which had been the original purpose of the research. This example suggests that paying attention to the politics of language can be important in a variety of contexts, whether in tourism employment as outlined above, or in questions of migration and integration (Sorgen, 2015), language policy and nation-building (Garibova, 2009).

When looking for simple strategies to working with different languages, it is also tempting to use online translation programmes, such as Google Translate (https://translate.google.com). These can certainly help to some extent with understanding material in other languages, but for research and fieldwork purposes they can also provide lots of pitfalls as the standard of grammar can be odd. Furthermore, the kind of language which is often being used in quite specific ways in research can be mis-translated very easily. When preparing a questionnaire, it is essential that the terms being used are precise and that they translate accurately. Leaving this to an online tool is rather risky. Online translation tools also tend to deal poorly with the kinds of text produced when interviews are transcribed. For example, running an extract from an interview about international student recruitment through Google Translate and asking it to translate from English into German resulted in a discussion of 'bumper numbers' (i.e. record levels of student recruitment) translated as having something to do with 'Autonummern' – i.e. car registration numbers. Thus, while it is possible to use such sites to help, they are not a reliable tool for detailed research and could in fact result in some very odd outcomes leading to inaccurate claims about what the participants had been discussing and the language that they had been using. Most professional bodies explicitly do not use machine translation because of these very issues. So although online translation tools can be a useful basic tool, they should always be used with extreme care and need to be used, if at all, in conjunction with other language skills.

Of course, it is possible to spend time learning the relevant language. Indeed it is almost essential if fieldwork is to last for a longer period, not least because even everyday life can be difficult without basic language skills (Watson, 2004). However, while some writers suggest that being able to speak the language of the research participants is the ideal situation, most researchers will at some time find it necessary or even desirable to work with an interpreter.

In the context of undergraduate research, finding an interpreter with suitable language skills and who you can afford to pay may be a real challenge and you might have to negotiate carefully about the expectations you have of the interpreter and that they have of you. Sarah Turner (2010) discusses aspects such as mutual expectations, cultural dynamics and variations in different contexts (in her case China and Vietnam) which are worth consideration in negotiating the relationship between interpreters and the researcher. Such negotiation is worth doing even if the interpreter is a fellow student. Interpreting is a challenging process for the person doing it, but it can also lead to frustrations between different members of a team if interpreters and their role in the research, including their relation to the other researchers, are not given consideration (Ficklin and Jones, 2009). Discussing beforehand such issues as how much interpreters will translate there and then for others in the team or whether they will just get on with the interview and translate key findings later can help immensely in smoothing the process. It also avoids leaving some team members feeling marginalized while others feel they are doing all the work. Having found an interpreter, however, the act of translation/interpretation is not merely a straightforward process of simply transmitting meaning from one language to another. Here two issues are worth considering: the approach to translation appropriate for the fieldwork; and the social relations between the interpreter, researcher and research participants, as well as the relation of each to the research.

Reporting on her research on domestic labour in Kenya and Tanzania, Janet Bujra (2006) outlines how she discussed carefully with the interpreters what kinds of terms were appropriate for asking people about domestic labour and people's social roles in order to understand the processes involved and, crucially, the meanings people ascribed to these practices. She also negotiated with the interpreter about the purpose of the research and what kind of translation was most suitable, agreeing that a rougher translation which preserved local terminology and slang was more appropriate than translating interviews and discussions into standard English or standard Swahili. This was a deliberate choice about the function of translation in the research. Often translation is assumed to be a direct transmission of meaning from one language to the other and the aim would be to make the equivalent meaning clear in the target language. For example, professional codes of conduct stress that the aim is to 'render, to the best of [the interpreter's] ability, a complete and accurate interpretation without altering or omitting anything that is stated. Interpreters shall not add to what is said nor provide unsolicited explanation' (Language Line Solutions, 2013). This approach is known as 'domesticating translation' (Venuti, 2012) and adherence to its principles is central to professional, reliable translation and interpretation. However, for Bujra, such a strict approach would have meant details of how people understood their own lives being lost. Thus she chose what can be termed 'foreignizing translation' (Venuti, 2012), where the aim instead is to get a sense of people's meanings in their own context, even if that does not produce a highly polished translated text. Neither of these strategies is necessarily right or wrong. Rather these choices indicate both the possibilities of meaning-translation and the impossibility of fully articulating all cultural references and meanings across languages and cultures (Smith, 1996; 2009).

It is also worth considering that translation is not only a process by which something of the original meaning is 'lost' but rather that by paying attention to translation issues, wider meanings and significance can be understood. Edwards et al. (2010) discuss a complex process by which meanings and cultural practices around particular forms of peat stacks in the Faroe Islands were understood in part because of realizing the limitations of direct translation and that what was self-evident about land-use interpretation in one context was far from clear in another.

Rather than assuming interpreters are neutral, almost-invisible transmitters of meaning between the researcher and research participants (Turner, 2010), it is worth considering their active role in the research. Would an outside interpreter bring less bias, or lack local knowledge? Does a local interpreter understand the issues better?

Might they introduce bias by tending to guide you to interview people they already know? Does their gender, class, ethnic or age position mean access to some participants is easier or more problematic? Remember that in many situations those who have the skills to act as interpreters may also be the more educated or affluent people in a society. Thus it is useful to consider how interpreters are involved in meaning-making in the research.

Twyman et al. (1999) discuss the use of interpreters in their study of society–environment links in rangelands management in Botswana. The British fieldworkers found the range of local languages too great and depended on high-level language skills supplied by their interpreters. Taking translation seriously within the research process, they argue that 'translation is a practice of intercultural communication [...] in which we understand other cultures as far as possible in their own terms but in our language' (p. 320). This neat formulation suggests translation involves 'mapping ideas and practices' between cultures (Twyman et al., 1999: 320). Rather than only recording answers to questions once they were translated into English, they taped and transcribed the whole process of communication back and forwards between the various research participants, including the translators. Analysing this revealed how the person interpreting often had to summarize roughly the meaning of what the interviewee said in order to let the interview proceed, but after the event the interviewer and the translator could discuss in more detail the ideas articulated. This revealed it was actually very difficult to 'map' the ideas and meanings of the research participants onto English language terms. In fact many interviewees were already talking in what was not their first language and used a variety of different language terms to communicate their attitudes and practices in land and livestock management. In writing up their research, Twyman et al. explored how meaning was mapped between and across cultures, noting particularly where this was problematic or provided new insights.

Awareness of language use and translation remains important even when you know and speak a language fluently. While being bilingual (or even multilingual) might make you more directly aware of the details of what is being discussed or alert you to how challenging it can be to translate meaning between different contexts, thinking consciously about the translation strategies used in research and the ethics and politics of translation remains important. As a simple strategy, being aware of and making clear the challenges of translating certain terms between languages can enrich the analysis. For example, in the analysis of interviews for her undergraduate dissertation Alicja Klek (2014) paid attention to the different nuances of the phrase 'to be home' in English and how migrants from Poland to the UK talked in Polish using the term 'być u siebie' ('to be at mine'). This allowed her to understand in more depth different meanings and processes related to home-making and belonging among the migrants. Likewise, Kathrin Hörschelmann (2002) describes choices made in her research on the cultural and social consequences of German unification. She interviewed in her native German, transcribed the interviews in German, and did the analysis on the original transcripts. Only when writing the final article in English did she translate quotations, but she tried to keep as much of the meaning and style of the original speech in her translated, written quotations, even leaving some terms untranslated and explained in footnotes (a 'foreignizing' approach rather than a 'domesticating' one in Venuti's terms).

Strategies that take translation seriously as a meaning-making process and as a political one are discussed by Martin Müller (2007). He calls for geographers to adopt critical approaches to translation and to understand the political and cultural impact of the loss of meanings for terms when they come to be translated between languages, particularly when they deal with key terms of geographical research such as home, nation, family, environment, power, inequality, community or citizenship. Too often apparently simple translations (typically into internationally standard English terms) become the dominant interpretations of situations which in fact require more nuanced understanding. Approaches such as those mentioned above can help to de-centre this process. Alternatively, in some contexts the action of translation and interpreting has itself become the focus for activist movements. One example surrounds the move towards activist interpreting at the World Social Forum by Babels (http://www.babels. org)(see Baker, 2013). It is fair to say, however, that there also remain considerable debates about the relative benefits and limitations of more 'activist' approaches to translation/interpreting. Such initiatives are perceived by some, particularly those involved in professional translation/interpretation, to be moves towards deskilling and casualization of a valuable profession which is key to facilitating effective cross-cultural communication (Boeri, 2008).

The examples in this section suggest it is sometimes possible to work in relatively simple ways with different languages, even in contexts where we are not fluent in the language, and through this to gain an insight into the cultures studied. However, if we want detailed understandings of the processes involved and of the meanings people attach to these processes, we need to pay attention to the possibilities and problems of translation and interpretation between languages. Furthermore, we must be aware of the politics of language use, especially where abilities to communicate in one language or another confer status or access to benefits or privileges, and of the possibility that translation itself can be a political act (Smith, 2009).

6.3 Working With Difference: Power Relations, Positionality and Beyond

Moving on to more general questions of difference, unequal power relations and the position of the researcher, we now consider a range of practical responses by geographers which illustrate the challenges and possibilities of cross-cultural research.

During my own doctoral research in the 1990s, whilst researching urban change in eastern Germany, I became aware of the unease many local residents felt about comparing their experiences to those of western societies. Claims about the 'victory' of the West after the collapse of communism and the practical need to adapt to western administrative and legal frameworks had served to devalue their own cultural and political experiences. In my research I tried to be open to the diverse experiences on which people were drawing to develop new community politics. At one meeting with a group which was developing a local heritage museum, members discussed the problems of industrial decline and unemployment facing their city. Then one person asked why I was interested in their city and what my home city in the UK was like. However, after my attempt to explain some similarities in the effects of deindustrialization and labour-market restructuring between Glasgow, in Scotland, and their city, several people rebuffed my arguments, claiming the situation in Glasgow was in no way like that in their own city.

Initially, I was embarrassed and felt I had lapsed into the problematic stance of what they saw as a 'typical westerner', comparing everything outside western society (usually unfavourably) to the 'normal' West, reflecting an 'ethnocentric' approach, as outlined above. On further reflection it seemed the group members were partially correct, since the pace of change in their city was greater than in Glasgow. However, what was more important was that I had set up a framework where I claimed to be interested in their experiences in and of themselves, without any claims to 'know best'. I had deliberately tried not to essentialize differences between 'East' and 'West', or between capitalism and communism, instead seeking connections as well as differences. Into this framework, I had then introduced a note suggesting what was new, difficult and often very painful for them, with many experiencing redundancy, was, on a global scale, not so unusual, special, or particular. My comparisons inadvertently belittled the severity of people's experiences, denying their individual significance and the particular social and political relations affecting their city. As I developed my analysis, it seemed to me that this moment was not just one where I made a 'mistake' in the fieldwork. The fact some group members found my comparisons unacceptable highlighted precisely the politics of naming processes, cultures and experiences as 'similar' or 'different' which were tied up with the negotiation of these post-communist transitions. Whilst this might seem like an old example, the points its raises remain valid for cross-cultural research today, as outlined by Monica Evans (2012) who, in relation to her research on fathers in Chile, outlines both the value of empathy in establishing relationships with research participants and the need for a very tentative approach which avoids being 'imperialistic' in projecting the self onto the lives of others.

Our research can never escape from the power relations shaping the situations in which we research. We must address these carefully and take account of them in the choices we make in our research practices as well as in the interpretations we develop in an on-going series of negotiations in the field and beyond. One strategy for addressing such inequalities is to work through the complex positionality[5] of the researcher, subjecting the research process itself to scrutiny and not assuming the researcher is a disembodied presence, removed from the research process. This does not mean adopting a self-centred view of research where 'other cultures and other people' become merely the 'exotic backdrops of authorial self-discovery' (Lancaster, 1996: 131). Nor does it mean assuming the researcher can know exactly how they are viewed by the research participants or can account for the significance of every element of their own identity in the research (Rose, 1997). Rather 'we must recognize and take account of our own position, as well as that of our research participants, and write this in to our research practice' (McDowell, 1992: 409) in ways that are sensitive to the difference our presence makes in the research, and how the process of research itself can shape social relations.

Many researchers provide detailed discussions of questions of positionality in cross-cultural research (Madge et al., 1997; Laurie et al., 1999; Sultana, 2007; Chattopadhyay, 2013). Tracey Skelton (2001: 89) provides a definition:

> By positionality I mean things like our 'race' and gender [...] but also our class experiences, our levels of education, our sexuality, our age, our ableness, whether we are a parent or not. All of these have a bearing upon who we are, how our identities are formed and how we do our research. We are not neutral, scientific observers, untouched by the emotional and political contexts of places where we do our research.

This means being aware of how aspects of our own identities are significant, or might change as we 'travel' (spatially or culturally) to different contexts. For Skelton, not having children as a young researcher marked her out as different from the women she interviewed about gender relations in Montserrat. Rather than avoiding their questions about this, she used this as a point of discussion, recognizing that some of the women felt luckier, or more mature than her. 'I found this a healthy way of letting the power I had in the interview context – I was the one asking questions – dissipate and shift into complex positions within the interviews' (Skelton, 2001: 91). To Skelton's discussion of individual positionality we can also add calls to consider in more detail the positionalities and relations between the full range of those involved in the research, including translators/interpreters and various research and field assistants whose roles are too often silenced (Ficklin and Jones, 2009; Turner, 2010).

At times, thinking about positionality leads researchers to question whether it is even appropriate to undertake studies where they are 'outsiders', especially where their outsider status might perpetuate the ways less powerful groups and cultures have been represented by those in more powerful positions, such as westerners representing the experiences of people in developing countries (Madge, 1994), or where those being researched might prefer their experiences are not opened up to public gaze (Barnett, 1997). Kim England (1994) withdrew from researching the lesbian community in Toronto, Canada, because she felt she was too much of an 'outsider' and the context of homophobia in Canada meant her trying to 'speak for' or 'give voice to' the lesbian community led to the danger of her colonizing the experiences of this group of women. She felt it more appropriate to leave research of the community to other lesbian women. However, Robina Mohammad's (2001) discussion of research on Pakistani women in Britain suggests that the need to consider positionality does not disappear where we appear to be 'insiders', since we are also partly 'outsiders' by the very fact that we are engaged in research. Likewise, other aspects of our own identities (dress, accent or education) can be markers of our difference as well as our similarities. Rather than her apparent 'insider' status allowing access to 'the truth', Mohammad suggests that within her study participants presented 'multiple truths'. The challenge was to understand 'which truth' was being told and whose interests were served by particular representations. Similarly, Chattopadhyay (2013: 154) presents a complex and nuanced narration of the insider and outside positions which she continuously negotiated during her fieldwork in the Narmada Valley in India, not only in more abstract and theoretical terms but in 'everyday acts' such as 'how I ate, how I sat […] and where I sat. […] These ways of presenting myself and interacting with my participants, however mundane, are not insignificant because they narrate my embodied situatedness as the researcher'.

A number of geographers engaged in cross-cultural research have argued that addressing positionality, while important, may not in itself be enough to tackle the fundamentally uneven and unequal social relations which structure the research process and which may indeed be the focus of the research. These researchers have explored the role that collaborative approaches (see also Chapter 13), working with individuals and groups to establish a shared set of goals and practices, can play in helping to move 'beyond positionality' and to develop research practices which do not simply involve reflecting on but also acting in ways which might challenge such inequalities. Examples include collaboration in the study of women's oral histories in India between Richa Nagar, Farah Ali and Sangatin Women's Collective (2003), Geraldine Pratt's long-term engagement with migrant communities in Vancouver, Canada (Pratt, 2007), and the collaboration between Red Thread (2000) and Linda Peake on issues of reproductive and sexual health among women in Guyana. Mistry et al. (2009) provide an example from physical geography of considering the opportunities and limitations of shifting from being 'top-down experts' to 'participatory facilitators' in their research relations with partners in Guyana. Usefully they also discuss the variety of often very practical issues that can affect the ability to collaborate, from different institutional requirements to different access to communication and resources.

These examples demonstrate a commitment not just to 'give voice' to those involved (thereby preserving the existing unequal social relations between the privileged researcher and what appear to be the otherwise silent research 'subjects'), but to work with people, to listen to their priorities, to engage with the politics and practices of social change, and to take seriously the challenges of not just noting difference but to work with it and across it. Often adopting participatory methods, collaborative approaches do not remove the need to address the challenges outlined in this chapter. Instead they are 'partnerships in which the questions around how power and authority would be shared cannot be answered beforehand, but are imagined, struggled over and resolved through the collaborative process itself' (Nagar et al., 2003: 369). Here multiple moves across borders and collaborations between and across sectors and diverse forms of difference might 'reimagine reciprocity' (Benson and Nagar, 2006: 581). Schenk (2013) suggests that there may be situations where more collaborative approaches are not possible, but argues an explicit approach to considering the relations between multiple members of what might be viewed as a research 'team' rather than an 'external' researcher and 'local subjects' might be possible even in situations characterized by conflict or emergencies.

To some extent, then, we are always involved in working with 'different' cultures and must negotiate the power relations of similarity and difference in our research whether these cultures are 'remote' or close at hand. As Heidi Nast (1994: 57) argues, 'we can never *not* work with "others" who are separate and different from ourselves; difference is an essential aspect of all social interactions that requires that we are always everywhere in between or negotiating the worlds of me and not-me'. The challenge is to address this in the research strategies we adopt and in the on-going social relations surrounding and permeating our research practice. As Saraswati Raju (2002) asks, 'We are different, but can we talk?'

6.4 (Re)Presenting 'Other' Cultures

Finally, it is important to consider how to represent the people and places studied in the field report or dissertation. Writing to thank those who have helped and, where appropriate, sending a copy of the report is a good start. If you worked in another language or your academic findings would be inappropriate (people might be interested in what you discovered, but not necessarily in your latest theoretical insight), a revised feedback report might be more appropriate, as might be giving a presentation, writing an article for the local newspaper, setting up a project website where people can access your findings or using social media to publicise the results. You might also seek to adopt more 'polyvocal' approaches to writing where the 'expert' researcher is de-centred in favour of a more collaborative approach (though this may be something you have to negotiate as being suitable for the demands of your particular institution if you are working on an assessed piece of work). Even when not producing a co-authored text, you can still ask for comments from participants on your analysis or may even develop the analysis in collaboration with them. These are often complex situations and it does not automatically follow that the research collaborators are right and you are wrong. You can instead work with that difference of interpretation to decentre your position as the apparently all-knowing researcher, compared to the research participants. You may need to discuss whether different kinds of analysis, interpretation or publications may be more important or appropriate for particular audiences (Nagar et al., 2003).

It is important to consider the language in which your research is to be written, though this may also be constrained by the requirements for work to be read and assessed in particular languages. More widely in geographical scholarship this question is important because of the global dominance of English-language publishing in academic writing (Bański and Ferenc, 2013). This affects which languages and concepts are more central to the development of geographical knowledges and which are marginalized (Garcia-Ramon, 2003). Your writing strategy might engage with this, for example by writing a bilingual text (Cravey, 2003). Such consideration applies to visual representations as well as to written texts, as the following example illustrates.

American anthropologist Kathleen Kuehnast (2000) had been researching the economic burdens faced by women in regions affected by farm privatization in Kyrgyzstan, a central Asian post-Soviet state. For the cover of her report she chose a photograph of 'a Kyrgyz elderly woman and her daughter-in-law, holding a baby, each dressed in the warm clothing of semi-nomadic herders' (Kuehnast, 2000: 105). She was surprised when some women in government jobs, whom she knew, were offended by this, feeling it presented their country in a poor light. After initially dismissing this as the unwillingness of higher-status women to address the problems facing nomadic women in their own country, Kuehnast wondered why else the photograph was problematic. Perhaps it failed to represent women's achievements in education and employment, buying in too strongly to the notion of women as victims of communist repression? Did it contradict important Soviet-era ideas of women as strong and competent workers? Alternatively, many of the women interviewed by Kuehnast saw in the western media images of glamorous women, which flooded the country after independence, an ideal of 'western', 'modern' or 'American' womanhood (with the 'right' clothes, make-up and leisured lifestyle). In this context, the portrayal of Kyrgyz women in traditional clothing and as working women may have seemed to illustrate too sharply the apparent 'failings' of Kyrgyz women in adopting suitably 'westernized' or 'modernized' gender identities.

Representation is fundamentally problematic and particularly so in cross-cultural research. The demands of 'analysis, writing up and dissemination of information often force us to detach ourselves, switch back to "western mode" to produce texts and develop "distance" to use information' (Madge, 1994: 95). As Kuehnast found, even analysis and representation which is meant to be helpful is not immune to being problematic for the research participants. Each researcher therefore must make her or his own choices, often resulting in pragmatic responses to the particular situation. Clare Madge (1994: 96), in her research on the Gambia, decided that some of the information she gathered talking to people who became her friends could not be included in her analysis, since to use that information would 'betray the trust of my friends'. Audrey Kobayashi (2001) opted to stop working on 'other' cultures outside Canada and instead to focus on working with Japanese-Canadians within Canada in collaborative activist research. Tracey Skelton at times felt it was impossible to write about her research on Montserrat without reproducing the unequal colonial relations she was trying to combat, so for some time she decided not to write on these topics. However, in the end she decided that 'as part of the politics of reflective and politically conscious feminist and/or cross-cultural research, we have to continue our research projects, we must publish and disseminate our research. If we do not, others without political anxieties and sensitivities about their fieldwork processes take the space' (2001: 95).

6.5 Conclusion

There are many challenges in working in different languages and different cultures and even experienced researchers do not always get it right. However, the key issue is to pay attention to the issues raised in this chapter as we plan and undertake fieldwork, and analyse and represent what we find, keeping in mind 'why we are doing it and what the research we do means to other people' (Skelton, 2001: 96). When done well, research in different languages and different cultures is incredibly enriching and challenges us to think about difference and diversity in productive and sensitive ways.

SUMMARY

- Cross-cultural research is challenging, enriching and rewarding.
- Concepts of cultural difference should inform such research, taking into account the fluidity of cultures, unequal social relations, the need to avoid ethnocentrism and to be open to hybrid cultural forms while keeping a balance between 'hopeful' and 'critical' approaches to difference.
- Simple strategies may address language differences but care needs to be taken with how attention to language use and the articulation of meanings might provide insight into the 'other' culture. Translation requires careful consideration of how meanings 'map' between cultures, the translation strategy being adopted, and negotiation of the roles of translators and interpreters as part of the research process.
- Addressing the power relations surrounding research, writing the positionality of the researcher into our research accounts, or adopting collaborative strategies can help reduce ethnocentrism or even begin to challenge unequal social relations. 'Outsiders' and 'insiders' should consider carefully their relation to the research. This may provide a variety of accounts without necessarily producing the single 'correct' answer to interpreting a particular situation.
- Choices about representation in written, verbal or visual formats should avoid reinforcing unequal power relations or stereotypes and be informed by the ethics and politics of the research.

Notes

1 'Other' is often used in cultural analysis with inverted commas, and sometimes an initial capital letter, e.g. the 'Other', or 'other' cultures, to imply that such cultures, social groups or societies may not be as different as is implied by cultural and social norms but that they may be constructed as such through a variety of social, cultural, political or economic processes.

2   'Ethnocentric' or 'ethnocentricism' relates to the prioritizing of one's own world-view and experiences, of one's own culture as the 'norm' against which others are measured for their 'strangeness', 'lack' of development, 'difference', or 'exoticism'. It often implies taking western experiences as the norm.

3   Drawing on the work of Edward Said (1978), 'orientalism' describes and critiques the set of ideas common among Europeans in their depictions of 'the Orient' during the era of imperialism. Such ideas emphasized both the apparent attractions and exotic nature of such people, places and cultures, and their supposed danger and lack of civilization, helping to 'justify' European colonial endeavours. The terms have subsequently been applied more widely to all colonial situations and also to contemporary representations of 'other' cultures, peoples and places.

4   'Diaspora' describes the dispersal or scattering of a population. It can also refer as a noun to dispersed or forcibly displaced populations such as the Black diaspora or Jewish diaspora. In more theoretical terms, the idea of diaspora, or diasporic communities, challenges notions of fixed connections between cultures, identity and place. The term 'transnational community' is used to describe social and cultural relations that transcend and escape the bounded spaces of the nation-state, possibly as a result of migration and diaspora or because the social group does not fit existing national boundaries (such as the Kurdish population).

5   Positionality can be understood as a consideration of how the relative position of the researcher may affect the process of the research. For example, information given by participants may depend on how the researcher is viewed in that particular context (threatening, insignificant, powerful). The researcher may also be relatively privileged by their positionality.

Further Reading

This guide to further reading identifies references for the key themes of the practicalities of fieldwork abroad, debates about cultural difference, translation strategies and negotiating power relations:

• Two articles by Nash (2000a, 2000b) discuss practical issues in undertaking independent fieldwork abroad. The first addresses establishing contacts, legal requirements for visas, collecting and exporting samples, health and safety issues, and training. The second considers budgeting and fundraising. The website of the Royal Geographical Society (UK) also gives a whole variety of links to practical sources about international fieldwork – http://www.rgs.org/OurWork/Fieldwork+and+Expeditions/Fieldwork+Expeditions.htm

• Dywer (2014), Skelton and Allen (1999) and McEwan (2008) explore current debates about contemporary cultural change. All provide useful overviews. Although it is an older source, Hall (1995) is a key discussion of debates about globalization, culture and difference which underpin recent debates.

• Smith (1996, 2009), Twyman et al. (1999), Bujra (2006), Müller (2007) and Turner (2010) all consider the issue of translation between different languages and of working with a translator. Each discusses how translation itself can become part of the focus for analysis.

• Websites of relevance to issues of translation include:

   o Google Translate – https://translate.google.com – probably the best known of the machine code translation sites available (though see the comments in this chapter about the limitations of such sites);

   o Linguee is an example of an online dictionary which covers a limited range of languages but which gives more detailed examples of the translation in context in both the source and the target language. Useful for understanding more of the nuances of translation – http://www.linguee.com;

   o An example of a professional code of ethics for interpreters and translators can be found at Language Line Solutions – http://www.languageline.co.uk/assets/Interpreter_Code_of_Ethics.pdf;

   o An example of 'activist' approaches to interpreting and translation can be found at "Babels.org" – http://www.babels.org

• Pratt et al. (2007), Nagar et al. (2003) and Chattopadhyay (2013) provide detailed examples of negotiating power relations, positionality and representation in cross-cultural research.

Note: Full details of the above can be found in the references list below.

References

Baker, M. (2013) 'Translation as an alternative space for political action', *Social Movement Studies*, 12: 23–47.

Bański, J. and Ferenc, M. (2013) '"International" or "Anglo-American" journals of geography?', *Geoforum*, 45: 285–95.

Barnett, C. (1997) '"Sing along with the common people": Politics, postcolonialism and other figures', *Environment and Planning D: Society and Space*, 15: 137–54.

Benson, K. and Nagar, R. (2006) 'Collaboration as resistance? Reconsidering the processes, products and possibilities of feminist oral history and ethnography', *Gender, Place and Culture*, 13: 581–92.

Boeri, J. (2008) 'A narrative account of the Babels vs. Naumann controversy: Competing perspectives on activism in conference interpreting', *The Translator*, 14: 21–50.

Bujra, J. (2006) 'Lost in translation? The use of interpreters in fieldwork', in V. Desai and R. Potter (eds) *Doing Development Research*. London: Sage. pp. 172–9.

Chattopadhyay, S. (2013) 'Getting personal while narrating the "field": A researcher's journey to the villages of the Narmada valley', *Gender, Place and Culture*, 20: 137–59.

Cloke, P. (2014) 'Self-Other', in P. Cloke, P. Crang and M. Goodwin (eds) *Introducing Human Geographies* (3rd edition). London: Routledge. pp. 63–81.

Collins, F.L. (2009) 'Transnationalism unbound: Detailing new subjects, registers and spatialities of cross-border lives', *Geography Compass*, 3: 434–58.

Cravey, A. (2003) 'Toque una ranchera, por favour', *Antipode*, 35: 603–21.

Driver, F. (2014) 'Imaginative geographies', in P. Cloke, P. Crang and M. Goodwin (eds) *Introducing Human Geographies* (3rd edition). London: Routledge. pp. 234–48.

Dudrah, R.K. (2002) 'Drum n dhol: British Bhangra music and diasporic South Asian identity formation', *European Journal of Cultural Studies*, 5(3): 363–83.

Dwyer, C. (2014) 'Diasporas', in P. Cloke, P. Crang and M. Goodwin (eds) *Introducing Human Geographies* (3rd edition). London: Routledge. pp.669–85.

Eco, U. (1993) *La ricerca della lingua perfetta nella cultura europea/The Search for the Perfect Language in the European Culture*. Rome: Laterza.

Edwards, K.J., Guttesen, R., Sigvardsen, P.J. and Hansen, S.S. (2010) 'Language, overseas research and a stack of problems in the Faroe Islands', *Scottish Geographical Journal*, 126: 1–8.

England, K. (1994) 'Getting personal: Reflexivity, positionality and feminist research', *Professional Geographer*, 46(1): 80–9.

Evans, M. (2012) 'Feeling my way: Emotions and empathy in geographic research with fathers in Valparaíso, Chile', *Area*, 44: 503–9.

Ficklin, L. and Jones, B. (2009) 'Deciphering "voice" from "words": Interpreting translation practices in the field', *Graduate Journal of Social Science*, 6: 108–30.

Garcia-Ramon, M.D. (2003) 'Globalization and international geography: The questions of languages and scholarly traditions', *Progress in Human Geography*, 27(1): 1–5.

Garibova, J. (2009) 'Language policy in post-Soviet Azerbaijan: Political aspects', *International Journal of the Sociology of Language*, 198: 7–32.

Gilroy, P. (1993) *The Black Atlantic: Modernity and Double Consciousness*. London: Verso.

Glass, M.R. (2014) 'Encouraging reflexivity in urban geography fieldwork: Study abroad experiences in Singapore and Malaysia', *Journal of Geography in Higher Education*, 38: 69–85.

Grundy-Warr, C. and Sidaway, J. (2008) 'The place of the nation-state', in P. Daniels, M. Bradshaw, D. Shaw and J. Sidaway (eds) *An Introduction to Human Geography*. Harlow: Prentice Hall. pp. 417–37.

Hall, S. (1995) 'New cultures for old', in D. Massey and P. Jess (eds) *A Place in the World? Places, Cultures and Globalization*. Oxford: Oxford University Press, Open University Press. pp. 175–213.

Hammersley, L.A., Bilous, R.H., James, S.W., Trau, A.M. and Suchet-Pearson, S. (2014) 'Challenging ideals of reciprocity in undergraduate teaching: The unexpected benefits of unpredictable cross-cultural fieldwork', *Journal of Geography in Higher Education*, 38: 208–18.

Hawthorne, T.L., Atchison, C. and LangBruttig, A. (2014) 'Community geography as a model for international research experiences in study abroad programmes', *Journal of Geography in Higher Education*, 38: 219–37.

Hörschelmann, K. (2002) 'History after the end: Post-socialist difference in a (post)modern world', *Transactions of the Institute of British Geographers*, 27: 52–66.

Ives, P. (2010) 'Cosmopolitanism and global English: Language politics in globalization debates', *Political Studies,* 58: 516–35.

Klek, A. (2014) *The Place-Making Process of Post-Accession Migrants in Dundee*. MA Dissertation. University of Dundee.

Kobayashi, A. (2001) 'Negotiating the personal and the political in critical qualitative research', in M. Limb and C. Dwyer (eds) *Qualitative Methodologies for Geographers*. London: Arnold. pp. 55–70.

Kuehnast, K. (2000) 'Ethnographic encounters in post-Soviet Kyrgyzstan: Dilemmas of gender, poverty and the Cold War', in H. de Soto and N. Dudwick (eds) *Fieldwork Dilemmas: Anthropologists in Postsocialist States*. Madison, WI: University of Wisconsin Press. pp. 100–118.

Lancaster, R.N. (1996) 'The use and abuse of reflexivity', *American Ethnologist*, 23(1): pp. 130–2.

Language Line Solutions (2013) *Interpreter Code of Ethics*. Available from http://www.languageline.co.uk/assets/ Interpreter_Code_of_Ethics.pdf (accessed 10 November 2015).

Laurie, N., Dwyer, C., Holloway, S.L. and Smith, F.M. (1999) *Geographies of New Femininities*. Harlow: Longman.

Levitt, P. (2007) *God Needs No Passport: Immigrants and the Changing American Religious Landscape*. New York: New Press.

Madge, C. (1994) 'The ethics of research in the "Third World"', in E. Robson and K. Willis (eds) *DARG Monograph No. 8: Postgraduate Fieldwork in Developing Areas*. Developing Areas Research Group of the Institute of British Geographers. pp. 91–102.

Madge, C., Raghuram, P., Skelton, T., Willis, K. and Williams, J. (1997) 'Methods and methodologies in feminist geographies: Politics, practice and power', in Women and Geography Study Group, *Feminist Geographies: Explorations in Diversity and Difference*. Harlow: Longman. pp. 86–111.

Marneros, S. and Gibbs, P. (2015) 'An evaluation of the link between subjects studied in hospitality courses in Cyprus and career success: perceptions of industry professionals', *Higher Education, Skills and Work-based Learning*, 5(3): 228–41.

Massey, D. and Jess, P. (1995) 'Places and cultures in an uneven world', in D. Massey and P. Jess (eds) *A Place in the World? Places, Cultures and Globalization*. Oxford: Oxford University Press, Open University Press. pp. 215–40.

McDowell, L. (1992) 'Doing gender: feminism, feminists and research methods in human geography', *Transactions of the Institute of British Geographers*, 16: 400–419.

McEwan, C. (2008) 'Geography, culture and global change', in P. Daniels, M. Bradshaw, D. Shaw and J. Sidaway (eds) *An Introduction to Human Geography*. Harlow: Prentice Hall. pp. 273–89.

Mistry, J., Berardi, A. and Simpson, M. (2009) 'Critical reflections on practice: the changing roles of three physical geographers carrying out research in a developing country', *Area*, 41: 82–93.

Mohammad, R. (2001) '"Insiders" and/or "outsiders": Positionality, theory and praxis', in M. Limb and C. Dwyer (eds) *Qualitative Methodologies for Geographers*. London: Arnold. pp. 101–17.

Müller, M. (2007) 'What's in a word? Problematizing translation between languages', *Area*, 39(2): 206–13.

Nagar, R. in consultation with F. Ali and Sangatin Women's Collective, Sitapur, Uttar Pradesh, India (2003) 'Collaboration across borders: Moving beyond positionality', *Singapore Journal of Tropical Geography*, 24(3): 356–72.

Nairn, K. (1996) 'Parties of geography fieldtrips: Embodied fieldwork', *New Zealand Women's Studies Journal*, 12: 88–97.

Nairn, K., Higgitt, D. and Vanneste, D. (2000) 'International perspectives on fieldcourses', *Journal of Geography in Higher Education*, 24(2): 246–54.

Nash, D.J. (2000a) 'Doing independent overseas fieldwork 1: Practicalities and pitfalls', *Journal of Geography in Higher Education*, 24: 139–49.

Nash, D.J. (2000b) 'Doing independent overseas fieldwork 2: Getting funded', *Journal of Geography in Higher Education*, 24: 425–33.

Nast, H. (1994) 'Opening remarks on "Women in the Field"', *Professional Geographer*, 46: 54–66.

O'Grady, R. (1975) Tourism, the Asian Dilemma. *Report of a Study of Asian Tourism* (Christian Conference of Asia, Jan–Jun 1975).

Phillips, R. (2014) 'Colonialism and postcolonialism', in P. Cloke, P. Crang and M. Goodwin (eds) *Introducing Human Geographies* (3rd edition). London: Routledge, pp. 493–508.

Potter, R., Binns, T., Elliott, J. and Smith, D. (2008) *Geographies of Development: An Introduction to Development Studies* (3rd edition). Harlow: Prentice Hall.

Pratt, G. (2007) in collaboration with the Philippine Women Centre of B.C. and *Ugnayan ng Kabataang Pilipino sa Canada*/Filipino-Canadian Youth Alliance 'Working with migrant communities: collaborating with the Kalayaan Centre in Vancouver, Canada', in S. Kindon, R. Pain and M. Kesby (eds) *Participatory Action Research Approaches and Methods: Connecting People, Participation and Place*. London: Routledge. pp. 95–103.

Raju, S. (2002) 'We are different, but can we talk?', *Gender, Place and Culture*, 9(2): 173–7.

Red Thread (2000) *Women Researching Women: Study on Issues of Reproductive and Sex Health and of Domestic Violence against Women in Guyana*. Report of the Inter-American Development Bank (IDB) Project TC-97-07-40-9-GY conducted by Red Thread Women's Development Programme, Georgetown, Guyana, in conjunction with Dr Linda Peake. available at http://www.hands.org.gy/download/wom_surv.htm (accessed 10 November 2015).

Rose, G. (1992) 'Geography as a science of observation: The landscape, the gaze and masculinity', in F. Driver and G. Rose (eds) *Nature and Science: Essays in the History of Geographical Knowledge*. London: IBG Historical Geography Research Group. pp. 8–18.

Rose, G. (1997) 'Situating knowledges: positionality, reflexivity and other tactics', *Progress in Human Geography*, 21(3): 305–20.

Said, E. (1978) *Orientalism*. Harlow: Penguin.

Schenk, C.G. (2013) 'Navigating an inconvenient difference in antagonistic contexts: Doing fieldwork in Aceh, Indonesia', *Singapore Journal of Tropical Geography*, 34: 342–56.

Simon, D. (2006) 'Your questions answered? Conducting questionnaire surveys', in V. Desai and R. Potter (eds) *Doing Development Research*. London: Sage. pp. 163–71.

Skelton, T. (2001) 'Cross-cultural research: Issues of power, positionality and "race"', in M. Limb and C. Dwyer (eds) *Qualitative Methodologies for Geographers*. London: Arnold. pp. 87–100.

Skelton, T. and Allen, T. (eds) (1999) *Culture and Global Change*. London: Routledge.

Smith, F.M. (1996) 'Problematizing language: Limitations and possibilities in "foreign language" research', *Area*, 28(2): 160–6.

Smith, F.M. (2006) 'Encountering Europe through fieldwork', *European Urban and Regional Studies*, 13(1): 77–82.

Smith, F.M. (2009) 'Translation', in R. Kitchin and N. Thrift (eds) *International Encyclopedia of Human Geography*. London: Elsevier. pp. 361–67.

Sorgen, A. (2015) 'Integration through participation: The effects of participating in an English Conversation club on refugee and asylum seeker integration', *Applied Linguistics Review*, 6(2): 241–60.

Sparke, M. (1996) 'Displacing the field in fieldwork: Masculinity, metaphor and space', in N. Duncan (ed.) *BodySpace*. London: Routledge. pp. 212–33.

Sultana, F. (2007) 'Reflexivity, positionality and participatory ethics: Negotiating fieldwork dilemmas in international research', *ACME*, 3: 374–85.

Thickett, A., Elekes, Z., Allaste, A.-A., Kaha, K., Moskalewicz, J., Kobin, M. and Thom, B. (2013) 'The meaning and use of drinking terms: Contrasts and commonalities across four European countries', *Drugs: Education, Prevention and Policy*, 20: 375–82.

Tobar, H. (2005) *Translation Nation: Defining a New American Identity in the Spanish-speaking United States*. New York: Riverhead Books.

Tremlett, A. (2009) 'Claims of "knowing" in ethnography: Realizing anti-essentialism through a critical reflection on language acquisition in fieldwork', *Graduate Journal of Social Science*, 6: 63–85.

Turner, S. (2010) 'The silenced assistant: Reflections of invisible interpreters and research assistants', *Asia Pacific Viewpoint*, 51: 206–19.

Twyman, C., Morrisson, J. and Sporton, D. (1999) 'The final fifth: Autobiography, reflexivity and interpretation in cross-cultural research', *Area*, 31: 313–26.

Valentine, G., Sporton, D. and Nielsen, K.B. (2008) 'Language use on the move: Sites of encounter, identities and belonging', *Transactions of the Institute of British Geographers*, 33: 376–87.

Venuti, L. (ed.) (2012) *The Translation Studies Reader* (3rd edition) London: Routledge.

Watson, E.E. (2004) '"What a dolt one is": Language learning and fieldwork in geography', *Area*, 36: 59–68.

ON THE COMPANION WEBSITE...

Visit **https://study.sagepub.com/keymethods3e** for author videos, chapter exercises, resources and links, plus **free** access to the following recommended articles:

1. **Panelli, R. (2008) 'Social geographies: Encounters with Indigenous and more-than-White/Anglo geographies', *Progress in Human Geography*, 32(6): 801–11.**

This review explores how paying attention to concepts and understandings of 'more-than-White' or 'Indigenous' geographies can offer ways to rethink and de-centre key concepts in social geography such as home, place and society–environment relations.

2. **Wright, M.W. (2009) 'Gender and geography: Knowledge and activism across the intimately global', *Progress in Human Geography*, 33(3): 379–86.**

The paper discusses the politics of knowledge production (whose knowledge counts, how is it heard, how is it represented) and of research-related activisms (what do we do about what we discover in research) through a focus on feminist research on the intersections of the local, the global and gender within the context of globalisation and social (in)justice.

3. **Glassman, J. (2009) 'Critical geography I: the question of internationalism', *Progress in Human Geography*, 33(5): 685–92.**

This article reflects critically on experiences at the 2007 International Critical Geographies Conference in Mumbai and on a related fieldtrip in order to discuss how, even in such critically-aware contexts, issues such as language, translation, power and positionality emerge as significant factors in how geographical knowledge develops.

SECTION TWO

Generating and Working with Data in Human Geography

7 Historical and Archival Research

Ruth Craggs

SYNOPSIS

This chapter describes different sorts of **historical evidence,** which can be anything from a document to an image, oral history interview, or a building or landscape. It stresses that historical sources are always **fragmentary** because not everything in the past leaves a trace, and not all traces produced are then preserved. It also demonstrates that historical evidence is also always **partial:** it represents the views, priorities and knowledge of those who produced it. It is also shaped by the subsequent visions and priorities of generations of governments, businesses, archive professionals, and individuals who have chosen to select, order and preserve certain historical sources (but not others). Therefore the chapter argues that historical evidence should be understood as **'socially constructed'**. In other words, historical evidence is not an objective record waiting to be uncovered, but is constructed through the cultural, political, economic and social contexts of its production and preservation. Therefore we need to conceptualize the historical **'archive'** not just as a (virtual or physical) location, but as a site of (selective) memory produced through, and reproducing, relations of power. The fact that historical evidence is fragmentary, partial, and socially constructed, means that careful **evaluation** and **analysis** are required in order to produce valid research.

The chapter provides a practical guide to developing historical research questions, finding sources, approaching archival collections, and analysing and presenting historical evidence. It is organized into the following sections:

- Introduction
- Developing historical questions
- Finding historical sources
- Approaching archival collections
- Analysing and presenting your data
- Conclusion

7.1 Introduction

'Enchanting, mysterious, seductive and addictive' are not words you might associate with historical research, and yet, these are precisely the terms used by Sarah Mills to describe archives in a recent article (2013a: 703). Although it can be time-consuming and challenging, work with historical sources can bring the researcher into contact with unique, ancient, and important materials. One might encounter, for example, the Scottish explorer and missionary of Africa David Livingstone's hat (which is in the Royal Geographical Society's collections), recordings of the first Beatles concert, or newspaper cuttings reporting Nelson Mandela's release from prison. Historical research can also provide privileged access to the everyday lives, hopes and fears of individual people through their personal letters, diaries, and photographs. You might hear the voices of the first Commonwealth migrants reflecting on their experience of moving to Britain, or read diaries of those caught up in the London blitz. The excitement of getting to see, touch, and examine these materials, of finding the missing piece of evidence which explains a question as yet unanswered, and of turning up something completely unexpected in the archives motivates many to conduct historical research.

Providing a window onto past lives, a historical perspective can also allow us to think differently about the present day. Whether your primary focus is historic or contemporary, there is a huge range of rich sources available through which it is possible to reconstruct lives and places past: oral history, documentary sources, diaries and letters, moving and still images, and historical objects. Geographers have even drawn upon landscape as historical text (see Duncan and Duncan, 1988; Duncan, 2004). In this chapter, I provide an introduction to a range of different historical sources available and provide guidelines for finding and using these sources. I also suggest ways of evaluating and analysing the evidence available.

Throughout, I highlight the need to approach these historic sources like any other form of data: carefully and critically. The past is not just there to uncover in the archives; historical sources provide only a fragmentary and partial record. Even records we might think of as objective, trustworthy and official – such as court records or census data for example – present an incomplete picture. Certain groups, particularly those who have historically been marginalized or less powerful – women, the lower classes, indigenous peoples – are often absent in these sources. When they are present in the record, it is often as they are viewed by (more powerful) others, rather than in their own words. Work, and public life, rather than domestic experiences (including domestic labour and paid work carried out in the home) also appear more often in the record (though see Blunt and John, 2014, on historical sources and methods for domestic practice). Historical research, therefore, entails not only finding relevant sources, but also working out what they can (and can't) tell you through careful analysis. It is the job of the researcher, therefore, not only to find the materials, but also then to spend time working out what they mean.

Section 7.2 considers how to ask and answer different sorts of historical questions. Section 7.3 provides an outline of the sorts of sources available to the researcher, and some strategies for locating them. Section 7.4 explores how to approach the archive, providing practical tips for visiting and working in archives, alongside some ways of conceptualizing archival research. Section 7.5 provides a summary of different approaches to analysing historical data, alongside some creative ways of presenting your research.

7.2 Developing Historical Questions

Historical research projects usually begin with a research problem or question. This might evolve from topics studied in class: poverty, travel, work, leisure or urbanisation.

Or they might respond to the questions asked by other researchers: often authors conclude their journal articles and books by suggesting future directions for research. Once you have settled on a question of interest, the first consideration must always be to investigate whether there are sources available with which to answer it. If the subject is relatively recent, then you may be able to speak to those involved (see Chapter 9 of this volume on semi-structured interviews and Riley and Harvey (2007) on the use of oral history in geography). Otherwise, you need to consider what other sources might be available.

Let's take an example: researching British women's experience of travel in the nineteenth and twentieth centuries. Where might relevant information be found? Travellers might have written diaries, letters, or published accounts of their experiences (Blunt, 2000; Keighren and Withers, 2011; 2012). Perhaps they took photographs, or sketched; more recent travellers may even have recorded films (Brickell and Garrett, 2013). Guides may have been written to help women to prepare for travelling abroad (Blunt, 1999). Perhaps these women applied for funding from travellers' societies, or gave lectures once they returned (Evans et al., 2013). Might they also crop up in the records produced by those who they met on the road? This brief example provides evidence of the wealth of different sorts of sources available, although it is important to note that any such records may be much more comprehensive for middle- and upper-class women who had the greater means through which to travel and also to record and disseminate their experiences.

The next stage is to consider what sorts of institutions might have collected these sorts of materials at the time, and might preserve them today. In the example above, alongside searches of published materials, such as autobiographies and travel narratives, one could also explore the institutional archives of organizations such as the Royal Geographical Society, involved in planning, funding and reporting men's and (later) women's travel, and the records of the UK Foreign and Colonial Offices, if they came into contact with British officials overseas. A good way of identifying helpful materials is to read others' research on similar topics and look at the sources they have used as a starting point. Might the same archival collection or type of source also be useful to you?

A second approach to developing a research project is to pinpoint an interesting source or collection of sources, and to consider what sorts of historical questions they suggest. In reality, research projects often develop through a mixture of both of these strategies: a topic of interest is identified, initial archival searches are conducted, and the sources uncovered lead to the refining of research questions (Ogborn, 2010). Sometimes searching an archive for answers to a particular question may alert you to a whole other set of questions that are far more interesting or important, provoking a change of direction. This means that it is important to have an initial look at the sources available early on in the process, and to be flexible when approaching the evidence available (see Lorimer, 2009, on archival research and 'make-do methods').

When developing a research project it is also important to be practical. A great set of sources can be useless if you can't access them, either because of their location, or because of the language they are written in. There are solutions to these issues – translation, reproduction of records, travel – but these are often time consuming and prohibitively expensive; they may also have methodological implications (see Chapter 6 of this volume for working with translators). You might find that all the evidence needed can be found in one place (more likely if you are concentrating your research on one sort of source – newspaper accounts, or tax returns, for example). However, often research will involve gathering together a range of different source materials in order to build up a fuller picture: Box 7.1 provides an example of such a project.

Box 7.1   Drawing on multiple archival sources

(Adapted from Livsey, 2014)

Tim Livsey's research used the University of Ibadan in Nigeria as a case study through which to explore the relationship between the built environment, colonial and nationalist politics, and development. His research involved drawing a wide range of archival and published sources together. The University of Ibadan's own institutional archives provided a crucial source. In addition, reports of official delegations visiting Nigeria were found in the Nigerian National Archive in Ibadan, whilst Nigerian newspapers (student papers, but also local and national titles) provided evidence for the reception of the university buildings once they were constructed and the behaviour of students. Personal papers, held in Ibadan, provided evidence for the experience and views of senior university staff, whilst published memoirs of former students – most notably the poet Wole Soyinka – provided a different view of the university. However, the university began as a British colonial project, therefore many important records were also found in UK archives. A published report on the state of higher education in West Africa prior to the university was found in a specialist library collection (The Institute of Education Library); UK parliamentary debates about the university were recorded in Hansard; and the National Archives and the Royal Institute for British Architecture (RIBA) – both in London – held correspondence with the university's architects. Photographs and other university records were uncovered in other specialist collections in the UK: the Royal Commonwealth Society Collection now housed at Cambridge University Library (see http://www.lib.cam.ac.uk/deptserv/rcs), and the former Rhodes House Library, Oxford (the collection is now housed in the Weston Library: http://www.bodleian.ox.ac.uk/weston).

Figure 7.1   A newly constructed university hall of residence, early 1950s.
Reproduced by kind permission of the Syndics of Cambridge University Library

7.3 Finding Historical Sources

As the example in Box 7.1 suggests, a huge variety of sources is available to the historical researcher. Books, published in the period that you are interested in, provide an accessible and important source that is easy to overlook. Fiction can provide evidence of historical landscapes and lives, as well as providing access to broader values and visions of the author and their readers. For example, Yi-Fu Tuan (1985) examined Arthur Conan Doyle's *Sherlock Holmes* novels in order to uncover Victorian ideas about the city, Tim Cresswell (1993) explored narratives of mobility in Jack Kerouac's *On the Road,* and Mandy Morris (1996) examined English gendered identity in Frances Hodgson Burnett's *The Secret Garden.* Travel narratives, guidebooks, poetry, political speeches and manifestos, as well as textbooks, encyclopedias and academic or popular journals, can provide clear evidence of the values and views of the past. Periodicals of learned societies or professions provide a particularly good source for examining the changing state of scientific, geographical and other forms of disciplinary knowledge (Philo, 1987); whilst similar publications by political and other interest groups provide an important record of contemporary political views (Keighren, 2013, provides a good summary of recent work in geography which has focused on the book and geographical knowledge). Other printed sources, such as parliamentary papers and Hansard, the edited record of the UK Parliament, provide a good insight into how the concerns of the period – health, welfare and the poor, empire, prisons – were discussed, understood, and acted upon (on prisons see Ogborn, 1995; on the poor law see Driver, 1993).

Specialist libraries hold collections pertinent to particular places: local studies libraries contain published and unpublished material about the town or county, whilst other institutions focus on particular sorts of source, such as newspapers (the British Library's Newspaper Library), or maps (The Royal Geographical Society, The British Library, The National Library of Wales). Others still bring together materials on particular issues, such as The Wellcome Library for the History of Medicine or The Women's Library, now housed at the LSE, which documents women's lives, focusing particularly on the UK and the political, economic and social transformations of the last 150 years. Containing over 60,000 books and pamphlets and 3,000 periodical titles, alongside over 500 archives and 5,000 museum objects, The Women's Library reflects a common trend: many libraries also hold significant and important archival collections. Other special collections also held at university libraries include the Foreign and Commonwealth Office Historical Collection, held at King's College London, the DH Lawrence collection at the University of Nottingham, and the Mass Observation Archive, held at the University of Sussex. If you are still deciding on a topic for your research, it is worth investigating the archival collections held by your institution or locally: they might provide the impetus for a project.

With the growth of online resources such as digitized records (of which more below), as well as online 'archives' such as Flickr, Instagram and Twitter (see Chapters 16 and 17), the physical space of a collection is also becoming less important than it once was. Nevertheless, much historical research still takes place in more traditional archives. Often the most well-known and largest archives in any country are those which hold the official record of government. In the UK, this is the National Archives, held at Kew, London, which is a huge, well-resourced collection open to all. These state records are particularly good at providing evidence of national policy in regard to health, education, housing and other forms of welfare, policing, law and justice, and foreign and colonial relations. Other official records also provide a good record of the interactions between the state and its people: for example through records of military personnel, immigration and emigration, tax, births marriages and deaths. However, official records are less good at recording different aspects of peoples' lives – their feelings, personal relationships, leisure activities, as well as anything – for example illegal activity – deliberately hidden from the official gaze (see Chapter 12 on researching emotion). Official records also sometimes omit materials which may be deemed incriminating to the authorities concerned (see Box 7.5).

Other collections, often smaller, less well-resourced or more difficult to access, can fill some of the gaps left in the official record. For example, the social research organization Mass Observation, founded in 1937 and active until the 1950s, aimed to provide a record of everyday life in Britain, or as they called it, an 'anthropology of ourselves'. The records of their team of writers, photographers and observers, as well as subsequent Mass Observation projects, provide evidence of attitudes to work, leisure, sexuality, religion, alcohol and many other issues, as well as observations of people's everyday practices. Other collections, such as that of the British Broadcasting Corporation (www.bbc.co.uk/archive/), the British Cartoon Archive (www.cartoons. ac.uk) and the British Film Institute (BFI) National Archive (www.bfi.org.uk/archive-collections), provide materials relating to popular culture.

Business archives, too, can provide different types of information to that held by the state; most obviously they can provide records of economic transactions, business practices, and networks of trade. But they may also offer insights into (geo)political structures and transformations. For example, the archives of Barclays Bank and the Bank of England have been utilized to explore processes of decolonization in different ways: employment policies and correspondence were drawn upon to examine the 'Africanisation' of staff in African bank branches, whilst the records of currency design held by the Royal Mint display the national imagery of newly independent countries (Decker, 2005; Eagleton, 2016). Thus business archives can tell us not only about business, but also about the social and cultural politics of race and representation. The Business Archives Council website provides useful information about researching business history (www.businessarchivescouncil.org.uk/).

The records of other institutions can also hold important materials. Individual schools, places of worship, special interest groups, societies, sports teams, political parties, campaign groups, and charities may have their own archival collections. For example, Sarah Mills (2013b; 2013c) has explored the historical geographies of youth, education and citizenship though the various records of the UK Scouting Association, and Gavin Brown and Helen Yaffe (2013) have explored anti-apartheid activism and solidarity in London through the 'Non-Stop against apartheid' archive (for a fantastic online archive of the UK Anti-Apartheid Movement, see also www. aamarchives.org).

Finally, the archives of individuals – from a shoebox in the loft to extensive collections formally catalogued – can also provide interesting and important insights. Often housed in domestic spaces, the experience of using such archives is very different, and they can afford different opportunities for engagement with the producer of the archives (Ashmore et al., 2012). They can tell you about 'the vernacular or private sphere of everyday life' but also provide insights that often stretch beyond the individual and the domestic (2012: 82–3).

How does one go about tracking down these archival collections? The UK National Archives provides a good starting point for research in the UK, providing useful catalogues and research guides (see Box 7.2 for details of these and other important catalogues). Another helpful place to begin is with secondary reading: others researching similar topics will provide details of the historical sources they have drawn upon in their work. Finally, if you have particular individuals, institutions or events in mind, it is worth doing a simple Internet search: in many cases obscure collections held by individuals or online archives turn up through this route (see for example the online photographs, diaries and present day reminiscences drawn upon to reconstruct 1960s expeditions in Craggs, 2011).

**Box 7.2   List of useful catalogues of archives
and online collections**

(Adapted from Ogborn, 2010)

Discovery, hosted by the UK National Archives, is a comprehensive online catalogue of archival records across the UK and beyond, from which you can search 32 million records, including the National Archives collections (9 million of which are in downloadable format), and the records of other archival collections (http://discovery.nationalarchives.gov.uk).

Research Guides are also provided by the National Archives (see http://nationalarchives.gov.uk/records). These provide helpful tips for beginning research, giving information about relevant archival holdings covering not only national records but also other repositories both in the UK and elsewhere, and provide practical information about how and where to access them.

History of Britain, a digital library of primary and secondary sources for the history of Britain, is available at British History Online (www.british-history.ac.uk).

Digital History Projects, which often provide good starting points for desk-based historical research, are listed at History Online (http://www.history.ac.uk/history-online/projects/list).

Vision of Britain provides historical maps, census information, and a statistical atlas for the UK (http://www.visionofbritain.org.uk/index.jsp).

Hansard, the edited record of what was said in the UK parliament, including votes, written statements and written answers to parliamentary questions, is searchable online from 1803 (http://hansard.millbanksystems.com).

Old Bailey Online is a fully searchable record containing details of nearly 200,000 criminal trials held at London's central criminal court between 1674 and 1913 (http://www.oldbaileyonline.org).

7.4 Approaching Archival Collections

Practical considerations

How should one approach the collections and materials themselves? This in part depends on whether your sources are online or in a physical repository (see Box 7.3 for some practical questions to consider prior to your visit). Whether your archive is virtual or physical, there are also methodological and conceptual issues to bear in mind.

Box 7.3   Before you visit

How will you access the archive? Collections have rules governing who they will allow to access their collections. National repositories such as the British Library, the National Library of Scotland or the UK National Archives are open to all but require proof of ID and address in order to issue readers' tickets. Others, such as those held by the Universities of Oxford and Cambridge, may also require evidence of your status as a student, and letters of affiliation or support from tutors. Some archives may charge for entry, particularly those in the global south (see Chapter 6 for other practical considerations for research overseas). Others may only be open by arrangement, and require you to contact the archivist ahead of time to schedule any visit. If you are using private archives, then any access will need to be negotiated with the owner, something that can take time and patience. Be aware that even large archives may not open every day. It is worth checking the up to date information on institutional websites prior to your visit.

  How will you access materials? Many archives have online catalogues, which you should familiarise yourself with before you go. In some places, you can order materials prior to your visit so that they are waiting for you when you arrive.

  How will you record what you find? Archives have different rules regarding how you can interact with their records. Although in some cases photocopying is allowed, this might be restricted to newer items or be very costly. Sometimes digital photography is allowed (without flash), but resist the temptation to photograph everything for reading later on (reading photographs for the first time on a computer screen is much more difficult than looking at the original documents). Consider whether you want to makes notes on a computer or by hand, and if the latter, be aware that pens are not allowed in many archives, so you will need to bring a pencil.

When using an archive, there are different ways of ensuring that you access the most relevant and important materials for your research. The catalogue (if it exists) is the most obvious place to begin. Try different strategies: searching (entering a range of key words) and browsing (moving from one record to others which are closely related, for example as part of a thematic series, or written by the same author) can often uncover different sets of material. It is important to be systematic, and keep detailed notes of the records examined. If you can't recall the reference codes then not only might you end up repeatedly looking at the same material, but your exciting findings will also be useless: these reference details are required in any project write-up so that others can use and verify your findings.

Although archival research is often cast as solitary work, to get the most out of any collection it is often important to talk to others there (Cameron, 2001). The archivists are experts in their materials and may well be happy to provide guidance and advice; particularly in the small and less well-documented collections this is an invaluable resource. If you are consulting personal or informal archives, building relations with the collection's owner may also be crucial to securing access and reproduction rights for material held there. As Ashmore et al. (2012) note, conversations with the producers of private archives can be an important part of the research process, providing insight, contextualization for materials, and spilling over into oral history interviews. Other researchers may also have helpful advice if they are researching similar topics. Given the importance of these relationships, it is often appreciated, particularly in smaller collections, if you keep in touch with those who have helped along the way, for example through offering copies of your research, or writing for their newsletters or websites. There might even be opportunities to work with archivists to catalogue material or curate an exhibition, collaborations that can be valuable for your own research, the archive, and the public (Mills, 2013c; see also Ashmore et al., 2012; Craggs et al., 2013).

Conceptual and methodological considerations

Alongside these very practical considerations, when working with historical sources, including archival collections, it is also important to ask some more conceptual questions about the evidence in front of you. For although in the larger collections it is possible to feel overwhelmed by the sheer volume of materials, many things are not available to consult in the archive. Not everything that happens in the past leaves a trace, and not all traces are then preserved (Ogborn, 2010). It seems an obvious point, but written records often pass more easily into the present; speech of all kinds, song and other sounds are more ephemeral (on spaces of speech, see Livingstone, 2007; on birdsong, see Lorimer, 2007). Everyday practices and experiences were often deemed too banal and commonplace to record, particularly in times when writing materials and skills were expensive and limited to a small minority of the population.

We've already seen that official records are dominated by the concerns of the state, whilst many other materials reflect the views and values of others who wielded economic, social or political power at the time of their production. These issues are true not only of the individual record, but also of archival collections as a whole. Collections, just as much as individual documents, are socially constructed; in other words, they are produced by the decisions of archive professionals and owners. Choices about selection, ordering and access are taken in relation to (changing) cultural, economic and political priorities (Craggs, 2008). In Francesca Moore's (2010) work on abortion, she noted that relevant material was to be found under headings such as 'Wisewomen', 'Herbalism' and 'Women's History' in local record offices. These cataloguing decisions may have reflected both the social and legal status of abortion when the material was collected, and shrewd actions by archivists, attempting to 'veil' these records in order that they could remain part of the historical record rather than being destroyed (Moore, 2010). It is crucial to consider how different priorities and contexts may impact on the evidence available to you, and to ask what these archival patterns, presences and absences might tell you (see Box 7.4).

Box 7.4   Questions to ask of an archive

- Who started the collection and why? What were their original aims?
- What were the political, cultural and economic contexts in which the materials were produce and assembled?
- Who owns and funds the collection now, and what might be their priorities?
- What does the collection aim to document?

(Continued)

(Continued)

- What selection decisions are made, and how do these impact on the evidence available? What has been kept, what has been thrown away? What has been withheld? And what was never collected?
- What ordering strategies are put in place? And how might these reflect and reproduce particular assumptions and narratives?
- How does you own experience and practice in the archive impact on the data you collect and how you interpret these?

Let us consider the case of the UK National Archives. This holds the records of government departments and agencies falling within this, the courts, the NHS, the armed services and many non-departmental public bodies. The records were produced in the day to day business of government, and state services, but they are now collected to ensure government openness and accountability, maintain a record of past actions, and support research (National Archives, 2012). However, although this archive is one of the largest in the world, it is by no means a complete record of government. Not all records are kept: once they are passed on, decisions are made as to what to deposit, with only a small percentage selected, with criteria for selection based on uniqueness and importance (National Archives, 2012). In addition, when material is deposited, it is embargoed for a number of years – until 2000 government records were closed for 30 years, though this has now been amended to 20 years, with provision to request access to even more recent documents through Freedom of Information. However, confidential or sensitive information is often closed for longer periods, or removed entirely (see Box 7.5). If you are reliant on digitized material, then current political concerns or popular topics may have been prioritized. Such is the importance of these structural factors in the archive that many authors have now begun to focus on the archive as the subject of their research, considering how collections produce certain visions of the world, construct partial histories, and reflect the institutional concerns of those that own(ed) them (Craggs, 2008). Archives should therefore be understood as sites of power and (selective) memory, rather than as places for the retrieval of some already formed objective past which is waiting to be discovered. These issues do not mean that archival sources are of no use, merely that they need to be treated carefully and critically, something which postcolonial scholars in particular have emphasized (Duncan, 1999; Burton, 2003; 2005). Ann Stoler (2002; 2009) for example, highlights the need to read colonial archives 'against the grain' – to uncover the silences, agency, and acts of resistance of marginalized colonial subjects hidden in the records – and 'along the grain' – to come to understand the categories and assumptions of the colonisers.

Box 7.5   The migrated archives

In the process of decolonisation, rather than being handed over to the new national governments in newly independent countries, many British colonial records were returned to London. Those returned were those that 'might embarrass Her Majesty's Government or local government; might embarrass members of the police, military forces, public servants, or others, e.g. police informers; might compromise sources of intelligence information; or might be used unethically by ministers in a successive government' (cited in Badger, 2012: 799–800). These records were not transferred to the UK National Archives, but kept at a separate facility at Hanslope Park, where they were not accessible to researchers. The existence of some of these migrated archives only became widely known in 2011 as a result of a legal case in which Kenyan citizens aimed to demonstrate the British Government had been complicit in torture in colonial Kenya in the 1950s.

The decision to migrate many records, and destroy many more (which were burnt in the months leading up to decolonisation), reflected both the political exigencies of the time (the materials were of ongoing use to the Foreign and Commonwealth Office in the era of decolonisation and the Cold War, and may have threatened the UK government's relationship with newly independent states) and what historian Richard Drayton (2013) has called 'historical narcissism': the desire to present a historical narrative in which Britain's imperial rule was both peaceful and fair.

The case of the Migrated Archives demonstrates several important issues which it is helpful to keep in mind when conducting archival research:

- Selection and release policies often reflect political priorities, as well as practical decisions about space, storage, and relevance.
- It is relatively easy for governments (and companies and organizations) to withhold historical records (see Curless, 2013).
- These decisions affect the evidence which historical researchers are able to draw upon.
- These decisions also impact the ability of people to hold governments, companies, and other organizations to account.

As users of archives, we also contribute layers of meaning to the documents we encounter, analyse and re-present (Cook and Schwartz, 2002), therefore it is also important to consider your own role in the construction of history. What is your relationship to the historical past that you are examining? What subjectivities and values do you bring to the research? How might the political, social, economic, geographical and cultural contexts of the present day from which you are researching affect your interpretations of the evidence? What assumptions might you be making? How might the evidence appear different if you were to select other records, or to view the same records from a different geographical or political perspective? It is likely that your own position in the world, and your own history, will affect what you decide to research, and how you view and interpret the evidence that you gather. It is important to be reflexive about your positionality, and to consider how this might impact your conclusions in your research. However, as many scholars have now come to accept, this should not be seen as problematic. Indeed, much of the most exciting work emerging over the last few years has been produced by scholar-activists who are deeply engaged in the politics and struggles of the communities that they research with (see Chapter 11 for participatory and activist research, and Cameron (2014), for archival activism; see also Bailey et al., 2009, for the impact of faith in archival research).

It is also important to reflect on the experience of being in the archive as a researcher (Burton, 2005). Sarah Mills (2012: 361) recounts 'a series of emotions and sensations: solitude, repetitiveness, the thrill of discovery and dejection of lost causes' that she felt in the archive, emotions that will be familiar to anyone who has worked with historical collections. These emotions can motivate and dishearten the researcher by turn, and thus are important to acknowledge from a practical and methodological perspective. As Mills suggests, they also intersect with ethical considerations: how to do people encountered in the archive justice and represent them fairly. Thus ethical practice in the archive involves not only being aware of 'issues surrounding privacy, confidentiality, consent, copyright and publication' when utilising personal and biographical information, but also a broader sense of connection and responsibility (Mills, 2012: 359).

7.5 Analysing and Presenting Your Data

Just like interview transcripts or soil samples, historical sources need analysis. In this section I suggest some ways in which historical sources might be approached and analysed. These brief comments are complemented by other contributions to this volume which provide more detail about analysing texts (Chapter 1), visual imagery (Chapter 15), online materials (Chapters 16 and 17) and qualitative data (Chapter 36).

Box 7.6   Approaching and analysing historical sources

- Is it genuine?
- Is it accurate?
- Who produced the source, and what difference might this make? (e.g. which individual/group wrote, painted, or otherwise constructed it? Who commissioned or funded it? Might the needs, views, and knowledge of the authors and commissioners shape the observations made?)
- Why was it produced? What was its intended use? (e.g. was it an official document meant to inform government policy, or a private note meant for the eyes of one person only? Was it meant to persuade, inform or mislead? What difference might this make to the language used, the level of detail, the sorts of information included?)
- Where did they produce it and how might this influence its contents? (e.g. in 'the field' or at home? From direct observation or from word of mouth?)
- Has the source been reproduced or translated? Are you looking at the original or a reproduction in a book? If you are reading published materials how do they relate to the originals? Are they in the same language? What role has the publisher and editor played in shaping the source? (See for example Withers and Keighren, 2011, on these questions.)
- How was the source consumed and by who? Who was its audience? How did they respond?

Box 7.6 sets out a set of basic questions to ask about your sources in relation to their production, qualities, and audiences. Beyond these simple questions, the sorts of analysis you choose depend on the sorts of questions you want to answer, as well as the sorts of data you have. Generally speaking, large collections of similar and comparable material (a run of newsletters, a set of diaries, records of business transactions for example), lend themselves to analysis that can provide descriptive overviews of the content of materials and highlight change over time. Content analysis – a method of counting elements (such as articles on particular topics or the incidence of words) – is one common method used in such circumstances, and researchers dealing with large sets of historical material often also use sampling to manage these materials. These methods are effective at answering 'who', 'what' and 'how' questions: 'Who was involved?' 'What was said or done?' 'How did things change?'.

Other, more qualitative forms of analysis – for example discourse analysis (see Chapter 36) – are better suited to answering questions about meanings, experiences and understandings, questions such as: 'Why were decisions made?' 'How were people and places described, understood and valued at different times?' 'How did people view and experience particular events?' These methods use some sort of coding or categorization (this could be as simple as colour-coding with highlighters) in order to draw out themes from the source material and examine the narratives they contain. These forms of analysis are an effective way of analysing sets of homogeneous material, but can also be utilized to compare disparate sources. For example, in the research about the University of Ibadan presented in Box 7.1, the correspondence, biographies, official reports, and newspaper accounts were analysed for evidence of the views about the university held by British officials, local elites, and the students and staff of the university. In turn, wider (overlapping and competing) discourses about the meaning and value of modernity and development in Nigeria could be discerned across the source material. Boxes 7.7 and 7.8 highlight the value of these different approaches. Box 7.7 summarizes some recent research about the Gemini News Agency and utilizes content analysis to explore who produced the agency's copy, and which countries were the focus of its coverage, whilst Box 7.8 illustrates the value of discourse analysis in order to explore the geopolitical discourses present in the popular weekly magazine the *Readers' Digest*. It is important to note that many studies – including those described here – effectively combine methods of analysis to provide both overarching descriptions and more detailed discursive case studies.

Box 7.7   Content analysis: The Gemini News archive

(Adapted with thanks from the PhD research of Ashley Crowson)

The Gemini News Service was a pioneering press agency, active 1968–2002. Emerging in the context of rapid decolonisation, it aimed to:

- tell the stories of people in the Global South ignored by the mainstream media;
- challenge the 'parachute' reporting of developing countries by Western correspondents, relying instead on reporters resident in the countries about which they were writing.

Ashley Crowson's research aimed to explore the extent to which Gemini was successful in achieving these aims. In order to answer these questions Gemini's original story ledgers, now held in the *Guardian*'s archive, were used. They record (most of) the stories that were sent to subscribers 1968–1997, including details of each published story's headline and author, date, subject category (e.g. news, economics, culture etc.) and the country/region that the story focused on. These red books contain records of approximately 16,850 stories. For the purposes of this analysis every fourth story was included in the sample, a total of 3,917 articles (or 23% of the total).

The analysis reveals a clear Commonwealth focus in stories filed, with 'non-western' countries featuring much more heavily than in the outputs of competitor agencies. It also shows Gemini had

(Continued)

(Continued)

relative but uneven success at using local journalists: over 90% of stories about India were produced by Indian journalists, with an overall total of 45% of stories produced by journalists from the country concerned. Figure 7.2 below, produced with ArcGIS software, represents these findings, and demonstrates the value of visual representations of your research findings – a presentation method which is relatively unusual in historical research (see section four of this volume for an introduction to visualising geographical data).

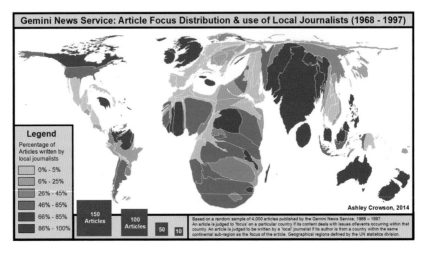

Figure 7.2   Cartogram showing number of articles focusing on each country and success in using 'local' journalists by country (Source: Ashley Crowson, 2014)

Box 7.8   Discourse analysis: *The Reader's Digest*

(adapted with thanks from Joanne Sharp, 'Publishing American identity: Popular geopolitics, myth and The Reader's Digest' (1993) and Condensing the Cold War: Reader's Digest and American Identity (2000)

The *Reader's Digest*, founded in 1922, had a circulation of over 16 million by 1991. In Joanne Sharp's research, she examined the ways in which discourses about America and the USSR were produced in the magazine's pages from its founding until 1990 – a period which included the Cold War between these two nations. Through a careful reading of the magazine's content – including not only the arguments presented but also the language used – she showed how narratives of American identity were produced in opposition to the characteristics and values of the USSR. She also demonstrated the ways in which the magazine produced imagined geographies of both the United States and the USSR that reflected these binary distinctions. For example, she quotes an article entitled 'Russian Winter', published in the *Digest* in 1984:

> Although some inhabited places are further north on the map and actually as cold, Russia's lumbering heritage of isolation and backwardness makes it more frozen. Russia remains psychologically the most northern of nations. (Feifer, 1984, cited in Sharp, 2000: ix)

> Overall, Sharp's research found that the *Digest* was overwhelmingly negative towards the Soviet Union. Of the 89 articles about the USSR produced between 1980–1990, only one showed any sympathy towards the Soviet viewpoint (Sharp, 1993). As a result of her discourse analysis of the magazine's articles, Sharp argued that:
>
>> A dualism is set up between the USA and the USSR so that a description of events in, or characteristics of, the USSR (totalitarianism, expansionism and so on) automatically implies that the opposite applies to the US (in this case: democracy, freedom…). The Soviet Union becomes a negative space into which The Reader's Digest projects all those values which are antithetical to its own ('American') values. It is not possible to have coexisting but different values in this system; always one set of values is right, the other exists in opposition and is thus wrong. By implication, the positive side of this value binary can be found to exist in the positive (conceptual and physical) space of America. (1993: 496)

7.6 Conclusion

Historical research is challenging but also exciting and important. In this chapter, I have provided a brief overview of some of the key conceptual, methodological and practical issues to consider in approaching historical topics. Focusing in particular on the archive, I have argued that we need to view the historical record as fragmentary and partial, but also powerful, and to be aware of these qualities when using these sources in our research. I have also offered some suggestions about how to analyse this material.

SUMMARY

- There is a huge range of historical sources available to researchers, from state archives to diaries, letters, photographs, art, film, music and sound, objects, buildings and landscapes. Whilst many of these are available online, many more still require a visit to a library, archive, or other repository – from museums to private homes.
- Historical sources and collections are fragmentary, partial and power laden.
- They offer important evidence for the historical researcher, but must be analysed both critically and ethically.
- Historical sources are also crucial in the present day, informing government policy and providing evidence for campaigns over rights to land and compensation.
- Archival research also offers opportunities for other collaborations that can benefit your research as well as the collection, and the public.

Further Reading

The edited volume *Practising the Archive* (Gagen et al., 2007) provides a series of reflections on **archival research**, including the use of urban plans, sound recordings, diaries, books, and theory.

Hayden Lorimer (2009) provides a very readable account of archival methods and the need to **improvise and make-do** in historical research.

Ann Stoler (2009) provides a detailed discussion of **power** in the archives and its implications for archival methods, from a postcolonial perspective.

Alison Blunt and Eleanor John's edited special issue of *Home Cultures* (2014) explores historical sources and methods for approaching **domestic practices**.

Laura Cameron's (2014) article provides a great discussion of **activist and participatory scholarship**, exploring historical research on aboriginal rights in Canada.

Sarah Mills (2013c) provides a good discussion of the opportunities for **collaboration** in the archives (see also Ashmore et al., 2012, and Craggs et al., 2013).

Mark Riley and David Harvey's (2007) special issue of *Social and Cultural Geography* provides a good introduction to the use of **oral history** in geography.

Note: Full details of the above can be found in the references list below.

Acknowledgements

Thank you to Ashley Crowson, Tim Livsey and Joanne Sharp for permission to reproduce their research here. Thanks also go to Miles Ogborn for allowing me to reproduce information from his chapter in earlier editions of this volume.

References

Ashmore, P., Craggs, R. and Neate, H. (2012) 'Working-with: Talking and sorting in personal archives', *Journal of Historical Geography* 38: 81–8.

Badger, A. (2012) 'Historians, a legacy of suspicion and the "migrated archives"', *Small Wars & Insurgencies* 23: 799–807.

Bailey, A., Brace, C. and Harvey, D. (2009) 'Three geographers in an archive: Positions, predilections and passing comment on transient lives', *Transactions of the Institute of British Geographers* 34: 254–69.

Blunt, A. (1999) 'Imperial geographies of home: British women in India, 1886–1925', *Transactions of the Institute of British Geographers NS* 24: 421–40.

Blunt, A. (2000) 'Spatial stories under siege: British women writing from Lucknow in 1857', *Gender, Place and Culture* 7: 229–46.

Blunt, A. and John, E. (2014) 'Domestic practice in the past: Historical sources and methods', *Home Cultures* 11(3): 269–74.

Brickell, K. and Garrett, B. (2013) 'Geography, film and exploration: Amateur filmmaking in the Himalayas', *Transactions of the Institute of British Geographers* 38: 7–11.

Brown, G. and Yaffe, H. (2013) 'Non-Stop Against Apartheid: Practicing solidarity outside the South African Embassy', *Social Movement Studies* 12: 227– 34.

Burton, A. (2003) *Dwelling in the Archive: Women Writing House, Home, and History in Late Colonial India.* Oxford: Oxford University Press.

Burton, A. (ed.) (2005) *Archive Stories: Facts, Fictions and the Writing of History.* Durham, NC Duke University Press.

Cameron, L. (2001) 'Oral history in the Freud archives: Incidents, ethics, and relations', *Historical Geography,* 29: 38–44.

Cameron, L. (2014) 'Participation, archival activism and learning to learn', *Journal of Historical Geography* 46: 99–101.

Cook, T. and Schwartz, J. (2002) 'Archives, records and power: From (postmodern) theory to (archival) performance', *Archival Science* 2: 171–85.

Craggs, R. (2008) 'Situating the imperial archive: The Royal Empire Society Library 1868–1945', *Journal of Historical Geography* 34: 48–67.

Craggs, R. (2011) '"The long and dusty road": Comex Travel cultures and Commonwealth citizenship on the Asian Highway', *Cultural Geographies* 18: 363–84.

Craggs, R., Geoghegan, H. and Keighren, I. (ed) (2013) *Collaborative Geographies: The Politics, Practicalities, and Promise of Working Together.* London: Royal Geographical Society (Historical Geography Research Series; no. 43).

Cresswell, T. (1993) 'Mobility as resistance: A geographical reading of Kerouac's *On the Road*', *Transactions of the Institute of British Geographers NS* 18: 249–62.

Crowson, A. (2014) 'News From Elsewhere: Journalism, Geopolitics and the Decolonisation of Knowledge'. Unpublished PhD thesis, King's College, London.

Curless, G. (2013) 'Covering Up the Dark Side of Decolonisation'. http://imperialglobalexeter.com/2013/12/10/covering-up-the-dark-side-of-decolonisation (accessed 11 November 2015).

Decker, S. (2005) 'Decolonising Barclays Bank DCO? Corporate Africanisation in Nigeria, 1945–69', *Journal of Imperial and Commonwealth History* 33: 419–40.

Drayton, R. (2013) 'The Foreign Office secretly hoarded 1.2m files. It's historical narcissism,' *The Guardian, Comment is Free.* www.theguardian.com/commentisfree/2013/oct/27/uk-foreign-office-secret-files (accessed 11 November 2015).

Driver, F. (1993) *Power and Pauperism: The Workhouse System, 1834–1884.* Cambridge: Cambridge University Press.

Duncan, J. (1999) 'Complicity and resistance in the colonial archive: Some issues of method and theory in historical geography', *Historical Geography* 27: 119–28.

Duncan, J. (2004) *The City as Text: The Politics of Landscape Interpretation in the Kandyan Kingdom.* Cambridge: Cambridge University Press.

Duncan, J. and Duncan, N. (1988) '(Re)reading the landscape', *Environment and Planning D: Society and Space* 6: 117–26.

Eagleton, C. (2016) 'Designing change: Coins and the creation of new national identities', in R. Craggs and C. Wintle (eds) *Cultures of Decolonisation: Transnational Productions and Practices, 1945–1970.* Manchester: Manchester University Press. pp. 222–44.

Evans, S., Keighren, I. and Maddrell, A. (2013) 'Coming of age? Reflections on the centenary of women's admission to the Royal Geographical Society', *The Geographical Journal* 179: 373–6.

Gagen, E., Lorimer, L. and Vasudevan, A. (eds) (2007) *Practising the Archive: Reflections on Method and Practice in Historical Geography.* London: Royal Geographical Society (Historical Geography Research Group Series; no. 40).

Keighren, I. (2013) 'Geographies of the book: Review and prospect', *Geography Compass*, 7: 745–58.

Keighren, I. and Withers, C.W.J. (2011) 'Questions of inscription and epistemology in British travelers' accounts of early nineteenth-century South America', *Annals of the Association of American Geographers*, 101: 1331–46.

Keighren, I. and Withers, C.W.J. (2012) 'The spectacular and the sacred: Narrating landscape in works of travel', *Cultural Geographies* 19: 11–30.

Livingstone, D. (2007) 'Science, site and speech: Scientific knowledge and the spaces of rhetoric', *History of the Human Sciences* 20: 71–98.

Livsey, T. (2014) "Suitable lodgings for students': modern space, colonial development and decolonization in Nigeria', *Urban History* 41(4): 664–85.

Lorimer, H. (2007) 'Songs from before shaping the conditions for appreciative listening', in E. Gagen, H. Lorimer, A. Vasudevan (eds) *Practising the Archive: Reflections on Method and Practice in Historical Geography.* London: Royal Geographical Society (Historical Geography Research Group Series; no. 40). pp. 57–73.

Lorimer, H. (2009) 'Caught in the nick of time: Archives and fieldwork', in D. DeLyser, S. Aitken, M.A. Crang, S. Herbert and L. McDowell (eds) *The SAGE Handbook of Qualitative Research in Human Geography.* London: Sage. pp. 248–73.

Mills, S. (2012) 'Young ghosts: ethical and methodological issues of historical research in children's geographies', *Children's Geographies* 10: 357–63.

Mills, S. (2013a) 'Cultural-historical geographies of the archive: Fragments, objects and ghosts', *Geography Compass* 7: 701–13.

Mills, S. (2013b) '"An Instruction in Good Citizenship": Scouting and the historical geographies of citizenship education', *Transactions of the Institute of British Geographers* 38: 120–34.

Mills, S. (2013c) 'Surprise! Public historical geographies, user engagement and voluntarism', *Area*, 45: 16–22.

Moore, F.P.L. (2010) 'Tales from the archive: Methodological and ethical issues in historical geography research', *Area* 42: 262–70.

Morris, M. (1996) '"Tha'lt be like a blush-rose when tha' grows up, my little lass": English cultural and gendered identity in *The Secret Garden*', *Environment and Planning D: Society and Space* 14: 59–78.

National Archives (2012) *National Archives Record Collection Policy*, available at http://www.nationalarchives.gov.uk/documents/records-collection-policy-2012.pdf (accessed 7 December 2015).

Ogborn, M. (1995) 'Discipline, government and law: Separate confinement in the prisons of England and Wales, 1830–1877', *Transactions of the Institute of British Geographers* 20: 295–311.

Ogborn, M. (2010) 'Finding historical sources' in N. Clifford, S. French, and G. Valentine (eds) *Key Methods in Geography*. London: Sage. pp. 89–102.

Philo, C. (1987) 'Fit localities for an asylum: The historical geography of the nineteenth century "mad business" in England as viewed through the pages of the Asylum Journal', *Journal of Historical Geography* 13: 398–415.

Riley, M. and Harvey, D. (2007) 'Talking geography: On oral history and the practice of geography', *Social and Cultural Geography* 8: 345–51.

Sharp, J. (1993) 'Publishing American identity: Popular geopolitics, myth and The Reader's Digest', *Political Geography* 12: 491–503.

Sharp, J. (2000) *Condensing the Cold War: Reader's Digest and American Identity*. Minneapolis: University of Minnesota Press.

Stoler, A. (2002) 'Colonial archives and the arts of governance', *Archival Science* 2: 87–109.

Stoler, A. (2009) *Along the Archival Grain: Epistemic Anxieties and Colonial Common Sense*. Princeton, NJ: Princeton University Press.

Tuan, Y-F. (1985) 'The landscapes of Sherlock Holmes', *Journal of Geography* 84: 56–60.

Withers, C.W.J. and Keighren, I. (2011) 'Travels into print: authoring, editing and narratives of travel and exploration, *c.*1815–*c.*1857', *Transactions of the Institute of British Geographers* 36: 560–73.

ON THE COMPANION WEBSITE…

Visit **https://study.sagepub.com/keymethods3e** for author videos, chapter exercises, resources and links, plus **free** access to the following recommended articles:

1. **Offen, K. (2013) 'Historical Geography II: Digital imaginations', *Progress in Human Geography,* 37(4): 564–77.**

This article explores ways in which geographers and others have made use of new digital technologies and media.

2. **Offen, K. (2014) 'Historical Geography III: Climate matters', *Progress in Human Geography,* 38(3): 467–89.**

This article reflects critically on experiences at the 2007 International Critical Geographies Conference in Mumbai and on a related fieldtrip in order to discuss how, even in such critically-aware contexts, issues such as language, translation, power and positionality emerge as significant factors in how geographical knowledge develops.

3. **Mayhew, R.J. (2009) 'Historical geography 2007–2008: Foucault's avatars – still in (the) Driver's seat', *Progress in Human Geography,* 33(3): 387–97.**

This article is useful as it summarizes the field of historical geography in this period. It provides good examples of the range of research areas that can be researched through archival research.

8 Conducting Questionnaire Surveys

Sara L. McLafferty

SYNOPSIS

In geography, questionnaire surveys have been used to explore people's perceptions, attitudes, experiences, behaviours and spatial interactions in diverse geographical contexts. This chapter explains the basics of why and how to carry out survey research.

This chapter is organized into the following sections:

- Introduction
- Questionnaire design
- Strategies for conducting questionnaire surveys
- Sampling
- Conclusion

8.1 Introduction

Survey research has been an important tool in geography for several decades. The goal of survey research is to acquire information about the characteristics, behaviours and attitudes of a population by administering a standardized questionnaire, or survey, to a sample of individuals. Surveys have been used to address a wide range of geographical issues, including perceptions of risk from natural hazards; social networks; coping behaviours among people with HIV/AIDS; environmental attitudes; travel patterns and behaviours; mental maps; power relations in industrial firms; gender differences in household responsibilities; and access to employment. In geography, questionnaire surveys were first used in the field of behavioural geography to examine people's environmental perceptions, travel behaviour and consumer choices (Rushton, 1969; Gould and White, 1974). Survey research methods quickly spread to other branches of human geography, and today they are an essential component of the human geographer's toolkit.

Questionnaire survey research is just one method for collecting information about people or institutions. When does it make sense to conduct a questionnaire, rather than relying on secondary data (see Chapter 30) or information collected by observational methods (see Chapters 11 and 13), for example? Survey research is particularly useful for eliciting people's attitudes and opinions about social, political and environmental issues such as neighbourhood quality of life, or environmental problems and risks. This style of research is also valuable for finding out about complex behaviours and social interactions. Finally, survey research is a tool for gathering information about people's lives that is not available from published sources (e.g. data on behaviours and attitudes in realms such as diet, health, and employment characteristics). In developing countries where government data sources are often out of date and of poor quality, questionnaire surveys are a primary means of collecting data on people and their characteristics.

Before embarking on survey research, it is critically important to have a clear understanding of the research problem of interest. What are the objectives of the research? What key questions or issues are to be addressed? What people or institutions make up the target population? What are the geographical area and time period of interest? These issues underpin how the survey is designed and administered. Surveys can be expensive and time-consuming to conduct, so both quality and type of information gathered are important.

Although each survey deals with a unique topic, in a unique population, the process of conducting survey research involves a common set of issues. The first step is **survey design**. Researchers must develop questions and create a survey instrument that both achieves the goals of the research and is clear and easy to understand for respondents. Second, we need to decide how the survey will be administered. Online questionnaires and telephone interviews are just a few of the many strategies for conducting surveys. Third, survey research involves **sampling** – identifying a sample of people to receive and respond to the questionnaire. This chapter provides a brief introduction to each of these issues, drawing upon examples from geographic research.

8.2 Questionnaire Design

Questionnaires are at the heart of survey research. Each questionnaire is tailor-made to fit a research project, including a series of questions that address the topic of interest. Decades of survey research have shown that the design and wording of questions can have significant effects on the answers obtained. There are well established procedures for developing a 'good' questionnaire that includes clear and effective questions (Groves et al., 2009).

Good questions are ones that provide useful information about what the researcher is trying to measure. Although this may appear to be simple, straightforward advice, it is often challenging to implement. Questions can range from factual questions that ask people to provide information, to opinion questions that assess attitudes and preferences. Writing good questions requires not only thinking about what information we are trying to obtain but also anticipating how the study population will interpret particular questions. Let's examine the following question: 'Are you concerned about environmental degradation in your neighbourhood?' This item raises more questions than it answers. What constitutes *concern*? What does *environmental degradation* mean? Do people understand it? How does each respondent define his or her *neighbourhood*? This question also presumes that there *is* environmental degradation, which could be considered 'leading' the respondent toward certain answers. Questions should be clear and easy to understand for survey respondents, avoid 'leading' questions, and they should provide useful, consistent information for research purposes.

One of the most important rules in preparing survey questions is to keep it simple. Avoid complex phrases and long words that might confuse respondents. Do not ask two questions in one. The question 'Did you choose your home because it is close to work and inexpensive?' creates confusion because there is no obvious response if only one characteristic is important. Jargon and specialized technical terms cause problems in survey questions. Terms like 'accessibility' or 'power' or 'GIS' are well known among geographers, but ambiguous and confusing for most respondents. Don't assume that respondents are familiar with geographic concepts! Define terms as clearly as possible and avoid vague, all-encompassing concepts. For example, asking people about their involvement in community activities – without specifying what kinds of activities and what level of involvement – is unlikely to produce useful responses. It is better to ask a series of questions about specific types of involvement rather than a single vague question. Finally, one should avoid negative words in questions. Words like 'no' and 'not' tend to confuse respondents (Babbie, 2013) (see Box 8.1).

Responses to survey questions are as important as the questions themselves. *Open-ended* questions allow participants to craft their own responses, whereas *fixed-response* questions offer a limited set of responses.

Box 8.1   Guidelines for designing survey questions

Basic principles:

- Keep it simple
- Define terms clearly
- Use the simplest possible wording

Things to avoid:

- Long, complex questions
- Two or more questions in one
- Jargon
- Biased or emotionally charged terms
- Negative words like 'not' or 'none'

Open-ended questions provide qualitative information that can be analysed with qualitative methodologies (see Chapter 36). Increasingly, geographers are using open-ended responses in questionnaire surveys as part of the broader shift towards qualitative methodologies. Open-ended questions have several advantages: respondents are not constrained in answering questions, they can express in their own words the fullest possible range of attitudes, preferences and emotions, and their 'true' viewpoints may be better represented. Relying on a mix of open-ended and fixed-response questions, Gilbert (1998) analysed survival strategies among working poor women and their use of place-based social networks. The fixed-response questions provided data on the demographic and household characteristics of the women and their social interaction patterns, while the open-ended questions offered detailed insights about women's coping strategies and life circumstances.

Fixed-response questions are commonly used in survey research, and the principles for designing such questions have been in place for decades. There are several advantages to fixed responses. First, the fixed alternatives act as a guide for respondents, making it easier for them to answer questions. Second, the responses are easier to analyse and interpret because they fall into a limited set of categories (Fink, 2013), and can be analysed quantitatively using simple percentages and a variety of statistical techniques. The downside is that such responses lack the detail, richness and personal viewpoints that can be gained from open-ended questions.

A simple type of fixed-response question is the factual question that, for instance, asks about age, income, time budgets or activity patterns. Responses may be numerical or involve checklists, categories, or yes/no answers. The key in framing these types of questions is to anticipate all possible responses. As in all phases of survey design, it is important to think about the kind of information needed for research as well as characteristics of the study population that might influence their responses. A 'don't know' or 'other' option is generally included to allow for the fullest range of responses. For numerical information (age, income), one must decide between creating categorical responses (e.g. 15–24, 25–34), or recording the actual numerical value. Creating categories involves a loss of information – a shift from interval to ordinal data[1] – but the categorical information may be easier to analyse. Also, for sensitive topics such as age and annual income, respondents are more likely to answer if the choice involves a broad category rather than a specific number.

Although factual questions appear straightforward, they often reflect an uneasy balance between the needs and views of survey administrators and those of respondents. Questions about 'race' are a good example. Race is a social construction that does not fit easily into the discrete response categories used in questionnaire surveys. In the census of the United States, the categories and options used to elicit information about race have changed over time reflecting changes in social understandings of race. The 1850 Census presented three options – 'white', 'black' and 'mulatto'[2] – for respondents to choose in identifying their race. In contrast, the 2010 Census provided 14 racial categories and permitted respondents to check multiple categories to identify themselves as mixed race. Even with this wider range of alternatives, in completing the recent census many people did not select a racial category, responding instead by writing down an ethnic identification.

Finding out about attitudes and opinions involves more complex kinds of fixed-response formats. In general, respondents are asked to provide a rating on an ordinal scale that represents a wide range of possible responses. The Likert scale presents a range of responses anchored by two extreme, opposing positions. For example, residents may be asked to rate the quality of the schools in their neighbourhood from 'excellent' to 'satisfactory' to 'poor'. The two extreme positions, 'excellent' and 'poor', serve as anchor points for the scale, and any number of alternative responses can be included in between (Box 8.2). It is best to use an odd number of responses – 3, 5 and 7 are common – so that the middle value represents a neutral opinion. Respondents often want the option of giving a neutral answer when they do not have strong feelings one way or the other. Odd-numbered scales give such an option, whereas even-numbered scales force the response to one side.

Another approach is to present the scale as a continuous line connecting the two anchors. Respondents are asked to draw a tick mark on the line at the location representing their opinion, and the distance along the line shows the strength of opinion. This gives maximum flexibility, but respondents are often confused about the process and it is difficult to compare results among respondents. Consequently most researchers work with fixed Likert scales.

Box 8.2   Examples of Likert-type responses

Please rate the quality of schools in your neighbourhood:

Excellent				Poor	(continuous)
Excellent		Satisfactory		Poor	(three-point scale)
Excellent	Good	Satisfactory	Fair	Poor	(five-point scale)

Attitudinal scales can be difficult to evaluate because there is no 'objective' standard for knowing whether or not a response is accurate. However, researchers can take several steps to improve validity. In general, it is better to offer respondents more possible answers than fewer – i.e. a five-point scale provides more information than a three-point scale. But as the number of categories increases, respondents lose their ability to discriminate among categories and the responses lose meaning. An intermediate number of categories (five or seven are commonly used) works best. Because responses often vary depending on how a question is worded, another good practice is to use multiple questions, with different wording and formats, to measure the same concept. By comparing responses across questions one can check if people give consistent responses. If so, the responses can be averaged or combined statistically to represent the underlying concept or attitude. This strategy was used in a study of the links between residents' perceptions of neighbourhood quality in Bristol and objective indicators of social deprivation (Haynes et al., 2007). To measure perceived neighbourhood quality, the authors asked respondents to evaluate levels of noise, pollution, friendliness, crime and social interaction in the local neighbourhood. Responses to these diverse questions were combined statistically to create composite measures that were correlated with area-based indicators of housing quality and social deprivation.

Questionnaires should also include a clear set of instructions to guide individual responses. For self-administered surveys – those that do not involve an interviewer, such as online and mail/email surveys – the questionnaire has to be self-explanatory. The instructions for respondents must be written in simple, direct language and be as clear and explicit as possible so that the questionnaire can be filled out without assistance.

Fixed-response questions work best in self-administered questionnaires, and the design and layout of the questionnaire are critically important. For questionnaires involving interviewers, the key is to have a clear and consistent set of instructions for interviewers to follow. There are well-tested guidelines for designing and formatting interviewer-administered questionnaires (Groves et al., 2009).

The final and critically important step in questionnaire construction is pre-testing (pilot-testing). In this phase, we test the questionnaire on a small group of people to check the questions, responses, layout and instructions. Are the questions understandable? Does the questionnaire allow all possible responses? Are the instructions clear and easy to follow? Is the questionnaire too long? Do any questions make respondents uncomfortable? Pre-testing often reveals flaws in the questionnaire that were not obvious to researchers. The questionnaire is then modified, and it may be pre-tested again before going to the full sample. Several pre-tests may be needed to achieve a well-designed questionnaire. For interview-based surveys, pre-testing has other benefits. It builds interviewing skills and helps interviewers develop confidence and rapport with respondents. In sum, pre-testing is an essential step in ensuring a successful questionnaire survey.

8.3 Strategies for Conducting Questionnaire Surveys

There are many strategies for conducting questionnaire surveys. Among the traditional methods are telephone surveys, face-to-face interviews and postal surveys. However, Internet, email, and social media-based survey research has expanded dramatically in the past decade, leveraging the enormous growth of digital communications networks. Survey strategies differ along many dimensions – from practical issues like cost and time, to issues affecting the quality and quantity of information that can be collected. Some survey strategies require the use of interviewers whereas others utilize self-administered questionnaires.

Face-to-face interviews

Face-to-face interviews are one of the most flexible survey strategies. They can accommodate virtually any type of question and questionnaire. The interviewer can ask questions in complex sequences, administer long questionnaires, clarify vague responses and, with open-ended questions, probe to reveal hidden meanings. The personal contact between interviewer and respondent often results in more meaningful answers and generates a higher rate of response. Interviews require careful planning. Interviewers need training and preparation to ensure that the process is consistent across interviewers. Thus, face-to-face interviews are generally the most expensive and time-consuming survey strategies. Another drawback is the potential for interviewer-induced bias. The unequal relationship between interviewer and respondent, embedded in issues of gender, race, ethnicity and power, can influence responses (Kobayashi, 1994).

Telephone interviews

Telephone interviews via landline or mobile phone are widely used in market research, although their role has declined with the growth of online communications. Telephone interviews combine the personal touch of interviews with the more efficient and lower-cost format of the telephone. In many places, firms can be hired to conduct telephone surveys, saving researchers the time and expense of training interviewers and setting up phone banks. Phone surveys, however, are generally limited to short questionnaires with fixed-response questions. Such surveys miss people who do not have telephones or screen their phone calls. Although phone surveys can be conducted via mobile or landline networks, the cost per call is generally higher via mobile phone, and it is difficult to link mobile phone survey respondents to a particular geographic location. Response rates are typically lower for mobile phone surveys. Finally, although the interviewer and interviewee are only connected remotely, issues of power and bias can creep into phone surveys.

Postal and 'drop and pick-up' surveys

Postal (or mail) surveys are self-administered questionnaires distributed in a post-out, post-back format. A stamped, addressed envelope is included for returning the completed survey, and reminder notes may be sent later to encourage people to respond. For interviewees, there is no time pressure to respond; forms can be completed at a convenient time. The main weakness of postal surveys is the low response rate. Typically, less than 30 per cent of questionnaires will be completed and returned, and those who respond may not be representative of the target survey population. People with low levels of education or busy lives are less likely to respond. Although widely used in the past, postal surveys have largely been replaced by online surveys.

A related strategy is the 'drop and pick-up' questionnaire. This involves leaving self-administered questionnaires at people's homes and picking the surveys up at a later date. The person dropping off the surveys can give simple instructions and a brief description of the survey effort. The personal contact in dropping off the survey gives response rates close to those for face-to-face interviews, but with much less time and interviewer training. Thus, the method combines the strengths of interview and self-administered strategies. This comes at a cost – the costs are substantially higher than are those for postal, online or telephone surveys, though still less than those for personal interviews.

Online surveys

Online surveys, distributed through email, Internet, and social media, are increasingly important tools for geographic research. While many online surveys have the same format as a standard questionnaire, online technologies also allow researchers to create 'intelligent', computer-assisted questionnaires that check and direct people's responses. Geographers Claire Madge and Henrietta O'Connor (2002) used an Internet questionnaire to find out how new parents use the Internet in acquiring information on parenting and in developing social-support networks. Every phase of their research methodology relied on the Internet. Respondents were solicited online: they volunteered for the project by clicking on a 'cyberparents' hotlink on a prominent parenting website. The web-based questionnaire included a series of fixed format questions with hyperlinks. Online interviews and discussion groups were then conducted to discover deeper levels of information.

Internet surveys like the one used by Madge and O'Connor have several advantages. They are inexpensive to administer, convenient for respondents, and facilitate automated data entry. They also provide low-cost access to geographically dispersed populations and physically immobile groups (Sue and Ritter, 2012). Another important advantage is that online questionnaires can include detailed colour graphics, such as maps, photographs, video clips and animations. In a survey of recreational use in the Norfolk Broadlands (UK), Bearman and Appleton (2012) embedded a Google Maps API within an online survey so that participants could digitize their preferred locations for recreational activity (Figure 8.1). Similarly, natural hazards researchers are using spatial videos of hazard-related damage to find out how people respond to and understand extreme natural events (Lue et al., 2014). This dynamic linking of GIS maps and questionnaires illustrates the many exciting possibilities of online and app-based survey research in geography.

On the negative side, distributing questionnaires online raises a host of sampling issues. Who are the respondents? Where are the respondents? Do they represent the target population? What types of people respond and don't respond to Internet surveys? Clearly people without access to email and the Internet will be left out of the sample. Although many questions remain, online surveys represent a significant innovation that increasingly dominates survey research.

Beyond the traditional questionnaire, Internet and mobile technologies are propelling the development of novel survey methods for examining people's behaviours and experiences during their everyday lives. Ecological Momentary Assessment (EMA) involves frequent sampling of respondents' perceptions and behaviours via mobile electronic communications devices. EMA technologies can include electronic diaries, questionnaires, and sensors that detect physiological responses such as physical activity or heart rate. Genevieve Dunton and colleagues used EMA to compare physical

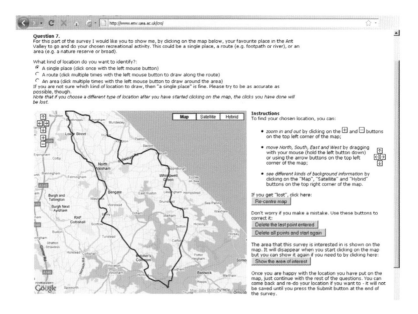

Figure 8.1 Screenshot of the user interface requesting input for an online survey of preferences for recreation activities in the Norfolk Broads area of East Anglia, England

Source: Bearman and Appleton (2012) Figure 1, page 162.

activity levels among children living in compact, walkable 'Smart Growth' communities and those in traditional, low-density suburban neighbourhoods (Dunton et al., 2012). Data describing physical activity-related behaviours and experiences were collected from children via mobile phone at specific intervals during the children's discretionary time. Although such real-time surveying of populations presents innovative research opportunities, it also raises many important ethical issues concerning privacy, confidentiality, and personal monitoring.

Each survey strategy has distinct advantages and disadvantages and the 'best' choice varies from one research project to another. Choosing a survey strategy involves weighing practical considerations, such as time and cost constraints, with research considerations, such as response rate, types of questions and the need (or lack of it) for interviewer skills. Frequently, the research context limits one's choice. Surveys in developing countries often rely on personal interviews (Awanyo, 2001); map-perception surveys often use computer-assisted questionnaires, 'smartphones', and the Internet. Regardless, researchers should note the limitations of the chosen method and attempt to minimize their effects.

8.4 Sampling

Sampling is a key issue in survey research because who responds to a survey can have a significant impact on the results. The sample is the subset of people to whom the questionnaire will be administered. Typically the sample is selected to represent some larger population of interest – the group of people or institutions that are the subject of the research. Populations can be very broad – e.g. 'all people in the UK' – or they can be quite specific, for example 'married women with children who work outside the home and live in Chicago'. Populations are bounded in time and space, representing a group of people or institutions in a particular geographical area over a particular time period. Effective sampling requires that this population of interest be clearly defined.

The first step in sampling is to identify the sampling frame – those individuals who have a chance to be included in the sample (Groves et al., 2009). The sampling frame may include the entire population or a subset of the population. Sometimes the design of the survey limits the sampling frame. For instance, in a telephone survey drawn from a telephone directory, the sampling frame only includes households that have (landline) telephones and whose telephone numbers are listed in the directory. Similarly, an Internet survey excludes people who do not have access to the Internet or do not use it. The resulting sample will be biased if those excluded from the sampling frame differ significantly from those included.

Sampling also involves decisions about how to choose the sample and sample size. Commonly used sampling procedures include random sampling, in which individuals are selected at random, and systematic sampling, which involves choosing individuals at regular intervals – e.g. every fourth house along a block. The former ensures that each individual has the same chance of being selected, whereas the latter provides even coverage of the population within the sampling frame. Sometimes the population consists of subgroups that are of particular interest – for example, different neighbourhoods in a city or ethnic groups in a population. If these subgroups differ in size, random sampling will result in the smaller subgroups being under-represented in the sample.

Stratified sampling procedures ensure that the sample adequately represents various subgroups. In stratified sampling, we first divide the population into subgroups and then choose samples randomly or systematically from each subgroup. Surveys that explore differences among groups or geographical areas often rely on stratified sampling. A study by Fan (2002) utilized stratified sampling to examine differences in labour-market experiences among three groups – temporary migrants, permanent migrants and non-migrants – in Guangzhou City in China. The sample consisted of more than 1500 respondents, stratified to represent not only the three migrant groups but also various occupational groups and districts within the city (Fan, 2002). Respondents were chosen randomly within each occupational and geographical group, with adjustments to ensure that the three migrant groups were appropriately represented.

Another important issue is how large a sample should be. Large sample sizes give more precise estimates of population characteristics and they provide more information for addressing the research problem. However, large samples also mean more questionnaires and more time and effort spent in interviewing and analysis. The cost of survey research increases proportionately with sample size. In choosing a sample size, analysts must trade off the benefits of added information and better estimates with the costs of administering and analysing the surveys.

One way to decide on a sample size is to focus on subgroups rather than the population as a whole. The sample must be large enough to provide reasonably accurate estimates for each of the subgroups that are being compared and analysed. Large sample sizes shrink quickly when divided into subgroups. For example, in analysing travel patterns by gender, urban/rural residence, and three categories of ethnic origin, there will be 12 (2x2x3) subgroups. An overall sample size of 100 yields just eight responses on average for each subgroup, which is too small to produce reliable subgroup estimates. To avoid this problem, the researcher should first identify the various subgroups and choose an adequate sample size for each. This is easy to do in a stratified sampling design. With other sampling procedures the issue is trickier. Small subgroups will be missed unless the overall sample size is large. Researchers should carefully assess the various subgroups of interest in choosing a sample size.

Sample size decisions also involve thinking about how much precision or confidence is needed in various estimates. Precision, the closeness in approximating a population value, always increases with sample size, but the improvements in precision decrease at larger sample sizes. When survey data are used to estimate a population proportion, improvements in precision diminish at sample sizes of 200–300.

It is also important to think about how the survey data will be analysed. Typically researchers use statistical procedures such as chi-square and analysis of variance (ANOVA) in analysing survey responses. These procedures require sample sizes of approximately 25 or more. Larger sample sizes are needed for multivariate statistical procedures such as multiple regression analysis and logistic regression. If separate statistical analyses will be performed for different subgroups in the sample, each subgroup must have a sample size that is sufficient for statistical analysis.

Finally, sample size decisions also involve very real budgetary and time constraints and these may well be beyond the researcher's control. In sum, there is no single answer to the sample size decision. The decision involves anticipating what the data will be used for and how they will be analysed, and balancing those considerations with the realities of time and money.

Sampling procedures are important because they can introduce various sources of bias into a research project. Sampling bias arises when the sample size is not large enough to represent accurately the study population or subgroups within it. More importantly, the sampling frame may be biased, as occurs in telephone or Internet surveys. Many survey procedures under-represent disadvantaged populations, including poor and homeless people and ethnic and racial minorities. Special efforts are needed to ensure that these groups are not excluded from the research project. Finally, non-response bias occurs when those who refuse to respond differ significantly from those who do respond. Non-response often correlates with age, social class, education and political beliefs, resulting in a sample that is not representative of the study population. Although non-response bias cannot be eliminated, its effects can be minimized with a good sampling design. Because survey results are often highly dependent on the characteristics of the sample, bias is a crucial issue in sampling and survey design.

8.5 Conclusion

Conducting questionnaire surveys involves a series of steps, including designing and pre-testing the questionnaire; choosing a survey strategy; identifying a sample of potential respondents; and administering the survey. These complex decisions are closely interconnected. The design of the questionnaire affects whether or not face-to-face interviews are needed. For many projects, financial constraints dictate the use of online or telephone surveys and relatively small sample sizes. Thus, in any survey project there is a continual give and take between different factors, framed by the goals and constraints of the research endeavour.

Questionnaire surveys have well-known limitations, as discussed at various points in this chapter. For geographical research, poorly worded questions, ambiguous responses and non-response bias are all issues that raise major concerns. Some geographers contend that survey information is of limited value, especially when compared to the rich and detailed information that can be gleaned from in-depth interviews and participant observation (Winchester, 1999). A more balanced view recognizes the strengths of questionnaire surveys – their ability to gather information from large samples, about large and diverse populations; their ability to incorporate both open and fixed questions and their use of trained interviewers to elicit information; and, finally, in the Internet era, their ability to reach widely dispersed populations with innovative, computer-assisted or app-based, graphically rich questionnaires. Despite their limitations, surveys remain the most efficient and effective tool for collecting population-based information.

Questionnaire surveys have a long history in geographic research, a history that continues to evolve as the discipline of geography changes. During the 1970s, survey methods facilitated the shift away from statistical analysis of secondary data towards behavioural and environmental perception research. During the 1980s and 1990s, survey methods became less popular as a result of a 'qualitative turn' in human geography. Today, as geographers search for a common ground between quantitative and qualitative methods, questionnaire surveys are playing an important role in innovative 'mixed' methodologies. As these developments unfold, questionnaire surveys will continue to provide a rich array of information about people's lives and well-being in their diverse geographical contexts.

SUMMARY

- Questionnaire surveys are useful for gathering information about people's characteristics, perceptions, attitudes and behaviours.
- Before embarking on survey research, clearly identify the goals and objectives of the research project. Decide what information you are trying to gather via the questionnaire survey.
- Survey research involves three key steps: designing the questionnaire, choosing a survey strategy and choosing the survey respondents (sampling).
- The questionnaire should be designed to acquire useful information about the research problem of interest. Questionnaires can include both open-ended and fixed-response questions. In either case, the questions should be clearly and simply worded and should avoid jargon and 'leading' questions.
- The types of survey strategies include face-to-face interviews, telephone surveys, postal surveys, drop and pick-up surveys and online surveys. Each has distinct advantages and disadvantages, and the choice among them depends on the type of questionnaire, desired response rate, and time and budgetary constraints.
- Sampling involves identifying the group of people to whom the questionnaire will be administered. The sample should be selected to represent well the target population and to minimize non-response bias. The sample size should be large enough to represent various subgroups in the population and to allow effective statistical analysis of results.

Notes

1   That is, a shift from simply seeking to collect numerical data that are comparable as these are based on the same interval (e.g. the measurement of age in years, or income in US dollars) to data that are more readily ordered, ranked or rated (e.g. age categories, or income bands).
2   The 'mulatto' category being used at the time to refer to a person who had one black and one white parent; a term that is now widely considered offensive.

Further Reading

A great deal has been written about questionnaire surveys in the social sciences. The following are just a few of the sources I have found useful:

- Bryman (2012) is a comprehensive, well-written book that provides an overview of questionnaire surveys and other social research methods including Internet and e-research approaches.

- Fink (2013) provide a very clear primer on how to conduct questionnaire surveys, focusing on 'how to' and practical advice.

- Groves et. al. (2009) provide an excellent, detailed, up-to-date discussion of survey research methods with an extensive bibliography. This book emphasizes methodological topics such as non-response bias, validity, sampling frames, questionnaire evaluation, interviewer bias, and ethical issues.

- Sue and Ritter (2012) offer a practical introduction to online survey research that adapts the well-established steps of survey research design for the online context.

- Survey Research Laboratory, University of Illinois at Chicago – Sites Related to Survey Research (www.srl.uic.edu/links.html). This is a comprehensive Internet site for survey research, including links to journals, organizations, sample questionnaires, software packages for analysing survey data, and sampling-related software and websites.

Note: Full details of the above can be found in the references list below.

References

Awanyo, L. (2001) 'Labor, ecology and a failed agenda of market incentives: The political ecology of agrarian reforms in Ghana', *Annals of the Association of American Geographers*, 91: 92–121.

Babbie, E. (2013) *The Practice of Social Research* (13th edition). Belmont, CA: Wadsworth.

Bearman, N., and Appleton, K. (2012) 'Using Google Maps to collect spatial responses in a survey environment', *Area*, 44: 160–9.

Bryman, A. (2012) *Social Research Methods* (4th edition). Oxford: Oxford University Press.

Dunton, G., Intille, S., Wolch, J., and Pence, M.A. (2012) 'Investigating the impact of a smart growth community on the contexts of children's physical activity using Ecological Momentary Assessment'. *Health and Place*, 18: 76–84.

Fan, C. (2002) 'The elite, the natives, and the outsiders: migration and labor market segmentation in urban China', *Annals of the Association of American Geographers*, 92: 103–24.

Fink, A. (2013) *How to Conduct Surveys: A Step-by-Step Guide* (5th edition). Thousand Oaks, CA: Sage.

Gilbert, M. (1998) '"Race", space and power: The survival strategies of working poor women', *Annals of the Association of American Geographers*, 88: 595–621.

Gould, P. and White, R. (1974) *Mental Maps*. Baltimore, MD: Penguin Books.

Groves, R.M., Fowler, F.J., Couper, M.P., Lepowski, J.M., Singer, E. and Tourangeau, R. (2009) *Survey Methodology* (2nd edition). New York: Wiley.

Haynes, R., Daras, K., Reading, R. and Jones, A. (2007) 'Modifiable neighbourhood units, zone design and residents' perceptions', *Health and Place*, 13: 812–25.

Kobayashi, A. (1994) 'Colouring the field: Gender, "race" and the politics of fieldwork', *The Professional Geographer*, 46: 73–9.

Lue, E., Wilson, J.P. and Curtis, A. (2014) 'Conducting disaster damage assessments with Spatial Video, experts, and citizens', *Applied Geography*, 52: 46–54.

Madge, C. and O'Connor, H. (2002) 'On-line with e-mums: Exploring the Internet as a medium for research', *Area*, 34: 92–102.

Rushton, G. (1969) 'Analysis of spatial behavior by revealed space preference', *Annals of the Association of American Geographers*, 59: 391–406.

Sue, V.M. and Ritter, L.A. (2012) *Conducting Online Surveys*. Thousand Oaks, CA: Sage.

Winchester, H.P.M. (1999) 'Interviews and questionnaires as mixed methods in population geography: The case of lone fathers in Newcastle, Australia', *The Professional Geographer*, 51: 60–7.

ON THE COMPANION WEBSITE…

Visit **https://study.sagepub.com/keymethods3e** for author videos, chapter exercises, resources and links, plus free access to the following recommended articles:

1. **Schuurman, N. (2009) 'Work, life, and creativity among academic geographers', *Progress in Human Geography*, 33(3): 307–12.**

This survey of academic geographers highlights stresses, challenges and opportunities in teaching, research, and academic life.

2. **Houston, S., Wright, R., Ellis, M, Holloway, S. and Hudson, M. (2005) 'Places of possibility: Where mixed-race partners meet',** *Progress in Human Geography,* **29(6): 700–17.**

This article explores how places and spaces of everyday life influence how, why, and where people of different races interact.

3. **Bailey, A. (2009) 'Population Geography: Lifecourse matters',** *Progress in Human Geography,* **33(3): 407–18.**

This report looks at how the transitions that occur through the lives of individuals strongly shape their mobility social interactions, health, and socioeconomic well-being.

9 Semi-structured Interviews and Focus Groups

Robyn Longhurst

SYNOPSIS

A **semi-structured interview** is a verbal interchange where one person, the interviewer, attempts to elicit information from another person by asking questions. Although the interviewer prepares a list of predetermined questions, semi-structured interviews unfold in a conversational manner offering participants the chance to explore issues they feel are important. A **focus group** is a group of people, usually between 6 and 12, who meet in an informal setting to talk about a particular topic that has been set by the researcher. The facilitator keeps the group on topic but is otherwise non-directive, allowing the group to explore the subject from as many angles as they please. This chapter explains how to go about conducting both semi-structured interviews and focus groups whether it be in-person or online.

This chapter is organized into the following sections:

9.1 Introduction

Talking with people is an excellent way of gathering information. Sometimes in our everyday lives, however, we tend to talk too quickly, not listen carefully enough or interrupt others. This applies to both talking with people in-person and online. Semi-structured interviews (sometimes referred to as informal, conversational or 'soft' interviews) and focus groups (sometimes referred to as focus-group interviews) are about talking with people but in ways that are self-conscious, orderly and partially structured. Krueger and Casey explain that focus-group interviewing (and we could add here, semi-structured interviewing) is about talking but it is also

... about listening. It is about paying attention. It is about being open to hear what people have to say. It is about being nonjudgmental. It is about creating a comfortable environment for people to share. It is about being careful and systematic with the things people tell you. (2000: xi)

Over the past few decades interesting debates have emerged in geography (especially amongst feminist geographers) about the utility and validity of qualitative methods, including semi-structured interviews and focus groups (Pile, 1991; Nast, 1994; Bennett, 2003; Crang, 2003, 2004, 2005; Schoenberger, 2007; McDowell, 2007; Davies and Dwyer, 2007; Longhurst et al., 2008; Secor, 2010; Hutcheson, 2013). Many geographers have moved towards what Sayer and Morgan (1985) call 'intensive methods' to examine the power relations and social processes constituted in geographical patterns.

Recently these 'intensive methods' have become even more 'intensive', or perhaps a more apt description is they have become 'performative' (Dewsbury, 2010; Pile 2010) with researchers using their own bodies as "instruments of research" (Longhurst et al. 2008). Performative methods focus on how different bodily practices involve multiple senses. For example, Longhurst et al. (2008) not only interviewed, but also cooked and ate with research participants. Duffy et al. (2011) conducted 'on the spot' interviews at a festival in order to feel the bodily rhythms of the sounds that surrounded them. Cain (2011) 'rummaged' with her participants through their wardrobes in order to elicit stories about clothing and bodies.

Geographers now employ a wide range of intensive or performative methods many of which are included in this volume. Semi-structured interviews, however, are probably still one of the most commonly used qualitative methods (Kitchin and Tate, 2000: 213). Focus groups are not as commonly used but they have become increasingly popular since the mid-1990s (see *Area*, 1996: Vol. 28 (2), which contains an introduction and five articles on focus groups).

Geographers have used focus groups to collect data on a diverse range of subjects. As early as 1988 Burgess et al. used focus groups (which they called 'small groups') to explore people's environmental values (Burgess et al., 1988a, 1988b). A decade later Miller et al. (1998) conducted focus groups (as well as surveys and ethnographic research) on shopping in northern London to explore the links between shopping and identity. Wolch et al. (2000) ran a series of focus groups in Los Angeles with an aim to find out more about the role played by cultural difference in shaping attitudes towards animals in the city. Skop (2006) examined the methodological potential of focus groups for population geography. Hutcheson (2013) conducted interviews and focus groups with families who relocated to another city after experiencing a significant earthquake and aftershocks in Christchurch, New Zealand. Hutcheson was acutely aware of the performative aspects of the interviews and focus groups and wrote notes about her participants' bodily reactions and facial expressions immediately after while sitting in her car.

Geographers have also used semi-structured interviews to collect data on an equally diverse range of subjects. Winchester (1999: 61) conducted interviews (and questionnaires) to gather a range of information about the characteristics of 'lone fathers' and the causes of marital breakdown and post-marital conflict in Newcastle, Australia. Valentine (1999) interviewed couples, some together, some apart, in order to understand gender relations in households. Johnston (2001) conducted interviews (and focus groups) with participants (or subjects – see McDowell, 1992: footnote 4, on the contested nature of the terms 'participant' and 'subject') and organizers at a gay pride parade in Auckland, New Zealand. Punch (2000) conducted interviews (and participant observation) with children and their families in Churquiales, a rural community in the south of Bolivia. Duffy et al. (2011) conducted short on-the-spot interviews in Australia with participants in a folkloristic parade who were performing songs and dance representing their Swiss-Italian heritage.

In this chapter I define briefly what I mean by semi-structured interviews and focus groups. These two methods share some characteristics in common; in other ways they are dissimilar. I also discuss how to plan and conduct semi-structured interviews and focus groups whether it be in-person or online via a medium such as Skype. This discussion includes formulating a schedule of questions, selecting and recruiting participants, choosing a location, transcribing data and thinking through some of the ethical issues and power relations involved in conducting semi-structured interviews and focus groups. Throughout the chapter empirical examples are used in an attempt to illustrate key arguments.

9.2 What are Semi-Structured Interviews and Focus Groups?

Interviews, explains Dunn (2005: 79), are verbal interchanges where one person, the interviewer, attempts to elicit information from another person. Basically there are three types of interviews: structured, unstructured and semi-structured, which can be placed along a continuum. Dunn explains:

> Structured interviews follow a predetermined and standardised list of questions. The questions are always asked in almost the same way and in the same order. At the other end of the continuum are unstructured forms of interviewing such as oral histories… The conversation in these interviews is actually directed by the informant rather than by the set questions. In the middle of this continuum are semi-structured interviews. This form of interviewing has some degree of predetermined order but still ensures flexibility in the way issues are addressed by the informant. (2005: 80)

Semi-structured interviews and focus groups are similar in that they are conversational and informal in tone. Both allow for an open response in the participants' own words rather than a 'yes or no' type answer.

A focus group is a group of people, usually between 6 and 12, who meet in an informal setting to talk about a particular topic that has been set by the researcher (for other definitions see Merton and Kendall, 1990; Greenbaum, 1993; Morgan, 1997; Stewart et al., 2006; Gregory et al., 2009). The method has its roots in market research. The facilitator or moderator of focus groups keeps the group on the topic but is otherwise non-directive, allowing the group to explore the subject from as many angles as they please. Often researchers attempt to construct as homogeneous a group as possible (but not always – see Goss and Leinback, 1996). The idea is to attempt to simulate a group of friends or people who have things in common and feel relaxed talking to each other. When Honeyfield (1997; also see Campbell et al., 1999) conducted research on representations of place and masculinity in television advertising for beer, he carried out two focus groups: one with five women, one with seven men. In both groups the participants had either met before, were friends or lived together as 'flatmates'.

Focus groups tend to last between one and two hours. A key characteristic is the interaction between members of the group (Morgan, 1997: 12; Bedford and Burgess, 2001; Cameron, 2005). This makes them different from semi-structured interviews which rely on the interaction between interviewer and interviewee. Focus groups are also different from interviews in that it is possible to gather the opinions of a large number of people for comparatively little time and expense. Recently I have been carrying out research on people's experiences of using Skype by running small focus groups on Skype. It is also possible to conduct focus groups via other software programmes such as Facetime or Google Hangout (see Madge and O'Connor, 2002, on 'on-line synchronous interviews', and Hanna, 2012, on using Skype as a 'research medium').

Focus groups are often recommended to researchers wishing to orientate themselves to a new field (Greenbaum, 1993; Morgan, 1997). For example, in 1992 I began some research on pregnant women's experiences of public spaces in Hamilton, New Zealand. There was no existing research on this topic so I wanted to establish some of the parameters of the project before using other methods. I did not know what words pregnant women in Hamilton used to refer to their pregnant bodies – tummies? stomachs? breasts? boobs? – therefore, it would have been difficult to conduct interviews. Focus groups provided an excellent opportunity to gather preliminary information about the topic (see Longhurst, 1996, for an account of these focus groups).

Both semi-structured interviews and focus groups can be used as 'stand-alone methods', as a supplement to other methods or as a means for triangulation in multi-methods research. Researchers often draw on a range of methods and theories. Valentine explains:

> Often researchers draw on many different perspectives or sources in the course of their work. This is known as triangulation. The term comes from surveying, where it describes using different bearings to give the correct position. In the same way researchers can use multiple methods or different sources to try and maximize their understanding of a research question. (2005: 112)

To sum up thus far, semi-structured interviews and focus groups can be used for a range of research, are reasonably informal or conversational in nature and are flexible in that they can be carried out in-person or online and can be used in conjunction with a variety of other methods and theories. It is also evident that semi-structured interviews and focus groups are more than just 'chats'. The researcher needs to formulate questions, select and recruit participants, choose a medium and/or location and transcribe data while at the same time remaining cognizant of the ethical issues and power relations involved in qualitative research. In the section that follows I address these topics.

9.3 Formulating Questions

Dunn (2005: 81) explains: 'It is not possible to formulate a strict guide to good practice for every interview [and focus group] context'. Every interview and focus group requires its own preparation, thought and practice. It is a social interaction and there are no hard and fast rules one can follow (Valentine, 2005). Nevertheless there are certain procedures that researchers are well advised to heed.

To begin, researchers need to brief themselves fully on the topic. Having done this it is important to work out a list of themes or questions to ask participants. People who are very confident at interviewing or running focus groups often equip themselves with just a list of themes. Personally, I like to be prepared with actual questions in case the conversation dries up. Questions may be designed to elicit information that is 'factual', descriptive, thoughtful, emotional or affectual. A combination of different types of questions can be effective depending on the research topic. Researchers often start with a question that participants are likely to feel comfortable answering. More difficult, sensitive or thought-provoking questions are best left to the second half of the interview or focus group when participants are feeling more comfortable. Box 9.1 contains a list of questions I drew up in order to examine large/fat/overweight people's experiences of place. This schedule could be used for semi-structured interviews or focus groups. Follow-up questions are in parentheses.

I would not necessarily ask these questions in the order listed. Allowing the discussion to unfold in a conversational manner offers participants the chance to explore issues they feel are important. At the end of the interview or focus group, however, I would check my schedule to make sure that all the questions had been covered at some stage during the interview or focus group.

It is important to remember that it can take time for participants to 'warm up' to semi-structured interviews and focus groups. If possible, therefore, it is worth offering drinks and food as a way of relaxing people although clearly this is not possible if you are conducting the research online. It is also useful at the beginning of a focus group to engage participants in some kind of activity that focuses their attention on the discussion topic. For example, participants might be asked to draw a picture, respond to a photograph or imagine a particular situation. This technique tends to be used more by market researchers but it can also prove effective for social scientists. Kitzinger (1994) presented focus group members with a pack of cards bearing statements about who might be 'at risk' from AIDS. She asked the group to sort the cards into different piles indicating the degree of 'risk' attached to each 'type of person'. Kitzinger (1994: 107) explains that '[s]uch exercises involve people in working together with minimal input from the facilitator and encourage participants to concentrate on one another (rather than on the group facilitator) during the subsequent discussion'.

Box 9.1   Semi-structured interview and focus-group schedule

Questions associated with Longhurst (1996)

- Can you remember a time in your life when you were *not* large/fat/overweight? (Tell me about that. How did people respond to you then?)
- Are there places that you avoid on account of being large? (Why? How do you feel if you do visit these places?)

(Continued)

(Continued)

- Are there places where you feel comfortable or a sense of belonging on account of your size? (Tell me about these places and how you feel in them.)
- In New Zealand there is a strong tradition of spending time at the beach. Do you go to the beach? (Explain. What is it like for you at the beach?)
- Describe your experience of clothes shopping. (Where do you shop? Are shop assistants helpful? Are the changing rooms comfortable? Do you ever feel that other shoppers judge you on account of your size?)
- When you shop for groceries or eat out in a public space, how do you feel? (Why?)
- Are there any issues concerning your size that arise at work? (What are these issues?)
- Do you feel cramped in some spaces? (For example, movie-theatre seats, small cars, planes?)
- Do you exercise? (If so, what do you do and where do you do it?)
- Have you made any modifications to your home to suit your size? (For example, altered doorways, selected particular furniture, arranged furniture in specific ways, modified bathroom/toileting facilities. Explain.)
- Do you imagine that your life would be different if you were smaller? (Explain.)
- Are there any issues that you would like to raise that you feel are important but that you haven't had a chance to explore in this interview/focus group?

9.4 Selecting and Recruiting Participants

Selecting participants for semi-structured interviews and focus groups is vitally important. Usually people are chosen on the basis of their experience related to the research topic (Cameron, 2005). Burgess's (1996, cited in Cameron, 2005: 121) study of fear in the countryside is a useful example of this 'purposive sampling' technique. When using quantitative methods the aim is often to choose a random or representative sample, to be 'objective' and to be able to replicate the data. This is not the case when using qualitative methods. Valentine (2005: 111, emphasis in original) explains that, unlike with most questionnaires, 'the aim of an interview [and a focus group] is *not* to be representative (a common but mistaken criticism of this technique) but to understand how individual people experience and make sense of their own lives'.

For example, if you were studying 'racial violence' you might anticipate interviewing and/or running focus groups with people from different ethnic groups, especially those thought to be involved in the violence. However, you might also want to examine the ways in which people's ethnic or racial identities intersect with other identities such as gender, sexuality, 'migrant status' and age in order to explore more fully the processes shaping racial violence. It is not only participants' identities that need to be considered, however, when conducting research. Valentine (2005: 113) makes the important point that 'When you are thinking about who you want to interview it is important to reflect on who you are and how your own identity will shape the interactions that you have with others'. She explains this is what academics describe as being *reflexive* or recognizing your own *positionality* (see England, 1994: 82; Moss, 2002; Bondi, 2003, on 'empathy and identification' in the research process; and Moser, 2008, on personality as the new positionality).

There are many strategies for recruiting participants for semi-structured interviews and focus groups. Some strategies work for both methods while others are more appropriate for one or the other. If you are recruiting participants for interviews it is common practice 'to carry out a simple questionnaire survey to gather basic factual information and to include a request at the end of the questionnaire asking respondents who are willing to take part in a follow-up interview to give their address and telephone number' (Valentine, 2005: 115). It is also possible to advertise for participants in local newspapers or on radio stations, requesting interested parties to contact you.

Alternatively, or in addition to these strategies, researchers are increasingly using social media (such as Facebook) and web-based surveys (such as SurveyMonkey) to reach out to potential participants in particular target groups. Special groups who share an experience or background can be identified and contacted via a survey link, email or text in an attempt to set up an in-person or online interview or focus group. For example, when conducting research on people's experiences of love and romance in the massively multiplayer online role-playing game (MMORPG) *World of Warcraft*, PhD student Cherie Todd invited players to respond to an online questionnaire and then asked if they would be prepared to be interviewed in-person if they lived locally or via Skype if they lived afar (Todd, forthcoming).

Using social media to recruit is to some degree replacing the earlier method of 'cold calling' which involves actually calling on people (usually strangers) to ask if they would be prepared to be interviewed. This can be a nerve-racking process because interviewers often get a high refusal rate.

As mentioned already, focus groups are often made up of people who share something in common or know each other. Group membership lists, therefore, can be a useful tool for recruiting. People who already know each other through sports clubs, online groups, community activities, church groups or work can make an ideal focus group. When I conducted focus groups on men's experiences of domestic bathrooms (a private space rarely discussed by geographers) I succeeded in enlisting (with the help of friends) four groups of men. The first group belonged to the same rugby club, the second were colleagues in a government department, the third were 'job-seekers' and the fourth were family/friends.

Another route useful for securing participants for focus groups is what Krueger (1988: 94) refers to as 'recruiting on location' or 'on-site recruiting'. I used this strategy to recruit first-time pregnant women to talk about their experiences of public places. Pregnant women were approached at antenatal classes, midwives' clinics and doctors' surgeries. These women 'opened doors' to me speaking with other pregnant women. Social scientists refer to this as 'snowballing': 'This term describes using one contact to help you recruit another contact, who in turn can put you in touch with someone else' (Valentine, 2005: 117).

9.5 Where to Meet

Not only is it necessary to decide how to select and recruit participants but also to decide where to conduct the interview or focus-group meeting. In the first instance you will need to decide whether to conduct it in-person or online. It comes as no surprise to most geographers that where an interview or focus group is held can make a difference. Ideally, the setting should be relatively neutral. I once made the mistake of helping to facilitate a focus group about the quality of service offered by a local council at the council offices. The discussion did not flow freely and it soon became apparent that the participants felt hesitant (understandably) about criticizing the council while in one of their rooms. However, it is worth noting that 'In most cases if you are talking to business people or officials from institutions and organizations you will have no choice but to interview them in their own offices' (Valentine, 2005: 118; but also see McDowell, 1997, on interviewing investment bankers in the City of London). Being in the environs you are studying can also prove useful. If you decide to conduct semi-structured interviews or focus groups online think which software package (e.g. Skype, Facetime, Google Hangout) might be most user-friendly and/or work most effectively for you and your participants.

It is not always possible to conduct interviews and focus groups in 'the perfect setting' but if at all possible aim to find a place that is neutral, informal (but not noisy) and easily accessible. For example, if you are conducting a reasonably small focus group it may be possible to sit comfortably around one computer screen or a dining-room table (see Fine and Macpherson, 1992, for an account of a focus group that took place 'over dinner'). Needless to say, if it is a larger focus group then multiple computer screens or a larger space will be required, perhaps a room at a school or club. The main consideration for both semi-structured interviews and focus groups is that interviewees feel comfortable whether in real or online space. It is important that the interviewer also feels comfortable (see also Chapter 2). Valentine (2005: 118) warns: 'For your own safety never arrange interviews with people you do not feel comfortable with or agree to meet strangers in places where you feel vulnerable'.

9.6 Recording and Transcribing Discussions

When conducting semi-structured interviews or focus groups it is possible to take notes or to audio/video record the discussion. I usually audio(record) the proceedings. If you are conducting the research online you can use either inbuilt software (Sound Recorder on Windows; QuickTime on OS X) or commercial applications to record the conversation. Recording allows me to focus fully on the interaction instead of feeling pressure to get the participants' words written in my notebook (see Valentine, 2005). Directly after the interview I document the general tone of the conversation, the key themes that emerged and anything that particularly impressed or surprised me in the conversation. Taking these notes, in a sense, is a form of data analysis (for information on qualitative data analysis, see Chapter 36; also Miles and Huberman, 1994; Kitchin and Tate, 2000).

It is advantageous to transcribe interviews and focus groups as soon as possible after conducting them (for how to code a transcript, see Chapter 36). Hearing the taped conversation when it is still fresh in your mind makes transcription much easier. Focus groups, especially large groups, can be difficult to transcribe because each speaker, including the facilitator, needs to be identified. In Box 9.2 is an example of a transcript from a focus group of young mothers who met to discuss their experiences of using various technologies as part of their mothering practices. Note the 'dynamism and energy as people respond to the contributions of others' (Cameron, 2005: 117). In this focus group excerpt one of the participants puts a question to other group members. Jasmine asks 'Who cares what Facebook does with users' photos?' This prompts a difference in opinion between participants. Note the various transcription codes: the starts of overlap in talk are marked by a double oblique //; pauses are marked with a dot in parenthesis (.); non-verbal actions, gestures and facial expressions are noted in square brackets; and loud exclamations are in **bold** typeface (for more detailed transcription codes see Dunn, 2005: 98).

Box 9.2   Transcription of a Focus Group

Lakin: I just got told that the photos you put on Facebook become Facebook property, even if you delete them off, they are still part of Facebook property. They've still got copies of them all //.

Jasmine: **But who cares**? What are they gonna do with them? **(.)**

Theressa: Well, I am not keen on that in a way [frowns]. That's pretty scary really.

Jasmine: But there are millions and millions and millions of photos on there.

Theressa: But imagine the worries about certain people getting hold of them. Like we would not let kids' photos at school go on the Internet. We won't let things like that happen because what if the wrong people get them. Just anything like that. If they are Facebook property anyone can go into the data base and get any photo of anything and do God knows what with it.

Source: Audio-tape excerpt from a focus group conducted by Robyn Longhurst in 2009 (see Longhurst, 2013, for a publication based on these data)

As this transcript illustrates, sometimes participants can disagree and data can be 'sensitive' (some mothers fear paedophiles on the Internet). It is not surprising, therefore, that there are numerous ethical issues to consider when conducting semi-structured interviews and focus groups (see also Chapter 3).

9.7 Ethical Issues

Two important ethical issues are confidentiality and anonymity. Participants need to be assured that all the data collected will remain secure under lock and key or on a computer database accessible by password only; that information supplied will remain confidential and participants will remain anonymous (unless they desire otherwise); and that participants have the right to withdraw from the research at any time without explanation. It is also sound research practice to offer to provide participants with a summary of the research results at the completion of the project and to follow through on this commitment. This summary might take the form of a hard copy or an electronic copy posted on a website (for example, the Department of Geography, Durham University provides reports on various research projects conducted by staff: see https://www.dur.ac.uk/geography/research/).

Focus groups pose a further complication in relation to confidentiality because not only is the researcher privy to information but also members of the group. Therefore, participants need to be asked to treat discussions as confidential. Cameron explains:

> As this [confidentiality] cannot be guaranteed, it is appropriate to remind people to disclose only those things they would feel comfortable about being repeated outside the group. Of course, you should always weigh up whether a topic is too controversial or sensitive for discussion in a focus group and is better handled through another technique, like individual in-depth interviews. (2005: 122)

Another ethical issue is that participants in the course of an interview or focus group may express sexist, racist or other offensive views. In an earlier quotation, Krueger and Casey (2000: xi) claim that researchers ought to listen, pay attention and be non-judgemental. Sometimes, however, being non-judgemental might simply reproduce and even legitimize interviewees' discrimination through complicity (see Valentine, 2005). Researchers need to think carefully about how to deal with such situations because there are no easy solutions.

Researchers also need to think carefully about how to interview or run focus groups in different cultural contexts (see Chapter 6). For example, 'First World' researchers investigating 'Third World' 'subjects' need to be highly sensitive to local codes of conduct (Valentine, 2005). In short, there is a web of ethical issues and power relations that need to be teased out when conducting semi-structured interviews and focus groups (see Law, 2004, on 'mess' in social science research). Feminist geographers in particular have made a useful contribution in this area (for example, see McDowell, 1992; Dyck, 1993; Katz, 1994; England, 1994; Gibson-Graham, 1994; Kobayashi, 1994; Moss, 2002; Bondi, 2003).

9.8 Conclusion

In this chapter I have outlined two qualitative methods – semi-structured interviews and focus groups – and how they can be employed in geographical research. Both methods involve talking with people in a semi-structured manner whether it be in-person on online. However, whereas semi-structured interviews rely on the interaction between interviewee and interviewer, focus groups rely on interactions amongst interviewees. Both methods make a significant contribution to geographic research, especially now that discussions about meaning, identity, subjectivity, emotion, affect, politics, knowledge, power, performativity and representation are high on many geographers' agendas. Critically examining the construction of knowledge and discourse in geography (see Rose, 1993) has led to an interest in developing methodological strategies that can be employed with a high level of reflexivity about the process of research. Semi-structured interviews and focus groups are useful for investigating complex behaviours, opinions, emotions and affects, and for collecting a diversity of experiences. These methods do not offer researchers a route to 'the truth' but they do offer a route to partial insights into what people do and think.

SUMMARY

- Semi-structured interviews and focus groups are about talking with people both in-person and online but in ways that are self-conscious, orderly and partially structured.
- These methods are useful for investigating complex behaviours, opinions, emotions and affects, and for collecting a diversity of experiences.
- Every interview and focus group requires its own preparation, thought and practice.
- There is a range of methods that can be used for recruiting participants, including advertising for participants, accessing membership lists (including Internet mailing lists), using social media, on-site recruiting and 'cold calling'.
- Interviews/focus groups ought to be conducted in a place or space where both participants and interviewer feel comfortable.
- When conducting semi-structured interviews or focus groups take notes and/or audio/video record the discussion.
- There is a web of ethical issues and power relations that need to be teased out when using these methods.
- Semi-structured interviews and focus groups make a significant contribution to geographic research, especially now that discussions about meaning, identity, subjectivity, emotion, affect, politics, knowledge, power, performativity and representation are high on many geographers' agendas.

Further Reading

There are numerous excellent books, book chapters and articles on semi-structured interviews (and interviewing more generally) and focus groups written by geographers and other social scientists. I have listed below some recommended published titles:

- Denzin and Lincoln's (2011) 4th edition of *The SAGE Handbook of Qualititive Research* provides a number of useful chapters including on interviews (e.g. chpt. 32 by Peräkylä and Ruusuvuori) and focus groups (e.g. chpt. 33 by Kamberelis and Dimitriadis). Other topics covered in *The SAGE Handbook* include ethics, politics, feminism, performance, technology and post qualitative research.

- Madge and O'Connor (2002) discuss their experience of conducting what they call 'semi-structured synchronous virtual group interviews' with a group of mothers. Using Internet technologies for research is gaining popularity as more possibilities open up for synchronous audio and visual communications.

- Cameron (2005) provides a geographer's perspective on focus groups, explaining the various ways they have been used, how to plan and conduct them, and how to analyse and present results.

- Valentine's (2005) chapter on 'conversational interviews' is highly readable and provides advice on whom to talk to, how to recruit participants and where to hold interviews. Valentine raises interesting questions about the ethics and politics of interviewing and alerts readers to some of the potential pitfalls that can occur in research.

- Dunn (2005) discusses structured, semi-structured and unstructured interviewing in geography, critically assessing the relative strengths and weaknesses of each method. His chapter provides advice on interview design, practice, transcription, data analysis and presentation. Like Valentine, Dunn has a useful guide at the end of the chapter to further reading.

Note: Full details of the above can be found in the references list below.

References

Area (1996) 28(2) 'Introduction to focus groups' by J.D. Goss and five papers on using focus groups in human geography by Burgess; Zeigler, Brunn and Johnston; Holbrook and Jackson; Longhurst; and Goss and Leinback.

Bedford, T. and Burgess, J. (2001) 'The focus-group experience', in M. Limb and C. Dwyer (eds) *Qualitative Methodologies for Geographers.* London: Arnold. pp. 121–35.

Bennett, K. (2003) 'Interviews and focus groups', in P. Shurmer-Smith (ed.) *Doing Cultural Geography.* London: Sage. pp. 153–62.

Bondi, L. (2003) 'Empathy and identification: Conceptual resources for feminist fieldwork', *ACME: International Journal of Critical Geography,* 2: 64–76. http://acme-journal.org/index.php/acme/article/viewFile/708/571 (accessed 12 November 2015).

Burgess, J. (1996) 'Focusing on fear: The use of focus groups in a project for the Community Forest Unit, Countryside Commission', *Area,* 28: 130–35.

Burgess, J., Limb, M. and Harrison, C.M. (1988a) 'Exploring environmental values through the medium of small groups. 1. Theory and practice', *Environment and Planning A,* 20: 309–26.

Burgess, J., Limb, M. and Harrison C.M. (1988b) 'Exploring environmental values through the medium of small groups. 2. Illustrations of a group at work', *Environment and Planning A,* 20: 457–76.

Cain, T.M. (2011) 'Bounded bodies: The larger everyday clothing practices of larger women', PhD thesis, Massey University, Albany, New Zealand.

Cameron, J. (2005) 'Focusing on the focus group', in I. Hay (ed.) *Qualitative Research Methods in Human Geography* (2nd edition). Melbourne: Oxford University Press. pp. 116–32.

Campbell, H., Law, R. and Honeyfield, J. (1999) '"What it means to be a man": Hegemonic masculinity and the reinvention of beer', in R. Law, H. Campbell and J. Dolan (eds) *Masculinities in Aotearoa/New Zealand.* Palmerston North: Dunmore Press. pp. 166–86.

Crang, M. (2002) 'Qualitative methods: The new orthodoxy?', *Progress in Human Geography,* 26: 647–55.

Crang, M. (2003) 'Qualitative methods: Touchy, feely, look-see?', *Progress in Human Geography,* 27: 494–504.

Crang, M. (2005) 'Qualitative methods: There is nothing outside the text?', *Progress in Human Geography,* 29 (2): 225–33.

Davies, G. and Dwyer, C. (2007) 'Qualitative methods: Are you enchanted or are you alienated?', *Progress in Human Geography,* 31: 257–66.

Denzin, N.K. and Lincoln Y.S. (eds) (2011) *The SAGE Handbook of Qualitative Research* (4th edition). Thousand Oaks, CA: Sage.

Department of Geography, Durham University, Projects: https://www.dur.ac.uk/geography/research/ (accessed 10 November 2016).

Dewsbury, J.D. (2010) 'Performative, non-representational, and affect-based research: Seven injunctions', in D. DeLyser, S. Herbert, S. Aitken, M. Crang and L. McDowell (eds) *The SAGE Handbook of Qualitative Geography.* London: Sage. pp. 321–34.

Dunn, K. (2005) 'Interviewing', in I. Hay (ed.) *Qualitative Research Methods in Human Geography* (2nd edition). Melbourne: Oxford University Press. pp. 79–105.

Duffy, M., Waitt, G., Gorman-Murray, A. and Gibson, C. (2011) 'Bodily rhythms: Corporeal capacities to engage with festival spaces', *Emotion, Space and Society,* 4, 17–24.

Dyck, I. (1993) 'Ethnography: A feminist method?', *The Canadian Geographer,* 37: 52–7.

England, K. (1994) 'Getting personal: Reflexivity, positionality and feminist research', *The Professional Geographer,* 46: 80–9.

Fine, M. and Macpherson, P. (1992) 'Over dinner: Feminism and adolescent female bodies', in M. Fine (ed.) *Disruptive Voices: The Possibilities of Feminist Research.* East Lansing, MI: University of Michigan Press. pp. 175–203.

Gibson-Graham, J.K. (1994) '"Stuffed if I know!": Reflections on post-modern feminist social research', *Gender, Place and Culture,* 1: 205–24.

Goss, J.D. and Leinback, T.R. (1996) 'Focus groups as alternative research practice: Experience with transmigrants in Indonesia', *Area,* 28: 115–23.

Greenbaum, T. (1993) *The Handbook for Focus Group Research.* Lexington, MA: Lexington Books.

Gregory, D., Johnston, R.J., Pratt, G., Watts, M. and Whatmore, S. (2009) *A Dictionary of Human Geography* (5th edition). Oxford: Blackwell.

Hanna, P. (2012) 'Using internet technologies (such as Skype) as a research medium: A research note', *Qualitative Research*, 12 (2): 239–42.

Honeyfield, J. (1997) 'Red blooded blood brothers: Representations of place and hard man masculinity in television advertisements for beer.' Master's thesis, University of Waikato, New Zealand.

Hutcheson, G. (2013) 'Methodological reflections on transference and countertransference in geographical research: Relocation experiences from post-disaster Christchurch, Aotearoa New Zealand', *Area*, 45: 477–84.

Johnston, L. (2001) '(Other) bodies and tourism studies', *Annals of Tourism Research*, 28: 180–201.

Kamberelis, G. and Dimitriadis, G. (2011) 'Focus groups: Contingent articulations of pedagogy, politics, and inquiry', in N.K. Denzin and Y.S. Lincoln (eds) *The SAGE Handbook of Qualitative Research* (4th edition). Thousand Oaks, CA: Sage., pp. 545–53.

Katz, C. (1994) 'Playing the field: Questions of fieldwork in geography', *The Professional Geographer*, 46: 67–72.

Kitchin, R. and Tate, N.J. (2000) *Conducting Research into Human Geography*. Edinburgh Gate: Pearson.

Kitzinger, J. (1994) 'The methodology of focus groups: The importance of interaction between research participants', *Sociology of Health and Illness*, 16: 103–21.

Kobayashi, A. (1994) 'Coloring the field: Gender, "race", and the politics of fieldwork', *The Professional Geographer*, 46: 73–9.

Krueger, R.A. (1988) *Focus Groups: A Practical Guide for Applied Research*. Thousand Oaks, CA: Sage.

Krueger, R.A. and Casey, M.A. (2000) *Focus Groups. A Practical Guide for Applied Research* (3rd edition). Thousand Oaks, CA: Sage.

Law, J. (2004) *After Method: Mess in Social Science Research*. London: Routledge.

Longhurst, R. (1996) 'Refocusing groups: Pregnant women's geographical experiences of Hamilton, New Zealand/Aotearoa', *Area*, 28: 143–9.

Longhurst, R. (2013) 'Using Skype to mother: bodies, emotions, visuality, and screens', *Environment and Planning D: Society and Space*, 31(4): 664–79.

Longhurst, R., Ho, E. and Johnston, L. (2008) 'Using "the body" as an "instrument of research": kimchi'l and pavlova', *Area*, 40: 208–17.

Madge, C. and O'Connor, H. (2002) 'On-line with e-mums: Exploring the internet as a medium for research', *Area*, 34: 102.

McDowell, L. (1992) 'Doing gender: Feminism, feminists and research methods in human geography', *Transactions, Institute of British Geographers*, 17: 399–416.

McDowell, L. (1997) *Capital Culture: Gender at Work in the City*. Oxford: Blackwell.

McDowell, L. (2007) 'Sexing the economy, theorizing bodies', in A. Tickell, E. Sheppard, J. Peck and T. Barnes (eds) *Politics and Practice in Economic Geography*. London: Sage. pp. 60–70.

Merton, R.K. and Kendall, P.L. (1990) *The Focused Interview: A Manual of Problems and Procedures* (2nd edition). New York: Free Press.

Miles, M.B. and Huberman, A.M. (1994) *Qualitative Data Analysis: An Expanded Sourcebook*. Thousand Oaks, CA: Sage.

Miller, D., Jackson, P., Thrift, N., Holbrook, B. and Rowlands, N. (1998) *Shopping, Place and Identity*. London: Routledge.

Morgan, D.L. (1997) *Focus Groups as Qualitative Research*. (*Qualitative Research Methods*: 16). Thousand Oaks, CA: Sage.

Moser, S. (2010) 'Personality: A new positionality?' *Area*, 40: 383–92.

Moss, P. (ed.) (2002) *Feminist Geography in Practice: Research and Methods*. Oxford: Blackwell.

Nast, H. (1994) 'Opening remarks on "women in the field"', *The Professional Geographer*, 46: 54–5.

Peräkylä, A. and Ruusuvuori, J. (2011) 'Analyzing talk and text', in N.K. Denzin and Y.S. Lincoln (eds) *The SAGE Handbook of Qualitative Research* (4th edition). Thousand Oaks, CA: Sage. pp. 529–43.

Pile, S. (1991) 'Practising interpretative geography', *Transactions of the Institute of British Geographers*, 16: 458–69.

Pile, S. (2010) 'Intimate distance: The unconscious dimensions of the rapport between researcher and researched', *The Professional Geographer*, 62: 483–95.

Punch, S. (2000) 'Children's strategies for creating playspaces', in S.L. Holloway and G. Valentine (eds) *Children's Geographies. Playing, Living, Learning.* London and New York: Routledge. pp. 48–62.

Rose, G. (1993) *Feminism and Geography: The Limits of Geographical Knowledge.* Cambridge: Polity Press.

Sayer, A. and Morgan, K. (1985) 'A modern industry in a reclining region: Links between method, theory and policy', in D. Massey and R. Meegan (eds) *Politics and Method.* London: Methuen. pp. 147–68.

Schoenberger, E. (2007) 'Politics and practice: Becoming a geographer', in A. Tickell, E. Sheppard, J. Peck and T. Barnes (eds) *Politics and Practice in Economic Geography.* London: Sage. pp. 27–37.

Secor, A.J. (2010) 'Social surveys, interviews and focus groups', in B. Gomez and J.P. Jones III (eds) *Research Methods in Geography.* Chichester: Blackwell. pp. 194–205.

Skop, E. (2006) 'The methodological potential of focus groups in population geography', *Population, Space and Place,* 12: 113–24.

Stewart, D.W., Shamdasani, P.N. and Rook, D.W. (2006) *Focus Groups: Theory and Practice* (2nd edition). Newbury Park, CA: Sage.

Todd, C.J. (forthcoming) '"Male blood elves are so gay": gender and sexual identity in online games', in G. Brown and K. Browe (eds) *The Ashgate Research Companion to Geographies of Sex and Sexualities.* Farnham: Ashgate.

Valentine, G. (1999) 'Doing household research: Interviewing couples together and apart', *Area,* 31: 67–74.

Valentine, G. (2005) 'Tell me about... using interviews as a research methodology', in R. Flowerdew and D. Martin (eds) *Methods in Human Geography: A Guide for Students Doing a Research Project* (2nd edition). Edinburgh Gate: Addison Wesley Longman. pp. 110–27.

Winchester, H.P.M. (1999) 'Interviews and questionnaires as mixed methods in population geography: The case of lone fathers in Newcastle, Australia', *The Professional Geographer,* 51: 60–7.

Wolch, J., Brownlow, A. and Lassiter, U. (2000) 'Constructing the animal worlds of inner-city Los Angeles', in C. Philo and C. Wilbert (eds) *Animal Spaces, Beastly Places.* London and New York: Routledge. pp. 71–97.

ON THE COMPANION WEBSITE...

Visit **https://study.sagepub.com/keymethods3e** for author videos, chapter exercises, resources and links, plus **free** access to the following recommended articles:

1. **Crang, M. (2003) 'Qualitative methods: Touchy, Feely, look-see?', *Progress in Human Geography*, 27(4): 494–504.**

This article is useful because it points readers to various articles and books devoted to qualitative methods in geography. It also discusses researcher positionality, and performative approaches, both highly relevant when thinking about semi-structured interviews and focus groups.

2. **Davies, G. and Dwyer, C (2007) 'Qualitative methods: Are you enchanted or are you alienated?', *Progress in Human Geography,* 31(2): 257–66.**

This article argues that methods such as semi-structured interviews and focus groups remain 'the backbone of qualitative methods in human geography' but there are transformations occurring in the way they are being used to construct and convey knowledge. Issues of agency, embodiment and emotion, being in nature, and the performativity of place are discussed.

10 Respondent Diaries

Alan Latham

SYNOPSIS

Diaries are pieces of autobiographic writing describing in a more or less systematic way a period of an individual's life. Diaries may be written for the express purposes of a research project, or they may be documents written at the time of a particular event, that the researcher later draws upon to understand that event. The basis of diary-based approaches is to gain a sense of the routines, rhythms, and texture of a person's life, or a part of their life, over a period of time. This chapter will concern itself with diaries written for the express purpose of a particular research project.

This chapter is organized into the following sections

- Introduction
- When are diaries useful?
- Different kinds of diaries
- The practicalities of organizing respondent diaries
- What kind of material might you expect? And what to do with it?
- The limitations of respondent diaries

10.1 Introduction

Geographers are often concerned with the everyday rhythms and textures of people's day-to-day lives. They often want to understand the spatial and temporal context within which particular social practices occur. They want to know, does a certain practice tend to occur at a certain time of the day? Or a certain time of the week? Or month? Or year? Geographers also often want to know the frequency and duration of particular social practices, and how these social practices are related to other events that are involved in structuring a person's daily, weekly, or monthly life path. To begin to understand British drinking patterns, for example, it would be crucial to have a sense of the ways the practice of drinking is related to the rhythms of the working week. Similarly, it would be difficult to make much sense of daily variations in traffic flow in a particular city without reference to the times most people begin and finish working, or when schools open and close. In many instances, research respondent diaries are a very productive way of generating such research material.

10.2 When are Diaries Useful?

There are many different ways to generate research material that explore the rhythms and textures of day-to-day lives. Interviews, questionnaires, participant observation, and focus groups can all provide insightful material. However, there are limitations to these methods.

Firstly, with interviews, questionnaires, and focus groups it is often unreasonable to expect people to reliably remember the frequency of activities that they carry out routinely and necessarily without much thought. Ask yourself, for example, how many times you have a used a bus, bicycle, or car this week? Then try and list the time and destination of every individual journey you have made in the past week. Unless you have an exceptional memory (and most people do not) – or really don't get out much – you will find it hard, if not impossible, to produce a reliable list without referring to your personal organizer or diary.

Secondly, a researcher might be interested in the texture of people's mundane interactions in certain kinds of spaces – public spaces, say, or within the workplace. Precisely because many of these interactions are fleeting, lacking in any obvious social consequence, or just plain routine, potential research respondents will generally struggle to say anything much about them. Respondents will rarely be able to cite specific instances of certain interactions, and without a concrete event to organize their account around will frequently reply in vague, and thus difficult to interpret, generalities. In other cases, people will simply not notice that they are involved in certain kinds of interactions at all. They will overlook the fact that they smile a hello at the newsagent cashier where they buy their morning newspaper, or that they always choose to sit at a particular place to eat their lunch, or that lunch always – or nearly always – consists of the same thing. (Indeed, people are often surprised when they realize the degree that their lives are entrained in particular routines and rituals of behaviour.) Participant observation might be one way of generating research material about these sorts of interactions. However, unless you follow an individual through a whole day, participant observation will be largely silent about how the observed interactions fit into people's wider routines and daily commitments. While some researchers have undertaken to follow their respondents through the routines of their day (see Laurier and Philo, 2003), this involves very significant commitments from both the researcher and the research respondent. And, if the respondent spends relatively little time actually involved in the activity that is the focus of research, this might prove to be an unproductive (if very rigorous) research strategy.

Diaries produced by those involved in the flow of the social practice the researcher is seeking to understand can offer ways around the two difficulties outline above. By asking people to note down when they are involved in a certain activity either whilst involved in it, or immediately afterwards, research is not so beholden to the capriciousness of memory. Diaries can thus produce more detailed, more reliable, and often more focussed accounts, than other methodologies. What is more, the creation of diaries by asking research respondents to attend to social practices or events with the view of reporting these to the researcher in effect allows the research respondent to stand in as a proxy for the researcher. The result is that the tracing out of someone's day, or week, or month, or whatever period of time does not require the constant presence of the researcher. Instead the diary allows the researcher to virtually accompany her or his research respondent as they go about their day-to-day routines without the intrusiveness and heavy time demands that actually physically shadowing a respondent would involve. In the same time that a strict participant observation-based research strategy might be able to gain data on just one respondent, a diary-based process might be able to obtain data from four or five diary respondents. (Although it is also worth noting that the detail of the observations gained from participant observation may well be of a greater quality.)

There are a number of further attractions to respondent diaries (these diaries are also know as solicited diaries; see Meth, 2003). Firstly, by asking people to attend to particular practices or events – whether that be journeys undertaken, people talked to, food eaten, or anything else – people become more aware of the practices that they are involved in. So, for example, in a study on interactions in public, the respondent may become more aware of the degree to which they are in fact involved with other people even though that involvement may not include explicit markers of engagement such as talk. Or, a food diary might make a diarist aware in a way that they had not previously been of the degree to which the pleasure of certain kinds of leisure activities – television watching, say – is bound up with the consumption of certain kinds of foods. Secondly, in a manner that few other research techniques can match, diaries can provide respondents with a chance to reflect upon their lives in a systematic and sustained way. Respondent diaries offer a chance for respondents to fashion, over a period of time, a narrative about their lives as told from within the perspective of their ordinary, day-to-day, lives. Diaries also offer a chance for respondents to reflect on the wider meaning of the events and activities reported in their diaries, and it gives the enthusiastic diarist opportunities to place the ordinary events recounted in their diary within a broader biographical canvas. So, respondents might explain the pleasure of a certain kind of food within a narrative about how their mother used to feed that food to them as infants. Or, the attraction of walking to work might be explained through the resonances such a walk has with childhood memories of walking to school. Thirdly, if the remit of the diary is left reasonably open, the events the diaries narrate and the manner in which the diarist describes them may well suggest avenues of interest and concern that the researcher had simply not thought about, or had considered irrelevant or trivial. Fourthly, as Felicity Thomas (2007) has highlighted in her work on HIV/AIDS in Southern Africa, diaries can provide an opportunity for respondents to explain and explore highly emotional and personally sensitive issues with a frankness and openness that face-to-face interactions might inhibit them from doing. And, fifthly, and finally, in reflecting upon their day-to-day lives, and the specific elements they have been asked to attend to, respondents may begin to offer folk theories and explanations that may help the researcher formulate their own theoretically literate accounts.

10.3 Different Kinds of Diaries

Geographers have used research respondent diaries to study a wide range of different themes. Examples of diary use range from the experience of urban pedestrianism (Middleton, 2009; 2010), practices of food consumption (Valentine, 1999), experiences of violence (Meth, 2003), the sociality of cafés, bars and other hospitality spaces (Latham, 2004; 2006), New Age spirituality (Holloway, 2003), the mobility of blind people (Cook and Crang, 1995), children's journeys to school (Murray, 2009b), consumer shopping decisions (Hoggard, 1978), the lives of street children (Young and Barrett, 2001), the relationship between fishermen and wildlife in Zimbabwe (McGregor, 2005), childhood in rural Bolivia (Punch, 2001), the lives of Eastern European migrants in London (Datta, 2011), the routines of homeless people (Johnsen et al., 2008) and the lives of women with HIV/AIDS (Thomas, 2007). This variety of topics indicates something of the versatility of respondent diaries as a research technique. The variety, however, also points to the diversity of ways in which respondent diaries can be (and have been) used by human geographers.

In some of the cases cited above the research relied solely on diaries. In other cases diaries were used together with a range of related methods such as participant observation and in-depth interviews. In yet other cases, the production of diaries was directly connected with follow-up diary-based interviews. There is also a great deal of variation in the kinds of diaries that were relied upon. In some cases, the researchers asked diarists only to attend to a very narrow range of parameters, while others simply asked diarists to describe what they felt was important, leaving the style and content of the diary entirely open to the respondent diarist's judgement. In some cases, respondents were not asked to write a diary at all but were asked to provide a photographic diary of their lives, or provided with a video camera and asked to make a video diary. So, rather than being a single, easily definable, method or technique, respondent diaries in fact represent a quite broad set of research techniques. That said, it is possible to break the respondent diary into five basic types.

Diary-Logs: a diary-log is simply a log-book where respondents are asked to note down as precisely as possible tightly defined details about certain key activities. Diary logs are useful for generating data where reliable quantifiable data are essential; for example, data about travel patterns or working hours. Diary-logs are highly prescriptive by design, providing little (or ideally) no scope for interpretation from the diarist (see Carlstein et al., 1978; Schwanen et al., 2008).

Written Diaries: A written diary is the form of diary most commonly associated with respondent diaries. A written diary – usually simply referred to as a diary – is a description of a period of a research respondent's life, written by the respondent, and commissioned by the researcher. The remit of a written diary can vary enormously. In some cases diarists are simply provided with a diary and asked to describe their day. In other cases, diarists are asked to focus only on certain kinds of activities, or activities that take place in particular places. A written diary may contain elements of a diary-log. For example, a researcher may be interested in the everyday movements of an individual and would like to use the material from the diary to produce a map of the diarist's daily movements. In this case the researcher may instruct the diarist to include very specific details about journeys undertaken, their timing, and purpose. Because the format of the diary is typically left open, the style, detail, focus, and depth of the diaries produced within a single diary-based project will often vary enormously from respondent to respondent.

Photographic Diaries: Photographic diaries differ from written diaries in that rather than relying on written accounts, research respondents are asked to describe or illustrate elements of their lives through the medium of photography. Typically respondents are given a disposable camera and asked to photograph that which the respondent feels is most relevant. Once the camera is full, the diary is understood to be finished. This obviously limits the scope of the diary. However, digital cameras offer a range of possibilities for widening the remit of the diarist – as they allow the diarist to take many more photographs, as well as allowing them to delete, edit, and retake photographs that they are unhappy with. In some cases, diarists might be asked to note down when and where each individual photograph was taken, and why the photograph was taken. In most cases, upon completion of the diary the diarist will talk through the photographs taken with the researcher. An advantage of photographic diaries is that they do not require any degree of literacy. A further advantage is that compared to a written diary in general photographic diaries require a lesser time commitment. Rather than having to compose a written diary entry, the diarist simply has to point a camera and take photographs. While photographic diaries represent a discrete form of diary production, in practice photographic diaries are often combined with written diaries.

Video Diary: Video diaries are a relatively new, and, at least within geography, a little explored form of diary production. Video diaries can be of two forms. They may involve the diarist simply talking to a static video recounting the events of a day. Or, as in the example of Murray (2009a), the video might be used as a device to record key elements of a respondent's day, for example a child's journey to school. The advantage of video is that through the ability to record significant blocks of time, and in catching movement, it provides an immediacy of context difficult to match in written and photographic diaries. That said, video diaries also have a number of disadvantages. They are more obtrusive than written or photographic diaries. The equipment necessary to make quality recordings is relatively expensive. And, the usefulness of the diary is very dependent on the ability of the diarist to competently use the equipment provided for them to produce a diary.

Diary, Diary Interview: Strictly speaking this is not another form of diary production. Following the lead of American ethnographers Zimmerman and Wieder (1977) many geographers who have used diaries as a research technique have treated diary writing as an iterative process (see Latham, 2006; Middleton, 2009; 2010). Research respondents are asked to write diaries. On completion of the diary, the researcher then undertakes an in-depth interview with the diarist based on the diary. In the diary interview the diarist is asked to lead the researcher through the diary. This allows the diarist to explain ambiguities in their written dairy. It also provides the diarist with an opportunity to reflect upon and – if they feel necessary – to expand on the accounts presented in the diary. Additionally, the diary interview offers the researcher a chance to ask about the wider context of the events presented in the diary. Researchers have the opportunity to explore the diarist's relationship with key actors present within the respondent's diary, and they can explore the extent to which the research respondent feels the events recorded in the diary are representative of the respondent's life.

10.4 The Practicalities of Organizing Respondent Diaries

Organizing the production of respondent diaries is an involved, drawn out, and far from straightforward process. That said, following a few simple rules can aid the smooth running of a respondent diary-based research project.

1. *Think carefully about what kind of information you want to generate from respondent diaries.* If your main interest is obtaining accurate details about when people engage in a certain activity, it may be superfluous to ask them to write about everything they do during a week. Similarly, if you are interested in the general texture of a person's day it may be inhibiting to demand that a diarist lists the exact time and date of everything recorded in the diary. In fact, it is a good idea to spend a few days, or better, a week filling out a diary in the manner that you intend to ask your respondents to. This will allow you to think about the details you would like your diarists to focus on, and the appropriate strategies that need to be adopted to facilitate this focus. Writing your own diary also allows you to gain a sense of the time demands that diary writing is likely to make on potential diarists.

2. *Think carefully about whom you want to recruit as diarists, and how you are going to recruit them.* The recruitment of diarists can be the most time intensive part of the research process. If you are fortunate you may have an existing research contact who will be able to provide willing diarists. More commonly you will have to think creatively about how to recruit diarists. The most reliable technique is to ask acquaintances if they know people who fit the profile of the kinds of respondent diarists you are wishing to recruit. Advertisements in local or community newspapers, listing magazines such as London's *Time Out*, or notices on Internet discussion boards, can also be effective. Similarly, do not be afraid to use wanted posters in places potential respondent diarists are likely to frequent. Once you have managed to recruit a number of initial diarists further recruitment through snowballing is generally effective.

3. *Think about the competencies of the people you are recruiting as diarists.* One of the great strengths of respondent diaries is that they draw on the narrative skills of those producing them. This, of course, demands that if you are going to ask a certain population of people to produce a diary that they have those skills. If you are working with a social group with a low level of literacy, written diaries may not be appropriate. Or, rather you may only be able to gain diary accounts from relatively highly educated, and privileged groups (although see Meth, 2003; Thomas, 2007). In this case, another form of diary keeping such as photographic diaries might be more appropriate.

4. *Provide diarists with a clear briefing of what you expect them to do.* Diarists need to have a good sense of what they are being asked to produce. They also need to have a reasonable sense of the purpose of the project to which they are contributing. Ideally the researcher should brief the diarist in person. This gives diarists the chance to clarify with the researcher just what they are being asked to do. Respondents should also be provided with a detailed instruction sheet. The instruction sheet should include information about who is undertaking the research, the institution the researcher is affiliated with, and contact details for both the researcher and others involved in supervising the research project. The instruction sheet should be firmly attached to the diary given to the diarist to complete their diary in.

5. *Provide your diarists with a notebook (or camera if doing a photographic diary) and a pen.* As the researcher is asking people to produce a diary the researcher must provide the diary. The diary should be robust, easy to carry around, and have enough pages for the respondent to complete the task asked of them. Also a pen should be provided to write the diary. Alternatively, you may choose to use a digital format

where people record their thoughts and activities on their mobile phones or computers using text or audio recordings and then share those with you electronically.

6. *Devise a straightforward procedure for returning completed diaries.* Getting diaries back from diarists can be surprisingly time consuming. The most reliable way of getting diaries returned is to pick them up directly from each diarist. This has the advantage that you can ask the diarist about how they found the diary writing process. If you are combining the diary with diary-interviews the diary pick-up also offers an opportunity to arrange a time for the interview. However, if you have a number of diarists writing at the same time, your diarists are very busy, or your diarists are dispersed over a large area, it may not be practicable to personally pick up each diary. In this case, you should provide diarists with a pre-paid self-addressed envelope and instruct them to return the completed diary by post.

10.5 What Kind of Material Might You Expect? And What to Do With It?

The kind of research material generated through respondent diaries is dependent on the instructions given to the respondent diarists.

In the case of a diary-log the diarist will have produced a set of responses that are easily assimilated into a quantitative database (see Schwanen et al., 2008). More open-ended diaries should be approached like any other set of qualitative data. As with a recorded interview, it is good practice to type out the text of the diary either into a word processing document or a qualitative research program such as NVivo (http://www.qsrinternational.com; Windows, Mac) or Ethnograph (http://www.qualisresearch.com; Windows). It is important to recognise that the quality and detail of diaries may vary enormously (see Figure 10.1). Do not treat shorter, less detailed, diaries simply as failures that should be ignored. Indeed, it may be tempting to organize any research account primarily around diary material generated from the most loquacious and personable diarists. This temptation should be resisted. While the longer and more detailed diaries may offer more obvious sources for quotation and illustration, the shortness of other diaries' might well point to equally important conclusions. Be prepared to recognise that there are multiple realities to any social situation, and work hard to construct research accounts that pay due respect to that.

In fact, in many ways the most challenging part of the analysis of diaries involves devising ways of (re)presenting diarists' accounts that respect the texture of the lives recounted in the original diary material. Especially if the reason for using respondent diaries is a desire to understand the rhythms and routines of people's day-to-day lives, it is important to attempt to produce research accounts that express something of the vitality of those lives. This does not simply mean the research accounts produced from respondent diaries should quote liberally from the original diaries – although that may well be one appropriate strategy. It also suggests the need to experiment with different ways of narrating research material (see Chapter 36 for suggestions). This could involve a range of different strategies, from exploring new ways of diagramming time-space (see Figure 10.2) to simply letting diary material speak for itself.

DAY FOUR   Weds   DATE: 11-2-09

Time of contact	Your geographical location at time of contact (place, town/city)	First name of Person contacted[a] (add G: if part of a group)	Relation of person to you (e.g. friend, partner, family member, relative, colleague)	Geographical location of person at the time of contact (place, town/city)	Where did this person live at the time of contact? (area, town/city)	Mode of contact (in person, landline, mobile phone, email, text, MSN messenger, or other – please specify)	Nature of contact (brief description of communication or activity)	How long have you known this person? (approximate months/years)
10 am	Work Teddington	Ed	friend	Kingston London	Kingston London	text	brief chat	3 mths
12.30	Work Teddington	Wayne	friend	Rogers Park Teddington	Rogers Park London	Mobile phone	Advice about camera	11 yrs
13.00	Cafe Teddington	Rebecca	Work colleague/friend	Cafe Teddington	?	In Person	Introduction chat	2 weeks
8pm –10pm	Home R/Park London	Ealee G Mylo G	Friend	Rogers Park London	Wimbledon, London	In person	Cell group discussion	18 mths 2 mths
		Jonni G	'' ''	'' ''	''	'' ''	'' ''	3 months
		Andy G	'' ''	'' ''	''	'' ''	'' ''	6 months
		Adam G	'' ''	'' ''	''	'' ''	'' ''	4 months
		Diane G	'' ''	'' ''	''	'' ''	'' ''	1 year
		Jay G	'' ''	'' ''	''	'' ''	'' ''	18 months

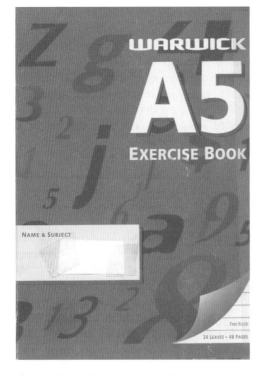

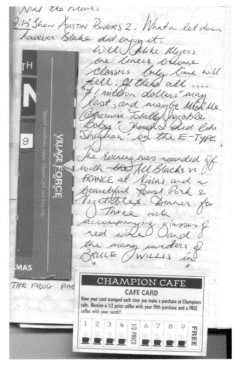

Figure 10.1   Diary excerpts (x3). Source: Author

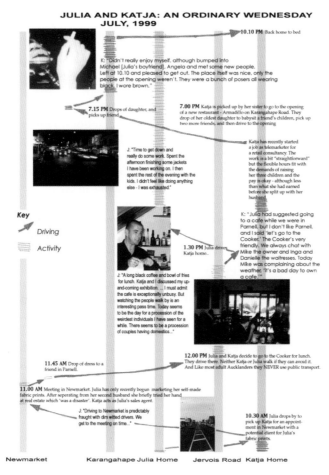

JULIA AND KATJA: AN ORDINARY WEDNESDAY JULY, 1999

10.10 PM Back home to bed

K: "Didn't really enjoy myself, although bumped into Michael [Julia's boyfriend], Angela and met some new people. Left at 10.10 and pleased to get out. The place itself was nice, only the people at the opening weren't. They were a bunch of posers all wearing black. I wore brown."

7.15 PM Drops of daughter, and picks up friend.

7.00 PM Katja is picked up by her sister to go to the opening of a new restaurant - Armadillo on Karangahape Road. They drop of her oldest daughter to babysit a friend's children, pick up two more friends, and then drive to the opening

J: "Time to get down and really do some work. Spent the afternoon finishing some jackets I have been working on. I then spent the rest of the evening with the kids. I didn't feel like doing anything else - I was exhausted."

Katja has recently started a job as telemarketer for a retail consultancy. The work is a bit "straightforward" but the flexible hours fit with the demands of raising her three children and the pay is okay - although less than what she had earned before she split up with her husband.

Key

Driving

Activity

1.30 PM Julia drives Katja home.

K: "Julia had suggested going to a cafe while we were in Parnell, but I don't like Parnell, and I said 'let's go to the Cooker.' The Cooker's very friendly. We always chat with Mike the owner and Inga and Danielle the waitresses. Today Mike was complaining about the weather, 'It's a bad day to own a cafe.'"

J: "A long black coffee and bowl of fries for lunch. Katja and I discussed my up-and-coming exhibition. ... I must admit the cafe is exceptionally unbusy. But watching the people walk by is an interesting pass time. Today seems to be the day for a procession of the weirdest individuals I have seen for a while. There seems to be a procession of couples having domestics..."

12.00 PM Julia and Katja decide to go to the Cooker for lunch. They drive there. Neither Katja or Julia walk if they can avoid it. And Like most adult Aucklanders they NEVER use public transport.

11.45 AM Drop of dress to a friend in Parnell.

11.00 AM Meeting in Newmarket. Julia has only recently begun marketing her self-made fabric prints. After seperating from her second husband she briefly tried her hand at real estate which 'was a disaster'. Katja acts as Julia's sales agent.

J: "Driving to Newmarket is predictably fraught with dim witted drivers. We get to the meeting on time..."

10.30 AM Julia drops by to pick up Katja for an appointment in Newmarket with a potential client for Julia's fabric prints.

Newmarket   Karangahape Julia Home   Jervois Road Katja Home

Figure 10.2 A time-space diagram based on respondent diaries

10.6 The Limitations of Respondent Diaries

Respondent diaries can produce research material that is enormously productive. However, it is important to stress the limitations of diaries as a research technique.

Firstly, respondent diaries make significant demands on both the researcher and the research respondent. For the researcher the logistics of finding appropriate diarists, of distributing the diaries, ensuring that diaries are completed in a manner that is aligned with the project's research aims, ensuring the retrieval of completed diaries, and arranging diary-interviews, should not be underestimated. Of course, all good research is demanding of a researcher's time and intellectual resources. More crucially, diaries make a much greater demand on a research respondent's time than more commonly employed techniques such as interviews or focus groups. What is more, respondent diaries require an on-going commitment from the diarist. This can make it difficult to recruit diarists. It may also mean that many groups who would make ideal diary subjects will refuse to write diaries due to the time constraints they are under.

Secondly, as has already been mentioned, diary keeping assumes a certain set of personal competencies. Researchers such as McGregor (2006), Meth (2003) and Thomas (2007) who have used respondent diaries in non-Western contexts, have stressed the need to recognise the extent that the diary is in all sorts of ways a Western technique of self-reflection deeply embedded in Western traditions of self-hood. More prosaically, producing a diary requires certain basic skills; an ability to write, or an ability to use a camera, an ability to keep track of the passage of time and so on. It might be assumed that these skills are generally distributed in most populations. However, even in highly educated societies such as the United Kingdom, the USA, Canada, or New Zealand there are wide variations in people's levels of literacy, capacity for self-organization, and so on. Even highly competent individuals may find completing a diary intimidating, especially if in their normal day-to-day life they are rarely called upon to produce self-directed blocks of handwritten text. Indeed, given the contemporary ubiquity of computer and keyboard use, it may well be necessary to devise on-line forms of diary keeping as handwriting becomes for many an archaic – and thus alien – technique of communication.

Thirdly, respondent diaries – whether written, photographic, or video-based – can produce enormous variety in the quality and depth of material generated. If one of the advantages of the diary technique is that they allow the respondent diarist to stand in for the researcher, the flipside of this is that many people are (sadly) rather poor observers and reporters. As with all qualitative techniques, it is tempting to focus on those respondents who produce the most detailed and compelling accounts. However, it is important to provide equal weight to the inarticulate and un-observant – not least because their accounts may well be more representative of the general state of affairs.

SUMMARY

This chapter has provided an introduction to the use of respondent diaries.

- Respondent diaries are diaries commissioned by a researcher to provide information on some aspect of each diarist's life.
- Respondent diaries can be an excellent way of generating material about the rhythms and routines of people's day-to-day lives.
- There are a number of different approaches to producing respondent diaries. The content of respondent diaries may be highly prescriptive, or left entirely up to the prerogatives of the diarists.
- As well as written diaries, geographers and social scientists have also used respondent photographic and video diaries. Video and photographic diaries have in some cases been combined with written diaries.
- Organizing respondent diaries is an involved and sometimes complicated process. The researcher must provide clear guidelines about how they want diarists to approach their diaries. The researcher must also provide a concise sense of the purpose of the research the diary is being solicited for.
- Diary-interviews often provide an excellent supplement to a written or photographic diary.
- Research material generated through respondent diaries can be analysed in a range of ways using standard social scientific analytic techniques.
- Research material generated through respondent diaries offers a range of distinctive opportunities for experimenting with different ways of diagramming the rhythms and textures of everyday life.

Further Reading

The following articles and books provide good examples of respondent diaries in practice:

- Alaszewski (2006) offers a comprehensive overview of the many ways diaries have been used in the social sciences.

- Latham (2006) provides an account of everyday urban sociality based around photo-diaries, written diaries, and diary interviews, and presents some interesting explorations of different ways to (re)present diary-based material. A detailed reflection on the development of the methods used to gather this material is available in Latham (2004).

- Meth (2003), in a reflection on the advantages of using respondent diaries, provides a compelling example of the capacity of diaries to generate research material about emotionally sensitive issues.

- McGregor (2005) provides a wonderful example of the shift in perspective respondent diaries can provide. A study of the relationships various groups have to the Nile crocodile in Zimbabwe, the respondent diaries produced by Tonga fishermen provide a striking counterpoint to the accounts offered by scientists and those campaigning for the protection of crocodiles.

- Middleton (2010) provides a thought-provoking study of the practice of walking to work and demonstrates the usefulness of diaries for conveying a sense both of the texture of an experience, and how that experience is related to wider routines.

- Zimmerman and Wieder (1977) is an account of how two ethnographers developed the diary-interview method as a way of obtaining observations about social spaces that the ethnographer would normally have difficulty gaining access to. This paper has been influential in the work of many geographers who have employed diary-based methods.

Note: Full details of the above can be found in the reference list below.

References

Alaszewski, A. (2006) *Using Diaries for Social Research*. London: Sage.
Carlstein, T., Parkes, D. and Thrift, N. (eds) (1978) *Timing Space and Spacing Time: Human Activity and Time Geography*, vol. 2. London: Edward Arnold.
Crang, M. and Cook, I. (1995) *Doing Ethnographies*. Norwich: Geobooks.
Datta, A. (2012) '"Where is the global city?" Visual narratives of London among East European migrants', *Urban Studies*, 49(8): 1725–40.
Hoggard, K. (1978) 'Consumer shopping strategies and purchasing activities,' *Geoforum*, 9: 415–23.
Holloway, S. (2003) 'Outsiders in rural society? Constructions of rurality and nature–society relations in the racialisation of English gypsy-travellers, 1869–1934', *Environment and Planning D*, 21(6): 695–715.
Johnsen, S., May, J. and Cloke, P. (2008) 'Imag(in)ing "homeless places": Using auto-photography to (re)examine the geographies of homelessness', *Area*, 40: 194–207.
Latham, A. (2004) 'Researching and writing everyday accounts of the city: An introduction to the diary-photo diary-interview method', in C. Knowles and P. Sweetman (eds), *Picturing the Social Landscape: Visual Methods and the Sociological Imagination*. London: Routledge. pp. 117–31.
Latham, A. (2006) 'Sociality and the cosmopolitan imagination: National, cosmopolitan and local imaginaries in Auckland, New Zealand', in J. Binny, J. Holloway, S. Millington and C. Young (eds), *Cosmopolitan Urbanism*. London: Routledge. pp. 89–111.
Laurier, E. and Philo, C. (2003) 'The region in the boot: Mobilising lone subjects and multiple objects', *Environment and Planning D*, 21(1): 85–106.
McGregor, J. (2005) 'Crocodile crimes: People versus wildlife and the politics of postcolonial conservation on Lake Kariba, Zimbabwe', *Geoforum*, 36(3): 353–69.

McGregor, J. (2006) 'Diaries and case studies', in V. Desai and R. Potter (eds), *Doing Development Research*. London: Sage. pp. 200–6.

Meth, P. (2003) 'Entries and omissions: Using solicited diaries in geographical research', *Area*, 35(2): 195–205.

Middleton, J. (2009) '"Stepping in time": Walking, time, and space in the city', *Environment and Planning A*, 41(8): 1943–61.

Middleton, J. (2010) 'Sense and the city: Exploring the embodied geographies of urban walking', *Social and Cultural Geography*, 11(6): 575–96.

Murray, L. (2009a) 'Looking at and looking back: Visualization in mobile research', *Qualitative Research*, 9(4): 469–88.

Murray, L. (2009b) 'Making the journey to school: The gendered and generational aspects of risk in constructing everyday mobility', *Health, Risk and Society*, 11(5) 471–86.

Punch, S. (2001) 'Multiple methods and research relations with young people in rural Bolivia', in M. Limb and C. Dwyer (eds), *Qualitative Methodologies for Geographers*. London: Arnold. pp.165–80.

Schwanen, T., Kwan, M-P. and Ren, F. (2008) 'How fixed is fixed? Gendered rigidity of time-space constraints and geographies of everyday activities', *Geoforum*, 39: 2109–21.

Thomas, F. (2007) 'Eliciting emotions in HIV/AIDS research: A diary-based approach', *Area*, 39 (1): 74–82.

Valentine, G. (1999) 'A corporeal geography of consumption', *Environment and Planning D: Society and Space*, 17: 329–51.

Young, L. and Barrett, H. (2001) 'Adapting visual methods: Action research with Kampala street children', *Area*, 33(2): 141–52.

Zimmerman, D. and Wieder, D. (1977) 'The diary: Diary interview method', *Urban Life* 5(4): 479–98.

ON THE COMPANION WEBSITE...

Visit **https://study.sagepub.com/keymethods3e** for author videos, chapter exercises, resources and links, plus free access to the following recommended articles:

1. **Davies, G. and Dwyer, C (2007) 'Qualitative methods: Are you enchanted or are you alienated?', *Progress in Human Geography*, 31(2): 257–66.**

An excellent overview of recent trends in qualitative work exploring how qualitative work in human geography increasingly seeks to understand the world's enchantments.

2. **Crang, M. (2005) 'Qualitative methods: There is nothing outside the text?', *Progress in Human Geography*, 29(2): 225–33.**

This article explores the ways qualitative research reaches beyond just text and language.

3. **Lorimer, H. (2005) 'Cultural geography: The busyness of being "more-than-representational"', *Progress in Human Geography*, 29(1): 83–94.**

A wonderfully evocative exploration of doing so-called non-representational theory.

11 Participant and Non-participant Observation

Eric Laurier

SYNOPSIS

Participant observation is a minimal method that, as its name suggests, involves participation in and observation of places, practices and people. It proceeds along two trajectories. The first trajectory is finding a setting where the social or cultural thing that the researcher wishes to study is happening and becoming intimate with the group that populates that setting. The second trajectory, which moves in parallel with the first, is that the researchers themselves change how they understand the setting and its inhabitants. By undertaking participant observation the researcher aims to gain a better understanding of people in places: to begin to identify what matters for that culture, how things are organized, and why locals do things in the distinctive ways that they do (but see Sultana, 2007, on dilemmas in the research setting). Geographers have used it to study communities, public spaces, institutions, embodied practices, game playing, illness, political action, online communication, everyday life and other spatial phenomena.

This chapter is organized into the following sections:

- What is participant observation?
- Legitimate peripheral participation
- From outsider to insider
- How to find a familiar thing interesting
- Changing your grip
- Strengths and weaknesses

11.1 What is Participant Observation?

Participant observation is perhaps the easiest method in the world to use because it is ubiquitous and we can all already do it. From the moment we are born we are, in various ways, observing the world around us and trying to participate in it. Children, acquiring language for the first time, listen to and watch what their parents are doing as well as how and when. They observe greetings and have greetings directed at them, and attempt to participate by, at first, looking, and later, waving and making sounds that approximate – and eventually become – hellos and goodbyes. It is, of course, not just children who use this method to acquire skills: air traffic controllers spend a great deal of time observing air traffic control and are gradually entered into the practical demands of directing planes as fully participating air traffic controllers. International migrants, finding themselves in foreign countries, have the massive task of observing a multitude of activities and exactly how they are done in order to participate in new cultures. Amongst other background knowledge they have to acquire the locals' taken-for-granted ways of getting everyday things done, such as greetings, ordering coffee, queuing for buses, making small talk, paying their taxes and so on.

Participant-observation is the foundation of ethnographic research, and is built from two familiar parts: observation and participation. Observation of spatial phenomena has been a central method for geography from its very outset, whether it be observing the movement of glaciers or the traffic flows of cities. With a reasonable vantage point, the right tools, and a table for recording data, researchers can begin to collect, compare and count the presence, movement, and features of individuals, groups and populations in defined spaces. Participation is a form of involvement in, or association with, a group, practice or event. For many students trained in the observational methods of the physical sciences the validity of their observations is based upon on distance and neutrality rather than participation. Combining observation with participation appears to risk destroying the quality of the data collected by this method. However, the power of participant-observation lies in its intimacy with, and grounded perspective upon, the places, practices and people studied. The knowledge acquired from this method is as much reliant on participation as it is on observation. To drive this point home, human geographers frequently reverse the terms of the title of the method and call it 'observant participation'.

Participant observation, as you are probably beginning to realize, requires us to reconsider the questions we can ask as geographers and to change our criteria for judging whether the method was adeptly executed or not. Typical indicators of 'scientific rigour' must be reconsidered for this approach, though this does not mean we abandon rigour (see Baxter and Eyles, 1997, for an excellent rubric of rigour in social geography). While ethnographers do not generally use hypothesis testing, we can change our perspective by being open to the unexpected, and question received opinions by employing critical reflexivity. While we cannot collect standardised data and keep our variables equal, we can identify patterns, show similarities, and uncover differences. In undertaking participant observation we can begin to collect *what* things are relevant to study for the situation we are participating in. We can explain to other researchers *why* those things are significant to the group or practice that we have been participating in. Most importantly of all, we can provide descriptions of *how* those ordinary and extraordinary things are accomplished by the people we are studying, with the equipment they have for living, situated in the varied environments that they inhabit. In sum, participant observation has strengths in describing the local *processes, practices, norms, values, reasoning, technologies,* and so on that constitute social and cultural lifeworlds.

To return briefly to the example of traffic flows, participant observation could be undertaken with any one of the groups that is involved in the production of traffic and which has different local understandings of traffic. You could spend time in the unfamiliar world of a traffic control room with engineers and planners to understand and be able to speak about how they manage traffic, and its disruptions (Gordon, 2012). You could also spend time in a more familiar world: accompanying commuters who drive through that traffic and begin to understand what their practices are, how they drive and why they take the routes that they do (Laurier et al., 2008).

Participant observation is a long-established method for engaging with familiar and unfamiliar lifeworlds not only in geography but also in anthropology and sociology. It is undertaken as a trajectory from one perspective to another and is most conspicuous in unfamiliar environments, such as a place, community, workplace or institution that you have not been part of before. Think of the effort and time required to do informative participant observation studies of UK motorway-traffic control (Gordon, 2012), social workers' urban outreach (Hall and Smith, 2013) or the French court system (Latour, 2010). Yet while it is a struggle to observe what is happening in these places, let alone participate in them, your findings will be of value by the very fact that how unfamiliar places are organized is not part of the everyday knowledge of members of society.

When participant observation is used in familiar settings such as shopping in the supermarket (Cochoy, 2008), walking along a coastal path (Wylie, 2005), the rave scene in Goa (Saldanha, 1999), or riding an elevator (Hirschauer, 2005), then the researcher has to work all the harder to notice the things that happen in these settings that they usually overlook. In other words, in familiar situations the problem is how to change the way we observe them from what phenomenologists called the 'natural attitude'. As participants in these common cultures, they have become so familiar to us that as ordinary members of society we no longer spend time noticing what we find relevant and irrelevant in them, what our assumptions are about them and how we inhabit them.

Gender is a classic example of a dimension of each society's common culture that is frequently studied through participant observation. For men, how it is that they go about orienting toward and producing themselves as men in the local cultures where they are located, has been learnt from their earliest years onwards and is profoundly taken for granted. A participant observation study in the UK or USA, for instance, could then select one of the many sites where gender has particular significance for the construction and performance of masculinity, such as a gym, and then spend time observing and participating in what it is that the male members do at the gym and how they go about it. Such a study also raises an abiding methodological problem for participant observation. Any individual researcher, according to their existing gendering by society, has both different claims to insider knowledge and limits to their participation. Age, race, (dis)ability, and a number of other social divisions that have been the focus of geographical research, pose similar challenges (and opportunities) for participant observation studies because the researcher's own perceived position in the society influences their surroundings and the actions of others (Sultana, 2007). Even for more permeable social groups, the extent to which the participant is able to approach or be a member of a group itself requires careful analysis.

11.2 Legitimate Peripheral Participation

We have established that a key assumption of participant observation is that observing social activity is predicated on participating, even in the most minimal of ways. However, my students often want to label the short studies that they undertake 'non-participant observation' because they worry that they were not participating in the activities that they were observing. A common case is when students have studied a city street and it seems to them that they simply sat on a bench taking notes and were not involved in the setting and its events. Are they then not non-participant-observers? No, because those very spaces accommodate our presence and give us a role; we change the spaces we are present in to greater or lesser degrees, even when we are seemingly passive. By our very presence we are participating members of the public, in a public space that provides persons using it with the right to do things like people-watching and to be watched by other people in return.

There are many other sites apart from public spaces which also allow us to participate in them in a minimal fashion while observing what happens. The education researchers Lave and Wenger (1991) have provided a useful term to make sense of the position that you can occupy in these settings. They call the involvement of researchers in these settings 'legitimate peripheral participation'. While many places allow people to sit at their edges without having to become full participants, they do so in different ways and proffer different rights (which are also useful features of the place to describe and analyse). Offices allow new employees to shadow their staff, classrooms allow trainee teachers to sit in on lessons, and Internet forums allow 'lurkers' (though this latter term hints at an incipient moral judgement of those that observe without participating more fully). From these peripheral perspectives there are informative findings that can be made about the cultures under study even if the researchers do not become members of the local cultures.

11.3 From Outsider To Insider

Having sketched out the possibility of legitimate peripheral participation to undertake limited participant observations of certain groups or activities, let us now turn to the aim that we began with: the outsider in their participation is becoming an insider and will change their perspective on the local world of the group that they are studying. That sounds rather grand; what it means more prosaically is that if you spend a couple of months working on a waste collection vehicle you will no longer look at the rubbish in the street in the same way, nor will you see neighbourhoods themselves with the same eye. You will have begun to acquire the rubbish collector's ways of perceiving, assessing, and acting upon the state of people's rubbish bins. As we noted above, there will very likely be limits to becoming a legitimate member of the groups you are studying. For becoming a rubbish collector, the limits of becoming an insider may be subtle and lie around whether the researcher is covert or overt, where they find themself positioned by members of the waste collection vehicle, whether they undertake ritual inductions, whether they have the abilities to undertake the job physically, and, of course, being able to opt out of the position as long-term employment when the research term is complete.

When undertaking participation, many researchers aim to follow a trajectory from 'outsider' to 'insider', even if they do run up against the limits of doing so. Those limits include qualities of the place, practice, or group that are open to analysis and these become part of the findings of the research. Despite there being no template for doing participant observation there are features in the course of its trajectory which, while not specifying what is to be done, will give you a sense of whether you are making any progress or not. You should be able to detect your own perspective changing as you begin to grasp how it is that people achieve the practices that they are doing. Why they do so may then also gradually become apparent. This is what Jack Katz (2001; 2002) calls 'from how to why'. More importantly, the group around you will change how they categorise you as you shift from a 'newby' or a 'beginner' to whatever they call a competent and/or accepted member or practitioner. It may be that you can become 'one of us' or that you are always only a partial member of the group as we noted earlier. Just how long it takes to become competent in or 'doing being' what you choose to study, and indeed whether it is possible to reach that state, will vary according to what it is and who you are already. Are you trying to serve coffee in a local café or become a marathon swimmer or the leader of a drug gang? Some roles are more feasible – and legal – than others, which should be carefully considered!

Should you choose to study a supermarket as a shopper or stacking the shelves as a member of the staff then it is not so demanding to become competent in these activities. However, with other communities of practice, such as urban gangs, dairy farmers or taxi drivers, it may take considerable time and effort before you will be recognized as an accepted member of their group. For whatever activity you decide to participate in, there will be different ordinary and expert ways in which you are instructed in how to do it, from the more formal (i.e. lessons, workshops, courses, rulebooks, etc.) to the informal (tips, jokes, brief chats). However long the journey, the participant observer makes notes at all points both to provide a record and to push themself to reflect on what they are experiencing, how people are doing the things around them that they do, and, significantly, the shifts in their own perspective on the phenomena that they are studying. It is at heart a process of becoming attentive and acquiring a new perspective on some small part of the world.

11.4 How to Find a Familiar Thing Interesting

In studying the *everyday* it is hard not to exercise a kind of professional scepticism that subverts the intelligibility of the everyday thing that we are able to see as participants in a shared culture. The danger is that in trying to find a new perspective we go too far and act like a Martian who has landed on Earth. Such a creature is without the shared methods that we have for making sense of the things in our local environment. A central insight from phenomenology is that the intelligibility of our environment is tied to the fact that we are participants in a shared lifeworld and that we perceive this lifeworld with the linguistic categories we have been given by the acquisition of language.

In doing participant observation in places where we are already competent inhabitants and usually take the appearance of things for granted, the solution to doing adequate descriptions of them is, then, not to import strange Martian labels for the things we see or hear or otherwise sense almost instantly. Instead we aim to notice, uncover, and analyse the categorisations of things that the locals do, and would use to describe their observations.

If you are a 'local' and thus an 'insider' you already have a substantial advantage in providing adequate descriptions of how and why things get done in the ways they get done. Yet you are also at the equal disadvantage of no longer noticing how such things get done because they are so familiar as to be taken-for-granted, *seen but unnoticed*, and you may never have attempted to be interested in them. Consider the ordinary social and cultural categorisations of things that locals use. In a student's notes on the gym we find, for instance, that 'a tired looking older woman' exchanged 'friendly' glances with 'another older woman', or that a 'teenage girl avoided the glance of another teenage boy'. The student is becoming attentive to how people use glances to do 'being friendly' or by not glancing back to do 'avoiding someone's attention'. The student begins to reflect on how, as a competent member of a shared place, she is able to see what someone else sees. That is, she is able work out what it was that the other person was looking at, not so much from pinpointing the exact direction of their look but by seeing what it is in the scene that they would glance at or who they would exchange glances with (e.g., the guy waiting for the exercise machine is glancing at his watch, two members of the same social category 'older women' are exchanging glances).

The challenges of a participant observation will be different if you have pursued participant observations of new and/or uncommon sets of skills. Geographers have researched diverse sub-communities and events, such as historical re-enactments (M. Crang, 1996), working in telebanking (Harper et al., 2000), Mexican women's labour conditions (Wright, 2001), marathon swimming (Throsby, 2013) and living among the mentally ill on the streets (Parr, 1998). For these more practically ambitious projects your adequacy as an observant participant turns on you having learnt things that other researchers cannot be expected to know. This certainly makes delivering 'news' easier since, unlike 'exchanging glances' or 'answering the telephone' or 'buying a newspaper', outsiders do not know how these more obscure, expert, secret, marginal or unfamiliar cultures are organized and what their local practices are, and what those local practices mean to them. Competence in performing the practices that constitute the setting that you have selected to investigate will be one way in which your findings from participant observation will attain a reasonable degree of adequacy. Your aim is then, as it was for the familiar, to make available for further description, analysis and, under some circumstances, judgment, the practices and events that members of those settings treat as unremarkable or routine.

11.5 Changing Your Grip

Recording what you notice at all stages of doing participant observation is vital. If you are an outsider, then you will notice many peculiar practices at the beginning, practices that you may forget were noticeable by the time you are an insider. At the beginning you have the perspective of 'any person' – who may well be the audience you wish to write your report for at the end of your fieldwork – and you may find yourself offered a perspective by the setting itself as an 'incomer', a 'tourist', a 'guest' etc. As I have hinted at earlier, if you are a young adult studying children's cultures or life in an old folks' home, then while your perspective and understanding will shift, you will never become a child again nor will you yet be in your old age. Without keeping a record of your own struggles to understand what people do, how they do it and why they do it, you are likely to forget the perspective that you began with. Consequently you will no longer appreciate what it is about local cultures that may seem odd, unfathomable, or otherwise inaccessible to other researchers. Writing good field-notes is an excellent solution to this quandary.

If you are already an insider, which is the case when studying many features of everyday life (though it may also be that you have knowledge of an expert, exotic, or esoteric practice), then the very process of recording should help you begin to see your taken-for-granted world afresh. Records, in the form mainly of notes, but also photographs and videos, will provide your materials for thinking with when it comes to writing-up your participant observation. Writing a powerful report on your participant observation requires documenting it as you go along. Not only will details slip away very quickly, your documenting process is also the beginning of your analysis.

Recording what you notice is not enough, of course; it needs to be coupled with techniques for helping the participant-observer open up their insider perspective for further inspection and reflection. The three techniques of note-taking for helping you get started that I will briefly describe here are: (1) bracketing phenomena (see Box 11.1); (2) describing practices through instructions (see Box 11.2); (3) breaching norms and rules.

Bracketing phenomena

This technique for note-taking works most effectively when trying to record details of the familiar. Your ambition is to suspend your taken-for-granted knowledge of the thing you are trying to observe. In the example in Box 11.1 it is the common recognition of a café as empty. The recognisable state of affairs [empty] is bracketed, quite literally. It could be something of a less familiar kind such as how a manager at an organization you have been studying has [the last word]. It could be, for skateboarders, doing an [ollie]. Once you have bracketed an ordinary phenomenon, your work as a participant-observer is to then begin to describe how this state of affairs is produced, how it is recognised, and start to trace out some of the implications this has for the group that uses or populates the site.

Be wary when bracketing familiar features of a world that you do not then adopt the perspective of our Martian Invader. As an 'insider' you already have a store of practical knowledge about how the things you are observing function. What you are trying to do is open up, inspect and analyse your insider knowledge.

Box 11.1   The café is [empty]

Staff can be present in the café but without customers the café is [empty]. What more is there to this easily recognizable state of affairs? There is: how the customer recognizes [empty] – which is bound up with the typical interior architectural construction of this café, and many others like it, which allows those entering to look around as they enter and see, at a glance, how busy the café is. It is bound up also with the expectations of time of day: 'this early hour' (e.g. around 7am) in this café known for its 'appearances as usual' on a weekday at this sort of time or as an environment of expectations. At the same time of day, an airport or flower market café is likely to be full.

(Continued)

(Continued)

What the customer makes of [empty] is related to their orientation to the 'awakening' of the day. That is, the reasons for empty-ness are temporally located – it is 'just opened'. A customer is not put off or curious about this observable empty-ness during the opening time, the way they would be were it to be observably empty at 1pm ('Why is it empty? Is the food bad? Are the staff rude? Is it expensive?'). (For more details see Laurier, 2008.)

Turning practices into instructions

Where some methods in geography provide minimal information on the processual nature of systems, participant observation excels at describing how people do things, step-by-step. A technique to try in note-taking is, then, to transform what you are trying to describe into instructions or sequences of actions that would allow someone else to redo the thing somewhere else. If you are moving from outsider to insider this will be all the easier because the people around will be showing you how to do it or may actually give you a list of instructions.

Box 11.2   Taking a Selfie

Taking 'a selfie' photograph: I pick up my smartphone and take a few selfies. I break what I have just done into instructions:

1 Look at some other people's selfies.
2 Examine your current appearance.
3 Sort out hair, straighten specs etc.
4 Activate your camera app.
5 Hold your phone in your out-stretched arm (this is peculiarly difficult and worth returning to).
6 Examine the background as to whether it captures the place you want to be seen in (or is a distraction because your facial expression is the main focus here).
7 Produce a face (but what face? why?).
8 Try and match the timing of pushing the fire button with the face you wish to take (this is also peculiarly hard without losing the face because you are concentrating. I ended up clicking the button three or four times.).
9 Examine the images produced on your camera.
10 Delete most of them.
11 Potentially redo the selfie on the basis of the earlier selfie.
12 Title the selfie (this leaves me thinking – what is this selfie doing, is it obvious, what else might be made of it?).

My initial list of instructions has not provided me with much by way of notes yet and what I want to do next is then extend the descriptions of each step, to consider how each step relates to the next and observe how others are doing these steps. Briefly here then, looking at other people's selfies I see a range of possible facial expressions people can make and consider which of these are ones that 'do' something – is that what we want to 'do', for example, looking cool or looking daft or looking amazed? How does this relate to where my 'selfie' is (this was already something that came up when I was writing down the steps to produce a selfie)? Am I at a major sports event, am I out with my friends, did my boyfriend just leave me? My appearance then should match what I am going to try and do with this selfie so people do not see me doing an 'amazed' expression because my boyfriend just left me. Or maybe they should. But then what is 'an amazed look' accomplishing and, as our analysis progresses, what does this tell us about the social worlds in which selfies are produced?

In producing instructions we are trying to avoid producing a definition of the practice, or trying to explain why it is important, or what it means. We are accepting that there is a practice in the world that other members know how to recognize and may well know how to do. In disassembling it into instructions we begin to open up the practical knowledge that is required to do it. Once we can put ourselves into the middle of the action we can begin to make visible what shared problems emerge in this practice. For the example of the 'selfie' in Box 11.2, this meant briefly exploring the shared problem of exploring what self-taken public faces look like in different situations, and then, working with the camera to try and capture a face that fits and that will mean what we intended it to mean for others.

These basic instructions that you write down on first brush can then themselves be considered for further features of the phenomenon that they have missed. What can I add to this instruction for it to make more sense? Was it important who did the action? What were the expectations of others? What are the criteria for success or failure? And so on. The additional questions and expansions that we make in response to rendering practices as instructions are ones that we will find from looking at other studies of the same or similar phenomena.

Participation by 'breaching' followed by observation

Breaching experiments arise out of classic studies in the 1960s where social scientists, borrowing from other radical elements of 60s culture, tried to disrupt social norms. They continue to serve a useful purpose for studying everyday settings because members of those settings use those very norms to make sense of the disruption.

Box 11.3   Queueing a sandwich

An exercise set for students by Nick Llewellyn at the University of Warwick was to disrupt the norms in supermarket checkouts by placing a sandwich on the floor at the end of a queue (line) in a busy supermarket. Students, having placed the sandwich (or similar ordinary item), would retire to

(Continued)

(Continued)

a distance and keep notes on how the sandwich was dealt with. Customers arriving at the end of the queue would stand behind the sandwich and move it along in front of them, while the queue itself moved along. They would look at the customers ahead who might shrug or look around for the customer who had left the sandwich behind. The sandwich almost always made it to the point of sale where the checkout operator and neighbouring customers would discuss its status, often looking slightly concerned or frustrated. There were many points of analysis that could be made about what happened. Here we can note briefly how the details of the place that an item from the supermarket is found in are used to make sense of it. The sandwich placed by the students was treated, not as a sandwich for buying or that had simply fallen onto the floor, but by examining its proximity to the queue, it was seen as a placeholder for a shopper who for whatever reason had had to depart the queue. Customers worked collectively to keep the absent shopper's place for them in the queue. In fact its removal was seen as a problem because it meant taking that person out of the queue.

The restrictions set in place by university ethics committees and procedures have made the breaching experiments of the 1960s harder to undertake. However mild forms of disruption, such as moving goods out of their usual locations in shops (see Box 11.3), which continue to cause concern and consternation, remain useful for revealing taken-for-granted norms and rules. The design of breaching experiments has to be done with care so that it allows a variety of responses to the breach, has to include a plan for follow-on responses from the participant-observer, and depending on the nature of the breaching, incorporate de-briefing elements for the members of the setting where it is administered.

In analysing what happens when the everyday practices in a particular place are disrupted we can look at, firstly, what the norms and rules are, and secondly, how these are used by members of different places. Planning the disruption will be the beginnings of this analysis because it forces us to change our perspective on how the place works and identify what its central features are that can then be breached. Breaching also brings to the surface the particular ways that members of settings maintain and repair the normal appearances of the specific practices that are being disrupted.

11.6 Strengths and Weaknesses

Human geography, like most other social sciences, engages with diverse dimensions of social difference such as power, class, race, identity and landscape. These form a complex back-drop for the concerns of actually doing participant observation, but they also saturate your position as researcher, others' views of you and your project, and many of your findings. While it is unlikely that you will solve problems that have dogged society and the social sciences for a century or more from your study of a drop-in centre, supermarket, or airport arrivals gate, the use of participant observation should enable you to ground these dimensions of difference through the shared practical problems of particular groups. It may be that in beginning to describe what happens at the local swimming pool your analysis begins to uncover how masculinity and femininity are manifest in the pleasures (and sufferings) of sports practices (Throsby, 2013). Or in a participant observation of a workplace you may be able to show how the staff of multinational fast food companies resist, subvert or conform to the scripts their company expects them to follow (Leidner, 1993).

Participant observation is by nature exploratory. Your results are unlikely to be wholly predictable or pre-determined by others' studies or theoretical accounts. Indeed, much of the insight gained in participant observation will be material you could not have imagined before doing your study. Thus, participant observation presents many opportunities for beginning (and seasoned) researchers who are willing to open their senses to the world around them, take note of even the most familiar of phenomena, and then use notes and other recordings to make sense of patterns, processes, and connections in social life.

The strengths of participant observation are:

- establishing what is important for particular groups or for specific practices;

- detailing how spatial activities and events are organized in terms that are recognisable to their members;

- understanding why specific groups or practices are carried out in the ways they are;

- uncovering taken-for-granted aspects of the everyday and holding them up for scrutiny and analysis;

- requiring no technical knowledge, yet it is challenging because it requires shifting to a new perspective.

The weaknesses of participant observation are as follows:

- It is not designed for generalizing beyond the event, group, or practice you are studying. This does not mean that it cannot be profitably compared with other studies nor become the basis a study using other methods to check its generalizability, but it does mean that participant observation's task is about findings that are specific and grounded.

- It is an exploratory method not suitable for hypothesis testing, though in another sense it should be about changing or challenging the assumptions that you began with.

- It does not permit standardization of the phenomena for comparison across multiple settings.

- Data are tied to the original researcher and can be challenging to represent and share.

Participant observation allows you to build detailed descriptions from the ground up that should be recognizable to the groups whose lives you have either entered into or already begun amongst. The kind of evidence that arises out of its detailed description allows your study to bring into view certain types of phenomena that are too complex for methodologies that seek and detect general features. While I have emphasized how participant observation is concerned with the details, it returns those details to bodies of studies and theoretical approaches (e.g. landscape, gender, children's geographies, more-than-representational theory, performativity, critical realism etc.). Like other methods in human geography, it is through articulating with and altering the ongoing work of these approaches that participant observation realizes it value.

SUMMARY

- Participant observation is a method that gathers local and contextualized knowledge of groups, events or practices.
- Insider knowledge can be gathered about places we are already familiar with (and thus are insiders already) or places that are at a distance from our current situation which require us to become competent in new practices, ways of understanding the world and local moral orders
- Note-taking and noticing are essential from the beginning to the end of any participant observation and can be aided by bracketing phenomena, turning things into instructions, and breaching experiments
- Participant observation supports the translation of know-how and exploring cultural difference and similitude.

Further Reading

The following books and articles provide good examples of participant observation in practice:

- Crang, P. (1994) is an example of participant observation used to examine how a waiter's work gets done and that by looking closely at this work we can learn about surveillance and display in workplaces.

- Harper et al. (2000) is based on two of the authors spending time working alongside the employees of new telephone-banking facilities and traditional banks and building societies.

- Parr (1998) successfully accesses and helps us understand a vulnerable group within the city using covert participant observation.

- Vannini (2012) is an ambitious, extensive and evocatively written participant observation of ferry travel on the West Coast of Canada. There is also a parallel website: http://ferrytales.innovativeethnographies.net (accessed 13 November 2015).

- Venkatesh (2009) is a popular participant observation study of black urban gangs. Venkatesh was originally held hostage by the gang before later being allowed to become a legitimate peripheral participant.

Note: Full details of the above can be found in the references list below.

References

Baxter, J. and Eyles, J. (1997) 'Evaluating qualitative research in social geography: Establishing "rigour" in interview analysis', *Transactions of the Institute of British Geographers,* 22: 505–25.
Cochoy, F. (2008) 'Calculation, qualculation, calqulation: Shopping cart arithmetic, equipped cognition and the clustered consumer', *Marketing Theory,* 8(1): 15.
Crang, M. (1996) 'Living history: Magic kingdoms or a quixotic quest for authenticity ?', *Annals of Tourism Research,* 23(2): 415–31.
Crang, P. (1994) 'It's showtime: On the workplace geographies of display in a restaurant in South East England', *Environment and Planning D: Society and Space,* 12: 675–704.

Gordon, R. J. (2012) *Ordering Networks: Motorways and the Work of Managing Disruption.* PhD Thesis, University of Durham, Durham.

Hall, T. and Smith, R. J. (2013) 'Knowing the city: Maps, mobility and urban outreach work', *Qualitative Research*, 14(3): 294–310.

Harper, R., Randall, D. and Rouncefield, M. (2000) *Organisational Change and Retail Finance: An Ethnographic Perspective.* London: Routledge.

Hirschauer, S. (2005) 'On doing being a stranger: The practical constitution of civil inattention', *Journal for the Theory of Social Behaviour*, 35(1): 41–67.

Katz, J. (2001) 'From How to Why: On luminous description and causal inference in ethnography (Part 1)', *Ethnography*, 2(4): 443–73.

Katz, J. (2002) 'From How to Why: On luminous description and causal inference in ethnography (Part 2)', *Ethnography*, 3(1): 64–90.

Latour, B. (2010) *The Making of Law - An Ethnography of the Conseil d'État.* Cambridge: Polity Press.

Laurier, E. (2008) 'How breakfast happens in the café', *Time and Society*, 17: 119–143.

Laurier, E., Lorimer, H., Brown, B., Jones, O., Juhlin, O., Noble, A., Perry, M., Pica, D., Sormani, P., Strebel, I., Swan, L., Taylor, A., Watts, L. and Weilnemann, A. (2008) 'Driving and "passengering": Notes on the ordinary organization of car travel', *Mobilities*, 3(1): 1–24.

Lave, J. and Wenger, E. (1991) *Situated Learning.* Cambridge: Cambridge University Press.

Leidner, R. (1993) *Fast Food, Fast Talk: Service Work and the Routinization of Everyday Life.* Berkeley, CA: University of California Press.

Parr, H. (1998) 'Mental health, ethnography and the body', *Area*, 30(1): 28–37.

Saldanha, A. (1999) Goa trance in Goa: Globalization, musical practice and the politics of place. Unpublished paper, 10th Annual IASPM International Conference, Sydney. http://nicocarpentier.net/koccc/Publications/Arunsydney.html

Saldanha, A. (2007) *Psychedelic White: Goa Trance and the Viscosity of Race.* Minneapolis: University of Minnesota Press.

Sultana, F. (2007) 'Reflexivity, positionality, and participatory ethics: Negotiating fieldwork dilemmas in international research', ACME 6(3): 374–85.

Throsby, K. (2013) '"If I go in like a cranky sea lion, I come out like a smiling dolphin": Marathon swimming and the unexpected pleasures of being a body in water', *Feminist Review*, 103(0): 5–22.

Vannini, P. (2012) *Ferry Tales: Mobility, Place, and Time on Canada's West Coast.* London: Routledge.

Venkatesh, S. (2009) *Gang Leader for a Day.* Harmondsworth: Penguin.

Wright, M. W. (2001) 'Desire and the prosthetics of supervision: A case of maquiladora flexibility', *Cultural Anthropology*, 16(3), 354–73.

Wylie, J. (2005) 'A single day's walking: Narrating self and landscape on the South West Coast Path', *Transactions of the Institute of British Geographers*, 30 (2): 234–47.

ON THE COMPANION WEBSITE…

Visit **https://study.sagepub.com/keymethods3e** for author videos, chapter exercises, resources and links, plus free access to the following recommended articles:

1. **Crang, M. (2003) 'Qualitative methods: Touchy, feely, look-see?', *Progress in Human Geography*, 27(4): 494–504.**

2. **DeLyser, D. and Sui, D. (2014) 'Crossing the qualitative-quantitative chasm III: Enduring methods, open geography, participatory research, and the fourth paradigm', *Progress in Human Geography*, 29(2): 294–307.**

3. **Coombes, B., Johnson, J.T. and Howitt, R. (2014) 'Indigenous geographies III: Methodological innovation and the unsettling of participatory research', *Progress in Human Geography*, 38(6): 845–54.**

12 Researching Affect and Emotion

Ben Anderson

SYNOPSIS

This chapter explores why affect and emotion are important, the challenges in researching them, and the ways geographers have responded to those challenges. It is not intended as a 'how-to' manual that would provide a series of ready-made and transferable techniques that you could simply apply to a research project 'on' the substantive geographies of this or that affect or emotion. Instead, it argues that affect and emotion pose particular challenges for research practice and those challenges must be constantly reflected upon over the course of a research project. Because of their intangibility and ephemerality, affect and emotion may appear difficult to research and study. The chapter details how these common challenges have been responded to by geographers in the context of a proliferation of interest in the topics and the coexistence of different theories of what affect and emotion are and do. Talk-based methods have been predominantly used in researching particular substantive 'emotional geographies', where emphasis is placed on people's expression of emotion. Both 'psychoanalytic geographies' and 'non-representational geographies' have experimented with methods and methodologies that start from the presumption that at least some of affective/emotional experience is inexpressible. In the case of non-representational geographies, this has involved less reflection on particular methods and more an attempt to cultivate a particular style of doing research – based on immersion in situations and experimentation. At the same time, and influenced by feminist methodologies in particular, geographers have researched with and through affect and emotion.
   This chapter is organized into the following sections:

- Introduction
- The challenges of researching affect and emotion
- Methods and theories of affect and emotion
- Non-representational geographies
- Conclusion

12.1 Introduction

Speaking at a makeshift refugee camp in Calais, France, a Syrian man fleeing war speaks obliquely of hope. He is one of approximately 16 million Syrians displaced by war. Asked by a reporter about daily life in the camp Ahmad replied:

Only the same story here. You wake up, we have breakfast, we smoke, we laugh, we joke, we eat again, and after that in the night we have a try [at escaping to England]. Every night.[1]

This weak hope, a hope he perhaps shares with others, folds into the other affects and emotions of forced exile, including estrangement and shame. Perhaps the hope of something better sustains him, for a while at least, in his displacement and exile, and as he and others make a daily life within conditions that, for many, would be unbearable and conditions Ahmad and others hope will only be temporary. I begin with hope in a refugee camp as an illustration of the starting point of this chapter: that affects and emotions are part of any and all geographies. There is not and cannot be a carefully circumscribed 'affective' or 'emotional' geography neatly separated from other topics, issues or concerns. In this case, the intimate and global geographies of forced displacement are impossible to understand without attention being paid to the hopes and the other affects and emotions that imbue the movements of people; the hopes of desperate, dangerous journeys across seas and lands; the fears and anxieties and hatreds intensified in media panics about 'immigrants'; the empathy for suffering at a distance that imbues and energizes networks of support in the midst of a refugee crisis, and so on.

If affect and emotion were once silenced, marginalized or occluded from geographical research (Anderson and Smith, 2001), then this is no longer the case. Over the past ten years, geographers have focused on the affective or emotive dimensions of an extraordinary array of geographies across different scales: domestic violence, war and military violence; contemporary financial capitalism in the aftermath of crisis; race and the contemporary forms of identity and belonging; austerity, and so on. Whilst affect and emotion are understood in quite different ways, a point I will elaborate on below, most work starts with a simple insight: how people feel and what causes these feelings are not natural and therefore not timeless or unchanging. Whilst the precise connections are understood in quite different ways, geographers increasingly presume that affective life shapes and is shaped by how societies are patterned and organized (although a key question becomes how the social 'gets into' affect and emotion if they may appear, for some people, to be natural and spontaneous). Given this, all research on affect and emotion is indebted to feminist scholarship that first disrupted the marginalization of affect and emotion under gendered forms of knowledge production and made room for considering how people relate to the world affectively and emotionally (see Rose, 1993). This work countered the neglect of emotional matters by disrupting the geo-historically specific divide between masculine rationality and feminine irrationality. The reason for researching affect and emotion can, then, be simply stated: it is through affect and emotion that people are connected to the world around them. Not only are affect and emotion central to what people attach to and belong to, but it is also, in part, through affect and emotion that people are touched by and caught up in larger events and processes. Think, for example, of how the European refugee crisis might be felt in a fleeting feeling of sympathy that energizes a desire to help and support, or how it might be felt in the weak hope of escaping an unbearable situation.

It is no longer the case, then, that researchers have to work hard to justify why affect and emotion are legitimate topics for geographical research. Whilst they may seem to be intangible, fleeting and ephemeral, it is now acknowledged by geographers that affect and emotion are important dimensions of how spaces are lived and experienced. The much harder task is to consider what types of things affect and emotion are, and how we might go about designing research projects and using particular methods in order to research them. What, then, are the key challenges to researching affect and emotion and how have methods and methodologies been experimented with to meet those key challenges? I'll start by summarising some of the challenges of researching the topics, using the terms affect and emotion interchangeably in the first section, before differentiating between how affect and emotion have been understood in subsequent sections.

12.2 The Challenges of Researching Affect and Emotion

It is not that affect and emotion are unresearchable; far from it. As we shall see, a wide variety of methods have and are being used and experimented with in research on and through affect and emotion; ethnography, interviews, diaries, theatre-as-method, focus groups, artistic practice, and so on. Indeed, the emphasis on affect and emotion as legitimate topics of inquiry has occurred alongside a new-found openness as to what counts as an appropriate method and how methods might be combined in specific research projects (see Shaw et al., 2015). Nevertheless, affect and emotion are presumed to pose particular challenges to conventional models of doing geographical research. However theorised, and I will come to some differences below, affect and emotion may appear to lack the observable material presence that other topics of geographical research can at least appear to have. As a class of phenomenon, it is difficult to fix and locate them. Consider the affects and emotions of a charged event such as a protest. Perhaps we find affect and emotion in the tone of an angrily shouted chant repeated by marchers, or perhaps in the calmness of a gesture of non-violence in the face of provocation. Or perhaps we find affects and emotions in the shared but background atmosphere of solidarity that participants might themselves only be barely aware of, or perhaps in the non-conscious relays between brain and body as participants are primed for action. Affect and emotion may be found in all of these and other places. As well as being difficult to fix and locate, particular affects and emotions are rarely clear and distinct and identifiable. Partly this is because, however theorised, they constantly mix and blur with other things. So, to return to our example of a protest, affects/emotions of solidarity, calmness and so on are mixed together with all the other things that make a protest a protest – the rhythms of walking, banners, gestures, shouting, policing tactics, and so on. But affects/emotions are rarely clear and distinct because something about them escapes names and other ways of fixing. How, for example, to characterise the atmosphere of a protest? How to characterise any one individual or group of protestors' affective attachments and investments in the protest or in the issue protested? They will change, they may not be accountable by that individual or group, and they may be multiple, contradictory, or fragmented. Perhaps they may also blur the lines between the geo-historically specific names we have for the different emotions – fear, hope, anxiety and so on – or not quite fit with any of the names that currently exist as part of our vocabularies for emotional experience. In other words, it is not only that affect and emotion are not easily identifiable, it is also that they may be indeterminate as they change, as they mix with other things, and as they blur with one another.

Now, we may want to pause and question how unique these challenges are to researching affect and emotion. Several of the challenges are generic to doing any kind of social research. Consider issues of causality: one of the key problems for research on affect and emotion is how they 'get into' particular places or spaces. Now it is assumed that there is nothing natural about how people feel and what people feel, attention has shifted to understanding what affects and emotions do in particular situations. Again, consider the example of the atmosphere of a protest. How might we understand what that atmosphere 'does'? Perhaps, enlivened by participation in the protest, people feel an intensified attachment to the issue? Perhaps, as they recall a sense of solidarity, they attend other protests in the hope of recreating an otherwise rare feeling? In each example the atmosphere 'does' something, but in ways that are difficult to definitively identify. How affects and emotions matter in particular situations and the differences they make are hard to trace in the midst of the tangle of other things that make up particular spaces or places. So, however theorised, affect and emotion are assumed to pose a set of challenges to how geographical research is planned, conducted and represented. But, as I have hinted above, there is not one single, agreed upon definition of affect or emotion, nor an understanding of how either 'gets into' spaces and places (and vice versa). The multiplicity of definitions and understandings is not unique to contemporary human geography, nor is it a recent phenomenon. Both terms have always been contested and always understood in diverse ways. McCormack's cautionary note is important to remember:

> Geographers are not always talking about the same thing when they talk about affect and emotion ... their conceptual, empirical, ethical and political emphasis will differ, and often profoundly so. (2006: 330)

Differences matter because how affect and emotion are theorised – what type of things they are taken to be – holds implications for the conduct of research and for the type of research material that can be generated. In the next sections I summarise what are generally taken to be the two main approaches to affect and emotion in contemporary human geography – 'emotional geographies' and 'non-representational geographies' – and draw out their implications for research. It is important to note, however, that this is in many respects an artificial divide that risks downplaying the many connections between approaches. The distinction is best thought of as one way of orientating within a fast moving field of research.

12.3 Methods and Theories of Affect and Emotion

I stressed in the introduction that all work on affect and emotion in geography is indebted to the feminist geographical scholarship which disrupted the dismissal of questions of affect and emotion as irrelevant, non-serious and trivial. What this work showed was that this dismissal and marginalisation were closely connected to the reproduction of masculine forms of knowledge production and authority (see Bondi, 2005; Rose, 1993). In the wake of these critiques, there are now numerous substantive 'emotional geographies' of a very wide range of issues. To illustrate the proliferation of work once it is presumed that emotions have a 'ubiquitous and pervasive presence'

(Bondi, 2005: 445), Pile (2010) provides a partial list of some of the emotions discussed in recent geographical work, including:

> ... ambivalence, anger, anxiety, awe, betrayal, caring, closeness, comfort and discomfort, demoralisation, depression, desire, despair, desperation, disgust, disillusionment, distance, dread, embarrassment, envy, exclusion, familiarity, fear (including phobias), fragility, grief, guilt ... (2010: 17)

The rationale for this expansion of interest in emotions is straightforward. As Bondi et al. (2005: 1) write at the start of their introduction to the edited collection *Emotional Geographies*, emotions 'inform every aspect of our lives'. The assumption is that it is through emotions that the world is made meaningful. Witness how, for example, Bondi et al. justify that emotions matter:

> ... clearly our emotions matter. They affect the way we sense the substance of our past, present and future; all can seem bright, dull or darkened by our emotional outlook. Whether we crave emotional equilibrium, or adrenaline thrills, the emotional geographies of our lives are dynamic, transformed by our procession through childhood, adolescence, middle and old age, and by more immediately destabilizing events such as birth or bereavement, or the start or end of a relationship. Whether joyful, heartbreaking or numbing, emotion has the power to transform the shape of our lives, expanding or contracting our horizons, creating new frissures or fixtures we never expected to find. (2005: 1)

So, emotions are subjective states through which human life is made meaningful. On this understanding, emotions are never solely personal (even if, for some people in some places at some times, emotions may be felt as intensely personal). Work in emotional geographies stresses that emotions are relational, and/or should be conceptualised relationally, because they cannot be reduced to discreet, intra-psychic, states of mind. Anderson and Smith (2001: 9) attend to both 'emotional relations' and how 'social relations are lived through the emotions'. Thien (2005: 450) argues for 'placing emotion in the context of our always intersubjective relations'. Bondi et al. (2005: 3) 'argue for a non-objectifying view of emotions as relational flows, fluxes or currents, in-between people and places rather than "things" or "objects" to be studied or measured'.

This conceptualisation of emotions poses a particular challenge for research: the aim is to simultaneously understand a) how the world is made meaningful through emotions and b) how those emotions are constituted through relations. In the main, 'emotional geographies' have been accompanied by the use of talk-based qualitative methods, biography and life history, as well as ethnography. The promise of these methods being that they allow for sensitivity to how emotions 'provide information about their [research subjects](changing) social worlds, their relation(ship)s with others and the "rules" and structures that permit specific behaviour, allowing/disallowing individuals from expressing particular feelings' (Bennett, 2004: 416). This is not to say, however, that talk-based methods and emotional geographies are synonymous. Nevertheless, talk-based methods tend to treat emotions as a particular kind of thing – a subjective state felt by an individual and about which individuals can develop some form of reflexive awareness of in an appropriately conducted interview. Never naively assuming that emotions are always directly speakable, what this work does so well is give space to participants' personal understandings of situations and, in turn, show how those personal feelings are constituted through various, contingent relations.

Let's give one example from quite early work – Gill Valentine's (1989) seminal work on women's fear of crime. The work was based on interviewing women in Reading, UK, on their perceptions and experiences of crime. By providing an emotionally safe space through a carefully thought through interview exchange, Valentine was able to elicit women's personal reflections on fear and violence. She revealed a 'vicious cycle' whereby 'women's fear of male violence does not just take place in space but is tied up with the way public space is used, occupied and controlled by different groups at different times' (1989: 389). Valentine's is a brilliant example of the use of interviews to generate a particular kind of research material – talk – that generates an in-depth understanding of lived experience and so challenges conventional understandings (in this case of how and why women use public space). The use of interviews to explicate people's emotional geographies resonated with the initial rationale for qualitative research methods – they allowed for sensitivity to the specificities and vagaries of lived experience in a manner that quantitative methods did not. Interviewing in particular was imbued with an ethical significance that expressed a particular affective or emotive relation with research subjects (see also Chapter 9). It was a means of listening and questioning with sensitivity and care in order to understand people's lifeworlds and their interpretation of those lifeworlds. Not always, but in most cases, interviews and other methods designed to explicate substantive emotional geographies involve researching with and through emotions. Consider Pain's (2014) research on the intimate geographies of domestic violence. Like Valentine, Pain (2014) uses interviews to understand the nature, effects and experience of fear over time. Pain attempted to navigate the tension between the need to speak about violence and the possible harmful effects of replaying trauma, through a combination of feminist methodology and counselling techniques. For Pain, feminist methodology enabled her to centre safeguarding the emotional and physical well-being of participants, as well as being reflexive about the ethics of research and the power relation between researched and researcher. Counselling techniques were also used for two reasons – both of which remind us that interviews are emotionally charged exchanges and interactions:

> First, counselling techniques help in learning how to listen and to encourage participants to talk in different ways about emotions, which can be hard to express in everyday language and conversation. Second, counselling techniques are founded on principles for interaction, such as empathy and congruence, that aim to foster care and minimise harm to participants and the counsellor (or, in this case, the interviewer); for example, being yourself, only being there for the other person and dealing with your own issues so they do not intervene. (Pain, 2014: 130)

Pain's is a good example of how research 'on' emotions necessarily involves reflecting on how relations between researcher and researched involves emotions. Research on substantive 'emotional geographies' influenced by feminist methodologies therefore typically involves attempts to expose and reflect upon how knowledge production is emotionally constituted, including the emotional investments or attachments to a particular topic or project, the emotional relations between researcher/researched in particular methods, and the emotional experience of conducting research (including 'writing up'). For example, Widdowfield (2000) highlights how her anger, upset, and distress at injustice in 'less-desirable-neighbourhoods' she researched in were challenged by a range of more 'positive feelings' during her research on lone parents in UK housing estates in Northern England. To give another example, Parr and Stevenson (2015) describe the emotional geographies of listening to 'witness talk' by the families of missing persons. They describe how their practice of interviewing was framed by an ethic of care and what they term 'positive regard' for the stories told. The practice results from Parr and Stevenson's (2015: 310) commitment to 'go beyond cataloguing and naming the various emotions that human absence generates (the liminal "ambiguous loss" at stake demands this), and instead recognize that neglecting family talk of the missing characterful "other" is central in producing painful aspects of loss for families'.

Research under the broad rubric of 'emotional geographies' has been concerned to allow participants to reflexively express emotions and understand how emotions are constituted in and through relations with others. If it is through emotions that people relate to the world, then talk-based methods are one means of explicating how people are attached to, and connected with, the world. Influenced by feminist methodologies, some work on emotional geographies has advocated researching through and with emotions rather than only 'on' emotions. This recognises that emotions are a constitutive part of all research practice and, consequently, folded into all practices of knowledge production.

Other approaches have, however, raised some questions about the emphasis on 'expressed emotions' in at least some substantive research on particular emotional geographies (these critiques include from feminist geographical research that pays attention to the non-human forces and events that, in part, constitute subjectivity, as well as the expressed emotions of subjects; see Colls, 2012). One emerging area of work that challenges the emphasis on 'expressed emotion' is psychoanalytic geographies concerned with the relation between the unconscious and space and influenced by a range of Freudian and post-Freudian theories (for a summary of psychoanalytic geographies see Kingsbury and Pile, 2014). Still somewhat neglected within human geography, psychoanalytic geographies have a particular relation with expressed emotions. Pile explains it as follows:

> While psychoanalytic geographies have tended to be suspicious of expressed emotions, this does not mean that they do not take them seriously. Instead, psychoanalytic geographies have tended to focus on desires and anxieties, phobias and pleasures – in the middle ground between inexpressible affects and expressed emotions ... (2010: 14)

This concern with the 'middle ground' has meant a methodological experimentation in the field of psychoanalytic geographers in order to move between the expressed and inexpressible whilst holding onto the sense that not all emotional experience can be expressed or can be fitted into pre-existing systems of meaning and sense. For example, Bingley (2003) combines in-depth listening with practical workshops involving various forms of sand-play in order to sensitively explicate and examine the conscious and unconscious relationship between self and other. Other work has stressed how psychoanalytic practices/principles of 'empathy' and 'identification' (Bondi, 2003) and 'unconscious communication' (Bondi, 2014) might supplement the listening practices that make up non-psychoanalytic methods in human geography. What this work does – as some of the partially connected work in the field of emotional geographies also does – is not presume that emotions can be expressed; it attempts to listen to and learn from their inexpressibility. This means much can be learnt from what any method appears to fail to do and appears to miss or overlook. For example, Proudfoot (2010) reflects on the difficulties and challenges in the process of interviewing football fans for a project on the emotional geographies of nationalism. He argues and shows that enjoyment disappears when articulated through speech. For this reason, Proudfoot stresses the need to attend to the enactment of enjoyment itself, as well as how subjects represent their enjoyment through discourse: so, tears, ecstatic chanting, and celebration. Proudfoot's point is a subtle one. He does not argue we should jettison talk-based methods per se. Rather, and as with other psychoanalytic methods, he advocates developing strategies within conventional methods that stay a while with the difficulty of researching affect and emotion. Instead of using alternative research practices as someone like Bingley (2003) does, Proudfoot practised what he terms 'asking awry' as a methodological strategy. So, rather than asking participants directly about 'enjoyment' and nationalism he asks indirect questions about style of play, team formations, and so on. A particular affect/emotion – enjoyment – is approached obliquely in recognition of some of the challenges of researching something elusive and ephemeral, but nevertheless profoundly important to the experience of space.

12.4 Non-Representational Geographies

Non-representational geographies share with psychoanalytic geographies a concern with the emphasis on 'expressed emotion' in some existing work and a scepticism about attempts to 'capture' and 'represent' emotions. The term 'non-representational theory' is an umbrella for a series of theories and theories that share a number of common concerns that centre around an attempt to take seriously what Lorimer (2005: 8) terms 'our self-evidently more-than-human, more-than-textual, multisensual worlds' (see also Thrift, 2007; Anderson and Harrison, 2010). Non-representational geographies are concerned, first and foremost, with doings – practices and performances – and how spaces are made through practical action. As Anderson and Harrison (2010: 2) put it, the emphasis is on 'the making of meaning and signification in the manifold of actions and interactions rather than in a supplementary dimension such as that of discourse, ideology, and symbolic order'. This directs attention to what people do and how people do it. Non-representational geographies are not, however, only concerned with people. In different ways, they are concerned with practical action (as it unfolds in dynamic situations) that always involves a range of more-than-human forces and objects. Non-representational work places an emphasis on how 'many different things gather, not just deliberative humans, but a diverse range of actors and forces, some of which we know about, some not, and some of which may be just on the edge of awareness' (Anderson and Harrison,

2010: 10). This emphasis on relations is complemented by a concern with events and how events in their indeterminacy and excessiveness reveal the contingency of orders.

These starting points mean that non-representational theories offer a different – but partially connected – account of affect and emotion from the other theories reviewed so far. Instead of emotions as a subjective state, the emphasis is on *affect as bodily intensities* that happen below the threshold of individual consciousness. Examples include hostility, as racism intensifies in an encounter between racially marked bodies (Swanton, 2010); the surprising force of a sexually charged glance (Lim, 2007); 'comfort' as bodies hold together (Bissell, 2008); or the feeling of participation in dance (McCormack, 2013). Happening through the body, affects are a twofold capacity a body has to affect and to be affected. Affect is distinguished from emotion on this account, in a way that it is not in some other traditions. Emotion is the becoming conscious of capacities to affect and be affected and their insertion into already existing webs of meaning and signification. At the same time, a trace of the excess of bodily intensities remains in emotions. As Massumi (2002: 35) puts it, emotion is the most intense expression of the capture of affect *and* an expression of affect's always ongoing escape.

The distinction between affect and emotion is a contentious one that has been subject to critique, defence and revision (see Thien, 2005; Pile, 2010; Anderson, 2014). Nevertheless, this distinction holds implications for research practice. As with non-representational geographies more broadly, the interest in affect has been accompanied by a two-fold challenge to conventional qualitative methods, particularly talk-based methods (a challenge shared, as we have seen, with some work in psychoanalytic and feminist emotional geographies). Let's consider interviews. Interviews are organised around a specific form of expression and interaction – talk – and, in most but not all approaches to interviewing, the presumption of the meaningfulness of the words spoken and/or the act of speaking. Words either serve to express how someone relates to a particular topic or they reveal the manner in which dominant structures are expressed and repeated. The first challenge issued by non-representational theory to qualitative methods centred on questions of the relation between spoken words and affective/ emotive experience. The argument was that much of affective/emotive experience either *was* not articulated or, more radically, *could* not be articulated. This critique was accompanied by calls to move beyond words/speech and find other modes of expression that might capture, evoke or dramatise feelings, emotions, and so on. Thrift (2000), for example, called for geographers to learn and draw from the performing arts and other traditions that work with bodies in order to research affective life, including 'street theatre, community theatre, legislative theatre' and 'forms of dance and music therapy, contact improvisation' (2000: 3). What else, though, do interviews presume? Critiques of the normalisation of interviews as method also argued that they presume the presence of a conscious subject who is able to express what they think and feel about the given topic of the interview. Much of affective/emotive life, so the critique of this presumption goes, happens outside of consciousness and may not be available for expression or articulation through talk in an interview setting.

As with any critique, there is an element of caricature here. We might think of how most 'how to' guides to interviewing advocate that the researcher pays attention to the tone, or manner, in which talk as a particular form of interaction happens. We might also think about how the unsaid is treated in interviewing as potentially revelatory of what matters to and has significance for participants. Rose (1997), for example, stresses the importance of attending to 'silences' when thinking about the enactment of reflexivity in reflecting on research encounters. Furthermore, interviews are recognised to be affect- and emotion-imbued and -animated encounters – whether in concerns about the power imbalances between interviewer and interviewee or in paying attention to the impact of the setting of the interview to how interview talk unfolds. Nevertheless, the critique of the reduction of world to word happened alongside calls for human geographers to experiment with a range of other methods that involved forms of expression in addition to talk. Affect-based research has been characterised, then, by a flourishing of methods. What these have in common is that they attempt to expand the types of material that are generated and the types of encounter that happen as part of research.

Let's look at two examples of research on and through affect whilst stressing, as Vannini (2015) does, that there is no one archetypal non-representational method. Rather, what affect-based research has in common is something closer to an ethos. The first example is Derek McCormack's (2003) encounter with dance movement therapy (DMT). For about 18 months, McCormack participated in DMT sessions at a centre in Bristol (through weekly, two hour drop-in sessions) with the initial aim of understanding DMT as one example of how techniques and practices were employed to work on affect and emotion. His problem was that stopping and reflecting on what was happening seemed to miss something important about participation in movement, much of which was non-conscious. In response to this failure of after-the-event talk-based methods to sense relations and practices of movement, McCormack tried to develop a different way of paying attention and being responsive to the 'wordless intensities' (2003: 493) of the practice as it was happening. His emphasis shifted to trying to find a way of 'becoming faithful to the relations and movements that played out through the enactment of the practice' (2003: 493). Practically, this involved McCormack participating in the practice as it unfolded and finding ways to become attentive to what happened and emerged as bodies moved together. He provides an example from one of the exercises he participated in:

> Everyone in the room is asked to get into pairs. One person of each pair must lead the other around the room. The person being guided must keep their eyes closed, and the exercise is to be nonverbal. With my right hand, I take my partner's left hand and hold it a little out in front of me, experimenting with different ways of holding. I find a position that is loose, yet hopefully influential, and very slowly, begin to guide this hand around the room, taking care that my partner does not bump into anyone, or anything else, and allowing the changes in direction and in the pressure of my hand to be as smooth as possible. (2003: 497)

McCormack's research practice is, then, at least threefold. It involves an immersion in the practice – research through and as participating in the exercises and other practices. It also involved McCormack attempting to cultivate attentiveness to the perhaps non-conscious dynamics between him and the other participants. This involves him interrupting the effort to search for a deeper or hidden meaning by '*going beyond or behind* either a surface understanding of or immersion in that event' (2003: 493, emphasis in original). And, looping back into the other two practices, it involved him learning a practice and art of slow, patient description that attempted to express or evoke something of the practice and of the event.

The second example is Kathleen Stewart's (2007) description of the 'ordinary affects' of contemporary America. Unlike McCormack's involvement in one demarcated space of practice, Stewart's work is a series of evocative descriptions of the ordinary encounters that make up everyday life in the contemporary USA. Presented as a series of brief vignettes combining close ethnographic detail, storytelling and critical analysis, Stewart's method is one of cultivating a kind of attentiveness to the disparate, fragmented experiences of everyday life. A typical example is a vignette on the ordinary practices of 'watching' that depends on Stewart's unhurried attention to how minor happenings compose everyday life:

> In the convenience store, there is an aggressively casual, noncommittal noticing – half furtive, half bored. If there's a checkout line, it's loose; people mill around waiting. They buy lottery tickets, cigarettes, junk food, and beer. There are those who buy a single giant can of cheap beer early in the morning … Differences of all kinds are noted automatically. There is irritation, sometimes amusement, or a hard-boiled, hard-hearted lack of interest in something someone else is doing. And sometimes there are displays of kindness – brief, flickering, half-made gestures that can be noted as a bright moment in the day, or ignored. (2007: 83)

Like McCormack, Stewart does not employ a single 'method' that is rigidly applied to the world by a researcher following a set of generic procedures. Neither McCormack's nor Stewart's practice can be translated into a useful 'how to' guide for someone starting a research project to simply follow and apply. Stewart starts by placing herself in the midst of and being open to the unfolding, inchoate dynamics of ordinary situations. As with so much research on affect and emotion, her work has routes in an ethnographic sensibility attentive to the ongoingness of lived experience and based on a commitment to learning from participants about their worlds (Herbert, 2000). Her and McCormack's practice might be best described, following Thrift (1996), as a style of observant participation in the sense that they attempt to develop a heightened attentiveness to what is happening (in ways that expand beyond the ocular emphasis implied by the term 'observant') and they do that through involving themselves in the eventfulness of situations as they unfold. Whilst different in the types of material generated and the encounters through which that material is made, there are a series of commonalities between Stewart's and McCormack's experiments. In neither case is the aim to be exhaustive of all possible affective and emotive experience. The aim is not to represent affect and emotion in a way that would fully capture the lived. Instead, the work is evocative or expressive in the sense that they attempt to generate a sense of the situation.

We might think, then, of affect-based research as being characterised by something close to an ethos that imbues diverse techniques (Dewsbury, 2009). Like anything as amorphous as an ethos, it is difficult to characterise. Stewart provides a summation of her practice that whilst not equivalent to all 'non-representational theories' gives a sense of the aim of much affect-based research. She describes her work as:

… an experiment, not a judgment. Committed not to the demystification and uncovered truths that support a well-known picture of the world, but rather to speculation, curiosity, and the concrete, it tries to provoke attention to the forces that come into view as habit or shock, resonance or impact. (Stewart, 2007: 1)

We see here how, in addition to rigour and accuracy, other ways of engaging with the world are valued – in Stewart's case speculation and curiosity – and how the aim is not to represent the world but to 'provoke attention'. In this sense, and resonating with the aim of some feminist-inspired emotional geographies work, Stewart's research works affectively, or at least attempts to. Dewsbury (2009) provides another summary of an ethos which, as with Stewart, emphasises the openness to being affected by situations that follow from a certain type of participation:

The idea is to get embroiled in the site and allow ourselves to be infected by the effort, investment, and craze of the particular practice or experience being investigated. Some might call this participation, but it is a mode of participation that is more artistic and, as with most artistic practices, it comes with the side-effect of making us more vulnerable and self-reflexive. It is not however an argument for losing ourselves in the activity and deterritorializing ourselves completely from our academic remit, but nor does it mean sitting on the sidelines and judging. Rather the move, in immersing ourselves in the space, is to gather a portfolio of ethnographic 'exposures' that can act as lightning rods for thought. It is then in those key 'times out' as we set upon generating inventive ways of addressing and intervening in that which is happening, and has happened, as an academic, that such a method produces its data: a series of testimonies to practice. (Dewsbury, 2009: 326–7)

We can draw out some commonalities in the ethos that characterises some affect-based research. The first is an expanded sense of what counts as 'empirical material' that follows from the emphasis on participation or immersion. Dan Swanton's (2010) work on racism in a northern town in the United Kingdom provides a wonderful example of this expanded sense of what counts as empirical material. Swanton spent time in various ordinary spaces in Keighley, and his empirical account is organised around affectively charged scenes in which racism intensifies in the midst of ordinary encounters before dissipating again. To present these scenes he juxtaposes fragments of conversation, interview quotes, and ethnographic descriptions. Second, this work values a sense of methodological experimentation, or what Dewsbury describes as an 'ethos of *stretching* the means by which research is done and *striving* to continue as experiments fail or always come short in the attempt' (Dewsbury 2009: 323; emphasis in original). One example here is Ruth Raynor's (2015) use of drama-based methods to understand how austerity is felt as it folds with the lives of women using a women's group in Gateshead, United Kingdom. Together with the women, Raynor devised a play that attempted to dramatise how austerity was lived and experienced. Alongside interviews, this involved a series of drama games and techniques that aimed to disrupt habits. Another example of this ethos of experimentation, this time using quite different methods, is provided by Spinney (2015) who worked with participants for a project on 'everyday mobilities' using new ways of moving. He used Galvanic Skin Response (GSR) sensors to measure affect understood as non-conscious physical bodily changes (whilst participants walked, cycled, and ran), together with video elicitation interviews, as ways to 'bring bodily data to the fore'.

What all this work shares, then, is a commitment to experimentation in the hope of producing a characterisation of the world of experience that resonates. None of these results are readily translatable into a set of prescriptions for conducting affect-based research that would mean you could read and apply a boxed summary. The emphasis is on a particular kind of ethos of immersion and experimentation that crosses between a set of methods that, in different ways, attempt to encounter and touch the non-conscious dynamics of affective life. What this has meant is that earlier calls for a rupture in how research is conducted have been supplemented by an attempt to imbue existing qualitative methods with what Latham (2003: 2000) calls 'a sense of the creative, the practical, and being with practice-ness'. One example is provided by Latham (2003) in his account of the everyday public life of Auckland, which integrates interview material with photographic and written diaries. His aim was to stay a while with the multiplicities of affective life as they happen. Latham's is an early example of a practice of supplementing or complementing the interview with other material, partly in order to sensitise interviews to the dynamics of affective life (see Dowling et al., 2015, for various examples). An example is David Bissell's (2014) use of in-depth, long interviews in his research on the affects of commuting, including the intensifications of stress at particular moments on daily commutes. He rationalises their use by emphasising how they afford a reflexive attunement to affective dynamics that may, ordinarily, have passed at or below the threshold of awareness. He writes:

> I wanted to use interviews as unpredictable, improvised encounters that potentially heighten an attunement to the volatile, unpredictable affective tensions that teeter on the threshold of perceptibility. In this respect the interview encounter is a self-reflective technique that solicits from the interviewee a heightened exposure to those subtle transformations that, when noticed, can create a cascade of backwards-tracing realisations. (Bissell, 2014: 193)

Here we see how a method that has been the mainstay of research on 'emotional geographies' is repurposed for affect-based research, in a manner that folds a concern with subtle bodily transformations with people's reflexivity (and so undoes any simple distinction between 'emotional geographies' and 'non-representational geographies'). As with Latham's supplementation of interviews, the result is an attempt to stay with the particular challenges of researching affect and emotion.

12.5 Conclusion

Research on affect and emotion has expanded over the past decade. Rather than being ignored, marginalised or downplayed, it is increasingly recognised by geographers that affect/emotion is a constitutive part of any and all geographies. In this sense, all current work on affect and emotion is indebted to feminist geographies which first disrupted their relegation to the realm of the trivial, marginal or unimportant. There is no consensus, however, on how they are theorised and should be researched, i.e. what kind of empirical material can and should be generated about and through them? Furthermore, in their apparent intangibility, ephemerality and changeability, they pose particular challenges to how conventional geographical research is conducted. The result has been the use of new methods and the supplementing/complementing of existing methods with the purpose of understanding how specific substantive affective and emotional geographies are constituted. There is not and could not be a single method for researching affect and/or emotion. In conjunction with theories that differ in their conceptualisation of what kinds of things affect and emotion are and do, techniques serve as ways of attuning to, disclosing and articulating different aspects of affective or emotional life.

SUMMARY

- There is no consensus about how to theorise affect and emotion or how to research them. Nevertheless, there is now widespread acknowledgement that affect and emotion matter to the constitution of particular geographies.
- Moving beyond claims that affect and emotion are unresearchable because of their ephemerality or intangibility, approaches differ as to whether their emphasis is on 'expressed emotions' or affect/emotion as something on the threshold of inexpressibility.
- The results of recent explorations and research have been seen in a proliferation of new and repurposed methods, alongside a reflection on different affective or emotive styles of doing research.

Further Reading

Davidson et al. (2005) bring together geographers and sociologies to present case studies and approaches at the intersection between emotions and concepts of space. New ways of approaching and appreciating established concerns such as gender, race, sexuality, mental health and aging are accompanied by explorations of deeper emotions such as death and loss, and these are placed in the wider context of environmental and cultural politics.

McCormack (2013) presents insights from critical theorists as focused on the domains of the moving body – from dance therapy and choreography through radio sports commentary. Geographies of experimental participation enable new ways of thinking and remaking maps of experience, using the central concept of the 'refrain'.

Vannini (2015) is an attempt to address the use of non-representational theory in research. It covers issues of data, evidence, methods, styles and genres of research and research presentation, and the ways that non-representational theory and perspectives influence these through the research process.

Widdowfield (2000) provides a personal account of undertaking research, and a more general consideration of the benefits or otherwise of including emotions in the research process. Some further opportunities to extend and structure the debate concerning the affective are also addressed.

Note: Full details of the above can be found in the references list below.

References

Anderson, B. (2014) *Encountering Affect: Capacities, Apparatuses, Conditions.* Farnham: Ashgate.
Anderson, B. and Harrison, P. (2010) 'The promise of non-representational theories', in B. Anderson and P. Harrison (eds) *Taking-place: Non-representational Theories and Geography.* Farnham: Ashgate. pp. 1–34.

Anderson, K. and Smith, S. (2001) 'Editorial: Emotional geographies', *Transactions of the Institute of British Geographers,* 26(1): 7–10.

Bennett, K. (2004) 'Emotionally intelligent research', *Area.* 36 (4): 414–22.

Bingley, A. (2003) 'In here and out there: Sensations between self and landscape', *Social & Cultural Geography* 4(2) 329–45.

Bissell, D. (2008) 'Comfortable bodies: sedentary affects', *Environment and Planning A* 40(7): 1697–712.

Bissell, D. (2014) 'Encountering stressed bodies: Slow creep transformations and tipping points of commuter mobilities', *Geoforum* 51: 191–201.

Bondi, L. (2003) 'Empathy and identification: Conceptual resources for feminist fieldwork', *ACME: An International Journal of Critical Geography* 2: 64–76.

Bondi, L. (2005) 'Making connections and thinking through emotions: Between geography and psychotherapy', *Transactions of the Institute of British Geographers* 30: 433–48.

Bondi, L. (2014) 'Understanding feelings: Engaging with unconscious communication and embodied knowledge', *Emotion, Space and Society* 10: 44–54.

Bondi, L., Davidson, J. and Smith, M. (2005) 'Introduction: Geography's "emotional turn"', in J. Davidson, L. Bondi and M. Smith (eds) *Emotional Geographies.* Farnham: Ashgate. pp. 1–16.

Colls, R. (2012) 'Feminism, bodily difference and non-representational geographies', *Transactions of the Institute of British Geographers* 37: 430–45.

Davidson, J., Bondi, L. and Smith, M. (eds) (2005) *Emotional Geographies.* London: Ashgate.

Dewsbury, J.D. (2009) 'Performative, non-representational, and affect-based research: Seven injunctions', in D. Delyser, S. Aitken, M. Crang and L. McDowell (eds) *The SAGE Handbook of Qualitative Geography.* London: Sage. pp. 322–35.

Dowling, R., Lloyd, K. and Suchet-Pearson, S. (2015) 'Qualitative Methods 1: Enriching the interview', *Progress in Human Geography* (online early). doi: 10.1177/0309132515596880.

Herbert, S. (2000) 'For ethnography', *Progress in Human Geography* 24(4): 550–68.

Kingsbury, P. and Pile, S. (2014) *Psychoanalytic Geographies.* Farnham: Ashgate.

Latham, A. (2003) 'Research, performance, and doing human geography: Some reflections on the diary-photograph diary-interview method', *Environment and Planning A* 35: 1993–2017.

Lim, J. (2007) 'Queer critique and the politics of affect', in K. Browne, J. Lim and &G. Brown (eds) *Geographies of Sexualities: Theory, Practices and Politics.* Farnham: Ashgate. pp. 53–68.

Lorimer, H. (2005) 'Cultural geography: The busyness of being "more-than-representational"', *Progress in Human Geography* 29: 83–94.

Massumi, B. (2002) *Parables for the Virtual: Movement, Affect, Sensation.* London: Duke University Press.

McCormack, D. (2003) 'An event of geographical ethics in spaces of affect', *Transactions of the Institute of British Geographers* 28(4): 488–507.

McCormack, D. (2006) 'For the love of pipes and cables: A response to Deborah Thien', *Area* 38(3): 330–2.

McCormack, D. (2013) *Refrains for Moving Bodies: Experience and Experiment in Affective Spaces.* Durham and London: Duke University Press.

Pain, R. (2014) 'Seismologies of emotion: Fear and activism during domestic violence', *Social and Cultural Geography* 15(2): 127–50.

Parr, H. and Stevenson, O. (2015) '"No news today": Talk of witnessing with families of missing people', *Cultural Geographies* 22(2): 297–315.

Pile, S. (2010) 'Emotion and affect in recent human geography', *Transactions of the Institute of British Geographers.* 35(1): 5–20.

Proudfoot, J. (2010) 'Interviewing enjoyment, or the limits of discourse', *The Professional Geographer* 62(4): 507–18.

Raynor, R. (2015) 'Dramatizing austerity: On suspended dissonance' (unpublished manuscript).

Rose, G. (1993) *Feminism and Geography: The Limits of Geographical Knowledge.* London: Polity.

Rose, G. (1997) 'Positionality, reflexivities and other tactics', *Progress in Human Geography* 21(3): 305–10.

Shaw, W., DeLyser, D. and Crang, M. (2015) 'Limited by imagination alone: Research methods in cultural geographies', *Cultural Geographies,* 22(2): 211–5.

Spinney, J. (2015) 'Close encounters? Mobile methods, (post)phenomenology and affect', *Cultural Geographies* 22: 231–46.

Stewart, K. (2007) *Ordinary Affects*. Durham, NC: Duke University Press.

Swanton, D. (2010) 'Sorting bodies: Race, affect, and everyday multiculture in a mill town in Northern England', *Environment and Planning A* 42: 2332–50.

Thien, D. (2005) 'After or beyond feeling? A consideration of affect and emotion in geography', *Area* 37(4): 450–6.

Thrift, N. (1996) *Spatial Formations*. London: Sage.

Thrift, N. (2000) 'Dead or alive?', in I. Cook, D. Crouch, S. Naylor and J. Ryan (eds) *Cultural Turns/Geographical Turns: Perspectives on Cultural Geography*. Harlow: Prentice-Hall. pp. 1–6.

Thrift, N. (2007) *Non-representational Theory: Space, Politics, Affect*. London: Routledge.

Valentine, G. (1989) 'The geography of women's fear', *Area* 21(4): 385–90

Vannini, P. (2015) 'Non-representational research methodologies: An Introduction', in P. Vannini (ed.) *Non-Representational Methodologies: Re-Envisioning Research*. London: Routledge. pp. 1–18.

Widdowfield, R. (2000) 'The place of emotions in academic research', *Area* 32(2): 199–208.

Note

1   Quoted in: '"I am strange here" Conversations with the Syrians in Calais', www.theatlantic.com/international/archive/2015/08/calais-migrant-camp-uk-syria/401459 (accessed 13 November 2015).

ON THE COMPANION WEBSITE...

Visit **https://study.sagepub.com/keymethods3e** for author videos, chapter exercises, resources and links, plus free access to the following recommended articles:

1. **Ballard, R. (2015) 'Geographies of development III: Militancy, insurgency encroachment and development by the poor', *Progress in Human Geography*, 39: 2.**

2. **Fuller, D. and Askins, K. (2010) 'Public geographies II: Being organic', *Progress in Human Geography*, 34: 5.**

13 Participatory Action Research

Myrna M. Breitbart

SYNOPSIS

Participatory action research seeks to democratize research design by studying an issue or phenomenon with the full engagement of those affected by it. It involves working collaboratively to develop a research agenda, collect data, engage in critical analysis, and design actions to improve people's lives and effect social change.

This chapter is organized into the following sections:

- What is participatory action research and why do it?
- Participatory action research principles
- Formulation of the research questions
- Research design and data-collection methods
- Some areas of challenge and concern
- With all the constraints, why do it?

13.1 What is Participatory Action Research and Why Do It?

Participatory research or, as it is more commonly known, **participatory action research** (PAR) involves the study of a particular issue or phenomenon with the full engagement of those affected by it. Its most distinguishing features are a commitment to the democratization and demystification of research, and the utilization of results to improve the lives of community collaborators and promote broader social change. In this sense, PAR is more than a methodology; it is a form of activism. Most definitions thus combine data collection, critical inquiry, and action. Where PAR departs most from other forms of social research that also focus on effecting change is in the *means* used to achieve this end. In PAR, the means by which data and new knowledge are co-generated and interpretations debated constitute a key part of the change process.

PAR was developed in part as a response to exploitative research and as a component of a much larger radical social agenda. There is a vast collection of writing on its history, theory and practice (see, for example, Fals-Borda and Rahman, 1991; Park et al., 1993; Greenwood and Levin, 1998; Reason and Bradbury, 2001; Bloomgarten et al., 2006; Kindon et al., 2007). Many participatory researchers situate their practice within a broader tradition of liberationist movements and early critiques of international development work that question the purposes, ethics and outcomes of social research conducted *on behalf* of other people. Fals-Borda and Paulo Freire, in particular, stress the role of participatory research in recovering knowledge 'from below' (Paulo Freire, 1970; Fals-Borda, 1982). Feminists have been especially active in promoting participatory research (see Maguire, 1987; Gluck and Patai, 1991; McDowell, 1992; Stacey, 1998; Kindon, 2003; Cahill, 2007a). They raise questions about who benefits from research and insist on the importance of building reciprocity by creating social spaces in which people can transform their own thinking and make meaningful contributions to their well-being, 'not serve as mere objects of investigation' (Benmayer, 1991: 160; Cameron and Gibson, 2005). PAR has also sought to address the root causes of social injustice and was a practice embedded in the Civil Rights and Black Nationalist movements (Bell, 2001).

William Bunge's 'Geographical Expeditions' in Detroit at the end of the 1960s and early 1970s provide early examples of the application of participatory data collection specifically within the field of geography. Bunge sought to elicit the perspectives of people living in poverty through creative data collection and analysis, creating 'folk geographers' of neighbourhood residents (Bunge, 1977; Bunge, et al., 2011). Bunge's work embraced some key concepts of PAR:

- The importance of field research as an educational tool.

- The power of knowledge when used to design social actions to make a difference in people's lives.

- The acceptance of professionals as both teachers and students.

In the years since Bunge's groundbreaking work, a number of geographers, environmental educators, planners and architects have employed participatory research techniques. Colin Ward pioneered participatory research with young people and the use of the built environment as a resource for critical learning through his work for the Town and Country Planning Association (Ward and Fyson, 1973; Ward, 1978; Breitbart, 1992, 2014; Burke and Jones, 2014). Roger Hart also took up this work in the 1970s. He and several additional researchers have continued to develop highly effective models for engaging children from multiple cultural backgrounds in genuine PAR projects around the world (Hart, 1997; Chawla, 2002; Driskell, 2002). In recent decades there has been a substantial growth of collaborative PAR projects involving children and youth (e.g. Breitbart, 1998; Wridt, 2003; Cope, 2009; Donovan, 2014; Flores Carmona and Luschen 2014; http://y-plan. berkeley.edu). The field of participatory research in geography with adults has also accelerated (e.g. Cahill, 2007b; Kindon et al., 2007; mrs. c kinpaisby-hill, 2011). As Pain explains in her review of the field, it is often the context-specific, place-based nature of this work that both seeks to uncover and draw upon local knowledge to improve physical and social environments (2004: 653). Many of these principles are further illustrated in the on-going work of the East St Louis Action Research Project (see Box 13.1).

Box 13.1　The East St Louis Action Research Project (ESLARP)

The East St Louis Action Research Project began in 1987 when a State Representative in Illinois challenged the president of the University of Illinois in Champaign-Urbana (UICU) to serve the research and education needs of this low-income neighbourhood that is more than 175 miles away from campus. What began as a series of research projects within the University's Architecture, Landscape Architecture, and Urban and Regional Planning programmes, eventually evolved into an on-going collaboration between many disciplinary departments at UICU and several community-based organizations in East St Louis (Reardon, 1997; 2005).[1]

In 1990, Kenneth Reardon, a professor of urban planning, discussed the prospect of doing research in East St Louis with Ms Ceola Davis, a 30-year community outreach worker. She greeted him with the following: 'The last thing we need is another university person coming to East St Louis to tell us what any sixth grader here already knows' (Reardon, 2002: 17). At the second meeting, according to Reardon, Ms Davis indicated on what terms she and others might be willing to work in partnership with university faculty and students. In what came to be known as the 'Ceola Accords', she laid down the following terms:

- Local residents, not folks from the university, would decide what issues would be addressed through the partnership.
- Local residents would be actively involved in all stages of the planning process.
- The university would have to agree to a six-month probationary period and, if passed, a minimum five-year commitment.
- The resident group wanted help incorporating as a non-profit organization (Reardon, 2002: 18).

The suspicion with which Ms Davis viewed outsiders coming to do research in her community is not unusual. Communities are often treated as laboratories, given no role in the research process and benefit little from the results of studies conducted within their borders. In the case of East St Louis, by the time Reardon arrived, there were already 60 reports sitting on the shelf that had not resulted in any improvements to the neighbourhood. The partnership that has since developed between the city and the university has nevertheless become one of the best examples of participatory research.

The project began with the collection of data used to produce a map of untapped resources and problems in the neighbourhood. The city of East St Louis was also compared with its suburbs utilizing photographs, landownership records, more GIS-produced maps of physical conditions, and census data. Patterns of uneven development, documented in a very real and accessible format, became the basis for further research and actions to produce change. School performance was evaluated, street clean-ups were begun and direct actions were organized to move the city to address its problems. Additional research and planning was undertaken to create recreational space for children, affordable housing and re-direct proposed light rail lines through the neighbourhood. Community partners have also made significant demands on the university, to the point of requiring them to create a satellite Neighborhood College for adult learners. East St Louis residents choose the curriculum, which includes courses on developing affordable housing, political economy and the 'ABCs of Organizing'. Still under way, with more than 30 community partners, this participatory action research combines technical and capacity building with attention to addressing real problems through a commitment to sustaining long-term projects and the constitution of interdisciplinary teams. In one collaboration, students and faculty from UICU partnered with residents to study and remedy the problem of food insecurity, the few and inadequate places in the neighbourhood that can be accessed to purchase quality and affordable food.

13.2 Participatory Action Research Principles

PAR is not a specific methodology with exact procedures, nor is it about data collection alone. In a participatory data collection process, great value is placed on the knowledge and full engagement of those who are conventionally the object of research (Park, 1993). For this reason, participatory research often relies on less formal data collection methods and seeks to foster a community's capacity to problem solve and design actions without having to depend solely on outside experts.

The most basic and distinguishing principle of participatory research is *sustained dialogue* between external and community researchers to produce a more complete understanding of a situation or environment. Caitlin Cahill points to the ways in which active participatory research which aims to bring about neighbourhood change and builds upon personal observation can open up many new avenues of inquiry and understanding. For example, her work with a racially diverse group of young women on the Lower East Side of New York produced greater clarity about the causes and complicated personal effects of gentrification as part of a global process of uneven development. It also challenged social exclusion by initiating a series of environmental actions and media projects designed to challenge stereotypes and representations of the young women held by many of the people moving into the neighbourhood (Cahill, 2007b).

PAR is about *sharing power* and involves a commitment to see that the outcomes benefit the community in measurable ways by building on its assets. In practice, this means that community members should be hired and trained, whenever possible, as research partners. They should have the power to define research objectives and offer opinions about data collection methods, utilizing their lived experience as a beginning basis for the investigation. Drawing from her project in New York, Cahill (2007a) examines how new knowledge was produced by the young women, and how the collective research capacity of the participants was built by beginning with their lived experience, and continuing with a process of 'learning by doing'. Cahill and the young women worked to create a 'community of researchers' with a real collective investment in the outcomes of the project (Cahill, 2007a).

Participatory research is often initiated outside the university, with academics taking on the role of consultant or collaborator (supplying information and technical skills). In this process, personal skills – such as the ability to facilitate conversation without dictating choice, along with an awareness of one's own biases and their impact on research – can be more important than technical skills. The research is clearly *not* participatory if you do *everything* – frame questions, conduct the research, and write up and disseminate the results. This is true even if you spend a lot of time with community members asking their opinions or having them review your findings.

Participatory inquiry can be used as a tool to build knowledge of community assets and strengths as well as to identify or address problems. For example, the Holyoke Community Arts Inventory, part of a broader study of urban revitalization strategies that use the arts to promote economic and community development, involved Holyoke artists in an exploration of their own cultural assets and a complicated search for the diverse meanings that residents give to the creative pursuits in their lives. The study challenged standard definitions of a 'creative class' by using PAR to uncover the hidden talents of local residents (Breitbart, 2013). Post-structural critiques of PAR focus on the importance of recognizing 'multiple representations of knowledge each with their own power,' and point out how community participants' views of themselves can be altered through the process of generating data and making these public (Cameron and Gibson, 2005). Because a central tenet of PAR is for community-based researchers to *represent themselves* rather than *be represented*, the PAR process also raises questions concerning who presents work at public meetings, conferences or in publications, and in what contexts and format this information is presented. Many academic researchers engaged in PAR co-present and co-author papers and action proposals (Breitbart and Kepes, 2007; Cahill and Torre, 2007; mrs. c kinpaisby-hill, 2008, 2011; Cope, 2009).

In general, the literature on participatory research provides considerably more details about the ideology and politics of the approach than about the research *process* (Reason, 1994). Yet it is the application of the above principles into all stages of the project that best distinguishes its practice. A great challenge in writing this chapter is the recognition that every participatory research project differs markedly in the research techniques that it employs.

13.3 Formulation of the Research Questions

In participatory research, the process of formulating a research topic ideally originates with the affected community. In this early stage, community members may form partnerships with faculty and students. It is not uncommon for community-based organizations to want to interview prospective collaborators. These initial meetings prior to data collection can be an important time when knowledge is shared and trust is built.

Data collection in PAR begins with the lived experiences of the collaborators and the education of researchers about the neighbourhood or region under study to create a shared base of knowledge. Over the past several decades, PAR practitioners have employed an ever-widening range of creative methods designed to reduce barriers to participation and increase the likelihood of generating useful and effective data, which often involves the design of training workshops with local researchers on the use of particular methodologies (Kindon et al., 2007; Percy-Smith and Thomas, 2010).

If the research question is undecided, exploratory methods may provide the information necessary to determine the questions, including brainstorming, focus groups, neighbourhood walks and photography storyboards. The Holyoke Community Arts Inventory, mentioned earlier, began with a focus group attended by city residents, who discussed how their involvement in arts-related activities currently impacts their lives. They also shared future visions of the city in the event that 'arts' and 'culture' were to play a larger role. Ideas generated during the discussion helped to define a research project, which focused on producing a community arts map and learning more about indigenous talent so that this information could be taken into account in current urban revitalization plans (Breitbart, 2013).

Activities such as brainstorming workshops, social mapping, and model construction provide effective ways for drawing out participants' existing knowledge. One project, which resulted in the *Youth Vision Map for the Future of Holyoke* (Massachusetts), began with a youth-led walk around the city. Participants photographed those aspects of the environment that they felt had potential and those in need of immediate attention. The images were then placed on a base map for later assessment and analysis (Breitbart and Kepes, 2007). Similar data-gathering techniques and resource inventories were employed in the East St Louis Action Research Project (Box 13.1), where residents used disposable cameras to identify important issues and sites in the neighbourhood with 'untapped' resources. In both projects, a preliminary gathering of information and opinions was used to help participants prioritize the issues they would like to study further.

13.4 Research Design and Data-Collection Methods

Once some preliminary sharing of knowledge and relationship-building takes place, participatory research requires decisions about what further information to collect and how to collect it. This can lead to discussions about how to train individuals who may be unfamiliar with the chosen research techniques. The choice of data collection methods in all stages of research is driven by the likelihood that they will produce relevant and robust data that connect directly to the research question(s). Data collection methodologies are also often chosen to ignite a process of personal and social change, and to nourish the critical and creative capacities of participants.

Participatory research requires a consideration of how the methods chosen will further community goals and demystify the research process. Attention is also given to encouraging maximum involvement and fostering a greater equalization of power among participants. Full participation in decision-making ideally requires discussions about everything, from the wording of questions to their sequencing. It also necessitates an acceptance of democratic process and full understanding of the decision-making structure and rules of discourse, no matter how young or old the participants (Hart, 1997; Percy-Smith, 2014). Some important things to consider in these deliberations are the range of skills that participants have to offer and the diversity of the group, including the mix of age, gender, and cultural or ethnic background. Issues of safety and access to equipment or transportation may also come into play. In almost all cases, more than one investigative approach is chosen so as to draw upon the widest range of skills.

Both quantitative (see Chapters 8 and 16) and qualitative data collection methods (see Chapters 7–10) are used in participatory research. Common qualitative data collection methods that go beyond brainstorming or photographic documentation, and are used with both children and adults, include interviewing, oral history, story telling, focus groups, drawing, social mapping and local surveys or environmental inventories. Since many of these data collection techniques are described elsewhere in this book, I will focus here on their specific use in participatory research. For example, in a recent project, Kimbombo, a theatre group in Holyoke, Massachusetts, was created within a community organization that provides GED (General Education Development high-school equivalency) and English as a second language training to Spanish-speaking residents. The project involved interviewing local students and culminated in the writing and performance of plays on issues the learners identified as important, such as domestic violence, diabetes awareness, and the acceptance of residents with HIV/AIDS. The resulting scripts and performances provided an opportunity for the community to control its own representation while simultaneously addressing critical health issues

(Breitbart, 2013). It is common in PAR for academics and community researchers, adults and youth, to do *paired* interviews and focus groups, as well as some oral history and group story telling. This variety of techniques enables partners to share perspectives and generate empirical data. The organic process of using these methods to come up with initial discussion questions, examine them, and then reformulate the questions and begin anew, contrasts with the more linear progression of typical social research. It reflects the fact that PAR can be a 'continuous educational process' that builds upon a deepening understanding of an issue or topic and does not necessarily begin and end with one project (Park, 1993: 15).

There are many examples in PAR of the use of *qualitative surveys*. For example, a PAR collaboration between students and faculty at the University of Utah and youth affiliated with the Mestizo Institute for Culture & Arts utilized a variety of qualitative methods to investigate the emotional and material impacts of the misrepresentation of immigrants resulting from negative stereotypes of immigrants (Box 13.2).

Box 13.2   Youth-driven reframing of immigration through Participatory Action Research

This participatory action research project was conducted by a diverse group of immigrant youth involved in the Mestizo Arts & Activism Collective, a university/community partnership developed at the University of Utah in collaboration with the Mestizo Institute for Culture & Arts in Salt Lake City that provides young people with a space to investigate their concerns and struggles. The project began with the everyday experiences of youth who personally, and through the media, had encountered animosity toward people who migrate to the US from other countries. This included particularly hurtful racist stereotypes about immigrant youth as 'illegal aliens' or criminals. The 'Dreaming of No Judgment' research and action project aimed at radically shifting the discourse and actions that often accompany such stereotypes.

Data collection began with a diverse group of Latino/Chicano, African-American, Asian, and bi-racial young people who felt that their views and experiences had been largely ignored by politicians and local residents. The youth shared personal stories of their misrepresentation and stereotyping. They also began to research the dominant anti-immigrant sentiments on a local website and within Utah's laws pertaining to immigration. The latter investigation included attendance at legislative hearings where the youth researchers became increasingly aware of the xenophobia prevalent in the views of legislators and the general public. They also identified what they felt was a general lack of understanding and empathy for the problems that immigrants face on a daily basis. The interview and storytelling uncovered white residents' fear of immigrants, which the youth compared to the fears that they and the larger adult immigrant community experience on a regular basis. The researchers began to question why one group's 'fears' should be given more authority and attention than another's and decided to use PAR to challenge negative representations of immigrants and set a new course for reframing the immigration debate. The continued sharing of personal stories and focus groups with a widening array of young people produced new knowledge about the emotional and economic impacts of stereotyping, including material struggles and the denial of access to opportunities such as education and jobs.

The reflection that youth from the Mestizo Arts & Activism Collective engaged in during this data collection process eventually helped them to design activist interventions that enabled them to share their findings and alter stereotypes. The *Dreaming of No Judgment* project resulted in the creation of a webpage to encourage more youth to begin to challenge stereotypes. To reach policy-makers, they also utilized theatre and literary arts to assert their sense of belonging and their rights. They created a performance piece entitled 'We the People', which they presented to the Utah State legislature to combat the own victimization and the more general misrepresentation of immigrants by politicians. Two articles written about the project provide an evocative and clear example of how a participatory action research process can produce questions that have real meaning for partici-pants, and how particular data-collection methods, once chosen by participants, enabled them to frame research questions and action goals that were of direct relevance to their lives (Cahill, 2010; Cahill, Quijada Cerecer and Bradley, 2010). Both the research process and the actions fostered some hope that the current situation can be transformed.

Visual methods of generating and representing data provide another important tool in PAR, especially for geographers (see Chapters 31 and 34). There are many ways to generate spatial representations, from sketches to annotated photographic essays (e.g. photovoice), and various forms of mapping (e.g. http://www.social-life.co/publication/atlas_social_maps/). *Social mapping* is a particularly useful low-technology data-collection tool that has been widely employed by geographers and architects who are engaged in participatory research as a repository for the raw data from environmental inventories. For example, in an on-going planning and housing project carried on in Guyana with an indigenous Amerindian population, Gabriel Arboleda utilized a variety of visual mapping tools to replace conventional forms of participation with a form of PAR designed to assess housing needs, prioritize households, and impact design (see Box 13.3).

Box 13.3 Participatory research on housing in the Guyana Hinterland

The Second Low-Income Settlement Program (LISP II) offered by a public institution to Amerindian indigenous communities in rural Guyana is a housing program that departs significantly from main-stream methods by engaging residents in participatory planning and implementation. The Guyana Hinterland has high poverty rates along with rich natural resources. The project, guided by Gabriel Arboleda, took place in eight villages and involved Amerindians in all phases of research and action: planning, design, administration, financing, and construction of the housing. More than 200 new homes were created and 60% of applicants' homes were upgraded (Arboleda, 2014).

Villagers began with an initial assessment of current housing to determine whether this was a priority. After agreeing that it was, they engaged in a collaborative mapping of their villages in order to determine the nature and extent of the housing problem. These data included detailed information about the dwellings, occupants, materials, water sources, toilets, communication devices, and other technology, such as lighting. While poverty and overcrowding were prevalent, the mapping revealed a strong desire and capacity on the part of Amerindians to 'act on their own behalf' (Arboleda, 2014: 206).

Empowered by these exercises, the *agency* of the villagers was achieved through the stages of participatory planning, design and construction that utilized local materials and built upon indigenous

(Continued)

(Continued)

building knowledge, negating the need for outside contractors. The PAR process enabled the bringing together of needs, problems and resources, and established agreements about such things as hiring local labour in lieu of cash, purchasing construction materials from villagers, prioritizing cultural preferences in design, and avoiding relocation. Villagers worked in diverse teams with a local builder or craftsperson to propose house designs, and final designs for each region drew upon these proposals. Village councils administered the program, including deciding on beneficiaries and procuring local materials. Drawing on cultural preferences, thatched houses were upgraded to metallic zinc roofs that included rainwater collection systems. Villagers were enabled to own their homes by contributing labour and materials rather than money. The PAR process, which elevated the importance of local knowledge, and identified cultural capital as a form of investment equity, in this way facilitated a dramatic and much-needed transformation of the living environment (Arboleda, 2014: 224).

Social mapping often begins with the acquisition of a large base map of the area under study, something local planning departments are often willing to share. In one planning exercise in Holyoke, Massachusetts, youth and adults were divided into small groups and enhanced the information on base maps by designing their own icons and using self-stick dots of varying colours to designate aspects of the environment that they considered important (for example, red dots for areas you like; green dots for locations you hang out in and enjoy; blue dots for places you avoid, etc.). This methodology can be used in paper formats or digitally with newly available open-source (free) customizable mapping programs and mobile phone apps (see Chapters 16 and 18).

Analysis of the geographical information on social maps compiled by multiple research partners often involves discussion of the relationship between the 'real' and perceived characteristics of an environment. In the early stages of participatory research, these contrasting spatial representations can be useful in identifying the problems or assets of an area from different perspectives based on participants' gender, age, socio-economic status, or ethnicity. This information can provoke debate and help decide on a focus for further research. In doing participatory research with residents of a neighbourhood in the Bronx, New York, for example, Roger Hart used templates representing such things as dangerous places and places used by teenagers alone to map the differing perceptions of the neighbourhood by diverse groups of residents. The resulting composite map used recordings of the conversations and analysis of spatial patterns to make recommendations for new types of recreational spaces (Hart et al., 1991).

It is helpful for researchers and participants to walk or drive around the study area with cameras prior to social mapping. Annotated photographs can then be added to the map to represent and categorize environmental data; again, this can be done in paper or digital form. There is a lot of room for creative interpretation in such a process. For example, during a collaboration between Hampshire College interns and youth from Nuestras Raices, a community development organization, co-researchers combed the city with papers, notepads, and cameras to collect data on sites they considered ecologically sound and those they felt were potentially dangerous. Finding inadequate a set of symbols already devised for the international Green Map project (www.greenmap.com), the young people produced a map of their findings that included Puerto Rican restaurants and locally owned businesses in their definition of 'green sites'. Health information compiled through neighbourhood interviews was also mapped, suggesting geographic patterns of illness such as asthma that necessitated further examination. The 'Holyoke Youth Green Map' was also used to begin to acquire vacant lots for future community gardens (Breitbart and Ferguson, 2002).

The *Youth Vision Map for the Future of Holyoke* resulted from an amalgamation of ideas drawn from many brainstorming sessions. It contains bright visual symbols, placed at various sites, of what modifications young people would like to introduce into the urban landscape. Examples include murals, a Store for Safety (to discreetly provide youth with safe-sex products), a Teen Café (to hang out in, access cheap food and listen to music) and a Youth Van (staffed by adults and youth, to bring youth information about health issues and youth programmes). In this case, because the data suggested several possibilities for action, a summit for all of the youth organizations involved in the data collection provided an opportunity for each to discuss what its priorities were. The results were shared and each group went home with its own project to work on (Breitbart and Kepes, 2007).

The data collection process in PAR has its own educative value, in part because it necessitates the pooling of information and perspectives among diverse research partners. The commonalities and differences in perspective among researchers that visual displays (such as social maps) present can generate fruitful discussions that imply avenues for further research or action. However, the meaning of information collected through social mapping is not always readily apparent, nor is it always possible to decide upon a focus for research based solely on a visual or graphic representation. Additional discussion among participants is necessary to uncover varied interpretations and help define priorities. This is often the case when more technically sophisticated forms of mapping, such as GIS (Geographic Information Systems), are used by geographers as a tool for data analysis in partnership with co-researchers. The time involved in training and the access required to purchase expensive technology limit its use. One partial remedy is for a PAR team with access to a geography department to ask for technical assistance with GIS to analyse their field-generated data, which has been employed in various ways (see Chapter 18 for examples of community GIS partnerships). Sarah Elwood has also developed many effective ways to generate a critical cartography and produce a powerful tool for community empowerment by combining GIS and PAR (Elwood, 2006). These examples demonstrate the challenges as well as possible solutions to bridging the 'digital divide' found in many communities.

There is also a long history and growing literature on participatory methods of data collection with children and youth that reflects the increased attention given to children's rights as citizens (see The Article 15 Project, http://crc15.org/; Driskell, 2002; Kindon et al., 2007; Percy-Smith and Thomas, 2010). Sparked by the pioneering work of the late Colin Ward (Ward and Fyson, 1973, Ward, 1978; Burke and Jones, 2014; Breitbart, 2014), many PAR projects are designed to encourage children and youth to become agents of change through the exploration and critical analysis of their neighbourhood environments. Data collection methods are creative and diverse, and often employ visual tools (e.g. drawing, photography, mapping, video story telling etc.) to communicate information and insights. The *YouthPower Guide* (Urban Places Project, 2000), a PAR project that involved Holyoke youth and university faculty and students, provides a step-by-step guide for engaging youth in community planning. With respect to data collection, it includes over 20 activities that can be completed in a one-to-two-hour session and that are all youth led. Some of the basic rules it sets out for brainstorming apply to work with adults as well. Y-PLAN (*Youth-Plan, Learn, Act, Now!*), an on-going educational and action research initiative based at the Center for Cities & Schools at the University of California-Berkeley, develops collaborative projects that utilize many spatial and mapping techniques to engage youth in community revitalization and the transformation of public spaces in the US and abroad. Y-PLAN's website provides several examples of work that employs PAR principles, as well as toolkits for initiating community-based research and action projects with youth (www.y-plan.berkeley.edu).

In general, whether working with youth or adults, PAR draws heavily upon a variety of social-research methods and minimizes approaches that are beyond the material or technical resources of the individuals involved. With quantitative data, this may mean the use of public data sources and descriptive statistics that help to tell a story, allow comparisons to be made, or identify causal relationships without requiring a significant level of quantitative expertise. In the ESLARP, census and other statistical data were put online by students and faculty, along with the software that allowed community residents to pose questions, plug in the data and produce a result in a number of different formats, including charts, diagrams and maps. Another strategy is to divide the labour so that while some researchers are collecting information, others are taking those data and putting them in a graphic or statistical form that allows them to be analysed.

Students who want to undertake participatory research should have some working knowledge of basic social-research techniques, as described in this book. A community-based learning course, where the reading and analysis of geographic theory and case studies are combined with work on a hands-on field project, is one effective way to gather experience using such techniques (see, for example, Meghan Cope, 2009). If data-collection methods are unfamiliar to community researchers, they, too, should be trained. This transfer of skills of geographic analysis was at the core of Bunge's work in Detroit, Elwood's GIS project in Chicago, and is currently under way in the ESLARP, where adult residents interested in acquiring social-research skills are offered free courses.

While PAR does not require everyone to be equally involved in all phases of research, there is a strong commitment to encouraging active participation. Utilizing a variety of research methods and a division of labour that consciously seeks to make use of the particular strengths of each collaborator is one way to assure widespread participation in the collection of information and its exchange.

13.5 Some Areas of Challenge and Concern

Participatory action research presents a number of challenges, some of which are inherent to the practice, and others of which are more systemic or structural. In terms of the former, the coordination of the different skills and levels of participation that each partner brings to the research project can present a real dilemma. In an insightful and honest account of a PAR project she supervized as part of a *Children's Urban Geographies* course, Meghan Cope (2009) points out how differing levels of student's personal commitment, experience, and ingrained local cultures can inhibit full involvement.

In these cases, academics must resist the urge to silence others by jumping in to fill the gaps in conversation, and employ good facilitation skills to encourage the participation of individuals whose ideas have never been seriously valued or who lack experience expressing their views. Devoting time to developing methods for building the collective capacity of a 'community of researchers' can also help, including the provision of tools for oral presentation (e.g. how to make outlines of main notes, use visual aids such as slides etc.) (Cahill, 2007c).

Other practical obstacles to engaging in PAR arise from the unpredictable aspects of all projects that attempt to address community-defined priorities. Priorities frequently change mid-stream due to political, personal, or socio-economic circumstances that necessitate a shift in focus. When my colleagues and I collaborated with our community partners to write a Housing and Urban Development Community Outreach Partnership Center (HUD-COPC) grant, many project ideas emerged from discussions. Once the grant to develop projects to address these priorities was funded a year later, massive cuts in public social services required us to re-order original priorities. Given such situations, it is incumbent on academics to accept the non-linearity of PAR and prepare ourselves and students for unexpected events and situations, such as when key participants may not agree among themselves or goals change (see also Pain, 2009). The social relationships that faculty build over time with individual community collaborators, and a deepening of trust to support frank communication and problem-solving, are key to adjusting successfully to inevitable unanticipated events. They also contribute in valuable ways to keeping community–university partnerships alive.

As more universities have come to recognize the importance of civic engagement, the issues of sustainability and reciprocity in partnerships with community-based organizations have gained increasing attention. PAR projects often involve adult community partners who are over-extended, lack money and resources and/or work several demanding jobs. Community organizations are quick to point out the time and effort it takes to train and supervize university students as well as the resource constraints that research partnerships can place on over-stressed non-profits (Bushouse, 2005). Students and faculty also face competing demands on their time, and work within a timeframe that differs markedly from the 'real world'. The fact that most colleges and universities operate on what I would call the 'Brigadoon' principle – disappearing at crucial times of the year, only to reappear and try to pick up where they left off after seasonal breaks – can make PAR projects very difficult to sustain. While academic calendars may preclude students from involvement past one semester, some faculty have the capacity, and many would say the *obligation*, to sustain projects over the long term with community partners. This may require a difficult renegotiation of teaching commitments, but it can also mean networking within and outside one's discipline to mentor new students and faculty into the process to insure a project's continuation. Participatory research is time-consuming and predicated on trusting relationships and a commitment to the project's duration. While it may be possible to incorporate such work into a semester-long class if project deadlines are short term, participatory processes are generally better suited to more flexible timeframes.

The writing required as part of a participatory research project may also be quite different from academic writing that would be done for a traditional course paper or a professional geography journal. Tensions can develop between 'practically-oriented' community partners who are looking for progressive change and 'theoretically-oriented' students or faculty, many of whom are depending on academic publications for their own career success (Perkins and Wandersman, 1997). Just as Ceola Davis made demands on Ken Reardon and his students in East St Louis, community collaborators can set down the criteria for investigation that academic partners must agree to. These demands can constrain the parameters of a project or redirect its focus. In pursuing community-based learning and research, Cahill and Fine also point out the often 'false distinction between college and community' when many of the students involved in PAR are either from the local communities they are working in, or have come from communities that experience similar struggles around issues such as affordable housing, food justice, education etc. (Cahill and Fine, 2014: 70). Additionally, because one project is not likely to address all the needs that stakeholders articulate, participatory research almost always begets more projects, soliciting the continued involvement of outside research partners. This issue of sustainability of the partnership presents its own challenges.

Restrictions placed on 'human subjects' research connected to Institutional Review Board (IRB) policies in higher education institutions in the US, and university and research council ethical guidelines and frameworks in the UK, present further challenges to students who seek to engage in PAR (Pain, 2004; Cahill et al., 2007; Dyer and Demeritt, 2009) (see also Chapter 3). These policies often require an articulation of research questions, methods, and goals *before* these have been negotiated with community partners. Academic critics of IRB policies, and the growing bureaucracy of ethical procedures more generally, point out that the principles and practices of PAR, informed by a long history of grass-roots organizing and critical race and feminist theory, seek to address unequal power relationships within society and between the university and the community. Yet human-subjects regulations can actually *reinforce* those unequal relationships and inhibit the exercise of agency on the part of community participants. An important on-going agenda results from these tensions and contradictions, suggesting the need to generate more effective ethical practices and a more active challenging and reformulation of IRB policies on college campuses (Cahill et al., 2007: 309). Balanced against these challenges is a growing tendency to hold colleges and universities more accountable as citizens of the cities, towns and region in which they reside (Bloomgarten et al., 2006). A recent compact negotiated between local colleges and community-based organizations in the city of Holyoke, Massachusetts, is meant to address many of the issues that arise in the context of partnership work, and may serve as a model for other institutions (http://www.holyokec3.org/index.php/campus-community-compact).

While many universities acknowledge the importance of instilling a sense of social responsibility in students, service-learning and community-based research are often viewed as co-curricular activities. Hampshire College and Kingsborough Community College in the US represent a minority of institutions of higher education that have woven community-based research and learning as a graduation requirement into the colleges' mission (https://www.hampshire.edu/academics/cel-2-requirement; Cahill and Fine, 2014). The goals of these academic programs are larger than enhanced student learning; they extend to the promotion of greater social justice through addressing issues of immediate concern to community partners.

The primary reasons why proponents continue to advocate for participatory research as an evolving method rest largely with this overt commitment to political aims. As several of the examples here reveal, PAR can at times challenge erroneous stereotypes and reveal the strengths and assets present in marginalized communities (Cahill, 2013). Because PAR is construed as more than a methodology, theory of knowledge, or pedagogy, much of its appeal to students and faculty in geography is its emphasis on improving environments and a belief that the research can contribute to struggles for social justice. While it is important to be realistic about what kinds of change can be achieved, and in what timeframe, many practitioners urge academics to persist and address its challenges (Klocker, 2012). This includes paying attention to differences in power and a consideration of how projects move from research alone to social action (Pain, 2004; Arboleda, 2014).

Many PAR projects must simultaneously try to meet divergent goals, such as the building of community and the development of a critical understanding of local assets and problems. In a recent PAR collaboration between Hampshire College students and public housing residents in a nearby city, particularly challenging issues were raised when both the students and the residents entered into the planning process with the city for housing renovations. The public housing residents had successfully mobilized politically to reverse an earlier decision by the city to tear down their project, which is conveniently located downtown. When the students were asked by a resident leader to assist in what was purported by the city to be a participatory process for planning renovations on the existing site, it soon became clear that participation meant very different things to different people. The mainly Spanish-speaking residents were not given adequate access to information, translation services, or a timeframe sufficient to aid their own research or enable them to contribute in meaningful ways to the decision-making process. Student collaborators also could only go so far in assisting in this research process given the structural barriers. They often felt inadequate as partners and uncomfortable as outsiders, when, as students of architecture and planning, they were assumed to have more knowledge and power than the residents. This kind of misapplication of the principles of participatory practice reminds us of Sherry Arnstein's original (1969) critique of the participatory processes put in place to address the inequities created by urban renewal in the mid-20th century. Her 'ladder of citizen participation' illustrates the many different forms that participation can take depending on whether the motivation is manipulation or the transfer of real power and control. Chatterton et al. (2007) also remind practitioners of PAR that meaningful social transformation does not generally come through particular participatory *techniques* but from working with community partners to understand and alter how larger policies function.

The problematization of PAR, as well as increasingly frank discussions of projects that have either failed or met serious challenges, emphasize the importance and the inherent complexity of doing this imperfect work. They represent an important commitment to improving transformative practices while also countering tendencies to over-romanticize participatory practice as the only type of research that can promote social change.

13.6 With All the Constraints, Why Do It?

For those of us who do participatory research, it is a humbling experience. After years of working with local neighbourhoods on collaborative projects, Randy

Stoeker confronted what he refers to as the 'haunting question' of how he can conduct this type of research so that it is both 'empowering and liberatory' (1997). He nevertheless urges students and faculty not to be so concerned about 'doing the right thing' that they become 'paralysed'. Community collaborators of the likes of Ceola Davis or the Amerindians of Guyana will always tell you when you have erred, and will usually give you the benefit of the doubt if you are honest, respectful and follow through. There are no 'pure' forms of PAR; only degrees of participative practice within real-world constraints.

The data that PAR can produce are also more likely to be useful, accurate and to lead to actions that address people's real needs and desires. In discussing the promise of civic engagement programs and their impacts on college and university students, Guarasci describes a graduate who characterized her collaborative research as a way to 'discover a second faculty' from within the community, from whom she learned a tremendous amount about neighbourhood issues (2014: 61). Augmenting a knowledge base with the new perspectives that community-based researchers add not only assures better quality scholarship and a greater likelihood of identifying and addressing issues of importance to a community-based constituency, it also has the capacity to change institutions of higher education. *Imagining America*, a national consortium of over 100 colleges and universities, aims to democratize civic culture by connecting scholars and practitioners in the arts, humanities, and spatial design with their local communities, utilizing practices drawn from community-organizing and PAR. The goal is to activate a critical but 'hopeful' re-imagining of the country and institutions of higher education to better serve community needs and generate new forms of community-based knowledge/talent. A further aim is to provide greater access to higher education for traditionally underserved populations (www.imaginingamerica.org).

Echoing the goals of Imagining America, Nadinne Cruz, consultant and former Director of the Haas Canter for Public Service at Stanford University, posed a thought-provoking question at a service learning conference: what if our institutions' exercise of social responsibility is to assess how adequate the knowledge of our faculty and students is for addressing critical issues in the world? In spite of the complexities, experience with participatory research suggests that we are more likely to address this challenge if we collaborate in methodology design, data collection, analysis and action with those individuals and groups most likely to gain from the investigative process.

SUMMARY

- In PAR the collaborative means by which data are co-generated, interpreted and used to design actions play a key role in social transformation.
- In PAR university researchers and their collaborators aim for more reciprocal relationships; they share knowledge, power, and a decision-making role.
- Participatory research begins with the lived experiences of the collaborators and the acquisition by outside researchers of basic knowledge about the place under study.
- Utilizing a variety of data collection methods, and a division of labour that consciously builds upon the strengths and experiences of each member of the team, serves to nourish the critical and creative capacities of all participants.
- PAR can help to address societal inequities and contribute to movements for greater social justice.

Note

1   The ESLARP project is currently coordinated by Dr. Howard Rambsy at Southern Illinois University Edwardsville through the Institute for Urban Research. Information about current projects can be found at https://www.siue.edu/graduate/iur/projects/eslarp/index.shtml.

Further Reading

The following books provide a range of examples of different forms of PAR:

- Park et al. (1993) begin with a foreword by Paulo Freire, one of the world's experts on participative research as a vehicle for personal and political transformation. Cases referenced here are drawn from the North American experience and address the relationships between power and knowledge, research methods and social action. An appendix identifies key organizations that promote participatory research.

- Reason and Bradbury's (2001) comprehensive collection of articles on action approaches to social science is directed at an academic audience. It is divided into four sections that address theories and methods of participatory research as well as the application of these approaches and the skills necessary for implementation. It also explores the role of universities in action research.

- Kindon et al. (2007) supply an inventive and comprehensive edited collection on participatory action research that includes a mixture of honest and critical reflection along with case studies from around the world. Chapters are relatively short and they are written in a highly accessible manner. There is also a clear emphasis on space and place, and a range of practical advice on the application of PAR in geography.

- Percy-Smith and Thomas (2010) provide an excellent review of some of the most recent theories and practices around children's participation in decision-making within family and community settings around the world and in a wide variety of circumstances. Besides including many different case studies, the book incorporates critical reflection on the part of young people and adults with the aim of better addressing challenges and improving participatory practices. The book concludes with new theoretical perspectives on how best to promote children and youth as active citizens.

- Burke and Jones (2014) focus on the prescient ideas and practices generated by British educator, architect and social anarchist, Colin Ward. Much of this work concerns the everyday involvement of children and youth in their own learning and the transformation and improvement of their living environments. Written from a cross-disciplinary perspective, each chapter takes the principles and practices promoted by Ward and presents historical and contemporary examples of their application.

Note: Full details of the above can be found in the references list below.

References

Arboleda, G. (2014) 'Participation practice and its criticism: Can they be bridged? A field report from the Guyana Hinterland' *Housing and Society*, 41(2): 195–27.

Arnstein, S. (1969) 'A ladder of citizen participation', *Journal of the American Institute of Planning*, 35(4): 216–24.

Bell, E. (2001) 'Infusing race into the US discourse on action research', in P. Reason and H. Bradbury (eds) *Handbook of Action Research: Participative Inquiry and Practice*. London: Sage. pp. 48–58.

Benmayer, R. (1991) 'Testimony, action research, and empowerment: Puerto Rican women and popular education', in S. Gluck and D. Patai (eds) *Women's Words: The Feminist Practice of Oral History.* New York: Routledge. pp. 159–74.

Bloomgarten, A. (2013) 'Reciprocity as sustainability in campus-community partnerships', *Journal of Public Scholarship in Higher Education*, 3: 129–45.

Bloomgarten, A., Bombardier, M., Breitbart, M., Nagel, K. and Smith, P. (2006) 'The Holyoke planning network: Building a sustainable college/community partnership in a metropolitan setting', in R. Forrant and L. Silka (eds) *Inside and Out: Universities and Education for Sustainable Development.* Amityville, NY: Baywood. pp. 105–18.

Breitbart, M. (1992) '"Calling up the community": Exploring the subversive terrain of urban environmental education', in J. Miller and P. Glazer (eds) *Words that Ring Like Trumpets.* Amherst, MA: Hampshire College. pp. 78–94.

Breitbart, M. (1995) 'Banners for the street: Reclaiming space and designing change with urban youth', *Journal of Education and Planning Research*, 15: 101–14.

Breitbart, M. (1998) '"Dana's mystical tunnel": Young people's designs for survival and change in the city', in T. Skelton and G. Valentine (eds) *Cool Places: Geographies of Youth Cultures.* London: Routledge. pp. 305–27.

Breitbart, M. (2013) *Creative Economies in Post-Industrial Cities: Manufacturing a (different) Scene.* Farnham: Ashgate.

Breitbart, M. (2014) 'Inciting desire, ignoring boundaries and making space: Colin Ward's considerable contribution to radical pedagogy and social change', in C. Burke and K. Jones (eds) *Education, Childhood and Anarchism.* London: Routledge. pp.175–85.

Breitbart, M. and Ferguson, B. (2002) 'Partnerships for social change: Community-based learning at Hampshire College', *New Village Journal: Building Sustainable Cultures*, 3.

Breitbart, M. and Kepes, I. (2007) 'The YouthPower story: How adults can better support young people's sustained participation in community-based planning', *Children, Youth and Environments*, 17: 226–53.

Bunge, W. (1977) 'The first years of the Detroit Geographical Expedition: Personal report', in R. Peet (ed.) *Radical Geography.* London: Methuen. pp. 31–9.

Bunge, W., Barnes, T. and Heynen, N. (2011) *Fitzgerald: Geography of a Revolution.* Atlanta: University of Georgia Press.

Burke, C. and Jones, K. (eds) (2014) *Education, Childhood and Anarchism.* London: Routledge.

Bushouse, B. (2005) 'Community non-profit organizations and service learning: Resource constraints to building partnerships with universities', *Michigan Journal of Community Service Learning*, 12(1): 32–40.

Cahill, C. (2007a) 'The personal is political: Developing new subjectivities in a participatory action research process', *Gender, Place, and Culture*, 14: 267–92.

Cahill, C. (2007b) 'Negotiating grit and glamour: Young women of color and the gentrification of the Lower East Side', *City and Society*, 19: 202–31.

Cahill, C. (2007c) 'Including excluded perspectives in participatory action research', *Design Studies*, 28(3): 325–40.

Cahill, C. (2010) '"Why do they hate us?" Reframing immigration through participatory action research', *Area*, 42: 152–61.

Cahill, C. (2013) 'The road less traveled: Transcultural Community Building', in J. Hou (ed.) *Transcultural Cities: Border Crossing and Placemaking.* London: Routledge. pp. 193–206.

Cahill, C. and Fine, M. (2014) 'Living the civic: Brooklyn's public scholars', in J. N. Reich (ed.) *Civic Engagement, Civic Development and Higher Education: New Perspectives in Transformational Learning.* Washington, DC: AC&U. pp. 67–72. Available from http://archive.aacu.org/bringing_theory/documents/4CivicSeries_CECD_final_r.pdf (accessed 16 November 2015).

Cahill, C., Quijada Cerecer, D.A. and Bradley, M. (2010) '"Dreaming of.....": Reflections on Participatory Action Research as a feminist praxis of critical hope', *Affilia: A Journal of Women and Social Work*, 25(4): 406–15.

Cahill, C., Sultana, F. and Pain, R. (2007) 'Participatory ethics: Politics, practices, institutions', *ACME: An International E-journal for Critical Geographies*, 6: 304–18.

Cahill, C. and Torre, M. (2007) 'Beyond the journal article: Representations, audience, and the presentation of participatory action research', in S. Kindon, R. Pain and M. Kesby (eds) *Connecting People, Participation and Place: Participatory Action Research Approaches and Methods.* London: Routledge. pp. 196–206.

Cameron, J. and Gibson, K. (2005) 'Participatory action research in a poststructuralist vein', *Geoforum*, 36(3): 315–31.

Chatterton, P., Fuller, D. and Routledge, P. (2007) 'Relating action to activism: Theoretical and methodological reflections', in S. Kindon, R. Pain and M. Kesby (eds) *Connecting People, Participation and Place: Participatory Action Research Approaches and Methods: Connecting people, participation and place.* London: Routledge. pp. 216–22.

Chawla, L. (ed.) (2002) *Growing Up in an Urbanising World.* London: Earthscan.

Cieri, M. (2003) 'Drawing on perception: Re-territorializing space and place from African-American perspectives'. Paper presented at the Association of American Geographers, New Orleans.

Cope, M. (2009) 'Challenging adult perspectives on children's geographies through participatory research methods: Insights from a service-learning course', *Journal of Geography in Higher Education*, 33(1): 33–50.

Donovan, G.T. (2014) 'Opening proprietary ecologies: Participatory action design research with young people', in G.B. Gudmundsdottir and K.B. Vasbø (eds) *Methodological Challenges When Exploring Digital Learning Spaces in Education*. Rotterdam: Sense Publishing. pp. 65–78.

Driskell, D. (2002) *Creating Better Cities with Children and Youth: A Manual for Participation*. London: Earthscan.

Dyer, S. and Demeritt, D. (2009) 'Un-ethical review? Why it is wrong to apply the medical model of research governance to human geography', *Progress in Human Geography*, 33(1): 46–64.

Elwood, S. (2006) 'Critical issues in participatory GIS: Deconstruction, reconstructions and new research directions', *Transactions in GIS*, 10: 693–708.

Fals-Borda, O. (1982) 'Participation research and rural social change', *Journal of Rural Cooperation*, 10: 25–40.

Fals-Borda, O. and Rahman, M. (1991) *Action and Knowledge: Breaking the Monopoly with Participatory Research*. New York: Apex.

Flores Carmona, J. and Luschen, K. (eds) (2014) *Crafting Critical Stories: Toward Pedagogies and Methodologies of Collaboration, Inclusion and Voice*. New York: Peter Lang.

Freire, P. (1970) *Pedagogy of the Oppressed*. New York: Seabury.

Gluck, S. and Patai, D. (eds) (1991) *Women's Words: The Feminist Practice of Oral History*. New York: Routledge.

Greenwood, D. and Levin, M. (1998) *Introduction to Action Research: Social Research for Social Change*. Thousand Oaks, CA: Sage.

Guarasci, R. (2014) 'Civic provocations: Higher learning, civic competency, and neighborhood partnerships', in J. Rich (ed.) *Civic Engagement, Civic Development, and Higher Education: New Perspectives on Transformational Learning*. Washington, DC: AAC&U. pp. 59–62.

Hart, R. (1997) *Children's Participation*. London: Earthscan.

Hart, R., Iltus, S. and Mora, R. (1991) *'Safe Play for West Farms': Play and Recreation Proposals for the West Farms Area of the Bronx Based Upon the Residents Perceptions and Preferences*. City University of New York: Children's Environments Research Group.

Kindon, S. (2003) 'Participatory video in geographic research: A feminist practice of looking?', *Area*, 35: 142–53.

Kindon, S., Pain, R. and Kesby, M. (eds) (2007) *Connecting People, Participation and Place: Participatory Action Research Approaches and Methods*. London: Routledge.

Klocker, N. (2012) 'Doing participatory action research and doing a PhD: Words of encouragement for prospective students', *Journal of Geography in Higher Education*, 36(1): 149–63.

Maguire, P. (1987) *Doing Participatory Research: a Feminist Approach*. Amherst, MA: Center for International Education, University of Massachusetts.

McDowell, L. (1992) 'Doing gender: Feminism, feminists and research methods in human geography', *Transactions: Institute of British Geographers*, 17: 399–416.

mrs c kinpaisby-hill (2008) 'Publishing from participatory research', in A. Blunt (ed.) *Publishing in Geography: A Guide for New Researchers*. London: Wiley-Blackwell. pp. 45–7.

mrs. c kinpaisby-hill (2011) 'Participatory praxis and social justice: Towards more fully social geographies', in V. Casino, M. Thomas, P. Cloke, and R. Panelli (eds) *A Companion to Social Geography*. Oxford: Blackwell, pp. 214–34.

Pain, R. (2004) 'Social geography: Participatory research', *Progress in Human Geography*, 28: 652–63.

Pain, R. (2007) 'Guest editorial: Participatory geographies', *Environment and Planning A*, 39: 2807–12.

Pain, R. (2009) 'Working across distant spaces: Connecting participatory action research and teaching', *Journal of Geography in Higher Education*, 33: 81–7.

Park, P. (1993) 'What is participatory research? A theoretical and methodological perspective', in P. Park and M. Brydon-Miller, B. Hall and T. Jackson (eds) *Voices for Change: Participatory Research in the U.S. and Canada*. Westport, CT: Greenwood Press. pp. 1–20.

Park, P. and Brydon-Miller, M., Hall, B. and Jackson, T. (eds) (1993) *Voices for Change: Participatory Research in the U.S. and Canada*. Westport, CT: Greenwood Press.

Participatory Geographies Working Group (2006) http://www.pygywg.org.

Percy-Smith, B. (2014) 'Reclaiming children's participation as an empowering social process', in C. Burke and K. Jones (eds) *Education, Childhood and Anarchism*. London: Routledge. pp. 209–20.

Percy-Smith, B. and Thomas, N. (eds) (2010) *A Handbook of Children and Young People's Participation: Perspectives from Theory and Practice.* London: Routledge.

Perkins, D. and Wandersman, A. (1997) 'You'll have to work to overcome our suspicions', in D. Murphy, M. Scammel and R. Sclove (eds) *Doing Community-based Research: A Reader.* Amherst, MA: LOKA Institute. pp. 93–102.

Reardon, K. (1997) 'Institutionalizing community service learning at a major research university: The case of the East St Louis Action Research Project', *Michigan Journal of Community Service Learning*, pp. 130–6.

Reardon, K. (2002) 'Making waves along the Mississippi: the East St. Louis Action', *New Village: Building Sustainable Cultures*, 3: 16–23.

Reardon, K. (2005) 'Empowerment planning in East St. Louis: A People's Response to the deindustrialization blues', *CITY*, 9(1): 85–100.

Reason, P. (1994) 'Three approaches to participatory inquiry', in N. Denzin and Y. Lincoln (eds) *Handbook of Qualitative Research.* Thousand Oaks, CA: Sage, pp. 324–39.

Reason, P. and Bradbury, H. (eds) (2001) *Handbook of Action Research: Participative Inquiry and Practice.* London: Sage.

Stacey, J. (1998) 'Can there be a feminist ethnography?', *Women's Studies*, 11: 21–7.

Stoeker, R. (1999) 'Are academics irrelevant? Roles for scholars in participatory research', *American Behavioral Scientist* 42(5): 840–54.

Urban Places Project (eds) (2000) *The YouthPower Guide: How to Make Your Community Better.* Amherst, MA: UMass Extension.

Ward, C. (1978) *The Child in the City.* New York: Pantheon.

Ward, C. and Fyson, A. (1973) *Streetwork: The Exploding School.* London: Routledge & Kegan Paul.

Wridt, P. (2003), 'The Neighborhood Atlas Project: An example of Participatory Action Research in Geography Education', *Research in Geographic Education*, 5: 25–47.

ON THE COMPANION WEBSITE...

Visit **https://study.sagepub.com/keymethods3e** for author videos, chapter exercises, resources and links, plus free access to the following recommended articles:

1. **Davies, G. and Dwyer, C. (2007) 'Qualitative methods: Are you enchanted or are you alienated?', *Progress in Human Geography*, 31(2): 257–66.**

Davies and Dwyer reflect on the links between qualitative methodologies, interpretative strategies, and movements away from 'certainty' as researchers begin to accept the 'textured nature of the world'.

2. **Blomley, N. (2008) 'The spaces of critical geography', *Progress in Human Geography*, 32(2): 285–93.**

Blomley examines debates surrounding where critical geography is located – in the academy/publications or in the community/sites of activist struggle – arguing for the importance of the latter and collective forms of knowledge production.

3. **Davies, G. and Dwyer, C (2008) 'Qualitative Methods II: Minding the gap', *Progress in Human Geography*, 32(3): 399–406.**

Davies and Dwyer raise many questions, and challenge several assumptions, regarding the way geographers conduct research in public/political domains, and articulate various 'publics', including through participatory engagement practices and the co-production of research.

14 Textual Analysis

Marcus A. Doel

SYNOPSIS

A text can be defined as anything that *signifies* something for someone or other. When you ponder a cloud-laden sky, or gaze upon a crepuscular landscape, or inspect your mole-ridden skin, or scan a beautifully crafted page such as this, then you are faced with so many texts: sky-text; earth-text; body-text; book-text. The world is full of texts. The world is fully textured. Conflating the notions of 'text' and 'texture' reminds us that the world's texts are irreducibly *material*, and that they do not necessarily need to be 'read' in the conventional sense in order to 'make sense.' For example, while a doctor's inquisitive eyes might peruse a patch of mole-ridden skin for signs of cancer, a lover's furtive eyes might scan the same expanse of skin for affects of pleasure. The skin-text can be taken up by both the optical eye that would impassively interrogate it for sense (meanings) and the haptical eye that would passionately caress it for sensations (feelings). Perhaps the world is *nothing but* texts and textures, *nothing but* sense and sensations. So, far from being highly prized artistic and literary artefacts, most 'texts' appear mundane and are scattered among the debris of everyday life (e.g. graffiti, footprints in the snow, and suchlike for human geographers; rock striations, tephra deposits, and suchlike for physical geographers). Consequently, this chapter sets out to challenge the bookish conception of 'texts' by showing not only how everything can come to signify something for someone or other, but also how texts give expression to social antagonisms.

This chapter is organized into the following sections:

- Introduction: Earth-writing for geographers
- Texts are not what you think
- Reading against the grain
- Just reading
- Mind that child!

14.1 Introduction: Earth-writing for Geographers

As he watched, the stars began to slide about, to realign themselves upon the black canvas of the sky as though to spell out some message for him ... What do they say, oletimer? he asked. What do the stars say? ... After a long silent time, the Indian said: They say the universe is mute. Only men speak. Though there is nothing to say.

Robert Coover, *Ghost Town* (1998: 83)

Human geography took a 'cultural turn' a couple of decades ago (see, for example, Cook et al., 2000), and since then virtually every part of the discipline – from Economic Geography to Medical Geography, and from Political Geography to Urban Geography – has become increasingly 'enculturated,' so that it is now not uncommon for geographers to study things like youth subcultures, cultures of nature, and cultural economy (Amin and Thrift, 2004). Meanwhile, cultural geography itself has changed from a relatively self-contained sub-discipline into one that embraces concerns that span the sciences, social sciences, and arts and humanities. As well as cultural geographies of nations, landscapes, and built environments, there are also cultural geographies of money, domestic appliances, and exotic fruit (Cloke et al., 2014; Horton and Kraftl, 2013). Basically, there are cultures of everything, just as there are histories and geographies of everything. Little wonder, then, that there are cultural texts as far as the eye can see. The world around us cries out to be read (Perec, 2010), but until the 'cultural turn' those cries often fell on deaf ears. So, whilst the *Empire of Culture* has made every nook and cranny of the world culturally significant, and the *Tyranny of the Signifier*[1] has made all of those nooks and crannies *mean something*, and despite some resistance here and there to the demand that the world should *make sense*, the dominance of significance and meaning seems to be largely accepted and generally unquestioned, much like the extraordinary power of clock-time, accounting conventions, and screws (Bartky, 2000; Fleischman et al., 2013; Rybczynski, 2000). Indeed, what is intolerable today is for something to *lack* significance, for something to *elude* meaning, and for something to *evade* sense. It is no longer tolerable for something to simply exist. Its existence must *mean something*. Its existence must *matter*. Even silence is now forced to speak louder than words!

One of the key legacies of this 'cultural turn,' and the expansion of 'cultural geography' to universal proportions, is the realization that the world is brimming with texts, the vast majority of which are not book-like: from the phallic claims of skyscraping architecture to the 'pedestrian utterances' of street-walkers down below (De Certeau, 1985); from the instantly obsolescent gibberish of social media to the 'uncreative writing' of joyful plagiarists (Goldsmith, 2011). The world is full of texts. Perhaps the world is *nothing but* texts and textures; *nothing but* sense and sensation. How apt, then, that our discipline should be called Geography: from the Greek, *gēo-graphia*, which means, *earth-writing*. Every Geographer is an earth-writer and an earth-reader. We in-scribe and de-scribe the world. We mark onto the world and into the world. We set down the world and set out the world. In short, the world is *constructed* and *transformed* by geo-graphical action; often with considerable violence (Pakenham, 1991; Weizman, 2007). As I gaze upon a glorious azure sky, I am reminded that even the clouds had to be rendered intelligible (Hamblyn, 2001).

Consequently, learning how to analyse texts has become a vital skill for every geographer. So, in this chapter I want to help you set out in the right direction. I will do this by focusing on four key issues. First, I want to alert you to the fact that 'texts' are almost certainly not what you think. Second, I want to encourage you to read – slowly, very slowly. As Dr Seuss famously cautioned: 'Take it slowly. This book is dangerous.' Third, I want to demonstrate that reading texts is nevertheless an easy feat to pull off – once you know the trick. Finally, I want to offer you an initial checklist for the effective analysis of texts.

Now, while it would be easy to make this introduction difficult – by filling it with technical jargon and edifying quotations – I want to keep things as simple as possible. I just want to provide you with the *flavour* of textual analysis and trust that as you progress through your studies you will acquire the taste for more. My message is simple. The worst that can happen is not that you read badly. The worst would be for you to prefer not to read at all. The photocopier once spared us from the effort of reading; just as the video-recorder once relieved us of the burden of watching TV. Nowadays, it is the PDF download, the search box, and the cut-and-paste key combinations that free us from the labour of reading. Why spend ages reading wonderful books like *This is Not a Pipe* (Foucault, 2008) and *Signs and Machines* (Lazzarato, 2014), when you can instruct your computer to transform an e-book, PDF or suchlike into an easily digested 'word-cloud' for instant visual consumption? With the blossoming of computer-assisted qualitative-data-analysis software (CAQDAS), such as NVivo (née NUD*IST), the *art of reading* – especially 'reading against the grain' (Eagleton, 1986) – is giving way to the *labour of coding* for automated computation (storing, sorting, searching, classifying, linking, mapping, and pattern-recognition of vast quantities of non-numerical, textual data), particularly for voluminous material, such as reams of interview transcripts or piles of policy documents or mountains of newspaper and magazine articles. You can download and computationally digest in a couple of hours what might otherwise take more than a lifetime to read. Here as elsewhere, as our culture of delirious acceleration approaches its zenith (Noys, 2014; Schivelbusch, 1993), the dog-end of reading may finally give out its last gasp as more and more texts become pre-digested for automatic consumption.

14.2 Texts Are Not What You Think

I suspect that the word 'text' will leave you cold. The word 'text' still tends to conjure up images of refinement, elitism, and distraction. Little wonder, then, that a predilection for texts is often assumed to pale into insignificance when compared to more pressing concerns like economic crises, regional conflicts, and global warming, all of which seem to demand evidence-based analysis and urgent action. For those who are lucky enough to escape the numerous horrors that stalk the modern world, a regard for texts can be a sign of distinction, a weapon of class war, and a way of distancing oneself from the trials and tribulations of everyday life. Texts are for the studious (to edify and enlighten) and the leisured (to entertain and distract), not for those who find themselves in alleyways, prisons, and shopping centres. Yet texts are indeed to be found in alleyways, prisons, and shopping centres (as well as in the rather palatial surroundings of my Grade II Listed Ivory Tower). Graffiti, for example, make for interesting reading (@149 St; Banksy, 2006; Lewisohn, 2011). For some, graffiti are merely signs of mindless vandalism: annoying, disrespectful, and perhaps even threatening. For others, graffiti demonstrate an attempt by marginalized and alienated people to reclaim the streets: kids, gangs, and wanna-be artists.

For still others, graffiti are anonymous messages hurled into the world: 'MAD 4 AEP,' 'Eat yourself fitter,' 'Do you want to play toilet tennis? Look left.' For others still, graffiti offer insight into long-lost worlds (Baird and Taylor, 2011). So, when one is invited to analyse texts, one should read graffiti, garbage, and carrier bags with the same rigour as one would read literature, legislation, and gene sequences. Indeed, I suspect that you are already well versed in reading the signs of everyday life: words, pictures, gestures, expressions, ambiences, street furniture, lipstick traces, etc. You will already be an accomplished analyst of worldly texts, with an acute sensitivity to both their sense and their sensation. Every kind of geography conveys *meaning* and *feeling*, even those that ostensibly mean 'nothing' and feel 'nothing,' such as the incessant noise of a city's soundscape (Attali, 1985), the blank regions of an explorer's map (Olsson, 2007), or the cold calculations of a trader's accounts (Hochschild, 2006).

As a point of departure, then, do not begin your analysis with a prejudice about what a geographer like you *should* and *should not* set out to read. The types of text you will need to take up will depend upon the particular context that you are interested in: British culture, urban culture, political culture, drug culture, neoliberal culture, pop culture, youth culture, etc. Be guided by your interests rather than by presuppositions about what counts – and does not count – as an acceptable text for a geographer. For instance, there is just as much to learn about cities from films, literature, and comic strips as there is from policy debates, planning documents, and numerical models (Clarke, 1997; Moretti, 1999; Ahrens and Meteling, 2010; Dittmer, 2014; Pratt and San Juan, 2014). I am interested in consumer culture, and my favourite cultural text is a two-page colour-chart produced by Dulux for a range of household paints (2000–2003). On the left-hand page there are 42 colours – with names like 'sheer amethyst,' 'blue topaz,' and 'rose lacquer' – arranged into three seemingly arbitrary groups. For over an hour my students failed dismally to work out what – if anything – these colours meant. When we could no longer stand the silence, I revealed the six-word solution that appears on the right-hand page of the colour chart: the first group of colours is 'Urban Discovery,' the second 'African Discovery,' and the third 'Oriental Discovery.' For the rest of the seminar they were able to draw out a seemingly endless series of interpretations about the geographical imagination of Western consumers that ranged from the ideological sentimentality of home to the colonial foundations of the discourse of 'discovery.' For example, the Urban colour scheme is predominantly made up of airy and watery blue-greys: cold, muted, and vapid. It is the perfect colour scheme for a superficial, artificial, and post-industrial world of anonymity, alienation, and interchangeability. Blue-grey is the colour of Capital (Lyotard, 1998). When one notes the names of these blue-greys – such as madison mauve™, city limit™, manhattan view™, plaza™, dot com, loft™, café latte™, platinum, and brushed steel™ – one can discern a highly gendered and class-specific lifestyle: single professional men immersed in the fast and furious world of e-commerce relish their cosmopolitan comforts (cf. Baudrillard, 1996, 1998).

Meanwhile, the African colour scheme is dominated by earthy and fiery colours: warm, grounded, and vibrant. While the Urban is alien, Africa is other: exotic, untimely, and tied to the earth. It is rendered through white heat™, turmeric™, fired ochre, bazaar™, ancient earth, raw umber, and beaten bronze™. Finally, the pastels of the Oriental colour scheme are mainly creams and greens. Yet rather than the *coldness* of the Urban (modernity) or the *heat* of Africa (tradition), they evoke a *calming* influence (nature): silken trail™, sea grass™, palm wood™, bamboo screen™, chi™, lotus blossom™, gold veil™, and eastern gold™. So, while the urban north/west and the African south pit the elements of air and water against those of fire and earth, the calming spirit of Nature wafts in from the Oriental east. Such is the ideologically loaded world according to Dulux.

I hope that this example has given you a flavour of how a few *divisions* – such as Urban/African/Oriental, North/South/East/West, worldly/spiritual, cold/hot/calm, and modernity/tradition/nature – can be arranged to organize a whole range of disparate materials into a functioning whole. So, try not to get bogged down in the details of a text (specific content; individual elements). Instead, seek out the divisions that structure those details into a configuration that apportions meaning, feeling, value, significance, visibility, etc. (general form; collective expressions). While these divisions are sometimes hard to pin down and extract, more often than not they hide in plain sight – and are therefore all too easily overlooked (Blonsky, 1985; Perec, 1999).

14.3 Reading Against the Grain

So, the world is overflowing with texts, and these texts are not simply collections of words fixed onto paper. From now on, think of a text in terms of its original Latin root: *texere* – weave. Every text is a *tissue of signs*, which can be folded, unfolded, and refolded in many ways, much like a handkerchief or a piece of origami paper. Dominant readings (i.e. powerful readings) leave deep creases (e.g. narrative ruts) and stubborn stains (e.g. threadbare clichés) in the tissue of signs, which can only be removed by reading otherwise – against the grain. As with washing and ironing, removing stubborn stains and deep creases from a text invariably requires a forceful reading, not least because of the violence that lodged them there in the first place. For instance, *you won't budge* the narrative ruts and threadbare clichés of 'capitalist realism' ('What really matters is the Economy, stupid!') or 'democratic servitude' (Citizens obediently 'Vote!' when summoned so to do by the State) *with a gentle laundering* of the dominant terms: a refreshed capitalism (Now with Merit!), a crisper democracy (Now with Tweets!), etc. They need to be *revolutionized* – blasted apart and replaced (e.g. Badiou, 2008; Dean, 2012; Derrida, 2009). Reading against the grain, then, is a counter-hegemonic *resistance movement* that challenges 'the powers that be' – not only in their rule of the Real, but also in their rule of the Symbolic and the Imaginary which sustains their rule of the Real. Hence the existence of Marxist readings, feminist readings, postcolonial readings, psychoanalytic readings, minoritarian readings, etc. Although do re-read *Reading Theory Now* (Dunne, 2013).

As a tissue of signs, a text is anything with a signifying structure. It is anything that leads one into decoding, exegesis, interpretation, and translation. Quite simply, a text is anything that *refers* meaning – and thereby the reader – elsewhere: to other texts, codes, situations, contexts, languages, expectations, institutions, etc. Needless to say, this process of referral can be prolonged infinitely and extended over innumerable domains, although in practice we tend to arrest this structure of dissemination in order to get things done (see Exercise 14.1).

Exercise 14.1   Signifying Nothing – A Flavour of Moral Geography

Do you believe me when I claim: (1) that a text necessarily refers you elsewhere; (2) that this process of referral is endless; (3) that it will lead you all over everywhere; and (4) that sense, meaning, and action are an interruption, rather than a culmination, of reading a text? No? Then consider this symbol, which the dominant 'powers that be' have let rip through the landscape like a virus: 🚭. Once you have read it, read it again: slowly — very slowly. Now, try to answer the following two questions. To *what* does 🚭 refer? To *what else* does 🚭 refer?

If you need some help, here is my starter for ten. First and foremost, 🚭 refers to 'No Smoking.' Second, it refers to a place and a time for 'no smoking' (the extent and duration of which remain unspecified). Third, it refers itself to you and to me: as readers of signs and users of space; as smokers and non-smokers; and as smoking non-smokers (i.e. passive smokers and free riders) and non-smoking smokers (i.e. 'good' smokers who keep their cigarettes unlit). Fourth, it refers above all to actions: to what people do. Specifically, they *do not* smoke. But 🚭 is much more than a simple statement of fact: *there is* no smoking here. It prescribes a required practice: there *must not* be smoking here. Smoking must be *made absent* and *kept absent* – by you and by me; and when all is said and done (or rather *not done*), *by force*, if necessary. Fifth, it thereby refers to all manner of expectations, rules, and sanctions that are meant to govern the relationship between people and place; to those people who devise, impose, and enforce them; and to the biopolitical sources of trust, expertise, authority, and ideology that legitimate them – such as law, politics, commerce, the media, the health professions, and academia. Sixth, if the sign 🚭 can manage to elicit practices of 'no smoking,' then it actually helps to create a non-smoking space. Seventh, does the prohibition against smoke and smoking occur at the sign, around the sign or from where one reads the sign? Eighth, does 🚭 still function as a sign of prohibition when there is no one there to refrain from smoking, when it appears upside down, when it is obscured by foliage or when it is used as an example of dissemination in a textbook? Ninth, although 🚭 seems to refer to 'no smoking' in general, it actually refers to particular cases of interpellation or hailing: 'Hey! You there' (Althusser, 2001). Finally, it is not entirely clear what 'no smoking' actually entails. Does 🚭 object to smoke, smoking or both? One can hold a lit cigarette without smoking it – as I often do when walking through 'no smoking' areas en route from one smoking place to another (just as people desist from cycling by dismounting and wheeling their bicycles through 'No Cycling' zones) – and one can smoke things other than cigarettes. One can even 'smoke' smokelessly thanks to the advent of newfangled e-cigarettes. In some especially miserable places, moves are afoot to ban this simulation of smoking, not only because the water vapour of e-cigarettes literally clouds the issue of enforcing a ban on *real* smoking, but also because e-cigarettes (like chocolate cigarettes and candy cigarettes before them) allegedly risk 're-normalizing' the consumption and enjoyment of real cigarettes. Much like the repeated rounds of simulation, counter-simulation, and dissimulation that have characterized the antagonistic game of one-upmanship between butter and margarine since the 1870s (Genosko, 2009), perhaps e-cigarettes will be compelled to distinguish their vapours from 'real' cigarette smoke via discolouration and malodourousness (just as cigarette packaging is increasingly called upon to look truly disgusting, much like the woeful packaging of supermarkets' 'value' products might be cynically designed to bring shame upon those consumers who debase themselves by being drawn to their bargain-basement prices – think about who could possibly be seduced by a tin of 'value' baked beans or a carton of 'value' orange juice, for example).

Although my starter-pack of ten is now empty, the explication of 🚭 remains inexhaustible. We will never be finished with the interpretation of texts and signs. We are hooked on reading. We are addicted to enigmatic signs that seduce us (Baudrillard, 1990). Nevertheless, we can gain great insight into how they are socially and spatially structured along the way. So let me restate my original question. Can you think of anything that does not lend itself to interminable interpretation and that does not lead into a labyrinth of spatial analysis? Hereinafter, treat all 'order-words' (which try to clamp everything down to their 'right-and-proper' place) as if they were 'pass-words' (in order to escape and take flight from this clamping down); and every 'world order' (i.e. large-scale clamp down) as if it were a 'world adrift' (Malabou and Derrida, 2004). Read against the grain.

Food, clothes and gadgets signify no less than written words, spoken language and The Human Genome (Barthes, 1993; Maines, 1999; Summers, 2001; Doy, 2002). Consequently, geographers can take virtually any artefact and attempt to draw out the meanings, values, dispositions, desires, knowledge, power relations, and practices that are encoded into it. Accordingly, what matters is not whether the artefact has a prominent place and obvious significance for the culture in question – such as Acts of Parliament, iconic landmarks, and famous works of art – but whether the artefact will enable you to access how that culture exists, experiences, and acts in the world. Think of *the* international language of business: is it English, the US dollar or the computer spreadsheet? Or think of *the* technology that holds our world together: is it telecommunication, synchronized clocks or the humble screw? To put it another way, what would have the greatest impact on our world: the disappearance of Acts of Parliament, iconic landmarks and famous works of art or the absence of money, spreadsheets, clocks and screws?

Since everything can be decoded, everything is a text. Everything opens out onto a social world. But a text is not simply something that purposefully carries a message or a meaning. Rather, a text is anything that *responds* to the call of an interpretative gesture. It is anything that leads someone to suppose – either implicitly or explicitly – that it might mean something. For example, take a landscape. On the one hand, this landscape can be read for the material traces of the various ways of life that have been encoded into it. So a landscape can be treated as the accumulated collective expression of innumerable modifications over time by a host of human and non-human agents: floras, faunas, technologies, political regimes, earth-surface processes, social formations, etc. Many people call this the 'Real.' No doubt you have gazed out from hilltops, studied maps, and processed data on computers in order to make sense of a landscape and extract the truth about its history and geography. On the other hand, a landscape may also function as an enigmatic blank on to which meaning and significance can be projected by different groups of people with very different kinds of interest. As with astrology, numerology and conspiracy theories, meanings and values can be read *into* things. In this way, a landscape can come to signify all kinds of things for all manner of reasons: nature, perfection, order, beauty, desolation, belonging, the sacred, wealth, alienation, eternity, woman, the future, the community, the nation, the People, etc. Many people call this the 'Imaginary' or the 'Symbolic.' Needless to say, while these kinds of representation usually achieve a certain consistency through habitual associations – often to the point of seeming natural, commonsensical, and self-evident to the groups whose interests they serve – they are nevertheless socially constructed and inevitably contested. For what appears to be natural and self-evident invariably turns out to be the product of a long and drawn-out social struggle (Schivelbusch, 1993). Like 'heterosexuality,' 'lifelong learning,' and 'working for a living,' things are invariably natural*ized* rather than natural. In other words, what counts as natural self-evidence is not so much a quality inherent in the world as an outcome of specific social settings. It is a social construct. The city, for example, can be represented as anything whatsoever: monumental, alien, desolate, erotic, chaotic, unruly, sacred, fun, a desert, a jungle, sociable, alienating, fleeting, eternal, fearful, artificial, second nature, a wilderness, an ocean … But these are not just representations without consequence since each elicits certain *practices* that impact upon the city, transforming both its built environment and its social life. By systematically studying these material traces and representations – by reading the real *and* imagined landscape – you will gain an insight into both the practices and the values that have shaped the relationship between people and place (Lefebvre, 1991). (See Exercise 14.2.)

Exercise 14.2   Signifying Self-evidence in Social Space

Look around the room that you are in. Note the people, the creatures, the plants, the objects, the ambience, the textures, the sensations and the dimensions. What do they tell you about who you are, your relationship to others and your place in society? What do they say about your social, cultural, political, economic and sexual status? Take any object whatsoever. Why is it there? Where did it come from? What connections does it establish between you and the outside world? Now think of the room that you are in as an ensemble, an assemblage or a 'machine for living.' What kinds of activity does it enable? And what kinds of activity does it restrict or even preclude? What is the focal point of the room – and what does that tell you about the way of life to which you have become subject and accustomed? Finally, think of three activities and three objects that would be *completely out of place* in this room. Now, imagine the way of life – the 'spatial practices' and the 'spatial representations' – that would make them as self-evidently in place as that focal point around which your entire existence is presently arrayed.

Most people use the word 'discourse' when they want to consider both the representations *and* practices of a particular social group. A discourse is a specific constellation of knowledge and practice through which a way of life is given material expression. It engenders a discourse-specific (i.e. partial and relative) incarnation of the world that tends to become both naturalized and taken for granted. When writers draw attention to these material and immaterial constellations of knowledge and practice, they usually do so in terms of the social and spatial power struggle between 'dominant discourses' on the one hand, and 'discourses of resistance' (or 'dominated discourses') on the other hand. Discourse analysis discloses how this constellation of knowledge and power is structured, and situates it within its appropriate social, cultural, and geo-historical context. For instance, think of the discursive conflict between adults and children, the rich and the poor, the colonizers and the colonized, and environmentalists and capitalists. Each draws upon a very different geographical imagination to frame and envision their take on the world (a framing and envisioning that is necessarily partial and selective, that necessarily 'distributes the sensible' (Rancière, 2004), making some things visible and other things invisible, some things audible and other things inaudible, some things thinkable and other things unthinkable, some things significant and other things insignificant, some things worthy and other things worthless, etc.), and each deploys a very different repertoire of spatial practices that jostle for supremacy.

A wonderful example of such a conflict is provided by Allen and Pryke's (1994) consideration of how foreign-exchange dealers, security guards, caterers, and cleaners inhabit 'the floors of finance' in the City of London. Although they all more or less occupy the same *physical place*, they each live and work in very different *social spaces*. For example, while the foreign-exchange dealers are hooked into the hyperactive 'global space of flows' through their computer screens, telephones, and social networks, the cleaners are expected to disappear without trace within the 'local space of place' after attending to the mundane materials that literally make up the floors of finance: carpet, wood, metal, glass, plastic, marble, etc.

Since the power relations between dominant and dominated discourses are asymmetrical and hierarchical, discourse analysis tends to be critical rather than neutral (see Chapter 36). For example, the discourses of debt, disaster, and resilience cry out for a critical discourse analysis (Dyson, 2006; Giroux, 2006; Klein, 2007; Graeber, 2011; Lazzarato, 2012; Neocleous, 2014). Critical discourse analysis shows not only how dominant discourses prevail and oppress through their powerful regimes of signs, but also how one might resist and overturn them. For instance, deconstruction, schizoanalysis, and forensic architecture take this subversive form of textual analysis in radical and refreshing directions (Derrida, 1988; Guattari, 2011; Weizman, 2011).

In summary, a geographical analysis of cultural texts and competing discourses needs to follow as rigorously as possible the spatial, temporal, and social traces of both real and imagined signifying structures: representations and practices. The expertise that you will need to draw upon will inevitably spin out, not only across the whole of human geography but also across many other disciplines. While this profusion is undoubtedly daunting, you should take comfort from the fact that the ability to make sense of the world is more likely to come from dogged reading than divine inspiration. Perhaps the best advice is to move slowly, to remain alert for possible connections, and to resist the trap of uncritically accepting common sense. So, you should try to enter into a photograph, artefact or document in the same way that you would enter into a film or a novel: they lead to worlds within worlds; social spaces within social spaces. But rather than passively submit to the line that has been laid out for you to follow by a hegemonic discourse, you should attempt to explore and survey systematically this world within a world for yourself. Take a landscape painting, an Act of Parliament or an advert for new housing. Or else a bridge, a nature reserve or a display of lucky thimbles. Ask:

- How, why, and for whom has it been constructed?

- What are the materials, practices and power relationships that are assumed by it and sustained through it?

- What codes, values, dispositions, habits, stereotypes and associations does it draw upon?

- What kind of personal and group identities does it promote? And how do they relate to other identities?

- What does it mean? What are its main structuring devices: oppositions, divisions, metaphors, illustrations, exemplars, etc.? And how do they over-determine and constrain the choice and arrangement of content?

- More importantly, what kind of work does it do? And for the benefit of whom?

- What has been included, excluded, empowered and repressed?

- What has been rendered visible, audible, thinkable, valuable, significant and expressible? And what has been rendered invisible, inaudible, unthinkable, worthless, insignificant and inexpressible?

- How might it be modified, transformed or deconstructed? How could this social space be inhabited differently?

Finally, since nothing ever comes alone, with what other assemblages does it fit and resonate? Are these assemblages synergetic or contradictory? How does it relate to the assemblages that interest, say, social geographers and population geographers? And since the work of contextualization and re-contextualization can be done indefinitely, there is never anything 'final' about following the manifold traces of textual analysis.

A good place for a geographer to start is with a map – not least because most people tend to assume that a well-prepared map, like a photograph, a measurement or a number, tells the truth. However, as with everything else, maps are made to serve particular purposes in specific social, cultural, economic, and political settings. They edit, transform, and remake the world in a way that suits the interests to hand. The 'ground truths' that they express are not only *constructed* truths, but also *partial* truths (Pickles, 2004; Wood and Fels, 2008; Brotton, 2012). This ineluctable partiality is one reason why geographers are forever engaged in *A Search for Common Ground* (Gould and Olsson, 1982) in the absence of *A Ground for Common Search* (Golledge et al., 1988). Both the ground and the common continually elude us, like a vanishing point on the horizon into which the sunny-side of *the* Truth is forever setting.

14.4 Just Reading

So far I have tried to clarify how you should *approach* texts: slowly, attentively, broadly, openly, and with an eye towards the struggles between competing discourses for representational and practical supremacy over social space. However, I have been asked by the editors of this book to do something strange: perhaps even pointless. Since this is a 'how to' book I have been asked to *instruct* you in – of all things – *reading*. But how could you have got to this chapter without already knowing how to read? What concerns readers like you is not learning *how* to read but *whether* to read, how *much* to read, and *what* to read. But you don't need me to address these questions because you have heard the answers repeated ad nauseam by teachers and tutors alike: 'You must read … as much as possible … of what is on your reading lists.' And the difficulties that you will face will probably have less to do with reading per se and more to do with the nature of the writing. Academic texts are notoriously dry, long-winded, boring, jargon-ridden, humourless, pompous, obtuse and turgid.

Now, while I get endless queries from students about how things should be *written* and *presented* – structure, flow, balance, objectivity, referencing, quoting, exemplification, contextualization, and especially length – I cannot recall anyone *ever* asking me how to read. This is odd: not only because many students experience an entirely understandable difficulty in reading 'academic' texts but also because reading in general is actually a demanding and skilled activity that requires training. Indeed, you should study techniques of reading in the same way as you would study any other qualitative or quantitative research technique, and believe it or not, there are countless ways in which to read – few of which boil down to reading word after word for page after page. I will *mention* just a few: hermeneutics, semiotics, psychoanalysis, schizoanalysis, structuralism, deconstruction, discourse analysis, frame analysis, conversational analysis, Marxist literary theory, feminist literary theory, and reader response. For the flavour of things, try some of Icon Book's comic-book series designed 'for beginners.' I especially enjoyed *Introducing Baudrillard*,

Introducing Cultural Studies, *Introducing Derrida*, *Introducing Postmodernism*, and *Introducing Semiotics*. For now, however, I just want to give you a feel for the power of innovative reading strategies (see Exercise 14.3).

Exercise 14.3   A Speedy Literature Review

A group of students have been struggling to perform an onerous but nevertheless important task: a so-called 'literature review.' Basically, rather than write an essay on something or other out there in the world, like the human slaughter industry or the joy of anaerobic digestion, they need to write about how geographers have written about something or other. Fifteen journal articles, several books, and four weeks later the reviews are in. Although they have read a vast amount they still find it hard to get a grip on what academics think about things. Imagine their dismay when I suggested this short cut after the event. Take any textbook – perhaps this one – and go and get a very old textbook on the same subject from the library (say something from the 1990s). Open their indexes and make three lists of words and phrases: (1) those that appear in both; (2) those that only appear in the old textbook; and (3) those that only appear in the new textbook. What are the dominant themes within each of the lists? Now, compare what has been preserved within the subject (list 1), what has been purged from the subject (list 2), and what has been added to the subject (list 3). What does this tell you about the changing interests of geography and geographers? Do these changes amount to progress? If you are really pressed for time, try the same procedure with the contents pages instead. This will give you a sense of how the subject is structured and the relative importance of various issues and debates.

When faced with the inconvenience of long indexes, feel free to restrict your textual analysis to a workable range of letters: say M–R. For those with a quantitative bent, try drawing inferences about the population of the entire index from this sample, jot down how confident you are about your inferences, and then compare your expected results with what is actually the case. This should alert you to the fact that quantitative analysis is a particular way of thinking rather than a fixation on numbers per se. If you extrapolate this way of thinking to vast amounts of textual data – such as Hansard's verbatim report of the proceedings of the UK's House of Commons and House of Lords, or William Shakespeare's entire *œuvre* (including the recipes), or every use of the phrase 'I feel' on social media over the last three months — then one can subject these 'populations' as a whole (these 'universes of discourse'), or 'samples' thereof (so-called 'corpora'), to the usual array of descriptive and inferential statistics in order to tease-out significant patterns in the data. (For example, given the feelings expressed on social media, where in the world do young women 'feel' especially happy or lonely or enraged?) This study of language through 'real-world' samples of text is called Corpus Linguistics, and it is flourishing in an environment characterized by an explosion of electronic data and the ability of computer applications such as WordSmith (http://lexically. net/wordsmith; Windows) to digest them.

14.5 Mind that child!

Let me bring this chapter to a close with an anecdote and a checklist. Apart from a few advertising slogans, the only phrase that I can remember from my childhood is: 'Mind that child!' It was written on the back of the ice-cream vans that toured the streets where I lived. Although there was nothing difficult about the words, I always felt unnerved by them. On every occasion I encountered the phrase, the anticipated pleasure of eating ice cream was tainted by anxiety over these enigmatic and sinister words. Now, believe it or not, it was only a couple of years ago that this childhood association between ice-cream vans and foreboding dissipated. I was about to overtake an ice-cream van when I noticed those words: 'Mind that child!' Suddenly, I realized that the words were addressed to *adults* rather than to children. They simply warned drivers not to run over children distracted by the pleasures of ice cream, rather than to warn *children* about the dangers of 'that!' – something so terrible that adults dare not even write its name. I had assumed that the statement 'Child: mind *that*!' was meant to warn me of evil car drivers who routinely ran children over (this was the moral panic par excellence of my infant years: soon to be followed by the horrors of sexually transmitted diseases, heroin abuse, and glue-sniffing; nuclear war was already passé, terrorism was still comedic, and fundamentalism didn't exist), but it always struck me as odd that one actually needed to be *in* the road in order to be able to read the warning – as if the whole thing were a terrible ruse to expedite the killing of children. Years later I would be reminded of these deadly ice-cream vans when I read about the Nazis' highly effective discourse of dissimulation: the slogan 'Arbeit macht frei' (Work makes free) appeared above the gates to the Auschwitz-Birkenau extermination camp; its gas chambers were presented as if they were showers; and the conversion of vans into mobile gas chambers at Chelmno and elsewhere were simply referred to as 'special vehicles' and often made to resemble furniture-removal vans (Lanzmann, 1985; Friedlander, 1995). The comeuppance is clear. When you start to analyse a text, do not take its message at face value or assume that it was meant for people like you. It almost certainly wasn't. Once you have worked out for whom the text was produced you will have gone a long way to reading it effectively.

So, here is a final checklist to get you going. Investigate *who* produced the text, *why* they produced it, *how* they produced it, and for *whom* they produced it (Du Gay et al., 1997). Sometimes all of this will have been consciously intended. On other occasions, however, it will have been unconscious and unintended. Investigate the *form*, *content* and *assumptions* of the text in question, paying just as much attention to what is *absent* from the text as to what is present. Set the text in appropriate *contexts*, and compare and contrast it with other relevant material: other texts, lifestyles, belief systems, practices, artefacts, etc. Investigate how they have been used and abused as mechanisms for articulating *power* and *resistance* by a range of people and groups. But above all, investigate what *work* these texts do: how do they impact upon and affect society and space? Once you have done that you can begin to approach the analysis of texts according to a wide variety of criteria that may interest you: power, knowledge, desire, truth, fidelity, meaning, exclusion, class, race, gender, sexuality, etc. And if anyone mocks your scholarly and geographical interest in *reading* texts, remember the story of Christopher Columbus's egg. In reply to a suggestion that *anyone* could have discovered America, Columbus challenged the guests at a banquet held in his honour to make an egg stand on end. When all had failed, Columbus did so by flattening one end with a sharp tap on the table, thus demonstrating that while others might follow, he had discovered the way.

SUMMARY

- 'Text' and 'textuality' have become key terms in geography, largely because of the 'cultural turn' in recent decades.
- The world is full of texts – texts that will lead you all over everywhere.
- Texts matter because they shape, inform, and constrain spatial practices and social spaces.

- It is important to distinguish between the producers and consumers of texts, and the form and content of texts.
- Power relations are encrypted into texts, and they invariably masquerade as being normal, natural and self-evident.
- 'Just reading' may seem easy. But reading *justly* is far from straightforward, especially when dominant discourses pervert our imaginary and warp our reality.
- There are many different kinds of textual analysis (e.g. discourse analysis, corpus linguistics, semiotics, schizoanalysis, and deconstruction).
- So, take time to read – and to learn how to read: critically, justly and against the grain of common (i.e. accepted, dominant and enforced) sense.

Note

1   It would be folly to try to explain the 'tyranny of the Signifier' here, but suffice to say that the Signifier is to sense and sensation, as Capital is to our life and our world. Everything must now submit to the power of the Signifier, just as everything must now submit to the power of Capital. Everything *must* 'mean' something, just as everything *must* be 'worth' something – even if that meaning is nonsense and that value is naught (Rotman, 1983). For us, the sun no longer sets on the empire of the Signifier, just as it no longer sets on the empire of Capital: 'From the moment that there is meaning there are nothing but signs. We *think only in signs*' (Derrida, 1997: 50). The empire of meaning, like the empire of money, has become fully globalized. The 'tyranny of the Signifier' is the key insight of structuralism (e.g. Ferdinand de Saussure's linguistics and Jacques Lacan's psychoanalysis), and the *turn away* from its despotism is the driving force behind post-structuralism, especially for deconstruction (Derrida, 1988), schizoanalysis (Guattari, 2011), symbolic exchange (Baudrillard, 1993), and other 'drift-works' (Lyotard, 1984, 2011) (Harland, 1987; Dosse, 1997; Dews, 2007; Howarth, 2013).

Further Reading

This guide includes an eclectic selection of examples of different forms of textual analysis:

- Blonsky (1985). A famous collection of essays, all devoted to the analysis of 'signs.'
- Darnton (2010). A wonderful account of the subversive power of 'viral' communication networks in a surveillance society long before our Internet age – eighteenth-century Paris.
- Du Gay et al. (1997). Everything you ever wanted to know about how the 'turn to culture' impinges on our lives using the once newfangled Sony Walkman as an example.
- Lazzarato (2014). The key production process for capitalism is the production of subjectivity, which is accomplished through a 'regime of signs' that ensures our social subjection and 'machinic enslavement' – we are the indebted subjects of finance capital.
- Perec (1999). An extraordinary writer, famed for his legendary novel, *Life: A User's Manual* (2008), crafts a glorious account of innumerable 'species' of spaces (from the Page, via the Bedroom and the Apartment, to the Country and the World). Question your teaspoons!
- Rybczynski (2000), in attempting to identify the most significant invention of the past millennium, uncovers the mind-boggling secret history of the screwdriver and screw. This is a glorious example of why one should never take anything for granted.
- Schivelbusch (1993) presents an astonishing world of social conflict and cultural struggle that surrounded the introduction of pepper, coffee, chocolate, tobacco and opiates into Europe. It is a remarkable study of the modernization and industrialization of culture.

Note: Full details of the above can be found in the references listed below.

References

@149 St. New York City Cyber Bench. http://www.at149st.com (accessed 11 December 2015).

Ahrens, J. and Meteling, A. (eds) (2010) Comics and the City: Urban Space in Print, Picture and Sequence. London: Continuum.

Allen, J. and Pryke, M. (1994) 'The production of service space,' Environment and Planning D: Society and Space, 12: 453–76.

Althusser, L. (2001) Lenin and Philosophy and Other Essays. New York: Monthly Review Press.

Amin, A. and Thrift, N. (eds) (2004) The Blackwell Cultural Economy Reader. Oxford: Blackwell.

Attali, J. (1985) Noise: The Political Economy of Music. Minneapolis, MA: University of Minnesota Press.

Badiou, A. (2008) The Meaning of Sarkozy. London: Verso.

Baird, J. A. and Taylor, C. (eds) (2011) Ancient Graffiti in Context. Abingdon: Routledge.

Banksy. (2006) Banksy: Wall and Piece. London: Century.

Barthes, R. (1993) Mythologies. London: Vintage.

Bartky, I R. (2000) Selling the True Time: Nineteenth-Century Timekeeping in America. Stanford, CA: Stanford University Press.

Baudrillard, J. (1990) Seduction. London: Macmillan.

Baudrillard, J. (1993) Symbolic Exchange and Death. London: Sage.

Baudrillard, J. (1996) The System of Objects. London: Verso.

Baudrillard, J. (1998) The Consumer Society: Myths and Structures. London: Sage.

Blonsky, M. (ed.) (1985) On Signs. Baltimore, MY: Johns Hopkins University Press.

Brotton, J. (2012) A History of the World in Twelve Maps. London: Allen Lane.

Clarke, D. B. (ed.) (1997) The Cinematic City. London: Routledge.

Cloke, P., Crang, P. and Goodwin, M. (eds) (2014) Introducing Human Geographies (3rd edition). Abingdon: Routledge.

Cook, I., Crouch, D., Naylor, S. and Ryan, J. (eds) (2000) Cultural Turns/Geographical Turns: Perspectives on Cultural Geography. Harlow: Prentice Hall.

Coover, R. (1998) Ghost Town. New York: Henry Holt.

Darnton, R. (2010) Poetry and the Police: Communication Networks in Eighteenth-Century Paris. Cambridge, MA: Harvard University Press.

Dean, J. (2012) The Communist Horizon. London: Verso.

De Certeau, M. (1985) 'Practices of space,' in M. Blonsky (ed.) On Signs. Baltimore, MY: Johns Hopkins University Press. pp. 122–45.

Deleuze, G. and Guattari, F. (1988) A Thousand Plateaus: Capitalism and Schizophrenia. London: Athlone.

Deleuze, G. and Guattari, F. (1984) Anti-Oedipus: Capitalism and Schizophrenia. London: Athlone.

Derrida, J. (1988) Limited Inc. Evanston, IL: Northwestern University Press.

Derrida, J. (1997) Of Grammatology. Baltimore, MY: Johns Hopkins University Press.

Derrida, J. (2009) The Beast and the Sovereign. Chicago, IL: Chicago University Press.

Dews, P. (2007) Logics of Disintegration: Post-Structuralist Thought and the Claims of Critical Theory. London: Verso.

Dittmer, J. (ed.) (2014) Comic Book Geographies. Stuttgart: Franz Steiner.

Dosse, F. (1997) History of Structuralism, two volumes. Minneapolis, MA: University of Minnesota Press.

Doy, G. (2002) Drapery: Classicism and Barbarism in Visual Culture. London: I. B. Tauris.

Du Gay, P., Hall, S., Jones, L., Mackay, H. and Negus, H. (1997) Doing Cultural Studies: The Story of the Sony Walkman. Milton Keynes: Open University Press.

Dunne, É. (2013) Reading Theory Now: An ABC of Good Reading with J. Hillis Miller. London: Bloomsbury.

Dyson, M. E. (2006) Come Hell or High Water: Hurricane Katrina and the Color of Disaster. New York: Basic Civitas.

Eagleton, T. (1986) Reading Against the Grain: Essays 1975–1985. London: Verso.

Fleischman, R. K., Funnell, W. and Walker, S. P. (eds) (2013) Critical Histories of Accounting: Sinister Inscriptions in the Modern Era. Abingdon: Routledge.

Foucault, M. (2008) This is Not a Pipe. Berkeley, CA: University of California Press.

Friedlander, H. (1995) The Origins of Nazi Genocide: From Euthanasia to the Final Solution. London: North Carolina University Press.

Genosko, G. (2009) 'Better than butter: Margarine and simulation,' in D. B. Clarke, M. A. Doel, W. Merrin and R. G. Smith (eds) *Fatal Theories*. Abingdon: Routledge. pp. 83–90.

Giroux, H. A. (2006) *Stormy Weather: Hurricane Katrina and the Politics of Disposability*. Boulder, CO: Paradigm.

Goldsmith, K. (2011) *Uncreative Writing: Managing Language in the Digital Age*. New York: Columbia University Press.

Golledge, R., Couclelis, H. and Gould, P. (eds) (1988) *A Ground for Common Search*. Santa Barbara, CA: Santa Barbara Geographical Press.

Gould, P. and Olsson, G. (eds) (1982) *A Search for Common Ground*. London: Pion.

Graeber, D. (2011) *Debt: The First 5,000 Years*. New York: Melville House.

Guattari, F. (2011) *The Machinic Unconscious: Essays in Schizoanalysis*. Los Angeles, CA: Semiotext(e).

Hamblyn, R. (2001) *The Invention of Clouds: How an Amateur Meteorologist Forged the Language of the Skies*. London: Picador.

Harland, R. (1987) *Superstructuralism: The Philosophy of Structuralism and Post-Structuralism*. York: Methuen.

Hochschild, A. (2006) *King Leopold's Ghost: A Story of Greed, Terror and Heroism*. London: Pan.

Horton, J. and Kraftl, P. (2014) *Cultural Geographies: An Introduction*. Abingdon: Routledge.

Howarth, D. R. (2013) *Poststructuralism and After: Structure, Subjectivity and Power*. London: Palgrave Macmillan.

Klein, N. (2007) *The Shock Doctrine: The Rise of Disaster Capitalism*. London: Penguin.

Lanzmann, C. (1985) *Shoah: An Oral History of the Holocaust*. New York: Pantheon.

Lazzarato, M. (2012) *The Making of the Indebted Man: Essay on the Neoliberal Condition*. Los Angeles, CA: Semiotext(e).

Lazzarato, M. (2014) *Signs and Machines: Capitalism and the Production of Subjectivity*. Los Angeles, CA: Semiotext(e).

Lefebvre, H. (1991) *The Production of Space*. Oxford: Blackwell.

Lewisohn, C. (2011) *Abstract Graffiti*. London: Merrell.

Lyotard, J.-F. (1984) *Driftworks*. New York: Semiotext(e).

Lyotard, J.-F. (1998) *The Assassination of Experience by Painting, Monory*. London: Black Dog.

Lyotard, J.-F. (2011) *Discourse, Figure*. Minneapolis, MA: University of Minnesota Press.

Maines, S. (1999) *The Technology of Orgasm: 'Hysteria,' the Vibrator, and Women's Sexual Satisfaction*. Baltimore, MY: Johns Hopkins University Press.

Malabou, C. and Derrida, J. (2004) *Counterpath: Traveling with Jacques Derrida*. Stanford, CA: Stanford University Press.

Moretti, F. (1999) *Atlas of the European Novel, 1800–1900*. London: Verso.

Neocleous, M. (2014) *War Power, Police Power*. Edinburgh: Edinburgh University Press.

Noys, B. (2014) *Malign Velocities: Accelerationism and Capitalism*. Alresford: Zero Books.

Olsson, G. (2007) *Abysmal: A Critique of Cartographic Reason*. Chicago, IL: Chicago University Press.

Pakenham, T. (1991) *The Scramble for Africa*. London: Weidenfeld & Nicholson.

Perec, G. (1999) *Species of Spaces and Other Pieces* (revised edition). London: Penguin.

Perec, G. (2008) *Life: A User's Manual*. London: Vintage.

Perec, G. (2010) *An Attempt at Exhausting a Place in Paris*. Cambridge, MA: Wakefield Press.

Pickles, J. (2004) *A History of Spaces: Cartographic Reason, Mapping and the Geo-Coded World*. London: Routledge.

Pratt, G. and San Juan, R. M. (2014) *Film and Urban Space: Critical Possibilities*. Edinburgh: Edinburgh University Press.

Rancière, J. (2004) *The Politics of Aesthetics: The Distribution of the Sensible*. London: Continuum.

Rotman, B. (1993) *Signifying Nothing: The Semiotics of Zero*. Stanford, CA: Stanford University Press.

Rybczynski, W. (2000) *One Good Turn: A Natural History of the Screwdriver and the Screw*. New York: Simon & Schuster.

Schivelbusch, W. (1993) *Tastes of Paradise: A Social History of Spices, Stimulants, and Intoxicants*. New York: Vintage.

Summers, L. (2001) *Bound to Please: A History of the Victorian Corset*. Oxford: Berg.

Weizman, E. (2007) *Hollow Land: Israel's Architecture of Occupation*. London: Verso.

Weizman, E. (2011) *The Least of All Possible Evils: Humanitarian Violence from Arendt to Gaza*. London: Verso.

Wood, D. and Fels, J. (2008) *The Nature of Maps: Cartographic Constructions of the Natural World*. Chicago, IL: Chicago University Press.

ON THE COMPANION WEBSITE…

Visit **https://study.sagepub.com/keymethods3e** for author videos, chapter exercises, resources and links, plus **free** access to the following recommended articles:

1. **Crang, M. (2005) 'Qualitative methods: There is nothing outside the text?',** *Progress in Human Geography*, **29(2): 225–33.**

Human Geography is largely made up of words (or squiggles, at least: signs, symbols, lines, shading, numbers, etc.), to which we entrust our truths, our sense, and our wisdom. The clue is in the title, as they say: geo-graphy or 'earth-writing.' But what happens when our trust in words is shaken to the core? And what happens when our qualitative methods try to reach out beyond words? When a photograph, for example, speaks louder than words, is there any way of escaping our prison-house of language? Aren't all of those other media and sensations little more than 'in other words'?

2. **Richardson, D.M. and Pyšek, P. (2006) 'Plant invasions: Merging the concepts of species invasiveness and community invasibility',** *Progress in Physical Geography*, **30(3): 409–31.**

Gregory Bateson's warning that 'there is an ecology of bad ideas, just as there is an ecology of weeds' is perfectly attuned to the way in which a truly noxious form of 'war-talk' is seeping into the political unconscious of our increasingly anxious age – the world-wide cult of 'resilience' being a prime example. The cultivation of war-talk as an ecology of bad ideas is beautifully illustrated by the discourse of 'invasion ecology,' with its notions of 'biological invasion,' 'invasional meltdown,' 'alien species,' and 'sleeper weeds.' The political unconscious of Physical Geography should invariably trouble and disturb Human Geographers.

3. **Caquard, S. (2013) 'Cartography I: Mapping narrative cartography',** *Progress in Human Geography*, **37(1): 135–44.**

Obviously, maps do so much more than merely re-present our vested interests in the world. (And by only ever re-presenting selective interests, they necessarily mis-represent by way of omission. They omit what lacks interest and they omit interests other than their own. And most interests tend to wane with time – except for the interests of the State, perhaps; and the interests of property, of course.) Maps make claims, make sense, and make do. One important thing that they do is 'tell tales,' and until very recently their fairly immutable form – being literally imprinted onto a material fabric, such as cloth or paper – meant that those tales had the implicit form: 'Once upon a time, long, long ago.' Most maps seem timeless, as if those pockmarks strewn across its surface were eternal – the main roads and battle sites, the pylons and the sewage works, the viewpoints and the railway stations, even the ruins and the scree. Boundaries, like all good fairy tales, especially tend to endure. But with the advent of digital and dynamic mapping, and the dissolution of mapping into the multi-media banality of everyday life, what new tales will they tell and whose interests will they serve?

15 Interpreting the Visual

Liz Roberts

SYNOPSIS

This chapter looks at the various ways geographers study visual images, as representations and as objects that people engage with in different ways. It shows how geographers are interested in images for different reasons – for what a painting says about a culture in a historical period of time, for how people use photographs of their family to create a sense of 'home' – and how they use different methods to find out about these. These methods include studying the representational content of images, asking people about how they produced or viewed an image, giving personal accounts or observing people looking at images. Geographers are, first and foremost, concerned with the different social and spatial effects of visual images.

Some key theoretical terms and debates that have been very influential to studying the visual image are introduced. These include:

- The idea of 'image as text' whereby images comprise a language of signs that can be decoded by the reader.
- The idea of 'intertextuality' – how meaning is shaped through relation to other texts, and how we interpret signs differently and from particular social positions.
- How images are products of their society. They are produced through – and reproduce – common ideas, knowledges and social relations held at the time of their production.
- Images as material objects – they 'do' things and have real effects. People use images to do things.
- How, when people interpret an image, they bring to it their knowledge and cultural references, as well as how images also engage the body. Viewing is embodied, affective and multi-sensory.

This chapter is organized into the following sections:

- Introduction
- Visual images as cultural texts
- Production and consumption
- Embodied and affective encounters with visual images
- Conclusion

15.1 Introduction

Geographers are required to analyse a range of visual images in their research, such as maps, paintings, photographs and graffiti, and in films, advertising, public art, websites and video games. There are many different approaches to the interpretation of these varied visual images, most of them informed by philosophical writing about vision, representation, subjectivity and the body. This chapter will focus on three important theoretical strands of visual theory and their associated methods, with examples of how these are being used by geographers. The first section looks at the image as 'text', something that can be 'read' or 'decoded' through understanding its signs, what they signify, and how they relate to other cultural texts. This has been important for geographers' understanding of landscape. The second looks at Gillian Rose's 'critical visual methodology', which considers the social effects of images through the sites of production, the content of the image, and consumption. This stresses the conditions that make viewing possible, the audience and the context of viewing as important to the image's interpretation. The last section looks at more recent geographical work that suggests visual image interpretation should be supplemented by consideration of non-representational and bodily aspects of viewing, particularly with more interactive visual media such as video games. This work is concerned with the centrality of the body to how we understand ourselves and encounter the world. The chapter concludes with some points on the ethics of interpreting visual images and related implications for interpretive methods such as our reflexivity: how we as geographers can reflect and be transparent about our own role (thinking about how our bodies, our identities and our predispositions come into play) in the interpretative approaches we use.

These three strands illustrate how geographers take images seriously, thinking about them as both shaped by the cultures in which they are produced, and as having real effects in the world rather than being simple reflections or representations of reality. In this way, they examine what images 'do' as well as what they 'say' about the world and highlight that images can reproduce and create social relations such as power inequalities both through the way they represent different people and places, and through the meanings they help to create about them. Their particular power lies in the way that they can make these meanings appear natural and pre-given. An example of this is the type of racial and gendered stereotypes reproduced in many films and video games. These simultaneously reproduce and naturalise existing social inequalities through the 'reality effect' they create. Much recent work in geography seeks to draw attention to the highly constructed nature of images like these, as well as more complex political ideas, through pointing out how they can be understood as part of wider discourses or ideologies in society. On the one hand, then, visual images help to reinforce the 'status quo' because of their referential status – they are commonly understood to transparently reflect the real world; on the other hand, they can also subvert such dominant meanings and act as forms of resistance. Some geographical work has focused on new forms of participatory visual art projects that have a political intention to do just this. For example, Loopmans et al. (2012) discuss the effects of two community-based photographic projects to subvert the negative meanings and representations of the deprived neighbourhoods depicted.

There are, then, multiple ways to think about visual images and explore different geographical concerns. Geographers use a range of theories and methods. However, they often do not focus on the visual methods they are using explicitly, and they do not always reflect on the implications for analysis of the methods they are choosing, or on their own role in the interpretation. It is therefore difficult to put forward an exact method that geographers are using to interpret visual images. They tend to use those ideas and approaches most appropriate to the images they are analysing. As Rose says, current interpretive approaches often encompass 'a mix of discourse analysis and semiology, with perhaps a dash of Lacan or Deleuze' (2012: 191), suggesting that interpretation is often a mixed bag of theories and methods. Rose's approach instead develops a more systematic methodology for social scientists to engage with visual images. However, she confesses herself, that rather than guidelines or rules, this approach instead raises a number of questions to be considered. In line with this, this chapter will raise critical questions for interpretation at the end of each section.

Before moving on to the first strand it is worth outlining some of the ways geographers are engaging with visual images and the reasons they are doing this. Rather than trying to give advice on how to choose images for analysis, which will depend on your research topic and questions, here examples are given of the types of visual images geographers choose and the methods they take to analyse them. Some geographers are interested in the geopolitical resonance of visual images, examining the techniques used in film and videogames that reproduce or subvert familiar scripts about racial or ethnic 'others' in simulated war contexts (Carter and McCormack, 2006: 228; Schwartz, 2006). Others are interested in the way important geographical topics are represented in different forms of visual image, for example counting the different types of images used in articles about climate change to understand how the issue is represented in news media (DiFrancesco and Young, 2011) or exploring different collaborative and participatory arts-science projects that challenge and re-think climate change science (Gabys and Yusoff, 2012; Gibbs, 2014). It should be clear from these few illustrations that there are many different reasons that geographers engage with visual images, and that their methods are varied and appropriate to their research concerns. Foremost, geographers are concerned to stress the real and powerful effects of visual images.

15.2 Visual Images as Cultural Texts

The power of imagery is clearly evident in how geographers interpret landscape. As part of the 'cultural turn' in the discipline, culture became increasingly important to how geographers understood landscape. The study of landscape shifted from the physical, material form of geographical areas to incorporate its representation in paintings, film, literary texts and photographs, as well as the 'desired, remembered and somatic spaces of the imagination and the senses' (Cosgrove, 2003: 249). Landscape was not 'out there,' distinct from the person seeing it, but also resided 'in the minds and eyes of beholders' (Cresswell, 2003). This made vision the primary way to 'get at' landscape, and meant it was necessary to think about how, exactly, we see landscape. How we see landscape was recognised as not just the biological process of vision. Cultural, imaginative and symbolic aspects were part of this process. The term 'visuality' was used to describe this difference. It referred to the way that how we are able to see is socially conditioned, determined by social norms and ideas. In landscape studies, this meant that if landscape could be understood as 'a social and cultural product, a way of seeing projected onto the land' (Cosgrove, 1984: 269) then it could also be read and decoded like a painting using iconographic approaches – in other words, land-scape as the study of the content of an image.

One way that this reading and coding can be approached is through Semiology, the study of signs in an image. Semiology, or semiotics, is also used to analyse paintings, advertisements, art, photography, TV, film, and web media. It works from the principle that images comprise of 'signs' – different components in the image – and that these carry meaning. Signs are a unit of language. Signs consist of both the 'signified' (the concept or object) and 'signifier' (word or image), with the 'referent' being the actual thing the sign relates to (Barthes, 1977; Rose, 2012) (see Figure 15.1).

The signifier does not have to resemble the signified, it just has to be recognised as carrying that meaning, e.g. the word 'dog' does not look like an actual dog and a four-leaf clover only arbitrarily symbolises luck (for further explanation see also Chandler, 2014). Signs, then, have a 'denotative' and 'connotative' meaning: they describe something denotatively and they have a culturally determined connotative meaning. Nowadays, social scientists recognise that there is no fixed link between the signs (the recognisable components in an image and their meanings) and their referents (their related objects in the world). Rather than inherent in an object or its representation, meanings were fixed and anchored only through the cultural systems of the interpreter. Meaning is determined through how they relate to other texts that the viewer is aware of. This is called 'intertextuality,' and refers to the meanings that visual images reference and recall for the viewer from other cultural texts.

To give an example, in order for us to make meaningful the story of the recent Disney-Pixar film *Brave* (2012) it helps for us to be aware of – to have in our 'inter-pretative repertoire' – previous Disney films and the trope of the 'Disney princess' and the romantic narratives that this entails – featuring the climactic conclusion in the form of a wedding – to help us recognise that *Brave's* Merida, as shown in Figure 15.2, with her wild ginger hair, active stance, and her bow and arrows indicat-ing that she can defend herself, is not a typical heroine. Unusually, she is fighting for her independence throughout the film, i.e. to remain unmarried/choose her own

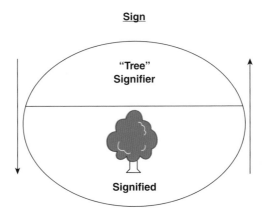

Figure 15.1   The components of a sign in its simplest form

husband. It also helps to have a knowledge of Scottish history and related cultural texts, to help us understand several visual puns in the film such as embarrassing incidents related to wearing kilts, for example, which subvert the stereotypical hyper-masculine Scottish hero who is courageous and at one with nature (for a video clip see https://www.youtube.com/watch?v=sKg_KVmvAFY). This particular image is reinforced in iconic films like *Braveheart* (1995) and *Rob Roy* (1995) and has been visually and symbolically referenced so prolifically across different cultural forms that likely the viewer does not need to have seen the original films for the types of meanings reinforced in them to permeate their interpretation of *Brave*. It is impor-tant to note, however, that these signifiers might mean different things for different audiences: an American audience might see a romantic, wild, almost mystical Scottish landscape, whilst a Scottish viewer might be offended by the (albeit well- and humuorously-intended) cultural stereotypes reproduced within the film landscape (actually, the extended run of *Brave* in a great deal of Scottish cinemas is suggestive that it was very well received by Scottish viewers).

Geographers are interested in the meaning created in visual images, through its signs and how they relate to other texts. Exploring the image's intertextuality in this way enables the analyst to say something about the broader discourses prevalent in society. It becomes possible to situate visual images as emerging out of their socio-political contexts, by linking common themes or visual motifs across other images and texts. DiFrancesco and Young (2011), for example, compare newspaper article texts with the images that accompany them to analyse the consistency of Canadian news media dis-courses on climate change, whilst historical geographers might use archived texts like letters or newspapers and cultural texts like contemporary novels to explore wider discourses circulating around historical visual images. Studying signs can tell us inter-esting things about how an advertiser is trying to sell a product or can reveal something about important ideas or social relationships at the time the visual image was pro-duced. Some geographers have argued that cultural representations in landscape

Figure 15.2   Merida from Disney Pixar's *Brave* (2012)

painting powerfully masked, aestheticized and erased unequal social relations by privileging one way of seeing, such as the class-based and gender hierarchies in 18th and 19th century British landscape painting, painted for and representing the perspective of the upper classes (Berger, 1972; Rose, 1993). Such images helped to reinforce ideologies, the sets of normative ideas that support relationships of dominance (Rose, 2012) such as the relationship between landowners and the workers on the land, or husband and wife at that time. Visual imagery of landscape, therefore, has the appearance of representing a prior reality, yet also powerfully constitutes that reality, shaping the ways we see and understand landscape. One task of geographers is to peel back and reveal the hidden social truths in landscape representations, through looking not only at the content of the image but also at the wider socio-historical practices at the time of its production.

Interpreting landscape as text that can be decoded reflects the ways in which the cultural, symbolic and imaginative aspects of vision inform how we see not only landscape, but also any visual image. Visual images can help to reproduce unequal power relations and so studying them can help to reveal these. Geographers can do this by interpreting the different aspects of the content of visual images such as representational features like their framing, perspective and positioning, as well as the signs and symbols in it. When interpreting the content of the image, then, we need to think about:

- What is the purpose of the image?

- What type of media is used – painting, mural, photography? How does this make a difference to the meanings created?

- What are the main components in the image? Which components (or signs) are privileged over others and why?

- What do they denote and connote?

- Which other texts are referenced in the image?

- What type of ideologies or stereotypes are reinforced or challenged?

Geographers have asked these sorts of questions not just about landscape painting, but also about public art sculptures, films, advertising, cartoons and newspaper illustrations, and magazine photographs. This approach has been criticised for only thinking about images (landscapes) as representations, only worth studying for what they 'say' rather than as objects that 'do' things.

15.3 Production and Consumption

Thinking about visual images as texts that can be read is nowadays only one part of studying visual images. Gillian Rose's (2012) 'critical visual methodology' asks geographers to think about the social effects of images through three sites: the site of production; the site of the image (its content); and the site of reception (its audience). Likewise, Hawkins writes about geography and visual art in terms of not only what an artwork 'means' but also how this is related to the context of its making and its audience:

...from 18th-century landscape painting to contemporary urban public art projects, and studies of photographs and other visual cultural materials, research proceeds from the tensions between the sites of art's production and consumption and those it represents or evokes. (Hawkins, 2013: 57)

For geographers, these interconnected sites are crucial to the interpretation of the visual image. Rose's methodology entails three criteria: acknowledging images have their own effects; that images have social conditions; and that studying them requires reflexivity by the viewer-interpreter, which means thinking about how we ourselves view images. Further to the three sites of the image, she is interested in the technologies and material qualities of an image as well as the social modalities – 'the range of economic, social and political relations, institutions and practices that surround an image and through which it is used and seen' (2012: 20). This framework is offered as a way to structure the broad and varied visual theories and methods available.

Not all geographers have engaged with this framework directly, although most take up one or two of the sites and criteria in their explorations of visual images. Having focused on the site of the image (its content) in the previous section, this chapter now turns to the sites of production and reception. It highlights the materiality of images, the importance of the process of making images and what they 'do', and finally considers how geographers are incorporating the audience into their analysis by acknowledging that the audience is not homogeneous – we are not all the same and we do not interpret images in exactly the same way.

Geographers are concerned with a number of aspects of the production of visual images. Film geographers explore the production locations, economies and spatialities of film industries, whilst studies of the video game industry have highlighted the importance of cultural differences in emerging industries and products (Izushi and Aoyama, 2006; Johns, 2006). Geographers working with visual arts focus on the process of creating visual images, finding it as important as the end product. They might interview or spend time with the creator(s) of the visual image, they might report on the piece's wider reception by critics etc. (Morris and Cant, 2006; Hawkins 2010) or take part in participatory art projects and interview participants (Mackenzie, 2004; Burk, 2006). Often they describe the content of the created visual or performance art piece in conjunction with these other methods. Importantly, geographical studies (or collaborations) with art are less concerned about the content of the visual image and think instead about the therapeutic, empowering, sensory and social aspects of making and experiencing art (Macpherson, 2008; Sharp, 2007). Hawkins (2013: 56) refers to this as a shift to thinking about the site of art as 'the "social site", including its community, as subject, material and audience'. Here, it is difficult to separate the site of production from that of consumption; the viewer does not gaze at the art-object from a distance (as is the traditional conception of art appreciation) but is actually part of the artwork; as they bring their own knowledge, memories and specificities to bear on it they produce the art. The viewer becomes an active participant in the artwork. This idea of the active viewer is evident in much geographical work on the audience of visual images.

Work on the site of consumption of the visual image seeks to both engage with different audiences of the image and to think more fully about the image beyond its visual content; they are concerned with what images 'do' as material objects, how people view, display and use them; and the role they play in people's lives, identities and social (e.g. family) relations. Images are displayed not only in galleries and cinemas, but also in peoples' homes, photo-albums and on different visual technologies that have a material status as physical objects in the world. More often than not, these days images are viewed on a screen, be it a TV, mobile phone or computer screen. This means the types of images viewed are either purely digital creations or are transformed by appearing on the screen, and because they can be interacted with or manipulated by a viewer/user (particularly in video games – discussed below). In what is referred to as 'convergence culture', images can now be viewed across a number of media, and so whether the meaning or reception of these images changes across them is also considered important (Rose, 2012). For example, they are often viewed at the same time as other images and texts within the screen so that their content is not the centre of the viewer's attention (see Figure 15.3). It is worth considering the different effects of images encountered through various material forms and screens/technologies and how the experience of viewing them and their meanings change.

To understand visual images as material objects, geographers have used interview-based and ethnographic methods to grasp exactly what the social practices related to different visual images consist of. Geographers have used interviews and time spent in family homes to establish how displayed photographs help construct family identities, act as 'prosthetic memories' that support or disrupt families' collective memories, and form part of visual-material displays in the homes of second- and third- generation migrants that constitute a sense of belonging to place (Rose, 2003; Tolia-Kelly, 2004; Roberts, 2012). Although not centrally concerned with the visual image, Spinney et al. (2012) conduct research on laptop use, one of the screens (alongside the growth in tablets and smartphones) through which visual images are increasingly viewed. Their research considers the 'always-on-ness' of the laptop, the way it is mobile and how users interact with laptops, that is bound up with the reproduction of specific domestic imaginaries of family and home, such as how people think they should behave in different rooms in the house. Their approach incorporates interviews, participant

Figure 15.3   Convergence of images through the screen

(*Source*: author)

diaries about laptop-use, and researcher notes from home visits. Importantly, a 'visual objects' approach acknowledges that images do things and have effects separate to their representational content.

A contrasting way that geographers explore the site of audiencing is through speaking to and observing different audiences. 'Audience studies' is predominantly used for film and TV and is often used in marketing. It became an increasingly important approach in light of growing recognition that the intended meaning of visual images was not always the meaning interpreted by the viewer. Different viewers interpret images differently from each other. Some accept the dominant meanings in a TV show, whilst others resist them. A great deal of writing about images involves the analyst giving themselves the job of decoding the image, highlighting the dominant or preferred meanings reproduced through its signs and establishing a resistant reading, revealing the real social practices that lie beneath it. This approach, however, assumes an imagined or ideal audience that constructs their reality through the same codes and meanings as the analyst, who is often from a particular social background, for example, white and middle-class (Nakassis, 2009). Recent audience studies aim to capture a mix of interpretations of a visual image, which may involve interviews, focus groups, and online analysis.

In geography, for example, those studying geopolitics have used the Internet Movie Database (IMDb – www.imdb.com) to analyse viewers' experience of film (Dodds, 2006). The website enables viewers to vote on, review, and discuss films. This allows access to a large number of viewers, unfolding discussions and insights into different opinions about a film. Dittmer (2011) uses IMDb reviews of superhero films to powerfully illustrate the way audiences engage with the concept of the 'American hero' in the superhero film genre. In the reviews, fans reproduce this discourse. However, it also comes up for renegotiation: one viewer notes that *Iron Man*'s military weapons are called Jericho, which 'sounds cool' until you think that 'the modern-day Jericho is the West Bank of the Palestinian territories, and the Stark's 'rockets' start to look a lot more loaded' (2011: p.126). The different audience responses illustrate awareness of contemporary political events and critique. Dittmer argues that they provide insight for popular geopolitics. It is possible, then, to conduct online ethnographies, spending time in online environments, tracking how an image travels across media, or analysing the discussion surrounding online images. Audiences can tell us things about the image we had not personally experienced, and consulting different audiences can highlight the implicit power relations and constructed-ness of an image, when what appears natural and self-evident to some viewers is unrealistic to others.

To explore the sites of production and consumption of the image we might ask:

- Who is the 'author' (artist, director, advertising agency) of the image and what is their intended meaning?

- What do different texts produced at the same time have to say about the visual image or how might they be likened to it? Are there similar visual motifs or discourses?

- What is the context of viewing for this image? Does it change?

- How might different audiences interpret the image's signs differently and why?

- How might different audiences' opinions about the visual image be accessed?

15.4 Embodied and Affective Encounters with Visual Images

Geographers have been keen to stress that the way we engage with visual images is multi-sensory. One criticism of the audience studies approach discussed above is that it leaves aspects of audiences' experience absent, such as the 'visceral thrills' of such films and how they are experienced by the viewer's body. Vision is an embodied process; rather than purely cognitive – involving the eyes and the mind – it engages the whole body. Ash refers to the 'affective materiality' of the visual image, by which he means that:

> [I]mages affect the whole body, on a series of biological, existential, and sensory levels... Biologically the image hits the skin and has physiologically demonstrable effects, such as increased heart rate or the raising of hair on the back of the neck. (2009: 2106)

Rather than thinking about the effects of an image based on its cultural and representational aspects discussed so far (that is, how effects are based on meanings, knowledges and contexts), this approach explores the non-representational and the types of feelings, movements and sensations that happen when viewing visual images. This kind of work has a focus on the centrality of the body in how we perceive the world. It argues that images like movies and paintings have an expressive materiality, where the image 'touches' the viewer through how it calls into being the fleshiness of the world. This is referred to as 'haptic' vision (Sobchack, 1992; Ash, 2009). Viewers draw upon their past unconscious, bodily experiences to flesh out an image. Unlike semiology, they do not decode the sign for 'shoe', for example, but instead anticipate the experience of the shoe through their body based on what they have previously experienced.

Video games are a particularly interactive and immersive technology for viewing visual images and as such they have inspired geographers to find new ways to study them, particularly for the non-representational bodily aspects described above. Whilst geographers are still concerned about the representational aspects of video games and their social effects (Power, 2007; Schwartz, 2009) they are especially concerned with the spatial effects and techniques used to construct these virtual spaces and how they affect the viewer's body:

> As the spatiality of the video game has evolved from simple two-dimensional to complex three-dimensional worlds, the importance of an affective experience to the player has become paramount. (Shaw and Warf, 2009: 1332)

Indeed, Ash (2009) has closely studied the mechanisms of *Call of Duty 4* to show how video games are worth studying not only as sites of representational content or discursive contexts, but also that they work to refigure and produce their own spaces. He first points out that how we usually navigate and orientate ourselves in the world, such as gauging the distance between things or whether something will break when dropped, is through touch (the haptic), which we can then approximate through visual cues. Whereas in video games, players navigate virtual worlds foremost through visual cues – the 'eye takes on a fully haptic function' (Ash, 2009: 2117) – and they receive haptic feedback, such as force feedback in the form of vibrations through the control pad.

However, in this game, the first person perspective on the screen limits the player's ability to access visual clues. The player's view is partial, and constantly shifting as their avatar moves through the space as part of the image, limited by the pre-determined moves designed by the game producer (see Figure 15.4). Within Figure 15.4, the effect is heightened through the use of point perspective to create depth so that the two buildings protrude into the centre from the edge of the screen, and the viewer has to intuitively move their avatar through the image to increase their visibility. Here, the hands come to stand for the eyes because it is necessary to press buttons or move a control stick to look around and move through the virtual game environment. The player's prolonged attention to the spaces of the game can actually reshape connections between the body and those spaces, which results not only in video game experiences being associated with news and cinema images, but also with how people experience real-world, such as urban spaces (reported often by fans on game forums; Ash, 2009).

Ash uses accounts of personal experience to analyse these video game effects, whilst others have used 'real time' communication whilst playing games and text functions on PC-based games to ask other players about their experiences, also tracing themes in multi-player discussions. Shaw and Warf (2009) stress that examining the 'constellation of affects' in video games should always be complementary to representational approaches rather than replacing them, and that all of the interconnections still need investigation. The affective and material aspects of the visual image become increasingly important with the increase in augmented and virtual reality technologies like Facebook's Oculus Rift and the Google Glass. These visual technologies require us to think differently about viewing and the spaces of visual images.

Whilst theories about 'material affectivities' have been used productively with video games, they are not limited to interactive visual technologies. In fact, geographers argue that all visual images are interactive and affective in different ways. Rycroft, for example, explores this in different forms of nonrepresentational art like the Op Art

Figure 15.4 First person video game perspective

(*Source:* http://mensuro-aero.com/blog/uav-grey-eagle-deployed-in-afghanistan-for-the-first-time)

paintings of Bridget Riley (Rycroft, 2005, see http://www.tate.org.uk/art/artists/ bridget-riley-1845). Significantly we might ask:

- How did you experience the image? How did it make you feel?
- Which non-visual aspects affect how you experience the visual image and in what ways?
- How do different technologies, e.g. a screen, change how we experience the image?
- What techniques are used to create bodily effects?

15.5 Conclusion

Geographers are interested in interpreting visual images in a number of different fields – geopolitics, film geographies, historical geography, artworks and family practices – and are using different theories and methodologies in this interpretation. Whilst Rose (2012) proposes a critical visual methodology that engages with three sites of the image (production, content, audience), most geographers explore only one or two of these, as relevant to their research topic and questions. This chapter has attempted to introduce and give examples of the breadth of geographical inter-pretations of the visual image and it is likely that further reading will be necessary to explore particular approaches in more detail. The questions at the end of each section should provide a good starting point, though, to begin to think about inter-preting a visual image and its role.

Geographers are beginning to acknowledge their own position as not only viewers and interpreters of the visual image, but also producers. Visual aids like maps, globes, models, slides and photographic illustrations are often crucial in geographical work. Increasingly, different methods for visually recording data are being used to capture complex aspects of everyday experience (see Chapter 38 on videography, for example). Using, analysing and creating visual images means that we have to think carefully about the types of meanings we are producing. Are we reproducing discourses that are bound up in unequal power relations? Can we be more transparent about our decisions to analyse particular images and the methods we use? In some instances there may be prac-tical ethical considerations, such as gaining permission to reproduce images or access them, or always letting people know when and what we are researching online. In other cases, there might be more theoretical ethical concerns, such as moral debates surround-ing the choice of particular images; some critics, for example, argue that holocaust images speak for themselves and it is not right to try to speak for them by interpreting them, whereas others argue that it is only through doing so that justice is sought. Studying images, then, requires also thinking about the role of the interpreter.

SUMMARY

- Geographers are concerned with the social and spatial effects of visual images.
- Interpreting the visual image encompasses three sites: contexts of production, the representational content and the audience/viewing space.

- Analysis explores the discourses, social constructs, and ideologies that images (re)produce.
- Recent analysis is concerned with what images do rather than say, for example, such as embodied responses and how people use images to construct identity or 'sense of place'.
- It is important to consider your own viewing position and ethical considerations when interpreting visual images.

Further Reading

- For a comprehensive overview of theory and methods for social scientists, **Rose** (**2012**, and see new edition published in 2016) takes a **'critical visual methodology'** approach that asks geographers to think about the social effects of images through three sites: the site of production; the site of the image (its content); and the site of reception (its audience). As well as some of the approaches discussed in this chapter, Rose also explores psychoanalysis, iconography, feminism and other significant theoretical developments in how images are understood and analysed. There is a web resource that accompanies the book.

- **Wylie (2007)** gives an in-depth account of how **the study of landscape** has been informed by different art criticism, cultural theory and developments in social and cultural geography, such as Cultural Marxism, phenomenology and non-representational theory.

- The relationship between **geography and art** is an expanding field with many geographers seeking to work with artists. **Hawkins (2013)** looks at the role of the visual within a broader conceptualization of installation and place-based art. In contrast **Rycroft (2005)** has focused on two-dimensional works that are abstract and have different kinds of effects to the landscape imagery that is the topic of much geographical writing on art.

- Work on the **materiality and affectivity** of visual images is a relatively new field in geography**. Tolia-Kelly (2004)** has drawn on geographical work about materiality to discuss photographs and other images in the home. **Ash (2009)** draws from work in visual and media studies on computer games, paying attention to the spatial implications of the viewing practices involved in playing computer games.

Note: Full details of the above can be found in the references list below.

References

Ash, J. (2009) 'Emerging spatialities of the screen: Video games and the reconfiguration of spatial awareness', *Environment and Planning A* 41: 2105–24.

Barthes, R. (1977) *Image, Music, Text.* London: Fontana.

Berger, J. (1972) *Ways of Seeing.* London: British Broadcasting Corporation, and Harmondsworth: Penguin.

Burk, A.L. (2006) 'Beneath and before: Continuums of publicness in public art', *Social & Cultural Geography* 6: 949–64.

Carter, S. and McCormack, D.P. (2006) 'Film, geopolitics and the affective logics of intervention,' *Political Geography* 25: 228–45.

Chandler, D. (2014) *Semiotics for Beginners: Signs.* http://visual-memory.co.uk/daniel/Documents/S4B/sem02.html (accessed 18 November 2015).

Cosgrove, D. (1984) *Social Formation and Symbolic Landscape.* London: Croom Helm.

Cosgrove, D. (2003) 'Landscape and European sense of sight: Eyeing nature', in K. Anderson, M. Domosh, S. Pile and N. Thrift (eds) *Handbook of Cultural Geography.* London: Sage. pp. 249–68.

Cresswell, T. (2003) 'Landscape and the obliteration of practice', in K. Anderson, M. Domosh, S. Pile and N. Thrif (eds) *Handbook of Cultural Geography.* London: Sage. pp. 269–81.

Daniels S (1989) 'Marxism, culture, and the duplicity of landscape', in R. Peet and N. Thrift (eds) *New Models in Geography: The Political-Economy Perspec-tive*, volume 2. London: Unwin Hyman. pp. 196–220.

DiFrancesco, D.A. and Young, N. (2011) 'Seeing climate change: The visual construction of global warming in Canadian national print media', *Cultural Geographies* 18: 517–36.

Dittmer, J. (2011) 'American exceptionalism, visual effects and the post-9/11 cinematic superhero boom', *Environment and Planning D: Society and Space* 29: 114–30.

Dodds, K. (2006) 'Popular geopolitics and audience dispositions: James Bond and the Internet Movie Database (IMDb)', *Transactions of the Institute of British Geographers, New Series* 31: 116–30.

Gabrys, J. and Yusoff, K. (2012) 'Arts, sciences and climate change: Practices and politics at the Threshold', *Science as Culture,* 21(1): 1–24.

Gibbs, L. (2014) 'Art-science collaboration, embodied research methods and the politics of belonging', *Cultural Geographies* 21: 207–27.

Hawkins, H. (2010) 'Turn your trash into … Rubbish, art and politics. Richard Wentworth's geographical imagination', *Social & Cultural Geography* 11(8): 805–27

Hawkins, H. (2013) 'Geography and art: An expanding field: Site, the body and practice', *Progress in Human Geography* 37: 52–71.

Hughes, R. (2007) 'Through the looking blast: Geopolitics and visual culture', *Geography Compass* 1(5): 976–94.

Izushi, H. and Aoyama, Y. (2006) 'Industry evolution and cross-sectoral skill transfers: A comparative analysis of the video game industry in Japan, the United States, and the United Kingdom', *Environment and Planning A* 38: 1843–61.

Johns, J. (2006) 'Video games production networks: Value capture, power relations and embeddedness', *Journal of Economic Geography* 6:151–80.

Loopmans, M., Cowell, G. and Oosterlynck, S. (2012) 'Photography, public pedagogy and the politics of place-making in post-industrial areas', *Social & Cultural Geography* 13(7): 699–718.

Mackenzie, F.D. (2004) 'Place and the art of belonging', *Cultural Geographies* 11: 115–37.

Macpherson, H. (2008) 'Between landscape and blindness: Some paintings of an artist with macular degeneration', *Cultural Geographies* 15: 271–69.

Morris, N.J. and Cant, S.G. (2006) 'Engaging with place: Artists, site-specificity and the Hebden Bridge Sculpture Trail', *Social and Cultural Geography* 7(6): 863–88.

Nakassis, C.V. (2009) 'Theorizing film realism empirically', *New Cinemas: Journal of Contemporary Film* 7(3).

Power M, (2007) 'Digitized virtuosity: Video war games and post-9/11 cyber-deterrence', *Security Dialogue* 38: 271–88.

Roberts, E. (2012) 'Family photographs: Memories, narratives, place', in O. Jones and J. Garde-Hansen (eds) *Geography and Memory.* New York and Basingstoke, UK: Palgrave Macmillan.

Rose, G. (1993) *Feminism and Geography: The Limits of Geographical Knowledge.* Cambridge: Polity Press.

Rose G (2003) 'Family photographs and domestic spacings: A case study', *Transactions of the Institute of British Geographers* 28: 5–18.

Rose, G. (2012) *Visual Methodologies: An Introduction Researching with Visual Materials*, 3rd edn. London: Sage.

Rycroft, S. (2005) 'The nature of Op Art: Bridget Riley and the art of nonrepresentation', *Environment and Planning D: Society and Space* 23: 351–71.

Schwartz, L. (2006) 'Fantasy, realism, and the Other in recent video games', *Space and Culture* 9: 313–25.

Schwartz, L. (2009) 'Othering across time and place in the suikoden video game series', *GeoJournal* 74: 265–74.

Sharp, J. (2007) 'The life and death of Five Spaces: Public art and community regeneration in Glasgow', *Cultural Geographies* 14: 274–92.

Shaw, I.G.R. and Warf, B. (2009) 'Worlds of affect: Virtual geographies of video games', *Environment and Planning A* 41: 1332–43.

Sobchack, V. (1992) *The Address of the Eye: A Phenomenology of Film Experience.* Princeton, NJ: Princeton University Press.

Spinney, J., Green, N., Burningham, K., Cooper, G. and Uzzell, D. (2012) 'Are we sitting comfortably? Domestic imaginaries, laptop practices, and energy use', *Environment and Planning A* 44: 2629–45.

Tolia-Kelly, D. (2004) 'Locating processes of identification: Studying the precipitates of re-memory through artefacts in the British Asian home', *Transactions of the Institute of British Geographers* 29: 314–29.

Wylie, J. (2007) *Landscape.* Oxford: Routledge.

ON THE COMPANION WEBSITE...

Visit **https://study.sagepub.com/keymethods3e** for author videos, chapter exercises, resources and links, plus **free** access to the following recommended articles:

1. **Tolia-Kelly, D.P. (2012) 'The geographies of cultural geography II: Visual culture',** *Progress in Human Geography*, 36(1): 135–42.

This paper situates recent work by geographers that focuses on visual culture within the broader context of cultural geography, especially with reference to material culture.

2. **Tolia-Kelly, D.P. (2013) 'The geographies of cultural geography III: Material geographies, vibrant matters and risking surface geographies',** *Progress in Human Geography*, 37(1): 153–60.

This paper discusses the importance of the political in material geographies which is central to most methodologies for the analysis of visual images and representations.

3. **Monmonier, M. (2006) 'Cartography: Uncertainty, interventions, and dynamic display',** *Progress in Human Geography*, 30(3): 372–81.

In geography, much theoretical work on the visual image stems from the study of maps. This paper touches on some of the issues I raise at the end of my chapter about an ethics of representation and responsible use of visualisations.

16 Using Geotagged Digital Social Data in Geographic Research

Ate Poorthuis, Matthew Zook, Taylor Shelton, Mark Graham and Monica Stephens

SYNOPSIS

As the Internet and associated information technologies have proliferated, online, geotagged digital social data have emerged as a key product of our social interactions, resulting in what some have called a 'data revolution'. Massive amounts and new kinds of digital data create ever more complex informational environments and the possibility to map and measure spatial activities, patterns, and processes. Digital social media may also be an object of study in and of itself. This chapter outlines some uses of web-based, geographically-referenced digital social data for geographical research. We also focus on the pitfalls associated with user-generated and social media data sources, and on the benefits of a mixed methods approach to these data. Primarily, digital social data may be mapped for visual analysis, but closer examination through qualitative methods can provide insights into particular people's perceptions and experiences of the world around them. Thus, while making maps is often the starting point for geographers working with this kind of research, it is rarely the end point. The collection, analysis and contextualization of digital social data therefore allow insight into larger questions about society.

This chapter is organized into the following sections:

- Introduction: Researching (with) digital social data
- The many forms of digital social data
- Making meaning from digital social data
- Conclusion

16.1 Introduction: Researching (with) Digital Social Data

Information always has a geography. It is created in places, it is used in places, it is changed and repurposed in places. Crucially, it also helps define how we understand and create places, making the study of the geographies of information – where it is, what it is and defines, who produces it and who is produced *by* it – fundamental to the study of human geography.

As the Internet and associated information technologies have proliferated, online, geotagged *digital social data* have emerged as a key product of our social interactions, resulting in what some have called a 'data revolution' (Kitchin, 2014). Not only are governments, corporations, and other large organizations producing, capturing and analysing massive amounts of data, so too are individuals, whether through open data-sets released by governmental agencies or through the direct 'scraping' (extracting data) of social media sites. This digital documentation of everyday life creates ever more complex informational environments and entanglements, including the possibil-ity to map and measure spatial activities, patterns, and processes via new data sources.

The proliferation of big and user-generated data makes a range of everyday social, economic and political activities more visible than was previously possible. For instance, topics that have hitherto been studied through more qualitative, resource-intensive approaches, such as regionally-specific expressions of religion (Zook and Graham 2010; Shelton et al., 2012; Wall and Kirdnark, 2012), language (Graham and Zook, 2013; Graham, Hale et al., 2014) and consumption habits (Zook and Poorthuis, 2014), as well as the more general questions of how and where events are discussed online (Crampton et al., 2013) and how places are represented and understood differ-ently by different people (Watkins, 2012; Graham, Zook et al., 2013; Power et al., 2013). Moreover, digital social data have proved crucial to pressing problems, such as responses to natural disasters (Crutcher and Zook, 2009; Goodchild and Glennon, 2010; Zook et al., 2010; Crooks et al., 2013; Shelton et al., 2014). The ephemerality and mundane nature of digital social data also mean that they can be used for less-traditional academic research, such as mapping the spatial distribution of references to zombies (Graham, Shelton and Zook, 2013), the retail price of marijuana (Zook et al., 2012), consumption of adult content, or identifying the places that drink the most alcohol (Zook, 2010; Zook and Poorthuis, 2014).

While one can use these data to understand a variety of social processes, so too can digital social media be an object of study in and of themselves, offering a key way to guard against the uncritical use of this new resource. For instance, studies of the biases built into these datasets, such as the over-representation of the perspectives of wealthy places (Graham and Zook, 2011; Graham, Hogan et al., 2014; Graham, 2014), urban dwellers (Hecht and Stephens, 2014) and men (Stephens, 2013), have proved to be fertile ground for identifying the gaps within such data. These issues are essential to keep in mind when using digital social data because these data are always selective representations of the world around us; and thus what is measured and mapped by digital social data are selective accounts of selective stories. As such, we should always be cautious in interpreting findings, and critical of any positivist claims regarding the neutrality of data, always grounding our work in broader concepts and methods, and most importantly an awareness of geographic context (for a wider discussion see Graham and Shelton, 2013).

The use of digital social data has become increasingly prevalent within geography research for a number of reasons: first, it is now relatively easy to gather and process this kind of data, especially from social media platforms. Second, much of what is captured by such data – particularly cultural and political markers – is not readily available in more conventional datasets, such as demographic statistics tracked by the national censuses. Third, such data tend to be 'big', allowing both macro-level understandings, as well as the potential to analyse smaller subsets of the data to bet-ter understand more specific processes. Fourth, the data tend to be produced and collected in real-time, allowing the inclusion of temporality in one's analysis, rather than relying on a static snapshot of social activity. Finally, these data allow us to understand some of the relational dimensions of social life, from whom we associate with to how we move through space, and how these things, among many others, are networked and connected.

This chapter outlines how one might utilize the massive amounts of web-based, geographically-referenced digital social data for geographical research. Because much of these data are user-generated and produced through social media platforms, we also focus on the pitfalls associated with such sources and the benefits of a mixed methods approach to these data. Not only can digital social data be mapped for visual analysis, they are also useful to use as a range of quantitative methods to understand relationships between different subsets of the data. In addition, closer, systematic readings via qualitative methods of social data provide insights of particular people's perceptions and experiences of the world around them. Thus, while making maps is often the starting point for geographers undertaking this kind of research, it is rarely the end point.

16.2 The Many Forms of Digital Social Data

One of the fundamental characteristics of digital social data is their variety. For the purposes of this chapter, we have chosen a particularly expansive definition, which we outline by defining each term in turn. First, *digital* refers to the way these data are collected and stored, as low cost sensors, widely prevalent computing capacity, and hard disk storage have created a situation where it is easy to collect and store many of the mundane happenings of everyday life. This collection is greatly aided by the ways that an increasingly larger amount of daily life is mediated by digital technologies.

Second, *social* highlights that many of the collected data represent aspects of human life – the habitual and relational interactions between family and friends – that hitherto were not extensively documented in ways that could be easily used by researchers. This stands in contrast to other kinds of transactions, most notably economic transactions, for which data were already likely to be produced, from retail purchases recorded by cash registers, the suppliers and customers of firms, and so on. In short, the nature of the data being collected has expanded into new realms of human geography and sociability. Together, *digital* and *social* are key defining characteristics of the data on which this chapter focuses.

It is also worth noting that two other terms – online and geotagged – often go hand in hand with digital social data, though we use these with some caveats. While much digital social data are 'online', the meaning of online can be quite complicated, ranging from fully accessible data on the Internet (a membership list or search result on the web that can be copied) to controlled access on the Internet (access via an application programming interface, or API, to selected parts of a social database, e.g. Twitter's API), to controlled access via social means (asking a provider for a copy of data), to datasets that are simply not, or at least very rarely, shared (e.g. Facebook transactions, mobile phone records). Although many researchers use the more accessible types of digital social data – as we do here for our case study – it is important to note that researchers are not limited only to these sources. Indeed, researchers selecting data not for their potential insight into pre-existing questions or concerns, but simply because they are available, is concerning and problematic.

The second ancillary term we highlight, *geotagged*, is used to make two key observations. The first is that geotagging – or the act of associating a given piece of digital social data with a particular location on the earth's surface – comes with an array of technical issues, such as the precision of measurement and accuracy of location. Much geotagged data come with digital latitude and longitude coordinates, but these points are collected with a variety of GPS receivers and/or Wi-Fi and mobile location technologies with different levels of accuracy. Moreover, some of the geotagged information comes in less precise forms, such as an unstructured text reference to an often incomplete city name. Thus, all geotagged data are not created equal. The second reason to make geotagged ancillary rather than primary is the difference between data where geography is explicitly the subject of the socialization (e.g. a geotagged Wikipedia article or an OpenStreetMap entry) and where geography is the references to location of a social event (e.g. a geotagged tweet or Foursquare check-in). The third and final motivation is to emphasize the wealth of non-geographic information – relational connections, contextual information, user characteristics, etc. – contained within digital social data, in order to encourage geographic researchers to think beyond the geotag and the overly simplistic approaches that are often taken with this kind of data (Crampton et al., 2013).

With these basic characteristics of digital social data in mind, it is useful to review the range of these digital social data sources that have been utilized in geographic research. Goodchild (2007) uses the concept of volunteered geographic information (VGI) to capture the novelty, at least at the time, of individuals posting geotagged information to the web on a range of topics. We wish to expand upon this idea of user-generated data by complicating the straightforward notion of 'volunteered' data, considering that much of the digital social data that are created on a daily basis run the gamut from *purposefully shared* to *reflexively distributed* to data collection that is *derived from users' actions, but not their conscious control*. In the first category of *purposefully shared*, we might place a number of crowdsourcing projects – e.g. OpenStreetMap, Wikipedia, crisis mapping – in which the participants are actively creating data about a topic and/or place as part of a larger group. Also included in this category of consciously created data are many forms of social media – e.g. Twitter, Facebook, Foursquare, Weibo – in which users create and share information with others. This type of data, however, also transitions into the *reflexively distributed* category, as the ways in and extent to which contributions are distributed to others and made available for research purposes are often opaque to the user. Social data primarily meant for a network of friends are shared reflexively as a matter of daily practice, even though the data can easily be repurposed by others in ways the creator may not have contemplated. The final category of user-generated data is that data which can be derived from users' actions, but are not the result of any conscious choice – either on purpose or by habit – to share such data. This includes mobile phone logs commonly held, but rarely distributed, by telecom operators for billing and system design uses, as well as a range of location-based services (LBS) deployed via smart phone apps that track the use and location of a mobile device. While this kind of tracking is 'authorized' by terms of service agreed to by the end user, these are rarely read and in any case soon forgotten.

Each type of user-generated data is accessed in specific ways; for example, many social media services make part of their data accessible via an API, while mobile phone or LBS data are only available to researchers within or working closely with a service provider, and lend themselves to particular types of research questions. This is tied to the specific variables, bias and other factors within each of these datasets that together inform a key decision for any researcher. Far from letting the data speak for themselves, good geographic research with digital social data requires a careful assessment of what any particular dataset contains, as well as the selection of substantive research questions that such data are capable of addressing. In this vein, we wish to also emphasize the value of 'small data', which need not be seen as opposed to the 'Big Data' phenomenon. For example, a key way to leverage a large amount of data, such as billions of tweets, is to extract much smaller datasets – even with a few dozen or hundred observations – that look more closely at localized events or outliers to the norm, which can provide insight otherwise not available.

16.3 Making Meaning from Digital Social Data

The process of working towards understanding the world does not change *fundamentally* when using digital social data instead of more conventional data sources, e.g. data derived from surveys, interviews, or a census. There are, however, some key differences and specific challenges associated with digital social data, such as the fact that available datasets are often much larger as well as much less structured than, for example, data published by the US Census Bureau. Often, just *opening* a data set derived from these sources can be a challenge, let alone conducting meaningful research. In the next sections we outline a series of steps that can help a geographer go from a raw data set to meaningful insights. Following specific procedures along the research process is especially important when using digital social data, because such data are often not scrutinized or cleaned in the same way as more conventional data sources. Digital social data are rarely purposefully designed to answer a specific set of scholarly or applied research questions (see also Chapter 30 on secondary datasets), and biases and quirks within the data are rarely examined or thoroughly documented. In other words, if we are not careful with our approaches to these data, we are likely to fall into the trap of 'garbage in, garbage out', and end up producing research that never succeeds in answering the questions it set out to, or even answering any questions at all.

Anatomy of data and avoiding overload

The first step towards producing meaningful analysis is identifying and understanding the dataset used to answer a research question. Each of these kinds of data comes with a particular set of advantages and disadvantages, as well as specific ways that the data can be accessed. In this chapter, we will focus specifically on the use of data from the social media platform Twitter, though the approach we lay out can easily be repurposed for other data sources.

In order to better understand data, it is important to review the different variables available within a given dataset. Figure 16.1 shows the different elements of a tweet that can be leveraged for geographical research, ranging from a geotagged location and timestamp to information about the user, such as their profile image, name and username; to textual and graphic content; relational connections seen through links to other profiles or websites in the content; and who a user follows and is followed by (although this is not shown in Figure 16.1) (see Crampton et al., 2013, and Graham, Hale, and Gaffney, 2014, for more detailed overviews of Twitter data). And while each individual tweet already has a multitude of dimensions – spatial, temporal, textual, etc. – it is the combination of these individual data points into collectives that holds much potential for employing digital social data in answering research questions. It is incumbent upon the researcher to understand both the range and nuances of these variables within different digital social data sources by reading documentation, and conducting background research and experimentation.

The wealth of digital social data sources, however, can also prove to be a significant challenge, even in just simply managing all of the data. For example, towards the end of 2013, Twitter users were sending over 500 million messages per day,[1] posing the issue of how to make sense of so many data points. Even opening relatively small datasets of this kind on a standard computer can be difficult; until 2007, Microsoft's Excel could only handle 65,000 rows! When confronted with such large data sets, it makes quantification and automation through powerful computers, software programmers, and clever algorithms seem almost necessary. While these approaches can certainly be useful in certain contexts, especially as it has become easier to set up systems, such as a *Hadoop cluster*, for working with large datasets, the goal of this chapter is to demonstrate the utility of conventional geographic research methods for analysing digital social data through a creative, conscious and critical (re-)combination.

Towards this end, the remainder of this chapter works with a relatively small dataset containing all geotagged tweets in the US mentioning 'grits' (a ground-corn porridge-like dish popular in the Southeastern United States) from June 2012 to September 2014, obtained from the DOLLY archive at the University of Kentucky (Zook and Poorthuis, 2013). These data consist of approximately 64,000 tweets. Considering that geotagged tweets represent only 2-3% of all tweets, this figure

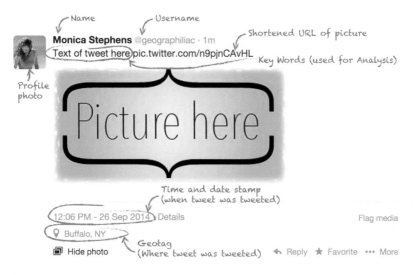

Figure 16.1 The anatomy of a tweet

represents a conversation of several thousand tweets per day and highlights the ability of digital social data to provide insight on a particular cultural phenomenon that would be unlikely to be the subject of other research approaches, such as official census records or nationwide surveys.

Identifying patterns in space and time

The possession of thousands of data points, however, does not necessarily translate into previously unforeseen insights, despite the claims of some of the biggest proponents of 'Big Data' (Anderson, 2008). Often, the simple mapping of the spatial distribution of these points (see Figure 16.2a) only mimics population distribution, a phenomenon seen in the now-ubiquitous, albeit poorly thought-out, 'animated ectoplasm maps' of Twitter activity (Field, 2014). Even when not mirroring population centres, these maps are also problematic in that when points overlap, one loses any sense of the actual number of points in a given location, obscuring the actual phenomena. One easy way to address this 'overplotting' is to make each point slightly transparent (Figure 16.2b). While this increases the legibility of the map somewhat, it is still largely representing population centres, as tweets for grits (as well as most things) are more likely to occur in places where people live.

For certain phenomena, this problem can be partially solved by creating 'heat maps' or density surfaces, widely known from weather maps. Though the results tend to be aesthetically pleasing and intuitive, the methods involved in creating density surfaces, such as kriging or kernel density estimation, assume a *continuous* surface. However, tweets, and most other human processes, are discrete: what happens in a city might be completely different from the directly adjacent countryside (see Galton, 2004; Longley et al., 2005). Hence, one should exercise caution in applying these techniques to geosocial media data.

Another common approach would be aggregating these individual data points to larger spatial units. For example, Figure 16.2c shows the number of tweets per county, and creates a much clearer spatial pattern without the problems of overplotting seen in Figure 16.2a. But it also creates a new problem: not all administrative units are the same size. In the United States, counties in the western portion of the country tend to be much larger. Not only will larger counties *ceteris paribus* have more tweets, they also stand out more visually on the map. For instance, Figure 16.2c over-emphasizes concentrations of grits-related tweeting in Los Angeles County, California, Clark County, Nevada, and Maricopa County, Arizona, due to these counties' large spatial footprint. But one advantage of working with point data like geotagged tweets is that we can define new areal units for aggregation, rather than relying on the more-or-less arbitrary census definitions. Figure 16.2d, shows a similar aggregation of tweets to polygons, but now the map uses a grid of identically-sized hexagonal cells (see Shelton et al., 2014), which diminishes the seemingly strong presence of large counties in the Southwest seen in Figure 16.3c, while simultaneously highlighting a more distinct concentration of tweets mentioning 'grits' in the Southeast.

Despite the fact that moving from the direct mapping of points in Figure 16.2a to the more informative mapping of density using hexagonal cells represents significant progress, the overall pattern still reflects population density, with large cities having more tweets than smaller towns or rural areas. In order to address this issue, which is again common across any range of phenomena, it is customary to *normalize* whatever is being measured by another variable that indicates size or population. Digital social data, however, can be normalized incorrectly. For example, normalizing the grits data based on the population of a state assumes that everybody in the state tweets at the same rate, leading to an over-representation of rural areas or places that

(a)

(b)

Figure 16.2   Simple but problematic mapping of digital social data

Figure 16.2 (Continued)

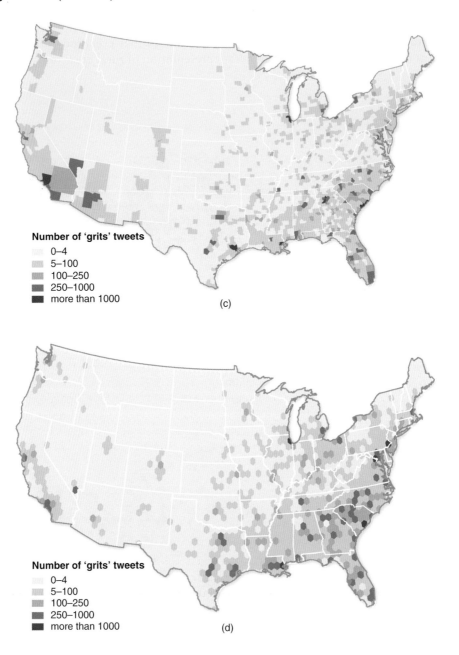

do not tweet frequently. A better strategy is to normalize by the total amount of digital social data activity (in this case tweets) in each area, i.e. using a 'tweeting population' rather than 'total population' to obtain a rate of 'tweets about grits' relative to the total amount of tweeting.

This leads to the question of what exactly does it mean that 3 out of every 100,000 tweets in a particular location mention grits and therefore, we use a slightly more complex measure called an odds ratio (OR):

$$OR = \frac{p_i/p}{r_i/r} \qquad [16.1]$$

where P_i is the number of tweets in area i related to phenomenon at hand and P is the total number of tweets related to the phenomenon. R_i is the number of 'population' tweets in area i and R is the grand total of that population. This measure corrects for differences in size, and has the additional advantage of allowing us to normalize by a sample population, rather than by the total population. In this case, we have used a 0.01% sample of the overall Twitter activity during the same time period (~180,000 tweets) derived from DOLLY (Zook and Poorthuis, 2013). Another advantage of this measure is that the odds ratio results in an easy-to-understand number, where a value of 1 for a particular areal unit indicates that there are exactly as many data points related to grits as one would expect based on the overall Twitter activity. Thus, any places with values less than 1 indicate that there are fewer 'grits' tweets than we would expect, and vice versa for values greater than 1. The resulting maps – seen using counties for aggregation in Figure 16.3a and the hexagonal cells in Figure 16.3c – that use this odds ratio measure reveal a very clear Southeastern 'grits' cluster that was only faintly distinguishable in Figure 16.2d above, and practically invisible in Figure 16.2a.

The odds ratio alone, however, confronts us with yet another issue: the problem of small numbers. If a given area has only a small number of total observations, thus making the denominator used for normalization small, the *variance* of the ratio is high, and thus might not be reliable. For example, one can see a littering of seemingly higher values for grits in less-populated areas of the Dakotas that are most likely an effect of a small number of observations. In order to address this issue, we instead calculate a *confidence interval* using the formula below, and then use the lower bound of that interval so that we know that, in this case, if we see a value of over 1 that value is *significant* as well (with 95% confidence).

$$OR_{lower} = e^{\ln(OR) - 1.96 * \sqrt{\frac{1}{p_i} + \frac{1}{p} + \frac{1}{r_i} + \frac{1}{r}}} \qquad [16.2]$$

This analysis is visualized in Figures 16.3b and 16.3d for both counties and hexagonal cells, and again demonstrates an ever clearer picture of 'grits' as a geographically specific cultural phenomenon associated with the American South, rather than simply an effect of population density. In effect, we have controlled for the significant amount of 'noise' in the dataset through the application of more geographically-contextualized quantitative analysis.

While Figure 16.3d makes it easy to discern a large cluster of grits-related tweets in the South, not all spatial patterns are this clear. Visual inspection of a map to identify clustering can be influenced by anything from colour choices to classification methods, making it important to combine this analysis with more statistical measures of spatial clustering. This differs from the previous analysis, in which we only considered the significance of the odds ratio for a single spatial unit (i.e. hexagon or county); now we turn to examining the significance of the odds ratio for that spatial unit in relation to all of its neighbouring areas. Conducted at a global scale, this analysis provides an indication of the degree of clustering for the entire spatial phenomenon, while on a local scale, it tells us which local clusters are significant. One way to do this is by calculating Moran's I (see Burt et al., 2009) and in this case, it comes as no surprise that the global pattern for grits tweets is clustered and the southeastern cluster shown visually

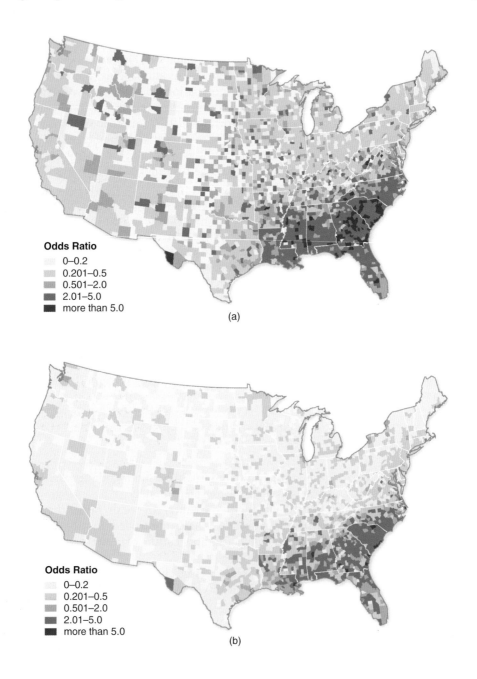

Odds Ratio
0–0.2
0.201–0.5
0.501–2.0
2.01–5.0
more than 5.0

(a)

Odds Ratio
0–0.2
0.201–0.5
0.501–2.0
2.01–5.0
more than 5.0

(b)

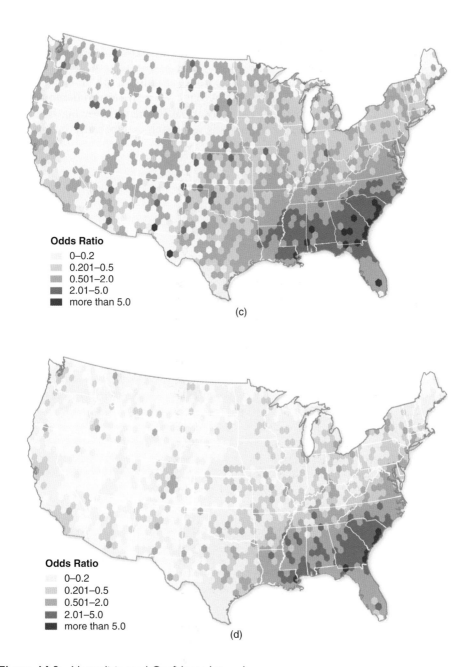

Figure 16.3 Normalizing and Confidence Intervals

is indeed a statistically significant cluster (Moran's I of 0.567) as well (see Figure 16.4a). One issue with cluster analysis is the findings of clustering are very much dependent on the size, shape, and placement of the areal units, i.e. the Modifiable Areal Unit Problem (see Openshaw, 1984; Burt et al., 2009, for an accessible discussion). Visually this is evident in Figures 16.4b–d, as variation in the size of the hexagons results in changes in cluster identification, such as the lack of high odds ratio clusters around Florida's larger cities in Figure 16.4c despite their presence with smaller hexagons in Figure 16.4d.

This exercise has identified the potential for normalizing phenomena-specific datasets by overall measures of social media activity. We now turn to demonstrating the potential for direct comparison of two subsets of this data, allowing for a side-stepping of the correspondence problem. This approach is especially useful when

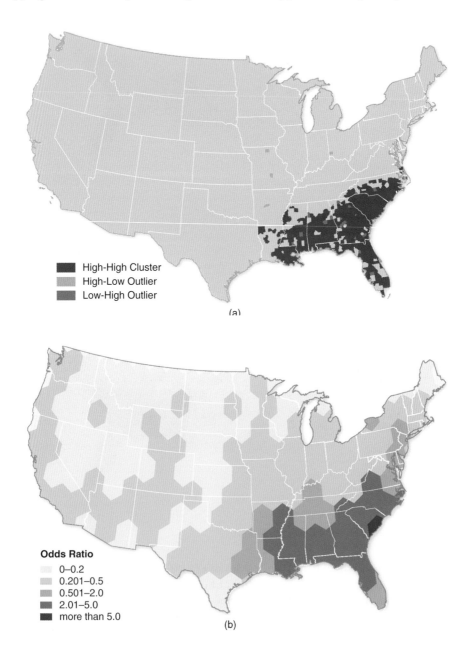

(a)

(b)

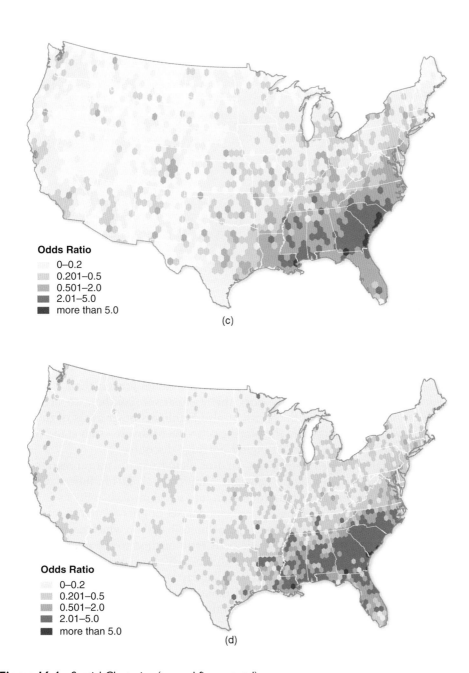

Odds Ratio
    0–0.2
    0.201–0.5
    0.501–2.0
    2.01–5.0
    more than 5.0

(c)

Odds Ratio
    0–0.2
    0.201–0.5
    0.501–2.0
    2.01–5.0
    more than 5.0

(d)

Figure 16.4 Spatial Clustering (two subfigures total)

comparing regional cultural differences (see Zook and Poorthuis, 2014, for an application to regional preferences in beer brands). Building on our earlier analyses, Figure 16.5 compares our existing dataset of tweets mentioning 'grits' with tweets mentioning 'oats'. The odds ratio is employed again, but now, instead of a 'total' pattern of tweets, we use oats-related tweets to normalize our values. In this analysis,

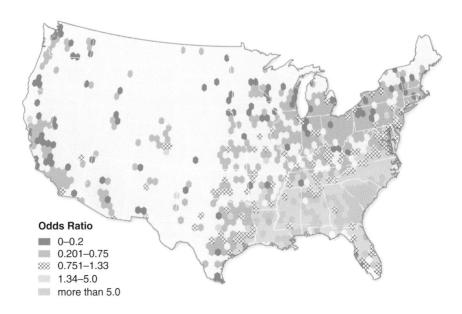

Figure 16.5 Grits vs. Oats, interaction between two datasets

values less than 1 signify a preference for oats, while values greater than 1 represent a tendency towards grits. Not only does this comparison continue to affirm our identification of a 'grits belt' in the South, it also highlights other areas of the country – an 'oats oval' stretching from the Northeast to the Midwest – that stand in stark contrast to the southeast in terms of digital porridge discourse.

As Figure 16.1 highlighted above, this kind of digital social data represents much more than is encapsulated in a simple geotag (Crampton et al., 2013). Indeed, variation over time can provide just as much useful insight into social media activity as variations over space. Rather than aggregating our original dataset of grits-related tweets to spatial units, we instead aggregate them to temporal 'units' based on which day of the week the message was sent, then apply the same odds ratio method to these temporal units in order to look at clustering in time. Figure 16.6 clearly shows that Saturday is by far the most popular day to talk about grits, with Sunday being significantly over an odds ratio of 1 as well. Thus, not only have we demonstrated significant clustering in space tied to grits' role as a regionally-specific cultural practice, but we are also able to identify that clustering in time suggests that grits – or least the tweeting about grits – is associated with leisure time. Of course, this opens up avenues for any number of follow-up research questions concerning the performativity of grits consumption as a marker of Southern culture, though this is tangential to the present chapter.

Understanding the context of digital social data

The final dimension of how digital social data can be leveraged for geographic research focuses on the social and spatial context in which they were created. Who produced the data and how do they move through space? To whom were they distributed

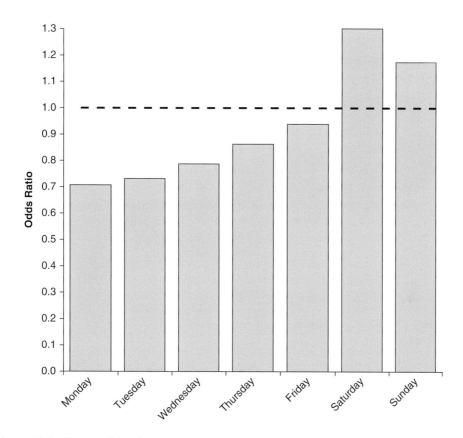

Figure 16.6   Temporal Variation

and what did the data signify? The ability to contextualize digital social data also represents a key difference between social media and other kinds of datasets (e.g. mobile phone records) that track movement and activity through space and time, but provide little information about what is transpiring in a particular place at a particular moment in time.

Thus a key avenue for analysis of digital social datasets is examining the relationships between individual users or individual messages. For example, on Twitter a user can have friends (those whom s/he follows) and followers (those who follow the user), and this information can be used to study how certain information travels between individuals (see Stephens and Poorthuis, 2015). It is also possible to identify relationships between places, based on visits or tweets made by the same person in these different places. Returning to our case study of 'grits' and its strong spatial clustering in the South, we might hypothesize that even those people tweeting about grits *outside* of the South are likely to have some kind of connection to the South. To examine this relationship, we began by looking for users who have tweeted about grits more than once – yielding a total of 8,958 users – and then drew a line from the tweet locations in chronological order. The resulting map (see Figure 16.7) clearly shows that there is a strong relational connection with the South for those

Figure 16.7   Relational 'gritspace' as seen through Twitter

who tweet about grits from other places, even for cities like Los Angeles that are quite distant in absolute space, as well as in terms of cultural identity. Indeed the gravity of grits appears quite strong because, of the users tweeting about grits from outside the South, approximately 55% of these also sent tweets from inside the cluster identified earlier in Figure 16.3d.

In addition to categorizing tweet content via a simple binary of the presence or absence of the term 'grits' one can examine the actual content of the tweets through a structured, qualitative analysis of the text much as one might do with interviews or secondary materials (see also Chapter 36). Many 'Big Data' researchers employ algorithmic sentiment analysis to examine this kind of content, perhaps influenced by the daunting task of reading so many tweets. Computers, however, tend to miss out on some of the important contextual clues that humans are more easily able to detect, especially when limited to 140 characters and contextualizing the relationship between a user's tweets over time and in different places. To provide an example of what might be done, Figure 16.8 highlights the case of grits tweets sent from Louisville, Kentucky, but focuses specifically on a set of five categories that we constructed from an examination and coding of each tweet. While these groupings are primarily for illustrative purposes – one can imagine a number of alternative subdivisions and refinements of this coding structure – they highlight the diversity of ways in which grits is used in tweets. '*Fandom*' tweets are referring to tweets in a loving (or hating) way, but do not necessarily indicate that a person is eating grits at that very moment. Similarly, '*craving*' tweets refer to a current craving for eating grits, while '*restaurant*' tweets are sent by people currently in restaurants, detailing the specific dishes they are eating. *Popular culture references* have little to do with eating: they are lyrics from rap songs by Ty Dolla $ign such as the phrase 'U gon' make them eggs cheesy with them grits' used to reference a particular cultural meme. In a similar way, the *expression* 'kiss my grits' is

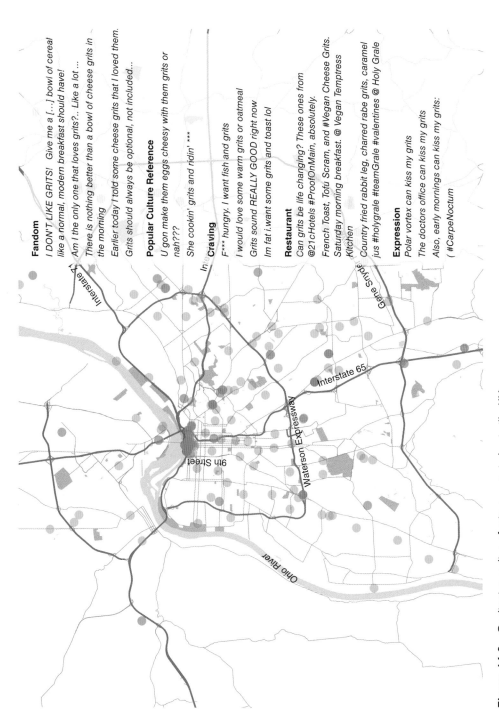

Fandom

I DON'T LIKE GRITS! Give me a [...] bowl of cereal like a normal, modern breakfast should have!

Am I the only one that loves grits?.. Like a lot ...

There is nothing better than a bowl of cheese grits in the morning

Earlier today I told some cheese grits that I loved them.

Grits should always be optional, not included...

Popular Culture Reference

U gon make them eggs cheesy with them grits or nah???

She cookin' grits and ridin' ***

Craving

F*** hungry. I want fish and grits

I would love some warm grits or oatmeal

Grits sound REALLY GOOD right now

Im fat i.want some grits and toast lol

Restaurant

Can grits be life changing? These ones from @21cHotels #ProofOnMain, absolutely.

French Toast, Tofu Scram, and #Vegan Cheese Grits. Saturday morning breakfast. @ Vegan Temptress Kitchen

Country fried rabbit leg, charred rabe grits, caramel jus #holygrale #teamGrale #valentines @ Holy Grale

Expression

Polar vortex can kiss my grits

The doctors office can kiss my grits

Also, early mornings can kiss my grits:

(#CarpeNoctum

Figure 16.8 Qualitative coding of grits tweets in Louisville, KY

a popular, family-friendly way of showing derision that finds its origins in the 1970s TV show *Alice*. As this brief review shows, qualitative methods can complement quantitative ones, and have much to offer in research on digital social data.

16.4 Conclusion

This introduction to the use of digital social data in geographic research is not meant to be exhaustive but rather to outline the potential and the shortcomings of this approach as well as provide an initial approach for research using these data. Therefore, emphasis has been on methodology, both on the nature and bias of the data, as well as possible approaches to working with them. In closing, however, we also wish to note that there are a number of larger questions associated with digital social data that go far beyond issues of how best to analyse and interpret the data.

For example, the processes behind the creation of both the means and motivations for documenting daily life that are manifest in digital social data are absolutely vital arenas for further research. What is the political economy of the creation and propagation of the geoweb and geosocial media and to whom and how do capital and profit accrue? How are we enrolled in the use of digital social data and how does their presence in our lives shape and discipline our everyday actions? How have representations and interactions with places – material, digital and hybrid – evolved with the advent of these records of daily life? Why and under what arrangements are individuals willing to take on the work-like tasks of collecting and categorizing data without remuneration and for the profit of others? How has the practice of privacy evolved and how do individuals and institutions negotiate understandings and regulations of what is in the public and private spheres?

There are many more questions like these to be asked and answered and we urge researchers not to lose sight of them. While more focused questions (such as our exploration of grits-space in this chapter) are extremely useful and absolutely necessary, they also provide the basis to pose questions not only about the practice and processes of digital social data but also to allow insight to larger questions about society.

SUMMARY

- The proliferation of big and user-generated data makes a range of everyday social, economic and political activities more visible than was previously possible, and the use of digital social data has become increasingly prevalent within geographical research.
- While one can use these data to understand a variety of social processes, digital social media can also be an object of study in and of themselves, offering a key way to guard against the uncritical use of this new resource.
- Key reasons for the growth and popularity of digital data use are: relative ease of collection of such data; novelty – much of what is reflected through digital data is not readily available in more conventional datasets; data tend to be 'big', allowing both macro-level understandings, as well as the potential to analyse smaller subsets of the data to better understand more specific processes; data tend to be produced and collected in real-time, rather than representing a static snapshot of social activity; and these data allow us to understand some of the relational networks and connections of social life.

- There are specific challenges associated with digital social data, especially the fact that available datasets are often much larger as well as much less structured than conventional data sets. Following specific procedures along the research process is important to account for the fact that such data are often not produced, collected and scrutinized in the same way as more conventional data sources; social data are rarely purposefully designed to answer a specific set of scholarly or applied research questions; and because biases and quirks within the data are rarely examined or thoroughly documented.
- Posing questions both about the practice and processes of digital social data also allows insight into some larger and emerging questions about society.

Note

1   https://blog.twitter.com/2013/new-tweets-per-second-record-and-how

Further Reading

- Goodchild (2007) coined the term 'Volunteered Geographic Information' (VGI) to describe the then-new production of geographic data online — often by individuals, not professionals. He compares these new data with more traditional citizen science and discusses the various technologies that enable this new data production. Written in 2007, it also serves as a testament to the initial excitement around this type of new data within Geography.

- boyd and Crawford (2012) critically reflect on what Big Data mean for science and knowledge production. Oft-cited, they stress that 'bigger' is not always better; that the context in which such data are produced matters; that the availability of data does not necessarily mean that it is ethical to use these in research; and that differences in access to big data might create new digital divides within the research community.

- Specific to Geography, Kitchin (2013) not only looks at the opportunities of Big Data for geographic research but also identifies a number of key challenges and risks. These range from the need for developing methods and techniques appropriate for such data to the danger of 'letting the data speak' for themselves.

- Crampton et al. (2013) attempt to address these challenges and critiques and think through how geographic research might look like when it does use digital social data but also remains cognizant of, or incorporates, the critical questions posed by boyd and Crawford (2012) and the risks identified by Kitchin (2013).

- Finally, Shearmur's (2015) editorial in *Urban Geography* serves as a reminder of the importance of more conventional data sources such as the Census to geographic research. While new forms of digital social data might reveal different or new things about society, that does not automatically mean that such data can replace census- or survey-type data.

Note: Full details of the above can be found in the references list below.

References

Anderson, C. (2008) 'The End of Theory: The Data Deluge Makes the Scientific Method Obsolete', *Wired.* http://archive. wired.com/science/discoveries/magazine/16-07/pb_theory (accessed 15 August 2015).

boyd, d., and Crawford, K. (2012). 'Critical questions for big data', *Information, Communication & Society* 15(5): 662–79. http://doi.org/10.1080/1369118X.2012.678878

Burt, J.E., Barber, G.M. and Rigby, D.L. (2009) *Elementary Statistics for Geographers.* New York, NY: Guilford Press.

Crampton, J. W., Graham, M., Poorthuis, A., Shelton, T., Stephens, M., Wilson, M.W. and Zook, M. (2013) 'Beyond the geotag: Situating "Big Data" and leveraging the potential of the geoweb', *Cartography and Geographic Information Science* 40 (2): 130–39. http://doi.org/10.1080/15230406.2013.777137

Crooks, A., Croitoru, A., Stefanidis, A. and Radzikowski, J. (2013) '#Earthquake: Twitter as a distributed sensor system', *Transactions in GIS* 17 (1): 124–47.

Crutcher, M. and Zook, M. (2009) 'Placemarks and waterlines: Racialized cyberscapes in post-Katrina Google Earth', *Geoforum* 40 (4): 523–34.

Elwood, S. (2014) 'Straddling the fence: Critical GIS and the geoweb', *Progress in Human Geography*. http://phg. sagepub.com/site/e-Specials/PHG_especial_intro.pdf (accessed 19 November 2015).

Field, K. (@kennethfield) (2014) 'I'm wondering when people will realise the animated ectoplasm twitter maps don't actually show anything http://t.co/SJVYLyBn1F' [Tweet]. 17 June. https://twitter.com/kennethfield/status/478775510386741248 (accessed 19 November 2015).

Galton, A. (2004) 'Fields and objects in space, time, and space-time', *Spatial Cognition and Computation*, 4(1): 39–68.

Goodchild, M. F. (2007) 'Citizens as sensors: The world of volunteered geography', *GeoJournal*, 69(4): 211–21. http://doi.org/10.1007/s10708-007-9111-y

Goodchild, M. F. and Glennon, J.A. (2010) 'Crowdsourcing Geographic Information for disaster response: A research frontier', *International Journal of Digital Earth* 3 (3): 231–41.

Graham, M. (2014) 'Internet geographies: Data shadows and digital divisions of labour', in M. Graham and W. Dutton (eds) *Society and the Internet: How Networks Of Information And Communication Are Changing Our Lives*. Oxford: Oxford University Press. pp. 99–116.

Graham, M. and Shelton, T. (2013) 'Geography and the future of Big Data: Big Data and the future of geography', *Dialogues in Human Geography* 3(3): 255–61. http://doi.org/10.1177/2043820613513121

Graham, M., and Zook, M. (2011) 'Visualizing global cyberscapes: Mapping user-generated placemarks', *Journal of Urban Technology* 18 (1): 115–32.

Graham, M. and Zook, M. (2013) 'Geographies: Exploring the geolinguistic contours of the web', *Environment and Planning A* 45 (1): 77–99.

Graham, M., Hale, S. and Gaffney, D. (2014) 'Where in the world are you? Geolocation and language identification in Twitter', *The Professional Geographer* 66(4). doi:10.1080/00330124.2014.907699

Graham, M., Shelton, T. and Zook, M. (2013) 'Mapping zombies', in A. Whelan, C. Moore, and R. Walker (eds) *Zombies in the Academy: Living Death In Higher Education*. Bristol: Intellect Press. pp. 147–56.

Graham, M., Zook, M. and Boulton, A. (2013) 'Augmented reality in urban places: Contested content and the duplicity of code', *Transactions of the Institute of British Geographers* 38 (3): 464–79.

Graham, M., Hogan, B., Straumann, R.K. and Medhat, A. (2014) 'Uneven geographies of user-generated information: Patterns of increasing informational poverty', *Annals of the Association of American Geographers* 104 (4): 746–64.

Hecht, B. and Stephens, M. (2014) 'A tale of cities: Urban biases in volunteered geographic information', in *Proceedings of the Eighth International AAAI Conference on Weblogs and Social Media*, Ann Arbor, Michigan, USA, June 1–4. pp. 197–205.

Kitchin, R. (2013) 'Big data and human geography: Opportunities, challenges and risks', *Dialogues in Human Geography*, 3(3): 262–67. http://doi.org/10.1177/2043820613513388

Kitchin, R. (2014) *The Data Revolution: Big Data, Open Data, Data Infrastructures and Their Consequences*. London: Sage.

Longley, P.A., Goodchild, M.F., Maguire, D.J. and Rhind, D.W. (2005) *Geographic Information Systems and Science* (2nd edition). New York: Wiley.

Openshaw S. (1984) *The Modifiable Areal Unit Problem*. Norwich: Geobooks.

Power, M. J., Neville, P., Devereux, E., Haynes, A. and Barnes, C. (2013) '"Why bother seeing the world for real?": Google Street View and the representation of a stigmatised neighbourhood', *New Media & Society* 15 (7): 1022–40.

Shearmur, R. (2015) 'Dazzled by data: Big Data, the census and urban geography', *Urban Geography*, 1–4. http://doi.org/10.1080/02723638.2015.1050922

Shelton, T., Poorthuis, A., Graham, M. and Zook, M. (2014) 'Mapping the data shadows of Hurricane Sandy: Uncovering the sociospatial dimensions of "Big Data"', *Geoforum* 52: 167–79.

Shelton, T., Zook, M. and Graham, M. (2012) 'The technology of religion: Mapping religious cyberscapes', *The Professional Geographer* 64 (4): 602–17.

Stephens, M. (2013) 'Gender and the GeoWeb: Divisions in the production of user-generated cartographic information', *GeoJournal* 78 (6): 981–96.

Stephens, M. and Poorthuis, A. (2015) 'Follow thy neighbor: Connecting the social and the spatial networks on Twitter', *Computers, Environment and Urban Systems* 53: 87–95.

Wall, M. and Kirdnark, T. (2012) 'Online maps and minorities: Geotagging Thailand's Muslims', *New Media & Society* 14 (4): 701–16.

Watkins, D. (2012) *Digital Facets of Place: Flickr's Mappings of the U.S.-Mexico Borderlands.* Unpublished MA thesis, University of Oregon Department of Geography.

Zook, M. (2010) 'The beer belly of America', http://www.floatingsheep.org/2010/02/beer-belly-of-america.html (accessed 18 November 2015).

Zook, M. and Poorthuis, A. (2014) 'Offline brews and online views: Exploring the geography of beer tweets', in M. Patterson and N. Hoalst-Pullen (eds) *The Geography of Beer.* Dordrecht: Springer. pp. 201–9.

Zook, M. and Poorthuis, A. (2013) 'DOLLY'. http://www.floatingsheep.org/p/dolly.html (accessed 18 November 2015).

Zook, M. and Graham, M. (2010) 'Featured graphic: The virtual "Bible Belt"', *Environment and Planning A* 42 (4): 763–4.

Zook, M., Graham, M. and Stephens, M. (2012) 'Data shadows of an underground economy: Volunteered Geographic Information and the economic geographies of marijuana'. Unpublished manuscript.

Zook, M., Graham, M., Shelton, T. and Gorman, S. (2010) 'Volunteered Geographic Information and crowdsourcing disaster relief: A case study of the Haitian earthquake', *World Medical & Health Policy* 2 (2): 7–33.

ON THE COMPANION WEBSITE...

Visit **https://study.sagepub.com/keymethods3e** for author videos, chapter exercises, resources and links, plus **free** access to the following recommended articles:

1. **Crampton, J.W. (2009) 'Cartography: Maps 2.0', *Progress in Human Geography*, 33 (1): 91–100.**

An early overview to the phenomenon of geotagged online social media that lay out what was and is novel about this kind of data used in geographic research.

2. **Elwood, S. (2010) 'Geographic information science: Emerging research on the societal implications of the geospatial web', *Progress in Human Geography*, 34 (3): 349–57.**

This article works to set geotagged social media into the larger theoretical contexts of critical GIS, public participation GIS, and power relations around data.

3. **Caquard, S. (2014) 'Cartography II: Collective cartographies in the social media era', *Progress in Human Geography*, 38 (1): 141–50.**

This article reviews how new data sources such as social media and ways of collecting data are introducing collective mapping practices and in so doing are changing the traditional roles of the state, the private sector and the individual.

17 Researching Virtual Communities

Mike Crang and Siti Mazidah Haji Mohamad

SYNOPSIS

This chapter asks what is distinctive about researching virtual communities. In so doing it troubles both those terms – arguing that 'virtual' may be an unhelpful descriptor, though it may throw into relief some equally problematic associations with the word 'community'. In the mid to late 1990s, researchers began to ask if a 'virtual community' could exist and if so what it might look like. This chapter highlights some of the baleful legacies left by those debates in how they defined virtual and community. Instead, we look at the actuality of virtual lives now, as we move from analyses of early adopters to mainstream formations. In so doing we track through some of the experiments in practice and analysis in the literature. We do so for four reasons: first to show the development of various concerns; second, to show the changing media and practices involved, and thus third, to show how we present inevitably a snapshot in a fast moving environment; and fourth, to suggest how some issues persist or mutate although the specific media may come and go. We do this in the context of a research issue that has moved from an esoteric interest to a mainstream phenomenon. We thus ask what new tactics or devices are called for to study 'virtual communities', how virtual communities change how we study social life generally, and what old techniques have been or can be modified to provide new insights.

This chapter is organized into the following sections:

- Thinking communities? Feeling, sentiment and social ties in a digital age
- Playing a role: Virtual game worlds and worlding virtual games
- Mediated social worlds
- Hybridised spaces of augmented communities
- Conclusion

17.1 Thinking Communities? Feeling, Sentiment and Social Ties in a Digital Age

With the rise of the Internet, largely in the last twenty years, there has been a huge concern with how this may change the nature of community. In the mid 1990s a spate of writing began to speculate whether digital communication media would erode certain forms of 'real world' community and/or create new alternative ones online. If the latter was the case, the debate was whether such communities would be 'real' and meaningful. Early approaches tended to characterise the advent of 'cyberspace' as an alternate realm that might result in the effacement of the 'real world'. The debate tended to be over the scale of such an effacement and whether it would result in positive or negative social change. It presented two distinct and apparently antagonistic domains. The 'virtual' was presented as ethereal and immaterial, which collapsed distinctions between spatial scales and was seen in terms of (accelerating or 'fast') time – promising instantaneous and real-time flows, even if actual systems tended to be more characterised by lag and buffering times. The virtual was seen as being constituted by trans-global interactions which undermined and reworked the pre-existing urban world of corporeal interaction, physical movement, and the traditional constitution of local life in physical places. The 'real' was then cast as spatially localised, materially embedded in place and occupying the time of the everyday. The effect is not only then to imagine the virtual as 'immaterial' but also to imagine the 'real' as possessing a Johnsonian solidity. In the words of one of the more apocalyptic prophets of the effects the result would be:

> tyranny of real time [where] the city of the past slowly becomes a paradoxical agglomeration in which relations of immediate proximity give way to interrelationships over distance. (Virilio, 1993: 10)

The virtual became associated with global, distantiated and fast. The local becomes the concrete, the human and the slow. The danger is not merely that this hyperbole reads off one kind of 'virtual world' but that it does so by mischaracterising the physical world too. As Thrift (2004) notes, we should be suspicious of the habitual coding of the everyday as small scale and local. Instead the everyday is where 'communities are far-flung, loosely-bounded, sparsely-knit and fragmentary. Most people operate in multiple, thinly-connected, partial communities as they deal with networks of kin, neighbours, friends, workmates and organizational ties' (Wellman, 2001: 227). Moreover, far from being bounded, 'communities consist of far-flung kinship, workplace, interest group, and neighbourhood ties concatenating to form a network that provides aid, support, social control, and links to other milieus' (Wellman and Hampton, 1999: 649).

In other words, many 'real world' communities are subtended by communication technologies – from paper letters to Voice Over Internet Protocol telecoms (like ooVoo, GoogleTalk and Skype). Telecoms have played a huge role in sustaining transnational communities among migrant workers or, as we shall see, international students and their families and friends at home (Vertovec, 2004; Uy-Tioco, 2007, Diminescu, 2008). Nor are such prosthetics confined to the distant, infrequent and remote. Many localised friendship groups are sustained by, and enacted through, applications such as instagram and whatsapp, while families' lives are connected and coordinated through a variety of devices from SMS messaging, to programmed recording functions and facetime apps (Morley, 2003; Wajcman et al., 2008).

Box 17.1   Tracing the social spaces of virtual communities in South Korea

Digital media were seen as antagonistic to the centralized and socially conservative media tradition-ally found in South Korea. As such they were the subject of periodic moral panics about young adults disconnecting from Korean norms of harmonious sociality. New communication technologies were feared to individualize human relationships and consequently destroy collective or familial affective relationships (Yoon, 2003: 328; Yoon, 2006).

However, researching the circuits of who messaged whom, how often, and why suggested that by offering new means of sharing intimacies (especially among young women), new media enhanced existing patterns of local sociality but without visible public arenas (Yoon, 2003: 339-40). Thus 'minihompys', as the home page feeds on smart phones were called, coopted the older sense of 'chon/cheong'. That latter term originally referred to the idea of relationship in terms of degrees of kinship separation but it now fits surprisingly well with the degrees of separation of friends and friends of friends on social media. To show commitment to these virtual bonds, users perform a *1-chon soonhwe*, or 'first degree connection tour', where they leave a message on their *1-chons* one by one as a gesture of courtesy (Choi, 2006: 177). Tracing the networks revealed a form of social space or 'cheong space' in which digital community practice appropriated tradition in identity formation (Yoon, 2003).

If real and virtual are entwined in producing communities that are now and have always been fractured, fragmented, and composed of the far-flung and near-by in dif-ferent measures, what might we say then of any effects of specific media on them? Gradually then what has emerged is research that examines information and commu-nication technologies not (at least simply) as a separate realm of cyberspace which is the domain of virtual communities, but prosthetic technologies which enhance and shape action over multiple times and distances. The research methods then are about how different technologies and media are used, to what ends and with what effects.

Some research has thus followed the role of media in the transmission, production and maintenance of feelings. New media render feelings more accessible to research – perhaps too temptingly so sometimes. Social media render the back and forth of social life perceptible (be that by academics, governments or more often corporations) through the digital traces – the data exhaust – they leave (see Chapter 16). Our banal social lives become digitally mediated and can be subject to quantitative encapsulation through lexical analysis. For instance, Alan Mislove and colleagues applied a word-rating system – scoring positive and negative connotations – to US-based geolocated tweets to produce stunning time lapse maps of the 'mood of the nation' (Mislove et al., 2010). Similar approaches have correlated emotional terms in tweets with stock mar-ket movements (Bollen et al., 2011) and yet so far the conclusions have been banal. The poetics and affective power of the visualisation have often been more powerful than the analysis. This kind of research renders apparent the transmission of affects and feelings to social life, but also its complicated and non-scalar nature. Complicated in the sense of the etymological root term 'pli' that is to fold or bend. Rather than just scraping the geolocations of tweets or photos to map the distribution of a pre-existing community in Euclidean space, tracing interactions by responses and reactions reveals how specific events in specific places bring together people from multiple locations to form more less transient or stable communities of interest (Crampton et al., 2013) that bend and distort senses of who is proximate and who is distant, with some physically nearby being bypassed and some remote others centrally involved.

Approaching the 'Big Data' of social media raises the question of quantifying the social life of a community. And yet a great deal of humanistic geographic work on emotions or affects has started from postulating them as unquantifiable. In contrast, Latour and Lépinay (2009) turn to Gabriel Tarde. At the start of the 20th century, Tarde argued that the problem of scientific study of society was not that it quantified, but that its measure (generally money) was wrong. He wondered about the possibilities of developing metrics for fame, charisma, or happiness by creating 'valuemeters' or 'glorimeters'. Latour and Lépinay note that social media now offer tools that indeed work through 'the calculation of authority, the mapping of credibility and the quanti-fication of glory' (2009: 29). Tracking influencers, followers, cross-postings and recirculations are all now common currency of both social judgment and research. Rendering the traffic of cultural life more perceptible, together with how different media render different actions more or less visible, now also alters how communities may be seen by their own members and by outsiders (be they researchers or other social actors). We should also be wary that in using such methods we are replicating *current* patterns of interaction, which those media are designed to serve; they tend to be conservative in repeating and reinforcing communication patterns and communica-tors that are currently dominant. As researchers we need to be clear that using social media as 'evidential' makes us 'not mere observers or utilisers of social media content but ... promoters of this infrastructure' where our analysis is going to be shaped by the affordances of the specific social media platform(s) we use (Wilson, 2015: 347).

17.2 Playing a Role: Virtual Game Worlds and Worlding Virtual Games

If there is a paradigmatic example of an apparently virtual community it is that created in online games. Emerging from role playing games conducted by groups around physical boards, the very first online games date back to the early 80s as so-called Multi-User Dungeons (MUD – later made to stand for the more generic Multi User 'Domains', losing the sword and sorcery content). By the 1990s the addition of graph-ical interfaces allowed Object Oriented MUDs (MOOs) which in turn evolved into both the increasingly popular Massively Multiplayer Online Role Playing Games (MMORPGs) and Virtual Worlds, that leave out roleplaying, and mixtures of both tailored for different ages – from Disney's Club Penguin for under 8s, who furnish their own virtual igloos, through to MovieStarPlanet for tweens who outfit their avatars, to variants of Minecraft, and MMORPGS such as World of Warcraft which on its own boasted over 5.5 million current players at the end of 2015 (http://www.statista.com/statistics/276601/number-of-world-of-warcraft-subscribers-by-quarter/). Even individ-ualized games have spawned communities of players with YouTube channels sharing moves and celebrating players' exploits. How might we research these apparently virtual communities? In this section we outline four different approaches that have been used. The first is iconographic, the second is using online analytics, the third is online ethnography and the fourth is a hybrid ethnography.

Iconographic analysis suggests there is little novel in the ideological content of many video games' depiction of racialised Others. We might see a not-so crypto-Orientalism in the conversion of Middle Eastern cities into backdrops and theatres of action, and of people into targets and victims, in 'first person shooter' games positioning the player as a western soldier (Leonard, 2003). Games such as *Full Spectrum Warrior: Ten hammers* use simulated news media coverage, in a CNN style, as framing devices (Höglund, 2008). The long running franchise of *Grand Theft Auto* plays on an American urban imaginary taken from TV and film – where Miami becomes Vice City in a homage to a 1980s cop show. It also trades on readily mobilized racialised and sexualised stereotypes to animate urban spaces based on Othering different neighbourhoods, associating race and crime, and where women all too often feature as prostitutes or victims (Atkinson and Willis, 2007).

A second approach is to use **online analytics** to look at the socialities and spatialities within online worlds. A start could be to actually map where the virtual characters or 'avatars' of players are in the simulated environment of the game, to look at patterns of activity (Börner and Penumarthy, 2003; Penumarthy and Börner, 2006). Studies then have mapped the patterns of play of characters through spatial game play analytics, often using software like Tableau (http://www.tableau.com/solutions/game-development) or this kind of functionality is currently being developed for the Unity game engine (https://unity3d.com/services/analytics) which can produce heat maps of areas which attract more players or where particular types of activities occur (for instance where people quit the game, where they pass through quickly or linger) or routes people take through the game environment. The techniques and visualisations are often the same as those used in real world geospatial analysis, though real world applications are less often focused on the likely movement of an orc horde (Drachen and Canossa, 2011; Drachen and Schubert, 2013). That said, the US Department of Defence and others have used examples of malware being introduced by hackers (a simulated plague that killed characters) to simulate likely real world reaction to biological weapons (Lofgren and Fefferman, 2007).

Other studies have done online ethnographies of the social life in diegetic spaces. Such then is to look at a game as working to enable sociality within a 'magic circle' that absorbs us and is walled off from everyday life (Copier, 2009). Within virtual worlds, the ability to shape the environment has been used by subcultural groups to articulate their identity; for instance in a study of a virtual Gorean community in the virtual world *Second Life* (www.secondlife.com), Bardzell and Odom (2008) note how the ritualized bodily performances and differentiated spaces of city and jungle, market place and tavern were used to enact the social stratifications. More generally one can analyse the online mechanisms used to perform (and play with) gender (Eklund, 2011) or racial identities (Monson, 2012). Within MMORPGS there can also be the development of dense socialities structured through online world mechanics – such as a proliferation of relatively small 'guilds' of likeminded players in World of Warcraft (Williams et al., 2006).

This would seem to confirm and amplify the sense of virtual spaces as isolated realms, but increasingly players break the magic circle both by para-game functions (with text messaging or communicative video interfaces of differing kinds between players on the edge of the game screen) (Ducheneaut et al., 2006: 284), and more channels of communication between players than just within the online world through YouTube channels playing game highlights, spin-off events, like 'Fantasy Fairs', 'Cosplay' festivals and meetings in real life (Copier, 2009). Indeed, the exchange of online gifts and goods breaks down any easy divide of online or virtual world versus real – with 'real' friendships made in online worlds. This link to outside worlds can be extended. There is a flourishing trade in virtual objects and skills whereby people use real money to buy virtual objects, magical swords or the like (Malaby, 2006). The term for this is 'Real Money Trading' and the scale of it is considerable. One company, Internet Gaming Entertainment, at its peak employed 500 people doing over a quarter billion dollars of business in trades by acting as a 'middleman for Western gamers eager to outsource the boring aspects of play to low-wage third worlders' (Salo, 2008; see also Kent, 2008). The result is an entire industry of sweat-shop gamers, in many low-income countries, but especially China, playing for hours in cramped conditions aiming not to 'win the game' but to accumulate specific arte-facts for sale – what is called 'Gold Farming'. Researching this could involve a hybrid ethnography tracing the outsourced geographies and back offices as in other globalized service sectors, alongside the (multiple, different) online and offline interactions.

Box 17.2   Hybrid Ethnographies: Where to research virtual communities spilling across different media and different material spaces?

Cockshut's (2012) participant observation on World of Warcraft for instance involved taking part with a self-defined group of players (a 'raiding guild') online, interviewing them around the planet via Skype and email, and looking at their posted video recordings of the game action. This chal-lenges the usual sense of participant observation – it entailed multiple places, and watched char-acters coming together in a simulated place, using media at every turn. And yet it also seems very familiar mixing interviews, observed behaviour and hanging out with people. In fact the new data possibility of using the computer logs was the one method that did not work since the detail there swamped the picture of how gamers behaved or related to each other. So researching virtual com-munities and online groups may well end up mean traveling to find members in the material and technical systems they use for more conventional interviews (Longan, 2015).

17.3 Mediated Social Worlds

The above suggests the intertwining of online communities and 'real world' ones – so that virtual communities tend to reinforce, not displace, other communities. A US survey found one-third of adults say that the Internet has improved their connections with friends 'a lot,' and nearly one-quarter that it has greatly improved their connec-tions with members of their family (Wellman et al., 2008). The effect has thus been one of reinforcing existing social networks. But within this, if we think about researching these networks, one might ask 'who' is doing what networking? Women have historically borne the primary responsibility in heterosexual couples for main-taining family networks and contacts (di Leonardo, 1987) – and this pattern has continued when this has moved online, with women primarily responsible for staying in touch with friends and family in 28% of married households, while men are the primary connector in just 4% of cases (Wellman et al., 2008: 23). Nor should we leap to the conclusion that families are fragmenting with each member connecting solely with their own social networks via different electronic channels. As Wellman et al. note 'multiple computers in the household does not necessarily lead family members to be in their own isolated technological corners' and 'married with children households with multiple computers have as much (if not more) shared time online with others than single-computer households' (Wellman et al., 2008: 16).

Some approaches see social networks as offering a natural experiment in social ties and habits data. A great many analysts have eyed the social graph of Facebook or other media greedily, and occasionally uncritically, as a source of large amounts of data. Indeed access to large phone company datasets allowed the mapping of Britain's regional communities of interaction according to landline calls (Ratti et al., 2010). These studies depend on getting access to these large datasets. But they are still partial – thus the UK telephone database of 22 million landlines is big but misses mobile traffic. Would that pattern be different? The honest answer is we do not know. So 'Big Data' does not mean complete data, and the usual caveats of selection, exclusion and the like apply (see Chapter 16).

However, these media are not simply reflecting behaviour or values but also part of the self-reflexive performance of those values. For instance, tastes (all those likes and favourites) are not simply being disclosed but performed for audiences (Lewis et al., 2008). People's postings and volunteered information on social media are influenced by the contexts in which they find themselves, where they make choices over their self-disclosure and self-censorship. Another research approach then is to see social media as a kind of performance arena. Drawing on the work of Erving Goffman, an approach called *symbolic interactionism* has focused on looking at how we perform identities. It argues we create a version of our self through performing an identity in a context for an audience. So at various times and in different places you may perform different identities – be that eager student, loving child, caring friend, lover, diligent worker and more. Goffman (2005) called this the regionalisation of performance in different locales. Some of these might be a front stage where we put on a specific performance and others a back stage where we might let the mask slip. So what you say to a tutor about the progress of your work may be different from what you say to friends in a coffee shop. More scrupulously though, that later 'back' stage may equally be a performance, of artfully disavowing the hard work you have done to make yourself appear less keen. These performances then are subject to group norms about what is acceptable, what will earn approbation and what reproach. Goffman thus spoke of 'facework' in terms not only facial expressions but as in the sense of not 'losing face' in interactions. In more fraught environments and for oppositional virtual communities, there is also actual censorship and the possibility of (extra-)legal sanctions for different sorts of self-presentation (Bamman et al., 2012).

Looking at online games, one can then see the echoes of this in the playing of roles. However, such environments are often pseudonymous where players are known by character names or player handles. Many social media by contrast are 'nonymous environments' (Zhao et al., 2008) in that they function by linking virtual and physical self-identities. But issues of self-presentation are just as applicable to them, where we choose to present different characteristics to a particular online audience. Some social media have options to differentiate audiences via account settings, but issues of virtual communities not aligning with usually separate identities performed in different contexts abound. For instance, many social disasters have come from 'frupervisors' (friends who are supervisors at work) seeing posts that conflict with accounts given of, say, illness. Many a teenager will not have parents as contacts and so forth. Virtual communities may also bring spatially discrete audiences into contact. New media can thus collapse social contexts.

In Siti Haji Mohamad's study of ethnically Malay Malaysian international students (Haji Mohamad, 2014) the focus was on how they used Facebook to remain in contact with friends and family at home but also deal with the way their behaviour as students in the UK might be judged by Malaysian standards. In this sense there were questions of whether their real world behaviour might change, how they might reconcile it with a virtual community that stretched across the globe and whether a globalised virtual community might actually be constituted among similarly mobile students in different locations through Facebook. To examine this, Siti used intermediaries to contact people, as well as asking around student societies, and then 'friended' various students. However, the simple graphing of networks and contacts did not really reveal how they were using the technology to manage their identity. To examine that, she assessed the use of different online technologies – Skype and Yahoo messenger, as well as Facebook. Then within Facebook, she documented what they posted, but also interviewed them about their choices, and looked at a range of factors: motivation for posting, self-reflexivity about their identity and audience, the collapsed contexts and the audiences with which they interacted, and if they used the features of Facebook to manage those audiences. She identified different strategies for managing audiences and performances. Some students selected who could see what, while others monitored their postings – being careful not to publish 'unsuitable' photos or updates – and worked to avoid being tagged in pictures by others depicting such. Still others reaffirmed their Malay Islamic identity through motivational and religious postings. In short, their identity performances were contextualised, spatialised and temporalised in manners that resonated with the varied everyday contexts. But a simple recording of what was posted or who was connected would hardly capture this active management of their presence in this virtual community.

Social media do not merely represent social life then but are the very fabric of sociality for some groups. Thus contemporary youth culture has evolved along with and using social media. The creation of youth culture online should and must be studied in order to understand the dynamics of youth culture generally – to not do so is to miss part of their lives.

Box 17.3   Ethics and online data

At first it may seem that analysing public postings and actions is easy and non-contentious ethically. This can be a flawed assumption when continual changes in online privacy settings mean people are not always aware of their own online exposure. However, even looking at some people's postings and pictures with their consent may well spill into watching other people's behaviour without their consent or even knowledge – when they are tagged in pictures, when third parties

(Continued)

(Continued)

share their materials and so forth. Just taking simple things like screen grabs may well reveal other users in terms of the information shown. Even if such data are then anonymised, they may well retain information that would subsequently allow identification if made public. The fact that data are accessible does not make their use ethical or in some cases legal (DeLyser and Sui, 2014). A variety of laws can prohibit the reuse of data for purposes other than they were originally intended for or their transport across borders – for instance, there are strict laws on moving personal data on EU citizens outside of Europe.

17.4 Hybridised Spaces of Augmented Communities

If we have thus far suggested that there are online forums which may sustain social interaction, some of which occurs virtually and some in the actual world, then we also need to consider how community and social action in the actual world are altered by the addition of virtual dimensions. We will do this through two examples. The first is to look at online gaming communities which transform the uses of concrete and specific places. The second is to look at hybrid games where the diegetic space of game play itself is fused with the physical world. In both cases we are pushing further with hybrid ethnographies of online communities, mixing interviews in real places with online data to think more about how these virtual elements may transform physical places. One term used to describe this is to think of places being augmented by layers of online data and media that overlay the physical place.

In talking about researching online games we touched on hybrid methods of tracking dispersed virtual communities of gamers. However, for some games and communities these are not just dispersed but concentrated in clusters of 'technosocial spaces' which mix the physical local space of physical association with distantiated actors. For example, in South Korea there are around 10,000 gaming cafés or *PC Bang* which each typically have up to 100 high spec work stations. About 50 per cent of them are in the capital region, and almost 25 per cent are located in central Seoul (Lee, 2005). One might expect such sites to have disappeared as residential broadband access increased. But they still thrive, as they foster specific temporalities of behaviour and rhythms of activities precisely as a third place of technosociality around gaming – allowing young adults to congregate and share online experiences. Sinchon, a university quarter in Seoul, developed a new urban consumption landscape based around these technospaces. They fostered a distinctive and spatially concentrated youth culture – which in turn became the object of 'moral panics' in the media regarding online addiction and social alienation as youngsters came together to play irrespective of the time of day. Researching 'virtual communities' of gaming here might well involve working through localized physical sites.

The second augmented environment is created by *hybrid-reality* or *locative games*. These are multiuser games generally played on smart phones that allow location awareness and data connections. These enable players to turn the city space into a game board. The first commercially released game was *BotFighters*, produced in Sweden in 2001. Based on the non-online game of Assassin, this was a game where players sought to tag each other in real space via mobile phone texts. It has been followed into numerous similar games where opponents track each other down in urban neighbourhoods as their mobile phones provide them with information on where other opponents are (Shirvanee, 2006). Many games have followed, some for individuals, some where players have to work as a team positioning themselves in the city, converting the urban fabric into a game board. Some now use augmented reality interfaces where looking through a smartphone camera shows players as zombies mounting attacks. The games 'create an imaginary playful layer that merges with the city space, connecting people who previously did not know one another via mobile technologies according to their movement in physical spaces' (de Souza e Silva, 2006: 272).

Box 17.4   Augmented reality

'Augmented reality' has been defined in contrast to virtual reality. Virtual reality attempts to simulate the look, shape and, sometimes, the feel, of physical objects and the tangible environment in online environments. Augmented reality instead overlays physical objects and environs with virtual objects and dynamic data. The world has, of course, always been overlaid with semiotic information, but the difference is the possibility for this to be dynamic, both in the sense of changing in real time and possibly individualized for different users. The information can then also be interrogated. For instance, many museums have been trialling interfaces, sometimes based on QR codes, the small squares of black and white patterns that can identify an object or display and link it to a database so a viewer can look up background details (like further details about the display, or who donated the object or other similar holdings). The process thus enriches the environment with greater information that is dependent not on the physical structure of the object, but on linking it with one or more databases. By being open to encounters with different objects and rendering the data interrogatable rather than being a fixed narrative, this technique has gone beyond the audio tour. Interfaces and processes for this have developed rapidly with smartphones using cameras to produce an image on screen that can be overlaid with information from databases. Applications have, for instance, enabled users to visualize nearby twitter activity in real time and real space.

Interest has emerged in connecting such locative media with social media – to offer location-based social media. Far from there being a conflict of virtual communities as distanciated or aspatial communities of interest contrasting with physical communities of proximity, much evidence now points to their confluence. The question arises as to whether there might be new digital scaffolding to support revivified forms of local community (Gordon and de Souza e Silva 2011). From people seeking to become e-mayor of their coffeeshop on foursquare, with 'checkins' and ratings gamifying the everyday, to local discussion groups on Facebook, a variety of new media are focusing on shared engagements with local places.

Furthermore, data have rapidly been adopted by local community activists as a tool. In other words, tools like GIS become not merely a means of representing a 'community' by geodemographics, but its use can become an active tool in fostering community through public participation and engagement. Local publics can be constituted through these means of representing themselves, often proffering their own data to contest, complement or complete official data (see also Chapter 13). These might be data enabled community formations almost as much as online games. Of course there are significant issues in terms of which communities thus were visible from official data and which are visible through generating their own data – and those that are invisible in both (for volunteered geographical information (VGI) see Chapter 16).

17.5 Conclusion

In the 1990s a vanguard of geographers, mostly moving from the study of media and communication, began taking new media seriously in terms of possible effects on community formation. The focus then was on the role of 'virtual' and online spaces as separate venues and alternative formations to 'physical world' communities. Such work pointed out the effect of media on dislocating, distantiating and refashioning social connections over time and space. New communities reconnected people separated over time (reanimating former social ties) and (re-)connected people currently distant from each other with shared interests. Initial studies and techniques focused on assessing whether these new formations were 'real' or 'authentic' and if they had limits in terms of commitment and engagement. They were measured against a model of real communities as materially practised, multi-dimensional, emotionally rich, temporally enduring and spatially bounded groups of people. What became clear was that 'real communities' were just as spatially complex as virtual ones, and that social relations mediated by online platforms were always materially practised, and could be as multi-dimensional, emotionally complex and temporally enduring as any others.

The study of digital communication and communities has thus moved to use hybrid methods looking at how the digital and material interact and hybridise each other. Some studies focus on patterns of online interaction and use, while others focus on the material embedding of actors. Moreover, as actors often use multiple media and move between them, research has tended to follow the same pattern. Indeed, geographical work has pointed to the rise of 'place' rather than its disappearance in how online communities work. Social networking sites can span vast distances but are also now deeply involved in shaping interaction and community in place. 'Real' world physical communities are now often subtended and enabled by online media. In addressing this, many conventional methods are used alongside the ability of online data to trace actions and offer quantitative summaries. However, using such data raises familiar and ethical issues of data partiality that any study needs to address.

SUMMARY

- There are significant debates over the emergence and characterisation of 'virtual communities', and what this means for research and for researchers. Research generally examines information and communication technologies not just as a realm of cyberspace, but as new technologies which enhance and shape action over multiple times and distances.
- One early and still growing kind of virtual community is that associated with online games. There are several ways these may be researched: iconographically, using online analytics, online ethnography, and through a hybrid ethnography.
- Social networking sites and social media form the other great area of 'virtual community'. There are different views as to how this does and does not characterise a virtual

community, how ordinary social life is increasingly mediated, and what this might mean for research. Those media are designed to serve *current* patterns of interaction, and thus they tend to be conservative in repeating and reinforcing communication patterns and communicators that are currently dominant. Using social media as 'evidential' further promotes the infrastructure of this communication, and analysis is going to be shaped by the affordances of the specific social media platform(s).

- It is also possible to consider 'virtual' communities in hybrid formations with location-based social networking and augmented reality media – how community and social action in place are altered by the addition of virtual dimensions. What results are hybrid ethnographies of online communities, mixing interviews in real places with online data to think more about how these virtual elements may transform physical places.

Further Reading

The following books and articles provide good examples of researching virtual communities:

- Driscoll and Gregg (2010) offer a corrective for anyone tempted to think an online ethnography is quick and easy – showing the complexities of engaging with online fora. They show how online ethnography challenges certain notions of presence but also repeats some elements of intimate understandings of how communities work.

- Goffman (2005) provides a wide ranging and accessible entry point into the many works by him, and to symbolic interactionism generally. Originally published in 1967 it shows its age occasionally, and accordingly has nothing to to say about mediated or digital encounters, but it is a classic in understanding how people's everyday interactions with each other are structured.

- Gordon and de Souza e Silva (2011) provide a series of examples of different social media fusing into and transforming daily life. There is less methodological discussion but this article is useful to get a sense of diverse kinds of communities, their locations and digital mediation.

- Hargittai and Sandvig (2015) have a series of more or less confessional essays by prominent researchers telling of pitfalls, triumphs and dilemmas.

- Longan (2015) offers an account of researching online community through looking at the materials and web sites involved, as well as through extensive interviews with the people who make it happen.

- Luh Sin (2015) provides an account of using social media while doing fieldwork and later to keep in touch with a case study while far away and the ethical conundrums it posed of what constituted fieldwork and what was shared with whom in what she terms the Field 2.0.

Note: Full details of the above can be found in the references list below.

References

Atkinson, R. and Willis, P. (2007) 'Charting the Ludodrome: The mediation of urban and simulated space and rise of the *flaneur electronique*', *Information, Communication & Society* 10 (6): 818–45.

Bamman, D., O'Connor, B. and Smith, N. (2012) 'Censorship and deletion practices in Chinese social media', *First Monday* 17 (3). http://firstmonday.org/article/view/3943/3169 (accessed 20 November 2015).

Bardzell, S. and Odom, W. (2008) 'The experience of embodied space in virtual worlds: An ethnography of a second life community', *Space and Culture* 11: 239–59.

Bollen, J., Mao, H. and Zeng, X. (2011) 'Twitter mood predicts the stock market', *Journal of Computational Science* 2 (1): 1–8.

Börner, K. and Penumarthy, S. (2003) 'Social diffusion patterns in three-dimensional virtual worlds', *Information Visualization* 2 (3): 182–98.

Choi, J.H.-j. (2006) 'Living in Cyworld: Contextualising Cy-Ties in South Korea', in A. Bruns and J. Jacobs (eds) *Use of Blogs*. New York: Peter Lang. pp. 173–86.

Cockshut, T. (2012) 'The Way We Play: Exploring the specifics of formation, action and competition in digital gameplay among World of Warcraft raiders', Doctoral Thesis, Durham University, Durham. http://etheses.dur.ac.uk/5931/ (accessed 20 November 2015).

Copier, M. (2009) 'Challenging the magic circle: How online role-playing games are negotiated by everyday life', in M. van den Boomen, S. Lammes, A.-S. Lehmann, J. Raessens and M. T. Schäfer (eds) *Tracing New Media in Everyday Life and Technology*. Amsterdam: University of Amsterdam Press. pp. 159–72.

Crampton, J.W., Graham, M., Poorthuis, A., Shelton, T., Stephens, M., Wilson, M.W. and Zook, M. (2013) 'Beyond the geotag: Situating "big data" and leveraging the potential of the geoweb', *Cartography and Geographic Information Science* 40 (2): 130–9.

de Souza e Silva, A. (2006) 'From cyber to hybrid: Mobile technologies as interfaces of hybrid spaces', *Space and Culture* 9 (3): 261–78.

DeLyser, D. and Sui, D. (2014) 'Crossing the qualitative-quantitative chasm III: Enduring methods, open geography, participatory research, and the fourth paradigm', *Progress in Human Geography* 38 (2): 294–307.

di Leonardo, M. (1987) 'The female world of cards and holidays: Women, families, and the work of kinship', *Signs* 12 (3): 440–53.

Diminescu, D. (2008) 'The connected migrant: An epistemological manifesto', *Social Science Information* 47 (4): 565–79.

Drachen, A. and Canossa, A. (2011) 'Evaluating motion: Spatial user behaviour in virtual environments', *International Journal of Arts and Technology* 4 (3): 294–314.

Drachen, A. and Schubert M. (2013) 'Spatial Game Analytics', in M. Seif El-Nasr, A. Drachen, A. Canossa and M. Schubert (eds) *Game Analytics: Maximizing the Value of Player Data*. Dordrecht: Springer. pp. 365–402.

Driscoll, C. and Gregg, M. (2010) 'My profile: The ethics of virtual ethnography', *Emotion, Space and Society* 3 (1): 15–20.

Ducheneaut, N., Yee, N., Nickell, E. and Moore, R.J. (2006) 'Alone together?: Exploring the social dynamics of massively multiplayer online games.' Proceedings of the SIGCHI conference on Human Factors in Computing Systems pp. 407–16.

Eklund, L. (2011) 'Doing gender in cyberspace: The performance of gender by female World of Warcraft players', *Convergence: The International Journal of Research into New Media Technologies* 17 (3): 323–42.

Goffman, E. (2005) *Interaction Ritual: Essays in Face-to-Face Behavior*. New York: AldineTransaction.

Gordon, E. and de Souza e Silva, A. (2011) *Net Locality: Why Location Matters in a Networked World*. Oxford: John Wiley & Sons.

HajiMohamad, S.(2014) 'Rooted Muslim cosmopolitanism: An ethnographic study of Malay Malaysian students' cultivation and performance of cosmopolitanism on Facebook and offline', Doctoral Dissertation, Durham University. http://etheses.dur.ac.uk/10871/ (accessed 20 November 2015).

Hargittai, E. and Sandvig, C. (2015) *Digital Research Confidential: The Secrets of Studying Behavior Online*. Cambridge, MA: MIT Press.

Höglund, J. (2008) 'Electronic empire: Orientalism revisited in the military shooter', *Game Studies* 8 (1). http://gamestudies.org/0801/articles/hoeglund (accessed 20 November 2015).

Kent, M. (2008) 'Massive Multi-player Online Games and the developing political economy of cyberspace', *Fast Capitalism* 4 (1). https://www.uta.edu/huma/agger/fastcapitalism/4_1/kent.html (accessed 20 November 2015).

Latour, B., and Lépinay, V.A. (2009) *The Science of Passionate Interests: An Introduction to Gabriel Tarde's Economic Anthropology*. Chicago, IL: Pricky Paradigm Press.

Lee, H. (2005) 'Multimedia and the hybrid city: Geographies of technocultural spaces in South Korea', Doctoral Dissertation, Durham University. http://etheses.dur.ac.uk/2727/1/2727_804.pdf (accessed 20 November 2015).

Leonard, D. (2003) '"Live in your world, play in ours": Race, video games, and consuming the Other', *Studies In Media & Information Literacy Education* 3 (4): 1–9.

Lewis, K., Kaufman, J., Gonzalez, M., Wimmer, A. and Christakis, N. (2008) 'Tastes, ties, and time: A new social network dataset using Facebook.com', *Social Networks* 30 (4): 330–42.

Lofgren, E. T. and Fefferman, N. H. (2007) 'The untapped potential of virtual game worlds to shed light on real world epidemics', *The Lancet Infectious Diseases* 7 (9): 625–9.

Longan, M.W. (2015) 'Cybergeography IRL', *Cultural Geographies* 22 (2): 217–29.

Luh Sin, H. (2015) '"You're not doing work, you're on Facebook!": Ethics of encountering the field through social media', *The Professional Geographer* 67 (4): 676–85.

Malaby, T. (2006) 'Parlaying value: Capital in and beyond virtual worlds', *Games and Culture* 1 (2): 141–62.

Mislove, A., Lehmann, S., Ahn, Y.-Y., Onnela J.-P. and Rosenquis, J. (2010) 'Visualisation of the Twitter Pulse of the Nation', http://www.ccs.neu.edu/home/amislove/twittermood (accessed 20 November 2015).

Monson, M.J. (2012) 'Race-Based Fantasy Realm', *Games and Culture* 7 (1): 48–71.

Morley, D. (2003) 'What's home got to do with it? Contradictory dynamics in the domestication of technology and the dislocation of domesticity', *European Journal of Cultural Studies* 6 (4): 435–58.

Mortensen, T.E. (2006) 'WoW is the New MUD: Social Gaming from Text to Video', *Games and Culture* 1 (4): 397–413.

Penumarthy, S. and Börner, K. (2006) 'Analysis and visualization of social diffusion patterns in three-dimensional virtual worlds', in R. Schroeder and A. Axelsson (eds) *Avatars at Work and Play*. Dordrecht: Springer. pp. 39–61.

Ratti, C., Sobolevsky, S., Calabrese, F., Andris, C., Reades, J., Martino, M., Claxton, R. and Strogatz, S.H. (2010) 'Redrawing the map of Great Britain from a network of human interactions', *PLoS ONE* 5 (12): e14248. http://doi.org/10.1371/journal.pone.0014248 (accessed 20 November 2015).

Salo, D. (2008) 'How the virtual gold trade works', *Wired* 16 (12) http://www.wired.com/2008/12/ff-ige-howto/ (accessed 20 November 2015).

Shirvanee, L. (2006) 'Locative viscosity: Traces of social histories in public space', *Leonardo Electronic Almanac* 14 (3). http://www.leoalmanac.org/wp-content/uploads/2012/07/Locative-Viscosity-Traces-Of-Social-Histories-In-Public-Space-Mapping-The-Emerging-Urban-Landscape-Vol-14-No-3-July-2006-Leonardo-Electronic-Almanac.pdf (accessed 20 November 2015).

Thrift, N. (2004) 'Driving in the City', *Theory, Culture & Society* 21 (4/5): 41–59.

Uy-Tioco, C. (2007) 'Overseas Filipino workers and text messaging: Reinventing transnational mothering', *Continuum: Journal of Media & Cultural Studies* 21 (2): 253–65.

Vertovec, S. (2004) 'Cheap calls: The social glue of migrant transnationalism', *Global Networks* 4 (2): 219–24.

Virilio, P. (1993) 'The third interval: A critical transition', in V. Andermatt-Conley (ed.) *Rethinking Technologies*. London: University Of Minnesota Press. pp. 3–10.

Wajcman, J., Bittman, M. and Brown, J.E. (2008) 'Families without Borders: Mobile phones, connectedness and work-home divisions', *Sociology* 42 (4): 635–52.

Wellman, B. (2001) 'Physical place and cyberplace: The rise of personalized networking', *International Journal of Urban and Regional Research* 25 (2): 227–52.

Wellman, B. and Hampton, K. (1999) 'Living networked on and off line', *Contemporary Sociology* 28 (6): 648–54.

Wellman, B., Smith, A., Wells, A. and Kennedy, T. (2008) *Networked Families*, 55. Washington, DC: Pew Internet & American Life Project.

Williams, D., Ducheneaut, N., Xiong, L., Zhang, Y., Yee, N. and Nickell E. (2006) 'From tree house to barracks the social life of guilds in World of Warcraft', *Games and Culture* 1 (4): 338–61.

Wilson, M.W. (2015) 'Morgan Freeman is dead and other big data stories', *Cultural Geographies* 22 (2): 345–9.

Yoon, K. (2003) 'Retraditionalizing the mobile: Young people's sociality and mobile phone use in Seoul, South Korea', *European Journal of Cultural Studies* 6 (3): 327–43.

Yoon, K. (2006) 'The making of neo- Confucian cyberkids: Representations of young mobile phone users in South Korea', *New Media and Society* 8 (5): 753–71.

Zhao, S., Grasmuck, S. and Martin, J. (2008) 'Identity construction on Facebook: Digital empowerment in anchored relationships', *Computers in Human Behavior*, 24: 1816–36.

ON THE COMPANION WEBSITE…

Visit **https://study.sagepub.com/keymethods3e** for author videos, chapter exercises, resources and links, plus **free** access to the following recommended articles:

1. Crang, M. (2005) 'Qualitative methods: There is nothing outside the text?', *Progress in Human Geography*, 29 (2): 225–33.

2. Lorimer, H. (2007) 'Cultural geography: Worldly shapes, differently arranged', *Progress in Human Geography*, 31 (1): 89–100.

3. Dwyer, C. and Davies, G. (2010) 'Qualitative methods III: Animating archives, artful interventions and online environments', *Progress in Human Geography* 34 (1): 89–97.

18 Critical GIS*

Matthew W. Wilson

*This chapter is based on 'GIS: A method and practice', which appeared in *Researching the City* (Sage, 2013).

SYNOPSIS

GIS is both a method and a practice. The use of the acronym 'GIS' may refer to a specific software package or a whole suite of visual technologies used to represent the Earth. GIS is not just a tool, nor is it just a software package. GIS may simply be a way to produce representations of your research results. Or, GIS will be used to conduct analyses of spatial phenomena. GIS may also be used as a prompt or an illustration to engage research subjects. This chapter outlines the various ways in which GIS may be used in research, providing examples to discuss the appropriateness of GIS as a method, while giving the basics for examining GIS as a practice. Critical use and applications of GIS at each stage of the research process involve fundamental decisions about how to observe, measure, analyse and represent the world. These fundamentals can be approached pragmatically, and provide necessary steps towards an engaged and responsible use of this technology in a critical framework.

This chapter is organized into the following sections:

- Introduction
- Criticality and GIS
- Preparing for research
- Doing Research
- Writing up Research
- Conclusion

18.1 Introduction

Geographic information systems (GIS) provide the means to not only study the world, in the traditional sense of gathering and analysing data about the world, but more broadly, also enable researchers to enact multiple modes of inquiry – ethnographic, quantitative, qualitative, critical, historical, and the more-than-representational (see Kwan, 2007). The use of the acronym 'GIS' may refer to a specific software package or a whole suite of visual technologies used to represent the Earth. GIS may stupefy your audience or excite them with visions of a finally realized Digital Earth. It may cause colleagues to raise an eyebrow ('is positivism alive and well?') or to congratulate you on doing something 'practical' or 'applied' ('you're making our discipline relevant!'). Perhaps no other 'method' within Geography would incite such varied responses, and yet, for many in and outside academe, GIS is *the* method produced by the geographical tradition. However, GIS is not just a tool, not just a software package. Rather, it motivates a method; it implies an inquiry and a perspective, a way to view and represent. Consideration of the use of GIS should motivate a methodology, that is, a study of the implications and *affordances* (opportunities) of such a method.

To place the method of GIS under interrogation, to study such implications and affordances, is, colloquially, to engage in critical GIS. Critical GIS is a tacking back-and-forth, between *using* GIS in radical or explicitly political ways and *situating* those GIS practices. GIS, understood from a critical perspective, is both a method and a practice. At the risk of dulling the impact of such a word, I employ 'critical' quite specifically. Beyond 'critical thinking', the use of the word 'critical' in critical GIS is akin to critique. As Jeremy Crampton so carefully argues, '[t]he purpose of critique is not to say that our knowledge is not *true*, but that the truth of knowledge is established under conditions that have a lot to do with *power*' (2010: 16, emphasis original).

The use of GIS and the study of the use of GIS imply such a critical approach in my work. In this chapter, I trace the various ways in which GIS may be used in research, providing some examples from my own work and that of others to discuss the appropriateness of GIS as a method, while providing the basics for examining GIS as a practice. In research I conducted on the use of spatial technologies to map community quality-of-life, GIS served as both the prompt in qualitative research with community members, and as a vehicle for analysing and visualizing the data created by community members (see Wilson, 2011a).

Throughout this chapter, I'll be making reference to a single map produced during my research in Seattle, Washington, USA. This research examined the use of mobile devices by a Seattle-based nonprofit in mapping quality-of-life concerns. Figure 18.1 depicts the ten neighbourhoods that participated in this survey project from 2004 to 2007. The graduated circles on each neighbourhood were produced by GIS to visually analyse the differences between the total features coded in each neighbourhood. Through typical steps of GIS use (data collection, preparation, analysis, and visualization), this map was not only used to address the central research question of the project (that is, how neighbourhoods participated in the production of digital spatial data), it was also enrolled in discussions with community members and nonprofit staff that participated in the four years of surveying. The map shows what seem to be greater participation in the Greenwood Phinney Ridge and the International District neighbourhoods, and begs two questions: why were more features coded and who participated in the survey activity?

Maps and GIS are objects that evoke and provoke. They do work beyond the desktop of the analyst, and critical geographers are well-placed for examining this work.

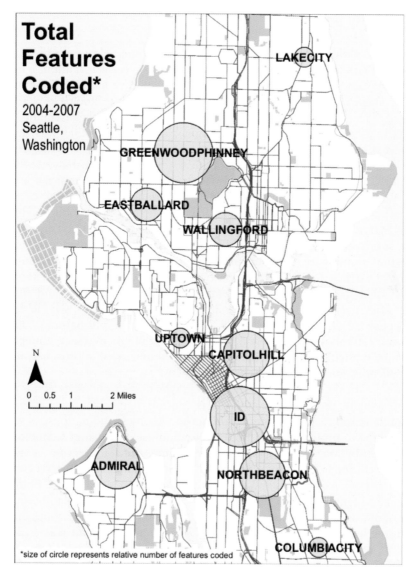

Total Features Coded*
2004-2007
Seattle, Washington

LAKECITY

GREENWOODPHINNEY

EASTBALLARD

WALLINGFORD

UPTOWN
CAPITOLHILL

N

0   0.5   1   2 Miles

ID

ADMIRAL
NORTHBEACON

COLUMBIACITY

*size of circle represents relative number of features coded

Figure 18.1 The ten Seattle neighborhoods that participated in the survey project, 2004–2007

I will use Figure 18.1 to illustrate my discussion of GIS as method and practice. This chapter is organized into five sections. Following a basic introduction of GIS as a method and practice (18.2), section 18.3 outlines necessary preparation when engaging in the use of GIS for human geography research, focusing particularly on conceptualization and formalization. In section 18.4, I discuss observation and analysis as the *doing* of critical GIS research. The writing up of this research – its representation – is presented in section 18.5, followed by a concluding section that outlines the affordances and limitations of critical GIS as method.

18.2 Criticality and GIS

Those working with GIS may reference 'GIScience' as the field in which scholarship about or with GIS takes place. GIScience is a relatively recent invention, with a contested history (Schuurman, 2000). A critical GIS perspective holds that the history of GIS development impacts the method and practice of such technologies. Early in the 1990s, geographers were debating what role GIS should play in the discipline (Macgill, 1990; Clark, 1992). Some felt that GIS might aid a discipline that was seen to be increasingly disparate in intellectual interests, that the technology would provide the glue to bind together modes of geographical inquiries (Openshaw, 1991, 1992; Dobson, 1993). Others were less optimistic – even antagonistic – and felt a focus on GIS was weakening the discipline, reducing the tradition to the accumulation of 'facts', as well as linking Geography to projects of violence and domination (Taylor, 1990; Taylor and Overton, 1991; Smith, 1992; Lake, 1993; Pickles, 1993). Through a series of interventions, an initiative labelled 'GIS and Society' sought to address this chasm by generating a multipart research agenda, eventually leading to the development of subfields like participatory and public participation GIS; spatial decision support systems and collaborative GIS; critical, feminist and qualitative GIS; and the spatial humanities (for overviews and edited collections towards this end, see Pickles, 1995, 2006; Sheppard, 1995, 2005; Craig et al., 2002; Dragicevic and Balram, 2006; Cope and Elwood, 2009; Bodenhamer et al., 2010). The use of GIS in research inherits from these various subfields, each of which marks a multiplicity of approaches, methodological challenges, and forms of public engagement.

Therefore, critical GIS attends to the ways in which this technology is not just software, but actually produces society – it is central to the planning, management, destruction, and reinvention of our neighbourhoods, cities, and nations, as well as to those who hope to profit from its proliferation and use. Geospatial technologies represent a multi-billion dollar global industry, and impact everyday experiences, embedded in consumer electronics and throughout the military-industrial complex to route goods/services/consumers and to direct predator drones and 'smart' bombs. They have become so central to daily life in advanced capitalist societies as to become invisible, part of our technological unconscious (Thrift, 2004). Spatialized codes direct how we search the Internet (Zook and Graham, 2007), how we use air travel (Budd and Adey, 2009), how we track urban quality-of-life (Wilson, 2011b), and how we interact in the living rooms of our homes (Dodge and Kitchin, 2009). Regardless of our awareness, systems for geographic information organize society. For instance, GIS is used toward a variety of urban functions, in community and urban planning (Talen, 2000; Elwood, 2002) as well as transportation planning (Nyerges and Aguirre, 2011), in public service delivery (Longley, 2005), and in disaster response (Zook et al., 2010).

Adopting a critical GIS perspective does not necessarily mean that you do not use GIS in research. You might be using GPS (global positioning systems), which could mean consumer-grade handheld-receivers or survey-grade receivers (distinguished by their levels of positional accuracy and attributal sophistication). Or it could indicate that you're using GPS trackers (that might clip to a human or other object of study) or the (assisted) GPS application on many mobile phones. You might also be using for-profit desktop software suites like ESRI's ArcGIS or Google Earth, or open-source software like QGIS. You might have heard about web-based GIS and map mashups, created using tools like Google MyMaps, ESRI's ArcGIS Online, or OpenStreetMap. You might have heard about something called 'qualitative GIS', and wonder where you might download such software (you can't . . . yet). This would not be the most appropriate venue for cataloguing all the systems, platforms, and software tools available (there are no doubt Wikipedia entries that serve this purpose). However, thinking of GIS as a method does not require you to turn off a critical sensibility. Instead, this approach exposes some preliminary questions and concerns that this chapter will now discuss: how do you express your understanding of the world in the computational structures of GIS? What are the specific considerations for spatial datasets? How do you represent the results of your work with GIS?

18.3 Preparing for Research

GIS are used in multiple ways to research spatial phenomena. As such, there can be no single 'step-by-step' process that works across all cases. Pavlovskaya (2004) uses GIS to understand the economics of everyday life in Moscow. Brown and Knopp (2008) use GIS in their study of queer oral histories in Seattle. Kwan (2008) uses GIS to visualize emotional geographies in Columbus, Ohio. Elwood (2006) works with community partners to use GIS in neighbourhood organizing and planning. In each project, GIS are used in different ways in the process of preparing for and doing research. As you prepare to do research using GIS, you will need to engage in the interrelated processes of *conceptualization* and *formalization*, understood as:

1. Considering the role of the map/GIS/data in research.

2. Formalizing both the elements to be included within GIS and the GIS practices.

Both conceptualization and formalization are important stages in GIS research. Even projects that may not seem to explicitly engage in conceptualization and formalization nonetheless must assume certain relationships between the objects represented by data and their material, physical manifestations: 'reality'. Fundamentally, these relationships are engaged through explicit discussions of the practices of conceptualization and formalization.

Conceptualization

It is helpful to begin with Nadine Schuurman's thoughts on the important disjunctures between conceptualization and formalization. In her research on GIS ontologies (understood simply as the categories that compose data), she writes:

> If we think of conceptualization as a cognate step toward understanding spatial processes and relationships, there remains a pressing need to express those relationships in a mathematical or formal notation as a precursor to coding them. (2006: 730)

Conceptualization in GIS research is a process by which a researcher not only makes sense of spatiality, but, to extend Schuurman, must also consider the relationships between the mapping technologies, the focus of the map, and the audience that views the map. In this sense, conceptualization is asking: what role does the map/GIS play in research?

- as a prompt/illustration for inquiry (e.g. a map used in a focus group discussion about neighbourhood improvements),

- as a vehicle for analysis (e.g. a GIS to compute the travel times of suburban residents),

- as a participative/collaborative object (e.g. a map made through the deliberations of community members to define a neighbourhood boundary dispute), or

- as a system for building visualizations (e.g. a GIS to model views from high points in the city).

The asking of this question underlines the entanglements of epistemology and method. Here it is useful to recall the arguments around quantitative geography in the 1990s (Lawson, 1995), more recently reinvigorated by qualitative GIS (Pavlovskaya, 2009): namely that methods need not presume or assume a singular way of knowing. Certainly, particular ways of knowing are enabled or made easier with the use of particular methods, but this need not be an automatic relationship. This is not to say that methods are, or can be, neutrally applied. Instead, it is an insistence on the creative role of the method – one that adapts and enrolls the epistemological stance of the researcher.

Conceptualization and formalization demand an attention to the entire research process, as part of your preparations. Figure 18.2 sketches an ordering of different aspects of research, beginning with conceptualization and formalization. During research preparation, it becomes important to consider the inputs/outputs of each aspect of research, as is true of many research endeavours. What is it that research is producing, at each stage? How will GIS be used? What are the requirements for these productions?

conceptualization ⟶ formalization ⟶ observation ⟶ analysis ⟶ representation

Figure 18.2   An ordering of different aspects of research

While these various stages of research are each important to consider as you prepare to use GIS, thinking about the analysis stage may be fruitful for understanding how to prepare. Despite common notions of GIS as a tool for spatial analysis, GIS need not be only understood as the vehicle for analysis in your study, and may actually precede analysis. Nor do GIS necessarily have to be used only for the representation of results from the analysis.

Figure 18.3 presents these three relationships of GIS to the analysis stage of research in Figure 18.2. In conceptualizing where GIS are used in the process of research, you will be better able to outline the entire project. By preceding analysis (relationship 'a'), GIS might be used as a prompt in an ethnographic project to

> (A) GIS precedes analysis
>
> (B) GIS is the analysis
>
> (C) GIS follows analysis

Figure 18.3   Three relationships of GIS to the analysis stage of research

engage research participants in a discussion of a local issue (perceptions of crime, effects of gentrification, historical preservation, etc.). The GIS acts as an illustration in a data collection effort, prior to analysis. GIS acting as the vehicle for analysis (relationship 'b') is perhaps the most prevalent assumption about the role of GIS in research (that a researcher collects data, feeds it into a GIS, and results appear!). Another prevalent assumption is that GIS are used as a final stage in research, following analysis (relationship 'c'), to create maps of the results of the study.

Formalization

Moving from conceptualization to formalization is the subject of much consternation in GIScience, particularly for those approaching the use of GIS from a critical perspective, as many of the concepts of critical geography can not be reduced to the computational vision of GIS (Schuurman 2006). In other words, as you prepare to do research, you may discover that the formalizations necessary for a particular line of inquiry are beyond the capacities of GIS. For instance, much of critical geography may explicitly reject the grid epistemology associated with GIS. In post-structural research, for example, there is an assumption that the categories of the study evolve over its duration (see Dixon and Jones, 1998). However, to use GIS is to make formal arrangements: What scale is utilized in your geographic research? How is place to be represented on the map? What spatial phenomena can be expressed in Euclidean space?

Once a researcher begins to formalize their conceptualizations central to their project, the affordances and limitations of GIS as method and practice will become apparent. There are a seemingly endless number of formal and formative factors to consider at this stage of the process, categorized here as:

1. data constraints

2. method constraints

3. system constraints

Of course, how a researcher makes decisions about these factors is directly dependent upon the relationship of GIS to analysis (see Figure 18.3). Therefore, 'data' may mean something different where GIS are used prior to analysis, compared to more ethnographic projects, where the GIS may operate as a prompt in discussions with research participants. 'Data', in an ethnographic project, may take the form of field notes and interview transcripts (see Chapters 9 and 11, this volume).

However, regardless of the relationship of GIS to analysis, a researcher needs digital data for GIS work. Prior to 'doing research' then, the researcher needs to consider how certain variables in the research project (race, class, gender, sexuality, property, home, public space, etc.) will be incorporated into the GIS and operationalized. As part of the process of formalization, operationalization entails making decisions about how to measure the variables selected by the researcher. For instance, a project interested in the spatiality of class in an urban area, might consider using income measures from a national census as one indicator of 'class'. These data are constrained by the spatial geometry of the enumeration areas; without which, the operationalization of 'class' using the spatial analytical capabilities of GIS would not be possible.

As you prepare to conduct work with GIS, consider how the method addresses the research question. It is tempting at the formalization stage to reconfigure the research question such that the available technical capability of the GIS can be directly enrolled. If the method available within GIS constrains the research question 'too much', per-haps the method is inappropriate. Relatedly, GIS are computer-based systems that have their own constraints – maximum data sizes, bandwidth restrictions, physical memory limitations, etc. – each of which may make the research question and the selected method untenable. These constraints are important to consider as a researcher pre-pares to conduct a study using GIS. Put directly, GIS are software with real limits.

Furthermore, not all data and methods are equally appropriate, and researchers will need to consider how data were created and manipulated as they formalize their pro-ject. Metadata is an important way of getting at appropriateness and is, very simply, data about data. GIS research relies upon metadata to understand basic issues of pro-jection and accuracy as well as more database-level concerns about attribute parameters and construction. If well-maintained, metadata provide a written record of the construction of spatial data as well as the manipulation of such data. Some meta-data are created through standards maintained by national bodies, such as the Federal Geographic Data Committee in the US (FGDC), while many datasets may lack prop-erly maintained metadata. As researchers prepare to use GIS, they may need to consult with those in charge of it to better understand how the data came to be. For instance, spatial data may only be intended for particular scales, where levels of accuracy are most appropriate. Metadata may assist you in understanding whether or not the data are appropriate for your research question.

As I prepared to create the map in Figure 18.1, I quickly realized the importance of understanding two key factors: (1) how the data categories were created and (2) how surveyors were instructed to collect information about urban quality-of-life. By speak-ing with those who created the software used in the survey and those who created the data categories, I better understood how the data were collected, and toward what ends, which meant that my project extended from the GIS and the map to many discus-sions with those involved in its production. Spatial data and maps have implications beyond the often narrow ways in which they are imagined. By addressing these two factors, the project evolved into an examination of the ways in which residents were trained to see their neighbourhoods as specific data objects (Wilson, 2011a) as well as how the data generated by the surveys were significant in different ways to different people and organizations (Wilson, 2011b).

As with many methods, the distinction between preparing for and conducting research is an artificial one. Conceptualization and formalization are integral aspects of the 'doing' of research. By enabling such separations, research runs the risk of bracketing assumptions from the research process, fixing them as neutral conditions for data collection, observations and analyses. Instead, researchers who use GIS should understand the processes of conceptualization and formalization as significant, contingent stages that make the research possible. They are therefore stages to be returned to throughout the research, to re-evaluate appropriateness and perhaps adjust parameters to better address the research question. This particular way of thinking through the 'doing' of GIS draws particularly upon qualitative GIS, as simultaneously both method and practice (Cope and Elwood, 2009).

18.4 Doing Research

Traditionally, the doing of research with GIS incorporates two aspects of research (see Figure 18.2) – observation and analysis. One must make observations and analyse such observations. Observations are dependent upon what can be measured, decisions that are made as part of the formalization stage, previously discussed. In what follows, I discuss different types of observations and related considerations. I then review a range of analytical approaches with GIS.

Observation

Broadly speaking, when *doing* GIS research you will be making observations. Depending on how a researcher views Figure 18.3 (for instance, if GIS are used as a vehicle for analysis), this section might also be titled 'data collection'. Data collection can be the most time-intensive aspect of GIS research. The lack of appropriate data can sink a research project that uses GIS, and data availability is commonly expressed as a major limitation for GIS-based research.

When locating data, there are three approaches to help guide the researcher. First, the required geographic area and scale should direct the data search (is this state or provincial data? metropolitan data? national data?). Second, the theme of the data should help isolate organizations that might collect and maintain such data (census, housing, natural resources, urban infrastructure, transportation, etc.). Finally, by contacting individuals or organizations with interests related to the research question, the sharing of datasets may be appropriate. More specifically, there are several questions to consider when selecting GIS data (see Box 18.1). As the researcher assembles data for analysis, these questions should help determine the appropriateness of the data for addressing the research question.

Box 18.1   Questions to consider when selecting GIS data

1   Are the data in a format that can be used directly in GIS (i.e. shapefile, geodatabase, kml)? If not, can the data be transformed into spatial data? Data tables that have fields like latitude and longitude, street address or administrative boundary may still be transformed into GIS data.
2   Are the spatial files' coordinate system parameters defined? In order to conduct spatial analysis, the appropriate projection needs to be defined.

(Continued)

(Continued)

3   What is the scale of the spatial data? Does it meet the needs of the research question? If the study is focused at the neighbourhood scale, datasets like sidewalks, building footprints, and green spaces might be appropriate.

4   At what summary level are the tabular data (by address, county, tract, postal code, state, etc.)? This will inform the kinds of spatial analysis that can be performed.

5   How recent are the data? When were the data created and most recently updated? The currency of a dataset may directly impact the ability to ask certain research questions.

6   What were the sources of the data? Data collected by governments may have specific timelines for when the data are updated.

7   What are the copyright requirements? Are there distribution restrictions or human subjects research guidelines? Some datasets may require the researcher to enter into privacy agreements, and may not be legally distributed or published.

(adapted from Stanford University Libraries, 2006)

In addition to acquiring existing spatial data, or tabular data with a spatial attribute, a researcher may also need to collect additional data in the field. This might be done through GPS, using handheld receivers that record data points and lines. Many of these devices have the capability of downloading spatial data directly into a standard format ready for GIS. In contemporary urban research, for instance, data observations may also be made using street addresses, which can be stored in a spreadsheet and then geocoded directly into the GIS.

Other types of observations with GIS may be made. If GIS are used as a discussion prompt with research participants (relationship 'a' in Figure 18.3), then 'observations' are likely ethnographic or qualitative. By working directly with research participants in using GIS, a researcher can better understand how spatial knowledge is constructed and made to do work. As such, GIS may be used directly in participatory action research (Elwood, 2009) to evaluate or interpret the information products created by GIS or to interrogate the categories enrolled in the process of observation itself.

Analysis

Just as the modes of observation depend upon how you formulate the role of GIS in your work, analysis is equally diverse in approach. By disentangling method from epistemology, it is possible to imagine how GIS might be used in interpretative as well as positivist modes of inquiry (Pavlovskaya, 2009). Whether understood as basic measurement, calculation, modelling, or as comparative, thematization, or affective, *analysis* is the aspect of research through which new knowledge is generated.

Chrisman (2002) organizes analytical operations with GIS from basic attributal operations (queries, categorical manipulations, and arithmetic procedures) to more advanced spatial analysis (overlays, buffers, surfaces, viewsheds, and networks; see Table 18.1 and refer to any of the more popular textbooks on the use of GIS for spatial analysis). Within GIS, these operations are often organized by spatial data model: vector or raster. Vector data are composed of points, lines and polygons that use geographic coordinates as vertices. Raster data are composed of grids that store geographic information cell-by-cell, where each cell corresponds to an area on the surface of the Earth. Due to the differences in the architecture of these spatial data models (as well as the differences in their philosophies; see Couclelis, 1992), certain analytical operations are connected to specific data models. For instance, networks are most efficiently analysed using vector data, while surfaces are often represented and analysed using raster data.

These modes of 'spatial analysis' lend themselves toward the computational, and certainly toward more quantitative data. Within the fields of qualitative GIS and the spatial humanities, scholars are exploring different modes of analysis that allow more interpretative moves within GIS and, thereby, qualitative data. The subfield of qualitative GIS is focused on the use of a mix of data types, quantitative and qualitative (Cope and Elwood, 2009). As a method, qualitative GIS builds reflexivity directly into the technical operation (see Knigge and Cope, 2006; Jung, 2009). The growing field of spatial humanities is similarly concerned with enrolling spatial techniques in humanistic inquiry (Bodenhamer et al., 2010). The spatial humanities thus envision working with the historical and the artifactual in GIS to bridge multiple ways of knowing (see Cooper and Gregory, 2011; Yuan, 2010).

Within more participatory uses of GIS, qualitative research methods like discourse analysis or grounded theory may be used (see Knigge and Cope, 2006; Wilson, 2009). Here, the emphasis is on analysing the GIS as a practice, as integrally part of the way the world is understood and experienced and not as a neutral bystander. For instance, the map produced in Figure 18.1 uses basic arithmetic to produce the graduated circles that represent the total number of records collected by neighbourhood residents. The GIS were used to analyse and represent the differences in the total number of records that residents collected. However, more important to this particular urban research project, the map produced in Figure 18.1 evocatively prompted residents to further discuss the significance of the differences across Seattle neighbourhoods – a discussion that was analysed using discourse analysis. Here, it is important to recognize GIS as part of, and not separate from, that discourse analysis;

Table 18.1   Some common analytical operations used with GIS and their descriptions

Analytical operation	Description
queries	retrieval from an attribute database based on characteristics of a data record (may be spatial characteristics)
categorical manipulations	reclassifying attribute data into new categories
arithmetic procedures	addition, subtraction, multiplication, and division between attribute fields and between records in an attribute database
overlays	examines how spatial phenomena are related, such as union (all features in both input data layers) and intersection (only those features in common in both input data layers)
buffers	examines the proximity between features
least-cost path	computes the path of least resistance (time, effort, cost, etc.) between two geographic points

the technical production of knowledge intervenes directly in the more interpretative/ discursive practice of making meaning.

18.5 Writing Up Research

After *doing* scholarly work (GIS or otherwise), you will typically need to engage in a process of 'writing up'. Practically speaking, the actual writing up of GIS research is an exercise in metadata construction. As discussed earlier, metadata are a written record of the creation and manipulation of data. By writing up research, you are extending the written record of spatial data, in ways that will assist future scholars in evaluating the research as well as continuing the study. When conducting GIS work, the write-up may involve authoring or updating the accompanying metadata for the datasets produced by the project, making the datasets accessible to the broader public or to the specific communities affected by such data collection, as well as creating representations of the results – often in cartographic form. In this section, I discuss these three considerations as part of 'representation', the last aspect of research in Figure 18.2.

Representation

As symbols that constitute reality, representations here are understood as maps or as data themselves. In the process of writing up, researchers engage in the production of these representations, producing maps for publication, documenting the representations in the form of metadata, or presenting findings to the research community or to the community impacted by such research. Due to the ways in which GIS data and maps evoke authority on a subject (Wood, 1992; King, 1996), the practice of representing with/through geographic information technologies demands considerable attention. These representations may travel, in that they may be enrolled into new knowledge projects (complementary or otherwise). Representation, thus, acts in excess of the agenda of the researcher. In other words, your projects may be taken up at some point by other researchers in ways that you cannot predict (and may not even agree with!).

As you complete your GIS projects, you may be requested to share their spatial data with others conducting work on similar issues. Therefore, it is important to carefully document data including information about the time period, contact information, data quality, spatial reference, attributes contained, and any additional information about how the data were collected or derived and manipulated. The practice of recording metadata allows the transfer of not just the data, but the contexts around which the data exist. This information proves of critical importance as researchers share and use others' data, to better understand the limitations or constraints around its use.

Cartographic representations are perhaps the most obvious result of GIS-based research. And while maps may take several forms (web-based, paper, mobile-device ready, etc.), there are general practices of map-making that should act as guides. Consider the audience, purpose, reproduction constraints (colour, black and white, digital, paper, etc.), and necessary annotations. Will there need to be a legend, a north arrow, a scale bar in your project? How will you indicate the source of the data, or the various participants in the project? Are there other visualizations that might aid the map? Yau (2011) covers a range of visualization techniques (many of which are entirely web-based) for both spatial (and non-spatial) data. The map is not merely decoration and should directly aid your reader in understanding the results of your project.

The map in Figure 18.1 was produced to engage Seattle residents as to the differences between the raw amount of data collected in their neighbourhoods. As but one symbol of the reality of this survey effort, this representation enabled residents to further discuss their experiences with the survey. It became an entry into more detailed discussions of residents' hopes for how these data were to be used and their concerns for how these kinds of data collections tend to justify specific urban policies. The 'writing up' of GIS work should involve such discussions of the work that representations do – by releasing the cartographic products back to the communities impacted by such representations.

18.6 Conclusion

Geographic information systems enable different ways of knowing and cannot be limited to a single method. GIS are both something one does, participates in, as well as something one uses. As media, GIS are both practice and method. In this chapter, I have described a range of concerns in both the doing of GIS and the paying of attention to how GIS practices permeate and produce society. Ultimately, GIS are technologies of *representation*. That maps are made, as a human endeavour, complicate and haunt cartography as science, exposing the subjectivity and selective interests of cartographic practice. And while these crises of representation should not justify a dismissal of GIS, they should inspire careful consideration – a healthy scepticism – of the limitations and affordances of GIS as a method of research.

That maps wield power is not lost on critical cartographers. As representations of space, maps produce territories. Maps, and the GIS that create them, insinuate themselves into everyday life, figuring our interactions with human and nonhuman others. They communicate expertise and authenticity, where none may exist. Maps may challenge the status quo or reinforce it. As Harris and Weiner (1998) write, GIS may be both empowering and disempowering. Marginalized groups can use GIS to stake claims to resources, advocate for changes in policies and planning, draw attention to alternative geographies and histories, and inspire more transparent and accessible decision-making in government. As these technologies can be opened up to collaborative or participatory projects, GIS have the potential to be a collective force as groups organize and direct its representational prowess.

Geographic information technologies are, nonetheless, risky investments in knowledge production. As an ocular technology, GIS specifically visions from an elevated perspective. The world is displayed as seen from nowhere and everywhere. From high above and disconnected, this neutralizing downward gaze serves to organize the world into layers of objects and actions. The risk of this disembodied visioning is the kinds of distancing that it enables – subjective experiences become bounded as objects. This ocularcentrism, with the organization of space through Cartesian perspectivalism, enables an *epistemology* (way of knowing) of the grid, which, as Dixon and Jones

(1998: 251) write, is 'a way of knowing that imposes itself upon and eventually becomes inseparable from those processes it helps to understand'. The use of GIS in research risks concretizing and elevating a specific ontology that is best analysed with GIS; the concern is for when the world *becomes* what GIS shows us it is, and little else.

In a variety of geographic research projects, GIS may simply be a means to an end, a method to analyse and visualize a set of spatial data. Here, I have attempted to situate this technology – to show that it did not just appear out of thin air and that each stage of the research process using GIS, with a critical perspective, is imbued with fundamental decisions about how to observe, measure, analyse, and represent the world. These fundamentals can be approached pragmatically – as is the goal for the use of this chapter – and such practices would take necessary steps toward an engaged and responsible use of this technology.

SUMMARY

- GIS is both a method and a practice. GIS is not just a tool, nor is it just a software package. GIS may simply be a way to produce representations of your research results; it can be used to conduct analyses of spatial phenomena; GIS may also be used as a prompt or an illustration to engage research subjects.
- GIScience is relatively recent, and has a contested history: geospatial technologies impact everyday experiences, and have become so central to daily life in advanced capitalist societies as to become invisible parts of our technological unconscious. Regardless of our awareness, systems for geographic information organize society. Critical GIS is a perspective encompassing *using* GIS in radical or explicitly political ways and *situating* those GIS practices.
- In preparing to use GIS, you will need to engage in the interrelated processes of *conceptualization* – considering the role of the map/GIS/data in research – and *formalization* of both the elements to be included within GIS as well as the practices with GIS. Doing the research with GIS involves *observation and analysis*. Writing up the research is a form of *metadata construction*, which may involve authoring or updating metadata for the datasets produced by the project, making the datasets accessible to the broader public or to the specific communities affected by such data collection, as well as creating representations of the results – often in cartographic form.

Further Reading

Cope and Elwood's *Qualitative GIS* (2009) has become the primary text for the theory, method, and practice of alternative GIS. Operating both conceptually, to shift the imagination of GIS in society, and practically, to produce new techniques around geographic representation, this edited collection sets the stage for a renewed criticality in GIScience.

Crampton's *Mapping* (2010) marks a coming-of-age of the critical cartography and critical GIS subfields. This text, targeted at students and faculty new to these ideas, presents the foundational concepts and origin stories for the practice of critical mapping scholarship. Here, Crampton brings the critical concepts of the GIS and Society agenda to bear on the emergence of the geoweb.

New to cartographic design? Start with *Making Maps*, by John Krygier and Denis Wood (2005). Krygier and Wood bring levity to the design decisions of map-making, beginning with basic questions around intent and audience and ending with more complicated decisions around colour, visual hierarchy, balance, and use of negative space.

If you are interested in learning more about the 'GIS wars' of the 1990s, the emerging GIS and Society agenda, and the perspective of GIScientists, then check out Schuurman's *GIS: A Short Introduction* (2004). Drawing forward her interventions in the field beginning in the late 1990s, Schuurman gives an overview of the key concepts and techniques of GIScience in a way that appreciates a diverse audience of human geographers and social scientists more generally.

Nathan Yau, the author of the www.FlowingData.com blog, pulled together *Visualize This* (2011) for a group of visualization enthusiasts who are increasingly looking to open-source, web-based technologies to present compelling graphics with 'Big Data'. The text presumes a kind of hacker sensibility and offers an entire chapter on web-based mapping techniques, with example scripts to be copied and pasted for your mapping projects.

Note: Full details of the above can be found in the references list below.

References

Bodenhamer, D.J., Corrigan, J. and Harris, T.M. (eds) (2010) *The Spatial Humanities: GIS and the Future of Humanities Scholarship.* Bloomington: Indiana University Press.

Brown, M., and Knopp, L. (2008) 'Queering the map: The productive tensions of colliding epistemologies', *Annals of the Association of American Geographers* 98 (3): 1–19.

Budd, L., and Adey, P. (2009) 'The software-simulated airworld: Anticipatory code and affective aeromobilities', *Environment and Planning A* 41: 1366–85.

Chrisman, N.R. (2002) *Exploring Geographic Information Systems* (2nd edition). New York: Wiley.

Clark, G.L. (1992) 'GIS – what crisis?', *Environment and Planning A* 24 (3): 321–2.

Cooper, D., and Gregory, I.N. (2011) 'Mapping the English Lake District: A literary GIS', *Transactions of the IBG* 36 (1): 89–108.

Cope, M., and Elwood, S.A. (eds) (2009) *Qualitative GIS: A Mixed Methods Approach.* London: Sage.

Couclelis, H. (1992) 'People manipulate objects (but cultivate fields): Beyond the raster-vector debate in GIS', *Lecture Notes in Computer Science* 639: 65–77.

Craig, W.J., Harris, T.M. and Weiner, D. (eds) (2002) *Community Participation and Geographic Information Systems.* New York: Taylor & Francis.

Crampton, J.W. (2010) *Mapping : A Critical Introduction to Cartography and GIS.* Malden, MA: Wiley-Blackwell.

Dixon, D.P., and Jones, J.P. III (1998) 'My dinner with Derrida, or spatial analysis and poststructuralism do lunch', *Environment and Planning A* 30 (2): 247–60.

Dobson, J.E. (1993) 'The Geographic Revolution: A retrospective on the Age of Automated Geography', *The Professional Geographer* 45 (4):431–9.

Dodge, M., and Kitchin, R. (2009) 'Software, objects, and home space', *Environment and Planning A* 41: 1344–65.

Dragicevic, S., and S. Balram (eds) (2006) *Collaborative Geographic Information Systems.* Hershey, PA: Idea Group, Inc.

Elwood, S.A. (2002) 'GIS use in community planning:A multidimensional analysis of empowerment', *Environment and Planning A* 34: 905–22.

Elwood, S.A. (2006) 'Beyond cooptation or resistance: Urban spatial politics, community organizations, and GIS-based spatial narratives', *Annals of the Association of American Geographers* 96 (2): 323–41.

Elwood, S.A. (2009) 'Integrating participatory action research and GIS education: Negotiating methodologies, politics and technologies', *Journal of Geography in Higher Education* 33 (1): 51–65.

Harris, T.M., and Weiner, D. (1998) 'Empowerment, marginalization, and "community-integrated" GIS', *Cartography and Geographic Information Systems* 25 (2): 67–76.

Jung, J.-K. (2009) 'Computer-aided qualitative GIS: Software-level integration of CAQDAS and GIS', in M. Cope and S.A. Elwood (eds) *Qualitative GIS: A Mixed-Methods Approach.* London: Sage. pp. 115–35.

King, G. (1996) *Mapping Reality: An Exploration of Cultural Cartographies.* New York: St. Martin's Press.

Knigge, L., and Cope, M. (2006) 'Grounded visualization: Integrating the analysis of qualitative and quantitative data through grounded theory and visualization', *Environment and Planning A* 38: 2021–37.

Krygier, J., and Wood, D. (2005) *Making Maps: A Visual Guide to Map Design for GIS.* New York: Guilford Press.

Kwan, M.-P. (2007) 'Affecting geospatial technologies: Toward a feminist politics of emotion', *The Professional Geographer* 59 (1): 2–734.

Kwan, M.-P. (2008) 'From oral histories to visual narratives: Re-presenting the post-September 11 experiences of the Muslim women in the USA', *Social and Cultural Geography* 9 (6): 653–69.

Lake, R.W. (1993) 'Planning and applied geography: Positivism, ethics, and geographic information systems', *Progress in Human Geography* 17 (3): 404–13.

Lawson, V. (1995) 'The politics of difference: Examining the quantitative/qualitative dualism in post-structuralist feminist research', *The Professional Geographer* 47 (4): 449–57.

Longley, P. (2005) 'Geographical Information Systems: A renaissance of geodemographics for public service delivery', *Progress in Human Geography* 29 (1): 57–63.

Macgill, S.M. (1990) 'Commentary: GIS in the 1990s?', *Environment and Planning A* 22 (12): 1559–60.

Nyerges, T.L., and Aguirre, R.W. (2011) 'Public participation in analytic-deliberative decision making: Evaluating a large-group online field experiment', *Annals of the Association of American Geographers* 103 (3): 561–86.

Openshaw, S. (1991) 'A view on the GIS crisis in geography, or, using GIS to put Humpty-Dumpty back together again', *Environment and Planning A* 23 (5): 621–8.

Openshaw, S. (1992) 'Further thoughts on geography and GIS: A reply', *Environment and Planning A* 24 (4): 463–6.

Pavlovskaya, M. (2004) 'Other transitions: Multiple economies of Moscow households in the 1990s', *Annals of the Association of American Geographers* 94: 329–51.

Pavlovskaya, M. (2009) 'Breaking the silence: Non-quantitative GIS unearthed', in M. Cope and S. A. Elwood (eds) *Qualitative GIS: A Mixed-Methods Approach*. London: Sage. pp. 13–37.

Pickles, J. (1993) 'Discourse on Method and the History of Discipline: Reflections on Dobson's 1983 Automated Geography', *The Professional Geographer* 45 (4): 451–5.

Pickles, J. (ed.) (1995) *Ground Truth: The Social Implications of Geographic Information Systems*. New York: Guilford.

Pickles, J. (2006) 'Ground Truth 1995–2005', *Transactions in GIS* 10 (5): 763–72.

Schuurman, N. (2000) 'Trouble in the heartland: GIS and its critics in the 1990s', *Progress in Human Geography* 24 (4): 569–90.

Schuurman, N. (2004) *GIS: A Short Introduction* (Short introductions to geography). Malden, MA: Blackwell.

Schuurman, N. (2006) 'Formalization matters: Critical GIS and ontology research', *Annals of the Association of American Geographers* 96 (4): 726–39.

Sheppard, E. (1995) 'GIS and society: Towards a research agenda', *Cartography and Geographic Information Systems* 22 (1): 5–16.

Sheppard, E. (2005) 'Knowledge production through critical GIS: Genealogy and prospects', *Cartographica* 40 (4): 5–21.

Smith, N. (1992) 'History and philosophy of geography: Real wars, theory wars', *Progress in Human Geography* 16: 257–71.

Stanford University Libraries (2006) *Guidelines for Finding GIS Data*. Stanford University, 16 March. https://lib.stanford.edu/gis-branner-library/finding-data-guidelines (accessed 21 November 2015).

Talen, E. (2000) 'Bottom-up GIS - A new tool for individual and group expression in participatory planning', *Journal of the American Planning Association* 66 (3): 279–94.

Taylor, P.J. (1990) 'GKS', *Political Geography Quarterly* 9: 211–2.

Taylor, P.J., and Overton M. (1991) 'Further thoughts on geography and GIS', *Environment and Planning A* 23 (8): 1087–90.

Thrift, N. (2004) 'Remembering the technological unconscious by foregrounding knowledges of position', *Environment and Planning D: Society and Space* 22: 175–90.

Wilson, M.W. (2009) 'Towards a genealogy of qualitative GIS', in M. Cope and S. A. Elwood (eds) *Qualitative GIS: A Mixed Methods Approach*. London: Sage. pp. 156–70.

Wilson, M.W. (2011a) '"Training the eye": Formation of the geocoding subject', *Social and Cultural Geography* 12 (4): 357–76.

Wilson, M.W. (2011b) 'Data matter(s): Legitimacy, coding, and qualifications-of-life', *Environment and Planning D: Society and Space* 29 (5): 857–72.

Wood, D. (1992) *The Power of Maps*. New York: Guilford Press.

Yau, N.C. (2011) *Visualize This : The Flowing Data Guide to Design, Visualization, and Statistics*. Indianapolis, IN: Wiley Publishing, Inc.

Yuan, M. (2010) 'Mapping text', in D. J. Bodenhamer, J. Corrigan and T. M. Harris (eds) *The Spatial Humanities: GIS and the Future of Humanities Scholarship*. Bloomington, IN: Indiana University Press. pp. 109–23.

Zook, M.A., and Graham, M. (2007) 'The creative reconstruction of the Internet: Google and the privatization of cyberspace and DigiPlace', *Geoforum* 38: 1322–43.

Zook, M.A., Graham, M., Shelton, T. and Gorman, S. (2010) 'Volunteered geographic information and crowdsourcing disaster relief: A case study of the Haitian earthquake', *World Medical & Health Policy* 2 (2): 7–33.

ON THE COMPANION WEBSITE…

Visit **https://study.sagepub.com/keymethods3e** for author videos, chapter exercises, resources and links, plus **free** access to the following recommended articles:

1. **Crampton, J. (2011) 'Cartographic calculations of territory', *Progress in Human Geography*, 35 (1): 92–103.**

Crampton places recent developments in political geography with critical cartography in conversation to begin to explore the possibilities of each to be furthered in both empirical objects of study and in theory.

2. **Elwood, S. (2010) 'Geographic information science: Emerging research on the societal implications of the geospatial web', *Progress in Human Geography*, 34 (3): 349–57.**

Elwood draws forward a critical perspective on GIS toward the emerging tools, data, and practices of the geoweb, linking scholars across the discipline in this new field of study.

3. **O'Sullivan, D. (2006) 'Geographical information science: Critical GIS', *Progress in Human Geography*, 30 (6): 783–91.**

O'Sullivan reviews progress in the GIS and Society agenda and, specifically, critical GIS, arguing that much of the 10-year-old agenda shared between critical geography and GIScience remains to be addressed.

19 Quantitative Modelling in Human Geography

Alan Marshall

Essentially, all models are wrong but some are useful.

George E. P. Box (1987: 424)

SYNOPSIS

Many of the social, economic and demographic characteristics of population display strong spatial patterns. For example, the distribution of poor health, mortality, migration, wealth and unemployment exhibits strong spatial patterns across and within countries and regions. Quantitative models are a valuable tool which enable Human Geographers to evaluate theories about the social processes driving spatial unevenness in population and socio-demographic characteristics. Whilst all models contain error, they also have the potential to reveal new understandings of spatial process, to challenge and develop existing theories, or to inform policy responses to particular social challenges.

This chapter introduces the use of quantitative models in Human Geography and whilst it cannot cover the full range of techniques within this exciting and vibrant sub-discipline, it does provide examples of a selection of models to illustrate the potential of quantitative methodologies. The chapter also considers the development and the debates around the use of quantitative models, how you might apply them in a research project, as well as future challenges and opportunities for quantitative Human Geographers.

This chapter is organized into the following sections:

- Introduction
- Debates around the use of quantitative models in Human Geography
- How do I use quantitative models in a research project?
- Quantitative models in Human Geography
- Challenges and opportunities: The importance of Big Data

19.1 Introduction

Quantitative models have long played a crucial and successful role in the Physical Sciences predicting aspects of the natural world, such as the movement of planets, with great accuracy. The successes of quantitative models in the Physical Sciences, in part, stimulated their use within the Social Sciences to investigate social phenomena. However, the successes of quantitative models in the Social Sciences are perhaps less clear cut. Importantly, the complexity and ephemeral and value-laden nature of social processes introduce important philosophical and methodological considerations that must be taken into account when applying quantitative models to explore social research questions. However, when applied and interpreted appropriately, quantitative models have the potential to deliver important insights into social processes and the mediating role of place.

Human Geographers have used quantitative models to answer a range of questions which consider how social processes and characteristics are influenced by place, including:

- Do the characteristics of an area influence an individual's health over and above their individual characteristics such as age, gender, social class, smoking status? For an example see Box 19.4.

- Are places becoming more or less segregated over time in the spatial distribution of social (e.g. wealth, economic activity) and demographic characteristics (e.g. ethnicity, age)? For an example see Box 19.2 or Lloyd (2014a).

- What factors predict population flows (residential moves, commuting, shopping) between areas? For an example involving flows of students in higher education see Singleton et al. (2012).

- How can we develop local estimates of population characteristics (e.g. specific disabilities) and behaviour (e.g. smoking) in the absence of collected survey data? For an example involving local estimates of smoking and excessive alcohol consumption see Twigg et al. (2000).

Quantitative models are valuable because they provide a simplified version of reality based upon a number of assumptions about a particular social process of interest, which can then be subjected to empirical and theoretical validation. They encompass a diverse set of often interlinked methods including statistical, mathematical or computational techniques which may or may not draw on empirical data (see Box 19.1). In Human Geography, quantitative models are usually applied to develop an understanding of the role of place within social processes. Fotheringham et al. (2000) identify Quantitative Geography as consisting of the analysis of numerical spatial data, the development of spatial theory or the construction and testing of mathematical models of spatial processes.

Box 19.1   Types of Quantitative Modelling and Examples of Their Use in Human Geography

Statistical models using numerical data

Techniques based on linear regression models (see also Chapter 31), such as multilevel models or geographical weighted regression models, are used to capture the relationships between a dependent variable and a number of explanatory variables using empirical datasets. For an example, see Box 19.4.

(Continued)

(Continued)

Mathematical models

Agent-based models are used to simulate the behaviour of a population of 'agents' within a complex system and to test hypotheses about particular processes. For example, an agent-based model might be used to investigate the impact of characteristics of place on health or residential segregation by ethnic group. For an example see Box 19.5.

Computational techniques using numerical data

Simulation models are used to generate synthetic populations for local areas with rich detail on their social, economic and demographic characteristics. Aggregate data on the population characteristics of local areas are used to guide this process. For example, see the work of Ballas et al. (2006) on the use of spatial microsimulation to explore health inequalities.

19.2 Debates Around the Use of Quantitative Models in Human Geography

The use of quantitative models in Human Geography has a long history. An early example is provided by Charles Booth's *Inquiry into the Life and Labour of People* conducted between 1886 and 1903 which quantified the spatial distribution of poverty and wealth by streets across London to illustrate stark spatial inequalities which others have subsequently shown still exist today (Orford et al., 2002). The Quantitative Revolution in Geography, from the late 1950s through to the late 1970s, saw the first concerted attempt to apply quantitative methodologies within the discipline. Like many paradigm shifts it was set in motion by frustrations with preceding modes of thought. Guided by logical positivism and the techniques of the Natural Sciences, with its emphasis on generalisation and the neutrality of knowledge, it sought to make the study of Human Geography more scientific (Robinson, 1998: 2)

The Quantitative Revolution was subjected to a number of criticisms which have had important implications for subsequent quantitative research in Human Geography. Many of the criticisms were directed at the positivist underpinnings of the approach (Peet, 1998). For example, it was claimed that the imposition of the Natural Sciences approach was inappropriate given the complexity and ephemeral nature of society and social systems. The claim that quantitative analysis offered an objective evaluation of social research questions received strong challenge and others argued that the application of quantitative methods lacked a consideration of human agency and structure, neglecting the values and meanings that make humans human and the capabilities that they possess (Cloke et al., 1991; Smith, 1998). Finally, some of the research of the Quantitative Revolution made the dangerous step of assuming that a statistical relation implied a causal relation.

The philosophical criticisms discussed above were largely accepted by quantitative Geographers; for example, Harvey recognised the danger of the inappropriate use of quantitative tools:

I believe that these tools have often been misapplied or misunderstood in Geography. I certainly plead guilty in this respect. If we are to control the use of these sharp tools in research we must understand the philosophical and methodological assumptions upon which their use necessarily rests. (1969: 7)

Box 19.2 compares two pieces of research on residential segregation by ethnic group during and after the Quantitative Revolution.

Box 19.2   Two studies of segregation by ethnic group, during and after the quantitative revolution

Farley, R. and Taeuber E. (1968) 'Population and residential segregation since 1960', *Science* 159 (3818): 953–6.

Simpson, L. and Dorling, D. (2004) 'Statistics of racial segregation: Measures, evidence and policy', *Urban Studies* 41: 661–81.

Farley and Taeuber's (1968) study explored the nature of population change for 'White' and 'Negro' populations in 13 US cities. Segregation is portrayed as a problem, without consideration of personal choices and positive aspects of congregation, and is measured using a dissimilarity index[1] in 1960 and 1965. The research makes little attempt to understand the social processes that cause segregation, numerical evidence is used to show that segregation exists, but there is no consideration of the cultural, social and historic reasons why it occurs.

Simpson and Dorling's (2004) research on segregation by race can be compared to Farley and Taeuber's (1968) study, to show the changing use of quantitative methodologies in Human Geography research. This study uses data from the 1991 and 2001 census as well as detailed data collected by Bradford council to assess claims of self-segregation of South Asian populations in Bradford in the wake of the 2001 disturbances (Cantle, 2001).

The paper begins by situating the research in the context of the 2001 'race riots' and the resulting social and political climate. The historical, cultural and economic reasons for segregation are considered and form the substantive theory that the use of quantitative techniques fits around. Statistics involving race are given detailed consideration: racial classifications are recognised as being products of society and as having the power to influence how people see themselves. The danger of misinterpreting statistical relations between ethnic groups and other characteristics as causational relationships is acknowledged. The study uses the same index of dissimilarity as Farley and Taeuber's (1968) research. However, there is more evaluation of the use of this technique including, for example, its inability to distinguish between enforced and voluntary segregation. Finally, Simpson and Dorling's research recognises the increased understanding that can be gained through the use of non-quantitative techniques. The findings of research using semi-structured interviews with estate agents are used to explore the role of estate agents in steering vendors to less racially mixed areas.

[1] The dissimilarity index ranges between 0 and 1. A value of 1 indicates complete segregation and a value of 0 indicates no segregation.

Separate from the philosophical criticisms of the Quantitative Revolution, a number of methodological weaknesses were identified. Many statistical techniques were simply borrowed in a 'cookbook fashion' from the statistical and econometric literature and applied to problems with a spatial dimension (Fotheringham et al., 2000). They were not always entirely appropriate for the study of place within social processes. For example, linear regression is a useful tool to explore how a dependent variable (e.g. an individual's income) is influenced by one or more explanatory variables (e.g. age, gender and education), but it assumes that observations are independent after accounting for the explanatory variables. However, people with similar characteristics (age, educational level) tend to cluster within similar neighbourhoods, and it is challenging, if not impossible, to include all these different individual characteristics within a model. Thus, it is plausible that the assumption of independence is violated in a linear regression that fails to fully account for the spatial clustering of individuals. The response to such criticisms was immensely positive because it led to the development of a bespoke set of quantitative models for spatial analysis.

19.3 How do I Use Quantitative Models in a Research Project?

The first step is to clearly define the research question of interest. There are different ways of arriving at a research question; the question might stem from a literature review of what is currently known in a particular area, or it might emerge from exploratory data analysis (see Chapter 29). Ideally, both processes should inform the research question you choose, as well as a topic that genuinely interests or concerns you.

Once you have decided on a research question, the next step is to consider whether there is data available for you to evaluate using a quantitative model. If such data do not exist then you have a number of options; you may choose to refine your research question to a form that can be explored with existing data; alternatively you might embark on your own data collection to obtain the information required to answer your research question. Finally, you might consider whether a quantitative model that doesn't necessarily require any data inputs (e.g. agent-based modelling; see also Chapter 21) might be appropriate.

The secondary data sources that might be used within a project can be divided into three main sources of census, survey and administrative data (see Chapter 28 and 29). Further information on these sources and how to access them is given in Holdsworth et al. (2014; see pages 41–6). Important questions to consider when evaluating data sources include whether you require data on areas or individuals, the time-frame for which data are required and whether your research question can be answered with a cross-sectional or longitudinal data source. More detail on these issues is given elsewhere (Holdsworth et al., 2013: 46–8).

A key decision concerns the type of quantitative model to adapt. This should be informed by the nature of the research question and the available data. Typically a mathematical model is developed to evaluate a particular question. So, for example, we might be interested in investigating the extent to which an individual's health is determined by their individual characteristics (age, wealth, whether or not they smoke) as opposed to the characteristics of the area in which they live (level of crime, food availability, deprivation). A regression-based model, such as a multilevel model (see Box 19.4), that predicts health based on an individual's age, gender and social class as well as the characteristics of the area in which they reside, offers one way to investigate such a question. Alternatively, if we are interested in exploring a 'what if' scenario, such as what might happen to a population if a school or factory were to open in an area, or if a particular policy were implemented, then a simulation might be more appropriate (see Chapter 28 and later in this chapter for further details on simulation). A key factor to consider here is whether your research question seeks to explain or describe a particular social process. For example, explaining why people move from one area to another is likely to require a different model from one that aims simply to estimate the extent of migration between areas.

The final step involves validating your model and drawing conclusions. This may involve a comparison of predicted results with observed empirical data. In addition, certain statistical models have assumptions that need to be considered carefully before drawing conclusions from your model. Finally, it is instructive to compare the results from your model with other research; qualitative, quantitative or theoretical. In what ways do your results confirm those in the literature and in what ways do you find different or new insights? Why do you think such differences and similarities result?

A selection of quantitative models developed, or used by, Human Geographers will be considered in the next section.

19.4 Quantitative Models in Human Geography

Multilevel models

Some research questions in Human Geography involve a hierarchy; for example, people are nested within neighbourhoods within larger areas such as districts or regions. If we are interested in understanding a particular characteristic, such as self-reported health, we might speculate that this hierarchy is substantively important; health might be influenced by both individual and area factors.

A wide body of research has demonstrated a clear health gradient according to individual income or other indicators of social position, with better health among the most affluent and worse health amongst the poorest. Similarly, we might expect that the lifestyle choices people make, such as exercise, smoking and excessive alcohol consumption, might all have an influence on their health.

At the same time, characteristics of areas are thought to influence health; so, in a hypothetical scenario, if we put two identical individuals in two areas with different characteristics we might expect their health might develop in different ways if area-based or environmental influences exist. These area influences might operate at local levels and could include the type of food on offer in a neighbourhood, the quality and accessibility of leisure and sporting facilities, the level of crime or the quality of the built environment, all of which have been linked to health outcomes (Duncan et al., 1995; Pickett and Pearl, 2001; Cummins et al., 2005; Diez Roux and Mair, 2010). Alternatively, national characteristics such as welfare provision, the extent of inequality or aspects of culture (Bobak and Marmot, 1996) might also exert an influence on an individual's health. Finally, as health is in part a social construct (Gatrell and Eliott, 2009), expectations of what it means to be in poor health are likely to be constructed through comparisons with neighbours, bringing another dimension through which place might influence health. Several studies have explored this issue by comparing self-reported measures of health with mortality between places. For example, Mitchell (2005) compares self-reported illness and life expectancy across all districts in Britain to show that for a given life expectancy the Scots are less likely to report a limiting long-term illness (LLTI) than the Welsh.

Thus, the literature on health determinants suggests that if we restrict our analysis to a single level of a hierarchy, we run the risk of drawing incomplete or worse, incorrect conclusions on the drivers of health inequality. If we simply analyse individual determinants of health using a linear regression (see Chapter 30) then we are likely to miss crucial aspects of the neighbourhood, regional or national context that influence individual behaviour and health. On the other hand, if we fit a model to predict why areas differ in terms of the health of the resident population using only area level information, we run the risk of making the 'ecological fallacy' (Robinson, 1950) should we try to transfer any conclusions to individuals.

Box 19.3   The ecological fallacy

In 1950, W.S. Robinson used the 1930 US Census to demonstrate that different results can be obtained if a dataset is analysed at the level of the individual and at the level of the area.

He calculated the correlation between the percentage of a state population who were Black and the percentage of the state population who were illiterate, revealing a correlation of 0.77, indicating that states with a higher proportion of Black people had a higher percentage of illiterate people. However, when Robinson calculated the individual correlation between illiteracy and being Black he found a much weaker correlation (0.20).

More strikingly, he also demonstrated that correlations could act in opposite directions at individual and area level. The correlation between foreign born and illiteracy equalled -0.53 at state level and 0.12 at individual level; so illiteracy was lower in states with high proportions of foreign born people, but foreign born people were more likely to be illiterate than natives.

Thus, Robinson argued convincingly that conclusions drawn about relationships between variables at aggregate levels cannot be extended to individuals.

Multilevel models have become an important aspect of geographical research addressing both the ecological and atomistic fallacies; they enable simultaneous evaluation of area and individual influences on a particular characteristic. Also known as 'Mixed' or 'Random' models, they extend a multivariate linear regression model (see Chapter 31) by partitioning the error term associated with parameter estimates to include a part attributable to each level of the hierarchy.

The simplest multilevel model is the *variance components model*, sometimes known as an 'empty model', which predicts a characteristic of interest (such as self-reported health or body mass index) with no explanatory variables, just a constant term relating to the mean of the characteristic of interest. The utility of the variance components model is that it enables us to apportion the variability observed in a particular characteristic to different levels of a hierarchy. For example, suppose we wish to determine how much of the variability in an individual's income is attributable to individual factors and how much is attributable to area characteristics, then a variance components model is specified below and displayed graphically in Figure 19.1:

$$y_{ij} = \beta_{0j} + e_{ij} \qquad \text{(Within area component)} \qquad [19.1]$$

$$\beta_{0j} = \beta_0 + U_{0j} \qquad \text{(Between area component)} \qquad [19.2]$$

Here, y_{ij} = the income of individual i in area j, β_{0j} = the mean income for the jth group, e_{ij} = error terms for the ith individual in the jth group which are assumed to be normally distributed with a mean of zero and a variance of σ_{e0}^2, β_0 is the grand mean income of all cases and U_{0j} the error associated with the jth group which is assumed to be normally distributed with a mean of 0 and a variance of σ_{u0}^2.

The variance components model can be extended to include covariates such as the age of an individual, giving a random intercept model, the specification of which is given in Equations 19.3 and 19.4. Here, x_i gives the value of this explanatory variable for individual i. The model is also illustrated in Figure 19.1:

$$y_{ij} = \beta_{0j} + \beta_1 x_i + e_{ij} \qquad [19.3]$$

$$\beta_{0j} = \beta_0 + U_{0j} \qquad [19.4]$$

Finally, explanatory variables can be allowed to vary at each level of the model in a random slope and intercept model as illustrated in Equations 19.5–7 and graphically in Figure 19.1.

$$y_{ij} = \beta_{0j} + \beta_{1j} x_{ij} + e_{ij} \qquad [19.5]$$

$$\beta_{0j} = \beta_0 + U_{0j} \qquad [19.6]$$

$$\beta_{1j} = \beta_1 + U_{1j} \qquad [19.7]$$

There are a number of advantages to using multilevel models. From a methodological perspective, by explicitly recognising hierarchical structure in the data, the standard errors (error due to sampling) of parameter estimates are less likely to be incorrectly estimated than in a multiple regression. Second, multilevel models are flexible and can be developed to include more than two levels; enabling, for example, investigation of individual, neighbourhood and regional determinants of a particular characteristic. More complicated cross-classified multilevel models allow for situations where individuals are not nested within a single area but might live in more than one location. Third, a temporal element can be readily accommodated within the multilevel model through a hierarchy of repeated observations within people, within neighbourhoods.

The use of multilevel models is linked to certain disciplines such as Human Geography and the Social Sciences. They are less commonly used in disciplines such as Economics where hierarchies might be handled in different ways, for example, by including dummy variables to capture the effect of living in each area. The advantage of multilevel models over such alternatives becomes increasingly clear as the number of areas in the model increases, making the inclusion of large numbers of dummy variables inefficient and interpretation challenging. Furthermore, multilevel models are particularly useful if we are explicitly interested in the effect of hierarchy, rather than treating it as a confounder.

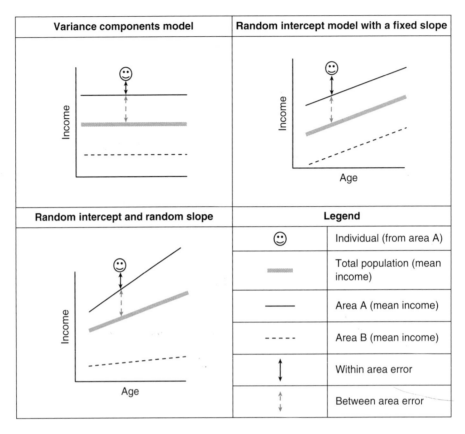

Figure 19.1 Multilevel models (Variance components model, Random intercept model, Random intercept and slope model)

Box 19.4　Wealth inequality and depression among older people

Case study: Marshall et al. (2014) 'Does the level of wealth inequality within a neighbourhood influence depression among older people?', *Health & Place* 27.

This paper considers two competing hypotheses about the impact of inequality in wealth within a neighbourhood: first, the wealth inequality hypothesis which proposes that neighbourhood inequality is harmful to health, and second, the mixed neighbourhood hypothesis which suggests that socially mixed neighbourhoods are beneficial for health outcomes. The paper considers whether the extent of inequality in house prices within a neighbourhood influences the prevalence of depression among older people using data from the English Longitudinal Study for Ageing (2002/3).

Middle Super Output Areas (MSOAs) are used as the neighbourhood geography. MSOAs are used in the dissemination of census data and were designed to improve the reporting of small area statistics. There are 7,193 MSOAs in England and Wales, with an average population of 7,200.

A multilevel model is fitted to predict whether a person has depression or not. The model includes individual correlates of depression including age, age squared (age squared is calculated as age*age. So if age=20 then age squared is equal to 20*20=400. By adding an age squared term we allow for the possibility that the relationship between our dependent variable (here depression) and an explanatory variable (here age) is non-linear. In other words an age squared term allows for the possibility that depression may increase (or decrease) at different rates at different ages, gender, marital status, economic activity, ethnicity, individual wealth, self-reported illness and educational qualification. The model also includes neighbourhood (MSOA) characteristics including median house price, level of deprivation (Indices of Multiple Deprivation 2004) and importantly for this paper a measure of area inequality in wealth based on the distribution of house prices (Gini coefficient). The model includes a random intercept which is allowed to vary across MSOAs.

A variance components model reveals that 10% of the variability in depression stems from the neighbourhood with the remaining variability attributable to the individual. The results are supportive of the mixed neighbourhood hypothesis: there is a significant association between neighbourhood inequality and depression, with lower levels of depression amongst older people in neighbourhoods with greater house price inequality after controlling for individual socio-economic and other area correlates of depression (e.g. deprivation). The association between area inequality and depression is strongest for the poorest individuals, but also holds among the most affluent. Thus the results are in line with research that suggests there are social and health benefits associated with economically mixed communities.

Geographically weighted regression

Suppose that we are interested in understanding the drivers of differences in the extent of self-reported illness across areas. One option is to perform a regression model where our dependent variable is the proportion of people with an illness in each area and a range of appropriate explanatory variables is used as predictors. We might expect, from previous research, to observe higher levels of self-reported illness in areas that have older age structures or that are deprived according to indicators such as unemployment, car ownership, tenure and wealth. We might include such variables in our regression model to test our expectations.

An assumption in a global regression model such as that described above is that the estimates of regression coefficients (which express the relationship between explanatory variables and the dependent variable) are the same in each area. However, it is equally plausible that in some situations the estimates of regression coefficients may vary over space. The problem is closely related to Simpson's paradox (see Figure 19.2), in which a relationship between two variables that is observed in two groups disappears, or is reversed, when the data are pooled.

As an example, car ownership is often used as a proxy for area deprivation and as such we might expect higher levels of illness in areas with low car ownership. Yet, the extent to which a car is needed differs depending on the quality of public transport, which varies spatially, especially across rural and urban areas. Thus spatial factors other than deprivation are related to car ownership and might complicate the relationship between car ownership and health. In Figure 19.2, we see the relationship between car ownership and % in poor health in areas comprising a set of neighbouring affluent urban areas (white dots) with good public transport and a group of neighbouring deprived rural areas (black dots) whose poor public transport

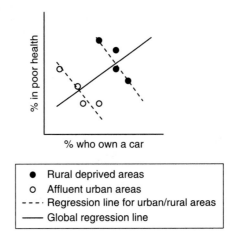

Legend:
- ● Rural deprived areas
- ○ Affluent urban areas
- ---· Regression line for urban/rural areas
- —— Global regression line

Figure 19.2 Simpson's paradox: comparing % in poor health and car ownership across and within different area types

makes car ownership very important. We see higher levels of poor health in the poor rural areas, perhaps as a result of the higher deprivation in such areas. Yet at the same time, levels of car ownership are very low in the affluent urban areas, not because of deprivation, but due to excellent public transport (and perhaps policies to deter car use – e.g. congestion charges). This combination of factors leads to a global relationship between % in poor health and car ownership that is positive and counterintuitive (areas with higher levels of car ownership have higher levels of poor health). However, within each area type, the relationship between heath and car ownership is negative as might expect (higher levels of car ownership are associated with lower levels of poor health). In such situations, referred to as spatial non-stationarity, traditional regression measures might give an incomplete, or worse misleading, perspective on the predictors of self-reported health over space.

One way of handling this challenge is to use geographically weighted regression which enables us to derive specific estimates of regression parameters for each observation (area) drawing on the data for this particular area of interest and those that are proximal to it.

In a global regression model (see Equation 19.8 below), y is the dependent variable, X is a matrix of explanatory variables and β provides the coefficients that capture the relationship between our explanatory variables (e.g. level of unemployment) and dependent variable (proportion with a self-reported illness). α gives the intercept term whilst ε gives an error term (difference between the observed and modelled values) which is assumed to be normally distributed with a mean of 0. We have one set of coefficients that are assumed to capture the relationship between dependent and explanatory variables in all areas.

$$y = \alpha + \beta X + \varepsilon \qquad [19.8]$$

A geographically weighted regression fits a different regression model for each area in the dataset, resulting in an *area-specific* set of coefficients reflecting relationships between dependent and explanatory variables in the region surrounding an area. The model below extends the global regression model by including a subscript, i, to indicate that there are a set of coefficient estimates for each area, i, in the dataset.

$$y_i = \alpha_i + \beta_i X + \varepsilon_i \qquad [19.9]$$

When fitting separate regression models for each area in the dataset, a key decision concerns which of the other observations (areas) to include in the regression. The more areas that are used, the closer the geographical regression becomes to a global regression. Yet, if a very small number of areas are used, the sampling errors around parameter estimates are likely to be large. An underlying assumption in geographically weighted regression is that proximal areas are likely to be more similar than distal areas in terms of the regression coefficients obtained. So, those areas that are closest to the area of interest are given the greatest weight. The weights range from 1 (for the area of interest) to 0 for the areas furthest away from the area of interest. A weight of 0.5 indicates that a particular area has half the influence within the regression model compared to the data for the area of interest itself.

Specification of a spatial kernel is an essential aspect of geographically weighted regression. It determines which observations are used in the estimation of regression parameters and the weight they are given. Figure 19.3 illustrates a spatial kernel around a particular observation; all those observations within the spatial kernel are used in the regression. However, observations nearer to the area of interest have larger weights (see the graph of the spatial weighting function in Figure 19.3) and so have more influence on the regression estimates.

Weights are determined by a spatial weighting function, and a number of different options exist. For example, a very basic spatial weighting function simply includes areas within the regression if they are within a certain distance (d) of the observation of interest:

$$w_{ij} = 1 \ \text{if} \ d_{ij} < d$$

and

$$w_{ij} = 0 \ \text{otherwise} \qquad [19.10]$$

Here, w_{ij} is the weight associated with area j, in the estimation of the geographical weighted regression model about area i. d_{ij} is the distance between area i and area j. In this simple model neighbouring areas that are included in the regression are given the same weighting (weight=1) as the area of interest.

However, we might intuitively expect areas to become less similar as the distance between them increases and require a weighting strategy that reflects this. Equation 19.10 provides one example of such a spatial weighting function (see Fotheringham et al., 2000, for more detail on this and alternative functions) which gives greater weights to observations close to the local observation of interest that diminish with distance:

$$w_{ij} = \exp\left(-\frac{d_{ij}^2}{b^2}\right) \qquad [19.11]$$

Here, w_{ij} and d_{ij} are defined as above whilst h is the bandwidth which determines how nearby locations are weighted and the degree of the decay in weights with distance from the area of interest. There are two main types of spatial kernels which are defined by fixed distance (see Equation 19.11) or by the number of neighbours (spatially adaptive kernels). Fixed distance kernels can be problematic in situations (e.g. rural areas) where data observations are sparse because they can lead to regressions involving a small number of observations. Spatially adaptive kernels expand in areas where data are sparse (see Figure 19.3) where it can seen that the black circle observation has a larger spatial kernel than the grey circle, reflecting the lower density of observations around the black dot. An example of a weighting function that gives a spatially adaptive kernel is below (Fotheringham et al., 2000).

$$w_{ij} = \left[1 - \left(\frac{d_{ij}}{h_i} \right)^2 \right] \text{ if } d_{ij} < h_i \text{ and } w_{ij} = 0 \text{ otherwise} \qquad [19.12]$$

Where h_i is the nth nearest neighbour from point i.

Spatial interaction models

Spatial interaction models examine flows between areas; these flows may comprise migrants, commuters, shoppers, goods or services. An early application of spatial interactions models used a 'gravity model' to understand migration between two areas

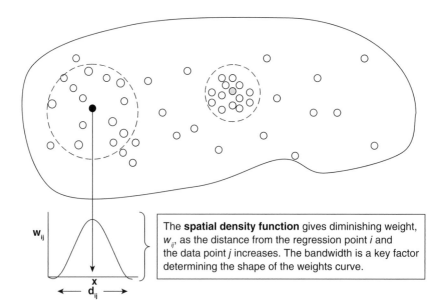

The **spatial density function** gives diminishing weight, w_{ij}, as the distance from the regression point i and the data point j increases. The bandwidth is a key factor determining the shape of the weights curve.

Figure 19.3 Example of a spatially adaptive kernel surrounding two areas; all the areas within the kernels are included within the regression for the two areas of interest with closer areas receiving greater weights

(Ravenstein, 1885; 1889) which is explained by two factors; the size or importance of each area, and the distance between the areas. Thus, the migration flows between places are understood in the same way as gravitational attraction between two celestial bodies. The model can be defined as:

$$F_{ij} = k\frac{P_i P_j}{d_{ij}} \qquad\qquad [19.13]$$

Where, F_{ij} = the number of flows between area i and area j, P_i and P_j are the population size (or importance) of areas i and j respectively, d_{ij} is the distance between i and j,[1] and k is a scaling factor which relates the number of flows (F_{ij}) to the ratio $\frac{P_i P_j}{d_{ij}}$.

The simple gravity model above can be modified to account for different types of flows which might be influenced to a greater or lesser extent by the distance or population size of each area. For example, distance might play slightly different roles in flows relating to residential moves, grocery shopping and commuting. In the model below, α, β and ω adjust the relationship between the explanatory variables (population size in, and distance between, each area) and the number of flows between areas. Figure 19.4 shows the effect of different values of ω on the number of flows between areas i and j. As ω increases, the influence of distance on flow becomes stronger, with a steeper decline in number of flows. The parameters α, β and ω are estimated empirically from observed data:

$$F_{ij} = k\frac{P_i^{\alpha} P_j^{\beta}}{d_{ij}^{\omega}} \qquad\qquad [19.14]$$

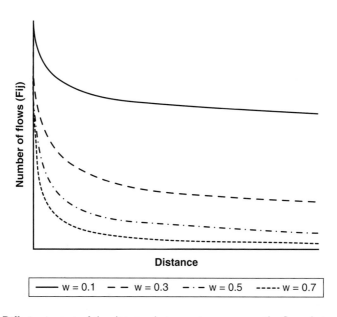

Figure 19.4 Differing impact of the distance between two areas on the flows between these areas

Whilst the gravity model has produced generally accurate estimates of flows, it doesn't offer any understanding or insight into why such flows occur. This criticism also applied to other models borrowed from other disciplines to understand spatial interactions, such as entropy maximising models. In response to these criticisms researchers began to develop models that attempted to explicitly model the decision-making process underlying a flow between two areas. The framework underlying such models involved viewing individuals as choosing a destination area from a set of possible destinations based on a decision making process which has the aim of maximising the utility gained from the move. Initially applications of such models did not take account of the impact of place on decisions around destination; for example, spatial disparities in house prices might mean that an individual living in Northern England has a more constrained choice of residential moves compared to an individual in London. However, subsequent research has explicitly included place and the associated constraints and opportunities within the modelling process. For further details see Fotheringham et al. (2000).

Simulation

One limitation of regression-based techniques, such as multilevel models, in identifying the impact of place on characteristics such as health, is that there are usually complex interrelations or feedback mechanisms linking both area and individual explanatory variables. Although other methodologies have been developed to address this shortcoming, these are, in general, poorly equipped to deal with complex situations where there are many dynamic interrelations among individuals and between individuals and their environment (Auchincloss and Diez Roux, 2008).

One response to this issue is a randomised controlled trial where samples of similar populations are placed in different neighbourhood environments. The subsequent social outcomes for each group are then tracked to assess whether differences develop. However, the ethical dimensions and cost of such schemes are usually prohibitive!

A more feasible alternative is to use agent-based modelling (ABM), a relatively modern technique that is intended to mimic complex systems (for more detail see Chapter 24). A computer model is used to simulate the behaviour of a population of 'agents' based on a set of inputs, assumptions and rules. Agents are assigned an initial condition which then changes over discrete time steps. Agents can flexibly interact with one another and with their environment with potential for feedback between individual and environmental attributes. Stochasticity is usually introduced to incorporate variability to agents' initial conditions and their interactions. Agent-based models have been used in a wide variety of settings such as crime (Malleson et al., 2010), residential segregation (Schelling, 1971), alcohol consumption (Giabbanelli and Crutzen, 2013), marriage and divorce (Hills and Todd, 2008) and access to public services (Harland and Heppenstall, 2012).

One of the key uses of agent-based models is to extend theory and to test hypotheses about particular processes; agent-based models enable a range of scenarios to be considered to identify the most salient areas of uncertainty, robustness and the identification of important thresholds (Epstein, 2008).

**Box 19.5   Schelling's Agent-based Model
of Segregation**

Schelling, T. (1971) 'Dynamic models of segregation', *Journal of Mathematical Sociology* 1.

In 1971, Schelling developed one of the earliest examples of an agent-based model to illustrate that spatial segregation of ethnic groups might plausibly develop from an interplay of individual choices relating to preferences for neighbours to be of the same ethnic group or an acceptance of neighbourhood mixing up to a certain point.

In Schelling's model agents are assigned to one of two groups and occupy cells of a larger rectangular space. Agents move according to the fraction of friends around them, where friends are those in the same group as a particular agent.

Schelling used his simulation to argue that whilst ethnic group segregation is linked to linguistic and economic factors as well as the practices of national and local organizations, it could also be the case that subtle individual choices and preferences might drive the spatial segregation of different groups (see http://ncase.me/polygons/).

An alternative use of simulation techniques is to derive synthetic populations for local areas. Microsimulation involves generating a list of individuals from a micro-dataset (a source containing information on individuals) that matches the known aggregate data for these areas as closely as possible (Dorling et al., 2005; Ballas et al., 2006). Williamson (2002) discusses a number of techniques for estimating spatially detailed population microdata.

19.5 Challenges and Opportunities: The Importance of Big Data

An exciting development for Human Geographers developing quantitative models is the arrival of 'Big Data', comprising an explosion in data from a range of relatively new sources such as financial transaction data, and Internet and mobile phone use, with the potential to deliver a valuable new understanding of society and the embedded role of place (see Chapter 14). For example, these new sources might offer insights into the evolution of opinions on social network sites, search trends from search engines, and consumer behaviour. One practical example of the use of Big Data with a geographical perspective is Google's analysis of search trends relating to flu. Initial analysis revealed very strong correlations between the frequency of Internet searches related to flu and actual diagnosed cases with the potential to predict rises of flu cases in particular areas, enabling health practitioners and policymakers to respond quickly to a flu epidemic (Ginsberg et al., 2009). More information on Google's analysis of trends in flu and access to historic data can be found at: http://googleresearch.blogspot.co.uk/2015/08/the-next-chapter-for-flu-trends.html.

Whilst the use of Big Data is undoubtedly an exciting avenue for new research, it has also been subject to challenge. For example, a subsequent research paper (Lazer et al., 2014) demonstrated that in the US, the Google flu trend analysis was predicting more than double the proportion of doctor visits for influenza-like symptoms compared to the Centers for Disease Control and Prevention. The researchers warn against the assumption that Big Data might substitute traditional data collection and analysis, pointing out that quantity of data does not sidestep issues of measurement and construct validity. One key issue in this instance is that search engines are continually evolving in order to improve their performance by, for example, autocompleteting search terms and suggesting results based on location and search history. Such processes are thought to influence what is actually searched for, thus complicating the comparison of search trends over time.

At the same time as new sources of Big Data are becoming available there has been a trend towards reduced collection of data by national governments with concerning implications for quantitative modelling. For example, the Census, the gold standard source of sub-national information on population characteristics, particularly in countries without population registration schemes, has been scaled back in many countries. The Belgian census has been discontinued and Canada has moved away from the collection of all but the most basic information in their most recent censuses. There are signs (at the time of writing) that the UK may follow suit for censuses beyond 2011 (Martin, 2005) with the possibility that such information might be collected exclusively online. A number of reasons are likely to be involved in the reduction of official statistics, with the cuts in public expenditure following the economic crisis that began in 2008 offering one explanation. The Radical Statistics group have played an important role in detailing cuts to official statistics and considering the implications of such cuts through a Reduced Statistics working group (http://www.radstats.org.uk/category/reduced-statistics).

Finally, a key, and enduring, challenge associated with any analysis of the role of place within a social process is the geographical scale of the areas in the analysis. The modifiable areal unit problem describes a situation where different scales or boundary specifications might lead to different conclusions around the role of place (Openshaw, 1977). An excellent account of the role of scale and the modifiable areal unit problem is provided by Lloyd (2014b).

SUMMARY

- Human Geographers have applied quantitative models to develop an understanding of the role of place within social processes.
- A quantitative model is a simplified version of a social process and consists of the analysis of numerical spatial data, the development of spatial theory or the construction and testing of mathematical models of spatial processes (Fotheringham et al., 2000).
- The complexity, ephemeral and value-laden nature of social processes introduces important philosophical and methodological considerations that must be taken into account when applying quantitative models to explore social research questions.
- A range of quantitative models has been developed that are tailored to investigation of the spatial unevenness of population and socio-demographic characteristics.
- Big Data offer both opportunities and challenges for quantitative human geographers who must also respond to reduced data collection by national governments.

Note

1   *dij*, the distance between areas i and j, can be calculated using Geographical Information Systems such as ArcGIS. The distance is usually based upon distance between the centroid of area.

Further Reading

Robinson (1998) and Fotheringham et al. (2000) provide a more detailed account of a wider range of quantitative models used by Geographers.

For those interested in multilevel models, Snijders and Bosker (2012) give an introduction to both basic and advanced techniques (see also the online resource below).

Lloyd (2014b) provides an excellent account of the role of spatial scale in geographical analysis and further details on methods such as geographically weighted regression and multilevel models.

Cloke et al. (1991) and Peet (1998) give a flavour of the philosophical debates around quantitative modelling in Human Geography.

Finally, many of the examples considered here relate to the influence of place on health. Much more detail on this theme can be found in the excellent *Health, Place, and Society* (Shaw et al., 2002).

For a more general account of the theoretical drivers and measurement of spatial unevenness in population see *Population and Society* (Holdsworth et al., 2013).

Note: Full details of the above can be found in the references list below.

Online Resources

Video – Alan Marshall discussing quantitative models in Human Geography and the aims of the chapter, visit https://study.sagepub.com/keymethods3e

Case study: using multilevel models to investigate predictors of Body Mass Index using MLWiN and the Health Survey for England. Data and practical details are available at:

http://discover.ukdataservice.ac.uk/Catalogue/?sn=6765&type=Data%20catalogue

Health, place and society (Shaw et al., 2002) available at:

http://www.sasi.group.shef.ac.uk/publications/healthplacesociety/health_place_and_society.pdf

References

Auchincloss, A., and Diez-Roux, A. (2008) 'A new tool for epidemiology: The usefulness of dynamic agent-based models in understanding place effects on health', *American Journal of Epidemiology* 168 (1): 1–8.

Ballas, D., Clarke, G., Rigby, J. and Wheeler, B. (2006) 'Using geographical information systems and spatial microsimulation for the analysis of health inequalities', *Health Informatics Journal* 12 (1): 65–79.

Bobak, M. and Marmot, M. (1996) 'East-West mortality divide and its potential explanations: A proposed research agenda', *British Medical Journal* 312: 421–5.

Box, G. E. P. and Draper, N. R. (1987) *Empirical Model Building and Response Surfaces*. New York, NY: John Wiley & Sons.

Cantle, T. (2001) *Community Cohesion: A Report of the Independent Review Team*. London: Home Office. http://resources.cohesioninstitute.org.uk/Publications/Documents/Document/DownloadDocumentsFile.aspx?recordId=96&file=PDFversion (accessed 22 November 2015).

Cloke, P., Philo, C. and Sadler, D. (1991) *Approaching Human Geography*. London: Chapman.

Cummins, S., Stafford, M., Macintyre, S., Marmot, M. and Ellaway, A. (2005) 'Neighbourhood environment and its association with self rated health: Evidence from Scotland and England', *Journal of Epidemiology and Community Health* 59 (3): 207–13.

Diez Roux, A.V. and Mair, C. (2010) 'Neighborhoods and health', *Annals of the New York Academy of Sciences* 1186: 125–45.

Dorling, D., Rossiter, D., Thomas, B. and Clarke, G. (2005) *Geography Matters: Simulating the Local Impacts of National Social Policies*. York: Joseph Rowntree Foundation.

Duncan, C., Jones, K. and Moon, G. (1995) 'Psychiatric morbidity: A multilevel approach to regional variations in the UK', *Journal of Epidemiology and Community Health* 49 (3): 290–5.

Epstein, J.M. (2008) 'Why model?', *Journal of Artificial Societies and Social Simulation,* 11 (4): 12. http://jasss.soc.surrey.ac.uk/11/4/12.html (accessed 22 November 2015).

Farley, R. and Taeuber, E. (1968) 'Population and residential segregation since 1960', *Science* 159 (3818): 953–6.

Fotheringham, A., Bunsden, C. and Charlton, M. (2000) *Quantitative Geography*. London: Sage.

Gatrell, A. and Eliott, S. (2009) *Geographies of Health: An Introduction*. Oxford: Blackwell.

Giabbanelli, P. and Crutzen, R. (2013) 'An agent based model of binge drinking among Dutch adults', *Journal of Artificial Societies and Social Simulation* 16 (2): 10.

Ginsberg, J., Mohebbi, M.H., Patel, R.S., Brammer, L., Smolinski, M.S. and Brilliant, L. (2009) 'Detecting influenza epidemics using search engine query data', *Nature* 457. doi:10.1038/nature07634

Harland, K. and Heppenstall, A.J. (2012) 'Using agent based models for education planning: Is the UK education system agent based?', in A. J. Heppenstall, A. Crooks, L. See. and M. Batty (eds) *Agent-Based Models of Geographical Systems*. London: Springer. pp. 481–97.

Harvey, D. (1969) *Explanation in Geography*. London: Arnold.

Heppenstall, A., Crooks, A.T., See, L.M. and Batty, M. (eds) (2011) *Agent-Based Models of Geographical Systems*. London: Springer.

Hills, T. and Todd, P. (2008) 'Population heterogeneity and individual differences in an assortative agent based marriage and divorce model (MADAM) using search with relaxing expectations', *Journal of Artificial Societies and Social Simulation* 11 (4): 5. http://jasss.soc.surrey.ac.uk/11/4/5.html (accessed 22 November 2015).

Holdsworth, C., Finney, N., Marshall, A. and Norman, P. (2013) *Population and Society*. London: Sage.

Lazer, G., Kennedy, R., King, G. and Vespignani, A. (2014) 'The parable of global flu: Traps in Big Data analysis', *Science*. 343: 1203–5.

Lloyd, C. (2014a) 'Assessing the spatial structure of population variables in England and Wales', *Transactions of the Institute of British Geographers*. doi: 10.1111/tran.12061

Lloyd, C. (2014b) *Exploring Spatial Scale in Geography*. London: Wiley.

Malleson, N., See, L., Evans, A. and Heptonstall, A. (2010) 'Implementing comprehensive offender behaviour in a realistic agent-based model of burglary', *Simulation* 88 (1): 50–71.

Marshall, A., Jivraj, S.., Nazroo, J., Tampubolon, G. and Vanhoutte, B. (2014) 'Does the level of wealth inequality within a neighbourhood influence depression among older people?', *Health & Place* 27: 194–204.

Martin, D. (2006) 'Last of the censuses? The future of small area population data', *Transactions of the Institute of British Geographers* 31: 6-18. DOI: 10.1111/j.1475-5661.2006.00189.x

Mitchell, R. (2005) 'Commentary: The decline of death–how do we measure and interpret changes in self-reported health across cultures and time?', *International Journal of Epidemiology* 34(2): 306–8.

Openshaw, S. (1977) 'A geographical solution to scale and aggregation problems in region-building, partitioning and spatial modelling', *Transactions of the Institute of British Geographers* NS 2: 459–72.

Orford, S., Dorling, D., Mitchell, R., Shaw, M. and Davey-Smith, G. (2002) 'Life and death of the people of London: A historical GIS of Charles Booth's enquiry', *Health & Place*. 8 (1): 25–35.

Peet, R. (1998) *Modern Geographical Thought*. Oxford: Blackwell.

Pickett, K. and Pearl, M. (2001) 'Multilevel analyses of neighbourhood socioeconomic context and health outcomes: A critical review', *Journal of Epidemiology and Community Health* 55(2): 111–22.

Ravenstein, E.G. (1885) 'The Laws of Migration', *Journal of the Statistical Society of London* 48 (2): 167–235.

Ravenstein, E.G. (1889) 'The Laws of Migration', *Journal of the Royal Statistical Society* 52 (2): 241–305.

Robinson, G. (1998) *Methods and Techniques in Human Geography.* London: Hodder.

Robinson, W.S. (1950) 'Ecological correlations and the behavior of individuals', *American Sociological Review* 15 (3): 351–7.

Sayer, A. (1985) 'Realism and geography'. In R. Johnston (ed.) *The Future of Geography.* London: Methuen.

Schelling, T. (1971) 'Dynamic models of segregation', *Journal of Mathematical Sociology* 1: 143–86.

Shaw, M., Dorling, D. and Mitchell, R. (2002) *Health, Place, and Society.* Harlow: Pearson Education.

Simpson, L. and Dorling, D. (2004) 'Statistics of racial segregation: Measures, evidence and policy', *Urban Studies* 41: 661–81.

Singleton, A.D., Wilson, A. G. and O'Brien, O. (2012) 'Geo-demographics and spatial interaction: An integrated model for higher education', *Journal of Geographical Systems* 14: 223-41. DOI 10.1007/s10109-010-0141-5

Smith, M. (1998) *Social Science in Question.* London: Sage.

Snijders, T. and Bosker, R. (2012) *Multilevel Analysis: An Introduction to Basic and Advanced Multilevel Modelling.* London: Sage.

Twigg, L., Moon, G. and Jones, K. (2000) 'Predicting small-area health-related behaviour: A comparison of smoking and drinking indicators', *Social Science & Medicine* 50 (7–8): 1109–20.

Williamson, P. (2002) 'Synthetic microdata', in P. Rees, D. Martin and P. Williamson (eds) *The Census Data System..* Chichester: John Wiley. pp. 231–42.

ON THE COMPANION WEBSITE…

Visit **https://study.sagepub.com/keymethods3e** for author videos, chapter exercises, resources and links, plus **free** access to the following recommended articles:

1. **Poon, J.P.H. (2005) 'Quantitative methods: Not positively positivist', *Progress in Human Geography*, 29 (6): 766–72.**

This paper provides a discussion of the philosophical underpinnings of quantitative methods within Geography.

2. **O'Sullivan, D. (2008) 'Geographical information science: Agent-based models', *Progress in Human Geography*, 32 (4): 541–50.**

This paper gives a flavour of the use of agent-based models within Geography, the challenges involved in using agent-based models and the ways in which agent-based models might be developed in future research.

3. **Curtis, S. and Riva, M. (2010) 'Health geographies I: Complexity theory and human health', *Progress in Human Geography*, 34 (4): 215–23.**

This paper provides an excellent account of the use of quantitative methods to understand the complex drivers of spatial health inequalities including the increasing role of interdisciplinary research strategies.

SECTION THREE

Generating and Working with Data in Physical and Environmental Geography

20 Making Observations and Measurements in the Field

Shelly A. Rayback

SYNOPSIS

In physical geography, fieldwork is central to how scientists learn about and understand the Earth. Working in the field encourages an integrative and iterative thinking process in which students and professionals alike must use observation, reasoning, synthesis and evaluation skills in order to develop a holistic picture of Earth's systems. In physical geography, there are a myriad of questions raised by scientists and many methods are used to address them. Considerations of spatial and temporal scale are essential in fieldwork design and how, when and where observations and measurements are made. Site selection must balance the research question with time and financial constraints. In addition, fieldwork planning and preparation must take into account issues of site suitability, feasibility and accessibility. Today, physical geographers are contemplating the longevity of field data by revisiting old field sites, leveraging old data and considering the potential uses of their data for future analyses.

This chapter is organized into the following sections:

- Introduction
- Importance of fieldwork
- Fieldwork design
- Site selection and planning
- Making observations and measurements
- Field data: Past, present and future
- Conclusion

20.1 Introduction

Ask any professional physical geographer what is their favourite aspect of their job and most individuals will reply without hesitation: fieldwork. Eagerly anticipated and actively enjoyed, fieldwork in physical geography is integral to how scientists learn about and understand the Earth. Fieldwork, by definition, takes place outside the classroom, laboratory or office in the dynamical setting of nature. While it includes making direct observations, taking measurements and samples, and recording data about the physical world, these physical acts are only part of a larger on-going cognitive process of the field researcher. It is in the field that the student and professional alike come face to face with the complexities and uncertainties of Earth's systems and processes. The researcher must make the mental leap from abstract concepts and theories to the real world by employing multiple problem solving skills such as observation, reasoning, synthesis and evaluation to the problem at hand. It is through the act of doing fieldwork that new knowledge is developed and old knowledge is tested, refined and occasionally rejected following direct observation of and experiential contact with the Earth.

Given this greater context, fieldwork, on the practical side, is the result of long hours of planning and preparation guided by multiple considerations including, the research question asked, appropriate methods and the researcher's interest. The primary research question shapes and drives the fieldwork and research project as a whole. There are an unlimited number of questions within each sub-field of physical geography that may arise from theoretically-driven reasoning or from more inductive thought processes. There are also an infinite number of methods that may be employed in field research. Some are closely linked to a particular sub-discipline in physical geography while others can be used to ask questions across multiple sub-fields. For example, dendrochronological methods have been used to answer questions relevant to both biogeography and climatology (see Chapter 22). Fieldwork may also be shaped by opportunities for training in or experience with particular field methods, including specific techniques and technologies. Some technologies, like GPS, are universally employed today by physical geographers to locate objects in space. Site selection, study design, and how field observations and measurements will be made are also important methodological choices that shape where and how fieldwork is carried out. Careful consideration of the fieldwork process and products will facilitate subsequent laboratory work and planned analyses, and may open up opportunities for future use of the collected field data in related studies. The researcher's motivation and interest in the topic under study are critical to the success of the fieldwork. When your clothes, tent and sleeping bag are damp from three days of driving rain, it is difficult to remain motivated if you do not feel passionate about your research topic. In this chapter, the importance of fieldwork within physical geography will be discussed, as well as critical issues around field-based studies including fieldwork design, site selection, observations and measurements, and the longevity of field data.

20.2 Importance of Fieldwork

Fieldwork is described as the crucible where science and scientists co-develop (Mogk and Goodwin, 2012). As an integrative process, it is where practitioners bring together content knowledge, observation, interpretation, analysis, experiment, and theory to form coherent, internally consistent interpretations of the landscape (Ernst, 2006). By using multiple investigative approaches and working hypotheses, physical geographers are continuously and simultaneously interrogating the Earth system from a reductionist/ analytical and a synthetic/integrative point of view (Mogk and Goodwin, 2012). However, fieldwork rarely exists in a vacuum. Rather, it works in tandem or is followed by analytical and experimental methods, physical, and computational modelling and the development and use of theory (Mogk and Goodwin, 2012). Thus, fieldwork is also part

Figure 20.1 Examples of undergraduate and graduate students carrying out fieldwork.
Photos by: Andrea Lini, Shelly Rayback and Beverley Wemple

of an iterative process in which field observations can raise new questions and lines of inquiry and in turn, theory, lab, and model results inform and require reinterpretation of field observations (Trop, 2000; Noll, 2003; Ernst, 2006). Thus the products of fieldwork contribute to the evolving nature of scientific knowledge within physical geography.

Fieldwork is one of the important spaces in which the training of the next generation of physical geographers occurs (Figure 20.1). It emphasizes engagement in 'authentic' activities and the practice of skills such as question asking, observations, ordering of experiences, representation, decision-making, setting priorities and communication (Neimetz and Potter, 1991; Carlson, 1999; Rowland, 2000). Many physical geographers point to their fieldwork experiences as formative spaces where they learned from and were incorporated into a collaborative network of colleagues and friends that have proven valuable over the duration of their careers. In addition, as the profession of physical geography becomes more diverse, including greater numbers of scientists who are also women or people of colour for example, the culture of fieldwork in physical geography is evolving and changing to meet and capitalize on new needs and talents (Bracken and Mawdsley, 2004).

20.3 Fieldwork Design

Through the examination of the Earth's physical and biological components and processes, physical geography seeks to understand resulting geographic patterns. In particular, physical geographers are interested in the interactions between the components and processes across space and time at multiple scales. However, while environmental changes occur over continuous, multi-scale spatio-temporal surfaces, individual studies must focus on a specific component or process at a scale of inquiry appropriate to the topic under investigation (Turkington, 2010). Specifically, to study a component or process both the phenomenon scale – the size at which the geographic structure exists and over which a geographic process operates – and the analysis scale – the size of units with which a phenomenon or process is measured – must be known or approximated (Montello, 2001). For example, for climatologists to detect and track the El Niño-Southern Oscillation (ENSO), atmospheric and oceanic measurements must be made across the equatorial Pacific to detect changes in atmospheric pressure and ocean temperatures and these measurements must take place over consecutive months.

Similar to space, consideration of temporal scale (from minutes to the geologic notion of deep time) is important in fieldwork design. Physical geographers must consider the length of time over which a process works and the rapidity with which Earth's components respond, as well as the frequency of measurement required to capture that process or response. In some cases, a palynologist may be able to remove a single, two metre-long sediment core from a lake to capture changes in regional vegetation communities that took place over a 10,000 year period, while a hydrologist may be required to take measurements at five minute intervals to capture an ephemeral pulse of nutrients into a stream following permafrost thaw on the arctic tundra.

An obstacle to both temporal and spatial scale considerations may be the available scale at which field data can be collected or have been collected by others (Montello, 2001). In some instances, the process or response under study operates at a scale different from that which is used to measure it. For example, biogeographers using Landsat imagery in fieldwork may be limited by the resolution of the pixels (30 x 30 m), as the scale is too coarse to capture finer-scale vegetation change, such as tree falls, within a stand. In field studies where time, money and effort are available, the researcher should consider taking measurements at finer-scales, leaving open the possibility of aggregating the data. Measurements made once every hour cannot be disaggregated into measurements made every minute. For the physical geographer who does not consider the temporal and spatial scales of inquiry in their fieldwork design, this neglect may result in unforeseen consequences for the degree to which their geographic information can be generalized and understood (Montello, 2001).

Broadly, physical geography field design can be categorized into four basic approaches including (1) sampling across space with controls, (2) space-for-time substitution, (3) long-term monitoring, and (4) experimental design. These approaches may be employed singly or in combination, depending on the research question, time, and financial constraints.

The first approach, **sampling across space with controls**, is characterized by the study of a target variable(s) across space with established independent control variables. Independent controls might include elevation, latitude, distance from a set point, and parent material. By establishing independent controls, the influence of the targeted variable on the physical or biological component, process, or response under study can be characterized. In conjunction with natural controls, a systematic or stratified sampling strategy (see Chapter 28) may be used to compare and contrast sites across space and determine the influence of the target variable on process and response (Turkington, 2010).

Using the **space-for-time substitution** approach, visible physical or biotic evidence on the landscape of different ages is used to understand environmental change through time. This approach assumes spatial and temporal variations are equivalent and that other controlling factors are not responsible for the change. This approach is also useful when general hypotheses about patterns and mechanisms are generated or trends sought. However, some problems arise when space is used as a surrogate for previous environments (Pickett, 1989). In particular, transient effects within systems may not be identified, and inheritance effects from previous environmental conditions or parent material may differ from the present context (Phillips, 2007). Biogeographers have used the space-for-time substitution approach when describing the process of succession, the changes in vegetation communities at a site over time. By observing the characteristics and composition of vegetation at similar sites, but at different time intervals following the same type of disturbance, biogeographers have inferred basic trajectories of change. However, recent research, particularly from long-term monitoring studies (see below), has shown the importance of transient effects (e.g. disturbance) and inheritance effects (e.g. land-use history) on vegetation change over time.

A third approach to field design, **long-term monitoring**, seeks to document and analyse evidence of environmental change at one site over long periods of time. The sites are often selected as representative of a particular landscape type or form (e.g. ecosystem, glacier) and they are monitored over time periods that are scaled to the processes under study (e.g. decades to centuries). In some cases, sites are selected for their natural state (e.g. Kluane Lake Research Station, Yukon, Canada, http://arctic.ucalgary.ca/kluane-lake-research-station; UNIS, Svalbard, Norway, http://www.unis.no/), while others are chosen because of, or despite, land-use history (e.g. Hubbard Brook Experimental Forest, Long Term Ecological Research site, New Hampshire, USA, http://www.hubbardbrook.org/). In yet other cases, sites are selected where large scale anthropogenic effects can be simulated and monitored over time in comparison to a control (H.J. Andrews Experimental Forest, Long Term Ecological Research site, Oregon, USA, http://andrewsforest.oregonstate.edu/).

Lastly, some field studies incorporate an **experimental design** approach to test specific hypotheses. Unlike laboratory studies in which all variables are controlled, experimental design studies take place in naturally-occurring environments. In these studies, one or a limited number of variables are manipulated, while other variables in the environment are controlled for to the greatest extent possible in a field. Like laboratory studies, experimental field studies use randomization of subjects or sampling units, as well as replication of treatments and controls. The International Tundra Experiment employs an experimental design approach in seeking to understand the response of circumpolar cold-adapted plant species and tundra ecosystems to changing environmental conditions and, in particular, to increasing summer temperature. The scientists use small open-topped chambers or greenhouses to manipulate temperature during the growing season (http://www.geog.ubc.ca/itex/).

20.4 Site Selection and Planning

Field site selection is guided by multiple considerations broadly organized around suitability, feasibility and accessibility. First and foremost, the field site must contain the object of study and/or be representative of the system or process under investigation (Turkington, 2010). Prior knowledge gleaned from the literature, from previous experience in the field, and reconnaissance visits (when and where possible) will help to inform the choice of site. Multiple technologies from topographic maps, to air photos, to Google Earth can aid in selecting appropriate sites, particularly ones that are remote. However, in some circumstances, field reconnaissance is the only way to determine the suitability of a site.

Fieldwork can be expensive and time consuming, so one must optimize and efficiently use each day in the field and every dollar spent. Thus, field time and funding must be balanced to achieve the desired results initially set out in the research question. A researcher may need to consider issues such as the timing, frequency, and duration of field visits. When a researcher goes into the field may be dependent upon a particular set of conditions under investigation or which control other variables or processes being studied. For example, a biogeographer may need to time her field investigations with the start of the growing season coincident with snowpack melt, or a fluvial hydrologist may need to go into the field directly following a weather event such as a rainstorm. Depending on the object, system, or process being studied, repeat field visits may be necessary. If the rate of change in the process is high or detailed sampling is needed to capture variability, then repeated field visits may occur on a daily, weekly, or monthly basis. Measurements of sediments in a glacially-fed stream may need to occur on a sub-daily to daily basis for the entire summer to capture diurnal and seasonal cycles in output. For studies of longer-term change, repeated annual visits over multiple years or decades may be warranted, such as those carried out at Long Term Ecological Research (LTER) sites in the United States or at the terminus of polar and temperate glaciers around the globe. In many instances, the duration of a field visit is less than a day or two, but in other cases a researcher must spend an entire field season at a site. The number of days in the field on field seasons is obviously limited by the amount of funding, but also by constraints on the degree program (e.g. a 2 year MSc degree) and/or funding deadlines.

Access to a site in the sense of (1) gaining permission to enter a site, and (2) being able to physically reach a site, is also an important consideration when selecting a field location. In the first instance, permission to conduct fieldwork on private and public land is required and the process of obtaining that permission may take weeks or months.At many institutions of higher education, requests to do fieldwork must pass through a risk management office to ensure that the university health insurance policy will cover the researcher for the work about to be undertaken. Despite these administrative hurdles, the act of not seeking permission from the land owner could result in the denial of requests for permission to do future fieldwork. Once permission to access the site is obtained, how one reaches the site and how long it takes to get there can be as simple as driving a car to a local beach or river meander and as complicated and time consuming as hiring a helicopter or pack mule and guide to reach a remote site.

A successful fieldwork season is built upon a foundation of thoughtful planning and preparation. The communication of the overall purpose of the research and the articulation of a clear set of fieldwork objectives are essential to building a collaborative work effort focused on accomplishing a longer-term goal and the set of tasks for the current field season (Laursen, 2011). Fieldwork should be prioritized according to the importance of each task to the overall project and the time needed to complete them. A strategy for making field measurements and collecting data and samples should be planned out ahead of time. Practise with field equipment and technology – a solid understanding of the limitations of each device may preclude some unwelcome surprises! When working in groups, it is recommended that the group practise together ahead of time, if possible, to facilitate consistent field practices, time management, and cooperation. The safety and welfare of the field participants should also be at the forefront of planning, too. Simple steps such as carrying a wilderness first aid kit, taking a first aid/CPR course, or working in groups can help when dealing with an injury or avoiding an unexpected encounter with wildlife. For those who do research in remote locations, fieldwork preparation courses (e.g. U.S. National Science Foundation Office of Polar Programs, http://www.nsf.gov; Swedish Polar Research Secretariat, http://www.polar.se/en; or the Royal Geographical Society, http://www.rgs.org) can prepare researchers to do work safely and efficiently in backcountry locations.

Once in the field, researchers must remain flexible. Periodic reevaluation of field objectives and how data and samples are collected must be made to facilitate successful fieldwork. It should be noted that professionals and new field researchers alike tend to have high expectations and over-program their field season (Laursen, 2011). However, most fieldwork takes longer than expected and, as not everything is controllable or can be anticipated, the ability to cope with the unexpected with a degree of patience and humour can salvage a project and reenergize a team. In addition, new ideas and interesting questions often arise from time in the field. Having a 'fertile mind', one that is open and prepared to make new observations and integrate these with extant knowledge to formulate new hypotheses and research directions, is critical to a rich fieldwork career (Mogk and Goodwin, 2012).

20.5 Making Observations and Measurements

In the Earth and natural sciences, some of the greatest scientific advances have come from direct observations of nature. Keen observers such as James Hutton, Louis Agassiz, Alfred Wegener, Charles Darwin, and Alfred Wallace produced insights on the natural world through their observations and interpretations which later led to the development of major scientific paradigms, such as geological time, plate tectonics, and evolution, that still hold true today (Mogk and Goodwin, 2012). Out of this work and to the present day, an epistemology grounded in direct observation and interpretation of natural processes and phenomena has emerged (Froderman, 1995). Thus, the making of observations is often the first step in fieldwork.

A reconnaissance visit to a potential field site is the first opportunity to make observations and a decision whether the site meets the study goals. A fieldworker may record notes in a notebook or electronically, capturing the salient features of the site via words, sketches, photographs, maps, and GPS points. Often this is an iterative process within a group as multiple perspectives, expertise and familiarity with a site are shared to develop an integrative first impression. Throughout the fieldwork component of the project, initial observations should be revisited by the group as data are collected and new ideas and interpretations emerge. However, the ultimate goal of reconnaissance visits is to determine whether the site contains the necessary phenomena or processes under investigation and whether you will be able to measure them and collect the necessary data. If the study is to be carried out at multiple field sites, it is important to determine whether there is enough similarity across sites to make appropriate comparisons. After a site visit, many field researchers will transcribe their field notes, making electronic and hard copies to ensure no information is lost or forgotten. Later, these field observations should be archived with the collected data and analysis for future use.

Given the time, money, and effort invested by researchers in collecting data in the field, it is important to consider carefully how those measurements will be made. In general, measurements are characterized by (1) their accuracy – the closeness of a measurement to the 'true' or expected value, and (2) their precision – the closeness of repeated measures to each other. In addition to increasing the sample size (see Chapter 32), accuracy and precision may be increased through consistent measurement practices. For example, to minimize measurement error, instruments should be calibrated before use, while observer error can be reduced through training and consistent field protocols. Any measurement problems should be rectified as soon as possible and a record made in the field notes to detail the timing and extent of the problem.

Based on the research question and opportunities presented by the field site, a sampling strategy to obtain the targeted measurements should be selected (see Chapter 28). The sampling strategy may be intensive or extensive based on the aims of the study. An extensive research design emphasizes highlighting pattern and regularity in data from which it is assumed that an underlying process or factor is the cause. In this case, large numbers of measurements are usually made in order to articulate generalizations from patterns in the data. Some form of random, systematic or stratified sampling allows the researcher to sample a large area thoroughly and then conduct a statistical analysis of the data. An intensive research design seeks to investigate a single case study or small number of studies in greater detail with the emphasis on hypothesis testing. More information on sampling strategies is available in Chapter 32. As discussed previously, temporal (frequency) and spatial considerations are also important when planning the distribution of measurements across space and time. The distribution of measurements across a landscape should be located so as to capture a measure of the size and variability of the geographic structure or process. The frequency of measurements must account for the process and response rate.

20.6 Field Data: Past, Present and Future

Today, physical geographers are increasingly considering the longevity of field data and their utility in addressing questions about the Earth's future. This consideration comes within the context of environmental change due to natural and anthropogenic forcing factors and the need to understand past conditions and predict future scenarios. In some subfields, like climatology and hydrology, long term datasets of atmospheric measurements or stream discharge have been recorded and maintained by national agencies for decades, which can be used by researchers to investigate change over time in the atmosphere or hydrosphere. In other instances where continuous long term data are not available, physical geographers have returned to old field sites and revisited old data sets to explore changes in natural processes and phenomena over time. For example, some dendrochronologists and palynologists are working to update archived field data and samples in order to extend the period of analysis up to the present.

In addition to exploring old data, physical geographers are also now considering the multiple future uses of the data they collect today. Recently collected field data that were originally intended to address a different set of research objectives are now being used in modelling, meta-analysis and synthesis studies. These larger projects often involve scientists from several institutions and countries with the goal of identifying and explaining larger continental, hemispheric, or global scale patterns. Finally, the archiving of field data is a growing concern in physical geography and in many cases, it is mandated by federal funding agencies. Currently, physical samples may be archived indefinitely in national and university research repositories, and electronic data may be uploaded and permanently stored on federally funded, publically accessible databases (e.g. National Oceanic and Atmospheric National Climate Data Center for Paleoclimatology, http://www.ncdc.noaa.gov/data-access/paleoclimatology-data).

20.7 Conclusion

In the 21st century, fieldwork continues to hold a central place in physical geography research. It is outside and within the landscape that physical geographers develop important intellectual skills to understand the Earth as a whole system, and not only as disconnected and disparate parts (Ireton et al., 1997). Physical geographers also have the ability to move fluidly and recursively from theory, to the field, to the lab and back, contributing to the evolving nature of scientific knowledge. It is this integrated and holistic view developed through fieldwork and viewed through the lens of space that places physical geographers in the unique position to be able to contribute effectively towards the ultimate goal of understanding and solving the pressing issues of global change.

SUMMARY

- Fieldwork is integral to physical geography and our holistic understanding of the Earth.
- For students carrying out fieldwork as a component in their research programme, it is essential to contemplate the theoretical foundation of the research question in relation to issues of temporal and spatial scale and the sampling site.
- Time and effort spent considering how to structure a rigorous approach to field design and sampling will be rewarded through the collection of quality data and the potential for a strong project outcome.
- Fieldwork remains one of the most enjoyable and rewarding aspects of research for physical geographers and will continue to be the focal point of many successful and informative research projects in the future.
- Field data can have longevity beyond the life of an individual research project.

Further Reading

- Gomez, B. and Jones, J.P. III. (eds) (2010) *Research Methods in Geography: A Critical Introduction.* Oxford: Wiley-Blackwell.

  Research Methods in Geography is a thorough introduction to research methods and techniques used in both physical and human geography. Each chapter introduces foundational concepts that focus upon major questions in the discipline today while addressing practical issues concerning data collection, analysis and interpretation.

- Montello, D.R. and Sutton, P. (2012). *An Introduction to Scientific Research Methods in Geography and the Environmental Sciences* (2nd edition). London: Sage.

  An Introduction to Scientific Research Methods in Geography and the Environmental Sciences is a holistic and comprehensive introduction to the process of doing research. The focus on novel topics such as scientific communication and the use of visualization are particularly noteworthy and appropriate for the next generation of geographers.

- Kastens, K. and Manduca, C.A. (eds) (2012) *Earth and Mind II: A Synthesis of Research on Thinking and Learning in the Geosciences*. GSA Special Paper 486.

  Earth and Mind II explores a common set of perspectives, approaches and values that geoscientists share through the themes of Time, Space, Systems and the Field. The value of this reading comes from its investigation of how we think about, teach and learn about the Earth and the contributions that this unique perspective has to offer the sciences as a whole.

Note: Full details of the above can be found in the references list below.

References

Bracken, L. and Mawdsley, E. (2004) 'Muddy glee: Rounding out the picture of women and physical geography field work', *Area* 36: 280–6.

Carlson, C. A. (1999) 'Field research as a pedagogical tool for learning hydrogeochemistry and scientific-writing skills', *Journal of Geoscience Education* 47: 150–7.

Ernst, G. (2006) 'Geologic mapping – where the rubber meets the road', in C.A. Manduca and D.W. Mogk (eds) *Earth and the Mind: How Geologists Think and Learn about the Earth*. Geological Society of America Special Paper 413. pp. 13–28.

Froderman, R. (1995) 'Geological reasoning: Geology as an interpretive and historical science', *Geological Society of American Bulletin* 107: 960–8.

Gomez, B. and Jones, J.P. III (eds) (2010) *Research Methods in Geography: A Critical Introduction*. Oxford: Wiley-Blackwell.

Ireton, M.F., Manduca, C.A. and Mogk, D.W. (1997) *Shaping the Future of Undergraduate Earth Science Education: Innovation and Change Using an Earth System Approach*. Washington, DC: American Geophysical Union.

Kastens, K. and Manduca, C.A. (eds) (2012) *Earth and Mind II: A Synthesis of Research on Thinking and Learning in the Geosciences*. GSA Special Paper 486.

Laursen, L. (2011) 'Field work: Close quarters', *Nature* 474: 407–9.

Mogk, D.W. and Goodwin, C. (2012) 'Learning in the field: Synthesis of research on thinking and learning in the geosciences', *Geological Society of America Special Papers 2012 Special Paper* 486: 131–63.

Montello, D.R. (2001) 'Scale in geography', in N. J. Smelser, and P.B. Baltes (eds) *International Encyclopedia of the Social and Behavioral Sciences*. Oxford: Pergamon Press. pp. 13501–4.

Montello, D.R. and Sutton, P. (2012) *An Introduction to Scientific Research Methods in Geography and the Environmental Sciences* (2nd edition). London: Sage.

Neimetz, J.W. and Potter, N. Jr. (1991) 'The scientific method and writing in introductory landscape development laboratories', *Journal of Geological Education* 39: 190–5.

Noll, M. (2003) 'Building bridges between field and laboratory studies in an undergraduate groundwater course', *Journal of Geoscience Education* 51: 231–6.

Phillips, J.D. (2007) 'The perfect landscape', *Geomorphology* 84, 159–69.

Pickett, S.T.A. (1989) 'Space-for-time substitution as an alternative to long-term studies', in G.E. Linken (ed.) *Long-Term Studies in Ecology: Approaches and Alternatives*. New York: Springer. pp. 110–35.

Rowland, S.M. (2000) 'Meeting of minds at the outcrop, a dialogue-writing assignment', *Journal of Geoscience Education* 48: 589.

Trop, J.M. (2000) 'Integration of field observations with laboratory modeling for understanding hydrologic processes in an undergraduate earth-science course', *Journal of Geoscience Education* 48: 514–21.

Turkington, A. (2010) 'Making observations and measurements in the field', in N. Clifford, S. French, and G. Valentine (eds) *Key Methods in Geography* (2nd edition). London: Sage. pp. 220–9.

ON THE COMPANION WEBSITE…

Visit **https://study.sagepub.com/keymethods3e** for author videos, chapter exercises, resources and links, plus **free** access to the following recommended articles:

1. **Mair, D. (2012) 'Glaciology: Research update I', *Progress in Physical Geography*, 36 (6): 813–32.**

This progress report highlights the importance of field work observations and measurements in glaciology to develop, test and refine models used to understand atmospheric melt-induced influences on the dynamics of the Greenland Ice Sheet.

2. **Meadows, M.E. (2012) 'Quaternary environments: Going forward, looking backwards?', *Progress in Physical Geography*, 36 (4): 539–47.**

In this progress report, Meadows makes a strong argument for the greater consideration and inclusion of a long-term temporal perspective in physical geography and other related physical and biological sciences, in particular, how this temporal perspective may be applied to issues of current environmental change.

3. **French, J.R. and Burningham, H. (2013) 'Coasts and climate: Insights from geomorphology', *Progress in Physical Geography*, 37 (4): 550–6.**

In this progress report, French and Burningham address how researchers must go through the process of questioning paradigms and theories with detailed studies in order to thoroughly test these assumptions. They also emphasize the need to examine not only local and regional influences, but also synergistic combinations, that might contradict ideas originally based on larger-scaled processes.

21 Making Observations and Measurements in the Laboratory

Scott A. Mensing

SYNOPSIS

There is a strong connection between fieldwork and laboratory work because many physical processes cannot be observed in the field but can only be discovered through laboratory analysis of samples collected in the field. Laboratories present a different working environment made up of specialized equipment that is critical for producing reliable and reproducible results. Geographers interested in understanding physical processes will often spend more time conducting research in a laboratory than in the field, and it is important to have a good working understanding of laboratory practices. This chapter describes the laboratory environment, discusses issues related to efficient and safe work within the laboratory, and reviews some of the common methodologies utilized by some of the sub-fields within geography.

This chapter is organized into the following sections:

- Introduction
- The laboratory environment
- Making observations and measurements
- Conclusion

21.1 Introduction

Laboratory methods are commonly intimately related to field methods. Many of the processes that are important in understanding the physical geography of the world cannot be observed in the field, but are only revealed through laboratory analysis of samples carefully collected in the field. Physical geographers often utilize specialized laboratory methods in their research to reconstruct both spatial and temporal changes in the physical world. For example, biogeographers and palaeoecologists interested in vegetation and climate change through time recover sediment cores from lakes, fens, bogs, and meadows and analyse a wide range of physical properties in the sediments, including pollen, organic matter, geochemical isotopes, charcoal and macrofossils (Faegri and Iversen, 1989). Dendrochronologists interested in reconstructing climate recover tree cores and measure ring widths, wood anatomy, and isotopes (Fritts, 1976). Geomorphologists and soil scientists interested in understanding processes that create landforms measure particle size and mineralogy, and hydrologists interested in understanding the water cycle and water resources measure water chemistry and quality (Goudie, 1990). Although laboratory experimentation is less common in geography than the biological sciences, fluvial geomorphologists can create flumes to observe and measure sediment transport to better understand processes controlling stream morphology (Church et al., 1998), and aeolian geomorphologists can use wind tunnels to examine the construction of features such as dunes (Dong et al., 2003).

Laboratories require dedicated workspaces with specialized equipment designed to measure small samples of physical materials. While some laboratories can serve multiple functions, most are designed for a specific purpose; therefore, it is important to know the capabilities of a laboratory when designing research. Furthermore, many laboratory analyses require specific amounts or types of material and it is important to understand these requirements prior to going into the field to collect materials. Once in a laboratory, all laboratories will have a series of safety procedures and protocols that you must follow while working in that lab, often including training prior to doing any work, so it is important to know these procedures and plan your time accordingly.

This chapter will describe the laboratory environment, discuss the issues inherent in working in a laboratory and review some of the particular methods geographers use to make observations and measurements in the laboratory environment.

21.2 The Laboratory Environment

Laboratories provide a controlled environment for collecting data on specific physical parameters. It is critical that these data be accurate and the processes be repeatable. For this reason, laboratories typically have a set of procedures in place to guide users in the appropriate use of equipment, methods for processing different types of materials, safe handling of dangerous substances or instruments, instructions on replacing consumables, and laboratory clean-up. Although the details in any particular lab will differ, there are some general guidelines that help one understand how to best conduct research in this environment.

Ideally, a laboratory is large enough to separate different functions that either require different equipment, or may lead to contamination of samples. For example, laboratories working with sediment cores may create separate spaces for 'dirty' and 'clean' tasks. Opening and examining sediment cores and processing sediments are relatively 'dirty' tasks, and ideally should take place in a separate room from the 'clean' space where microscopy is done. In some cases, where laboratories have sufficient space to isolate each of the different tasks, sediment description, sediment processing, and microscopy are all isolated to different rooms. To further reduce the potential for cross-contamination, sometimes samples collected domestically are isolated from samples collected internationally. Space constraints do not always allow the physical isolation of different activities, and in these cases a laboratory can organize its work to isolate different functions temporally by only performing one type of task at a time in the laboratory, and cleaning thoroughly between different tasks. In this way, even a small constrained laboratory can maintain good practices for reducing cross-contamination of samples and keep a clean and organized workspace.

All laboratories are designed for a specific function and therefore will have a suite of instruments particular for that function. While some instruments may be common in many different types of laboratories, it is important to recognize that even common instruments may come in many different forms, and it is important to know exactly what your need is to ensure that a laboratory has the right instruments to answer a specific question. A good example of this is a microscope. While microscopes are common to many labs, there are many different types and qualities of microscope. Some microscopy uses samples on glass microscope slides (e.g. pollen, diatoms) and requires a bright-field light microscope with magnification of 100x up to 1000x. In other cases, a dissecting scope is required (e.g. dendrochronology, macrofossils, charcoal) with magnifications from 10x to 50x. For mineralogical analysis, a petrographic scope with polarized light is essential. For very specialized work, sometimes a scanning electron microscope is needed. In each case, microscopes have many components of differing quality, and the better the quality (and typically the greater the cost) the higher the resolution that can be achieved. Microscopes used in a teaching laboratory are often not of sufficient quality for doing research.

In contrast, most laboratories also contain unique pieces of equipment that serve a very specific purpose, and though not necessarily used on a regular basis, are critical at a certain point in an analysis. For example, a **hydrometer** measures density of liquid and is required for certain flotation methods. A **spectrophotometer** is required for analyses of relative lightness or darkness of a liquid, a measure in peats that can be used to identify periods of greater or less oxidation (inferred as periods of drier or wetter climates). A **sedigraph** is able to rapidly and efficiently determine particle size of soils or sediments, producing results far more quickly and precisely than older mechanical methods that only required inexpensive equipment common to many laboratories. Some equipment is both expensive and only used irregularly, and therefore it is not uncommon for some instrumentation to be shared within institutions. This requires learning and following the protocols of the laboratory where the instruments are held, and usually reimbursing the costs of consumables or the time of technicians. However, such an approach may be much more economical than purchasing and supporting a specialized piece of equipment. In some countries, national laboratories have been set up exactly for this purpose. In the United States, the National Science Foundation has established the National Lacustrine Core Facility (LacCore) to supports lake sediment research in the field and laboratory, archiving of cores and data, and distribution of samples.

Some equipment is so specialized that samples are routinely sent off to a governmentally supported or commercial laboratory for analysis. This is particularly true for radiometric age dating (^{14}C, ^{210}Pb, ^{10}Be, ^{137}Cs), which is a standard method required for palaeoecologic reconstructions. These laboratories will provide protocols for pretreating samples for shipment. In some cases, the laboratories' analytic costs are based on the level of pre-treatment that you provide. Geochemical analyses (e.g. oxygen isotopes $\delta^{18}O$) are also routinely used by physical geographers to measure past changes in climate. While some researchers maintain their own lab, it is not uncommon to send samples to a professional lab.

Even common equipment can be more complicated than would initially appear. For example, test tubes come in a wide range of shapes, sizes and materials, and it is essential, before making a selection, to know what they will be used for. Sieves come in multiple sizes and materials, and for very fine size fractions, many researchers use filter cloth rather than traditional metal sieves. While metal sieves might require a mechanical shaker for efficient sieving, cloth filters may require a vacuum system to assist in sieving material efficiently. While some labs may come equipped with vacuum lines, simple faucet aspirator vacuum systems have been designed to run off of a faucet and can be set up inexpensively in any laboratory.

The accuracy and precision of instruments also need to be considered. Accuracy relates to the closeness to the true measure of something. Precision refers to the ability of the instrument to repeat the same measure. Balances, common in many labs for weighing samples, provide a good example of the accuracy of an instrument. It is not uncommon for a laboratory to have several balances; a relatively inexpensive balance for weighing heavy samples (≤500 g) to an accuracy of 0.01 g, and an analytical microbalance for very small objects suitable for accurate measurement of samples to 0.0001 g. Many laboratory analyses require measuring very small differences between samples and it is essential to have instruments capable of highly accurate measurements in order to obtain meaningful results. On the other hand, not all measures need to be of such high accuracy and it is important to know when such accuracy is essential in order to use the correct instrument.

While safety is always a concern when working in laboratories, some types of work require specific safety equipment. Any use of chemicals requires a fume hood where all sample processing takes place. Safety equipment such as gloves, eye protection, and laboratory coats is essential when working with any chemicals. Similarly, used chemicals must be properly stored and disposed of. All institutions have a set of safety protocols that must be rigorously followed to comply with local and federal laws. Noncompliance with regulations can have very serious legal consequences, and it is the researcher's responsibility to be aware of proper handling of dangerous substances. In some instances where excessive dust is produced, adequate ventilation and breathing apparatus must be provided. Open flames from Bunsen burners or very high temperatures from furnaces are also hazards. Furnaces may reach 1000° C and require specialized safety equipment for placing and removing samples. When particularly dangerous procedures are being done it is advisable to have a partner always present in case of emergency. Emergency contact numbers should always be posted in a visible location and emergency safety equipment readily available. Most labs working with chemicals will have an eye-wash system and often safety showers in case chemicals are accidentally spilled on a worker. Kits for cleaning up spills or specialized first aid kits are also mandatory in laboratories with dangerous chemicals. Even in safe labs, sharp tools are commonly used and glass beakers break. Specific 'sharps' containers should be available to ensure that hazardous wastes are kept separate from non-hazardous wastes. It is critical that all personnel using a laboratory receive adequate safety training before using the lab and that certificates of training be made available for inspections.

Maintaining an organized lab and reducing clutter are part of keeping a lab safe. All substances need to be properly labelled. If a material is found in an unmarked container, it must be assumed to be hazardous. Laboratory benches are meant for working and not storage and should be kept clear when not in use. Contamination of samples can be a problem with many types of analyses and keeping both the workspace and materials clean is critical. In some cases where cleanliness is particularly important, a temporary working surface (such as aluminium foil) can be laid on the benches and changed between different tasks. In laboratories where the potential for contamination between different analyses can potentially ruin the sample, temporary working surfaces are changed between each new analysis. Each person is responsible for the quality of their own work, and in labs used by multiple people, if there is a concern for contamination with samples, it is a good practice to wash all containers, tools and surfaces prior to use.

21.3 Making Observations and Measurements

Similar to good practices in the field, it is a good practice to record all steps in an analysis in a laboratory notebook that remains in the laboratory. This maintains a record of all activity in the laboratory and can be an essential reference when analysing the data at a later date. This notebook should follow standardized naming conventions for samples and include every step made during a procedure – even mistakes. If data have outliers, it is important to be able to determine whether these represent true variation, or resulted from an error in the processing. It is common to enter data directly in an electronic form; however replication of these data as hard copy in a permanent notebook is a backup procedure that often proves to be well worth the time and effort.

The goal of laboratory analysis is to obtain high quality, reproducible data often from multiple samples. Many laboratory procedures are simple but repetitive. Others may require multiple technical steps that must be done carefully and in a specific order. In both cases, the need for extended attention to detail can lead to making errors in reading instruments, transposing numbers when recording data, and skipping or forgetting critical steps. One remedy to this is to develop a set of procedures and error checks that are applied as you work. For example, it is not uncommon for some analyses to have to weigh a container before adding material, reweigh it after the material has been added, and then continue to reweigh it following different procedures. An error checking routine can be added to your data that lets you know if a sample weight exceeds the previous weight – typically a sure sign that a number has been entered wrong or possibly the sample has been entered in the wrong place.

Before using an instrument to take a measurement, make sure that the instrument has been calibrated. Some equipment has calibration systems built in (e.g. balances, magnetometers) and others require calibration against some standard (e.g. pH meters). Be aware that instruments may drift over time creating a bias in the data. In such instances, regular calibration may be necessary. Never be afraid to throw away data that you suspect contains errors. It is better to redo the analysis than to work with bad data.

Replication is an important practice in laboratory work. When using an instrument to make a measurement, a common practice is to take multiple repeated measurements, and then average these for the actual measure. No matter what the quality of the instrument, measures may vary between samples due to atmospheric conditions, variation in the sample, or variation in the instrument. Never assume that the first measure is correct. Most instruments will record more significant digits than can be accurately measured. The last digit is often thought of as a 'rounding' digit and can be recorded (for rounding) or disregarded. It is important to know the limitations of the equipment you are using so that you do not publish data at a level of accuracy beyond what the instrumentation was designed for.

All procedures should be documented so that the work can be replicated. It is not uncommon for some analyses to be so time consuming that a team of researchers may work on the same analysis. In such cases, it is imperative that a suitable number of replicates are analysed by all team members and the data compared to cross-validate the work and ensure that the results are consistent. Only after the methods have been shown to be applied consistently by each team member across the full breadth of variation of samples should team members continue individually. It is also important to create consistent naming conventions for data files, since consistent organization of the data is as important as consistent processing of samples. Multiple data files are often created in laboratory data analysis and simple and consistent file naming conventions are also critical in laboratory work so that data can be transferred and shared without creating errors.

Many different types of materials are brought into laboratories for analysis. An incomplete list of materials commonly analysed by physical geographers includes soil, sediment, water, plant matter, tree-cores and ice cores (see Chapter 22). An equally diverse array of analyses are done on these materials, such as particle size, humification, charcoal, pollen, diatoms, macrofossils, chemistry, geochemical isotopes, radiometric isotopes, wood anatomy, dust, organic content and mineralogy. Many analyses are destructive, meaning that they consume a portion of the material that was collected in the field. Before an analysis is planned, and ideally before field work even begins, it is important to know what types of analyses will be done on the material being collected and whether that analysis is destructive or non-destructive. In the case of destructive analyses, the researcher must know beforehand how much material will be consumed by the analysis, to ensure that sufficient material is gathered from the field to meet the needs of all of the different proposed analyses.

Sub-sampling of material also has different constraints. For any material collected, the outer portion of the samples is always more likely to be contaminated by foreign substances than the interior of the sample. It is important to know whether a specific analysis is more or less affected by potential contamination. For example, when working with lake sediments or peats, a common first step is to measure the water and organic content and bulk density. These measures are less sensitive to contamination than other analyses, such as isotopes, pollen or diatoms. In this case, the outer portion of a core may be safely subsampled for the wet/dry analysis, whereas subsamples for isotopes, pollen and diatoms should only be taken from the centre of the core to minimize the potential for contamination.

A well-designed study will have material remaining after all sub-samples have been taken and will have a plan for archiving these materials for future use. New laboratory methods are regularly being developed or improved and, considering how difficult it is to collect materials initially, a common practice is to save half of the original material and to archive this for future use. The form of archiving differs by material. Materials such as sediment or ice cores must be permanently refrigerated or frozen, whereas soils, tree rings and plant matter may only need to be kept in a cool, dry space. In some cases, national and international repositories have been created for archiving physical samples. In addition, data centres have been established to accept most of the different data types produced by laboratory analyses, and, once the results have been published, it is a good practice to submit the analytical data to these centres for sharing with the broader research community.

A few examples of laboratory methods from the different sub-disciplines within Geography will demonstrate the range of options that may be considered and the type of choices to be made in the laboratory.

Biogeographers and palaeoecologists may investigate vegetation and climate history reconstructed from lake or peat sediment cores (Birks and Birks, 1980). In the laboratory, they may begin by describing the sediments using non-destructive methods such as using instruments for creating a high resolution image, measuring magnetic properties, and using an X-ray fluorescent scanner to measure elemental content. They may then sub-sample for destructive analyses to further describe the sediments, including taking a sample for measuring percent water, percent organic material, percent inorganic material and bulk density. These analyses would require use of fine balances for weighing, an oven for drying the sample, a furnace to combust the sample, and a desiccator for storing samples between steps. Separate sub-samples may then be taken for pollen, to reconstruct vegetation history; diatoms, to reconstruct lake aquatic history; charcoal to reconstruct fire history; and geochemical isotopes for an independent measure of vegetation or lake history. These processes may require a fume hood for chemical analyses, bright-field microscopes for pollen and diatom analysis, dissecting microscopes for charcoal analysis, and potentially sending samples off to a professional laboratory for isotopic analysis. Sediment cores require obtaining radiometric dates as well, and materials would have to be identified and pre-processed for sending to a specialized laboratory for radiometric dating. The pre-processing differs for each different radiometric method, so the researcher would likely need to contact that laboratory to learn what procedures they follow. Since it is common for multiple researchers to take responsibility for each different analysis, it is important for the team to develop a systematic sampling strategy before the work begins to avoid sample bias and determine an appropriate sampling resolution.

Geomorphologists may collect rock or mineral samples to address questions regarding the age of glacial moraines or the formation of a feature such as an alluvial fan. Advances in ^{10}Be dating techniques have provided an opportunity for geomorphologists to improve our understanding of the timing of glacial advances (Bentley et al., 2006). This work requires collecting rock samples in the field, followed by a complicated and time-consuming series of chemical analyses in the laboratory before a sample can be submitted for ^{10}Be dating. This processing requires a specialized lab. Soils samples collected along a transect on an alluvial fan would then require analysis of sediment particle size to determine rates of soil development or energy level at different points in the fan. In the laboratory, sediment analysis may be done with mechanical methods (sieves and settling tubes) or with a more specialized sedigraph if that piece of equipment is available.

Dendrochronologists recover tree cores in order to reconstruct past climate, study tree growth, and determine the response of trees to climate (Fritts, 1976). For climate reconstruction, tree cores must be placed into core holders for processing. These are typically made of wood and a woodshop with power tools is necessary to make the core-holders. Sanding of the cores is also essential, and the initial steps are also typically done with power tools. Laboratories with power tools require particular requirements for safety training and proper dust ventilation. Once samples have been sanded, further analysis moves to a clean room with binocular microscopes for ring counting. Dendrochronologists have a formalized method of sample notation that is written onto each core so that the data are directly available on the sample. Specialized equipment is required for measuring ring-widths. Tree-cores can then be subsampled for isotopic analysis or wood anatomy. Once analyses have been completed, the samples themselves can be archived for potential additional analysis at a later date.

21.4 Conclusion

While geography can be considered as a field-oriented science (Petch and Reid, 1988), many important discoveries of geographic processes can only be made by analysing samples in the laboratory. Laboratory work is often quite specialized and efficient research requires knowledge of available instrumentation, protocols and specifications. It is critical to have knowledge of laboratory procedures prior to conducting field work to ensure that field samples are collected properly and in sufficient quantity to be able to conduct later laboratory analyses. It is rare for one researcher to have all the necessary equipment in their laboratory, and collaboration between researchers, or use of multiple laboratories, are not unusual. Well-designed laboratory procedures can be critical to obtaining successful research results.

SUMMARY

- Laboratory analysis generally is based on samples collected in the field, and an understanding of laboratory protocols is essential in designing fieldwork.
- Laboratories are typically designed for specialized functions, and instrumentation, protocols and safety procedures will differ between laboratories.
- Due to the wide range of potential analyses performed in laboratories, there is a wide range in types and quality of instruments for similar functions, and obtaining good results requires selecting the right laboratory equipment and procedures.
- Laboratories often have legally mandated safety procedures and no research should be initiated without proper training.
- The goal of laboratory research is to obtain high quality reproducible results, and this requires rigorous attention to detail and consistent application of standardized procedures.
- Many laboratory analyses consume samples, and projects should be designed to provide adequate material for all analyses plus archiving of extra material for future analyses.

Further Reading

Birks and Birks (1980) provide a comprehensive review of both field and laboratory methods used in palaeoecologic studies. The chapters on laboratory methods cover many techniques used within geography.

Massart et al. (1993) Many laboratory texts are specialized for subfields within chemistry and medicine. This book, while also somewhat specialized, goes into greater detail on subjects of accuracy and reliability and provides interesting chapters on collecting and analysing data in relation to water quality, air pollution and environmental modelling.

DiBerardinis et al. (2013) Originally written for industry, this book describes many of the criteria around designing labs that make it possible to conduct scientific investigations in a safe and healthy environment

Note: Full details of the above can be found in the references list below.

References

Bentley, M.J., Fogwill, C.J., Kubik, P.W. and Sugden, D.E. (2006) 'Geomorphological evidence and cosmogenic $^{10}BE/^{26}Al$ exposure ages for the Last Glacial Maximum and deglaciation of the Antarctic Peninsula Ice Sheet', *Geological Society of America Bulletin* 118: 1149–59.

Birks, H.J.B. and Birks, H.H. (1980) *Quaternary Palaeoecology*. Baltimore: University Park Press.

Church, M., Hassan, M.A. and Wolcott, J.F. (1998) 'Stabilizing self-organized structures in gravel-bed stream channels: Field and experimental observations', *Water Resource Research* 34: 3169–79.

DiBerardinis, L.J., Baum, J.S., First, M.W., Gatwood, G.T. and Seth, A.K. (2013) *Guidelines for Laboratory Design: Health, Safety and Environmental Considerations*. Oxford: Wiley.

Dong, Z., Liu, X., Wang, H. and Wang, X. (2003) 'Aeolian sand transport: A wind tunnel model', *Sedimentary Geology* 161: 71–83.

Faegri, K. and Iversen, J. (1989) *Textbook of Pollen Analysis* (4th edition). New York: Wiley.

Fritts, H.C. (1976) *Tree Rings and Climate*. New York: Academic Press.

Goudie, A. (1990) *Geomorphological Techniques* (2nd edition) London: Unwin Hyman.

Massart, D.L., Dijkstra, A. and Kaufman, L. (1993) *Evaluation and Optimization of Laboratory Methods and Analytical Procedures*. Amsterdam: Elsevier Science.

Petch, J. and Reid, I. (1988) 'The teaching of geomorphology and the geography/geology debate', *Journal of Geography in Higher Education* 12: 195–204.

ON THE COMPANION WEBSITE…

Visit **https://study.sagepub.com/keymethods3e** for author videos, chapter exercises, resources and links, plus **free** access to the following recommended articles:

1. **Meadows, M. (2014) 'Recent methodological advances in Quaternary palaeoecological proxies', *Progress in Physical Geography*, 38 (6): 807–17.**

This progress report provides a detailed account of many new materials and methods that contribute to reconstructing past environments and generally require some level of laboratory analysis to extract paleoecologic data.

2. **Lowe, D. (2008) 'Globalization of tephrochronology: New views from Australasia', *Progress in Physical Geography*, 32 (3): 311–35.**

Tephrochronology is an important sedimentalogical dating technique widely used in paleoecology and this article describes recent advances in laboratory methods and provides examples of how this tool has been used.

3. **Furley, P. (2010) 'Tropical savannas: Biomass, plant ecology, and the role of fire and soil on vegetation', *Progress in Physical Geography*, 34 (4): 563–85.**

This progress report details some of the soil chemical properties used around the world to help explain controls on vegetation type, distribution, and management that require some knowledge of laboratory analytic methods to collect soil data.

22 Getting Information from the Past: Palaeoecological Studies of Terrestrial Ecosystems

Laura N. Stahle and Cathy Whitlock

SYNOPSIS

In this chapter, we examine some of the approaches and data used to reconstruct terrestrial ecosystem change. Research employing these methods comes under the auspice of ecological biogeography, which is concerned with the factors that have led to the current distribution of plants on the landscape. Palaeoecology is the field of biogeography that utilizes environmental proxy data to investigate the response of species, populations, communities and ecosystems to past environmental change. Palaeoecological research usually concerns ecosystem dynamics during the Quaternary period (the last 2.58 million years), and in particular over the last 21,000 years, which includes the last glacial maximum (LGM) and the Holocene (the current interglacial period that began 11,700 years ago). Environmental proxy data are sensitive to ecosystem change at particular temporal and spatial scales and describe aspects of the environment to which the organism or physical property is sensitive. We outline the approaches taken to reconstruct terrestrial ecosystem change and review some of the common proxy data used to reconstruct past terrestrial ecosystem change, including tree rings, pollen, plant macrofossils, packrat middens, and charcoal and dung fungal spores, as well as supporting palaeolimnological information from diatoms, chironomids, and lake-sediment lithology, geochemistry and mineralogy. Proxy records can be used alone, but fuller understanding is achieved by combining and comparing multiple proxy data from single or groups of sites. Alternatively, developing geographic networks of records from a single proxy type, such as charcoal or pollen data, provides insights about patterns of environmental change over broad spatial scales.

This chapter is organized into the following sections:

- Introduction
- Palaeoecological approaches and data
- Types of proxy data in palaeoecological studies
- Multi-proxy records and global syntheses
- Conclusion

22.1 Introduction

The present is considered key to understanding the past, but historical and prehistorical data conversely play a critical role in informing both the present and the future. The past provides information on the historical range of variability, improves understanding of the natural and anthropogenic legacies that shape the present landscape, and reveals the sensitivity of eco-systems to a range of environmental conditions. Knowledge of the past comes from many sources. Real-time and near real-time data are acquired from field and remote instruments; aerial photography and satellites provide information about the past few decades. Historical data come from centuries of human observation and measurement. Information about the period before observational and instrumental data is obtained from environmental proxy evidence, such as those preserved in tree rings and lake sediments (Figure 22.1).

Palaeoecology is the field of biogeography that utilizes such proxy data to investigate the response of species, populations, communities and ecosystems to past environmental change. The field is multidisciplinary, drawing on techniques from biogeography, palae-ontology, geochemistry, and archaeology in order to reconstruct past environmental conditions, biotic responses, and climate and nonclimatic forcings. While palaeoecology has been applied to marine settings and even the cryosphere, our focus is on examining reconstructions of the terrestrial biosphere.

Most palaeoecological research concerns ecosystem dynamics during the Quaternary period (the last 2.58 million years), when vast ice sheets cyclically expanded and con-tracted in response to small variations in the configurations of the Earth's orbit around the sun and the internal feedbacks those variations precipitated (Figure 22.2). Of this body of literature, most attention is given to the last 21,000 years of history, starting with the time of the last glacial maximum (LGM) and extending through the current interglacial period, the Holocene, which began 11,700 years ago. In this chapter, we describe some of the types of data used in terrestrial palaeoecology. Our examples come from work in which we have been involved and published studies that we admire for their detail or approach. Our coverage of this topic is not meant to be comprehen-sive in concept or geography, and for that we refer to other helpful references.

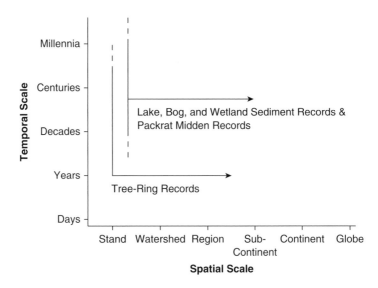

Figure 22.1   Temporal and spatial scales of proxy data discussed in text. The primary scale that these proxy data operate at is shown with solid lines. The dashed lines indicate scales that can be achieved in certain unique cases.

Global chronostratigraphical correlation table for the last 2.7 million years
v. 2010

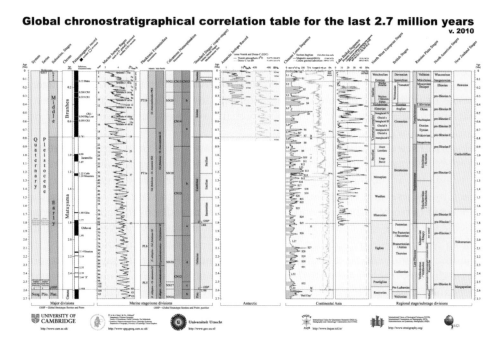

Figure 22.2 The International Commission on Stratigraphy's International Stratigraphic Chart for the Quaternary Period. (www.stratigraphy.org/upload/QuaternaryChart1.JPG)

22.2 Palaeoecological Approaches and Data

Palaeoecological reconstructions draw on a variety of datasets that, because of their sensitivity to climate or environment, can be used as evidence of past environmental conditions. Such proxy data can be temporally discontinuous information, as in the case of plant remains preserved in packrat middens or fossil vertebrates, providing a 'snapshot' of conditions and biota during a particular period. Ideally, they are continuous or nearly-continuous time series, such as tree-ring records or lake-sediment profiles, in order to examine ecological change on fine temporal scales. Some stratigraphic proxy data span hundreds of thousands to millions of years, as in the case of trace atmospheric gas records entombed in Antarctic ice cores or foraminiferal data from long marine cores, whereas other time series offer seasonal to annual resolution for the last few centuries, such as tree-ring data.

Palaeoecological studies that describe vegetation history generally span centuries to millennia and utilize fossil pollen, plant macrofossils and other plant remains. Chronologies for such records come from incremental age models, such as counting annually laminated lake sediments, or from radiometric dating methods, primarily radiocarbon (^{14}C) dating of terrestrial organic remains and lead-210 (^{210}Pb) dating of recent sediments. Radiocarbon dating is by far the most widely used technique for developing late-Quaternary chronologies, and age determinations are made on small fragments of terrestrial plants, charcoal particles, and sometimes on organic lake sediments. Calibrated radiocarbon years incorporate the effect of variations in radiocarbon production in the atmosphere, the impact of climate cycles, storage in different carbon reservoirs, and the effects of human activity. Age models are mathematical constructs that use a sequence of individual dates from a sedimentary record to develop an interpolated chronology for the entire time series. The veracity of radiometric dates and the age model can sometimes be determined by comparison with independent time-stratigraphic markers, such as a volcanic tephra of known age, the presence of pollen from an non-native plant species with a known invasion history, variations in palaeomagnetic field strength that have been dated elsewhere, or other historic events with a detectable signature.

Historical sciences, like palaeoecology, often answer questions through an iterative process of testing multiple working hypotheses. Plausible hypotheses (explanations) are formulated at the outset of the study, and the data are used to evaluate the merits of each hypothesis and reject those that do not stand up to scrutiny. Some hypotheses are rejected outright, others are modified in light of new discoveries, and new hypotheses emerge during the course of an investigation. In the case of vegetation history, research and testable hypotheses have focused on reconstructing ancient ecosystems (including the individuals, populations, and communities therein) and their response to past changes in climate, disturbances, and human activity. The motivation is that understanding past ecological interactions will provide insight into those occurring at present and likely in the future. Broad-scale investigations are often concerned with understanding the hierarchy of climatic and nonclimatic drivers that have shaped ecosystems along a variety of spatial and temporal scales.

Sediment cores from natural lakes and wetlands are the best source of information on the history of terrestrial environments. Fossils, geochemistry and other proxy data preserved in the layers of sediment provide a record of conditions in the watershed through time. Such records begin when the lake was first formed, and they end with the uppermost layer deposited in the current year. Most natural lakes and wetlands were created by ice recessional processes following the LGM, but lakes can also be formed following volcanic eruptions or as a result of fluvial processes, landslides, and coastal aggradation. Boreal wetlands are features of deglaciation but wetlands can also be formed by coastal, fluvial and other geomorphic processes that dam waterways or impound natural springs. Selection of a lake or wetland site for study is based on decisions about whether the vegetation, geological substrate, and climate of the location are representative of a particular region. A good research question in palaeoecology has importance beyond the boundaries of the study or site (i.e. it is one that addresses a timely scientific question or that proposes to look at old findings with a new, possibly transformative approach). In addition, such a research question can be answered by thoughtful selection of study sites, critical examination of multiple palaeoecological proxy, and a plan for using site results to gain broader inference.

22.3 Types of Proxy Data in Palaeoecological Studies

In this section, we discuss different sources and types of palaeoecological proxy data and provide examples of studies that have employed them. Of particular importance are the temporal and spatial scales that each type of proxy data addresses (Figure 22.1). Temporal scales include annual, decadal, centennial and millennial. Spatial scales range from metres to many hundreds of kilometres.

Proxy Data of Terrestrial Ecological Change

Tree-Ring Data

Most temperate tree species (i.e. those growing between 25–65° latitude) produce annual growth rings. Tree rings can be sampled using a simple coring device that extracts a pencil-width core of wood from the bark to the pith of a tree. Cores are prepared for analysis by simple sanding and polishing procedures. Rings are sequentially counted to provide the age of a tree. The varying width of rings over time provides information about the environmental and climate history of a locale. By analysing the pattern of ring widths, insight can be gained into the conditions under which the tree was subjected over the course of its life. The ring patterns are analogous to barcodes that can be matched to nearby trees. It is by matching or 'crossdating' the distinct pattern of ring widths between living and dead trees that a tree-ring chronology can be constructed. Dating by tree rings is called 'dendrochronology'. Crossdating is a straightforward concept but complicated by differences in growth rates among individual trees, local factors and the occasional problem of missing rings because of poor growing conditions. It is important to sample many trees in an area to overcome these issues.

Dendrochronological data cover a timespan of centuries to millennia at annual-to-interannual resolution. Tree-ring studies have been conducted on six continents. Tree-ring data are utilized to make ecological and climatological inferences about the past. These data have also been used to study the history of fire – this will be discussed later in the chapter.

Dendroecology is the application of tree-ring analysis to ecological questions. One major area of research in this subdiscipline is forest disturbance dynamics. This research area seeks to reconstruct and understand forest dynamics as driven by external disturbances both biotic (e.g. insect outbreaks) and abiotic (e.g. fire, flood, windthrow). Investigations often focus on identifying the timing and spatial extent of tree-growth releases or suppressions and cohort establishment events.

Dendroclimatology is the study of past climate using tree-ring data. Climate records are developed by comparing ring-width patterns with modern climate data to build a calibration function. The instrument-based calibration is then used to convert the tree-ring record from the earlier, pre-observational period and estimate variations of the climate variable back in time. A major strength of tree-ring climate proxy data is that reconstructions can be replicated across large spatial networks. The North American Drought Atlas contains a 2005-year-long record of yearly drought estimated from 835 tree-ring chronologies (Cook et al., 2004; http://iridl.ldeo.columbia.edu/SOURCES/.LDEO/.TRL/.NADA2004/pdsiatlashtml/pdsiviewmaps.html). These data show that long-lived severe droughts have been a feature of the North American climate for centuries and had negative consequences for some pre-industrial societies (e.g. Stahle et al., 1998; Munoz et al., 2014).

Pollen Data

The assemblages of pollen grains in lake, wetland and bog sediments preserve the record of past vegetation and data derived from pollen studies can be used to provide an indication of the response of vegetation to climate and environmental changes as well as to human impacts over thousands of years. Pollen is produced by angiosperms (flowering plants) and gymnosperms (seed-producing plants), and not surprisingly, wind-pollinated species that produce a lot of pollen each year are more abundant in the sediments than insect-pollinated species. Pollen falls on the surface of a body of water and becomes incorporated in the sediment and cores can be extracted from the sediment accumulated at the bottom of a lake, wetland or bog. The temporal scale of a pollen record depends on how fast sediment accumulates in a lake or wetland basin. Typically records of vegetation change derived from pollen analysis have centennial-scale resolution. However, greater temporal resolution can often be achieved if the research questions require it. The spatial resolution of pollen records varies from site to site and is often not well resolved. The size of sample lakes and wetlands strongly influences pollen source area and smaller (< 0.5 ha) sites are preferred to provide vegetation histories at the watershed scale (Ritchie, 1987).

In order to perform pollen analysis, sediment samples are taken at regular intervals in the lake/wetland/bog sediment core, and these samples are treated with a variety of chemicals to remove all the constituents except the pollen grains (see Chapter 21). The residue of pollen is mounted on glass slides and examined under the microscope at magnifications of 400–1000x. Pollen grains (generally between 25-100 microns in size) are identified by comparison with modern reference material and published atlases. Typically, 300–400 pollen grains are tallied for a given sample in the core, and it can take a trained analyst two or three hours or more to 'count' a sample. The ability to assign a pollen grain to a particular plant taxon is variable and the taxonomic resolution limits interpretation in some cases. For example, grass pollen cannot be identified below the taxonomic level of family (Poaceae), so it is not possible to determine whether the grass pollen comes from alpine, steppe or riparian species. Most pollen grains are securely identified to the level of genus or family but species identifications are often inferred by phytogeography. The presence of seeds, needles and other plant remains in the core also provides species identifications in cases where pollen cannot. A typical pollen record from temperate latitudes will include about 50 different pollen types from trees, shrubs and aquatic plants.

Pollen counts at each level in the sediment core are converted to percentages and accumulation rates, and changes in the proportion of different taxa through time are the basis for interpreting past vegetation. Because pollen does not have 1:1 relationship with the plants that produce it, modern studies are used to interpret past pollen assemblages. Modern pollen information comes from the surface sediments of lakes or pollen traps set out by researchers. The number and quality of surface pollen studies vary from place to place. In North America and Europe hundreds of surface samples have been collected and provide excellent calibration for the interpretation of pollen data through time (www.neotomadb.org).

As an example of the application of pollen data to understand past vegetation and climate, Williams et al. (2006) produced a synthesis of the late-Quaternary vegetation history in northern and eastern North America, examining changes across different levels of ecological organization from individual taxa to biomes. Broad-scale features of vegetation history emerged by comparing the records from multiple sites, and these could be compared with site-specific features that described local responses at individual sites. Different aspects of past vegetation dynamics were revealed by individual pollen time series, pollen maps, dissimilarity measures and estimates of temporal rates of vegetation change. The synthesis suggested that distribution and composition of vegetation were relatively stable during the LGM and during the mid- to late Holocene (last 6000 years). This stability was in contrast to the rapid changes that occurred during the late-glacial period to early Holocene transition (14,000–6000 years ago) and in the last 500 years. The history of particular pollen types suggested that the dominant tree species behaved independent of each other in their response to past climate change, and shifts in range were attributed to changes in regional moisture patterns, causing west-to-east responses and changes in temperature evidenced by south-to-north shifts. Some of the common plant communities today developed in the early Holocene, but there are also vegetation types common to the late-glacial period (e.g., *Picea–Cyperaceae–Fraxinus–Ostrya/Carpinus*) that no longer exist. The study is an excellent example of how vegetation changes across multiple spatial and temporal scales and the use of pollen data to reconstruct the distribution, composition and structure of plant communities over time.

Plant Macrofossils

Macroscopic remains of plants, including seeds, leaves, needles and fruits, can be found in most deposits suitable for pollen analysis (Birks, 2013). The best sites for these types of records are wetland sites or the littoral margins of lakes within steep catchments, where the opportunity for slopewash or stream input to deliver organic detritus is increased. Species identifications provided by these remains complement pollen-based interpretations by providing greater taxonomic resolution and also confirming the local presence of particular plants (unlike their pollen which can be transported long distances). In wetland deposits, the identification of bryophytes has offered useful information on temperature and hydrologic conditions (Mauquoy and van Geel, 2013). While the majority of plant remains in lake sediments are from aquatic taxa, the less common occurrence of terrestrial plant macrofossils has helped resolve important points in vegetation history (Jackson and Weng, 1999).

Packrat Middens

In semiarid regions, packrat midden data provide important vegetation information from semi-arid regions where lake-sediment pollen records are rare (Betancourt et al., 1990). The middens of packrats (21 different species of *Neotoma*) and other nest-building mammals are composed of plant material cemented into caves and rock crevices by urine. In dry settings, these nests are preserved for thousands of years, entombing plant remains of past vegetation as a series of cemented layers. The preservation of plants in midden deposits is excellent, and the remains have been used to reconstruct the local plant communities and population dynamics as well as make broader inferences about climate through the analysis of isotopic signatures and stomata density. Vegetation reconstructions depend on understanding the foraging area and dietary preferences of packrats, as well as the factors that may have led to differential preservation of the remains also posing challenges for interpretation (Finley, 1990; Elias, 2013).

In the American Southwest, plant assemblages in packrat middens that date to the LGM indicate large downslope shifts in the biogeographic range of subalpine conifers, such as *Pinus longaeva* and *Pinus flexilis*, as a result of cool humid conditions. Midden studies have also been used to trace the migration history of conifer species during the Holocene. A recent study of remains of Utah juniper (*Juniperus osteosperma*) in packrat middens documents its spread into Wyoming and Montana during the Holocene (Lyford et al., 2003). Utah juniper became established first in northeast Utah in the early Holocene about 9000 years ago (Figure 22.3 a–h). In mid-Holocene the species had advanced into central Wyoming and southern Montana.

Because of the wide distribution of packrat middens in the region, it was possible to study the dispersal, landscape structure, and climate variability that governed the spatial and temporal patterns of the juniper establishment (Lyford et al., 2003).

Fire

Two forms of primary proxy data have been used to study the history of fire on the landscape: tree-ring data and charcoal particles from lake, bog, small hollow and wetland sediments. Tree-ring data are widely used to reconstruct the timing and occasionally the extent of past fires. Fires that are not severe enough to kill trees often leave distinctive scars that can be used to determine the exact calendar year a fire occurred. These 'fire scars' are preserved within tree-ring sequences. Fire-scar chronologies have been compiled at local-to-regional scales in many regions including western North America and southern South America (Swetnam, 2002; Veblen et al., 2003; Falk et al., 2011). The fire-scar network for western North America includes over 800 fire chronologies spanning centuries. This network shows a close correlation between drought years and years with extensive fire activity (Swetnam, 2002). Fire-scar networks can be created and analysed at a range of spatial scales, from trees and stands to subcontinents, revealing different patterns and processes at different scales (Falk et al., 2011). In forests where few fire scars are present, stand-age analysis has been used to reconstruct the dates of past fires. This method requires the dating of a large number of trees in an area in order to determine when even-aged cohorts of trees became established. Pulses of tree regeneration occur after fire and other disturbances such as insect outbreaks, thus determining the age of tree cohorts in a forest allows researchers to determine the timing of forest disturbances.

The fire-scar network in western North America shows that large-scale climate teleconnections strongly influence fire activity on a sub-continental scale. A study by Kitzberger et al. (2007) employed both dendroclimatological methods and a large network of fire-scarred trees in the western United States in order to examine the links between large-scale climate teleconnections and fire activity. Three climate indices – El Nino Southern Oscillation (ENSO), Pacific Decadal Oscillation (PDO), and Atlantic Multidecadal Oscillation (AMO) – were reconstructed using tree-ring chronologies from the US and Mexico in the case of ENSO and PDO, and Finland, France, Italy, Jordan, Norway, Russia, Turkey, and the US in the case of AMO (Kitzberger et al., 2007). The fire-activity record was composed of individual fire chronologies from 238 sample sites. Synchronous fire activity across the sample sites was compared to the climate indices. The study found the warm phase of AMO has synchronized fire activity over multidecadal timescales in the western US for the last 500 years (Figure 22.4; Kitzberger et al., 2007).

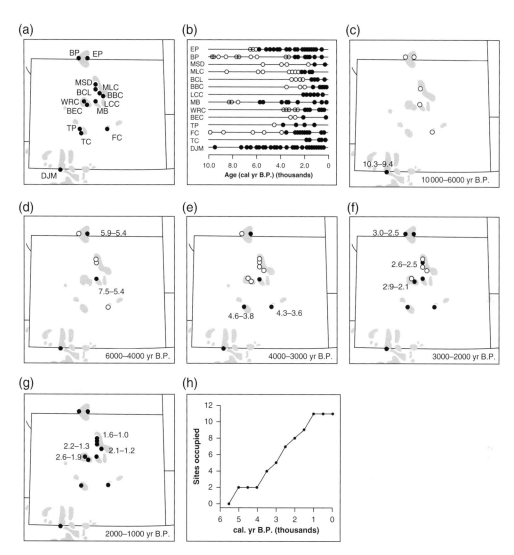

Figure 22.3   The location of woodrat-midden records of Utah juniper invasion in Wyoming and adjacent regions. Calibrated dates given in thousands of years before present (ka). (a) Locations of woodrat-midden study sites (black circles) and modern distribution of Utah juniper (grey circles) in Wyoming and adjacent states. (b) Holocene records of presence (filled circles) and absence (open circles) of Utah juniper macrofossils from woodrat middens at 14 sites. (c) Map of presence (filled circles) and absence (open circles) of Utah juniper macrofossils in woodrat middens dating between 10 and 6 ka. (d) Presence and absence of Utah juniper macrofossils between 6 and 4 ka (e) Presence and absence of Utah juniper between 4 and 3 ka. (f) Presence and absence of Utah juniper between 3 and 2 ka. (g) Presence and absence of Utah juniper between 2 and 1 ka. (h) Number of study sites occupied by Utah juniper as a function of time in the mid-to-late Holocene.

Source: Lyford et al., 2003: 576/7. © 2003 by the Ecological Society of America

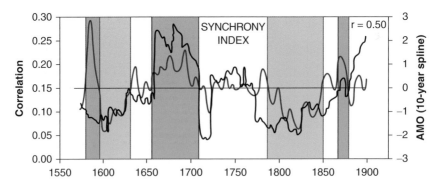

Figure 22.4   The indices of fire synchrony reconstructed from tree-ring data (50-year moving correlations between selected regions, black line) compared with a 10-year spline of reconstructed Atlantic Multidecadal Oscillation (AMO; grey line). Dark and light shaded areas indicate periods of low and high AMO, respectively. The association of AMO and continental-scale synchrony of fire across western North America is consistent with the role of AMO in determining patterns of drought across the western U.S. Fire was strongly synchronous across western North America during 1650–1770 (except 1710–1725) and after 1880 and weakly synchronous during 1550–1649 and 1750–1849. The highest degree of fire synchrony occurred from 1660–1710, coincident with the longest and warmest phase of the AMO during the past five centuries. Conversely, the lowest degree of fire synchrony occurred from 1787–1849, coincident with the coldest phase of the AMO.

Source: Kitzberger et al., 2007: 546. © (2006) National Academy of Sciences, USA

The second palaeofire proxy data are the charred pieces of wood, leaves and grass deposited and preserved in anaerobic environments such as bogs, lakes and wetlands (Brown and Power, 2013). The charcoal from sediment cores from these depositional environments is used to reconstruct fire activity over thousands of years. This method of fire-history research is based on extracting the charcoal from contiguous intervals of sediment cores and examining this under the microscope. Generally two different size classes of particles are analysed: microscopic and macroscopic charcoal. Microscopic particles are counted from pollen slides at a magnification of 400x, while macroscopic pieces greater than 125 microns (μm) are isolated from the sediment and analysed at 40x magnification. Microscopic charcoal can travel long distances before settling in a lake, and it is not typically analysed at continuous intervals. Macroscopic charcoal comes from fires within a radius of < 20 km of the lake (Higuera et al., 2009). It is analysed in every sample throughout a core and used to reconstruct local-scale fire episodes (Whitlock and Larson, 2001). Macroscopic charcoal data are converted to charcoal accumulation rates (number of particles cm^{-2} yr^{-1}). As with pollen and tree-ring data, modern studies are used to calibrate and inform the interpretation of charcoal abundance in the pre-observational period.

Analysis of charcoal is undertaken in order to determine and describe the fire regime of a particular area of study. A fire regime describes the characteristics of fire (frequency, size and severity) and its role in a particular ecosystem. A suite of climate, fuel, and landscape variables are required for fire to occur and spread, but their relative importance changes across spatial and temporal scales. Recent advances in statistical methods of charcoal analysis have added to our understanding of past fire regime characteristics (e.g. Higuera et al., 2009; Kelly et al., 2011). Figure 22.5 summarizes the most commonly reconstructed fire regime metrics, including charcoal accumulation rates, background charcoal and mean fire return intervals.

In a regional-scale study from the South Island, New Zealand, McWethy et al. (2009) explored changes in fire activity after Polynesians arrived on the island ca. 1280 CE. Prior to this time, New Zealand was uninhabited by humans and fire was

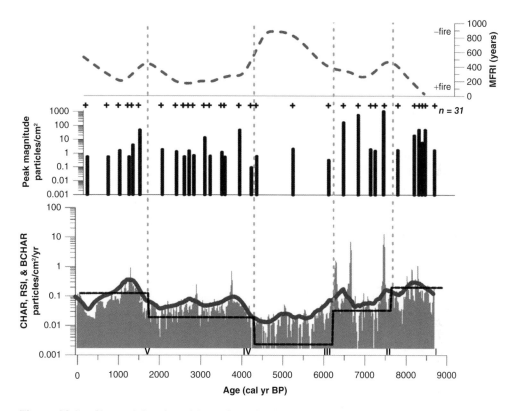

Figure 22.5 Charcoal data from Morris Pond, Utah, USA, over the past 9,000 years. This figure summarizes the typical metrics of a fire regime that charcoal data have been used to reconstruct. The top panel shows the Mean Fire Return Interval (MFRI). This metric reflects the frequency of fire through time at the site. The second panel shows statistically significant peaks in fire activity and the associated magnitude of the peaks. These peaks are interpreted as local (within 1–3 km) fire episodes. Peak magnitude is a measure of the total charcoal deposition for a fire episode and often reflects the type and amount of vegetation that was burned. The bottom panel shows Charcoal Accumulation Rate (CHAR), Background Charcoal (BCHAR), and the regime shift index algorithm (RSI). CHAR is the measure of the number of charcoal particles per cm^2 per year. Background charcoal (BCHAR) is the slowly varying trend in CHAR and changes in this metric often represent variability in the abundance of fuel or biomass at the site (for example, forests produce more charcoal than grasslands or tundra). The RSI was used to identify statistically significant changes in BCHAR and delineate fire regime zones. The dashed vertical lines indicate fire regime zones derived RSI.

Source: Morris et al., 2013: 30. © 2012 University of Washington. Published by Elsevier Inc.

infrequent and ecologically insignificant. The approach of this study was to investigate regional trends in biomass burning by reconstructing local-scale fire activity at many watersheds across a climatic gradient from high to low precipitation. The study also examined climate to determine if temperature anomalies may have contributed to flammability in the decades after human arrival. In order to reconstruct watershed-scale fire activity, high-resolution macroscopic charcoal was analysed in continuous samples through the sediment core. Individual records showed very low levels of fire prior to human arrival, a period of high fire activity, representing one to a few fires, followed by low fire activity until the time of European arrival. Comparison of several charcoal records across the South Island revealed the

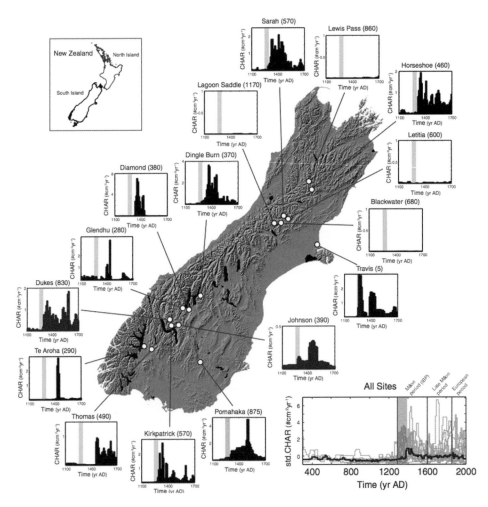

Figure 22.6 Local-scale charcoal records from 16 sites on the South Island, New Zealand. Most records have a clear peak in fire activity within 200 years of human arrival, implicating anthropogenic burning as the ignition source of these fire events.

Source: McWethy et al. 2010: 21344. © The National Academy of Sciences of the USA

regional patterns of fire activity over the last 1,000 years (Figure 22.6). A period of high fire activity lasting a few decades was evident in all but the wettest and highest elevation sites, although the dates of this period varied. This Initial Burning Period marked the arrival of Māori to the island and the rate of deforestation at this time suggests deliberate and targeted use of fires (Perry et al., 2012). Following the Initial Burning Period, Māori maintained low level fires to sustain vital resources. The study found that fires were not associated with anomalous summer temperatures, and thus the influence of climate on the fire activity was negligible. This study is a good example of using charcoal analysis to test hypotheses of the drivers (human vs. climatic) of fire activity.

Dung Fungal Spores

Dung fungal spores in lake, bog and wetland sediments are a proxy data source that has rapidly gained use in the last decade (Gill et al., 2013). These fungi are a type of coprophilous fungi that grow on animal dung. They require herbivore digestion to complete their life cycle, producing spores on the dung of mammals. Palaeoenvironmental reconstructions that incorporate dung spore analysis hold promise as a proxy for the presence and perhaps abundance of large herbivores. These animals can have strong effects on ecosystems by maintaining vegetation openness and patchiness, removing material that would otherwise fuel landscape fire, dispersing seeds, and physically disturbing soil and recycling nutrients (Rule et al., 2012).

The dung fungal spores of the genera *Sporormiella*, *Sordaria* and *Podospora* are the three most reliable indicators of large herbivore activity (Baker et al., 2013). The spores are transported to lakes and other depositional environments by slope-wash. The relationship between the amount of dung spores in sediments is influenced by both the abundance of dung in the watershed as well as its distance from the lakeshore. The methodology for quantifying dung spore source area and large herbivore population size is still developing. Reconstruction of herbivore densities has been performed using accumulation rates (number of spores/cm^2/year), and percentage in relation to the pollen sum. The use of accumulation rates is thought to be a superior approach as relative percentage does not provide a proxy for herbivory (plant consumption) that is independent from vegetation changes (Baker et al., 2013).

A number of recent palaeoecological studies have used *Sporormiella* spores to investigate the question of the timing and impacts of megafaunal decline (e.g. Robinson et al., 2005; Davis and Shafer, 2006; Gill et al., 2009; Rule et al., 2012). A study from northeast Australia examined the population collapse of megafauna that occurred around 40 thousand years ago shortly after people first arrived on the continent. This study reconstructed vegetation, fire and *Sporormiella* between 3,000 and 130,000 years ago with a special emphasis on the period between 39,000–43,000 (Rule et al., 2012). *Sporormiella* declined markedly between 40–41,000. This decrease was followed by an increase in charcoal and the pollen of grass and sclerophyll shrubs that tolerate frequent fires (Figure 22.7). This transition in vegetation occurred in the absence of major climate perturbations and thus it appears the mass extinction of megafaunal triggered major changes to vegetation and ecosystem functioning.

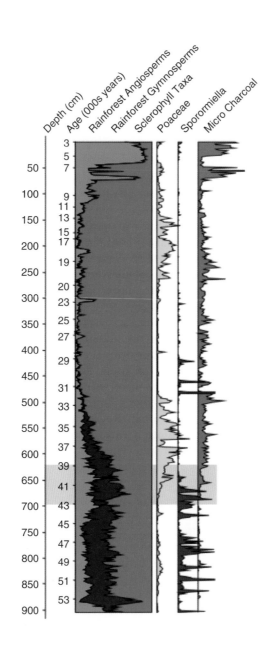

Figure 22.7 Pollen, charcoal and *Sporormiella* diagrams for Lynch's Crater. The interval during which *Sporormiella* declined and charcoal first increased is shaded grey. A large shift in vegetation from a rainforest to sclerophyll assemblage is evident in the millennia following *Sporormiella* decline and charcoal increase.

Source: Rule et al., 2012: 1484. © Science

Supporting Limnologic Proxy Data

Diatoms

Diatoms are algae in the division Bacillariophyta that occur in almost all aquatic environments (Jones, 2013). The siliceous skeleton or frustule of these microscopic unicellular organisms is identifiable to the species level and provides information on aquatic conditions that can be tied to changes in nutrients, water temperature, pH, and light penetration. Quaternary diatom remains have proven useful as indicators of local limnologic conditions and used to reconstruct past lake-level changes, water chemistry variations, and human disturbances of lake ecosystems. They are found in a wide range of aqueous or subaqueous environments as benthic (bottom dwelling), epiphytic (attached to plants), or planktonic (free-floating) and thus occupy a wide variety of niches. A key issue in diatom analysis is the accuracy with which diatom assemblages in sediments reflect the composition of the source communities and habitats from which they are collected. To this end, most analyses rely on comparison of the composition of assemblages, as well as the physical limitations of indicator taxa (Korhola, 2013).

Diatom analysis has provided perhaps the strongest case for acid rain effects from industrialization. In a classic study, Battarbee et al. (1984) showed that changes in the assemblages of diatoms preserved in lake sediments provided evidence of 19th and 20th lake acidification in northwestern Europe and North America. The strong relationship between diatom occurrence and water pH allows reconstruction of past pH levels with remarkable precision. Although long-term acidification is a natural process for lakes in areas of resistant base-poor bedrock, diatom analyses of sediments spanning the last 150 years indicated rapid and unprecedented acidification.

In studies where diatom records have been developed at several sites, the limnologic reconstructions often show considerable variability that reflects local site-specific variability superimposed on climate change (Fritz and Anderson, 2013). In some cases, the variability is a result of the changes in vegetation, particularly the early colonization of plants following deglaciation and their impact on catchment nitrogen cycling (Fritz et al., 2004). In other locations, the differences reflect the influence of different substrates and the mineralogy of the parent rock on lake nutrients, pH, and water clarity (Bigler et al., 2002).

Chironomids

Diptera: Chironomidae (non-biting midges) are a large taxonomic group of insects that live in most aquatic or semiaquatic habitats during their larval stages (Walker, 2013). The exoskeletal remains that are sloughed during larval molting accumulate in the sediments of lakes, and the well-preserved chitinous head capsules are identifiable to species by comparisons with specimens of extant species. As with diatoms, information on the modern biology and habitat of individual taxa is the basis for environmental reconstructions. Palaeoecological studies have focused on the development of quantitative calibration models (e.g. transfer functions) for the reconstruction of past environmental parameters based on the composition of the chironomid assemblage (Walker, 2013). This calibration is then applied to the time-series of chironomid remains to interpret past conditions. Some of the issues that vex specialists include misinterpreting past assemblages because modern samples are generally taken in summer (whereas the sedimentary record preserves the year-round assemblage); concerns about taphonomy (processes by which biological material is differentially deposited and preserved) and the redeposition of fossil remains from different locations in the lake, thus misinterpreting the assemblage; differences in taxa abundance related not to the environment but to differences in the duration of larval life cycle and number of generation and larval stages; and differential preservation of more robust head capsules thus distorting the record (Brooks et al., 2010; Velle and Heiri, 2013).

Chironomid records have been studied in a variety of locations to reconstruct climate change, land-use and other human activities. For example, they have been used to examine the ecological consequences of eutrophication of temperate lakes; monitor water pollution related to industrial inputs and airborne pollutants; record acidification trends in lake sediments; and assess changes in water salinity through time. Because air temperature influences processes of emergence, swarming and settling, the statistical correlation between chironomid assemblages and July air temperature is often stronger than that of surface water temperature (Massaferro et al., 2009). Available records indicate that chironomids can be successfully used to reconstruct temperature changes, especially if chironomid analysis is embedded in multiproxy, or multisite, studies.

Lithology, Geochemistry and Mineralogy

Examination of the sediments, independent of their biotic constituents, provides information on the history of the watershed and the lake as an important context for palaeoecological reconstructions. A variety of non-destructive tools, including colour reflectance, digital imaging, radiography and X-ray imaging and CT scanning, are used in initial descriptions (Kemp et al., 2001; Hodder and Gilbert, 2013). Further analysis of the inorganic fraction of sediment, through mineralogical and geochemical analysis, has been used to identify changes in inorganic inputs to the basin that may relate to changes in erosion, windiness, pollution inputs and nutrients (Last, 2001). Variations in the authigenic (locally formed) carbonate content of the sediments can be a good measure of chemistry and pH changes related to water temperature. The organic component of the sediments is produced by different types of biota in from the lake and watershed (Meyers and Ishiwatari, 1993). Organic matter in lake sediments comes from the detritus of terrestrial plants to aquatic algae and bacteria, each with different chemical signatures. The original composition of organic matter may be further altered by biotic and abiotic processes, and the degree of alteration also contains important palaeoenvironmental information (e.g. the degree of water-column mixing).

The isotope compositions of authigenic and biogenic carbonates and diatom silica are commonly examined to better understand changes in temperature, precipitation patterns, evaporation and the carbon cycle. Fluctuations in the isotope composition of authigenic or biogenic minerals are mainly a function of long-term changes in the balance between precipitation and evaporation as well as relative contributions between surface water and groundwater. Interpretation of isotope data from the various components within a lake sediment core requires a detailed knowledge and modern calibration of the processes that control and modify the signal; this must be determined for an individual lake system to establish the relationship between the measured signal, the isotopic composition of the host waters, and climate.

22.4 Multi-Proxy Records and Global Syntheses

Comparing multiple proxy data from a single site provides opportunities to reconstruct past environmental change within the whole watershed. A good example of this type of approach comes from Yellowstone National Park where a 9400-year-old core from Crevice Lake was analysed for pollen, charcoal, geochemistry, mineralogy, diatoms and stable isotopes to develop a nuanced understanding of Holocene environmental history (Figure 22.8; Whitlock et al., 2012). The pollen data indicated that the watershed supported a closed *Pinus*-dominated forest and low fire frequency prior to 8200 (calibrated) calendar years before the present (cal yr BP; where present = CE 1950), followed by open parkland until 2600 cal yr BP, and open mixed-conifer forest thereafter. Charcoal data suggested that fire activity shifted from infrequent stand-replacing fires initially to frequent surface fires in the middle Holocene and stand-replacing events in recent centuries. Low values of $\delta^{18}O$ were evidence of high winter precipitation in the early Holocene, followed by steadily drier conditions after 8500 cal yr BP. Carbonate-rich sediments before 5000 cal yr BP implied warmer summer conditions than after 5000 cal yr BP. High values of molybdenum (Mo), uranium (U), and sulphur (S) indicated anoxic bottom-waters before 8000 cal yr BP, between 4400 and 3900 cal yr BP, and after 2400 cal yr BP. Diatom assemblages suggested well-developed spring conditions and water-column mixing through much of the Holocene, but also revealed a period between 2200 and 800 cal yr BP with strong summer stratification, phosphate limitation, and oxygen-deficient bottom waters. Together, the proxy data implied wet winters, protracted springs, and warm, effectively wet summers in the early Holocene and less snowpack, cool springs, and warm dry summers in the middle Holocene. In the late Holocene, the region experienced extreme changes in winter, spring, and summer conditions, with particularly short springs and dry summers and winters during the Roman Warm Period (~ 2000 cal yr BP) and Medieval Climate Anomaly (1200–800 cal yr BP). Long springs and mild summers occurred during the Little Ice Age (500–100 cal yr BP) and these conditions persist to the present. Although the proxy data indicate effectively wet summer conditions in the early Holocene and drier conditions in the middle and late Holocene, summer conditions were governed by multi-seasonal controls on effective moisture that operated over different time scales.

Whereas multi-proxy comparison allows for in-depth analysis of a single site or a group of nearby sites, combining the records from a single proxy such as charcoal or pollen data affords the opportunity to reconstruct palaeoenvironmental history over a very large spatial scale. Continental- to global-scale syntheses of late glacial and Holocene fire activity and vegetation distribution have been compiled and are the subject of ongoing research. Global-scale syntheses of fire activity have been compiled and analysed by the Global Palaeofire Working Group (www.gpwg.org; Daniau et al., 2012). The Global Charcoal Database (GCD v2) contains almost 700 sedimentary charcoal records from six continents (Figure 22.9). The general global pattern of biomass burning over the last 21,000 years shows a widespread increase associated with the transition from cold glacial to warm Holocene climates (Figure 22.10; Power et al., 2008; Daniau et al., 2012; Marlon et al., 2013). The influence of temperature on fire activity at this large scale is pervasive.

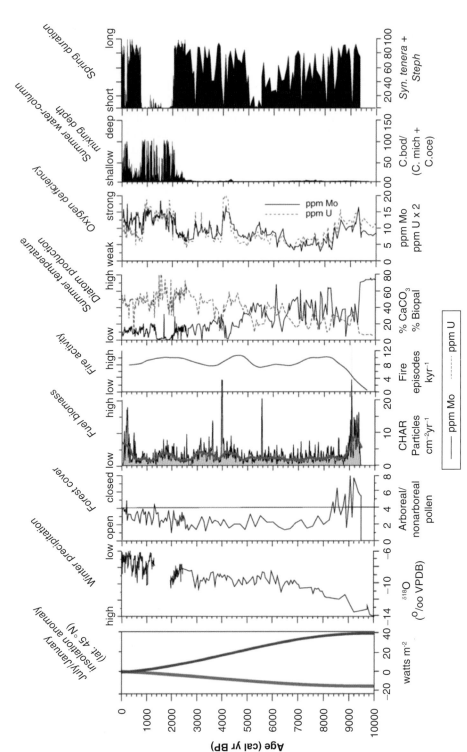

Figure 22.8 The summary of the environmental proxy at Crevice Lake over the last 9400 cal yr BP plotted with July and January insolation anomalies

Source: Whitlock et al., 2012: 99. © 2012 Elsevier B.V.

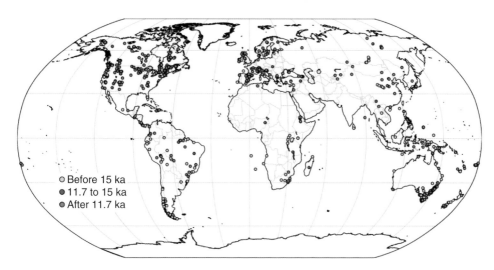

Figure 22.9 The location of charcoal records in the Global Charcoal Database version 2 (GCD v2) maintained by the Global Palaeofire Working Group (www.gpwg.org/gpwgdb.html)

Source: Daniau et al., 2012: 4. © 2012. American Geophysical Union.

The Palaeovegetation Mapping Project (known as BIOME 6000: Prentice and Webb, 1998) was developed to create fully-documented pollen and plant macrofossil data sets for 6750 and 21,000 cal yr BP, and to construct global maps of biomes for these time periods based on plant functional types and biomes. The BIOME 6000 database is publically available and updated with new datasets (www.bridge.bris.ac.uk/resources/Databases/BIOMES_data). The most recent version of the BIOME 6000 database (v4.2) has records for 11,166 modern sites, 1794 sites at 6750 cal yr BP, and 318 sites at 21,000 cal yr BP (Figure 22.11). Palaeovegetation datasets have been utilized to train and test palaeovegetation models (Figure 22.12; Kaplan et al., 2003; Prentice et al., 2011; Levavasseur et al., 2012) and to model past climate (Figure 22.13; Cheddadi et al., 1996; Ferrera et al., 1999; Bartlein et al., 2011).

Important conclusions have been drawn from these synthetic efforts. They demonstrate that at a global scale fire is controlled largely by temperature (Marlon et al., 2013). Thus, while prehistoric humans likely influenced fire activity in certain locales, globally climate is the primary control. These syntheses are also important for data-model comparisons. Many vegetation and climate models operate at large spatial scales (for example at 0.5-1.0° grid cells) and in order to make meaningful comparisons, palaeoenvironmental reconstructions need also to be compiled at a large spatial scale. One critique of these global-scale syntheses is that there is an uneven distribution of data with a more dense concentration of sites in North America and Europe than in the other continents, which may skew interpretation. Thus, it is important to support continued palaeoecological studies in Asia, Australia, South America and Africa and incorporate them into global databases.

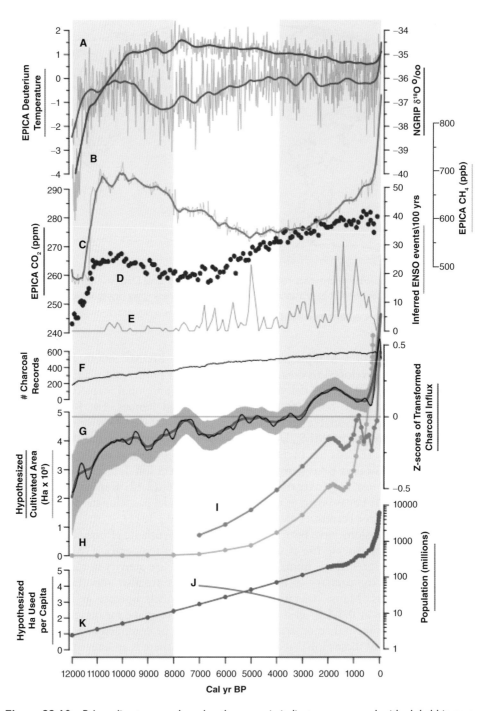

Figure 22.10 Palaeoclimate records and anthropogenic indicators compared with global biomass burning reconstructed from sedimentary charcoal records. Globally fire was low at the beginning of the Holocene and increased during the Holocene (panel G). This trajectory is consistent with the global increase in temperature through the glacial-interglacial transition (panels A and B)

Source: Marlon et al., 2013: 18. © 2013 Elsevier Ltd.

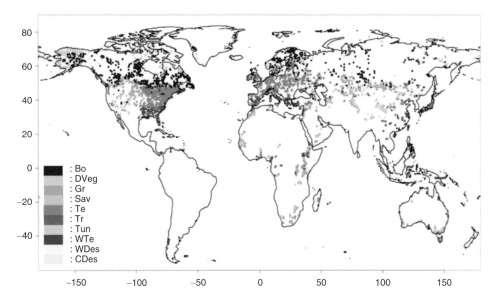

Figure 22.11   Pollen data collected in the BIOME 6000 database for the modern period (0 cal yr BP). In the legend, 'Bo' stands for boreal forests, 'DVeg' for desert vegetation, 'Gr' for grasslands and dry shrublands, 'Sav' for savannas and dry woodlands, 'Te' for temperate forests, 'Tr' for tropical forests, 'Tun' for tundra, 'WTe' for warm-temperate forests, 'WDes' for warm deserts and 'CDes' for cold deserts. Data from Prentice and Webb (1998), available online at www.bridge. bris.ac.uk/resources/Databases/BIOMES_data

22.5 Conclusion

The study of terrestrial palaeoecology has seen stimulating developments during recent years and it has become one of the most dynamic areas of biogeographical research. Geographical insights can be achieved by employing the appropriate types of data to answer specific, well-formed research questions. The application of independent dating, multivariate analyses, the establishment of databases, and the increasing quantitative precision of proxy reconstructions have broadened the scope of the discipline and helped extend our understanding of ecosystem dynamics beyond modern observations. Studies range from understanding the selective pressure and genetic make-up of individuals, to reconstructing populations, communities and ecosystems across the entire globe. Many of the proxy data sets allow simultaneous examination at multiple spatial and temporal scales; combinations of data-data, data-model and inter-model comparisons allow better understanding of the hierarchy of biophysical drivers of past ecosystem change.

Proxy data operate at different spatial and temporal scales (Figure 22.1), and while their interpretation is firmly grounded in uniformitarianism, the strongest proxies are those that are well grounded in modern empirical and observational research so that the constraints on present distribution are well understood. This information is essential to calibrate and refine interpretation of the past. With proxy data, there is also the opportunity to explore past conditions that may have no analogue in the present. The examples of novel ecosystems in the past offer some of

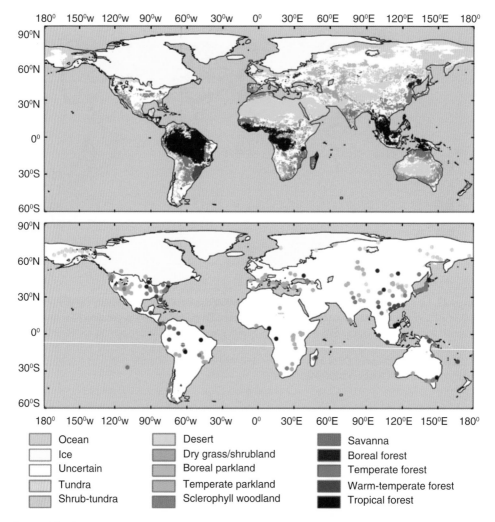

Figure 22.12 A comparison of simulated biome distribution at the last glacial maximum (LGM, 21,000 years ago) by the Land Processes and eXchanges (LPX) model (top panel) with LGM biomes inferred from pollen and plant macrofossil records compiled by the BIOME 6000 project (bottom panel)

Source: Prentice et al. 2011: 993. © 2011 New Phytologist Trust.

the most intriguing areas of research as we explore ecosystem responses in the future under projected climate change.

Most areas of palaeoecology are moving from an inductive data-gathering stage to more deductive approaches that generate and explore testable hypotheses about past environments. Individual records lie at the heart of any interpretation, but data composites and mapped summaries are capturing the dynamics of ecosystem change at broad temporal and spatial scales. Geography, with its interest in pattern and process at different scales, has the capacity to contribute greatly to palaeoecology. Collaborative research

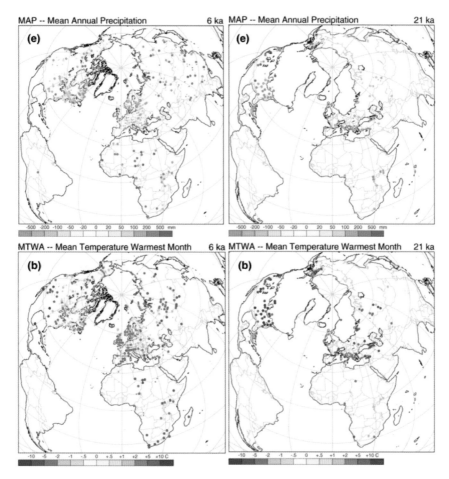

Figure 22.13 Mean annual precipitation and mean temperature of the warmest month at 6,000 and 21,000 cal yr BP reconstructed from fossil pollen data

Source: Bartlein et al. 2011: 792–793. © The Author(s) 2010.

and exchange among biogeographers and palaeoecologists, climate and ecosystem modellers, molecular geneticists, archaeologists and historians, and conservation biologists and land-use managers, will continue opening new doors of inquiry and relevance.

SUMMARY

- We examine some of the approaches and data used to reconstruct terrestrial ecosystem change. Research employing these methods comes under the auspice of ecological biogeography, which is concerned with the factors that have led to the current distribution of plants on the landscape.
- Palaeoecological research usually concerns ecosystem dynamics during the Quaternary period (the last 2.58 million years), and in particular over the last 21,000 years, which includes the last glacial maximum (LGM) and the Holocene (the current interglacial period that began 11,700 years ago).

- We outline the approaches taken to reconstruct terrestrial ecosystem change including tree rings, pollen, plant macrofossils, packrat middens, charcoal and dung fungal spores, as well as supporting palaeolimnological information from diatoms, chironomids and lake-sediment lithology, geochemistry and mineralogy.
- Proxy records can be used alone, but fuller understanding is achieved by combining and comparing multiple proxy data from single or groups of sites. Alternatively, developing geographic networks of records from a single proxy type, such as charcoal or pollen data, provides insights about patterns of environmental change over broad spatial scales.

Further Reading

Elias, S.A. and C.J. Mock (2013) *Encyclopedia of Quaternary Science.*

This encyclopedia is the most comprehensive and up-to-date overview of Quaternary science available. It contains 357 broad-ranging articles authored by well-respected researchers from around the world.

Last et al. (2001) *Tracking Environmental Change Using Lake Sediments: Volume 2: Physical and Geochemical Methods.*

Smol et al. (2010) *Tracking Environmental Change Using Lake Sediments: Volume 3: Terrestrial, Algal, and Siliceous Indicators.*

These reference books detail the most important indicators used by palaeoecologists and palaeolimnologists, including pollen analysis, plant macrofossils, charcoal, diatoms, stable isotopes, geochemistry, lithostratigraphy and mineralogy.

Delcourt, H.R. and Delcourt, P. (1991) *Quaternary Ecology: A Paleoecological Perspective.*

Birks, H.J.B. and Birks, H.H. (2004) *Quaternary Palaeoecology.*

These volumes discuss the approaches by which Quaternary terrestrial environments can be reconstructed from fossils and sediments. They provide in-depth explanation of the methods and assumptions for working with a variety of proxy data.

Note: Full details of the above can be found in the references list below.

References

Baker, A.G., Bhagwat S.A. and Willis, K.J. (2013) 'Do dung fungal spores make a good proxy for past distribution of large herbivores?', *Quaternary Science Reviews* 62: 21–31.

Bartlein, P.J., Harrison, S.P., Brewer, S., Connor, S., Davis, B.A.S., Gajewski, K., Guiot, J., Harrison-Prentice, T.I., Henderson, A., Peyron, O., Prentice, I.C., Scholze, M., Seppa, H., Shuman, B., Sugita, S., Thompson, R.S., Viau, A.E., Williams, J. and Wu, H. (2011) 'Pollen-based continental climate reconstructions at 6 and 21 ka: A global synthesis', *Climate Dynamics* 37: 775–802.

Battarbee, R.W., Thrush, B.A., Clymo, R.S., Le Cren, E.D., Goldsmith, P., Mellanby, K., Bradshaw, A.D., Chester, P.F., Howells, G.D. and Kerr, A. (1984) 'Diatom analysis and the acidification of lakes [and discussion]', *Philosophical Transactions of the Royal Society B* 305: 451–77.

Betancourt, J.L., Van Devender, T.R. and Martin, P.S. (1990) *Packrat Middens: The Last 40,000 Years of Biotic Change.* Tucson, AZ: University of Arizona Press.

Bigler, C., Larocque, I., Peglar, S.M., Birks, H.J.B. and Hall, R.I. (2002) 'Quantitative multiproxy assessment of long-term patterns of Holocene environmental change from a small lake near Abisko, northern Sweden', *The Holocene* 12: 481–96.

BIOME 6000 'Palaeovegetation Mapping Project', www.bridge.bris.ac.uk/resources/Databases/BIOMES_data (accessed 24 November 2015).

Birks, H.H. (2013) 'Plant macrofossils – introduction', in S.A. Elias and C.J. Mock (eds) *Encyclopedia of Quaternary Science* (2nd edition). London: Elsevier. pp. 593–612.

Birks, H.J.B. and Birks, H.H. (2004) *Quaternary Palaeoecology* (reprint of 1980 edition). Caldwell, NJ: Blackburn Press.

Brooks, S.J., Axford, Y., Heiri, O., Langdon, P.G. and Larocque-Tobler, I. (2010) 'Chironomids can be reliable proxies for Holocene temperatures: A comment on Velle et al., 2010', *The Holocene* 22: 1482–94.

Brown, K.J. and Power, M.J. (2013) 'Charred particle analyses', in S.A. Elias and C.J. Mock (eds) *Encyclopedia of Quaternary Science* (2nd edition). London: Elsevier,. pp. 716–29.

Cheddadi, R., Yu, G., Guiot, J., Harrison, S. P. and Prentice I.C. (1996) 'The climate of Europe 6000 years ago', *Climate Dynamics* 13: 1–9.

Cook, E.R., Woodhouse, C.A., Eakin, C.M., Meko, D.M. and Stahle, D.W. (2004) 'Long-term aridity changes in the Western United States', *Science* 306: 1015–18.

Daniau, A.L., Bartlein, P.J., Harrison, S.P., Prentice, I.C., Brewer, S., Friedlingstein, P., Harrison-Prentice, T.I., Inoue, J., Izumi, K., Marlon, J.R., Mooney, S., Power, M.J., Stevenson, J., Tinner, W., Andric, M., Atanassova, J., Behling, H., Black, M., Blarquez, O., Brown, K.J., Carcaillet, C., Colhoun, E.A., Colombaroli, D., Davis, B.A.S., D'Costa, D., Dodson, J., Dupont, L., Eshetu, Z., Gavin, D.G., Genries, A., Haberle, S., Hallett, D.J., Hope, G., Horn, S.P., Kassa, T.G., Katamura, F., Kennedy, L.M., Kershaw, P., Krivonogov, S., Long, C., Magri, D., Marinova, E., McKenzie, G.M., Moreno, P.I., Moss, P., Neumann, F.H., Norstrom, E., Paitre, C., Rius, D., Roberts, N., Robinson, G.S., Sasaki, N., Scott, L., Takahara, H., Terwilliger, V., Thevenon, F., Turner, R., Valsecchi, V.G., Vanniere, B., Walsh, M., Williams, N., and Zhang, Y. (2012) 'Predictability of biomass burning in response to climate changes', *Global Biogeochemical Cycles* 26, GB4007.

Davis, O.K. and Shafer, D.S. (2006) 'Sporormiella fungal spores, a palynological means of detecting herbivore density', *Palaeogeography, Palaeoclimatology, Palaeoecology* 237: 40–50.

Delcourt, H.R. and Delcourt, P. (1991) *Quaternary Ecology: A Paleoecological Perspective*. Dordrecht: Springer.

Elias, S. (2013) 'Plant macrofossils methods and studies: Rodent middens', in S.A. Elias and C.J. Mock (eds) *Encyclopedia of Quaternary Science* (2nd edition). London: Elsevier. pp. 674–83.

Elias, S.A. and C.J. Mock (2013) *Encyclopedia of Quaternary Science* (2nd edition). London: Elsevier. Available via http://www.sciencedirect.com/science/referenceworks/9780444536426 (accessed 24 November 2015).

Falk, D.A., Heyerdahl, E.K., Brown, P.M, Farris, C., Fulé, P.Z., McKenzie, D., Swetnam, T.W., Taylor, A.H. and Van Horne, M.L. (2011) 'Multi-scale controls of historical forest-fire regimes: New insights from fire-scar networks', *Frontiers in Ecology and the Environment* 9: 446–54.

Ferrera, I., Harrison, S.P., Prentice, I.C., Ramstein, G., Guiot, J., Bartlein, P.J., Bonnefille, R., Bush, M., Cramer, W., von Grafenstein, U., Holmgren, K., Hooghiemstra, H., Hope, G., Jolly, D., Lauritzen, S.E., Ono, Y., Pinot, S., Stute, M. and Yu, G. (1999) 'Tropical climates at the Last Glacial Maximum: A new synthesis of terrestrial palaeoclimate data. I. Vegetation, lake-levels and geochemistry', *Climate Dynamics* 15: 823–56.

Finley, Jr. R.B. (1990) 'Woodrat ecology and behavior and the interpretation of paleomiddens', in J.L. Betancourt, T.R. Van Devender and P.S. Martin (eds) *Packrat Middens: The Last 40,000 Years of Biotic Change*. Tucson, AZ: University of Arizona Press. pp. 28–42.

Fritz, S.C. and Anderson, N.J. (2013) 'The relative influences of climate and catchment processes on Holocene lake development in glaciated regions', *Journal of Paleolimnology* 49: 349–62.

Fritz, S.C., Juggins S. and Engstrom, D.R. (2004) 'Patterns of early lake evolution in boreal landscapes: A comparison of stratigraphic inferences with a modern chronosequence in Glacier Bay, Alaska', *The Holocene* 14: 828–40.

Gill, J.L., McLauchlan, K.K., Skibbe, A.M., Goring, S., Zirbel, C.R. and Williams, J.W. (2013) 'Linking abundances of the dung fungus Sporormiella to the density of bison: Implications for assessing grazing by megaherbivores in palaeorecords', *Journal of Ecology* 101: 1125–36.

Gill, J.L., Williams, J.W., Jackson, S.T., Lininger, K.B. and Robinson, G.S. (2009) 'Pleistocene megafaunal collapse, novel plant communities, and enhanced fire regimes in North America', *Science* 326: 1100–3.

Global Palaeofire Working Group (2013) 'Global Charcoal Database version 2', www.gpwg.org (accessed 24 November 2015).

Higuera, P.E., Brubaker, L.B., Anderson, P.M., Hu, F.S. and Brown, T.A. (2009) 'Vegetation mediated the impacts of postglacial climate change on fire regimes in the southcentral Brooks Range, Alaska', *Ecological Monographs* 79: 201–19.

Hodder, K.R. and Gilbert, R. (2013) 'Paleolimnology: Physical properties of lake sediments', in S.A. Elias and C.J. Mock (eds) *Encyclopedia of Quaternary Science* (2nd edition). London: Elsevier. pp. 300–12.

Jackson, S.T. and Weng, C. (1999) 'Late Quaternary extinction of a tree species in Eastern North America', *Proceedings of the National Academy of Sciences* 96: 13847–52.

Jones, V.I. (2013) 'Diatom introduction', in S.A. Elias and C.J. Mock (eds) *Encyclopedia of Quaternary Science* (2nd edition). London: Elsevier: pp. 471–80.

Kaplan, J.O., Bigelow, N.H., Prentice, I.C., Harrison, S.P., Bartlein, P.J., Christensen, T.R., Cramer, W., Matveyeva, N.V., McGuire, A., Murray, D.F., Razzhivin, V.Y., Smith, B., Walker, D.A., Anderson, P.M., Andreev, A.A., Brubaker, L.B., Edwards, M.E. and Lozhkin, A.V., (2003) 'Climate change and Arctic ecosystems II: Modeling, palaeodata-model comparisons, and future projections', *Journal of Geophysical Research* 108: 8171.

Kelly, R.F., Higuera, P.E., Barrett, C.M. and Hu, F.S. (2011) 'A signal-to-noise index to quantify the potential for peak detection in sediment-charcoal records', *Quaternary Research* 75: 11–17.

Kemp, A.E.S., Dean, J. and Pearce, R.B. (2001) 'Recognition and analysis of bedding and sediment fabric features', in J.P. Smol, H.J.B. Birks and W.M. Last (eds) *Tracking Environmental Change Using Lake Sediments*, Vol. 2. Amsterdam: Springer. pp. 7–22.

Kitzberger, T., Brown, P.M, Heyerdahl, E.K., Swetnam, T.W. and Veblen, T.T. (2007) 'Contingent Pacific–Atlantic Ocean influence on multicentury wildfire synchrony over western North America', *Proceedings of the National Academy of Sciences* 104: 543–48.

Korhola, A. (2013) 'Diatom methods: Data interpretation', in S.A. Elias and C.J. Mock (eds) *Encyclopedia of Quaternary Science* (2nd edition). London: Elsevier. pp. 489–500.

Last, W.M. (2001) 'Mineral analysis of lake sediments', in J.P. Smol, H.J.B. Birks, HJB and W.M. Last (eds) *Tracking Environmental Change Using Lake Sediments*, Vol. 2. Amsterdam: Springer. pp. 42–81.

Last, W.M., Smol, J.P. and Birks, H.J.B. (2001) *Tracking Environmental Change Using Lake Sediments: Volume 2: Physical and Geochemical Methods*. Dordrecht: Springer.

Levavasseur, G., Vrac, M., Roche, D.M. and Paillard, D. (2012) 'Statistical modelling of a new global potential vegetation distribution', *Environmental Research Letters* 7: 044019.

Lyford, M.E., Jackson, S.T., Betancourt, J.L. and Gray, S.T. (2003) 'Influence of landscape structure and climate variability on a late Holocene plant migration', *Ecological Monographs* 73: 567–83.

Marlon, J.R., Bartlein, P.J., Daniau, A-L., Harrison, S.P., Maezumi, S.Y., Power, M.J., Tinner, W. and Vannière, B. (2013) 'Global biomass burning: a synthesis and review of Holocene paleofire records and their controls', *Quaternary Science Reviews* 65: 5–25.

Massaferro, J.I., Moreno, P.I., Denton, G.H., Vandergoes, M. and Dieffenbacher-Krall, A. (2009) 'Chironomid and pollen evidence for climate fluctuations during the Last Glacial Termination in NW Patagonia', *Quaternary Science Reviews* 28: 517–25.

Mauquoy, D. and Van Geel, B. (2007) 'Plant macrofossil methods and studies: Mire and peat macros', in S.A. Elias and C.J. Mock (eds) *Encyclopedia of Quaternary Science*. Amsterdam, Netherlands: Elsevier Science. pp. 2315–36.

McWethy, D.B., Whitlock, C., Wilmshurst, J.M., McGlone, M.S. and Li, X. (2009) 'Rapid deforestation of South Island, New Zealand, by early Polynesian fires', *The Holocene* 19: 883–97.

McWethy, D.B., Whitlock, C., Wilmshurst, J.M., McGlone, M.S., Fromont, M., Li, X., Dieffenbacher-Krall, A., Hobbs, W.O., Fritz, S.C. and Cook, E.R. (2010) 'Rapid landscape transformation in South Island, New Zealand, following initial Polynesian settlement', *Proceedings of the National Academy of Sciences* 107: 21343–48.

Meyers, P.A. and Ishiwatari, R. (1993) 'Lacustrine organic geochemistry: An overview of indicators of organic matter sources and diagenesis in lake sediments', *Organic Geochemistry* 20: 867–900.

Morris, J.L., Brunelle, A., DeRose, R.J., Seppä, H., Power, M.J., Carter, V. and Bares, R. (2013) 'Using fire regimes to delineate zones in a high-resolution lake sediment record from the western United States', *Quaternary Research* 79: 24–36.

Munoz, E., Schroeder, S., Fike, D.A. and Williams, J.W. (2014) 'A record of sustained prehistoric and historic land use from the Cahokia region, Illinois, USA', *Geology* 42: 499–502.

Neotoma 'Paleoecology Database'. www.neotomadb.org (accessed 24 November 2015).

Perry, G.L., Wilmshurst, J.M., McGlone, M.S., McWethy, D.B., and Whitlock, C. (2012). 'Explaining fire-driven land-scape transformation during the Initial Burning Period of New Zealand's prehistory', *Global Change Biology* 18: 1609–21.

Power, M.J., Marlon, J.R., Ortiz, N., Bartlein, P.J., Harrison, S.P., Mayle, F.E., et al. (2008) 'Changes in fire regimes since the Last Glacial Maximum: An assessment based on a global synthesis and analysis of charcoal data', *Climate Dynamics* 30: 887–907.

Prentice, I.C. and Webb III, T. (1998) 'BIOME 6000: global paleovegetation maps and testing global biome models', *Journal of Biogeography* 25, 997-1005.

Prentice, I.C., Harrison, S.P. and Bartlein, P.J. (2011) 'Global vegetation and terrestrial carbon cycle changes after the last ice age', *New Phytologist* 189: 988–98.

Ritchie, J.C. (1987) *Post-glacial Vegetation of Canada*. Cambridge: Cambridge University Press.

Robinson, G.S., Burney, L.P. and Burney, D.A. (2005) 'Landscape paleoecology and megafaunal extinction in southeastern New York state', *Ecological Monographs* 75: 295–315.

Rule, S., Brook, B.W., Haberle, S.G., Turney, C.S.M., Kershaw, A.P. and Johnson, C.N. (2012) 'The aftermath of mega-faunal extinction: Ecosystem transformation in Pleistocene Australia', *Science* 335: 1483–6.

Smol, J.P., Birks, H.J. and Last, W.M. (2010) *Tracking Environmental Change Using Lake Sediments: Volume 3: Terrestrial, Algal, and Siliceous Indicators*. Dordrecht: Springer.

Stahle, D.W., Cleaveland, M.K, Blanton, D.B., Therrell, M.D. and Gay, D.A. (1998) 'The lost colony and Jamestown droughts', *Science* 280: 564–7.

Swetnam, T.W. (2002) 'Fire and climate history in the western Americas from tree rings', *PAGES* Magazine 10, 6–8.

Veblen, T.T., Kitzberger, T., Raffaele, E. and Lorenz, D.C. (2003) 'Fire history and vegetation changes in northern Patagonia, Argentina', in T.T. Veblen, W.L. Baker, G. Montenegro and T.W. Swetnam (eds) *Fire and Climatic Change in Temperate Ecosystems of the Western Americas*. New York, NY: Springer-Verlag. pp. 265–95.

Velle, G. and Heiri, O. (2013) 'Chironomid records; postglacial Europe' in S.A. Elias and C.J. Mock (eds) *Encyclopedia of Quaternary Sciences* (2nd edition). London: Elsevier. pp. 386–97.

Walker, I.R. (2013) 'Chironomid Overview', in S.A. Elias and C.J. Mock (eds) *Encyclopedia of Quaternary Science* (2nd edition). London: Elsevier. pp. 355–60.

Whitlock, C. and Larsen, C.P.S. (2001) 'Charcoal as a fire proxy' in J.P. Smol, H.J.P. Birks and W.M. Last (eds) *Tracking Environmental Change Using Lake Sediments: Volume 3 Terrestrial, Algal, and Siliceous Indicators*. Dordrecht: Kluwer Academic Publishers: pp. 75–97.

Whitlock, C., Dean, W.E., Fritz, S.C., Stevens, L.R., Stone, J.R., Power, M.J., Bracht-Flyr, B.B., Rosenbaum, J.R., Pierce, K.L. and Bracht-Flyr, B.B. (2012) 'Holocene seasonal variability inferred from multiple proxy records from Crevice Lake, Yellowstone National Park, USA', *Palaeogeography, Palaeoclimatology, Palaeoecology* 331: 90–103.

Williams, J.W., Shuman, B.N., Webb III, T., Bartlein, P.J. and Leduc, P.L. (2006) 'Late-Quaternary vegetation dynamics in North America: Scaling from taxa to biomes', *Ecological Monographs* 74: 309–34.

ON THE COMPANION WEBSITE...

Visit **https://study.sagepub.com/keymethods3e** for author videos, chapter exercises, resources and links, plus **free** access to the following recommended articles:

1. **Schreve, D. and Candy, I. (2010) 'Interglacial climates: Advances in our understanding of warm climate episodes', *Progress in Physical Geography*, 34 (6): 845–56.**

Major advances in our understanding of the climate, duration and stratigraphy of Quaternary interglacials have occurred over the decade. This review details palaeoenvironmental evidence contained within British interglacial deposits and their correlation with interglacial episodes recorded in marine and ice core records, allowing us to understand in greater detail how northwest Europe responded to different periods of climate warming. This information is important to understanding the evolving climate of the Holocene.

2. **Meadows, M.E. (2012) 'Quaternary environments: Going forward, looking backwards?', *Progress in Physical Geography*, 36 (4): 539–47.**

A longer-time perspective is needed in order to better understand contemporary and near-future global environments. This progress report reviews how an understanding of environmental dynamics over extended time periods is now incorporated into science dealing with predictions of future climate change by the IPCC consortium, how possible analogues for a warmer future are still vigorously explored and how information on past environments may better inform an understanding of contemporary ecosystem processes and influence the future management of biodiversity in protected areas.

3. **Hessl, A. and Pederson, N. (2013) 'Hemlock Legacy Project (HeLP): A paleoecological requiem for eastern hemlock', *Progress in Physical Geography*, 37 (1): 114–29.**

Eastern North American forests have effectively lost two major tree species (American chestnut and American elm) in the last 100 years, and two more, eastern and Carolina hemlock, will be functionally extinct over much of their ranges within a couple of decades. This progress report describes a community-based approach to salvaging palaeoenvironmental archives that could serve as a model for collections from other important species currently threatened by exotic forests pests and pathogens (e.g. whitebark pine, ash). The approach calls for building connections between scientists, students, environmental NGOs, and land managers focused on old-growth forests.

23 Numerical Modelling: Understanding Explanation and Prediction in Physical Geography

Stuart N. Lane

SYNOPSIS

This chapter introduces the use of numerical modelling for understanding environmental systems. Numerical models are used extensively throughout society, and hence by physical geographers, for dealing with situations where the researcher is 'remote' from what they are studying: where the events that are of interest have occurred in the past (e.g. reconstruction of past climates); may occur in the future (e.g. patterns of inundation associated with future flood events); or where they are occurring now, but cannot be measured or studied using other methods. The potential of numerical models aside, the problems of modelling environmental systems mean that models are commonly scientifically wrapped crystal balls, whose predictions must be treated with caution at best and scepticism at worst.

This chapter is organized into the following sections:

- Introduction: Why model?
- Fundamental aspects of environmental modelling
- What a model can and cannot do
- Conclusion

23.1 Introduction: Why Model?

In *Winnie-the-Pooh*, A.A. Milne demonstrates clearly the basic reason why we need to consider using numerical models. Pooh Bear finds a jar labelled 'hunny' with a yellow substance in it. Being a 'good' scientist, he cannot be sure that it is 'hunny' until he has done a proper scientific experiment to test that his hypothesis is valid. This has to be grounded in direct observation of what is in the jar and involves him tasting it. However, he cannot be sure that it is 'hunny' until he has tasted all of it, right to the bottom of the jar as, according to Pooh bear, his uncle had once said that someone once put cheese in a 'hunny' jar for a joke. The view that *all* jars labelled 'hunny' contain 'hunny' has already been proven false, and so each individual jar that Pooh bear finds has to be subject to Pooh-type assessment, in which the entire contents of the jar are consumed. If we apply this to environmental systems, we find ourselves visiting a series of great environmental disasters that have been created because as a society we have been unwilling to accept certain evidence (e.g. that of a jar labelled 'hunny' with a yellow substance in it) until there is definitive observational evidence that confirms that a partial observation or a theory is correct. The best example of this is provided by depletion of stratospheric ozone concentrations. The potential that chlorofluorocarbons might lead to long-term ozone depletion was demonstrated in the early 1970s (Molina and Rowland, 1974). It was not until the direct observation of a Spring ozone hole over Antarctica in 1985 (Farman et al., 1985) that this theoretical notion was accepted and subsequent environmental policy was developed. The same scenario has emerged in relation to global climate change: we have a theoretical basis to expect that atmospheric greenhouse gas accumulation has the potential to change climate; much of the critique of this theory is based around the fact that there is no demonstrable evidence to confirm this hypothesis (e.g. Michaels, 1992). We will not do anything about it until we have actually observed it.

The problem with this view is also illustrated by Milne. When Piglet goes to visit his heffalump trap, he finds what looks like a heffalump. Being only a small animal, and with the trap being very deep, he needs to get closer to see if the heffalump (Pooh bear with an empty 'hunny' jar stuck on his head) is indeed a heffalump. This mirrors classic observation-based science, in which we search for a better understanding of an apparent phenomenon, through more in-depth observation. However, if Piglet discovers that what he thinks is a heffalump is actually a heffalump then, being a small animal, he is likely to be in a lot of trouble as the heffalump may attack him. We do not want, through more intensive investigation based upon observation, to discover problems with potentially serious consequences (e.g. heffalumps, ozone holes, global warming, severe species loss, serious organic pollution). However, sometimes, we find ourselves facing an unresolvable circularity, where what we think is a matter of concern can only be confirmed as such through observation of the very phenomena we wish to avoid.

The numerical model is one of the tools that the geographer might use to break out of the circularity that Pooh bear and Piglet find themselves in. It provides a tool for investigating things that are inaccessible. Inaccessibility arises because we are commonly interested in: (i) environments in the past, before records began, or where environmental reconstruction may be unreliable or impossible; (ii) environments in the present, from where we cannot obtain measurements, perhaps because those environments are inaccessible, whether because they are remote, or too large to measure, or too small to measure; and (iii) environments in the future, where we are concerned about the possible impacts of decisions made now for future generations. This chapter seeks to introduce the basics of environmental modelling as part of geographical enquiry, but also to reflect upon the challenges and problems that result more generally with environmental modelling. The first section of the chapter introduces the fundamentals of environmental modelling, in terms of conceptual, empirical and physically-based approaches. The second section seeks to evaluate models critically, by considering what a model can do and what it cannot do.

23.2 Fundamental Aspects of Environmental Modelling

There are two distinct approaches to mathematical modelling: (i) empirical, or data-laden; and (ii) physically-based, or theory-laden. However, as this section will note, these two distinct approaches are not actually that distinct. They are end members of a spectrum of modelling approaches (Odoni and Lane, 2010) and they actually share fundamentally the idea that there is a resilient and defensible conceptual model of the system that is being considered.

The conceptual model

Without a conceptual model of the system that is being modelled, it is impossible to develop either an empirical or a physically-based mathematical model. Conceptual models involve a statement of the basic interactions between the components of a system (see also Chapter 24). If we think about the understanding of past climates, we can start to think about a simple conceptual model for climate change. Empirical evidence has shown that the Earth's climate has fluctuated between periods that were much colder than present (glacial periods) and periods that were slightly warmer than present (Figure 23.1). Why has this occurred? The initial conceptual model might assume that it is due to an external forcing, and we know that a good candidate in this respect is the nature of the Earth's rotation around the sun (see Imbrie and Imbrie, 1979), which varies over a number of different time-scales. There is good support for some of this cyclical behaviour being explained by orbital forcing (e.g. Imbrie and Imbrie, 1979), but there is also much evidence to question the conclusion that this is the only explanation (e.g. Broecker and Denton, 1990). It is well-established that processes internal to the earth-atmosphere system may exert an important conditioning role upon the way in which these external forcing factors affect climate. This is where we can develop a simple conceptual model to illustrate what form this system might take.

A system is made up of components that are connected together by links. Flows between components are driven by processes in a way that depends upon both the components and the nature of the links. Thus, in the case of glacial cycles, we may start to build a system by considering three components: (i) albedo which relates to the way

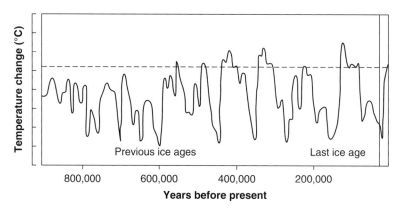

Figure 23.1   Globally averaged temperature change over the last million years. The dashed line indicates (approximately) the 1880 temperature.

Source: Houghton et al., 1990: Figure 7.1a

in which a surface reflects incoming solar radiation; (ii) temperature; and (iii) ice-sheet growth. As a first approximation, these may be linked together (Figure 23.2). The links are specified very simply as positive and negative: (i) a negative link between albedo and temperature, which reflects the fact that as albedo (earth surface reflectivity) goes up, more short-wave radiation will be reflected and hence temperature will go down; (ii) a negative link between temperature and ice-sheet growth, which reflects the fact that as temperature goes down, ice-sheet growth will go up; and (iii) a positive link between ice-sheet growth and albedo, which reflects the fact that as ice-sheet growth goes up, so albedo goes up as ice cover tends to reflect more short-wave radiation than other types of land cover. Put more generally, a positive link involves the effect variable responding in the same direction as the cause variable; a negative link involves the effect variable responding in the opposite direction to the cause variable.

Figure 23.2a illustrates a very important property of feedback, which occurs in all environmental systems and which arises from the combined interaction of these links between components: as temperature goes down, ice-sheet growth goes up; as ice-sheet growth goes up, albedo goes up; and as albedo goes up, temperature will go down. Thus, the net effect of the system in Figure 23.2a is positive feedback, where an initial reduction in temperature would be amplified through the ice-sheet growth and albedo links to result in further temperature reduction. Positive feedback causes a system to change or evolve.

In Figure 23.2b, precipitation is introduced as an additional and important component of glacier growth and decay: precipitation is required to add mass to an ice sheet. Precipitation levels will be governed by temperature, as this controls evaporation. For ice-sheet growth, the global oceans act as the major water source. As temperature goes down, precipitation will go down, due to less evaporation (i.e. the link is positive as both changes are in the same direction). If precipitation goes down, ice-sheet growth will go down. Thus, if we examine the links from ice-sheet growth to albedo to temperature to precipitation and back to ice-sheet growth we have a negative feedback: as ice-sheet growth goes up, albedo goes up, temperature goes down, precipitation goes down, and hence ice-sheet growth goes down. Negative feedback is a self-limiting feedback, which causes a system to resist change.

With these basic ideas about positive and negative feedback, we have a means of understanding the dynamics of glacier growth and decay in the system in Figure 23.2. If negative feedbacks dominate in a system, then the system will be maintained in the state that it is in, either as a glacial period or as an interglacial period. If positive feedbacks dominate, then we may have change, which may be rapid, and which may cause a system to evolve, either from a glacial to an interglacial or from an interglacial to a glacial. In general, positive feedback is not maintained in perpetuity. Rather, it continues until limiting factors slow or stop the feedback, or the state of the system evolves to allow other negative feedbacks to develop. In the presence of some sort of external forcing, the effects of that forcing will depend upon whether the system displays positive feedback and hence the forcing causes change, or negative feedback, in which the forcing is absorbed by the system. In relation to Milankovitch forcing (Figure 23.2c), the orbital driving factors are either enhanced or reduced by the system to which they are applied. In practice, the system is much more complex than this. Feedbacks may be delayed and simple components like 'temperature' and 'precipitation' need to be disaggregated in relation to glacier growth; e.g. enhanced winter

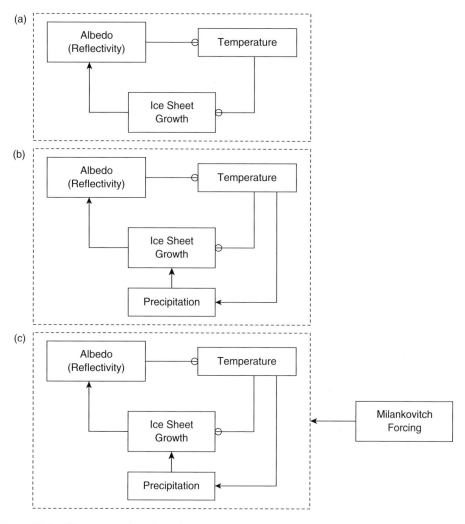

(a)

(b)

(c)

Figure 23.2   Three examples of simple systems that are relevant to glacial cycles

snowfall (increased precipitation) and reduced summer melt (lower temperatures) are generally considered to be the most conducive to glacier growth.

Thus far, we have only considered the environment as changing through time. However, the environment has a critical spatial and vertical dimension. Thus, feedbacks do not simply operate through time, but also through space: if you change one part of the environment, feedbacks may not be limited to just that part of the environment, but may also affect processes operating in other places in the environment. For instance, in the case of glacial cycles, it is well established that ice-sheet growth in the high latitudes can have major environmental impacts at both the mid- and the low latitudes. As the ice sheets extended to lower latitudes than present, cold polar air also extended to lower latitudes, causing the mid-latitudinal zones where warm and cold air mix and the westerly jet stream to move equatorwards. In the American south-west, this caused substantial increases in precipitation during the last glacial maximum (e.g. Spaulding, 1991). Similarly, as more water was locked up in the ice sheets during the last glaciation, and with changes in atmospheric and oceanic circulation, the tropical latitudes were somewhat cooler and drier than today. This has resulted in suggestions that tropical plant and animal communities retreated into refugia, although this is contested, with suggestions that continuous forest cover was maintained throughout glacial periods (e.g. Colinvaux et al., 2000). This provides an important reminder of the connected nature of the environment, and the way in which changes in one part of the environment can affect other parts of the environment.

Conceptualization, then, is the process of proscribing the system that is to be modelled. In turn, this conceptualization is heavily dependent upon how we perceive the environment (Odoni and Lane, 2012). However sophisticated the theoretical content of our models, they are also influenced strongly by the mixture of personal experience, community practice, data and observations that we encounter and formulate through our day-to-day practices as scientists (Lane, 2012). The dependence of modelling upon the modeller has important implications for the status of models as pieces and producers of knowledge, and the practices needed to guarantee trust in models and these are addressed below. Restricting the focus upon conceptualization, the key point is that when modelling the environment it is crucial to consider: (i) the proper representation of the spatiality of processes, especially because many processes are driven by gradients that are spatial (e.g. the slope of a hill controls the velocity with which water moves over it); (ii) vertical gradients of process (e.g. flow at the bed of a river is generally faster than at the water surface); (iii) the way the system evolves through time in response to the operation of these two- and three-dimensional processes; and (iv) feedbacks themselves manifest as change in the process drivers in response to the operations of processes themselves; e.g. as water moves over a hillslope, it may cause erosion (or deposition) and change the slope, so that processes at future times operate in different ways. The conceptual model allows us to specify (i) through (iv) and hence: (a) what is excluded explicitly in the model; (b) what is included, but represented in a simple way; and (c) what is excluded. As we will see later, this means that models are rarely generic representations of whole systems, but truncated representations of reality, where model predictions are partly defined by how the modeller builds the model, how the model is closed (Lane, 2001).

With a conceptual model identified, the question becomes how should the model be built? Should the model be built from observations or data such as through statistical association or machine learning, that is as an **empirical model**? Or should it be **physically-based,** that is built around a set of laws or rules that can be partially derived from laws (e.g. by arguing that Newtonian physics applies to the problem being studied and then using derivatives of Newton's conservation laws)?

Empirical approaches

Mathematical models based upon empirical approaches involve making a set of observations of a number of phenomena and then using these observations to construct relationships amongst them. Thus, this approach is heavily dependent upon either field or laboratory measurements to provide the data necessary to construct the models. In the most extreme case, the form of the conceptual model may be driven entirely by the relationships between data that can be identified on the basis of the properties of those data using artificial intelligence or machine learning (e.g. Licznar and Nearing, 2003).

Central to empirical approaches are statistical methods. The key assumption is that one or more forcing (independent) variables cause changes in a response (dependent) variable. A good example of this is the eutrophication of shallow lakes. Eutrophication is a natural process associated with a progressive increase in lake primary productivity as lakes are ineffective at removing accumulated nutrients. Research in the 1970s (e.g. Schindler, 1977) established that the key limiting nutrients for eutrophication are phosphates rather than nitrates, as is commonly assumed, because certain primary producers are capable of fixing atmospheric nitrogen. Figure 23.3 illustrates this for Barton Broad, using a dataset when no submerged plants (macrophytes) are present: as the total phosphate concentration of the water column rises, so the level of primary productivity increases. Thus, we have a conceptual model in which we assume that levels of eutrophication are determined by levels of phosphate in a given lake ecosystem. By synthesizing a set of data from different lake ecosystems, Cullen and Forgsberg (1988) were able to build simple statistical rules that allow levels of chlorophyll-a, an indicator of the level of primary productivity in the system, to be predicted from phosphate concentrations. In practice, eutrophication is also affected by predator–prey interactions and seasonality effects, and more complex models can be developed for predicting the level of eutrophication in a lake by using more than one forcing variable (e.g. Lau and Lane, 2002). This is not a straightforward task if the forcing variables are themselves related to one another or, as is commonly the case with ecological data obtained for a single ecosystem (e.g. a lake), individual observations are correlated through time or across space. Additional information on empirical modelling of quite complex environmental systems is provided in Pentecost (1999).

The use of an empirical approach to modelling requires a number of assumptions. First, the empirical relationship must have a justifiable theoretical basis in the form of an appropriate conceptual model. Whilst the use of statistical methods allows assessment of how good the model is (e.g. by assessing the goodness of fit of the empirical relationship – see Chapter 32) this is not normally a sufficient test of a model: empirical relationships may be spurious, and hence have a poor predictive ability. They may also be strongly affected by individual observations, especially at the extremes of a variable, which can affect the model's predictive ability in relation to extreme situations. Unfortunately, many of the environmental issues of most concern are those that are connected with extremes (e.g. floods and droughts).

Second, many empirical models perform poorly when used to predict beyond the range of observations upon which they are based. For instance, Cameron et al. (2002) compare traditional (using probability distributions and kinematic routing) and machine learning (neural network) approaches to flood forecasting. The latter is based upon using artificial intelligence methods to construct an empirical relationship between forcing parameters (e.g. upstream rainfall) and the key response variable (e.g. water level). Cameron et al. (2002) found that these models are only good at predicting water levels associated with patterns of forcing parameters that have happened before. Questions over temporal validity also appear in relation to spatial validity. In the case of lake eutrophication, there is a good reason to assume that lakes are generally phosphate limited, but research has shown (Kilinc and Moss, 2002) that in certain situations lakes can be nitrate limited. Thus, a generalized model (e.g. Figure 23.3) will not hold everywhere. There are two important implications of this problem: empirical models do not always hold through time; and they often transfer poorly to different places. At the root of these problems is the poor generalizability of empirical models. It is often argued that this is because they do not necessarily have a good physical basis, as they are grounded upon statistical interactions rather than fundamental physical processes.

Third, however the relationship is constructed, it is important to remember that empirical relationships involve statements of uncertainty. The scatter around the line in Figure 23.3 means that, for a given phosphate loading, there is a range of possible levels of eutrophication. The greater the scatter, the greater the uncertainty, and it is necessary to provide predictions with an uncertainty attached to them, normally in the form of a standard deviation (i.e. prediction ± standard deviation) or confidence limits about a regression fit (dotted lines in Figure 23.3). When predictions from more than one empirical model are combined, it becomes especially important to propagate the uncertainty associated with each prediction. For instance, it is common to estimate the sediment load produced by a catchment from the product of an estimated discharge (predicted on the basis of an empirical relationship between discharge and the more readily measured parameter, water level) and an estimated suspended sediment concentration (predicted on the basis of an empirical relationship between suspended sediment concentration and discharge). Both of these have uncertainties that will tend to magnify error when those uncertainties are combined. In some situations, this uncertainty can be propagated mathematically (e.g. Taylor, 1997). However, as the assumptions required for mathematical propagation of error are commonly violated,

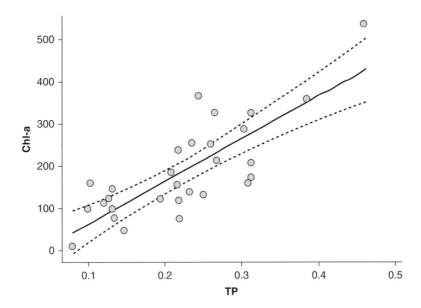

Figure 23.3   The relationship between chlorophyll-a loading (a measure of the level of eutrophication) and total phosphate concentration for periods of macrophyte absence in Barton Broad, Norfolk

Source: Lau (2000)

it is more common to propagate error statistically using techniques such as Monte Carlo analysis (e.g. Tarras-Wahlberg and Lane, 2003).

Physically-based numerical models

Empirical approaches have a physical basis insofar as they have some sort of conceptual model that justifies the form of the relationship developed, except in the extreme cases of machine learning. Physically-based numerical models take this one stage further by using the conceptual model to define links between fundamental physical, chemical and occasionally biological principles, which are then represented mathematically in computer code. At the spatial scale of the natural environment, we are fortunate in having a number of key principles, largely deriving from Newtonian mechanics: (i) rules of storage; (ii) rules of transport; and (iii) rules of transfer. Rules of storage are based upon the law of mass conservation: matter cannot be created or destroyed, but only transformed from one state into another. Thus, in relation to the prediction of flood inundation extent, and assuming that evaporation and infiltration losses are negligible, an increase in river discharge must result in one or more of: (i) an increase in flow velocity; (ii) an increase in water level; or (iii) transfer of water onto the floodplain. Mass conservation underpins almost all numerical models in some shape or form. However, it is rarely sufficient to predict how a system behaves. Thus, in the case of flood inundation, how an increase in discharge is divided between changes in flow velocity, water level and floodplain transfer depends upon what is conventionally labelled a force balance, and which is normally based upon further Newtonian mechanics if they apply: for example, every body continues in its same state of rest or uniform motion unless acted upon by a force. In the case of a river, this partitioning involves consideration of the pressure gradients and potential energy sources which drive the flow and the loss of momentum, due to friction at the bed and due to turbulence, which slows the flow: this is a rule of transport. For instance, in the simplest of terms, with a rougher bed, and subject to the shape of the river channel, an increase in discharge is less likely to lead to an increase in flow velocity than an increase in water level. Finally, rules of transfer allow for the possibility that chemical reactions cause a change in the state of an entity. For instance, phosphate bound to aluminium or iron may become soluble in a eutrophic lake, and hence available for fuelling eutrophication, if there are oxygen deficits sufficient for reducing conditions to develop.

Following from the discussion of conceptual models above, the key to the operation of these rules is to allow feedback between them and the parameters that describe them. For instance, in the case of the river example above, an increase in discharge was noted to lead to more of an increase in water level for a rougher bed than a smoother bed. However, if water level rises, the effect of bed friction will be reduced, making it easier to translate increases in discharge into increases in velocity. Here we see one of the fundamental advantages of numerical models over empirical models: they are more likely to capture the dynamics of the system through incorporating feedbacks between system parameters, something that our consideration of conceptual models identified as being crucial. However, they can only do that in so far as the relevant processes and feedbacks have been correctly included when the model is conceptualized.

Stages in the development of a physically-based numerical model

Stages in the development of a mathematical model are shown in Figure 23.4 for the case of a model of eutrophication processes in a shallow lake (from Lau, 2000). This introduces a number of important components in model building. First, it demonstrates the dependence upon a proper **conceptual model**. As noted above, the conceptual model is 'closed' as the boundaries of the system that the model will address must be defined. Ideally, all relevant processes will be included, and the closure will not exclude processes that might matter. In practice, processes are excluded for two reasons: (i) if they do not matter for the particular system being studied; or (ii) if there are limitations on the possibility of including a particular process. Situation (i) can arise for a number of reasons. First, the geography or the history of the problem may allow processes to be excluded (i.e. they may not be occurring in a particular place or at a particular time, and so can be ignored). Second, there may be time-scales of space-scales that are not relevant to the system that is being considered. For instance, if we wish to model the spatial patterns of flood inundation over large lengths of river, it is not necessary to include a sophisticated treatment of turbulence, a short time-scale aspect of the process, as uncertainties in other aspects of the model will tend to dominate. This reflects the common situation that time and space are coupled (e.g. Schumm and Lichty, 1965), where processes of interest over large spatial scales are commonly associated with longer time-scales. Unfortunately, numerical modelling faces a serious challenge when processes couple across time and space scales, and a fine resolution of process representation (in space and time) is required in order to get an adequate system representation. Situation (ii) follows from (i) when there are limits placed upon process representation because of limits to computation, but also more generally where either the process knowledge is missing, or process representation is difficult. For instance, in eutrophication modelling, there is a major difficulty in coupling the aggregate behaviour of phytoplankton to the species-specific behaviour of an individual fish, in relation to its life cycle. This situation is commonly dealt with by using a simpler version of a process's effects. In the case of lake eutrophication, the system may be driven by both nutrient limitations and food chain interactions. The latter are controlled by the presence of bottom growing vegetation which act as refugia for zooplankton that graze upon algae. It is not necessary to model plant life cycles in most shallow lakes as these are relatively straightforward functions of seasonality. Thus, they may be dealt with using simple parameterizations (see below).

The second component of model building involves taking the conceptual model, identifying appropriate process rules and transforming these into a **simulation model** that can solve the equations. This can be the most difficult stage of model development, as many equations do not lend themselves to easy solutions. This is well illustrated for the case of predicting the routing of floods through a drainage network. The discharge entering the network varies as a function of time. In simple terms, the rate at which it moves through a part of the network depends upon the water surface slope in that part of the network (steeper slopes mean faster flows). Thus, there is a spatial dependence. The water surface itself will evolve as the water moves through the network. This leads to two basic problems. First, the combined space–time dependence of flood routing means that the dominant equations are partial differentials (as they contain derivatives in time and space). This is common to almost all environmental models, as we are interested in how

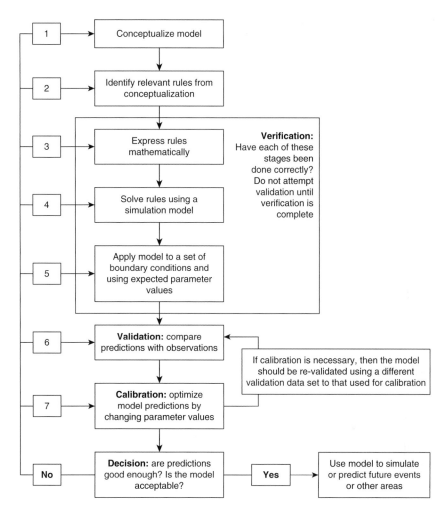

1. Is the model properly conceptualised? Are there enough processes represented? Are there too many processes represented? Are the correct processes represented?
2. Have the correct rules been identified? Are the rules properly specified? What assumptions have been introduced here? Are they acceptable?
3. Is the mathematical expression of the rules correct? Have any unsupportable assumptions been introduced during expression?
4. Are the rules to be solved correctly? Is the computer program free of programming error? Is the numerical solver sufficiently accurate? Is the model discretisation robust?
5. Are the boundary conditions properly specified? Is it the error in the boundary conditions that is affecting the model? Is the poor model performance due to a lack of necessary boundary conditions?
6. Are the observations being used to validate the model correct? Are the observations equivalent to model predictions? Are they representative of model predictions? Where (in space and time) is the model going wrong when compared with observations and predictions?
7. Has the calibration process produced realistic parameter values? Does the model produce more than one set of parameter values that are equally good at predicting when re-validating the model? Is the model overly sensitive to parameterisation?

Figure 23.4 A general approach to model development

Source: Lau (2000)

things move through space. Moving through space takes time, and hence all models should contain both space and time. Partial differentials are very difficult to solve. Second, all models require some form of initial conditions. In this case, we need starting values for water level and discharge throughout the drainage network. We will also need to know some combination of discharge, velocity and depth at the inlet and/or the outlet. Hence, models have a crucial dependence upon the availability of data to initialize them.

Third, a solution to the governing equations commonly requires us to **introduce parameters**. This may be because during the conceptualization process, we chose either to exclude certain processes, or to represent them in a simplified way. In the algal modelling described above, lake vegetation effects were treated in a simple way (a presence or absence conditioned by seasonality). Flood routing commonly ignores the lateral and vertical movements of water, and flow turbulence. Whilst later/vertical movements and turbulence may affect the routing of discharge, their direct inclusion in a model may make a solution impossibly time consuming if we are interested in a drainage-network-scale analysis. Whilst equations can have a good physical basis, as they are simplified, new terms can appear whose physical basis is less certain and, most commonly, whose field or laboratory measurement is especially difficult. Parameterization can also be required in situations where a process has been excluded from a model, with its effects being represented through one or more parameters. A good example of this in flood routing studies is the use of a roughness parameter that not only represents the effects of friction at the bed upon flow hydraulics but also turbulence. The process of parameterization is rarely conducted independently from data. Rather, a model is optimized by changing parameter values such that the differences between data and model predictions are minimized. This can result in parameters taking on values that are very different from what they might appear to be if they can be measured in the field. For this reason, they may be called *effective* parameters, ones required to allow a model to predict known behaviour, rather than derivable from fundamental process analysis. Importantly, this causes us to question the supposed generality of models as the results of parameterization may not be guaranteed to hold beyond the range of conditions for and the location at which parameterization has been undertaken. That is a model can only be effective once some data are available to parameterize it.

Model assessment involves two important stages: **verification** and **validation**. Verification is the process by which the model is checked to make sure that it is solving the equations correctly. This may involve debugging the computer code, doing checks upon the numerical-solution process and undertaking sensitivity analysis. The latter may be used to make sure that the model behaves sensibly in response to changes in boundary conditions or parameter values. Validation is the process by which a model is compared with reality. This normally involves the definition of a set of 'objective functions' that describe the extent to which model predictions match reality (see Lane and Richards, 2001). This is where there can be confusion with regard to validation and parameterization. Once an objective function has been determined, model parameters may be adjusted to reduce the magnitude of the error defined by the objective function, that is a specific approach to parameterization known as 'calibration'. It may involve a more radical redevelopment of the model through the incorporation of new processes or alternative treatment of existing processes.

Figure 23.6 shows default and optimized predictions of chlorophyll-a concentrations obtained for the eutrophication model described opposite (Figure 23.5), as applied to

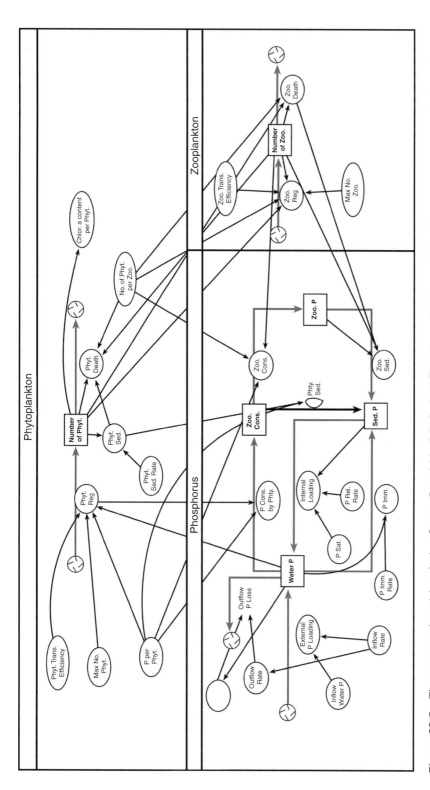

Figure 23.5 The conceptual model applied to Barton Broad, Norfolk, England. This shows that the model had three major components: the algal component (phytoplankton), a nutrient component (phosphorous) and an algal grazing component (zooplankton). There were then interactions both within and between these three major components. For instance, phytoplankton are re-generated according to nutrient availability and die naturally. They are also grazed by zooplankton. As phytoplankton are generated, die and are grazed, so phosphorous is moved around the various components in which it can be stored.

Source: Lau (2000)

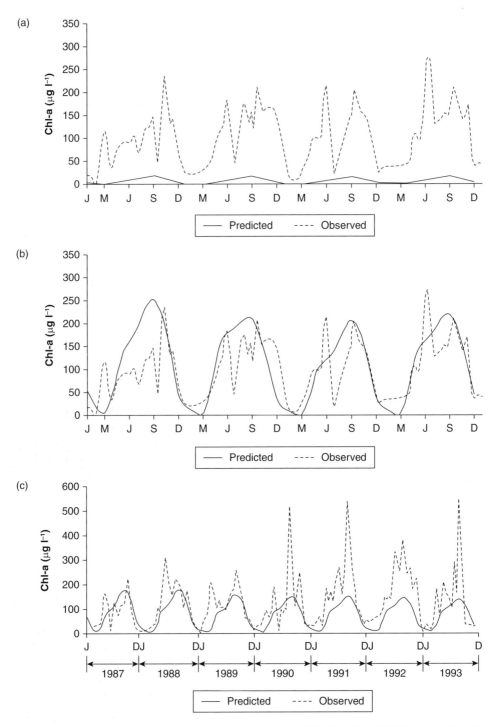

Figure 23.6 The default predicted (23.6a), optimized using 1983–1986 data (23.6b) and predicted using 1987–1993 data (23.6c) and chlorophyll-a concentrations for Barton Broad

Source: Lau (2000)

Barton Broad. Figure 23.6a shows the default predictions and Figure 23.6b shows the optimized predictions, in which parameterization was used to maximize the goodness of fit between the model and independent data. Unfortunately, Figure 23.6b is not an example of validation, as we have no evidence that the fit between the model and the measurements is due to anything other than the fact that we have forced the model to fit the measurements, what Oreskes et al. (1993) call 'forcing empirical adequacy'. To claim, from Figure 23.6b, that the model is validated, is tantamount to cheating. Modellers get around this problem through one of a number of means. First, it may be possible to do a split test, in which some data are not used in model parameterization, but held back for independent use for validation. An example of this is shown in Figure 23.6c, in which the model optimized on one period of data (Figure 23.6b for 1983 to 1986 data) is used to predict a second period, from 1987 to 1993, which is then compared to data to establish the model's validity. This demonstrates why the optimization undertaken for 1983 to 1986 does not constitute validation: when the model is applied in a predictive mode (1987 to 1993), it is clear that there is a progressive divergence between observations and model predictions and the model is not capturing some part of the system. An optimized model is not necessarily a validated model.

Second, validation can be undertaken by taking a model that has been parameterized for one location and, under the assumption that the parameters can be transferred, applying it to a second location in order to provide independent validation data.

The main purpose of validation is to place confidence in the extent to which a model can be used to simulate or to predict beyond the range of conditions for which it is formulated. As we noted in the introduction, this is the core purpose of a model: to extend the bounds of space and time. Here we face some serious difficulties, and these cause us to look very critically at deterministic mathematical models and what they can and cannot do.

23.3 What a Model Can and Cannot Do

The above examples illustrate four basic reasons why we might wish to develop numerical models and, connected to each of these, four reasons why models might not deliver what we hope. The first relates to the need to understand whole systems. This relates to the idea that there are potentially many processes that might operate in a system, that these connect with one another through both positive feedbacks (which amplify change) and negative feedbacks (which slow or even stop change), and that these processes operate over space and through time. Designing either field or laboratory experiments to understand such systems is a serious challenge. The numerical model provides the opportunity to integrate processes together in order to understand whole-system behaviour. The problem here is that numerical models are different from field or laboratory experiments in that they are subject to 'closure'. Models require assumptions to be made about what does and does not matter. For instance, climate models vary in the extent to which they include feedbacks associated with core components of the earth system (e.g. vegetation feedbacks). As noted above, they also involve simplifications of the included processes. Also in climate models, the effects of clouds upon the energy and water balance of the atmosphere have to be dealt with using simplifications as the spatial scale of cloud physics is typically smaller than that at which models are operating. Thus, models are closed representations of reality and a model is partly dependent upon how the modeller sets up the model (i.e. defines the closure), that is how they perceive the system that they are modelling and how they develop a conceptual model on this basis. This is similar to the way in which the experimental design associated with fieldwork (e.g. choice of field site, measurement methods) and laboratory work (e.g. definition of controlled experimental conditions) partly determines the results that are obtained (Lane, 2001).

The second important role of a model relates to system understanding through sensitivity analysis. This is where the model is used in the same way as a laboratory experiment, and it may even be referred to as 'numerical experiments': model predictions derived from models with different structures, process representations or parameter values are compared in order to understand which processes and parameters have most effect upon the system. This tells us which aspects of a system are most important for further research, for fieldwork, or for laboratory analysis. In a three-dimensional numerical model of flow over a rough gravel bed, model predictions were highly insensitive to changes in bed roughness, but highly sensitive to bed surface structure (Lane et al., 2002; Figure 23.7). Bed roughness affects the momen-

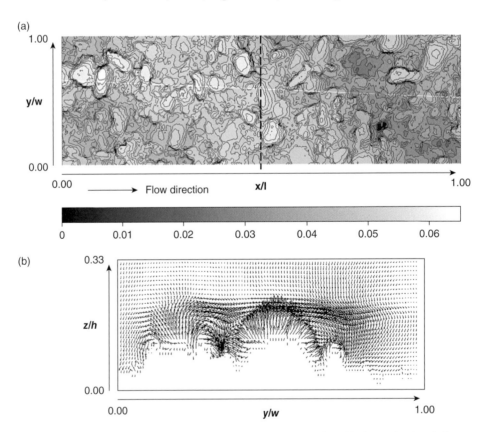

Figure 23.7 Model predictions of flow over a rough gravel-bed surface: (a) shows the digital elevation model and (b) shows velocity vectors in a cross-section taken along the dashed line in (a) (i.e. vectors show the cross-stream and vertical components of velocity); w is width, h is depth and l is channel length; y is position along the section, x is position downstream and z is position in the vertical

Source: Based on Lane et al. (2002)

tum balance at the bed. Bed-surface structure affects both the momentum balance and mass conservation (i.e. flow blockage). The poor sensitivity to bed roughness demonstrates that the effect of bed-surface structure will not be properly represented through a bed-roughness term, and requires us to reconsider how bed-surface structure is represented in models of flow over gravelly surfaces. The problem with this type of sensitivity analysis is similar to the whole system problem. Do the results hold in a more general sense, or are they a product of the nature of the model as it has been developed and as it is being applied? Unfortunately, this is not a simple question to answer. In the gravel-bed river-flow model, the treatment of turbulence is known to be inappropriate for this kind of flow. Turbulence has a major effect upon momentum transfer at the bed. Thus, whilst the conclusion that blockage affects associated with bed-surface structure matter does hold, the extent to which this is the case must remain uncertain as turbulence effects are not being included. This means that you have to take care in inferring real world behaviour from the results of a numerical experiment.

The third role of a model is the one that is of most interest to environmental managers: the prediction of system properties either at future time periods or for locations where reliable measurements cannot be made. This rationale allows us to break out of the 'hunny' jar and heffalump problems described in the introduction by using technology (a properly validated numerical model) to try to make statements about what might happen before they actually happen. We live with this type of modelling on a day-to-day basis. For instance, short-term weather forecasting is based upon regional predictive models of weather systems. Flood warning systems are commonly driven by a predictive model that indicates where water is expected to reach for a range of river discharges, and which can be used to issue warnings to potentially affected properties when a particular discharge occurs. The main issue here is the extent to which trust can be placed in a model's predictions as a result of uncertainties in model predictions. These uncertainties can be classified into seven broad headings (Table 23.1).

Looking across Table 23.1, it is possible to identify three core issues that we must evaluate when considering models as a strategy within physical geography. First, uncertainty will be an endemic characteristic of all modelling efforts. Attempts to reduce uncertainty are tempting, through the adoption of more sophisticated modelling approaches, or the improved specification of boundary conditions. Unfortunately, research has shown that attempts to deal with uncertainty tend to introduce yet more uncertainty, in the way that Wynne (1992) conceptualizes the dynamics of science as being the generator of uncertainty rather than the eliminator of it. For instance, Lane and Richards (1998) argue that three-dimensional models of river flow are required in tributary junctions in order to represent properly the effects of secondary circulation upon flow processes. Lane and Richards (2001), having used a three-dimensional model for this purpose, demonstrate the significant new uncertainties that have emerged from: (i) the difficulty of specifying inlet conditions in each of the tributaries in three dimensions; (ii) problems of designing a numerical mesh that provides a stable numerical solution that minimizes numerical diffusion; (iii) uncertainties over the performance of a roughness treatment in three dimensions; and (iv) problems with finding an appropriate turbulence model. The continual creation of uncertainty keeps the science of this sort of modelling alive. However, when models need to be put to practical use for forecasting or process understanding, obvious questions

Table 23.1 Model uncertainties

Type of uncertainty	Explanation
Closure uncertainties	This relates to uncertainties that arise because certain processes have been included or excluded during model development. A common strategy to deal with this is to include processes, test their effects, and to ignore or simplify them if they seem to have only a small effect upon model predictions. The problem with this is that the effect of a process can often change as the effects of other processes (i.e. system state) change. Thus, it is impossible to be certain that a given process will always be unimportant. Note that problems of closure are not just associated with models, but are an inherent characteristic of all science (see Lane, 2001). In general, closure uncertainties are easy to spot in any model application. We often forget that the same criticism of closure applies equally to almost all other aspects of scientific method.
Structural uncertainties	These arise because of uncertainties in the way in which the model is conceptualized in terms of links between components. A good example of this is whether or not a component is an active or a passive component of a model. For instance, in a global climate model, the ocean may be assumed to be a passive contributor to atmospheric processes: it acts as a heat and moisture source, but does not itself respond to atmospheric processes. This type of structural uncertainty will define limits to model applicability. For instance, the ocean has a high specific heat capacity, and so will respond slowly to atmospheric processes. Thus, specification of the ocean as a passive source of heat and moisture is acceptable if the model is being applied over a timescale (e.g. days) when the ocean can assume to be in steady state. As with closure uncertainties, structural uncertainties limit a model's applicability and remind us of the importance of evaluating a model in relation to the use to which it will be put. Any model can be criticized on the grounds of structural uncertainties without a purpose-specific evaluation.
Solution uncertainties	These uncertainties arise because most numerical models are approximate rather than exact solutions of the governing equations. Commonly, a numerical solution involves an initial guess, operation of the model upon that initial guess, and its subsequent correction. This continues until operation of the model on the previous corrected guess does not alter by the time of the next guess. This process can result in severe numerical instability in some situations, which is normally easy to detect. However, more subtle consequences, such as numerical diffusion associated with the actual operation of the solver, can be more difficult to detect. Following guidelines in relation to good practice can help in this respect.
Process uncertainties	These arise where there is poor knowledge over the exact form of the process representation used within a model. A good example of this is the treatment of turbulence in models of river flow. Turbulence can have an important effect upon flow processes as it extracts momentum from larger scales of flow and dissipates it at smaller scales. Most river models average turbulence out of the solution, but then have to model the effects of turbulence upon time-averaged flow properties. Turbulence models vary from the simple to the highly complex and it can be shown that different turbulence treatments are more or less suitable to different model applications. Thus, the process representation in a model, as with structural aspects of a model, needs careful evaluation with reference to the specific application for which the model is being used.

Type of uncertainty	Explanation
Parameter uncertainties	These arise when the right form of the process representation is used but there are uncertainties over the value of parameters that define relationships within the model. Particular problems can arise when parameters have a poor meaning in relation to measurement. This can arise in two ways. First, some parameters are difficult to measure in the field as they have no simple field equivalent. Second, during model optimization, parameters can acquire values that minimize a particular objective function, but which are different to the value that they actually take on the basis of field measurements. A good example of this is the bed roughness parameter used in one-dimensional flood routing models. It is common to have to increase this quite significantly at tributary junctions, to values much greater than might be suggested by the shape of the river or the bed grain-size. In this case, there is a good justification for it, as one-dimensional models represent not only bed roughness effects but also two- and three-dimensional flow processes and turbulence through the friction equation. Roughness is therefore representing the effects of these other processes in order to achieve what Beven (1989) labels 'the right results but for the wrong reasons'.
Initialization uncertainties	These are associated with the initial conditions required for the model to operate. They might include the geometry of the problem (e.g. the morphology of the river and floodplain system that is being used to drive the model) or boundary conditions (e.g. the flux of nutrients to a lake in a eutrophication model).
Validation uncertainties	Given the above six uncertainties, a model is unlikely to reproduce reality exactly, and validation is required to assess the extent to which there is a reasonable level of agreement. However, validation data themselves have an uncertainty attached to them. This is not simply due to possible measurement error, but also when the nature of model predictions (their spatial and temporal scale, the parameter being predicted) differs from the nature of a measurement. A commonly cited example of this is validation of predictions of soil moisture status in hillslope hydrological models, when point measurements of soil moisture status (in space and time) are used to validate areally integrated predictions. This creates problems for modelling in two senses. First, apparent model error may actually be validation data error. Second, if validation data are then used for model optimization (and note that data used for model optimization should not then be used for validation), uncertainty will be introduced into model predictions as the data that the model are optimized to may be incorrect. Following Beven (1989) this means that we may get the wrong results for the wrong reasons.

emerge as to whether or not numerical models are little more than computationally intensive crystal balls.

Second, Table 23.1 emphasizes that when models are to be used in an applied sense, it is vital that the associated uncertainty is communicated along with those model predictions. This is necessary to avoid false faith being placed in model predictions. This requires methods for: (i) determining what the uncertainty is; (ii) representing the uncertainty in a manner that has meaning to the user of a model's predictions (as well as the modeller themselves); (iii) communicating uncertainty (e.g. Stephens et al., 2012) and (iv) persuading those that use model predictions to accept both uncertainties in model predictions and uncertainties in determining the uncertainty of model predictions (Lemos and Rood, 2010). Before commenting on these, it is worth remembering the four-fold division introduced by Wynne (1992) of different types of uncertainty. Wynne argues that all uncertainties can be given one of four labels: (a) risks, or quantifiable uncertainties; (b) unquantifiable uncertainties; (c) uncertainties that we are ignorant about, but which we may find out about through further experience or investigation; and (d) indeterminacies, or uncertainties that cannot be determined through any form of investigation, prior to them happening. Thus, the determination of uncertainty is largely about the determination of risk, or quantifiable uncertainty, and the representation of that uncertainty accordingly. However, as uncertainty is communicated, it is vital to include unquantifiable uncertainty, ignorance and indeterminacy. As an example, consider an estimation of the uncertainty in patterns of floodplain inundation. Sensitivity analysis can be extended to uncertainty analysis through assessing the implications of model uncertainties (e.g. Table 23.1) for model predictions. As Binley et al. (1991) have demonstrated, we need formal methods for doing this (e.g. General Least-squares Uncertainty Estimation, or GLUE). This is not a straightforward task as a result of the large number of parameters that we are typically uncertain about. Nonetheless, uncertainty bands can be determined in the form of probabilities of floodplain inundation for events with different return periods (e.g. Romanowicz et al., 1996). These provide a partial account of uncertainty. Actual predictions of flood inundation will be affected by unquantifiable uncertainties (e.g. the difficulties of determining likely run-off generation given a particular combination of antecedent moisture conditions and rainfall patterns), ignorance (e.g. incorrect assumptions built into the model, such as that extreme flood events only occur when a catchment is fully saturated) and indeterminacy (e.g. aspects of floodplain management, such as culvert maintenance and repair, that cannot be determined in any realistic way before an event occurs). Unfortunately, we are not good at communicating these uncertainties *and* accepting them as a normal aspect of the science of environmental management. The latter is compounded by the problem that uncertainty is measured on a continuous probability scale whereas decisions over floodplain management have to be made on a discrete scale involving action (e.g. improve floodplain defences, refuse flood insurance protection) or no action (do not improve floodplain defences, allow flood insurance protection).

Third, we need to be sensitive to those who are the ultimate judge of model predictions and for whom the inevitable errors in what a model predicts translate into real day-to-day impacts. One of the best examples of this are the online flood maps of the Environment Agency of England and Wales (https://www.gov.uk/check-if-youre-at-risk-of-flooding) which are accessible to anyone and which identify 'indicative floodplains', which are overlain on basic Ordnance Survey maps, allowing anyone to find out whether or not they live in a floodplain. However, these maps are bound up with the process of modelling and the way in which flooding is represented in those models in complex ways. The models follow standard methodologies for predicting flood inundation using one-dimensional and two-dimensional models. Such models are commonly evaluated using contingency tables (e.g. Table 23.2) that evaluate, predict, and model inundation for a particular flood event. In Table 23.2, we would hope that the diagonal cells marked with the * would add up to 100 per cent. They never do, and values of 70 to 80 percent are commonly considered acceptable. Thus, a model which is an acceptable model overall may actually have specific locations within the model that are wrong. However, it is commonly the specific locations of the models that are used in decision-making: solicitors doing searches for potential buyers commonly report on whether or not a property is in the indicative floodplain; insurance companies may change premiums to reflect a possible flood risk. Thus, model predictions at the scale of the individual property, a scale at which the models are known to be wrong, may have a material impact upon the value of that property and those who live in it. Second, the model predictions actually indicate the area that is going to be flooded during an event with a particular return period (with and without defences). This is not the same as somewhere that *could* be flooded.

Both of these issues have resulted in the flood maps, and thus the models that underpin them, becoming contested (e.g. Porter and Demeritt, 2012). This dispute is aided by both: (i) the growing ease with which model predictions, especially those with a spatial content, can be disseminated using online methodologies and in an era of growing freedom of information; and (ii) the growth within society more generally of distrust of both experts and expertise. Model uncertainties are starting to take on a material form, giving those with expertise that is traditionally excluded from the modelling process the means of questioning and interrogating the model predictions and models to which they are increasingly subject. The externalization of model predictions is increasingly lending them up to different kinds of scrutiny and will change the way in which we use models as part of the process of making difficult decisions (Lane, 2014).

The final role for modelling is for simulation, in which we ask 'what if?' questions about environmental behaviour. In many senses, this is the combination of the above three reasons for using numerical models: use a representation of the whole system (reason 1), in which we vary system parameters (sensitivity analysis, reason 2), to make predictions (reason 3) that allow us to improve environmental management. The need to do this was emphasized in the introduction: awaiting confirmed observational evidence from a noisy system means that significant damage may have occurred before action can clearly be justified. Evidence suggests that despite this being one of the most

Table 23.2 An example contingency table for flood-risk assessment

	Predicted flooded	Predicted as not-flooded
Observed flooded	43%*	12%
Observed not-flooded	13%	32%*

powerful reasons to use models in environmental policy development and decision-making, it remains remarkably difficult for evidence from models to lead to major policy shifts (e.g. Weaver et al., 2013).

23.4 Conclusion

Harré (1981) identified three different roles for scientific investigation: (i) as formal aspects of methodological investigation; (ii) to develop the content of a theory; and (iii) in the development of technique. Table 23.3 applies these to numerical models in an attempt to demonstrate what models can and cannot do. In practice, this sort of classification should be treated as a fuzzy one, and this table is included as a basis for discussion, perhaps in a tutorial. For instance, following A1 in Table 23.3, models may be used to explore the characteristics of a naturally occurring process, but the extent to which this can be done will depend upon the confidence that can be placed in a model and, as noted above, there will always be uncertainty in the 'natural' characteristics that the results from a model may suggest. Similarly, strictly, models may not be used to provide negative results (A6), as we never know whether or not the null result is due to the model as constructed or a real characteristics of the system that the model is being used to represent. However, a model may provide an indication of a null result that causes us to look elsewhere in order to find additional supporting evidence. This discussion of both A1 and A6 emphasizes a key theme in the use of models: a model's effectiveness largely depends upon the ability of a modeller to engage with a broad spectrum of methods, including those that are field- and laboratory-based. This is where *Winnie-the-Pooh* again provides us with key guidance. During the flood, Pooh bear finds a 'missage' in a bottle. Being unable to read, he has to get to Owl. Surrounded with water he uses a classic piece of reasoning (i.e. a model): if a 'missage' in a bottle can float, then a bear in a 'hunny jar' can float. Having got the skeleton for his model, he develops his model by trying various positions on the jar, until he finds one that is stable. Here we see the crucial iteration between model development and empirical observation as Pooh bear tries to get his model just right. He now has an optimized model, which he proceeds to validate by successfully floating off to Owl. Finally, he demonstrates the transferability of his model by going to rescue Piglet, with Christopher Robin, in an upside-down umbrella. This requires much less development and represents the common transition of a model, through time, from being a developmental piece of science to a practical part of technology. Unfortunately, Pooh bear gives us little guidance as to when a model as a piece of science is sufficient for us to make it a practical piece of technology. However much we, as modellers, might debate the philosophical and methodological aspects of what we do, we cannot escape the fact that the believability of a model is no longer solely the domain of the academic or the policy-maker.

There is a bigger picture here, in relation to the role that we expect of models more generally. Our collective understanding of the past is that it is full of surprises, things we wish to avoid. Models become a means of making the future predictable. Futures that are predictable can be avoided and so controlled. It is not surprising

Table 23.3 What models can and cannot do

Models as formal aspects of method		Can a numerical model help?
A1	To explore the characteristics of a naturally occurring process	☑ The classic role of numerical simulation – asking 'what if?' questions
A2	To decide between rival hypotheses	☑ Used as part of sensitivity analysis in which rival hypotheses are tested using a model
A3	To find the form of a law inductively	☒ Generally not possible as laws are required to get a model to work, so generating laws from model predictions runs the risk of circular argument (but what is a law?)
A4	As models to simulate an otherwise unresearchable process	☑ A critical function of models and perhaps where they are most powerful but also most problematic (are a model's results a reflection of reality or a reflection of the way the model has been set up?)
A5	To exploit an accidental occurrence	☑ But more to understand possible accidental occurrences that might occur (e.g. what if there was a dam break?)
A6	To provide negative or null results	☒ A major problem for models – is the negative or null result a true property of the system, or simply because a model has been used?
Models in the development of the content of a theory		
B1	Through finding the hidden mechanism of a known effect	☒ Can't find hidden things if they are not included in the model
B2	By providing existence proofs	☑ Models can provide corroboration of things observed using other methods
B3	Through the decomposition of an apparently simple phenomena	☒ Not possible
B4	Through demonstration of underlying unity within apparent variety	☑ Identification of general patterns
In the development of technique		
C1	By developing accuracy and care in manipulation	☒ Not possible
C2	Demonstrating the power and versatility of apparatus	☒ Not possible

Source: After Harré (1981)

that models, then, increasingly permeate our day-to-day existence in ways that we rarely appreciate. However, there are occasions when our reliance on models emerges, often during times when they unsettle, undermine or simply fail to deliver the secure and predictable futures that we wish. During such occasions, not only do models come under scrutiny, but so do the very knowledge principles and practices of which they form a part, as we struggle to come to terms with the uncertainty that is endemic to any understanding of the future. This makes it very exciting to be a modeller in the twenty-first century, as it is one of those areas of scientific endeavour that increasingly needs the kind of interdisciplinary working that is so characteristic of geographical research.

SUMMARY

- Numerical models are used by physical geographers for dealing with situations where the researcher is 'remote' from what they are studying: where the events that are of interest have occurred in the past (e.g. reconstruction of past climates); may occur in the future (e.g. patterns of inundation associated with future flood events); or where they are occurring now, but cannot be measured or studied using other methods.
- Without a conceptual model of the system that is being modelled, it is impossible to develop either an empirical or a physically-based mathematical model. Conceptual models involve a statement of the basic interactions between the components of a system.
- The way that the environment is modelled needs careful consideration, so that feedbacks are properly incorporated.
- Empirical approaches have a physical basis in so far as they have some sort of conceptual model that justifies the form of the relationship developed. Physically-based numerical models take this one stage further by using the conceptual model to define links between fundamental physical, chemical and occasionally biological principles, which are then represented mathematically in computer code.
- Model assessment involves two important stages: verification and validation. Verification is the process by which the model is checked to make sure that it is solving the equations correctly. Validation is the process by which a model is compared with reality.
- Modelling has an important role in understanding the operation of environmental systems. This is achieved by using models to integrate multiple processes over many time-scales; to undertake experiments of sensitivity; to predict outcomes; and through simulation to ask 'what if?'-type questions.
- Models and the modelling process demonstrate many of the fundamental aspects of scientific research. The transition from scientific model to a useable technology for environmental management requires that it is transferrable between case studies, and reflects the need for qualitative as well as quantitative judgements.
- Models are increasingly subject to scrutiny, which is changing the way that they are used in decision-making.

Further Reading

- Beven (1989) is a very useful introduction to critical thinking in relation to numerical models. Beven introduces a set of ideas that challenged an emerging paradigm regarding the power of numerical models and provides a critical framework for evaluating the role of modelling in this case for the hydrological sciences, but with implications for the modelling of environmental systems more generally.

- Rather like the Beven paper, Anderson and Bates (2001) is useful because it brings together a very wide range of theoretical and methodological perspectives in relation to numerical modelling, albeit around the hydrological sciences.

- Kirkby et al. (1992) is a good general introduction to numerical modelling in physical geography. It is especially strong on the way modelling is done and has some easy but useful examples of models that can be coded to illustrate principles of model building.

- Beven (2000) is definitely the best book around on modelling in hydrology in general, with excellent coverage of material specific to hydrology but illustrating environmental modelling in general. Likewise, Huggett (1993) is very effective on conceptual modelling across the environmental spectrum and how to apply conceptual models using a range of modelling techniques.

- Jakeman et al. (1993) is useful for understanding modelling over a range of spatial scales and especially at the global scale.

Note: Full details of the above can be found in the references list below.

References

Anderson, M.G. and Bates, P.D. (eds) (2001) *Model Validation: Perspectives in Hydrological Science.* Chichester: John Wiley & Sons.

Beven, K.J. (1989) 'Changing ideas in hydrology: The case of physically-based models', *Journal of Hydrology*, 105: 157–72.

Beven, K.J. (2000) *Rainfall-runoff Modelling: The Primer.* Chichester: Wiley.

Binley, A.M., Beven, K.J., Calver, A. and Watts, L.G. (1991) 'Changing responses in hydrology: Assessing the uncertainty in physically-based model predictions', *Water Resources Research*, 27: 1253–61.

Broecker, W.S. and Denton, G.H. (1990) 'What drives glacial cycles?', *Scientific American*, 262: 42–50.

Cameron, D., Kneale, P. and See, L. (2002) 'An evaluation of a traditional and a neural net modelling approach to flood forecasting for an upland catchment', *Hydrological Processes*, 16: 1033–46.

Colinvaux, P.A., De Oliveira, P.E. and Bush, M.B. (2000) 'Amazonian and neotropical plant communities on glacial time-scales: The failure of the aridity and refuge hypotheses', *Quaternary Science Reviews*, 19: 141–69.

Cullen, P. and Forgsberg, C. (1988) 'Experiences with reducing point sources of phosphorous to lakes', *Hydrobiologia*, 170: 321–36.

Farman, J.C., Gardiner, B.G. and Shanklin, J.D. (1985) 'Large losses of total ozone in Antarctica reveal seasonal CLOx/Nox interaction', *Nature*, 315: 207–10.

Harré, R. (1981) *Great Scientific Experiments: Twenty Experiments that Changed Our View of the World.* London: Phaidon Press Limited.

Houghton, J.T., Jenkins, G.J. and Ephramus, J.J. (1990) (eds) *Climate Change: The IPCC Scientific Assessment.* Cambridge: Cambridge University Press.

Huggett, R.J. (1993) *Modelling the Human Impact on Nature.* Oxford: Oxford University Press.

Imbrie, J. and Imbrie, K.P. (1979) *Ice Ages: Solving the Mystery.* London: Macmillan.

Jakeman, A.J., Beck, M.B. and McAleer, M.J. (1993) *Modelling Change in Environmental Systems.* Chichester: John Wiley & Sons.

Kilinc, S. and Moss, B. (2002) 'Whitemere, a lake that defies some conventions about nutrients', *Freshwater Biology*, 47: 207–18.

Kirkby, M.J., Naden, P.S., Burt, T.P. and Butcher, D.P. (1992) *Computer Simulation in Physical Geography.* Chichester: John Wiley & Sons.

Lane, S.N. (2001) 'Constructive comments on D. Massey Space-time, "science" and the relationship between physical geography and human geography,' *Transactions of the Institute of British Geographers*, NS26: 243–56.

Lane, S.N. (2003) 'Environmental modelling', Chapter 12 in A. Rogers and H. Viles (eds) *The Student's Companion to Geography.* Oxford: Blackwell.

Lane, S.N. (2012) 'Making mathematical models perform in geographical space(s)', in J. Agnew and D. Livingstone (eds) *Handbook of Geographical Knowledge.* London: Sage, pp. 228–46.

Lane, S.N. (2014) 'Acting, predicting and intervening in a socio-hydrological world', *Hydrology and Earth System Sciences*, 18: 927–52.

Lane, S.N. and Richards, K.S. (1998) 'Two-dimensional modelling of flow processes in a multi-thread channel', *Hydrological Processes*, 12: 1279–98.

Lane, S.N. and Richards, K.S. (2001) 'The "validation" of hydrodynamic models: Some critical perspectives', in P.D. Bates and M.G. Anderson (eds) *Model Validation: Perspectives in Hydrological Science.* Chichester: John Wiley & Sons. pp. 413–38.

Lane, S.N., Hardy, R.J., Elliott, L. and Ingham, D.B. (2002) 'High resolution numerical modelling of three-dimensional flows over complex river bed topography', *Hydrological Processes*, 16: 2261–72.

Lau, S.S.S. (2000) 'Statistical and dynamical systems investigation of eutrophication processes in shallow lake ecosystems', PhD thesis, University of Cambridge.

Lau, S.S.S. and Lane, S.N. (2002) 'Biological and chemical factors influencing shallow lake eutrophication: A long-term study', *Science of the Total Environment*, 288: 167–81.

Lemos, M. C. and Rood, R. B. (2010) 'Climate projections and their impact on policy and practice', *WIREs Climate Change*, 1: 670–82. doi: 10.1002/wcc.71

Licznar, P. and Nearing M.A. (2003) 'Artificial neural networks of soil erosion and runoff prediction at the plot scale', *Catena*, 51: 89–114.

Michaels, P.J. (1992) *Sound and Fury: The Science and Politics of Global Warming*. Washington, DC: Cato Institute.

Molina, M.J. and Rowland, F.S. (1974) 'Stratospheric sink for chlorofluoromethanes: Chlorine atom-catalysed destruction of ozone', *Nature*, 249: 810–2.

Odoni, N. and Lane, S.N. (2010) 'Knowledge-theoretic models in hydrology', *Progress in Physical Geography*, 34: 151–71.

Odoni, N. and Lane, S.N. (2012) 'The significance of models in Geomorphology: From concepts to experiments', in K.J. Gregory and A.S. Goudie (eds) *Handbook of Geomorphological Knowledge*. London: Sage. pp. 154–74.

Oreskes, N., Shrader-Frechette, K. and Belitz, K. (1994), 'Verification, validation, and confirmation of numerical models in the Earth Sciences', *Science*, 263: 641–6.

Pentecost, A. (1999) *Analysing Environmental Data*. Harlow: Longman.

Porter, J. and Demeritt, D. (2012) 'Flood-risk management, mapping, and planning: The institutional politics of decision support in England', *Environment and Planning A*, 44: 2359–78

Romanowicz, R., Bevan, K.J. and Tawn, J. (1996) 'Bayesian calibration of flood inundation models', in M.G. Anderson, D.E. Walling and P.D. Bates (eds) *Floodplain Processes*. Chichester: John Wiley & Sons. pp. 333–60.

Schindler, D.W. (1977) 'Evolution of phosphorous limitation in lakes', *Science*, 195: 260–2.

Schumm, S.A. and Lichty, R.W. (1965) 'Time, space and causality in geomorphology', *American Journal of Science*, 263: 110–19.

Spaulding, W.G. (1991) 'Pluvial climatic episodes in North America and North Africa – types and correlations with global climate', *Palaeogeography, Palaeoclimatology and Palaeoecology*, 84: 217–27.

Stephens, E. M., Edwards, T. L. and Demeritt, D. (2012) 'Communicating probabilistic information from climate model ensembles – lessons from numerical weather prediction', *WIREs Climate Change*, 3: 409–26. doi: 10.1002/wcc.187

Tarras-Wahlberg, N.H. and Lane, S.N. (2003) 'Suspended sediment yield and metal contamination in a river catchment affected by El Niño events and gold mining activities: The Puyango river basin, southern Ecuador', *Hydrological Processes*, 17: 3101–23.

Taylor, J.R. (1997) *An Introduction to Error Analysis: The Study of Uncertainties in Physical Measurements* (2nd edition). Sausalito, CA: University Science Books.

Weaver, C. P., Lempert, R. J., Brown, C., Hall, J. A., Revell, D. and Sarewitz, D. (2013) 'Improving the contribution of climate model information to decision making: The value and demands of robust decision frameworks', *WIREs Climate Change*, 4: 39–60. doi: 10.1002/wcc.202

Wynne, B. (1992) 'Uncertainty and environmental learning: Reconceiving science and policy in the preventive paradigm', *Global Environmental Change*, 2: 111–27.

ON THE COMPANION WEBSITE...

Visit **https://study.sagepub.com/keymethods3e** for author videos, chapter exercises, resources and links, plus **free** access to the following recommended articles:

1. **Peel, M.C. and Bloschl, G. (2011) 'Hydrological modelling in a changing world', *Progress in Physical Geography*, 35: 249–61.**

This paper illustrates the challenges of applying models to predict the future and how modelling methods need to be adapted to do this.

2. **Hessl, A.E. (2011) 'Pathways for climate change effects on fire: Models, data, and uncertainties', *Progress in Physical Geography*, 35 (3): 393–407.**

This paper provides a case-study of how mathematical models can be used to inform a particular problem: wildfires.

3. **Odoni, N.A. and Lane, S.N. (2010) 'Knowledge-theoretic models in hydrology', *Progress in Physical Geography*, 34 (2): 151–71.**

This paper thinks through the relationship between models and data/observations and so reflects upon the status of geographical knowledge produced by modelling.

24 Simulation and Reduced Complexity Models

James D.A. Millington

SYNOPSIS

The process of simulation modelling iterates through system conceptualization, data collection, model construction, evaluation, and model use, demanding continual reflection on the part of the modeller. Simulation models, a product of the simulation modelling process, are *representations of reality* that couple theory with data to dynamically represent processes, interactions and feedbacks (potentially across space). For example, spatial simulation models allow examination of feedbacks between spatial patterns and processes. Advances in computing mean that many modelling environments and programming languages/libraries are now available which enable relatively quick production of simple simulation models for investigating geographical systems. This chapter includes examples of how to use these readily available tools, providing an introduction to simulation modelling for geographers with a focus on Reduced Complexity and Agent-Based Models (ABM). The importance of *conceptualization* of key processes to represent in a model, and careful consideration for how they should be *operationalized in code* (via equations and/or rules), is also discussed.

This chapter is organized into the following sections:

- Introduction: Conceptualizing the geographical world
- Reduced complexity models in physical geography
- Simulating pattern-process feedbacks
- Developing your simulation model

24.1 Introduction: Conceptualizing the Geographical World

You may have read this chapter before, but repetition with new initial conditions is often a useful way to explore and understand a concept or system (although deciding when to stop and do something else is also important!). In trying to understand the geographical world, we often resort to simplified representations of it that we can experiment with. These simplified representations are often called models, and include the conceptual models we carry around in our brains and that enable us to navigate our way through the world without needing to observe and comprehend precisely every detail of what is going on around us. For example, imagine you are walking along a busy city street. Your understanding about how humans move and interact with one another (based on previous observations of other people's movements), combined with observations of the locations and rates of movement of the people on the particular street you are on, provides a mental model that allows your brain to anticipate the dynamics of the crowd and help you to avoid collisions. In a similar way, more formal simplified representations of the world can help us to understand and anticipate dynamics of physical systems over geographical scales that are harder to comprehend and observe given human sensory and cognitive capacities. For instance, in computer-simulation modelling, a conceptualization of target phenomena – such as erosion and deposition in a landscape over several hundred years – is specified in computer code that is then executed in a computer (i.e. simulated) to produce output that allows the logical consequences of the conceptualization to be examined. Whereas walking down the street your brain combines your conceptual model of human movement (e.g. walking humans move at about 5 km/hr, on the horizontal plane and not vertically, etc.) with observations of individual people to anticipate the dynamics of a particular crowd, a computer simulation of landscape evolution might combine a mathematical model of erosion and deposition with input data for a particular landscape to anticipate landscape dynamics. Beyond the difference in scales of processes represented by each (seconds and minutes walking down the street, years and decades in the landscape model), a key difference between our brains using mental models and computers simulating formal models is that we can use the computer to simulate many instances of the same system to explore how different input data and other variations produce different outcomes. In the real world we only get one chance to walk down a given street at any given point in history (and who knows who we'll bump into), but by representing pedestrians on the street in a computer simulation (e.g. Torrens, 2012) we might explore, for example, how different initial conditions (e.g. people starting in different places or moving at different speeds) result in different flows of pedestrians. Similarly, computer simulation model models of landscape evolution offer the chance to examine how different initial landscape states result in different patterns and trajectories of change (e.g. Wainwright, 2008). Asking such 'what if...?' questions with simulation models is not restricted only to initial conditions, but can also be used to investigate alternative representations of processes (what if some pedestrians *want* to bump into another? How do different vegetation species influence sediment deposition differently?), and through repeated use, experimentation and reflection, simulation modelling allows us to improve our understanding of the world being represented.

Computer simulation models and modelling differ from other forms of model and modelling discussed elsewhere in this book. Quantitative statistical models (see Chapters 19 and 28) make assumptions about the form of relationships between measured variables and then estimate parameters (numeric values) to describe those relationships. As such, although statistical models can quantify relationships in observed measurements they are heavily dependent on data which do not necessarily represent dynamics of change well. In contrast, analytic models are driven primarily by theory and represent system dynamics and change through mathematical equations or expressions (e.g. based on differential equations). Although these models allow exact solutions based on their formulation (and therefore clear links to underlying theory), those formulations (and therefore the theory represented) are heavily constrained by mathematical possibilities and therefore may not represent as many influences on dynamics as might be desired. The advent of digital computers and the growth of their processing power now provide an alternative to these two older approaches (which could feasibly be executed by hand and brain power alone). Computer simulations models are able to use statistical relationship or mathematical expressions to represent the world, but can also use other means that have less restrictive assumptions. For example, analytic and statistical models tend to simplify representation of the world by aggregating system elements (people, animals, grains of sand etc.), assuming individuals are identical and uniform (analytic) and/or by representing them with population-level summaries (statistical). In contrast, computer simulation models now allow representation of the dynamics of interactions between disaggregated, differentiated and discrete individual elements (Bithell et al., 2008). Such discrete-element techniques are increasingly being applied to better understand how broad, general patterns in environmental and social systems are generated as a result of specific, local interactions between individual elements. Although the appropriate use of models in geography has a long history, the advent of discrete-element approaches only brings yet more questions and debate about their possibilities and limits, how they should be used (e.g. for prediction or for understanding?), and how their output can and should be interpreted (Clifford, 2008; Millington et al., 2012).

Although the discrete representation of system elements is possible in a computer simulation model (but not required), another form of *discretization* is always necessary. Whereas relationships found or expressed by equations are usually continuous in nature (e.g. differential equations describe rates of change between theoretically infinitesimal intervals) the digital nature of computers means that models implemented using them require discretization *of time and space*. Space must be split into discrete areas that are assumed to be internally homogeneous in all characteristics (e.g. a lattice or grid of cells), and time must be split up into discrete steps ('timesteps', in which change is assumed to happen between but not within). Consequently, to represent change and dynamics over space and time, computer simulation requires that the same calculations be repeated over and over again for each of these discrete chunks of space and/or time in order that interactions and the change between them can be represented. Each repeated calculation is known as an *iteration* and demands that to represent a change in time and space, the code that the computer executes must be written so that it loops over on itself (Box 24.1).

Box 24.1   Loops and NetLogo

As we'll see throughout this chapter, 'loops' are important for representing change across space and through time in computer simulation. This chapter also challenges you to develop and use computer simulation models yourself to understand geographical systems, and provides example models for you to explore in the freely available modelling software NetLogo (Wilensky, 1999). As your first challenge, see if you can understand and implement the NetLogo code shown below. Download NetLogo (from http://ccl.northwestern.edu/netlogo/), install it on your computer (Windows or Mac), open the NetLogo programme and then type the code below into the Code tab (make sure it is *exactly* the same as shown below).

```
to go                                                  ;; line 1
    let population 4                                    ;; line 2
    let growth-rate 2                                   ;; line 3
    while [population < 1000]                           ;; line 4
    [                                                   ;; line 5
        set population (population * growth-rate)       ;; line 6
        print population                                ;; line 7
    ]                                                   ;; line 8
end                                                     ;; line 9
```

This code demonstrates how a 'while' loop can be used to simulate the exponential growth of a population that starts with a size of two individuals (where in the code is this initial value specified?). To execute the code in NetLogo (after you've typed it in), go to the Interface tab and type 'go' at the bottom of the screen where is says 'observer>', then hit enter. You should see the value of the population as it grows; each timestep is printed on the screen (if not check you have typed the code correctly; computers are stupid and will only do exactly what you tell them so make sure the code is correct!).

Here's what the computer does when it reads the code. First, the size of the population is checked (line 4). If less than 1000, the code between the second set of square brackets will be executed (lines 5–8). The size of the population is checked again (line 4). If the less than 1000, the code between the second set of square brackets will be executed again. This continues until the population is no longer less than 1000 (i.e. while the expression population < 1000 is true). In each iteration of the loop, population doubles (lines 3 and 6) and the current value of the population is printed (line 7). Inherently in this code we assume that each iteration of the loop is an advance in time of one unit (i.e. a single 'timestep'). How many timesteps are simulated (and why)? What is the last population value printed (and why)? Loops can work equally well across discrete areas of space and can also be nested within themselves so that they self-repeat (think what you would see if you held up two mirrors facing one another). This concept of nested loops – known as recursion – is often used in computer programming. Once you've tried the code above, have a run through the tutorials that come with NetLogo to learn more about how to use this flexible modelling environment (in Netlogo go to the Help menu then 'NetLogo Manual' then click Tutorial #1 on the left).

The idea of loops and looping seems to pervade simulation modelling and we'll see three types of loop in this chapter. First, are the loops of computer code that execute the same commands or calculations over and over again (Box 24.1) and which are well visualised through flow charts (as we will see below; see also Figure 24.4). Second are feedback loops in the real-world geographical systems that we might aim to represent with our models, which we will consider those further in Box 24.3. The third loop in computer simulation we will consider here is the process of modelling itself, from model conceptualization, through data collection, model construction, model evaluation and model use. We will look at how that first type of loop is related to the other two loops in the final section of the chapter which provides advice for how you might go about using and developing computer simulation models yourself. Before those sections however, we'll consider in more detail one particular type of simulation model used in physical geography.

24.2 Reduced Complexity Models in Physical Geography

Reduced complexity models (RCMs) are simulations models used by geomorphologists to represent processes and change at 'intermediate' scales, generally 1s – 100s km^2 in extent and 10s – 100s years in duration. Such scales are intermediate between finer scale representations of physically-based models to understand processes of fluvial sediment transport and deposition, and broader scale representation of landscape evolution models to understand longer term impacts of climate and land-use change on channel dynamics (Brasington and Richards, 2007). These intermediate scales are also those that are potentially most useful, for example, to river and environmental managers (Stott, 2010). All computer simulation modelling of physical systems must negotiate the trade-off between representational detail on the one hand and the computational resources needed to simulate that detail across time and space on the other. The trajectory of much simulation modelling in geomorphology and hydrology through the late twentieth century followed the reductionist-deterministic perspective towards evermore detailed and smaller-scale quantitative representation of physical processes using empirical or theoretical relationships in the form of mathematical equations (e.g. Navier–Stokes equations; see Reddy, 2011). In these physically-based models, increased representational detail demands increased numbers of equations, parameters and calculations, in turn requiring increased computational resources. If larger spatial extents or temporal durations are to be simulated (e.g. up to the 100s km^2 and 1000s of years of 'intermediate' scales), computational resources increase yet further. Despite continuing rapid increases in computational power in the late twentieth and early twenty-first centuries, the need and motivation to investigate processes and change at 'intermediate' scales required a new approach with a reduced level of representational detail commensurate with available resources. Hence, the growth in use of 'reduced complexity' models with their relatively more simple representation of the laws of physics. For example, by relaxing some assumptions of equations determining fluid flow, fluvial geomorphological RCMs are able to provide rapid solutions to calculations of water depth and velocities (Coulthard et al., 2007).

Simplified equations are only one aspect of RCMs however, the other being the adoption of lattice structures (e.g. grids) to represent space discretely (e.g. Figure 24.1). Using a lattice of discrete elements ('cells'), each of which corresponds to some area of land (which is assumed to be internally homogeneous), a more general or holistic representation of catchments and landscapes is possible. For example, while many hydrological and geomorphological modellers were developing ever-more refined equations to represent processes at finer scales, other geomorphologists interested in how landscapes are shaped over long time periods and large spatial extents (many thousands of years and hundreds of square kilometres) have used landscape evolution models (LEMs; see Tucker and Hancock, 2010). To investigate how the shape of the landscape changes – due to the entrainment, transport and deposition of sediment as water flows from points of higher elevation to lower – LEMs use rules to determine which route water (and sediment) takes as it moves from one cell to another adjacent cell. Similarly, aeolian geomorphologists have experimented with the use of 'cellular automata'-type approaches, in which each cell is assumed to be in a discrete state, to simulate the creation of dune formations (e.g. Baas, 2002). By exploiting the computational efficiencies of a lattice structure, albeit with a finer spatial resolution (i.e. cells represent smaller areas of land surface), and combining it with simpler versions of the equations used in fine-scale physically-based models or entirely different conceptual abstractions (e.g. 'slabs' of sand instead of individual grains; see the DECAL model below), RCMs are able to efficiently represent processes and change at intermediate scales. The lattice or cellular structure of these models means that they are often referred to as cellular models, while their simplified representation of physical processes means that they are often seen as useful for explanation or exploration (i.e. examining 'what if...' scenarios for management) rather than prediction (where physically-based models offer greater precision). In future, combinations of different types of models, exploiting the different strengths of the various approaches, may become more prevalent (Nicholas et al., 2012; see Box 24.2).

Possibly the 'original' RCM was developed to investigate river form and process feedbacks in fluvial geomorphology (Murray and Paola, 1994). One of the primary assumptions of this model, and an important example of representational simplification, is that water is assumed to flow only in a pre-defined downstream direction (i.e. no eddies back upstream will be represented). This simplification allows the iterative application of rules that determine water and sediment movement between cells, starting from the row of cells at the upstream of a simulated reach and finishing at the downstream end (before returning to the upstream end to begin the next model iteration). From this ground-breaking model, many other cellular models have been developed and applied across a range of fluvial environments (Coulthard et al., 2007, 2002; Van De Wiel et al., 2007; Nicholas et al., 2012). Cellular models have also been developed to examine dynamics of aeolian systems. For example, Baas (2002) outlines how a 3-D cellular model enables the representation of aeolian sand entrainment,

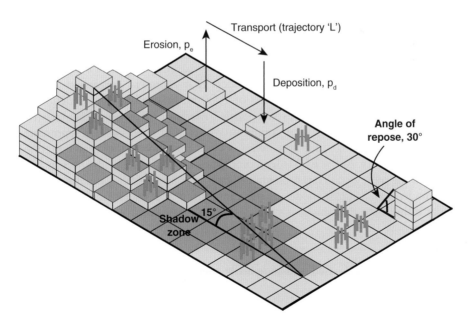

Figure 24.1   Illustration of the representation of simulated sand erosion, transport and deposition in the DECAL model (after Baas 2002, Figure 3). Flow charts describing the scheduling of the model can be found in Baas (2002) and Nield and Baas (2008).

transport and deposition processes by representing the topography (height) of sand dunes using stacked 'slabs' of sand on a lattice (Figure 24.1). Sand transport is simulated by moving slabs across the lattice between adjacent cells, with rules determining movement dependent on wind strength (with slabs in short stacks in the lee of taller stacks being less likely to be moved) and avalanching due to steep slope angles (to maintain an angle of repose at around 33°). Vegetation is represented dynamically in the model, with rules specifying locations and rates of growth dependent on dune height (i.e. number of sand slabs stacked in a cell). In turn, the presence and growth of vegetation in cells limits movement of sand slabs through those cells around that part of the lattice, and thereby influencing dune height and subsequent vegetation growth. Thus, feedbacks between vegetation and sand erosion, transport and deposition are represented and the dynamics of system interactions and dune morphology can be efficiently simulated. Nield and Baas (2008) used a model based on the structure described above, now named DECAL (Discrete ECogeomorphic Aeolian Landscape model), to show how multiple landforms (including Barchan and parabolic dunes) can be produced by the model which is built on simple local rules for movement of sand slabs between cells (aeolian entrainment, deposition, avalanches) and for vegetation growth. For example, if avalanching is only simulated in a given timestep between adjacent cells if the difference in height of slabs between those cells is greater than some predefined value. Thus, through the DECAL model, we see some of the primary characteristics of reduced complexity models: the ability to efficiently simulate the dynamics (due to feedbacks) of geomorphological systems by using rules of interactions between cells in a lattice structure.

Although the term 'Reduced Complexity Model' has been confined mainly to geomorphology, other sub-disciplines of physical geography have adopted similar cellular modelling approaches that enable simulation over larger spatial extents and temporal durations. In terrestrial plant ecology for example, models range from those that represent individual trees and their interactions with one another (e.g. competition for light and other resources) and their environment over areas on the order of 0.1 km^2 (Liu and Ashton, 1995; Pacala et al., 1996) to models that represent succession-disturbance dynamics across landscapes on the order of 1000 km^2 (Scheller and Mladenoff, 2007; Millington et al., 2009). At these broader scales, models usually use a cellular structure and represent plants at species- or community-level, and processes are represented by rules for vegetation change (due to succession or disturbance) rather than equations for growth and mortality. As for RCMs in geomorphology (Box 24.2), advances in computing are enabling more detailed representation over ever larger scales, and the primary use of the broader-scale cellular models is for exploration and understanding of landscape dynamics rather than prediction of system states at particular points in time or space.

Box 24.2   Reduced complexity models in geomorphology

The label 'Reduced Complexity Model' (RCM) is something of a paradox (Brasington and Richards, 2007). To see this we first need to understand that the term 'reduced' in this context is relative to the perspective that a 'standard' level of complexity in geomorphological models is the computational fluid dynamics (CFD) approach which uses physically-based partial differential equations (e.g. Navier-Stokes equations; see Lane et al., 1999). RCMs are simplified (or have 'reduced complexity') relative to CFD models in terms of the physical representation of fluid flows and processes of erosion, transport and deposition of solid material. The benefit of such simplification is that RCMs have vastly reduced computational demands compared to CFD models, enabling simulation of longer time durations and greater spatial extents that are more relevant to environmental management (i.e. 'intermediate' scales of 1–100s km^2 and 10s–100s years). However, the simplified representation of RCMs does not preclude the modelling issues of process conceptualization, parameterization, spatial and temporal resolution, and validation faced by developers of more complicated models. The reduced representation fidelity of RCMs can potentially lead to physically-inconsistent results and difficulties reproducing empirical observations (Coulthard et al., 2007). Future increases in computational power may mean that CFD approaches are possible at the scales of study that RCMs are now seen to be valuable for, but in the meantime it has been suggested that use of both in a hybrid, hierarchical manner might prove fruitful (Nicholas et al., 2012). The general view is that the promise of RCMs is in their use for explanation of system dynamics rather than prediction of static states: to understand system behaviour and change (Coulthard et al., 2007), to think differently and challenge assumptions about forms and processes (Odoni and Lane, 2011), and to investigate emergent properties of geomorphological systems (Brasington and Richards, 2007; Murray, 2007). Given that emergence lies at the heart of complexity theory (e.g. Harrison, 2001) this final point further compounds the paradoxical nature of the label 'reduced complexity model' – what is simple and useless for some, is complex and useful for others.

24.3 Simulating Pattern-Process Feedbacks

In geography we are often interested in areal differentiation and spatial patterns, such as how plant and animal species are differentially distributed across the earth's surface or how river channels vary in their planiform shape. Spatial patterns such as these are interesting in and of themselves and identifying and mapping spatial distributions and patterns was at the heart of early geography. Although still important, contemporary geography is often interested as much in the processes producing observed patterns as in the patterns themselves, sometimes simply to better understand, but at other times to be able to predict or make more informed management recommendations or decisions. Furthermore, as our understanding of the processes producing spatial patterns has improved, so we have often become interested in how the patterns themselves influence and modify or change the processes producing them. Consequently, an important reason simulation modelling has become popular in disciplines dealing with space (such as geography, landscape ecology, geomorphology, etc.) is because they provide a means to examine feedbacks between temporal processes and spatial patterns (see Box 24.3).

Box 24.3   Spatial simulation of pattern and process

Computer simulation models provide a means to represent spatial patterns and processes of change for investigation and experimentation: representation of processes may be modified, alternative measurements of patterns tested, and each done repeatedly. Providing numerous examples that

(Continued)

(Continued)

readers can explore for themselves using freely available software, O'Sullivan and Perry (2013a) present and discuss spatial simulation for investigating pattern and process in geographical and ecological systems. In particular, O'Sullivan and Perry (2013a) suggest that three processes underlie the majority of spatial simulation models currently available. The three spatial processes are:

- *Aggregation/Segregation:* Aggregation and segregation are two sides of the same coin, the former driven by the tendency of similar elements to group together in space, and the latter by the tendency of dissimilar elements to separate in space. If elements are unable to move in space, aggregation may occur as they change their attributes to become more similar to their local neighbours, thereby likely becoming more dissimilar to elements farther away. A primary means to represent this process is through iterative local averaging, in which values at a location are updated through time as the mean of spatially-local neighbours' attributes.
- *Mobile Entities and Random Walks:* A spatial walk is a succession of 'steps' (i.e. movements), each of which moves an entity from one location in space to another. Spatial walks might be random (the direction and length of each step are random) or influenced by attributes of the environment the entity is moving through, by attributes of the entity itself, or even by other entities moving through space. Examples include individual animals herding or flocking or pollutants moving through an environment. In some cases the 'walking' entity may change the environment or influence the walks of other entities and, in turn, this may reciprocally influence the entity's walk.
- *Spread:* Spread processes include diffusion, growth and percolation, and refer to the movement of material or phenomena in a more aggregated form than that considered in a spatial walk. For example, the diffusion of a gas through a vacuum from a point source results in the gas becoming evenly distributed across a space, but this process could also be thought of at an atomistic level as being the aggregate result of random walks of all the individual gas particles. Growth, in the context of spread, refers to expansion at a common boundary or front. A prime example is the spread of fire across a landscape, leaving burned land behind as it moves into unburned areas. Percolation shares many characteristics of diffusion and growth, but the emphasis here is on how the environment through which a material or phenomenon moves influences spread (rather than the characteristics of the material or phenomenon itself).

O'Sullivan and Perry (2013a) suggest that these processes might provide the 'building blocks' from which to start simulating a wide variety of spatial patterns. They provide example NetLogo models to demonstrate this which you can explore yourself (see the link on the companion website).

At its most fundamental, feedback is information about the state of an entity in a system communicated from that entity to another entity in the system. Commonly, feedback is understood to form reciprocal links between system entities, known as feedback loops. Thus, we meet the second kind of loop in simulation modelling. Positive feedback loops reinforce trajectories of change, whereas negative feedbacks act to stabilise and reduce change. For example, to illustrate how positive feedback loops can lead to areal growth or spatial clustering of vegetation, consider the relationship between vegetation and soil in semi-arid environments. In these environments, higher plant densities facilitate greater infiltration of water into the soil than at lower plant densities (HilleRisLambers et al., 2001). Where soil is devoid of plants, falling rain will barely infiltrate and will run off across the surface until it reaches a location where it can infiltrate more readily. Because plants facilitate infiltration this is likely to be near existing patches of vegetation, thus increasing soil moisture availability in the vicinity of the patch. Consequently, conditions for plant establishment are better near existing patches of vegetation, leading to the growth of the vegetation patch. This further raises plant density facilitating yet greater infiltration. In this feedback loop the information about soil (the original entity) causes a change in the extent of vegetation patches (the second entity), which in turn provides information which causes a change in the soil (original entity) at the periphery of the vegetation patch. This can be demonstrated by a simple simulation model (online Model 24.1 – download the code, examine it and test it yourself in NetLogo). The important point to highlight here is that changes are being caused at the periphery of an entity – soil conditions change at the periphery of the vegetation patch and in turn the periphery of the vegetation patch moves. It is because changes due to feedback do not occur at exactly the same point in space, but rather in rather spatially adjacent positions, that allows the areal growth of the vegetation patch.

Across larger spatial areas occupied by a single vegetation patch, HilleRisLambers et al. (2001) found that this positive feedback loop alone, without any spatial heterogeneity in the environment (e.g. slope), can lead to spatial patterns of vegetation patches alternating with bare soil. This can be demonstrated by a second simulation model (online Model 24.2 – try it yourself in NetLogo) in which raindrops fall randomly across an area and then run overland in random directions (due to lack of slope) in a random walk process (see Box 24.3). As the rainwater runs overland it infiltrates into soil, increasing the moisture available for plants to grow, with infiltration rates influenced again by plant density. Even with random rainfall, the modifications to infiltration rates due to plant density (as in the Model 24.1) result in clustering of plant patches across the simulated space. This second model is therefore similar is some ways to the DECAL model described above, in which sand is randomly entrained by wind, transported and deposited depending on the location of other sand (i.e. more likely to be deposited in the lee of higher piles of sand). Again, a simple simulation model (online Model 24.3) can demonstrate how randomness with some simple rules of interaction between landscape elements can lead to spatial pattern, because the processes implied by the interactions are dependent on existing spatial patterns.

In ecology, this concept of processes being shaped by their history (via spatial pattern) is known as the 'memory' of the process (Peterson, 2002). Feedback loops are created if the process has memory – the attributes of entities distributed across space record information about previous events (changes in state) caused by the process. For example, in landscapes that experience frequent fire, a mosaic of vegetation patches of varying age (i.e. due to varying time since last burn) can be produced (Figure 24.2). At landscape scales (e.g. 10–10,000 km^2) multiple factors influence how a fire spreads, including wind, physical relief and vegetation cover. Wind can be highly variable between fire events, whereas physical relief varies very little between individual events. Consequently, memory – and therefore feedbacks between spread and spatial pattern – have been conceptualized as being contained in vegetation flammability as a function of time since the last fire (Peterson, 2002). Peterson (2002) demonstrated this using a cellular model of vegetation growth and fire spread (online Model 24.4 is a NetLogo version of this model for you to try yourself). Debate continues about the importance of memory for wildfire in real world ecosystems, particularly in Mediterranean regions (Piñol et al., 2005; Keeley and Zedler, 2009). However, a positive feedback loop is known to exist between landscape homogeneity and fire spread (Loepfe et al., 2010). Homogeneous landscapes are characterized by fewer, larger patches of similar vegetation, meaning that fire spreads equally well through large areas of the landscape. A fire that spreads through a large area further homogenizes the landscape vegetation pattern (by burning the same area and likely more), in turn facilitating spatially consistent fire spread in future events. In contrast, landscapes with greater spatial heterogeneity in land uses (i.e. more, smaller patches) compose a negative feedback loop with fire spread because greater spatial variation in vegetation means fire spreads non-uniformly spatially, producing further heterogeneity.

There are, of course, many simplifications and assumptions in the models described above. However, as we saw for RCMs, such assumptions (allied with a cellular structure) enable rapid simulation and exploration of system dynamics. Furthermore, when used in an appropriate manner, such models can provide an experimental toolkit to examine hypotheses about which processes are most important for producing patterns observed in the real world. For example, the 'Pattern-Oriented Modelling' (POM) approach has been advocated for using simulation models to understand ecological systems (Grimm et al., 2005; Grimm and Railsback, 2012). The POM approach examines different representations of alternative hypothesized processes influencing individual elements in the models (e.g. sand or vegetation in the examples above) and how they combine to reproduce patterns observed in the real world at

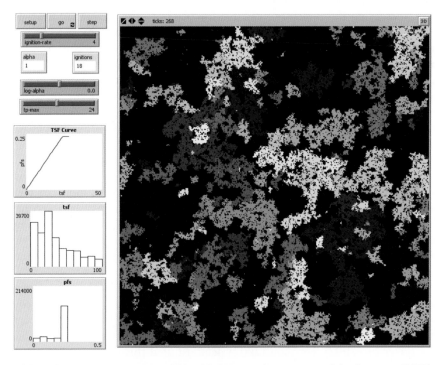

Figure 24.2   Spatial patterns produced by a NetLogo implementation of the Peterson (2002) model. This model is available as Model 24.4 from the book website.

different levels of organization. As an experimental approach, different model structures or parameters (e.g. probabilities of sand entrainment due to wind in the DECAL model) are examined systematically to see how outputs vary. One of the keys to POM is that support is provided for the different model structures (or parameters) only if they are able to reproduce *multiple* patterns observed in the real world, preferably at different levels of organization. For example, a model structure simulating the relationship between vegetation and soil in semi-arid environments may represent processes acting on individual plants (as online Model 24.2 does), but comparison of model output should be not only at that level but also the landscape level (e.g. clustering of vegetation) and also non-vegetation variables (e.g. soil moisture). Thus, careful and systematic use of such relatively simple simulations models can help advance our understanding of geographical systems and move such models 'from animations to science' (Grimm and Railsback, 2013a; and see below).

Before we turn to the use of these models in your own research, a final set of feedbacks that simulation models are currently being used to explore is those between human activity and environment processes. In particular, a form of simulation known as agent-based modelling (often abbreviated to ABM) can couple explicit representation of individual actors, their attributes, interactions and decisions with cellular representations of the physical environment to explore how the pervasive influence of human activity interacts with environmental processes. For example, Wainwright and Millington (2010) describe two models that link human activity with the environmental processes described above. The CybErosion model (Wainwright, 2008) links a Landscape Evolution Model with an agent-based model that represents human and animal agents and their reciprocal influence on soil erosion over several hundred years. The SPASIM model links an agent-based model of contemporary agricultural decision-making (Millington et al., 2008) with a cellular model of Mediterranean-type vegetation succession and fire disturbance (Millington et al., 2009) to explore the reciprocal impacts of land use/cover change and wildfire regimes. As with RCMs, agent-based modelling is still relatively new for investigating geographical systems and offers great promise (Heppenstall et al., 2012), but will potentially produce new loops of interactions that will need to be to be negotiated (Hacking, 1995; Millington et al., 2011).

24.4 Developing your Simulation Model

Deciding where to start when embarking on an environmental modelling project has been likened to the age-old Chicken-or-Egg dilemma; do you start by collecting data to then inform the construction of your model, or do you start with a conceptual model implemented in code and then collect data to identify the parameter values needed for the model to represent the real world (Mulligan and Wainwright, 2013)? One way of breaking such an infinite loop (or circular reference) might be to envisage the development of simulation models as a *process of modelling*, a series of yet more feedback loops (Figure 24.3). The steps of the modelling process (as I conceptualize it) are:

- Identification of Objectives;
- System Conceptualization;

- Data Collection;
- Model Construction;
- Evaluation; and
- Model Use.

A brief overview of each step is provided below, but further discussion can be found in Mulligan and Wainwright (2013) and Grimm and Railsback (2013b:Chapter 1); a succinct example of model development can be found in O'Sullivan and Perry (2013b: Chapter 8).

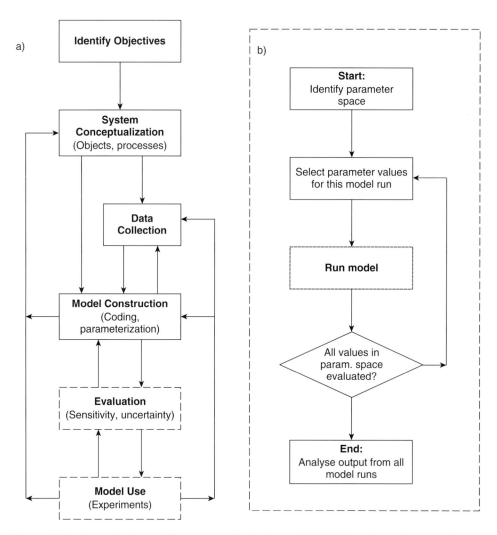

Figure 24.3 Loops in the modelling process. Evaluation and Model Use steps in the general modelling process (a) may have the process for multiple model runs (b) embedded within them. In turn, the 'run model' step in (b) may be represented by a further flowchart for a particular model (e.g. Figure 24.4)

Identify Objectives

Before stepping into the modelling process (loop), it is important to highlight that encompassing the entire processes is the need for a clearly defined objective for the modelling project. Models are sometimes distinguished between their use for prediction and explanation (e.g. Perry and Millington, 2008) and the debate about how models should or can be used in geography has a long history (Clifford, 2008). Whereas data-driven modelling approaches like statistical models are more likely to aim to predict system states, simulation modelling is often used to improve explanation of a process or phenomena. If prediction is the ultimate goal of a simulation model it may be at a lower level of precision (e.g. a more coarse scale) than other forms of modelling. Others may take a different view, but my emphasis here is on the use of simulation for explanation and improving understanding. Therefore, before starting out on the development of your simulation model you should have a good idea about either the phenomena you wish to represent via simulation (e.g. dune forms) or the processes (e.g. vegetation-fire feedbacks) you wish to better understand. In either case, (prediction or explanation) you will ideally also have information about patterns and forms observed in the real world against which you can compare the consequences of simulated processes (e.g. frequency-size distributions, landscape pattern metrics, ranges of variability, and so forth).

Conceptualization

From problem identification, we will step into the modelling process (loop) at the conceptualization stage – largely because simulations are essentially abstracted representations of the world used to explore and understand patterns and processes (Odoni and Lane, 2011), but also because when learning how to develop a simulation model it might be better to identify the possibilities and limitations of such tools before spending a lot of time and effort collecting data (see also Box 24.4). Regardless, conceptualization is a key stage in any modelling endeavour, and in geography in particular the issue of identifying model or system boundaries for representation has been long recognized (Richards, 1990; Lane, 2001; Brown, 2004). 'Bounding' or 'closure' of a model is an important step in geographical modelling because the real world is open, in the sense that there is free flow of energy, material and information, but computer models are 'closed' in the sense that a boundary must be defined beyond which the model cannot represent such flows. Model 'closure' thus involves deciding which processes will be represented explicitly within the model, which processes will not be represented in the model but will be parameterized or provide 'boundary conditions', what the spatial extent and resolution of the model will be, what the initial state of the modelled environment will be, and when the simulation will stop. For example, consider the DECAL model described above (and the simple version available online). The model represents sand transport in the wind direction but not variations due to turbulence, it specifies wind direction and strength (boundary conditions) but not the atmospheric conditions producing that wind, and it models an arbitrary area or space and length of time. Similarly, general circulation models for examining possible climate change must decide which processes to simulation, what area (global or regional) and resolution (what grid-size for atmospheric representation) and usually do not represent processes causing changes in atmospheric composition (e.g. greenhouse gas emissions are model boundary conditions specified by scenarios of future human activity). Further decisions in model conceptualization are what objects will be represented and what processes or relationships will be represented by parameters within the model. For example, in the DECAL model 'slabs' of sand are represented, not individual grains, and deposition in non-shadow regions is probabilistic (e.g. see parameter p in the simple version online) avoiding the need to represent fine-detailed processes of aeolian entrainment. Finally, variables must often be defined to be able to 'measure' model output so that it can be compared to empirical observations. For example, in models of vegetation growth and disturbance by fire, fire sizes must be recorded (and output to the user) so that simulated and observed frequency-size distributions can be compared (Millington et al., 2009). Model description protocols, such as the Overview, Design concepts, and Details (ODD) protocol developed for individual and agent-based simulation models (Grimm et al., 2010), can be useful to aid model conceptualization and construction.

Data Collection

Data may be required for several different aspects of model development – to establish parameters, to provide boundary conditions (e.g. scenarios of change), to establish empirical patterns against which to compare model output (e.g. pattern-oriented modelling). The particular data required will vary from project to project depending on the subject, and methods for collecting data are covered in detail in several other chapters in this book (e.g. Chapters 30–32).

Model Construction

This step refers to the process of converting the conceptual model into code that a computer can then execute to simulate, often thought of as 'computer coding'. In the past this step of the modelling process required detailed knowledge of computer programming and how computers actually function to achieve their calculations. However, recently several 'modelling environments', programming languages and libraries have been created that simplify coding (e.g. by taking care of things like memory allocation and providing functions to automate certain tasks) so that those interested in modelling geographical systems can do so much more efficiently. The choice of modelling environment depends in part on the objectives of the modeller (see *Identify Objectives* above), their programming skills, and the characteristics of the modelling environment (or programming language). The benefits and constraints of several modelling environments and languages for developing simulation models of geographical systems depend on factors such as ease of use, speed of execution and types of representation of spatial interactions (these are discussed further on the companion website). Example online models accompanying this chapter have been written in code that can be implemented in the freely available modelling environment, NetLogo (Wilensky, 1999). Designed for the creation of individual- and agent-based models, but also useful for developing geographic simulation models more generally, NetLogo is a good place to start for geographers learning simulation modelling because of its flexibility and simple syntax (see Box 24.1). These characteristics mean many people are now using this environment to develop simple, abstract models (Railsback and Grimm, 2012; O'Sullivan and Perry, 2013a), but as model complexity increases computational overheads (a consequence of the architecture needed to keep the programming language simple) mean that NetLogo models can run slowly and alternatives may need to be sought. Whether using NetLogo or another environment, e.g. PCRaster (http://pcraster.geo.uu.nl; Windows, Linux), MASON (http://cs.gmu.edu/~eclab/projects/mason/; Cross-platform) or language

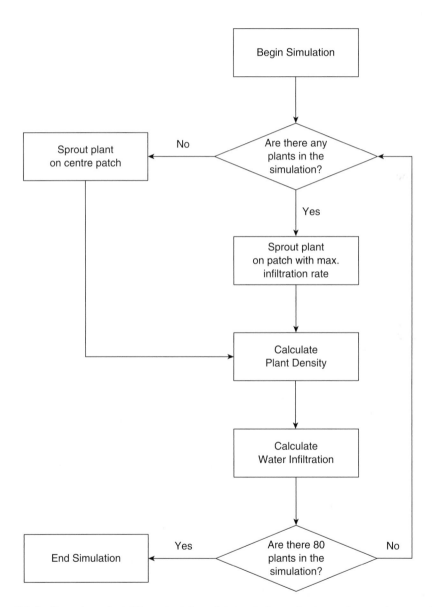

Figure 24.4 Flow chart describing scheduling of the code for a simple areal vegetation growth model (online Model 24.1, NetLogo code for which is available on the companion website). The 'Calculate Plant Density' step could in turn be represented by another flow chart.

(e.g. Python), some competence in programming will be needed. Introductory materials (including books, websites, Internet forums, other programmers) are available and should be sought out and used!

An important foundation for learning to program for simulation modelling is the ability to think computationally so that effective algorithms can be written. An algorithm is a step-by-step procedure that enables calculations and can be as simple as the population growth example in Box 24.1, or contain multiple algorithms to produce an entire model (e.g. the example models provided online). Because algorithms are step-by-step procedures, and often demand looping, they can be thought of as the coded equivalent of a flow chart. Indeed flows charts can be very useful when developing and presenting algorithms and models as they provide a visual means to trace the sequences of calculations or events required (Figure 24.3a and Figure 24.4). Other techniques may also be required in model construction, for example the use of regression modelling to establish relationships to be represented in the simulation model (Millington et al., 2013) but these techniques are beyond the scope of this chapter.

Evaluation

Once a model has been constructed in code ready for execution in the computer, several forms of evaluation are needed which may send a modeller right back to the previous step in the modelling process (model construction) or earlier! Most generally, two forms of evaluation can be distinguished: verification and validation (Oreskes et al., 1994; Rykiel, 1996). Verification can be thought of as ensuring the model matches the intended conceptual model (by comparing output against expectations – 'building the model correctly'), whereas validation can be thought of as ensuring the model matches processes in the real world (by comparing output to empirical observations – 'building the correct model'). Validation is important if the objective of the modelling is prediction, but the fundamental difficulties of 'true' validation of (closed) models of real-world (open) geographical systems are well known, if often forgotten (Oreskes et al., 1994). Verification is more important when the objective of the modelling is explanation or understanding, as aspects of it can even overlap with the *Model Use* step. The three forms of verification that are often useful (particularly for finding 'bugs' or errors in your model code) are: model exploration, sensitivity analysis and uncertainty analysis (Malamud and Baas, 2013). Model exploration is a form of model testing that involves playing with the model code to check it does what you expect by comparing alterations in code to expected outcomes in model output. As well as helping to identify bugs it can help to explore the appropriateness of the chosen model conceptualization by identifying unrealistic or empirically impossible system dynamics. Sensitivity analysis examines how variations in parameters or constants in a model influence outputs (e.g. what percentage change in model output does a given percentage change in parameter X produce?). Uncertainty analysis is very similar, but whereas sensitivity analysis is interested solely in the relationship between input and output, uncertainty analysis demands that the uncertainty in the input value is accounted for (e.g. via a probability distribution for the input value). Both sensitivity and uncertainty analyses are useful for identifying aspects of the model structure that are more or less important (or even irrelevant and therefore unnecessary) for understanding model outputs by tracing how different parameter or constant values influence processes and dynamics. They are also useful for identifying bugs in model code (when variations are unfeasible) and each can examine input values independently (vary values one at a time) or jointly (vary multiple values in a given model run). The importance of output for evaluation highlights the thought needed to establish which aspects of the modelled system will be 'measured' and reported via model code. Visualization of model output is also often important and the use of plots of variables, images and movies of model dynamics can be helpful for model evaluation.

Model Use

As with model evaluation, model use depends on the objectives of the modelling. Whereas for predictive purposes a model may be used to simulate the outcomes of different scenarios (of boundary conditions) for decision-making or management in particular instances, for explanatory purposes a model may be used to identify model structures that produce observed patterns (as in pattern-oriented modelling) or explore system dynamics more generally (e.g. existence of location of limits or thresholds). For example, some simulation models of wildfire model physical processes of ignition and combustions to represent fire behaviour as accurately as possible so that firefighters can anticipate where a fire will spread and therefore how to tackle it (e.g. Anderson et al., 2007). In contrast, ecologists may be more interested in understanding relationships between vegetation succession and fire as an ecological disturbance, and therefore will represent processes differently to understand dynamics more generally (Peterson, 2002). However, a common aspect to both cases is the execution of the model multiple times (possibly up to thousands of times), each execution being known as a model run. Multiple model runs may use the same initial conditions to identify the range of likely outcomes (predictive mode) or parameters may be varied systematically in 'simulation experiments' to explore system dynamics (Peck, 2004; Railsback and Grimm, 2012). As we saw in the examples above, the ability to run a simulation model multiple times enables experimentation in systems that would not be possible empirically due to scale issues and exploration of alternative trajectories of change into the future. In NetLogo the 'BehaviorSpace' is useful for specifying multiple model runs with alternative parameter values (Peterson, 2002).

Key to appropriate use and interpretation of a model is understanding how output is related to input via simulated processes. Questions will often arise about those connections during *Model Use* (and *Construction* and *Evaluation* steps), due to mismatches between model output and empirical data or theoretical expectations, and modellers will often find themselves returning to the *Conceptualization* step for consideration of alternative models. In returning to conceptualization, we realize that modelling has the potential to be a never-ending process. This may seem frustrating (see Box 24.4), but ultimately it is simply a reflection of the use of the scientific method to continually refine our understanding. Critics of simulation models may see them as incomplete representations of reality that have little bearing on our understanding of the real world (Goering, 2006; Simandan, 2010), but they fail to see the inductive and hermeneutic value of the process of modelling for understanding (Kleindorfer et al., 1998; Peck, 2008) and how models can be more or less useful (Box, 1979) or reliable (Winsberg, 2010) for understanding the world. Furthermore, innovative participatory modelling means such advantages may be appreciated by non-scientists and scientists alike (Lane et al., 2011; Souchère et al., 2010). Ultimately, all models (including maps and mental models) are simplifications of the real world and do not provide perfect representations of it. However, to the extent that we gain trust in a simulation model through the modelling process (e.g. by confronting it with data and our theoretical expectations), it allows us to explore the structures needed to produce observations or expectations and how much scope for different outcomes there might be.

Box 24.4   Tips for developing a simulation model

Algorithms are step-by-step procedures at the heart of computer simulation models. Although simple algorithms describing the process of developing a simulation model in a step-by-step fashion are possible (e.g. Figure 24.4), modelling is as much an art as a science and no single set of instructions is appropriate in all circumstances. Identifying the key real-world objects, interactions and processes to represent, and carefully considering how they should be appropriately operationalized in computer code, demands imagination and experience as much as theory and knowledge. This experience must be gained personally, but a few tips might save the novice some heartache and time:

- Start with as simple a model (i.e. representation) as possible and build from there. This has the benefit of both adhering to Ockham's Razor (Wainwright and Mulligan, 2013) and will help to keep tests of your patience to a minimum (but accept that your patience *will* be tested).
- Be clear about your objectives right from the start. This will help keep you on track (and sane) when you find yourself wrestling with the implementation of an algorithm or playing curiously for hours with the behaviour of some interesting minor aspect of your model. Being clear about objectives will also help with your (simple!) system conceptualization.
- In your system conceptualization, be clear about your model boundaries (and the constraints they impose), what your key objects are (multiple of which produce patterns), what the vital processes are (that cause changes in states of objects) and how these are all related. In turn, this will allow you to understand what parameters, constants and variables you need and what data might be needed to provide values for them. For example, for the code in Box 24.1 we can see the following:

    o Object: Population
    o Process: Growth of the population
    o Initial conditions: Initial population size (in this case 4)
    o Parameter: Growth rate

- When using a new modelling environment or language, take a little time to learn about how it is designed to be most efficient and get yourself up to speed on the built-in commands and data structures (often known as 'primitives'). These will save time both in coding up your model and executing it in the computer by increasing processing speed.
- Do ask advice of those who have been where you are now. Look at how others have conceptualized similar systems and solved similar coding problems, and use the Internet as the prodigious programming resource it is (there *will* be someone who has had a similar programming problem and posted it on an Internet forum somewhere).
- Don't expect things to work first time. Be patient, remember your objectives and keep going!

SUMMARY

As with all types of model, simulation and reduced complexity models are simplified representations of reality that can be useful in different ways depending on the objectives of the modeller. As simplified representations they share some features of their targets but not all, the choice of which can be likened to writing an essay. Much like modelling, when writing an essay (or this book chapter!) decisions must be made about the importance of different aspects of a subject in deciding the emphasis and time spend discussing them. In both essay writing and modelling, aspects of the target/subject need to be weighted by their importance based on previous understanding (literature), data, resources (computing power/word limit) and the objectives of the modeller/writer and what they want to explore. In simulation modelling, decisions are needed about what objects and processes will be included within the model boundary and how they are coupled, with the choice and weighting determined by the aim (e.g. explain or predict). For the simulation models discussed here, the focus has been on explanation and the exploration of system dynamics and feedbacks rather prediction of (static) states at particular points in time or space. My emphasis has been on models for explanation rather than prediction and I have stressed the utility of simulation for dynamically representing feedback loops. Developing your first simulation model, deciding what to include and leave out and working out how represent the real world in computer code, will likely be challenging. But understanding the need and utility of the nested loops discussed here – iterated execution of computer code to represent feedbacks in the geographical world through a continually reflexive modelling process – will be helpful, as should (re-)reading book chapters like Millington (2016).

Acknowledgements

I'm grateful to Nick Clifford, Andreas Baas (who provided Figure 24.1), Jacob White and Brynmor Saunders for comments and assistance with this particular manuscript, and to numerous other colleagues for their support during the loops of my own modelling career.

Further Reading

An excellent textbook covering a wide range of modelling approaches for physical geographers, including the process of modelling, is provided by Wainwright and Mulligan (2013). For more specific treatment of modelling geomorphological systems, see Brasington and Richards (2007) and Malamud and Baas (2013). Wainwright and Millington (2010) provide examples of how agent-based modelling might be combined with geomorphological models, whereas the textbook by Grimm and Railsback (2012) gives a comprehensive introduction to agent-based and individual-based modelling. Finally, as discussed throughout the chapter, O'Sullivan and Perry (2013a) explore the use of simulation models for investigating pattern and process in geographical and ecological systems, offering multiple examples for use in NetLogo (Willensky, 1999) and proposing a series of fundamental 'building block' models on which many geographical systems might be examined.

Note: Full details of the above can be found in the references list below.

References

Anderson, K., Reuter, G. and Flannigan, M.D. (2007) 'Fire-growth modelling using meteorological data with random and systematic perturbations', *International Journal of Wildland Fire* 16: 174–82.

Baas, A.C.W. (2002) 'Chaos, fractals and self-organization in coastal geomorphology: Simulating dune landscapes in vegetated environments', *Geomorphology* 48: 309–28.

Bithell, M., Brasington, J. and Richards, K. (2008) 'Discrete-element, individual-based and agent-based models: Tools for interdisciplinary enquiry in geography?', *Geoforum* 39: 625–42.

Box, G.E.P. (1979) 'Robustness in the strategy of scientific model building', in R.L., Launer and G.N. Wilkinson (eds) *Robustness in Statistics*. New York, NY: Academic Press. pp. 201–36.

Brasington, J. and Richards, K. (2007) 'Reduced-complexity, physically-based geomorphological modelling for catchment and river management', *Geomorphology* 90: 171–7.

Brown, J.D. (2004) 'Knowledge, uncertainty and physical geography: Towards the development of methodologies for questioning belief', *Transactions of the Institute of British Geographers* 29: 367–81.

Clifford, N.J. (2008) 'Models in geography revisited', *Geoforum* 39: 675–86.

Coulthard, T.J., Hicks, D.M. and Van De Wiel, M.J. (2007) 'Cellular modelling of river catchments and reaches: Advantages, limitations and prospects', *Geomorphology* 90: 192–207.

Coulthard, T.J., Macklin, M.G. and Kirkby, M.J. (2002) 'A cellular model of Holocene upland river basin and alluvial fan evolution', *Earth Surface Processes and Landforms* 27: 269–88.

Goering, J. (2006) 'Shelling Redux: How sociology fails to make progress in building and empirically testing complex causal models regarding race and residence', *Journal of Mathematical Sociology* 30: 299–317.

Grimm, V., Berger, U., DeAngelis, D.L., Polhill, J.G., Giske, J. and Railsback, S.F. (2010) 'The ODD protocol: A review and first update', *Ecological Modelling* 221: 2760–8.

Grimm, V. and Railsback, S.F. (2012) 'Pattern-oriented modelling: A "multi-scope" for predictive systems ecology', *Philosophical Transactions of the Royal Society B: Biological Sciences* 367: 298–310.

Grimm, V. and Railsback, S.F. (2013a) *Individual-based Modeling and Ecology*. Princeton: Princeton University Press.

Grimm, V. and Railsback, S.F. (2013b) 'Introduction',in V. Grimm and S.F. Railsback (eds) *Individual-based Modeling and Ecology*. Princeton: Princeton University Press. pp. 3–21.

Grimm, V., Revilla, E., Berger, U., Jeltsch, F., Mooij, W.M., Railsback, S.F., Thulke, H.H., Weiner, J., Wiegand, T. and DeAngelis, D.L. (2005) 'Pattern-oriented modeling of agent-based complex systems: Lessons from ecology', *Science* 310, 987–91.

Hacking, I. (1995) 'The looping effects of human kinds', in D. Sperber, D. Premack, and A.J. Premack (eds) *Causal Cognition: A Multidisciplinary Approach*. Oxford: Oxford University Press. pp. 351–83.

Harrison, S. (2001) 'On reductionism and emergence in geomorphology', *Transactions of the Institute of British Geographers* 26: 327–39.

Heppenstall, A.J., Crooks, A.T., See, L.M. and Batty, M. (2012) *Agent-Based Models of Geographical Systems*. London: Springer.

HilleRisLambers, R., Rietkerk, M., van den Bosch, F., Prins, H.H. and de Kroon, H. (2001) 'Vegetation pattern formation in semi-arid grazing systems', *Ecology* 82: 50–61.

Keeley, J.E. and Zedler, P.H. (2009) 'Large, high-intensity fire events in southern California shrublands: Debunking the fine-grain age patch model', *Ecological Applications* 19: 69–94.

Kleindorfer, G.B., O'Neill, L. and Ganeshan, R. (1998) 'Validation in simulation: Various positions in the philosophy of science', *Management Science* 44: 1087–99.

Lane, S.N. (2001) 'Constructive comments on D Massey – "Space-time, 'science' and the relationship between physical geography and human geography"', *Transactions of the Institute of British Geographers* 26: 243–56.

Lane, S.N., Bradbrook, K.F., Richards, K.S., Biron, P.A. and Roy, A.G. (1999) 'The application of computational fluid dynamics to natural river channels: Three-dimensional versus two-dimensional approaches', *Geomorphology* 29: 1–20.

Lane, S.N., Odoni, N.A., Landström, C., Whatmore, S.J., Ward, N. and Bradley, S. (2011) 'Doing flood risk science differently: An experiment in radical scientific method', *Transactions of the Institute of British Geographers* 36: 15–36.

Liu, J. and Ashton, P.S. (1995) 'Individual-based simulation models for forest succession and management', *Forest Ecology and Management* 73: 157–75.

Loepfe, L., Martinez-Vilalta, J., Oliveres, J., Piñol, J. and Lloret, F. (2010) 'Feedbacks between fuel reduction and landscape homogenisation determine fire regimes in three Mediterranean areas', *Forest Ecology and Management* 259: 2366–74.

Malamud, B.D. and Baas, A.C.W. (2013) 'Nine considerations for constructing and running geomorphological models', in A.C.W. Baas (ed.) *Quantitative Modeling of Geomorphology* (Treatise on Geomorphology, vol. 2). San Diego: Academic Press. pp. 6–28.

Millington, J.D.A. (2016) 'Simulation and Reduced Complexity Models', in N. Clifford et al. (2016) *Key Methods in Geography*. London: Sage pp. 119–59.

Millington, J.D.A., O'Sullivan, D. and Perry, G.L.W. (2012) 'Model histories: Narrative explanation in generative simulation modelling', *Geoforum* 43: 1025–34.

Millington, J.D.A., Romero Calcerrada, R. and Demeritt, D. (2011) 'Participatory evaluation of agent-based land-use models', *Journal of Land Use Science* 6: 195–210.

Millington, J.D.A., Romero-Calcerrada, R., Wainwright, J. and Perry, G.L.W., (2008) 'An agent-based model of Mediterranean agricultural land-use/cover change for examining wildfire risk', *Journal of Artificial Societies and Social Simulation* 11 (4): 4. http://jasss.soc.surrey.ac.uk/11/4/4.html.

Millington, J.D.A., Wainwright, J., Perry, G.L.W., Romero Calcerrada, R. and Malamud, B.D. (2009) 'Modelling Mediterranean landscape succession-disturbance dynamics: A landscape fire-succession model', *Environmental Modelling & Software* 24: 1196–1208.

Millington, J.D.A, Walters, M.B., Matonis, M.S. and Liu, J. (2013) 'Filling the gap: A compositional gap regeneration model for managed northern hardwood forests', *Ecological Modelling* 253: 17–27.

Mulligan, M. and Wainwright, J. (2013) 'Modelling and model building', in J. Wainwright, and M. Mulligan (eds) *Environmental Modelling: Finding Simplicity in Complexity*. Chichester: Wiley. pp. 7–26.

Murray, A.B. (2007) 'Reducing model complexity for explanation and prediction', *Geomorphology* 90: 178–91.

Murray, A.B. and Paola, C. (1994) 'A cellular model of braided rivers', *Nature* 371: 54–7.

Nicholas, A.P., Sandbach, S.D., Ashworth, P.J., Amsler, M.L., Best, J.L., Hardy, R.J., Lane, S.N., Orfeo, O., Parsons, D.R. and Reesink, A.J. (2012) 'Modelling hydrodynamics in the Rio Paraná, Argentina: An evaluation and intercomparison of reduced-complexity and physics based models applied to a large sand-bed river', *Geomorphology* 169: 192–211.

Nield, J.M. and Baas, A.C. (2008) 'Investigating parabolic and nebkha dune formation using a cellular automaton modelling approach', *Earth Surface Processes and Landforms* 33: 724–40.

O'Sullivan, D. and Perry, G.L.W. (2013a) *Spatial Simulation: Exploring Pattern and Process*. Chichester: Wiley-Blackwell.

O'Sullivan, D. and Perry, G.L.W. (2013b) 'Weaving it all together', in D. O'Sullivan and G.LW. Perry (eds) *Spatial Simulation: Exploring Pattern and Process*. Chichester: Wiley-Blackwell. pp. 229–64.

Odoni, N.A. and Lane, S.N. (2011) 'The significance of models in geomorphology: From concepts to experiments', in K.J. Gregory and A.S. Goudie (eds) *The SAGE Handbook of Geomorphology*. London: Sage. pp. 154–173.

Oreskes, N., Shrader-Frechette, K. and Belitz, K. (1994) 'Verification, validation, and confirmation of numeric models in the earth sciences', *Science* 263: 641–6.

Pacala, S.W., Canham, C.D., Saponara, J., Silander, J.A., Kobe, R.K. and Ribbens, E. (1996) 'Forest models defined by field measurements: Estimation, error analysis and dynamics', *Ecological Monographs* 66: 1–43.

Peck, S.L. (2004) 'Simulation as experiment: A philosophical reassessment for biological modelling', *TRENDS in Ecology and Evolution* 19: 530–4.

Peck, S.L. (2008) 'The hermeneutics of ecological simulation', *Biology & Philosophy* 23: 383–402.

Perry, G.L.W. and Millington, J.D.A. (2008) 'Spatial modelling of succession-disturbance dynamics in forest ecosystems: Concepts and examples', *Perspectives in Plant Ecology, Evolution and Systematics* 9: 191–210.

Peterson, G.D. (2002) 'Contagious disturbance, ecological memory, and the emergence of landscape pattern', *Ecosystems* 5: 329–38.

Piñol, J., Beven, K. and Viegas, D.X. (2005) 'Modelling the effect of fire-exclusion and prescribed fire on wildfire size in Mediterranean ecosystems', *Ecological Modelling* 183: 397–409.

Railsback, S.F. and Grimm, V. (2012) *Agent-Based and Individual-Based Modeling: A Practical Introduction*. Princeton, NJ: Princeton University Press.

Reddy (2011) http://dx.doi.org/10.1007/978-90-481-8702-7_37

Richards, K.S. (1990) '"Real" geomorphology', *Earth Surface Processes and Landforms* 15: 195–7.

Rykiel, E.J. (1996) 'Testing ecological models: The meaning of validation', *Ecological Modeling* 90, 229–44.

Scheller, R.M. and Mladenoff, D.J. (2007) 'An ecological classification of forest landscape simulation models: Tools and strategies for understanding broad-scale forested ecosystems', *Landscape Ecology* 22: 491–505.

Simandan, D. (2010) 'Beware of contingency', *Environment and Planning D: Society & Space* 28: 388–96.

Souchère, V., Millair, L., Echeverria, J., Bousquet, F., Le Page, C. and Etienne, M. (2010) 'Co-constructing with stake-holders a role-playing game to initiate collective management of erosive runoff risks at the watershed scale', *Environmental Modelling & Software* 25: 1359–70.

Stott, T. (2010) 'Fluvial geomorphology', *Progress in Physical Geography* 34 (2): 221–45.

Torrens, P.M. (2012) 'Moving agent pedestrians through space and time', *Annals of the Association of American Geographers* 102: 35–66.

Tucker, G.E. and Hancock, G.R. (2010) 'Modelling landscape evolution', *Earth Surface Processes and Landforms* 35: 28–50.

Van De Wiel, M.J., Coulthard, T.J., Macklin, M.G. and Lewin, J. (2007) 'Embedding reach-scale fluvial dynamics within the CAESAR cellular automaton landscape evolution model', *Geomorphology* 90: 283–301.

Wainwright, J. (2008) 'Can modelling enable us to understand the role of humans in landscape evolution?', *Geoforum* 39: 659–74.

Wainwright, J. and Millington, J.D.A. (2010) 'Mind, the gap in landscape-evolution modelling', *Earth Surface Processes and Landforms* 35: 842–55.

Wainwright, J. and Mulligan, M.(eds) (2013) *Environmental Modelling: Finding Simplicity in Complexity*. Chichester: Wiley.

Wilensky, U. (1999) 'NetLogo' http://ccl.northwestern.edu/netlogo (accessed 25 November 2015).

Winsberg, E. (2010) *Science in the Age of Computer Simulation*. Chicago: Chicago University Press.

ON THE COMPANION WEBSITE...

Visit **https://study.sagepub.com/keymethods3e** for author videos, chapter exercises, resources and links, plus **free** access to the following recommended articles:

1. **O'Sullivan, D. (2008) 'Geographical information science: Agent-based models', *Progress in Human Geography*, 32 (4): 783–91.**

This article reviews the range of approaches that agent-based simulation models embody, considering their implications for representation of geographical processes and patterns.

2. **Stott, T. (2010) 'Fluvial geomorphology', *Progress in Physical Geography*, 34 (2): 221–45.**

This paper provides an overview of reduced-complexity geomorphological modelling for river and catchment management and how this approach relates to others.

25 Remote Sensing and Satellite Earth Observation

Martin J. Wooster, Thomas Smith and Nick A. Drake

SYNOPSIS

Earth Observation (EO) is the term generally used to describe the study of the Earth from space via the techniques of remote sensing. Such 'environmental remote sensing' can be defined as the approach of acquiring information on the Earth from a distance, via measurements of naturally occurring or artificially generated electromagnetic radiation (EMR) that has interacted with the targeted Earth environment(s). Earth orbiting satellites are now the most commonly used platform from which to collect remote sensing data of the Earth, but aircraft-mounted instruments or even ground-based sensors are also used. A range of techniques, from simple one line equations to highly complex 'retrieval algorithms' described in tens of thousands of lines of computer code, can be used to create new datasets and information from pre-processed remotely sensed imagery and data, revealing a host of information on key Earth System variables. The timings of the main developments illustrated in this chapter demonstrate the comparatively young age of the Earth Observation discipline, and this coupled with its high technological dependence, mean that its capabilities, particularly in terms of satellite capabilities, remain rapidly evolving.

This chapter is organized into the following sections:

- Introduction
- Remote sensing fundamentals
- Earth observation missions and uses
- Development of EO missions, sensors, and platforms
- Image processing for earth observation
- Mission to planet Earth
- Conclusion

25.1 Introduction

Earth Observation (EO) generally refers to the study of the Earth from space using the techniques of remote sensing. Remote sensing has long been defined by phrases such as 'methods of obtaining information about an object through the analysis of data collected by instruments not in physical contact with that object', which is in direct contrast to 'on site' or *in situ* measurements. More recently, the term remote sensing has specifically become associated with the exploitation of electromagnetic (EM) radiation, either visible 'light' or EM radiation in other wavelength regions, and their use in 'sensing' the properties of an object under investigation. This is analogous to the way that the human vision system 'senses' visible light, using it to construct colour images of the world around us which are interpreted further by the human brain to extract the required information on our environment. Earth Observation replaces the human vision system with instrumentation that can, if desired, measure categories of electromagnetic radiation other than visible wavelength light, for example thermal infrared radiation that carries information related to the temperature of the emitting object. Earth Observation methods target one or more of Earth's land surface, water or atmospheric environments, though many of the techniques used to process and interpret the collected electromagnetic radiation signals are common across these application areas. Many of the techniques are in fact also very similar to those used in other remote sensing applications, for example for astronomy or planetary science.

Remote sensing is often classified into either active approaches, when the electromagnetic radiation signal is first emitted by the sensing instrument before interacting with the object under study, or passive methods, where naturally emitted light such as sunlight or thermal infrared radiation emitted by the object itself is measured. An overall definition of Earth Observation could therefore be formulated as 'the approach of acquiring information on the Earth from a distance, via measurements of naturally occurring or artificially generated electromagnetic radiation that has interacted with the targeted Earth environment(s)'.

25.2 Remote Sensing Fundamentals

Since Earth Observation is focused around measurements of electromagnetic radiation, a basic understanding of the subject is central to effective use of EO methods and remotely-sensed data in general. Electromagnetic radiation is generated by natural mechanisms occurring all around us, such as changes in atomic and molecular energy levels, as well as by more extreme circumstances such as nuclear reactions (e.g. in our Sun). In the 1860s the British physicist James Clerk Maxwell conceptualized electromagnetic radiation as perpendicular electric and magnetic fields that vary in magnitude in a direction perpendicular to the direction of the field propagation, and this concept is often applied to understand many of the processes exploited in environmental remote sensing methods. The wavelength of an electromagnetic wave is taken to be the distance between consecutive wave crests, and the wave propagates forward at the speed of light (c) which is a constant 300 million meters per second in a vacuum (Figure 25.1a). The full electromagnetic spectrum covers all possible wavelengths of electromagnetic radiation, and is subdivided into different wavelength regions; for example visible EM radiation covers the wavelength range 0.38–0.75 µm (with a µm being one millionth of a meter) and the infrared radiation range 0.7–1000 µm. The infrared region itself is further subdivided, for example the near infrared (NIR) region is the region at wavelengths just longer than red wavelength light (~ 0.75 - 1.0 µm) and substantial amounts of NIR electromagnetic radiation still arrive at the Earth from the Sun. By contrast, in the longer wavelength thermal infrared (TIR) spectral region (~ 3 – 1000 µm) the source is generally thermal emission from the Earth itself. Measurements made at these latter wavelengths can thus be converted into measurements of TIR 'brightness temperatures' which are related to (but different from) the kinetic temperature of the emitting surface. Further regions of the electromagnetic spectrum exist at wavelengths shorter (e.g. ultraviolet) and longer (e.g. microwaves) than the visible and infrared region respectively (Figure 25.1b).

For a variety of reasons, such as their differing ability to penetrate Earth's atmosphere, the different wavelength regions of the electromagnetic spectrum are not used equally for Earth Observation. Regions of the electromagnetic spectrum that penetrate the atmosphere quite effectively are called 'atmospheric windows', and it is at these wavelength regions where most remote sensing of the Earth's land and surface water is conducted. Regions of the electromagnetic spectrum outside of these windows may be targeted to provide information more related to the makeup of Earth's atmosphere itself, for example its humidity, temperature, pressure and the concentration of certain atmospheric gases. Of fundamental importance to Earth Observation is the fact that the different wavelength regions of the electromagnetic spectrum can be used to probe very different Earth system properties. In terms of passive remote sensing, the Sun is the source of electromagnetic radiation in the UV, visible, near-infrared and middle-infrared part of the spectrum (becoming increasingly weak and increasingly negligible beyond around 3.5 μm), whereas at wavelengths longer than around 3 μm any measured EM radiation increasingly originates from the thermal emissions of the Earth itself (Figure 25.2). Most active remote sensing occurs in the microwave region of the electromagnetic spectrum (radar), but laser-based instruments working usually in the near infrared or shortwave infrared region (lidar) are also increasingly used.

When electromagnetic radiation meets a component of the Earth system (e.g. vegetation or soil on the land, water in the ocean, gases in the atmosphere) it can interact with the material, generally at an atomic or molecular level. These interactions vary according to the electromagnetic radiation wavelength, and particular wavelengths of electromagnetic spectrum energy can either be reflected (or 'scattered'), transmitted or absorbed. In active remote sensing the measurement made by the observing instrument

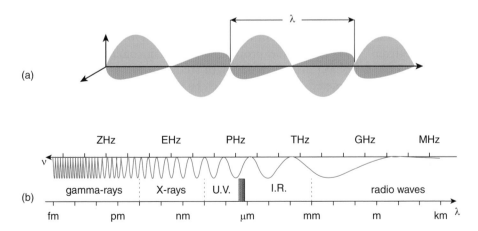

Figure 25.1 Electromagnetic waves (a) and electromagnetic radiation (b)

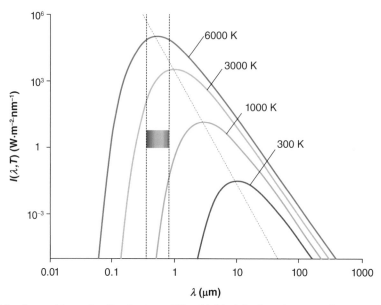

Figure 25.2 Spectral intensity distribution of Planck's black-body radiation as function of wavelength for different temperatures

essentially relates to the proportion of an artificially generated and transmitted EM radiation pulse that is received back at the sensor from the illuminated area, whereas in passive remote sensing the EMR being measured is naturally occurring (reflected sunlight or thermally emitted radiation from the Earth) and analysis is often based on an examination of how various atmospheric or surface processes have changed the amount of sensed EMR from what otherwise would have been recorded at particular wavelengths. It is generally the identification of characteristic 'spectral signatures' present at particular wavelengths or wavelength regions that allows the various targeted Earth system properties to be probed. For example, brightness temperatures measured at various wavelengths in the TIR spectral region allow information on the Earth's surface and atmospheric temperatures to be derived. Similarly, remotely sensed measurements of EMR made at visible to shortwave infrared wavelengths are often converted into a 'reflectance spectrum' that depicts the proportion of the incoming solar radiation reflected by the targeted object at each wavelength studied, from which the type and certain properties of the targeted material can often be inferred (Figure 25.3).

Rather than recording continuous spectra of the type shown in Figure 25.3 however, most Earth Observation instruments actually make multi-spectral measurements, meaning measurements at a few different wavebands (ranges of wavelengths), rather than across a continuous range of wavelengths. More complex hyperspectral instruments typically make measurements at many dozens, sometimes even thousands, of different wavebands, each very narrow in terms of their wavelength range. The more detailed the spectral measurements are, often the more precise the information extractable from the measured spectra can be. For example, an instrument measuring only in one 'red' (e.g. 0.62–0.75 μm) and one near infrared (e.g. 0.78–1.0 μm) waveband and observing vegetation may simply be able to discriminate between live and dead plants,

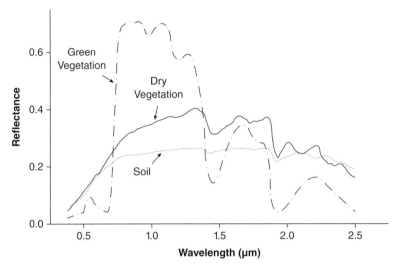

Figure 25.3 Spectral reflectance of green vegetation, dry vegetation, and soil (data taken from Clark et al. 1999)

whilst an instrument measuring at a few different wavebands across the visible and near infrared spectral regions might also be able to separate broad vegetation types (e.g. samples from grasslands from those of woodlands). By contrast, data collected by a hyperspectral instrument may be able to be used to estimate detailed biophysical-properties, such as leaf pigment concentrations. Whilst the former basic capabilities may not seem that useful in a laboratory setting, where the human eye might be able to do just as well or better, they become very powerful if the ability to collect even simple multispectral measurements is combined with an imaging capability, resulting in a system that can be used to provide information quite rapidly and repeatedly over large areas, even across our entire planet. This can be used to examine, for example, trends in deforestation or the response of crop growing regions to drought. Hyperspectral imaging systems similarly provide a very powerful capability that can potentially be used to discriminate individual plant species, based on for example the derived biophysical characteristics.

25.3 Earth Observation Missions and Uses

As already stated, remote sensing instruments are commonly mounted on aircraft, and there also exist many handheld instruments used to collect 'close-to-target' data. However, the remainder of this chapter focuses on observations made from Earth orbiting satellites, which are by far the most commonly used platform now used to collect remote sensing data of the Earth. There are in fact hundreds of EO satellites that have been launched by commercial companies, governments and space agencies such as the National Aeronautics and Space Administration (NASA) and the European Space Agency (ESA), each acquiring data that can be used to extract information about the Earth system. They often do so continuously, for many years non-stop, and are therefore often termed Earth Observation 'Missions' because their successful operation involves much more than just the orbiting satellite itself. The capabilities offered by these missions are often not fully appreciated, since many people's most common contact with satellite 'Earth Observation' is through the type of 'true colour' imagery used in applications such as Google Earth and Bing Maps. These types of imagery appear very similar to aerial photography, since they are collected at the same visible wavelengths as standard colour photography, but now taken from a satellite orbiting hundreds of kilometres above the Earth. However, as we have already discussed, there exist remote sensing imaging instruments that operate at wavelengths outside of the visible part of the electromagnetic spectrum, providing information that standard photography cannot hope to probe, and also in addition to imaging instruments there are sensors that acquire important information only along particular profiles or at individual points.

As an example of EO profiling techniques, the CALIPSO satellite, which is a joint mission between NASA and the French space agency CNES, uses active remote sensing with a spaceborne lidar to illuminate a narrow strip below the satellite, with some of the laser light being scattered back to the instrument from airborne particles and clouds, providing insights into the role that these play in regulating Earth's weather, climate, and air quality (Figure 25.4a). Conversely, the Japanese Greenhouse gases Observing SATellite (GOSAT) uses passive remote sensing approaches targeting individual points on the Earth to measure both reflected sunlight and thermally emitted radiation, including at wavelengths where carbon dioxide and methane in the atmosphere absorb strongly. By assessing how much electromagnetic radiation is arriving at the sensor in these particular wavelength regions, EO 'products' related to the 'total atmospheric column' amounts of CO_2 and CH_4 at the targeted locations can be provided (Figure 25.4b).

Much of the data from the type of 'scientific' EO missions highlighted above, and the more commonly available high to moderate spatial resolution (i.e. pixel sizes of 10 m to 1000 m) Earth observation multispectral 'imagers' such as Landsat, the Moderate Resolution Imaging Spectroradiometer (MODIS), the Sentinel-2 MSI and the Sentinel-3 Sea and Land Surface Temperature Radiometer (SLSTR), are provided free to users via online databases (e.g. http://ladsweb.nascom.nasa.gov/ and https://scihub.copernicus.eu/dhus/), either as relatively 'raw' remotely sensed imagery, preprocessed observations (e.g. spectral reflectance values and/or infrared brightness temperatures that still require subsequent processing by experienced users to extract the necessary information (e.g. forest cover maps or maps of sea surface temperature), or as 'higher level products' already converted to geophysical information about the Earth using computer-based algorithms (essentially linked sets of mathematical equations coded into software). In addition to these largely scientific EO missions operated by national and international space agencies, there are increasing numbers of commercial satellites that – for a price – will provide users with high spatial resolution imagery, where objects as small as 20 cm can be discriminated. This type of very spatially detailed imagery is often used by television broadcasters, by government agencies interested in rapid mapping, and for planning responses to swiftly changing situations such as natural disasters (Figure 25.5).

The decreasing cost and increasing availability of both scientific and commercial EO datasets, along with the provision of already processed 'EO products' and the

(a)

(b)

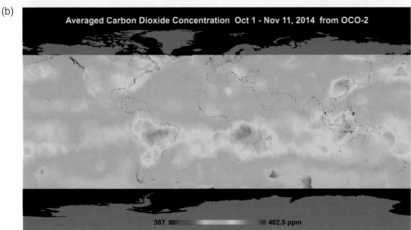

Figure 25.4   Example of Earth Observation profiling techniques from (a) the CALIPSO satellite and (b) averaged Carbon Dioxide Concentration from the Greenhouse gases Observing SATellite (GOSAT)

improving power of desktop computers and image processing/spectral analysis software, means that use of EO data previously restricted to a limited number of specialists is now open to many. This has led to the increasing use of such datasets in a wide range of scientific studies and environmental monitoring programmes, beyond the traditional remote sensing satellite application of weather forecasting, and opened up a very wide variety of applications across both physical and human geography. The following section explains some of the background to how this very diverse set of capabilities has come about.

(a)

(b)

(c)

Figure 25.5   High-resolution QuickBird imagery from (a) before and (b) after the 2004 Boxing Day tsunami in Banda Aceh, Indonesia; and (c) reconstruction by 2009

25.4 Development of EO Missions, Sensors, and Platforms

A brief look at the history of remote sensing instruments and platform development demonstrates the rapid technological advances made over the last half-century or so, which has led to the current wide use of and reliance on EO satellite technology in many areas. Photography was of course the first form of remote sensing, developed in the 1820s and 1830s, but it was not until 1858 that the first aerial photograph was taken from a balloon. Colour photography followed in the 1930s and Sputnik-1, the first Earth-orbiting satellite, was launched in 1957 by the USSR. By this time it was already clear that a spacecraft orbiting the Earth could provide an excellent vantage point for photographic military intelligence gathering, and the first satellite intelligence photography was obtained by the US in 1960. At that time, and for more than the next decade, these early 'Earth Observation' systems relied on exposed film capsules being re-entered into Earth's atmosphere and caught by aircraft over the Pacific Ocean. Only by relying on highly advanced version of traditional photographic technology could these early EO missions provide the spatial detail required for intelligence gathering purposes (Figure 25.6a), for example counting numbers of 'enemy' planes and tanks, but applications such as weather forecasting can make do with far worse spatial detail, and the first civilian Earth Observation mission targeting this type of meteorological application was in fact also launched in 1960 carrying a modified television camera providing black and white ('panchromatic') Earth imagery (Figure 25.6b).

After the initial development of Earth-orbiting missions targeting primarily military and meteorological needs, it took little more than a decade for sensors and satellites to improve such that 1972 saw the launch of the first in the series of the much more capable Landsat missions, this arguably being the most important single series of remote sensing satellites ever in the field of environmental Earth Observation. The Landsat series continues until this day, adding ever-more precise multispectral imaging capabilities, with Landsat-8 launched in 2013 carrying the 8-waveband Operational Land Imager (OLI) that records multispectral imagery with a spatial resolution of between 15 m and 30 m, and the two waveband Thermal InfraRed Sensor (TIRS) that records in the longwave infrared spectral region to provide 100 m spatial resolution data relayed to Earth surface temperatures. Multispectral imagery

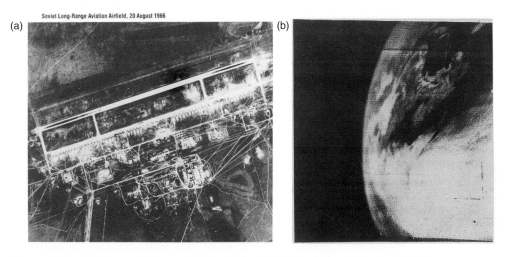

Figure 25.6   (a) Corona photograph from 1966 of Soviet Long-Range Aviation Airfield and (b) TIROS imagery

Figure 25.7   Landsat composite imagery of crops, forests, and cities

from Landsat is often used to create colour composites (Figure 25.7), reflecting different properties of the Earth surface that can be interpreted visually, and also has one of its main applications as land cover and land use mapping, where the reflectance spectra of each pixel are used to categorise it as a particular land cover or land use type (e.g. bare soil, forest, water, cropland etc.).

Like Landsat, most EO missions are placed into low-Earth orbit, typically lower than ~ 1000 km above the Earth's surface and usually circling approximately north-south or south-north whilst the Earth rotates east-west underneath. In this way, these type of near 'polar orbiting' EO missions can often provide coverage of the entire Earth surface over a period of hours, days or weeks, depending on the area covered by each individual acquired scene or dataset. The main alternative approach to this is to use a geostationary orbit, where the satellite platform is located almost 36,000 km above the equator, such that the spacecraft takes exactly 24 hours to revolve around its orbit. The satellite therefore appears stationary to an observer on Earth's surface, and this means that the temporal resolution of the data (the time between acquisition of consecutive images of the same area) is limited only by the time it takes to acquire each image, so very rapidly updated information can be provided. Geostationary orbits are commonly employed for meteorological satellite missions, such as the European Meteosat, where data of the Earth disk containing Africa and Europe are provided every 15 minutes for the tracking of weather and other rapidly changing phenomena such as vegetation fires (Figure 25.8). However, the large distance from the Earth required for geostationary orbit means that the sensors typically provide images of quite low spatial resolution, with each 'ground pixel' typically covering an area of 3 km × 3 km or more. A single geostationary satellite sensor also cannot view the entire Earth, but instead views only the single Earth hemisphere continuously presented to it, and the acquired imagery suffers significant distortion close to the edges of the viewed 'Earth disk'. Typically a ring of four or five geostationary satellites is used to provide almost continuous coverage of most of the Earth, apart from high latitudes due to the aforementioned edge distortion, and this presents a superb capability when analysing clouds and other rapidly changing phenomena, such as the temperature of the land surface.

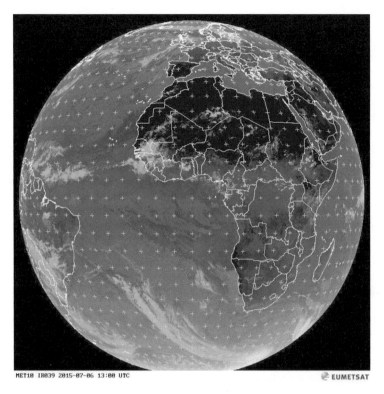

MET10 IR039 2015-07-06 13:00 UTC   ⌒ EUMETSAT

Figure 25.8 Example of geostationary meteorological satellite imagery from European Meteosat

25.5 Image Processing for Earth Observation

Before useful information can be gleaned from the 'raw' data that spaceborne Earth observing sensors capture, it is often necessary to perform corrections to remove image distortions and artefacts. Geometric correction of the data, to account for the effect of Earth curvature and rotation, as well as the movement of the spacecraft itself, is usually required to map images to a fixed and known coordinate system. Atmospheric correction is also often used to try to minimise the effects of gases and particles in the atmosphere on the measured signal. Imagery that has undergone 'pre-processing' for both geometric and/or atmospheric correction is now routinely available for download without the user having to perform these operations themselves.

Pre-processing can provide geo-corrected imagery with pixel-values containing information on the amount of electromagnetic energy emitted or reflected by the Earth in a particular waveband. A range of techniques, from simple one-line equations to highly complex 'retrieval algorithms' described in tens of thousands of lines of computer code, may be used to create new datasets and information from this pre-processed imagery, revealing a host of key Earth System variables, for example ocean temperatures, vegetation cover, surface height and atmospheric profiles of water vapour, ozone and carbon dioxide. An example of the simplest kind of algorithm would be image thresholding, where pixel values above or below a specified threshold in a particular spectral band are classified as a particular feature. Using imagery taken in a thermal infrared waveband for example, such thresholding may be used for the identification of cold (high-altitude) cumulonimbus clouds (which are quite good indicators of precipitation), or for detecting Earth surface 'hotspots' caused by active volcanoes or burning wildfires. Slightly more complex 'image arithmetic' operations can be used to combine information from multiple images of the same area taken at different times, or from multiple wavebands in a single image, via basic arithmetic equations. A straightforward example is change detection, where information from two images of the same area acquired at different times in the same waveband are subtracted from one other, emphasising differences that have occurred between the image acquisitions. Another example is the Normalised Differential Vegetation Index (NDVI), which uses spectral reflectance's (σ) calculated simultaneously from red (R) and near-infrared (NIR) waveband measurements (i.e. σ_{NIR} and σ_R) to provide a metric that can be related to the presence and photosynthetic state of vegetation:

$$\text{NDVI} = (\sigma_{NIR} - \sigma_R) \, / \, (\sigma_{NIR} + \sigma_R) \quad\quad\quad\quad [25.1]$$

The NDVI exploits the characteristic spectral reflectance signature of healthy vegetation, which peaks at near-infrared wavelengths and has a minimum at red wavelengths due to photosynthesis. The two forms of 'image arithmetic' can be combined, for example to examine changes in an area's vegetation over time by subtracting the NDVI image obtained at one date from the another (Figure 25.9).

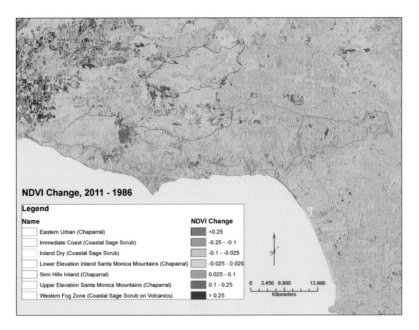

Figure 25.9 Changes in NDVI from Landsat over the Santa Monica Mountains between 1986 and 2011

Another popular form of image analysis that users can quite easily perform themselves is image classification, which is generally used to group or 'cluster' pixels with similar spectral signatures into particular land cover classes, thus generating a classified map of the imaged region. Semi-automated approaches allow users to 'train' these types of clustering algorithm to automatically recognise spectral signatures of key land cover types, for example water, croplands, forests, and particular soils or minerals. However, being ever more precise in the classification detail (e.g. moving from simply discriminating vegetation from bare soil to trying to classify different types of vegetation), often results in lowered classification accuracy. Classification accuracy assessment using data that were not used in the training stage is therefore an important part of any classification procedure.

Before the turn of the last millennium, most remote sensing datasets had to be converted into useful information about the Earth by the users themselves, using algorithms such as those above, and often far more complex techniques as well. However, since the late 1990s a whole series of mostly polar orbiting satellites have been launched specially targeting 'Earth System Science', initially as part of NASA's Mission to Planet Earth (MTPE) but also with major contributions by the European Space Agency and others. In a similar manner to how meteorological satellites are focused on providing remote sensing observations for the specific purpose of weather forecasting, this endeavour was largely focused on studying the signatures and drivers of global climate and environmental change, both natural and human induced. As such, these particular EO Missions were often targeted at addressing specific measurement issues, such as the changing temperature of the ocean surface, the percentage of forest cover, or the concentration of atmospheric greenhouse gases, and often their data were provided to users in the form of already processed 'EO products' dedicated to these topics, rather than the type of multi-spectral imagery typically provided by Landsat that users must generally process themselves as described above. This meant that environmental modellers and other users non-expert in the detail of remote sensing methods could still readily exploit the information provided by these sensors.

25.6 Mission to Planet Earth

A key instrument of the 'Mission to Planet Earth' era is the Moderate Resolution Imaging Spectroradiometer (MODIS), one of the most widely used Earth Observation tools ever deployed. MODIS was carried on two NASA satellites (called Terra and Aqua) launched in 1999 and 2002 respectively, and lasted well beyond its design lifetime (indeed both instruments are still operating well at the time of writing). MODIS collects continuous imagery of the Earth in 36 individual spectral bands over an imaging width ('swath') of 2300 km, and these data are routinely processed using complex algorithms to deliver multiple different EO products, reflecting various properties of Earth's land surface, ocean and atmosphere (Figure 25.10).

Terra's orbit around the Earth is timed so that by day it passes from north to south across the equator in the morning, while Aqua passes south to north over the equator in the afternoon. The MODIS instruments on the Terra and Aqua satellites

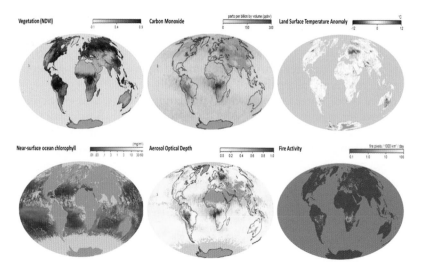

Figure 25.10   Examples of Earth Observation satellite products

view the entire Earth's surface every one to two days, and the data and products MODIS has provided have led to fundamental insights into many Earth System properties and processes, and to an increased ability to measure and even forecast global environmental changes, including those required by policy makers to help ensure well-founded decisions are made concerning the protection of the Earth's environment. Users may explore near real-time MODIS imagery and processed products using the Worldview online tool (https://earthdata.nasa.gov/labs/worldview/). NASA's Giovanni (http://giovanni.gsfc.nasa.gov/giovanni/) provides another online tool, giving users powerful capabilities to actually analyse these types of EO dataset, even those covering the entire planet, through a relatively simple web-based interface and without ever having to download the data themselves (Figure 25.11). Google Earth Engine is a further example of online data display and analysis, this time focused on exploitation of Landsat imagery (https://earthengine.google.org/).

In terms of European programmes, at the time of writing (2015) Europe's Copernicus programme has started to launch the first in a series of the many Sentinel satellites to be operated over the next few decades, many of which are targeted at the same kinds of environmental phenomena as the Terra and Aqua satellites and the other MTPE missions, but with a further specific 'operational' requirement for their data and information, beyond the kind supplied by 'one off' missions aimed more at research and methodological development. Data from many of the European Sentinel missions are to be routinely fed into a series of highly complex computer models that will be used to routinely monitor the state of Earth's surface, oceans and atmosphere, in the same way as meteorological models do now for the weather, and certain of these will also provide short-term forecasts of key environmental phenomena, for example of air quality, providing a new capability for early-warning of significant changes.

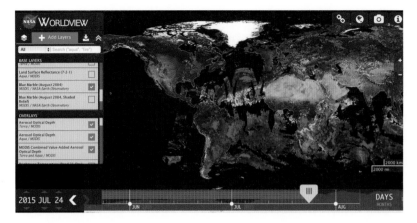

Figure 25.11 NASA's Worldview website and example of data layers

25.7 Conclusion

The timings of the key developments illustrated in this chapter hopefully demonstrate the comparatively young age of the Earth Observation discipline, and this – coupled with its high technological dependence – means that its capabilities, particularly in terms of satellite remote sensing, remain rapidly evolving. This swift advance means that textbooks can often fail to contain information on the most recent developments in platforms, instruments, EO products, algorithms and applications. Such detail can, however, often be gained through study of the relevant scientific journals, from the publicity material from Space Agencies and centres such as the UK NERC National Centre for Earth Observation (NCEO), and from appropriate Internet resources such as the excellent NASA Earth Observatory (http://earthobservatory.nasa.gov/).

SUMMARY

- Earth Observation (EO) generally refers to the study of the Earth from space using the techniques of remote sensing.
- Earth Observation can be defined as the approach of acquiring environmental information about the Earth from a distance, via measurements of naturally occurring or artificially generated electromagnetic radiation (EMR) that has interacted with the targeted Earth environment(s).
- Earth orbiting satellites are the most common platform now used to mount the instruments that collect remotely sensed data of the Earth, but aircraft and even ground-based stations are also used.
- A range of techniques, from simple one line equations to highly complex 'retrieval algorithms' described in tens of thousands of lines of computer code, may be used to create new information from the pre-processed imagery and data provided by satellite EO, revealing a host of information on key Earth System variables.
- The timings of the key developments demonstrate the comparatively young age of the Earth Observation discipline, and this coupled with its high technological dependence, mean that its capabilities, particularly in terms of satellite remote sensing, remain rapidly evolving.

Further Reading

- Rees, G. (2012) *Physical Principles of Remote Sensing* provides a detailed account of how remotely sensed observations are made and used.

- Jensen, J.R. (2007) *Remote Sensing of the Environment: An Earth Resource Perspective* provides a nice overview of the history of remote sensing and spaceborne sensors.

- Warner, T.A., Foody, G.M., and Nellis, M.D. (2009) *The SAGE Handbook of Remote Sensing* provides an overview of remote sensing and analysis techniques.

Note: Full details of the above can be found in the references list below.

ON THE COMPANION WEBSITE...

Visit **https://study.sagepub.com/keymethods3e** for author videos, chapter exercises, resources and links, plus **free** access to the following recommended articles:

1. **Aplin, P. (2005) 'Remote sensing: Ecology', *Progress in Physical Geography*, 29 (1): 104–113.**

   This article reviews applications for remote sensing and ecology.

2. **Wooster, M. (2007) 'Remote sensing: Sensors and systems', *Progress in Physical Geography*, 31 (1): 95–100.**

   This article reviews a range of new satellite Earth Observation (EO) sensors and their applications.

3. **Song, C. (2013) 'Optical remote sensing of forest leaf area index and biomass', *Progress in Physical Geography*, 37 (1): 98–113.**

   This article focuses on the use of optical remote sensing in mapping leaf area index (LAI) and abovegrand biomass for forests.

26 Digital Terrain Analysis

Peter L. Guth

SYNOPSIS

This chapter will introduce you to terrain analysis, which combines information about the earth's surface topography with other geospatial data to describe, visualize, or model the landscape. In today's work almost all terrain analysis is digital, using digital elevation models (DEMs). Terrain analysis can be done for basic scientific description, which falls under geomorphometry, a specialized branch of geomorphology, or it can be used to guide decision makers, most notably for military terrain analysis, which recognizes the vital role played by the physical aspects of the battlefield.

This chapter is organized into the following sections:

- Introduction
- Working with digital data
- Morphometric approaches in landscape
- Applying and interpreting analysis
- Comparison with RS and GIS
- Conclusion

26.1 Introduction

Terrain analysis employs elevation data, usually in conjunction with other geospatial information, to describe the landscape, for basic visualization, modelling, or to support decision making (Mayhew, 2009; Wilson and Deng, 2009; DoD, 2010; 2012). This chapter will get you started on using terrain analysis for your research. While terrain analysis can create tables, scatterplots, or histograms, the primary product will almost always be a map. The two reasons to do terrain analysis – to explore data and see relationships, and then communicate results to others – can be considered part of telling a story. Terrain analysis is the same as any other study in geography, or indeed any intellectual endeavour, and differs only in the questions asked and the data employed.

Terrain analysis ranges from largely qualitative, typified by military terrain analysis, to sophisticated numerical computations in geomorphometry. Military terrain analysis is 'the collection, analysis, evaluation, and interpretation of geographic information on the natural and man-made features of the terrain, combined with other relevant factors, to predict the effect of the terrain on military operations' (DoD, 2010). The same principles would apply to engineering site analysis or the selection of locations for economic development, where users seek to understand how the terrain affects and limits human activity. Geomorphometry is 'the science of topographic quantification; its operational focus is the extraction of land-surface parameters and objects from digital elevation models' (Pike et al., 2009: 4).

Terrain analysis grew out of air photo interpretation (Way, 1973; 1978), and some users prefer the term digital terrain analysis to emphasize the importance of computers to current usage. Other than speeding up execution to make possible exhaustive computations not previously realistically feasible, the use of computers and digital data has not led to many fundamental conceptual changes. What computers can do, however, is allow users to create 'results' without real understanding of the underlying processes and data. If you want to avoid GIGO (garbage in, garbage out), the need to really understand both earth processes and digital data remains. Digital terrain analysis is really a specific application of geographical information systems (GIS, see Chapter 37). You must understand the properties of the digital data, the characteristics of the algorithms applied to it, and finally the peculiarities of the software you use.

Every geography student would benefit from the simpler tools and techniques exemplified by military terrain analysis, which look at the landscape to understand what created the landforms and how that influences human activities. Much of your geographic education seeks to enable you to look at the earth, whether in the field, on a map or on a satellite image, and visualize the characteristics of that part of the world. You will know that this goal has been reached when you interpret the background landscape in the movies you watch, and ask yourself where the filming took place. Much of this simple terrain analysis can take place in Google Earth, but you must clearly understand the limits of that software. Google Earth satisfies most people most of the time, using the KISS (keep it simple, stupid) software design principle. This limits the number of options and choices, and allows users to jump in and use the software with minimal training and no reference to documentation. If you want to pursue more sophisticated terrain analysis, you will need to understand a great deal more background information to interpret the results, and you will also need a lot of training or to read a lot of documentation.

Exercise 26.1

Figure 26.1 shows five global maps. As a geography student you should be able to explain the spatial patterns shown on each map. While you should discuss a number of factors, for terrain analysis you would concentrate on how the absolute elevations and the local variations shown in Figure 26.1a contribute to the variations in the other maps. Most terrain analysis will probably operate at a much larger scale covering a small area with much more detail, but will seek the same understanding of relationships.

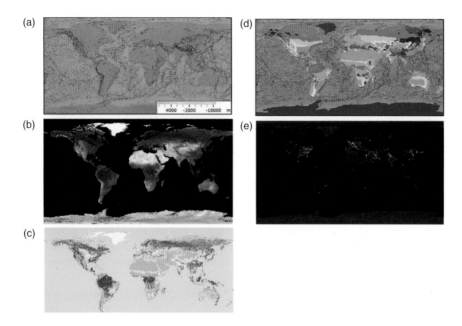

Figure 26.1   Global terrain analysis: (a) topography; (b) cloud free composite satellite imagery; (c) global land cover; (d) Köppen-Geiger climate classification; and (e) night lights

26.2 Working with Digital Data

Terrain analysis has evolved along with improved depictions of topography, and moved from a qualitative focus to allowing sophisticated quantitative work. The first depictions of topography used hachures (Imhof, 2007) – the depiction of the mountains in the American Southwest was famously described as 'abrupt ranges or ridges, looking upon the map like an army of caterpillars crawling northward' (Dutton, 1886: 116). While hachures have an artistic elegance, unless both the map-maker and the analyst exercise incredible care, they allow only very simple, qualitative assessments of the landscape. The move to contour lines required greater education for the user, but simple templates allowed rapid computation of slope based on the spacing of contour lines. With quadrangle maps having standard scales and contour line spacing, the templates could be standardized and an analyst could compute slope anywhere on the map. The move to digital topography meant that computer applications could rapidly compute slope, and a host of additional parameters, for every point on the map.

Digital topography can be stored in four formats: random points which can be arranged in triangulated irregular networks (TINs), contour lines, gridded digital elevation models (DEMs), or point clouds. No significant data sets have been produced as TINs. The Terrain Analysis Programs for Environmental Science (TAPES) program initially used digitized contours (Wilson and Gallant, 2000) but that format has since been eclipsed by grids. While grids might require more storage than TINs or contours, the low cost of storage means that the ease of processing with simple algorithms makes grids the overwhelming choice for mapping agencies and commercial firms making DEMs. A new shift to using point clouds, mostly from lidar on land but also from sonar underwater, is currently underway.

Maune (2007) defined DEM as a generic term for gridded topography or bathymetry, with two important variants. A DEM is often considered to be a 2.5-D surface, since there can only be a single elevation value at any location, meaning that caves and overhanging cliffs are not allowed. In mathematical terms this means the DEM represents a continuous, single valued function with the elevation (z value) as a function of the x and y values. A digital surface model (DSM, also called the first return surface) contains the highest point, and includes buildings, vegetation, and power lines. A digital terrain model (DTM, or bare earth surface) shows only the land surface and removes vegetation and cultural features. Older DEMs at what are now small and medium scale could not easily differentiate between DSMs and DTMs; their scales generalized features anyway. Most geomorphometry will use the DTM, but many users will want the cultural and vegetation information in the DSM. A difference grid computed with map algebra, which goes by several names including height above ground (HAG), will largely reflect the vegetation heights.

Point clouds can create TINs, but should be considered a separate category because of the huge data volumes, high point density (surveys commonly now produce > 10 pts/m^2) and the fact that they create true 3-D data sets. Current DEM production with lidar point clouds (Renslow, 2012) can easily produce both a DSM and DTM from a single survey, and these can be used for traditional terrain analysis. In addition to the huge increase in the size of the data sets to cover reasonable areas, the lidar DEMs show fine details which might be overkill – do we need to see every gully and boulder, or is a degree of smoothing more appropriate? We are still working on the answer to that question, as well as the degree to which we want to do terrain analysis with the original point cloud instead of the derived DTM.

Box 26.1   Key DEM characteristics

- Projection: geographic or projected like UTM.
- Horizontal datum: most will be WGS84 or a virtual equivalent like NAD83, and will hopefully be handled automatically by the GIS software.
- Vertical datum: most terrain analysis and GIS programs care most about relative elevations and local changes and can ignore vertical datums, but they will be increasingly important in dealing with sea level rise where small vertical changes have amplified horizontal impact. If collecting GPS data, you must know whether the elevations have been shifted from the native ellipsoidal height.
- Scale or data spacing: scale for DEMs can use the scale at which they can match a printed map, such as 1:24K or 1:250K, but the most useful characterization is the point spacing in either meters or arc seconds.
- Vertical resolution: elevations are usually stored either as 16 bit integers, in which case they can only record elevations to the nearest meter, or as 32 bit floating points, which can have a misleading precision but are probably only accurate to the nearest decimetre or centimetre.
- DTM (bare earth) or DSM (highest reflective surface).

Gridded DEMs have used two philosophies for the spacing. They can be created on a regular Cartesian grid from a particular map projection, most commonly Universal Transverse Mercator (UTM) or geometrically similar projections like UK OS or US State Plane (which has 126 variants, at least one per state). The map projections are defined over small regions over which earth curvature can be ignored, and equations can use a single grid spacing. Problems arise when the size of the analysis region increases and earth curvature cannot be ignored, or when the region of interest spans the boundary between adjacent zones. While GIS software can reproject and merge DEMs across zone boundaries, almost all DEM providers opted to use geographic coordinates with spacing in arc seconds or arc minutes. When the USGS created the National Elevation Dataset (NED; Gesch et al., 2002) and a web interface for downloading the data, they called it the Seamless Server to emphasize that data could be merged natively.

The best freely available DEMs with widespread coverage occur at global scale and medium scale, and almost all use geographic coordinates. Scale is subjective and has changed over time; before the rise of lidar data, many people would have considered the medium scale data to be large scale. A common scale is 3", which is often called either 90 m or 100 m data because the y spacing (longitude) is very close to the 90 m (about 92 m). The x spacing goes from 92 m at the equator to about 60 m at 50° latitude. The US NASA and NGA have released data at this scale, some in conjunction with European and Japanese space agencies. The availability of free mapping data depends on government philosophies, with the United States and Canada being the most open. Because of the uses of digital data for military terrain analysis, other countries regard mapping data as state secrets. Between these extremes, governments can sell data, or only expensive commercial data will be available.

Probably the single most important DEM for terrain analysis is the Shuttle Radar Topography Mission (SRTM) (Guth, 2006). This space shuttle mission collected a consistent, medium scale DEM that covers most of the earth's land surface. The earliest version suffered from data voids ('holes') in mountainous terrain, water bodies, and some deserts, and led to several hole filling versions. The most common format for this data uses HGT files which have no metadata and cannot be renamed because software must get the location from the file name. SRTM data are so important that most software can read the files, a testament to the revolutionary view of Earth's topography provided by the mission.

Box 26.2   Free DEMs and lidar point clouds

- Small scale (postings > 500 m, in arc seconds or arc minutes) with global coverage.

  - ETOPO5/ETOPO1: http://www.ngdc.noaa.gov/mgg/global/global.html – Land surface and underwater.
  - SRTM30 plus: http://topex.ucsd.edu/WWW_html/srtm30_plus.html – Land surface and underwater.
  - GMTED2010: http://topotools.cr.usgs.gov/GMTED_viewer/ – Spacings at 30", 15", and 7.5" for land surface only.

(Continued)

(Continued)

- Medium scale (postings 10-100 m, all in arc seconds except for UK OS Terrain)

  o SRTM version 3.0: http://earthexplorer.usgs.gov/ This covers the land surface south between 60°N and 54°S with 1" and 3" spacing.
  o ASTER GDEM: http://asterweb.jpl.nasa.gov/gdem.asp. Despite its nominal 1" spacing, GDEM is generally 'worse' than SRTM. You should probably only get it if the SRTM has many holes (mountains or deserts), or your region is north of the limits for the SRTM mission.
  o USGS NED: http://nationalmap.gov/viewer.html has full coverage of the United States at 1" and 1/3", and increasing coverage at 1/9".
  o UK OS Terrain 50: https://www.ordnancesurvey.co.uk/business-and-government/products/terrain-50.html – 50 m spacing, on the UK grid.
  o Canadian CDEM: https://www.nrcan.gc.ca/earth-sciences/geography/topographic-information/free-data-geogratis/11042 – Geotiff at 12", 6", 3", 1.5" or 0.75" spacing.

- Small scale (point clouds or grids with postings < 2 m). You can get LAS or LAZ files, and in some cases grids.

  o USGS: http://earthexplorer.usgs.gov/
  o NOAA Coastal Explorer: http://www.csc.noaa.gov/dataviewer/#
  o OpenTopography: http://www.opentopography.org/

Florinksy (2012) discussed the challenges of computing many derived geomorphic parameters for what he calls spheroidal equal angle grids, where the spacing between grid postings is in arc seconds, or arc minutes, or arc degrees. The best of these, such as the USGS NED, Canadian CDEM, SRTM, or ASTER GDEM, in fact use an ellipsoidal rather than spheroidal earth model, so that alternative names like latitude/longitude grids, arc second DEMs, or the geographic coordinate system (GCS) used by ArcGIS might be more appropriate. Of the terrain analysis software discussed by Wood (2009) here, only MICRODEM and River Tools use algorithms that correctly account for DEMs with geographic spacing. At the scale of these DEMs, the geometry can locally be viewed as quasi rectangular, with different spacing in the x and y directions. The y spacing is effectively constant (very small differences due to the ellipsoidal shape of the earth), and the x spacing varies with latitude. Equations for computations on the DEM which use the x and y spacing as equal in meters (e.g. Hengl and Reuter, 2009; Florinksy, 2012) can be revised to make the x and y spacing different, and to make the x spacing a function of latitude computed once and then looked up to perform computations.

Most GIS programs will either perform very poorly with the geographic DEMs, or refuse to process them. The solution is to reproject the DEM to UTM coordinates or a similar system. This will interpolate new elevations at new locations. At best it will create no noticeable changes, and at worst it will raise low points and lower the peaks – there is no way the reinterpolation process will improve on the source DEM.

Wood (2009) classified eight geomorphometry packages, all of which can be used for n-basic terrain analysis, showing their placement in a ternary diagram with an axis for general GIS, geomorphometry, and hydrology. Wood's own LandSerf has not been updated since 2009, and the support forum is closed and might have problems with current Java implementations. TAS has been renamed to Whitebox. Two of the programs, ArcGIS and RiverTools, are commercial and you should choose them only if you already own them or know how to use them; otherwise you should explore one of the free options listed below. They will all produce equivalent results for basic terrain analysis, and if you want to pursue advanced capabilities beyond what can be covered in this chapter, your choice will probably be dictated by which program performs the operations you require.

The choice of software to use depends on two factors: what your institution has available, and what you want to do. If you have educational institution access and learned how to use a package in class, that would be the best starting point. All of these programs will do basic terrain analysis, but for specialized analysis many of these will outperform ArcGIS for specific operations. ArcGIS is a generalized commercial tool, and the others have been programmed by scientists for teaching or research.

Box 26.3   Suggested terrain analysis programs (free except for ArcGIS)

- ESRI ArcGIS: the commercial gold standard for GIS. http://www.esri.com (Windows).
- qGIS with GRASS: http://www.qgis.org/en/site/ and http://grass.osgeo.org/ – qGIS provides a Graphical User Interface to work with GRASS (Windows, Mac).
- SAGA: http://www.saga-gis.org/en/index.html (Windows).
- ILWIS: http://52north.org/communities/ilwis (Windows).
- Whitebox: http://www.uoguelph.ca/~hydrogeo/Whitebox/index.html (Windows, Mac).
- MICRODEM: listed last since I might not be the most impartial judge of its capabilities. http://www.usna.edu/Users/oceano/pguth/website/microdem/microdemdown.htm

---

Except for MICRODEM, none of these programs will handle DEMs in a geographic projection very well, and you will have to reproject the DEMs into a coordinate system like UTM before use. If you stick to the common data formats listed above, you can mix and match the programs to take advantage of their different capabilities. Most modern GIS programs can reproject data on the fly, and will correctly let you mix and match data sets, but in this list only ArcGIS and MICRODEM will do that and the others require that you ensure that all your data have the same projection before beginning analysis.

Box 26.4   Additional free data for terrain analysis

- Satellite imagery, Landsat 8 has global coverage with 30 m resolution
- Land cover showing several dozen categories at scales from 30 m (US) to 1 km (global), and can show changes over time
- Vector roads and other cultural data, from OpenStreetMap
- Night lights, which show population density
- Köppen-Geiger Climate classification

The help file for MICRODEM has descriptions of these data sets and download instructions (download link: http://www.usna.edu/Users/oceano/pguth/microdem/win32/microdem.chm)

26.3 Morphometric Approaches in Landscape

The hillshade or shaded reflectance map probably provides the single most useful graphic for terrain analysis, as shown in Figure 26.2. Most programs will require creating the hillshade as a separate map layer, and require you to merge two layers for a display like Figure 26.2a, but MICRODEM will create it on the fly for display, and display the elevation while roaming. Figure 26.2a shows the combination of elevation and hillshade, which works best as a stand-alone figure. When the elevation map is used as a base for other map layers, as in Figures 26.2a and 26.2c, the greyscale hillshade provides relative relief and highlights ridges and valleys, leaving colour for the overlaid layers which can be vector data (Figure 26.2b) or raster data (Figure 26.2c). Figure 26.2d shows the same area displayed with elevations alone, and neither the coloured nor the greyscale version comes close to the detail in Figure 26.2a. In addition, the greyscale version is too dark for combining with other geospatial data as in Figure 26.2b or c. ArcGIS will create hillshades from geographic coordinate DEMs, but they will be too dark and noisy for use; you must use a UTM DEM.

Figures 26.2b and c show the power of simple, qualitative terrain analysis using GIS visualization. The town of Harpers Ferry lies at the confluence of the Potomac and Shenandoah Rivers. The rivers and the folded Appalachian Mountains determine the routes of the roads and railroads, and Figure 26.2b explains why settlement development

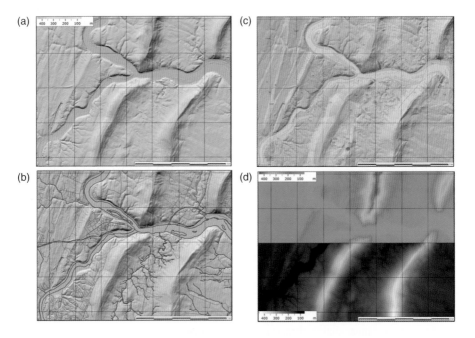

Figure 26.2   Harpers Ferry, West Virginia, from a 1/3" NED DEM: (a) hillshade combined with colours for elevation; (b) greyscale hillshade with TIGER roads, streams, and railroads in cyan; (c) greyscale hillshade combined with geologic map (Southworth et al., 2000); (d) elevations displayed with colour and greyscale

occurred where it did, and goes a long way towards explaining the military importance of the arsenal at Harpers Ferry during the American Civil War. The folded terrain also explains the two bloodiest battles of the war, which occurred just to the north at Antietam and Gettysburg when the South attempted to invade the North via the Great Valley between the prominent folded mountains. Figure 26.2c shows how the terrain reflects the underlying geology.

Once you move beyond the hillshade, the number of potential geomorphometric variables increases dramatically. Many are local, or focal operations in the GIS terminology, since they require a point's elevation and that of its immediate neighbours. The most important are the slope or steepness, and the aspect or downhill direction. These are the components of the first derivatives of the elevation surface, which is a vector with slope the magnitude and aspect the direction. There is a large literature on the best algorithms to use to compute slope, but all the proposed algorithms correlate strongly and the differences degenerate into philosophical definitions: is a peak flat, since at least at one point the tangent surface will be horizontal, and in a valley, do you want the slope along the valley or perpendicular to it? Other than these singular points, the regional slope patterns do not depend on the algorithm used. The values will depend slightly on the algorithm, and perhaps more on the DEM used. In general, slope increases as the DEM spacing decreases due to smoothing generalizations in the medium and small scale DEMs. The SRTM contains radar speckle which causes slopes in the flatter regions to be gentler than those computed from cartographic DEMs like NED or CDEM, while the SRTM is gentler in the steeper areas. This should not present problems if you stick with the same DEM source and scale for comparing landscapes.

The DEM also shows ridges and valleys with curvature, the second derivative of elevation or the rate of change of slope, whether the land surface is convex or concave. This derivative can be done in the horizontal or vertical direction (profile or plan curvature), and a number of additional variants have been proposed (Schmidt et al., 2003; Olaya, 2009; Minár et al., 2013). As the second derivative, curvature greatly amplifies small imperfections in the original DEM, and the DEM should probably be filtered before computing curvature (Hengl and Evans, 2009). The distribution of curvature values, with most values very close to 0 with small tails of both positive and negative curvature, means that you will probably have to adjust the colour scheme to show the results on a map.

Zonal characters require a computation region about the point, and as a result require the analyst to pick the size of the region based on the characteristics of the landscape and the data spacing of the DEM. This operation requires a region large enough for valid statistics, while hopefully remaining relatively homogeneous. The most common of these parameters are the fours moments of the elevation, slope, and two curvature distributions (Evans, 1988). Others include relief, and upward and downward openness (Yokoyama et al., 2002). Many of the parameters that have been proposed actually measure slope and closely correlate with it. A wide variety of geomorphic parameters are available (e.g. Florinsky, 2012; Olaya, 2009). Figure 26.3 shows maps with a number of parameters. Some show fine details, but others – like relief – change very slowly because the difference between the highest and lowest points in the specified region will only exist when the moving box that computes relief encounters new extreme values.

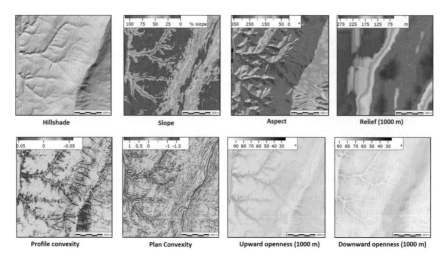

Figure 26.3 Maps of part of the 1/3" NED DEM near Harpers Ferry, United States. Five maps use focal operations (hillshade, slope, aspect, and two convexities), while three use zonal operations (relief and two openness measures) which require definition of a 1000 m region.

Box 26.5   Key terms

- Hillshade/shaded reflectance is probably the most useful and versatile map display for terrain analysis, and for use as a base map to show the relations of topography to other geographic data.
- Slope: can be measured in degrees or as a percentage (100 times the tangent of the slope angle, so a 100% slope corresponds to a 45° slope). For human operations slope determines mobility, but it also affects erosion and mass movement.
- Focal operations work on a point and its immediate neighbours.
- Zonal operations work on a region about the point, and the analyst must specify the region size.
- Viewsheds (or 'weapons fans' for military terrain analysis) show what terrain has line-of-sight with a particular point, and has uses for cell phone coverage or land preservation around parks and other locations.
- Hydrologic operations can extract drainage basins and stream channels from the DEM, and compute Strahler orders (stream hierarchy).

A current focus in geomorphometry attempts classification of landforms from the DEM. Combining several of the geomorphometric parameters leads to a virtually unlimited number of possible classifications. MacMillan and Shary (2009) discuss some of this history, including simple zonal calculations that can categorize points as pits, peaks, ridges, valleys, passes, and more complex schemes based on several curvature methods. Approaches to these classifications require (1) the analyst to set limits for the categories for each parameter, which might depend on the character of region analysed and the properties of the data, or (2) the use of object-based image analysis with programs which seek coherent regions in the data and grow them to find a reasonable balance between the number of size of the regions. Iwahashi and Pike (2007) presented

Figure 26.4 Landform classification performed in SAGA, exported as a Geotiff to MICRODEM, and exported as KMZ to Google Earth. The irregular white collar is a result of reprojecting the UTM data from SAGA into the geographic projection required for Google Earth

a classification based on defined categories of slope, local convexity, and surface texture (the 'grain', or horizontal spacing of features, and the hardest of the three to quantify), with impressive maps which SAGA can create (Figure 26.4). The GEO-Object-Based Image Analysis (GEOBIA) most commonly uses the commercial eCognition software (e.g. Drăgut and Blaschke, 2006; Drăguț and Eisank, 2012). Attaching meaning to the classifications represents the current challenge in geomorphometry.

If you want to investigate terrain analysis beyond the capabilities of the turnkey software packages, you have several options: (1) scripting in ArcGIS with Python, (2) extending one of the open source programs like SAGA, WhiteBox, GRASS, or qGIS, (3) using R with GRASS or SAGA, or (4) using MATLAB. Do not try to reinvent the wheel, and recognize that if you use common data formats like Geotiff for grids and shapefiles for vectors, you can ingest data from other programs, and export these for further processing or display.

26.4 Applying and Interpreting Analysis

Military terrain analysis uses the acronym OCOKA, which stands for 'observation and fields of fire, cover and concealment, obstacles, key terrain, and avenues of approach' (US Army, 1990). This mnemonic helps the analyst to provide a mental checklist, and to consider the most common factors. For most geographers, experience replaces the explicit checklist, but the goal remains to understand and explain the landscape.

Terrain analysis benefits from the largest monitor you can get, or multiple screens. In addition to the map on screen, you might have to open a web site to download

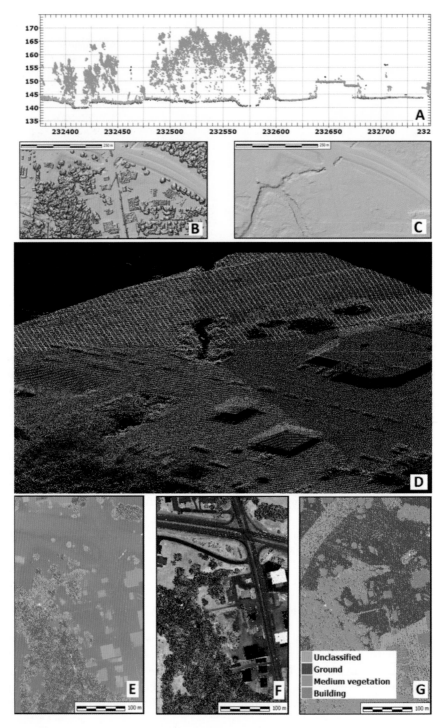

Figure 26.5 Four views of a lidar point cloud from Keene, New Hampshire. (a) east-west one meter thick slice; (b) one meter DSM created from the data; (c) one meter DTM; and (d) 3-D view of the point cloud; (e) maps showing points coloured by elevation; (f) return intensity, and (g) classification.

data or look up ancillary information, and a word processor to report your findings. Terrain analysis produces maps, and almost always your map will be disseminated via a PowerPoint presentation or a word processor document, where it might find its way into a periodical or even a book like this. For those sources, you want the map itself to use as much of the screen or the page as possible. My bias is to always include a scalebar, except on global maps likes those in Figure 26.1 where the scale varies too much to allow a meaningful scalebar, and to put in only as much legend as needed to show important features that are not obvious by standard symbology or context. If you must produce maps for a traditionalist who insists on a north arrow, full legend, and location map, that's how you must produce your maps, but pay attention to the amount of unnecessary white space many of those maps produce. Your job is to show the detail to understand the map area, and that requires maps as large as you can create.

Figure 26.5 shows a lidar survey, and how it can be visualized for terrain analysis. Lidar records three distinct parameters that can be displayed: elevation (Figure 26.5e), return intensity (26.5f), and a classification based on the geometric relationships on neighbouring points (26.5g). The display also show slices through the point cloud (26.5a), and in addition to blue roofs, this slice shows two utility lines. While the slice allows easy quantitative measurement, a 3-D interactive view can enhance qualitative assessment (26.5d). The point cloud can also generate a DSM (26.5b) and DTM (26.5c), which will provide the input for geomorphometric modelling. The lidar classification (26.5a, 26.5d, and 26.5g) requires post processing and is not always done. The latest specification for the point clouds adds a number of categories for utility transmission lines, because of the capability to monitor line status and vegetation encroachment with rapid corridor surveys, but the addition will add other options like military terrain analysis where power lines present significant obstacles for helicopter operations.

26. 5 Comparison with RS and GIS

Terrain analysis, especially as used by the military, predates both geographical information systems and modern remote sensing. While some purists will continue to rely largely on a printed paper map, almost all terrain analysis now uses digital tools and data whether the users refer to the process as digital terrain analysis or not.

Effective terrain analysis requires two extensions for ESRI's ArcGIS software (Spatial Analyst and 3D Analyst). While this may be largely a marketing strategy to reduce the apparent cost of the product, it shows that at least some knowledgeable people do not consider the tools needed for terrain analysis to be an essential component of the GIS toolbox. However, I think most users would say that digital terrain analysis simply uses parts of the GIS toolbox to automate and improve techniques that formerly relied on topography maps and airborne imagery.

Terrain analysis relies on remote sensing for base maps, and derived data sets like land cover or vegetation indices. Even the digital topography almost always comes from remote sensing imagery, whether stereo imagery (visible, near infrared in the case of the ASTER GDEM, or interferometric radar for SRTM) or the new lidar point clouds.

26.6 Conclusion

Terrain analysis is already possible on smart phones and tablets. Imagery and maps can be taken into the field, and GPS sensors overlay your position on the imagery as you move. On a cross country vacation, I can load my tablet with satellite imagery, geologic maps, and topographic maps in standard GIS formats like GeoTIFF or GeoPDF, and perform terrain analysis from the airplane window or while hiking. For someone whose initial field mapping was entirely analog, this capability astounds me, but does not change the need to appreciate the landscape and relate the features to fundamental processes. At the moment taking a full GIS software package for terrain analysis would require something a little larger than a pocket, but that will soon change, and terrain analysis in the field will be the norm.

A second trend will move some terrain analysis into the cloud, with only a browser or other light software on your computer. It's already possible to do satellite remote sensing on the Internet, and web services like slope maps and viewsheds can also be called from GIS programs. This removes the need to download and manage data, and to do the computations locally, especially if the cloud can use multiple processors to do the computations in parallel. I see three cautions for the full adoption of this: (1) cost of the cloud services, (2) inability to access the cloud for critical uses in the field, where from military or scientific users, and (3) likely limitation to only the more common functions.

As a geographer, terrain analysis is in your future, and in fact will be part of the online experience for all people using web programs like Google Earth and the programs that will be ubiquitous on our handheld devices.

SUMMARY

- Modern terrain analysis uses digital terrain, supplemented with other geospatial data, to describe and interpret the landscape.
- Terrain analysis started with printed maps, but now uses DEMs and is starting to use lidar point clouds.
- Terrain analysis uses GIS software as a tool to achieve a particular outcome.
- The hillshade or shaded relief map is the most useful terrain analysis product, and should be considered by all geographers (and many others) to provide a base map when the hills and valleys have influenced what the map should depict.
- More advanced terrain analysis uses focal and zonal parameters computed from the DEM, using geomorphometric algorithms.

Further Reading

- Way (1973, 1978) provides a classical description of terrain analysis with air photos and topographic maps. The principles still apply, and will help your understanding of digital geospatial data.

- The US Army (1990) Field Manual is out of date, but provides an understanding of how militaries use terrain analysis and the combination of field work and map analysis that can help guide planning and decision making. A Google search for FM 5-33 turns up several places to download a PDF, but no official government web sites.

- The papers in Maune (2007) summarize the uses and properties of gridded DEMs. Since most terrain analysis currently uses DEMs, this book provides the best starting place to understand the data and their limitations. Chapter 8 briefly discusses user applications in flood insurance, wetlands, forestry, utility corridors, coastal management, transportation, disaster and military operations.

- The papers in Renslow (2012) cover topographic lidar, including data collection, processing, and applications of the data. As usage proliferates of very large scale DEMs from lidar, and the point clouds themselves, this topic becomes vital. Look for a second edition, as the field explodes in popularity. Chapter 10, with 24 authors, discusses applications in forestry, corridor mapping, flood mapping, building extraction and reconstruction, airport surveys, coastal and hydrological monitoring and natural hazards.

- The papers in Hengl and Reuter (2009) provide an introduction to geomorphometry, with descriptions of eight software packages. Section 3 has ten chapters covering applications for mapping soils, vegetation, mass movements and landslides, ecology, hydrological modelling, meteorology and precision agriculture.

- The papers in Smith et al. (2011) discuss geomorphological mapping. Chapter 8 by Smith covers the visualization, interpretation, and quantification of landforms.

- Florinsky (2012) discusses digital terrain analysis in soil science and geology. His LandLord software is only available for researchers doing collaborative research.

Note: Full details of the above can be found in the references list below.

References

DoD (2010 amended through 15 November 2015) *Department of Defense Dictionary of Military and Associated Terms*, Joint Publication 1-02. http://www.dtic.mil/doctrine/new_pubs/jp1_02.pdf (accessed 26 November 2015).

DoD (2012) *Geospatial Intelligence in Joint Operations*: Joint Publication 2-03, http://www.dtic.mil/doctrine/new_pubs/jp2_03.pdf (accessed 26 November 2015).

Drăgut, L. and Blaschke, T. (2006) 'Automated classification of landform elements using object-based image analysis', *Geomorphology* 81: 330–44.

Drăgut, L. and Eisank, C. (2012) 'Automated object-based classification of topography from SRTM data', *Geomorphology* 141: 21–33.

Dutton, C.E. (1886) 'Mount Taylor and the Zuñi Plateau', in *Report by the Director of the United States Geological Survey*, U.S. Government Printing Office. pp. 111–98.

Evans, I.S. (1998) 'What do terrain statistics really mean?', in S.N. Lane, K.S. Richards and J.H. Chandler (eds) *Landform Monitoring, Modelling and Analysis*. Chichester: Wiley. pp. 119–138.

Florinsky, I.V. (2012) *Digital Terrain Analysis in Soil Science and Geology*. Kidlington: Elsevier.

Gesch, D.B., Oimoen, M., Greenlee, S., Nelson, C., Steuck, M. and Tyler, D. (2002) 'The national elevation dataset', *Photogrammetric Engineering and Remote Sensing* 68: 5–11.

Guth, P.L. (2006) 'Geomorphometry from SRTM: Comparison to NED', *Photogrammetric Engineering and Remote Sensing* 72: 269–77.

Hengl, T. and Evans, I.S. (2009) 'Mathematical and digital models of the land surface', in T. Hengl, and H.I. Reuter (eds) *Geomorphometry: Concepts, Software, Applications* (Developments in Soil Science Series). Kidlington: Elsevier. pp. 31–63.

Hengl, T. and Reuter, H.I. (eds) (2009) *Geomorphometry: Concepts, Software, Applications* (Developments in Soil Science Series). Kidlington: Elsevier.

Imhof, E. (2007) *Cartographic Relief Presentation*. Redlands: ESRI.

Iwahashi, J. and Pike, R.J. (2007) 'Automated classifications of topography from DEMs by an unsupervised nested means algorithm and a three-part geometric signature', *Geomorphology* 86: 409–40.

MacMillan, R.A. and Shary, P.A. (2009) 'Landforms and landform elements in geomorphometry', in T. Hengl and H.I. Reuter (eds) *Geomorphometry: Concepts, Software, Applications* (Developments in Soil Science Series). Kidlington: Elsevier. pp. 227–254.

Maune, D.F. (ed.) (2007) *Digital Elevation Model Technologies and Applications: The DEM Users Manual*. Bethesda: American Society for Photogrammetry and Remote Sensing.

Mayhew, S. (2009) *A Dictionary of Geography*. Oxford: Oxford University Press.

Minár, J., Jenčo, M., Evans, I.S., Minár Jr., J., Kadlec,M., Krcho, J., Pacina, J.,Burian, L. and Benová, A. (2013) 'Third-order geomorphometric variables (derivatives): Definition, computation and utilization of changes of curvatures', *International Journal of Geographical Information Science* 27: 1381–402.

Olaya, V. (2009) 'Basic land surface parameters', in T. Hengl and H.I. Reuter (eds) *Geomorphometry: Concepts, Software, Applications* (Developments in Soil Science Series). Kidlington: Elsevier. pp. 141–69.

Pike, R.J., Evans, I.S. and Hengl, T. (2009) 'Geomorphometry: A brief guide', in T. Hengl and H.I. Reuter (eds) *Geomorphometry: Concepts, Software, Applications* (Developments in Soil Science Series). Kidlington: Elsevier. pp. 3–30.

Renslow, M.S. (2012) *Manual of Airborne Topographic Lidar*. Bethesda: American Society for Photogrammetry and Remote Sensing.

Reuter, H.I., Hengl, T., Gessler, P. and Soille, P. (2009) 'Preparation of DEMs for geomorphometric analysis', in T. Hengl and H.I. Reuter (eds) *Geomorphometry: Concepts, Software, Applications* (Developments in Soil Science Series). Kidlington: Elsevier. pp. 87–140.

Schmidt, J., Evans, I.S. and Brinkmann, J. (2003) 'Comparison of polynomial models for land surface curvature calculation', *International Journal of Geographical Information Science* 17: 797–814.

Southworth, S., Brezinski, D.K., Orndorff, R.C., Logueux, K.M. and Chirico, P.G. (2000) *Digital Geologic Map of the Harpers Ferry National Historic Park*. US Geological Survey Open-File Report OF-2000-297. http://ngmdb.usgs.gov/Prodesc/proddesc_34293.htm (accessed 26 November 2015).

Smith, M., Paron, P. and Griffiths, J.S (eds) (2001) *Geomorphological Mapping: Methods and Applications* (Developments in Earth Surface Processes). Kidlington: Elsevier.

US Army (1990) *Terrain Analysis: Field Manual*. FM5-33.

Way, D.S. (1973) *Terrain Analysis: A Guide to Site Selection Using Aerial Photographic Interpretation*. Stroudsburg, PA: Dowden Hutchinson & Ross.

Way, D.S. (1978) *Terrain Analysis: A Guide to Site Selection Using Aerial Photographic Interpretation* (2nd edition). New York: McGraw Hill.

Wilson, J.P. and Deng, Y. (2009) 'Terrain analysis', in K.K. Kemp (ed.) *Encyclopedia of Geographic Information Science*. London: Sage. pp. 465–8.

Wilson, J.P. and Gallant, J.C. (eds) (2000) *Terrain Analysis: Principles and Applications*. Abingdon: Wiley.

Wood, J. (2009) 'Overview of software packages used in geomorphometry', in: T. Hengl and H.I. Reuter, (eds) *Geomorphometry: Concepts, Software, Applications* (Developments in Soil Science Series). Kidlington: Elsevier. pp. 257–68.

Yokoyama, R., Sirasawa, M. and Pike, R.J. (2002) 'Visualizing topography by openness: A new application of image processing to digital elevation models', *Photogrammetric Engineering and Remote Sensing* 68: 257–65.

ON THE COMPANION WEBSITE…

Visit **https://study.sagepub.com/keymethods3e** for author videos, chapter exercises, resources and links, plus **free** access to the following recommended articles:

1. **Caquard, S. (2014) 'Cartography II: Collective cartographies in the social media era', *Progress in Human Geography*, 38 (1): 141–50.**

This paper looks at ways to mobilize citizenry to create maps and digital data, which can serve as the inputs for terrain analysis. Perhaps typical for human geographers, this is a thought piece with no figures.

2. **Gillespie, T.W., Willis, K.S. and Ostermann-Kelm, S. (2015) 'Spaceborne remote sensing of the world's protected areas', *Progress in Physical Geography*, 39 (3): 388–404.**

This paper surveys the current satellites, which provide a critical input data for digital terrain analysis. The focus is on their use for protected areas, but the principles apply to any regions.

3. **Roche, S. (2014) 'Geographic Information Science I: Why does a smart city need to be spatially enabled?', *Progress in Human Geography*, 38 (5): 703–11.**

Human terrain analysis is one of the hardest subjects for quantification in GIS because the data are fuzzy and hard to collect. This paper explains how social media can help map urban areas, but without showing any maps or other figures.

27 Environmental GIS

Thomas W. Gillespie

SYNOPSIS

Environmental GIS refers to the use of Geographic Information Systems in environmental science and management. One of the most fundamental aspects of environmental GIS for geographic research is to collect, analyse, and create new layers with environmental data and compare results with other GIS datasets to test hypotheses in science. This can be done over different spatial scales (landscape, regional, and globally) and temporal scales (present, past, and future) with an amazing array of environmental datasets that are freely available. There are a number of common spatial analyses used in environmental research that are readily available in GIS software packages like ArcInfo and QGIS. There are a number of commonly used geospatial analyses used in environmental GIS such as measurements, overlays, proximity analyses, connectivity analyses, spatial modelling, and visualization. Environmental GIS is a fundamental part of environmental management used by national and local governments, businesses, and non-profits organizations for decision-making, governance, and planning. The use of environmental GIS will continue to increase in the future.

This chapter is organized into the following sections:

- Introduction
- Collecting environmental GIS data
- Environmental GIS datasets
- Environmental analyses
- Applied environmental GIS
- Conclusion

27.1 Introduction

GIS has become a fundamental part of geography and environmental science. In essence, GIS is simply georeferenced spatial (point, line, and area) data with attributes in a digital format. It has evolved with advances in computer speed and storage space, algorithm development and visualization techniques. Today GIS is a computer system designed to import, store, analyse, and visualize all types of geographical data. Environmental GIS focuses on the collection, analysis and visualization of geological, biological, chemical, climatic and natural resource datasets. The strength of GIS is that you can formulate and test theories, which can result in real world products that can be used by scientists, decision makers and the general public.

There are a number of commonly used terms in GIS that should be made clear to fully understand the applications and limitations of GIS. Point, line, and area data are sometimes referred to as node(s), arc(s), and polygon(s) respectively. Georeferenced point, line, and area data are collectively referred to as a **vector data** structure with similar data structure to a computer drawing program. In contrast, **raster data** are rectangular or square-based cells in a gridded system of rows and columns similar to an image from a digital camera. Each cell contains information on location and a value related to some environmental attribute that has been created for a region and has generally been collected from an airplane or satellite. Both vector and raster data structures are referred to as layers or coverages and both are commonly used in environmental GIS analyses. Most GIS is done with commercial ESRI software that includes ArcView for basic GIS analyses and ArcInfo for more advanced analyses. However, there are a number of open source GIS software programs like QGIS, that can be used for scientific analyses and contain commonly used GIS operations (see Chapter 26 Box 3). The first step is to get data into a GIS and this is done in three general ways. One can collect field data on an environmental theme of interest (i.e. soils, pollution, plants) to create new GIS layers, digitize data not currently in a GIS format, or download environmental GIS layers. The second step is to undertake spatial analysis in GIS to test hypotheses and theories with GIS layers. Finally, GIS provides statistical and visual data for scientists and decision-makers who need accurate and clear environmental information. This chapter reviews ways to collect, analyse, and apply environmental GIS data of interest.

27.2 Collecting Environmental GIS Data

The strength of GIS is that you can formulate and test hypotheses and theories in geography and environmental science, which can result in real world products that can be used by scientists, decision makers, and the general public. The first step is getting data into a GIS. This can be done a number of ways with field data collection, digitizing or geocoding existing environmental data, or downloading environmental GIS products.

Collecting field data on physical features (temperature, soil type), biodiversity (species, vegetation), and environmental themes (water chemistry, natural resources) is generally done at the point level with Global Positioning System (GPS) units to identify locations and a standard method to collect environmental information or attributes (see Chapter 20). In many ways this type of standard and repeatable collection of GIS data is the most important part of research and based on a hypothesis that is to be tested. The vast majority of environmental GIS data are samples of a population (e.g. temperature from climate stations, species inventories) rather than an exhaustive or complete inventory of environmental themes. All practitioners of environmental GIS should try to collect at least 30 locations in the field to truly understand the challenges of collecting field data and creating high quality GIS layers. For environmental GIS, one of the most important aspects is the exact location of the feature. It is easiest to collect point data and more complicated to collect line and area data in the field with GPS units. As a geographer, remember that you will need a standard and repeatable field collection method to create your GIS layers and if you do excellent work people will want to use and test your work in the future. GIS is one standard and repeatable method that can be used to store and distribute this data.

The second way to get data into a GIS is to use resources that are currently not in a georeferenced format. This can include a host of resources such a paper maps, historic records, aerial photographs, databases, or numerical data. The object is to get these data into a digital spatial database. First, you need to identify the data you need for your analyses and determine if they should be digitized as point, line, or area data. Point features are easiest to geocode by simply entering coordinates of latitude and longitude into a spreadsheet with attribute information; this can be imported into GIS. Line and area features are easiest to enter by tracing maps on screen (heads-up digitizing) or classifying georectified imagery to create environmental GIS data layers. Imagery in Google Earth can be digitized as point, line, or area in Keyhole Markup Language (KML) and easily be converted into files for use in GIS software. Point features such as tree locations, line features such as trails, and area features such as lakes or vegetation types, can all be collected with digital imagery. Remote sensing imagery from airborne and spaceborne sensors can also be used to quantify geographic metrics.

Finally there has been a significant increase in the number of environmental GIS datasets and analyses that have been done in science with point, line, and area data that would be impossible for one person to collect in their lifetime. Thus there is sometimes a need to use existing environmental GIS datasets to test hypotheses in environmental science and geography.

27.3 Environmental GIS Datasets

It is important to identify the best environmental GIS datasets to allow you to examine your environmental interest across temporal and spatial scales. This should also make you an expert on what GIS data exist for your environmental theme, your study area, and the quality of the data. It is important to critically examine the data you are using and this process begins by reading the metadata files to better understand the applications and limitations. Since digital GIS has been rapidly evolving over the last 30 years, there are a number of public access environmental datasets that currently exist. Scientists, governments, and non-governmental agencies have created and updated an amazing array of large environmental datasets, and set up portals online where GIS data can be freely downloaded (Table 27.1). Below I provide information on high quality environmental GIS datasets related to environmental science across different spatial and temporal scales (see Chapter 28).

There are a number of scientists that have devoted their lives to collecting standard and repeatable data on a number of environmental themes. Indeed, scientists are encouraged to manage and share their GIS data with the scientific community. A landscape example of high-resolution GIS point data is inventories of all plants > 1 cm in diameter at breast height that have been mapped in a 50 ha plot of tropical rainforest every five years since 1980 in Panama (http://ctfs.si.edu/webatlas/datasets/ as an additional resource) (Figure 27.1a). Regional examples of polygons are GIS datasets estimating organic carbon storage to 3 m depth in soils of the northern circumpolar permafrost region (Hugelius et al., 2013; Figure 27.1b). There are also some truly amazing global

Table 27.1 Useful environmental GIS datasets and links (URLs checked 27 November 2015)

Source	Environmental GIS data	Link
USGS	Topography, geology	http://ned.usgs.gov
EPA	Pollution, water	http://www.epa.gov/geospatial
Fish and Wildlife	Species, habitats	http://www.fws.gov/gis/index.html
US National Maps	US GIS datasets	http://nationalmap.gov/viewer.html
European Environ. Agency	Environmental datasets	http://www.eionet.europa.eu/gis/
UN Environ. Agency	Environmental datasets	http://ggim.un.org
Map of Life	Species distributions	http://www.mappinglife.org
IUCN	Threatened species	http://www.iucnredlist.org/technical-documents/spatial-data
NASA	Remote sensing	http://gcmd.nasa.gov
WorldClim	Past, present, and future climate	http://www.worldclim.org
Landscan	Population and age structure	http://web.ornl.gov/sci/landscan/
HYDE	Past land cover and population	http://themasites.pbl.nl/tridion/en/themasites/hyde/

datasets such as Hansen et al. (2013) forest cover dataset. This online dataset contains forest cover, loss, and gain in 30 m pixels at a global spatial scale for 2000 to 2012 from Landsat imagery (Figure 27.1c). There has also been an increase in citizen scientists' databases such as Ebird (www.ebird.org), which contains empirical location and density data on bird species in near real time. For your research, it is important to identify scientists that share your interests and passion in environmental research and identify any GIS datasets they produce.

National, state or province, and local governments have created an astounding array of GIS layers over the last 30 years. In general, these government agencies seek to create maps with at least 85% accurately. Within the United States, the US Geological Survey, US Environmental Protection Agency, and the US Fish and Wildlife Services have some of the most accurate and widely used environmental datasets. In Europe, the European Environment Agency provides GIS data on a wide range of environmental themes for Europe. Datasets include layers on geologic (substrate, faults, earthquakes), natural resources (water, forest types), and chemical (air and water pollution) features. The spatial and temporal resolution of some of these datasets are impressive such as the US National Ice Centre – Antarctic Daily Ice Extents which proves daily ice extent maps for the area around Antarctica in KML format so that it can be observed in Google Earth daily.

There are a number on non-governmental international institutions that provide regional and global datasets on environmental themes such as the United Nations Environment Agency that covers a huge range of physical geography and environmental topics. Their website has links to over 300 vector and raster datasets on physical geography (land cover, ecology, land use, climate). One of the best sites for biodiversity data is the Map of Life that contains data on terrestrial vertebrate and fish species for over 150 million point-occurrence records from the Global Biodiversity Information Facility (an intergovernmental warehouse of digitized species data), expert range maps from the International Union for Conservation of Nature, and regional presence/absence checklists from the World Wildlife Fund (Jetz et al., 2012).

(a)

(b)

(c)

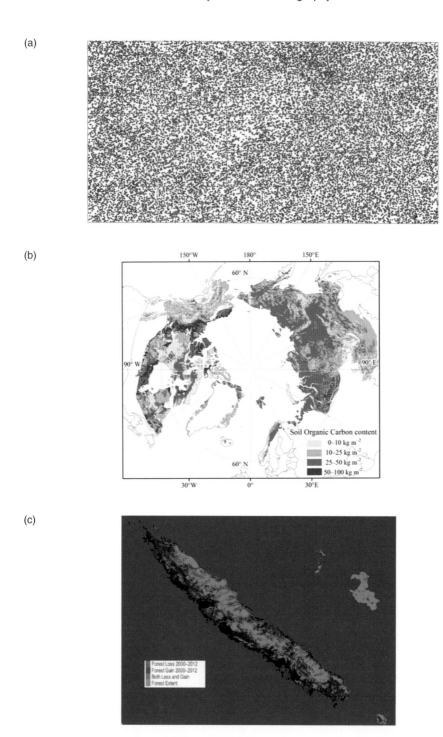

Figure 27.1   Examples of environmental GIS datasets for (a) trees > 10cm diameter at breast height in a 50ha plot on Barro Colorado, Panama; (b) circumpolar organic carbon storage; and (c) forest cover in New Caledonia from 2000 to 2012

Remote sensing has considerable potential for environmental information at a site, landscape, continental, and global spatial scale (Turner, 2014). Remote sensing from satellites offers an inexpensive means of deriving complete spatial coverage for large areas in a consistent manner that may be updated regularly. Although the collection and processing of remote sensing data is complex, there are a number of websites with processed remote sensing imagery that can be used in a GIS (see Chapter 25). Indeed, the National Aeronautics and Space Administration and the European Space Agency provide an impressive amount of ready to use GIS data (Figure 27.2). This may be superior to traditional GIS estimates of temperature, precipitation, and wind based on interpolation among widely dispersed climate stations in isolated regions. Estimates of primary productivity, carbon and vegetation phenology are available back to 2000 at a 1km resolution. Topography data have also been a fundamental component environmental GIS (see Chapter 27). Topography data are usually collected from digitized elevation maps but 90 m and 30 m elevation and topography data are available at a near global extent due to the Shuttle Radar Topography Mission. Recently, Google Earth Engine has brought together a massive dataset of satellite imagery dating back over 40 years and made it available online with tools for scientists and researchers to mine this warehouse of data to detect changes, map trends and quantify differences on the Earth's surface. These spatial explicit remote sensing GIS datasets should be useful for research on any environmental theme or region of the world.

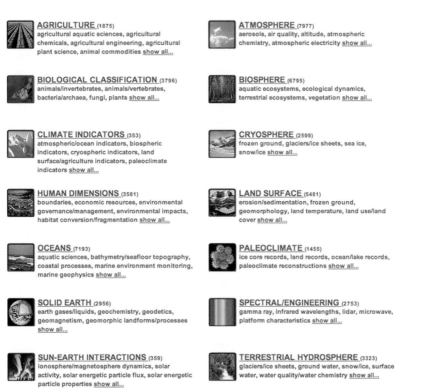

AGRICULTURE (1875)
agricultural aquatic sciences, agricultural chemicals, agricultural engineering, agricultural plant science, animal commodities show all...

ATMOSPHERE (7977)
aerosols, air quality, altitude, atmospheric chemistry, atmospheric electricity show all...

BIOLOGICAL CLASSIFICATION (3796)
animals/invertebrates, animals/vertebrates, bacteria/archaea, fungi, plants show all...

BIOSPHERE (6795)
aquatic ecosystems, ecological dynamics, terrestrial ecosystems, vegetation show all...

CLIMATE INDICATORS (353)
atmospheric/ocean indicators, biospheric indicators, cryospheric indicators, land surface/agriculture indicators, paleoclimate indicators show all...

CRYOSPHERE (2599)
frozen ground, glaciers/ice sheets, sea ice, snow/ice show all...

HUMAN DIMENSIONS (3581)
boundaries, economic resources, environmental governance/management, environmental impacts, habitat conversion/fragmentation show all...

LAND SURFACE (5481)
erosion/sedimentation, frozen ground, geomorphology, land temperature, land use/land cover show all...

OCEANS (7193)
aquatic sciences, bathymetry/seafloor topography, coastal processes, marine environment monitoring, marine geophysics show all...

PALEOCLIMATE (1455)
ice core records, land records, ocean/lake records, paleoclimate reconstructions show all...

SOLID EARTH (2956)
earth gases/liquids, geochemistry, geodetics, geomagnetism, geomorphic landforms/processes show all...

SPECTRAL/ENGINEERING (2753)
gamma ray, infrared wavelengths, lidar, microwave, platform characteristics show all...

SUN-EARTH INTERACTIONS (359)
Ionosphere/magnetosphere dynamics, solar activity, solar energetic particle flux, solar energetic particle properties show all...

TERRESTRIAL HYDROSPHERE (3323)
glaciers/ice sheets, ground water, snow/ice, surface water, water quality/water chemistry show all...

Figure 27.2 NASA Portal for environmental remote sensing datasets that can be used in GIS (Data Sets link on http://gcmd.nasa.gov)

There are a number of GIS databases that reconstruct past physical and cultural features of the world. Most geographers are interested in physical and cultural data that go back to the beginning of the Pleistocene, Holocene, or Anthropocene (see Chapter 22). The best physical historic datasets are the ones associated with climate (temperature and precipitation) over the Holocene or Anthropocene. These GIS maps are based on hindcasting models of past climate and provide 1km to 100km data on climate from the last glacial maximum to present. For instance, climate data from the last inter-glacial (~120,000–140,000 years BP), last glacial maximum (~21,000 years BP), and mid-Holocene (~6000 BP) are available at a 1 km pixel resolution from WorldClim (http://www.worldclim.org/).

One of the most important aspects of environmental GIS is spatial datasets used to map predictions in the future. The most common predictive GIS layers examine changes in climate such as changes in temperature and precipitation. The International Panel on Climate Change provides global climate change scenarios from a diversity of climate change models. The GIS data on the impact of climate change range from 1 km to 100 km resolution and provide data up to 2100. These climate scenarios are widely used to model the impacts of climate change on select environmental themes.

Although environmental GIS focuses mainly on physical, chemical and biological patterns and processes, relationships with humans are of fundamental interest and importance in geography. At a global scale, LandScan provides population density and demography data at a 1 km resolution (http://web.ornl.gov/sci/landscan/). At a regional scale, there are a number of countries that provide free census data on humans in GIS formats. There are also a number of historic datasets available at a global spatial scale. The recent development of the History Database of the Global Environment (HYDE 3.1) dataset provides a spatially explicit database of human-induced global land-use change over the past 12,000 years (Klein Goldewijk et al., 2011). HYDE also provides population density data at regular intervals from 12,000 B.P. to the present.

Geographers interested in environmental GIS should undertake a review of which GIS datasets currently exist for their environmental themes and collect and assess the accuracy and quality of existing GIS datasets.

27.4 Environmental Analyses

The most important aspect of environmental GIS is to test hypotheses and theories related to environmental themes of interest. Once you have your environmental layers in a GIS software program make sure that all your GIS data are in the same projection then identify the spatial scale of your analysis. After this a number of hypotheses can be tested, accepted or rejected like any other in science. Once your data are in a GIS, the strength is the layer stacking and repeatable analyses that can be done to test hypotheses and theories. For environmental themes (physical, biological, chemical, natural resources, and human) GIS layers can easily be compared to your data to create results (Figure 27.3). The strength of your analyses, however, will depend on the quality of the data and the power of your statistical analyses (see Chapters 30, 31 and 32). Finally and most importantly, one needs to have good hypotheses or research questions that can be analysed in a GIS and there is clearly no lack of important environmental issues (drought, extinction, pollution) that need to be examined in geography and environmental science.

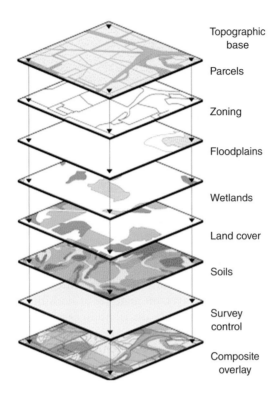

Figure 27.3 Examples of environmental GIS data layers

All GIS software has standard spatial analyses tools (Box 27.1). Geospatial analysis is a large field. For a comprehensive geospatial analyses review, visit online resources like Geospatial Analysis Online (de Smith et al., 2015; http://www.spatialanalysis online.com). However, there are a number of commonly used spatial analyses in environmental GIS. Indeed, most research that uses some form of GIS in top tier research journals like *Science, Nature,* or *Proceeding from the National Academy of Science* uses relatively simple GIS analyses to test hypotheses and answer research questions, the most common of which are measurements, overlays, proximity analyses, connectivity analyses, and predictive modelling.

Box 27.1   Key GIS software

- ArcGIS is the industry standard for GIS software. Most universities will have an ArcGIS license. This is a very powerful GIS software program. http://www.esri.com/
- QGIS is a free and open source GIS software program. QGIS provides viewing, editing, and analysis capabilities very similar to ArcGIS. http://www.qgis.org/

(Continued)

(Continued)

- GRASS GIS is a free and open source GIS software program used for geospatial data management and analysis, image processing, graphics and maps production, spatial modelling, and visualization. http://grass.osgeo.org
- Clark Labs TerrSet (which includes the IDRISI GIS) provides a host of utilities and procedures to optimize, analyse and visualize vector data, raster data, and remote sensing imagery. Student licenses are less than $100 US. http://www.clarklabs.org

Once your data are in a GIS it is possible to select, measure, and calculate statistics for select features and layers. All GIS software provides the user with tools to query or select features in turn to identify points, lines, areas, and cells of interest. This is usually done by selecting attributes based on their names (e.g. and, or, not), on their number value (e.g. >, <, =), or using simple math (i.e. +,−, *, /). These types of queries are regularly done to select features of interest and create new GIS layers. Within GIS it is possible to calculate descriptive statistics of each feature or layer and results from queries. Thus for each feature or layer, an analysis of both the central tendency that describes the centre of the attribute data distribution (e.g. mean, median, mode) and dispersion (e.g. count, range, variance, standard deviation) can be easily calculated for all point, line, area, and celldata in a layer or after a query (Bolstad, 2008). One of the most powerful aspects is to examine how environmental data change over time and space. All point, line, area, and celldata can be examined over time to see if there is no change, significant increases, or a decrease in distribution and density. You can also examine how environmental themes change over spatial scale by examining your GIS over different spatial scales. These measurements of GIS layers are an important first step in analysing your environmental GIS data.

Overlays are one of the most common and powerful analyses in GIS and environmental GIS. Sometimes all one needs is the simplest analysis to test hypotheses and theories. Examples include overlapping species richness, rarity, and threat patterns for mammals, birds, and amphibians at 100 km pixel resolution to identify if diversity patterns of different taxa overlap at a global scale (Figure 27.4) (Orme et al., 2005). This overlay found that geographic patterns of species richness generally overlap, but geographical areas with a high concentration of threatened species from one taxonomic group like amphibians do not necessarily overlap with other taxonomic groups like birds and mammals. Overlays can also be done by selecting features that are hypothesized to be related to an environmental theme and identifying where the features overlap, where there is no overlap, combing features of interest (union), or removing features that are not of interest (clip). Calculating area of overlays also provides important metrics because estimates of area are one of the central aspects in environmental GIS. This can also be undertaken for point and line data within polygons. Thus using overlays identifies patterns and infers processes of environmental themes.

Distance or proximity analyses are also a simple and powerful way to analyse GIS data. The distance between points, lines, areas, and cells is very useful for identifying correlations between two sets of data to identify relationships (e.g. cell phone towers and incidence of cancer). There are a number of simple distance metrics (buffers, interpolation) that can be calculated. A buffer is a region enclosing a point, line, or

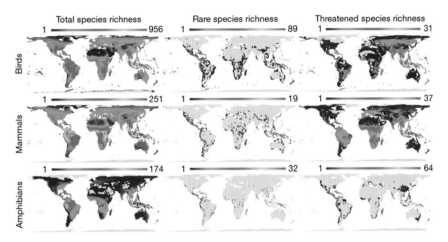

Figure 27.4 Total species richness, rare species richness, and threatened species richness for birds, mammals, and amphibians (Data taken from Orme et al., 2005)

polygon at a specified distance, usually based on straight-line distances from the selected features. Buffers around a location provide data for impacts around a location. Interpolation is another spatial analysis that is commonly done in environmental GIS. Interpolation is a method of estimating point or pixel values at unsampled locations based on the known values of closest points or pixels. If you have point data it is possible to create gradients or contour maps of your samples and adjust these models based on other GIS layers. For instance, temperature data are commonly interpolated to produce new GIS layers of temperature over a wide geographic area. However, caution should be taken when undertaking such interpolation analyses because geological, biological, and chemical interpolations are generally associated with a number of natural features and accuracy depends heavily on sample size and dispersion of samples (Heuvelink, 1998).

Connectivity is the state or extent of being connected or interconnected. There are a number of common connectivity analyses that are used in environmental GIS such as least cost path and network analyses. Least cost path is the calculation of the shortest distance between two points or the path of least resistance based on parameters provided in GIS layers. These types of analyses are commonly done for hydrology with surface topography, land cover, and geology layers. Water is one of the most important environmental themes and there are a host of GIS methods that model the movement of water across the landscape and region. Lines form networks of linear features that can be used to predict water movement with a high degree of accuracy. In particular, if a watershed area is known along with estimates of permeability of the substrate, then estimates of stream flow can be made or modelled based on the observed or predicted amounts of precipitation across a topographic grid. There are a number of software programs, like the Hydrologic Modelling System, that are designed to simulate the complete hydrologic processes of dendritic watershed systems. However, most GIS software contains modelling tools like the Simulator for Water Resources in Rural Basins (SWRRB), Environmental Policy Integrated Climate (EPIC), and Groundwater Loading Effects of Agricultural Management Systems (GLEAMS) that are for specific environmental settings.

Modelling in GIS using distribution modelling, ecological niche modelling, or spatial modelling has been growing at a striking rate in the last 20 years and provides both estimates of feature distributions over space together with estimates of bias in the models. Species distribution models are based on presence, absence, or abundance data from museum vouchers or field surveys and environmental GIS layers to create probability models of species distributions within landscapes, regions and continents (Franklin, 2010). Distribution modelling software, such as Maxent (http://www.cs. princeton.edu/~schapire/maxent/) (Windows, Mac), uses a general purpose machine learning technique called Maximum Entropy Modelling to create maps of predicted probabilities of a species or feature. Maxent is easy to download and can model the predicted distribution of species or features based on presence data (point data). Most environmental layers used in these species distribution models have been based on raster data (e.g. 10 m, 1 km) that can easily be collected over different spatial and temporal scales. There has also been an increase in the incorporation of spaceborne remote sensing data on climate, vegetation and land cover that has great potential to improve models of species or features over different spatial scales. Thus if you have point location data of a feature it should be possible to model the predicted distribution of the feature with Maxent.

Finally, visualization is useful for helping people learn effectively and to communicate your research results. Although we learn verbally or with text, visualizations of environmental GIS clearly support the adage that a picture is worth a thousand words. Thus the power on a simple GIS map, figure, or image should not be underestimated. Environmental GIS datasets can be displayed in the traditional two-dimensional maps or three-dimensional formats to highlight important results or relationships. Most maps are two-dimensional and thematic or apply colour theory techniques such as rainbowing (gradients of colours in the rainbow) or heat (gradient of heat intensity in shades of red) based on attribute values. Three-dimensional maps of environmental themes are also powerful ways to display your data. All GIS software has standard tools that can be used to examine how to display your data most effectively. However, it is important to remember that all maps generated from a GIS should be easy to understand, spatially accurate, and will need appropriate legends, scale bars, and place names.

27.5 Applied Environmental GIS

Environmental GIS is used extensively in planning, policy and governance around the world and is based on a number of environmental laws. In the United States, the National Environmental Policy Act passed in 1977 requires that any development on federal lands requires federal agencies to prepare environmental assessments and environmental impact statements or environmental impact reports. These environmental reports need to identify the impacts of development on an array of environmental themes (e.g. air, water, sound, native habitats, traffic). In particular, the environmental impacts of the proposed action have to identify any adverse environmental impacts, reasonable alternatives to the proposed action, the relationship between short- and long-term impacts, and any irreversible impacts on resources. GIS is well suited to manage and communicate environmental themes and ideas in environmental impact reports. Although they are not necessarily hypothesis-driven, they commonly use environmental GIS analyses in order to inform decision makers concerning single, multiple and cumulative impacts. Environmental GIS is a multi-billion dollar industry and GIS is used extensively in planning and governance of the environment.

27.6 Conclusion

The most important aspect of environmental GIS is to ask important research questions and test hypotheses. Then your results can be scaled up and tested across different temporal (present, past and future) and spatial scales (landscape, regions and globally). This should make you an expert on your chosen environmental topic. First and foremost, identify the environmental theme of interest to you and hypotheses you would like to test. Try to collect environmental data in the field, collect relevant and high quality environmental GIS datasets, and enter them into a GIS software program. Think about the geospatial analyses and statistical tests you will need to test your hypotheses or research questions. Undertake the analyses and create excellent maps that can be useful for scientists, resource managers, and the general public. As an example, it is important to understand the current status of terrestrial protected areas around the world. One could collect and overlay the current distribution of all terrestrial protected areas and examine the past, present, and future estimates of climate to identify protected areas that have experienced or will experiences high and low extremes in temperature and precipitation over time. One could also test the hypothesis that global protected areas are in regions that have historically had low densities of human populations since 12,000 BP. This could result in on-line products and maps that can be used by resource mangers for the conservation of biodiversity. However, your interests will be different and you need to outline the spatial scale of your analysis, accuracy of GIS data (applications and limitations), and statistics that will be used. There are an overwhelming number of environmental issues that need to be researched and the future of environmental GIS depends on you for environmental science, justice, and management in the future.

SUMMARY

- Environmental GIS focuses on the collection, analysis and visualization of geological, biological, chemical, climatic and natural resource datasets.
- Environmental GIS can be done over different spatial scales (landscape, regions and globally) and temporal scales (present, past and future) with GIS datasets.
- Scientists, governments and non-governmental agencies have created an amazing array of large environmental datasets.
- There are a number of commonly used geospatial analyses used in environmental GIS such as measurements, overlays, proximity analyses, connectivity analyses, spatial modelling and visualization.
- Environmental GIS is also used extensively in planning, policy and governance.

Further Reading

- Berg, L.R. and Hager, M.C. (2009) *Visualizing Environmental Science* provides a nice overview of contemporary environmental issues in science and geography.

- Bolstad, P. (2008) *GIS Fundamentals: A First Text on Geographic Information Systems* provides an excellent overview of GIS and analysis techniques.

- Longley, P.A., Goodchild, M.F., Maguire, D.J. and Rhind, D.W. (2001) *Geographic Information System and Science* provides a nice overview of GIS and science applications.

Note: Full details of the above can be found in the references list below.

References

Berg, L.R. and Hager, M.C. (2009) *Visualizing Environmental Science* (2nd edition). Hoboken, NJ: John Wiley and Sons.

Bolstad, P. (2008) *GIS Fundamentals: A First Text on Geographic Information Systems* (3rd edition). Minnesota: Eider Press.

de Smith, M., Longley, P. and Goodchild, M. (2015) 'Geospatial Analysis Online', http://www.spatialanalysisonline.com (accessed 27 November 2015).

Franklin, J. (2010) *Mapping Species Distributions: Spatial Inference and Prediction.* Cambridge: Cambridge University Press.

Hansen, M.C., Potapov, P.V., Moore, R., Hancher, M., Turubanova, S.A., Tyukavina, A., et al. (2013) 'High-resolution global maps of 21st-century forest cover change', *Science* 342: 850–3.

Heuvelink, G.B. (1998) *Error Propagation in Environmental Modelling with GIS.* Boca Raton, FL: CRC Press.

Hugelius, G., Bockheim, J.G., Camill, P., Elberling, B., Grosse, G., Harden, J.W., Johnson, K., Jorgenson, T., Koven, C.D., Kuhry, P., Michaelson, G., Mishra, U., Palmtag, J., Ping, C.-L., O'Donnell, J., Schirrmeister, L., Schuur, E.A.G., Sheng, Y., Smith, L.C., Strauss, J. and Yu, Z. (2013) 'A new data set for estimating organic carbon storage to 3 m depth in soils of the northern circumpolar permafrost region', *Earth System Science Data* 5: 393–402.

Jetz, W., McPherson, J.M. and Guralnick, R.P. (2012) 'Integrating biodiversity distribution knowledge: toward a global map of life', *Trends in Ecology and Evolution* 27: 151–9.

Klein Goldewijk, K., Beusen, A., Van Drecht, G. and De Vos, M. (2011) 'The HYDE 3.1 spatially explicit database of human-induced global land-use change over the past 12,000 years', *Global Ecology and Biogeography* 20: 73–86.

Longley, P.A., Goodchild, M.F., Maguire, D.J. and Rhind, D.W. (2001) *Geographic Information System and Science.* Abingdon: John Wiley & Sons, Ltd.

Orme C.D.L, Davies R.G., Burgess M., Eigenbrod F., Pickup N., Olson, V.A., Webster, A.J., Ding, T-S., Rasmussen, P.C., Ridgely, R.S., Stattersfield, A.J., Bennett, P.A., Blackburn, T.M. Gaston, K.J. and Owens, I.P. (2005) 'Global hotspots of species richness are not congruent with endemism or threat', *Nature* 436: 1016–19.

Turner, W. (2014) 'Sensing biodiversity', *Science* 346: 301–2.

ON THE COMPANION WEBSITE...

Visit **https://study.sagepub.com/keymethods3e** for author videos, chapter exercises, resources and links, plus **free** access to the following recommended articles:

1. **Gaston, K.J. (2006) 'Biodiversity and extinction: Macroecological patterns and people',** *Progress in Physical Geography*, **30 (2): 258–69.**

Gaston (2006) focuses on the study of broad geographic patterns in the size and structure of species assemblages in the US. Gaston is one of the foremost authorities on macroecological patterns and local, regional and global extinction.

2. **Foody, G.M. (2008) 'GIS: Biodiversity applications', *Progress in Physical Geography*, 32 (2): 223–35.**

This is an excellent and timely review of GIS applications for biodiversity. The article covers important issues related to GIS, remote sensing, and conservation.

3. **Richardson, D.M. and Pysek, P. (2006) 'Plant invasions: Merging the concepts of species invasiveness and community invasibility', *Progress in Physical Geography*, 30 (3): 409–31.**

This is simply one of the best reviews about the impacts of invasive plants on biodiversity and the environment. The article syntheses a number of very important topics that are still important to this day.

28 Models and Data in Biogeography and Landscape Ecology

George P. Malanson and Benjamin W. Heumann

Since all models are wrong the scientist cannot obtain a 'correct' one by excessive elaboration. On the contrary following William of Occam [s]he should seek an economical description of natural phenomena. Just as the ability to devise simple but evocative models is the signature of the great scientist so overelaboration and overparameterization is often the mark of mediocrity.

George E.P. Box (1976: 792)

SYNOPSIS

Models and data are the bases of science. We focus on specific types of models, especially simulations, and provide a starting point for the exploration of existing data for studies in biogeography and landscape ecology. The primary use of models in biogeography and landscape ecology is to test hypotheses and develop exploratory models that can potentially open the way to new theory. We examine the importance of resolution and scale, model hierarchy, and spatial and temporal choices for model building. We also examine examples of model approaches and complexity (empirical, phenomenological and mechanistic models) and examples of modelling techniques (species distribution models, biogeochemical cycle models). When building models it is important to develop conceptual and quantitative frameworks and consider model evaluation (verification and validation) and model communication. An array of organismal, landscape, soil, terrain, surface hydrology, biogeochemistry, climate, and ocean data and databases, which can be used to build models, are available. Modelling is essential to the advance of biogeography and landscape ecology, and physical geographers need to 'model-up' in the areas of mathematical and computer-programming skills that are essential for explicitly specifying and evaluating meaningful theory.

This chapter is organized into the following sections:

- Introduction
- Modelling with a focus on simulation
- Examples of models
- Building models
- Data
- Conclusion

28.1 Introduction

Models and data are the bases of science. Here, we will focus on specific types of models, especially simulations, and provide a starting point for the exploration of existing data for studies in biogeography and landscape ecology. The existing literature has good introductions to modelling and particular platforms for implementation. Among the platforms (with a focus on introductory texts), R (http://www.r-project.org) may be the most general and versatile, but it is particularly well suited to empirical (especially statistical) modelling (more definitions below) (see Bolker, 2008; Stevens, 2009; Gardener, 2014). For many simulations, the STELLA platform (http://www.iseesystems. com/ (Windows, Mac) is more useful because it has a graphical interface that doubles for design and programming. It is well suited to modelling population and ecosystem processes of inputs and outputs (Grant and Swannack, 2008). Better suited to spatially explicit modelling is the NetLogo platform (https://ccl.northwestern.edu/netlogo/), which is designed for agent-based models in a cellular framework (Railsback and Grimm, 2012; O'Sullivan and Perry, 2013); several more advanced (efficient but requiring more skill) agent-based modelling platforms exist (e.g., Repast, repast.sourceforge. net (Windows, Mac)). We introduce broad concepts, and follow-through with one or more of the above, or similar platforms, which will be necessary for moving modelling into practice.

28.2 Modelling with a Focus on Simulation

Purposes

Models are abstractions or representations of our concepts about the world. While we will present more general models for context, we will focus on simulation models as a means to explore and test our concepts, or, in some cases, to use them to guide action. Data are fundamental to models in that they are used to build and/or test them, even if only loosely.

The primary use of models in biogeography and landscape ecology is to test hypotheses. We can represent concepts of the world in specific form in the structure of a model, sometimes using data, and then test whether the hypotheses might be right by comparing the outcomes of the model to data. Basic statistical analyses exemplify this type of model. However, we can build more complicated models in simulations that we can also use to test hypotheses either formally or informally. Here we use simulation as a tool that operationalizes one or more models in a computer program, and most of what we discuss is about simulation models and modelling. Either way, the models are the basis for advancing our concepts.

Another use for models is more exploratory. Without specific hypotheses, and perhaps representing a conceptual world not specifically representing the one we are in, exploratory models can be informative about relations that are difficult to observe. Exploratory models can potentially open the way to new theory. An example is the development of ideas about self-organized criticality, which leads to regular patterns at levels of organization that we do not usually observe, that were developed largely through exploratory modelling (see Bak, 1999). Less abstract, Larsen et al. (2014) explained how exploratory modelling could help derive first order explanation, or causation, from complex and complicated systems. They developed a deductive-inductive loop between models and observation (more accurately, we think, a hypothetico-deductive process) in which the exploratory part of modelling is in comparing multiple models (Figure 28.1). This loop echoes the loop in model building, mentioned below, identified by Grant and Swannack (2008) and Railsback and Grimm (2012). It is also a loop that moves from empirical models to phenomenological models and, in the inner loop, to mechanistic models (all discussed below) in turn.

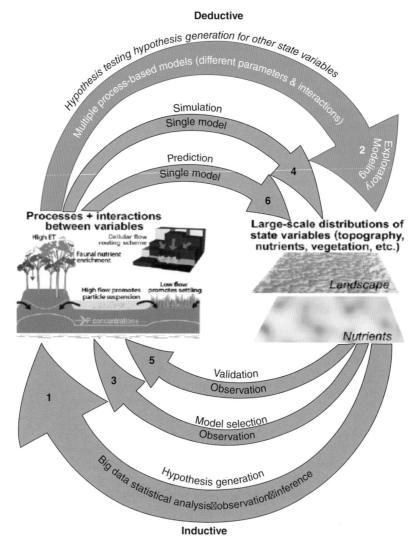

Figure 28.1 Larsen et al. (2014) described an inward-spiralling dialectic between modelling (deductive) and data (inductive)

One often thinks of using models for prediction and to guide decision-making and action, but this type of modelling is less common in biogeography and landscape ecology. In other environmental sciences models are more commonly used as a basis for action – such as weather and hydrological models – but the similar uses of models in landscape ecology and biogeography as might be applied in conservation have been limited because, it seems, there are greater uncertainties – especially if human behaviour must be modelled.

Broad modelling concepts

Resolution and scale

The resolution of a model is a choice, and the answer must fit the purpose of the modeller. The first issue is the ecological (also biogeographical/biological/system) resolution. What do we want to know and in what detail – and so at what level do we need to observe and model? Second are spatial and temporal resolution, which are linked in models although not often explicitly. Conceptually, ecological, spatial and temporal resolution are inextricable. The three are often discussed as questions of 'scale,' which has been defined so variously that it is now a shorthand for a vague linkage of the three resolutions.

Hierarchy

The ecological resolution at which one models has become a fundamental choice for environmental scientists. The importance of ecological resolution and how it can be addressed have been explored numerous times, and we will briefly synthesize our approach.

Hierarchy theory has developed in landscape ecology (Allen and Starr, 1982) and can be applied in biogeography. The basic proposition is that one must consider three levels of space and/or time for a given ecological resolution. The ecological resolution, or scale, is that of the relationship one wishes to explore. First, this focal scale must be chosen. The focal scale will differ with the question asked, and one might find in the process of modelling that the initial scale chosen is wrong for a given question. Second, a higher scale – in space a larger area and in time a longer period – is chosen, which constrains what occurs at the focal scale. Third, the lower scale – smaller spatial and temporal units, truly higher resolution – is chosen for the processes that create the relations at the focal scale. In many cases the upper and lower levels of the hierarchy are not modelled explicitly, but all three levels can be addressed in a model. Consider trees on a hillslope. If we wish to examine hypotheses about the changing dominance among species we might choose our ecological scale as the individual tree, with the focal scale being an area the size of a tree and the time-step of a year (since trees outside the tropics usually have an annual cycle). The higher scale would be that of the hillslope, encompassing the populations of the observed species, over a generation or average lifespan of the species. The lower level would be spatially within a tree (perhaps representing its carbon accumulation in its parts) and at a complementary time scale (perhaps daily for carbon balance). The dynamics of the individual trees and where they establish, grow, and die happen at the focal scale, but are based on the processes at the finer scale while constrained by the overall populations and environment at the larger scale.

Spatial and temporal choices

Part of choosing the ecological scale will be the choice of whether the model is to be spatially and temporally lumped or explicit. Simulations are temporally explicit and can be spatially explicit. Although the hierarchical approach as condensed above assumes a spatially and temporally explicit model, some models are created for single spatial and temporal units at the focal scale. Such spatially 'lumped' models treat the focal phenomena as comprised of a single unit of space and time. That could be the hillslope of trees at a moment of observation relative to its environment. Simulations could change the number of trees in each species over a time period, which could be a year, decade, century, ..., without reference to their locations. This spatially lumped approach has been the basis for most statistical modelling and much simulation, but in landscape ecology and biogeography, where spatial relations are core questions, spatially explicit simulations are more informative.

Within spatially explicit simulation, the choice of scale is the choice of extent and of spatial resolution. If space is represented continuously with partial differential equations, resolution is not the issue, but extent still matters. If space is represented discretely, as with a grid of cells, then both extent and resolution must be determined. The spatial resolution is the smallest spatial unit as represented in the model (although it can change across the spatial or temporal extent of a given model). The resolution should be chosen for the question, so that the phenomena are apparent at that scale (i.e., the resolution is not smaller than the organism) and are not averaged out across space (i.e., the area contains too wide a variety of organisms). The extent needs to be large enough to encompass the phenomena of interest (e.g., the populations of trees on the hillside) but not so large as to incorporate confounding processes (e.g., the valley-bottom populations). Many models applied in biogeography are pseudo-spatially explicit in that they represent an extent by a grid of cells, but wherein the cells are independent of one another, and these have their uses.

Temporal resolution has issues similar to spatial resolution. First, the difference between continuous and discrete representations of time, represented by differential and difference equations, respectively, embody differences in ecological resolution as well. Differential equations are usually solved for a single point in time, but can be used in simulations through time – but in the latter are often solved using numerical methods that discretize the equations, albeit with shorter time steps than any phenomena would require. Discrete time steps, represented by difference equations, usually choose a temporal scale to match phenomena, such as a daily or annual cycle.

Where both spatial and temporal scale are explicit in a discrete simulation (originally identified for partial differential equations – but for the numerical solutions of these, which are discrete) their connection is constrained. The Courant-Friedrichs-Lewy Criterion (Courant et al., 1967) holds that for a given spatial resolution (i.e., grid cell size), the process being simulated cannot skip over a cell in a single time step; otherwise it would not be influenced by the environment of that cell, which in fact it passes through (and vice-versa).

Types of models

The types of models used also have a hierarchy. Models come in a range of types and complexity. Even the simplest linear model:

$$Y = ax + b \qquad [28.1]$$

can be a statement of a relationship, a basis for testing that relationship, or a prediction for specific variables in a given time and place. This simple model could also be embedded in a more complex model with other equations as a simulation. Given our focus on simulations, we will examine such embedding, but work to keep this simple relationship as a touchstone.

Given the range of what could be covered by 'model,' a basic differentiation of approaches can aid in interpreting existing models, choosing among them, or designing new ones. Reynolds et al. (1993) defined three levels of model approaches and complexity: empirical, phenomenological and mechanistic. The three are not sharply differentiated, but provide a useful framework. Bolker (2008), perhaps with justification, assumed that models are empirically based and differentiates between phenomenological and mechanistic approaches. He too notes that the same model could be either, depending on context.

Empirical models

Empirical models describe relations between variables without knowing how or why a relationship exists. Input and outputs or independent and dependent variables can be observed, and the observed connection described, without much understanding. Simple statistical models that describe a correlation and a direction, such as in Equation 28.1, are empirical. This type of model is ubiquitous in biogeography and landscape ecology. In biogeography, the most repeated and trusted model of this type is the species area relationship:

$$S = cA^z \qquad [28.2]$$

where S is species richness, A is area, and c and z are constants based on observation (i.e., they are empirically-based). The log-linear relationship is revealed in observations. For growth of a tree, Reynolds et al. (1993) give an empirical equation based on growing-degree-days:

$$W(t) = k_1 \log(D) + k_2 \qquad [28.3]$$

where $W(t)$ is growth over period t, D is the sum of growing degree-days (GDD), and k are empirical constants.

Phenomenological models

Phenomenological models include some conceptual understanding. We believe that we have an understanding of the processes of birth and death that underlie the logistic population model:

$$dN/dt = rP\,(1 - N/K) \qquad [28.4]$$

where N is population, r is the growth rate (births – deaths per unit time) and K is carrying capacity. Herein the phenomenon of carrying capacity is represented although not directly observed and the individual births and deaths are not specifically tallied. However, this relationship is based on earlier observations. Reynolds et al. (1993) give a similar equation for tree growth:

$$dW/dt = r(Wmax - W), \text{ which integrates to } W(t) = \int r(Wmax - W)\, dt \quad [28.5]$$

This type of model can be solved for a single t or used in a simulation to examine change through time. Another modelling approach blends empiricism and phenomenological levels in structured equation modelling. In these models the direction of cause and effect through multiple connections can be chosen to represent hypotheses and then the statistical outcome assessed.

Mechanistic models

Mechanistic models attempt to fully represent our understanding of cause and effect. Most models that are described as mechanistic do not embody relationships but explicitly represent the transfer of matter or energy, or sometimes information, among compartments, or for populations, explicit births and deaths (from Kendall et al., 1999):

$$dJ(t)/dt = B(t) - R(t) - M(t) \qquad [28.6]$$

where J is the density of juveniles, B is the birth rate, R is the rate of change from juvenile to adult, M is the mortality rate, and t is time.

Reynolds et al. (1993: 131) defined mechanistic models as those that 'decompose a system into its component parts and describe the behaviour of the whole through the interaction of those parts,' with explicit reference to hierarchy theory. They contrast phenomenological and mechanistic models by explaining that the phenomenological are at the level of the focal system and derive from a general understanding, while the mechanistic models use phenomenological equations from a lower hierarchical level to represent the focal level and then generalized back to a phenomenological representation. They illustrate a mechanistic model with interacting root, shoot, carbon pool, and nitrogen pool compartments as well as a set of parameters.

Deterministic and stochastic models

In building models another primary differentiation is whether to model stochastically or deterministically. Some simulations combine both. Stochastic models include randomness in some way, often in an attempt to represent the noise in observations or assumed unobservable processes that cannot be modelled deterministically (such as turbulence). Stochastic models depend on statistical relationships. Deterministic models ignore the noise and represent the average relations in a system. They represent relations that have little noise and can be captured in direct mathematical equations. Simulations often combine deterministic and stochastic elements: the average relation is computed, and then some stochastic variation is added.

In computer simulations, the code determines the outcome, and so these are naturally deterministic. Simulations become stochastic by adding random numbers, drawn by a random number generator, primarily in two ways. First, computations can take place or not (lines of code executed or not) depending on whether a random number meets a threshold. The threshold is often set by the probability that an event will occur. Second, the random number can draw from a distribution and be used directly in an equation. Deterministic modelling works best for systems that have well-understood mechanistic processes for which their initial conditions are known (Newtonian orbital mechanics). Stochastic modelling works better in most cases in biogeography and landscape ecology. For example, the distribution of seeds dispersed from a parent by wind depends on wind turbulence, the exact size and shape of the seed, and how and when it is released. We do not know enough to model the process deterministically, and seeds do not all end up in the same place. What we do not know is represented by random numbers.

28.3 Examples of Models

Empirical: species distribution models

Among the most widely used empirical models in biogeography and landscape ecology today (or among all models ever in these fields) are species distribution models (sometimes called niche models) (Franklin, 2010), and we expand on them accordingly. Niche was defined by Hutchinson (1957) as a hypervolume with the dimensions being environmental conditions and resources describing the needs and tolerances of species to persist. Species distribution models use a GIS approach to relate species to their environment using geo-spatial data. Geo-located species observations and geo-spatial environmental data are used to extract and build species-environment relationships that quantitatively define Hutchinson's hypervolume. Some of the variables may be related to resources that the species uses, such as precipitation for plants, others may be limiting factors, such as date of last freeze, and others may be indirect indicators of either, such as topographic slope. Once the species-environment relationships are known, they can be projected using the geo-spatial environmental layers to map species distributions.

Approaches to modelling species distributions based on the niche include statistical and machine learning approaches. Most approaches, especially those based on traditional statistical methods, require presence and absence data, which is often problematic because true absence data are difficult to obtain. The lack of a species observation does not necessarily mean that the species is absent. An alternative approach to the more typical presence/absence model is the presence-only model.

The most well-known presence-only model is Maxent, which is based on the principle of maximum entropy (http://www.cs.princeton.edu/~schapire/maxent/). This stand-alone software was written using java and will run on a wide variety of platforms (Windows/Mac/Linux). Maxent uses presence data to model the niche hypervolume using a machine-learning optimization algorithm. This algorithm seeks to find the most uniform distribution (i.e., maximum entropy) given the environmental constraints. The Maxent solution is deterministic, although randomized partitioning of data for testing and training will yield varying results. Another advantage of Maxent is that it can combine linear and non-linear functions to build complex species-environment relationships that may be more realistic then assumed linear or non-linear functions. The species-environment relationships are constructed as percent likelihood of species occurrence. These relationships can then be mapped using the geo-spatial environmental data with a resulting likelihood of occurrence map. Maxent has been used in over one thousand peer-reviewed journal articles and has been shown to be among the best species distribution model, especially for small datasets. The Maxent software package also provides users with an easy to use graphical user interface.

Despite Maxent's advantages, there are several problems pertaining to species distribution modelling (see Austin, 2007). First, niche-based species distribution modelling requires a clear definition of the niche concept to link niche theory with modelling outcomes, but the approach uses where species are now (their realized niche) to define where they could be (their fundamental niche). More recently 'mechanistic niche models' are being developed that take aspects of species physiology and/or phenology into account – at least for modelling response to climate change (Kearney and Porter, 2009). Soberon (2007) provided a good discussion on interpretations of different niche concepts.

Second, environmental variables are selected. As is the case with any computational model, the software will find a solution, logical or not. Pre-selecting variables may introduce bias into the model, but including all possible variables may produce spurious results. Furthermore, inclusion of correlated variables may alter species-environment relationships and ultimately the modelled distributions, but how to select among correlated variables is not certain.

Third, model assessment, especially assessment of presence-only models, can be perplexing. For example, the Maxent software uses the receiver-operating characteristic (ROC) to compare the false positive and true positive rates. It is desirable for the resulting model to have much higher true positive than false positive rates. The lack of true absence data complicates the definition of false results. If the definition of a true absence is lacking, then how can a false positive be determined? In this situation, the absence data are pseudo-absences drawn from randomized background locations. Therefore the model is comparing the true positives with random results. While this may seem logical in its similarity to many statistical approaches, this formulation of the ROC makes it sensitive to the distribution of the species compared to the extent of the study area. Wide-spread or generalist species will inherently have a distribution closer to random than rare or specialist species. Furthermore, the extent of the study area relative to the species distribution will also affect the ROC results. Therefore, the ROC can only serve as an indicator of model accuracy, rather than a definitive assessment. An alternative approach is to partition validation data prior to modelling, and compare the model output to this truly independent data. In the end, model assessment requires both qualitative and qualitative analysis to ensure that the results are reasonable and logical.

Species distribution modelling is a powerful approach to examining the relationship of species and the environment, and a useful tool for conservation. However, these models can produce unrealistic results unless careful attention is made towards the interpretation of niche theory, variable selection, and model assessment.

Phenomenological: Forest stand, cellular automata, and agent-based models

While the basic logistic population models are core to ecology, more complicated models that use such relations can also be phenomenological. One of the most widely used class of phenomenological models are forest stand simulations (Botkin, 1993; Shugart, 1998). These models simulate the growth of individual trees based on phenomenological relations, in turn based on empirical observations. For example, the growth of trees is based on observations of the maximum sizes of old trees and a growth function that would reproduce this growth for ideal conditions, which is then modified by a parabolic function of assumed adaptation to climate based on observations of the range limits of the species (again, the realized niche). The consequences of the assumptions are problematic, and other assumptions would produce significantly different outcomes (e.g., Malanson et al., 1992). Population dynamics can represent individuals, but the basics of the logistic model are often embedded. Botkin (1993) used this type of model to examine effects of increasing CO_2 on forest dynamics 35 years ago, and they are still on the front lines of simulating ecological responses to climate change.

These forest stand models were originally developed as spatially lumped and run for a single hypothetical or representative location. They then were used to represent areas by implementing runs independently on a grid of cells in a pseudo-spatially explicit (see Malanson, 1996) simulation. The models became increasingly spatially explicit by adding one or another spatial process that linked cells in a grid, but remained phenomenological (e.g., light and shade in ZELIG and seed dispersal in MOSEL; see Urban et al. (1991) and Hanson et al. (1990) respectively). For example, dispersal was modelled phenomenologically by assuming distance decay functions for the movement of seeds because useful dispersal data for empirical modelling are rare and the process is too complex and complicated for mechanistic modelling.

In an effort to become more spatially explicit but computationally tractable, another class of phenomenological models was borrowed that reduces the ecological resolution. In their most basic form, cellular automata change the state of a cell depending on the states of its neighbours (Hogeweg, 1988). While more can be done by increasing the size of the neighbourhood and the strength of each cell's effect, models that are still called CA are often much less automatic in that they incorporate stochasticity (a true CA is deterministic). Many CA are focused on response to habitat fragmentation (e.g., Kupfer and Runkle, 2003) or ecotones (e.g., Zeng and Malanson, 2006). CA are well suited to examine the feedbacks between spatial patterns and processes. A further development is the use of agent-based models (ABM) or multi-agent systems (MAS) (Parker et al., 2003). The underlying grid of a CA is maintained, but agents are defined as entities that interact with the environment of each cell but also move across the grid, and they can potentially change their behaviour. While the essence of ABMs was explicit in Hagerstrand's (1965) simulation of the diffusion of innovation (developed earlier in his thesis), ABMs per se were not developed in biogeography and landscape ecology for decades. At their most simple, they are a step beyond CA in that the agents are separate from the underlying environment and are mobile (Malanson and Cramer, 1999), whereas more developed ABMs have agents that alter their environment, learn, and move. Feedbacks spread among the agents directly and through the environment. For example, Bennett and Tang (2006) simulated the movement and learning process for elk migrating in the Yellowstone National Park area wherein elk could learn and remember resource distributions and affect their herd. Some ABMs are less ambitious, however, and learning is not common. Smith-McKenna et al. (2014) built on Zeng and Malanson's (2006) CA by representing the trees as agents that could grow into adjacent cells; while a modest modelling advance, they provided a platform to explore fundamental theoretical problems, perhaps the better use for CA. While we have focused on ABMs focused on other species, they are particularly suited to including humans as decision makers in a feedback loop with landscapes (Boone and Galvin, 2014), although their interface with other simulations can be difficult (Yadav et al., 2008).

Mechanistic: BioGeochemical Cycles models

Where Reynolds et al. (1993) set a framework for mechanistic models, Running (Running and Coughlan, 1988; Running and Gower, 1991; Running and Hunt, 1993) developed a simulation for trees, later extended for other plant forms, that computed the components the former called for, notably photosynthesis and respiration, as the basis for areal net primary productivity. Mechanistic simulations that capture ecological processes were developed for fundamental flows of matter and energy, such as in photosynthesis, but these are difficult to scale up to biogeographic and landscape questions. Some do this by a combination of simplifying assumptions. The BioGeochemical Cycles (BGC) class of models did this by simplifying the representation of plants while still calculating photosynthesis. A model covers a large, lumped area so that flows among areas would be negligible and can be ignored in the model. The original BGC explicitly assumed a 'big leaf and computed a single process of photosynthesis for an area based on an average leaf area index for that extent. The exploration of various representations of leaf area can link process to pattern (Cairns and Malanson, 1998; Cairns, 2005). The more useful applications have, however, developed some spatially explicit connections across a grid of cells (starting with a pseudo-spatially explicit version (Band et al., 1991), particularly for water (Band et al., 1993) but also fire (Keane et al., 1996)).

28.4 Building Models

Grant and Swannack (2008) outline and describe four steps for modelling: creating a conceptual model, creating a quantitative model, model evaluation, and model application (including communication). Although all the sub-steps they describe are needed, we will focus on a further subset. However, they and others note that this is a cyclic process, with return to the concept state following the evaluation stage repeatedly (e.g., Railsback and Grimm, 2012). Grant and Swannack (2008) also provided a useful chapter on 'common pitfalls' at each step.

The conceptual model

The first step is to define a system to model. This step includes deriving a specific problem from our general objectives and determining what parts of the system to include. The relations among those parts need to be specified. This stage is usually focused on creating a diagram of the relations and/or compartments and processes that are to be modelled, which becomes a guide for the computer program.

The quantitative model

At this stage the conceptual diagram is written in computer code (note that some systems use a graphic interface in the programming step, e.g. STELLA), but the emphasis is on quantification. Quantification includes selecting the scales, specifying the form and the parameters of the equations (and also Boolean 'if…, then…' conditions for alternatives) and setting the order in which the steps will be executed. The form of the equations may be determined by the conceptual model or may be a form that is derived empirically, as when a regression equation (linear or nonlinear? used deterministically or stochastically?) is used. Parameters can be chosen from theoretical relationships (division by distance squared), observations, or best guesses (the latter often tested and changed in iterations of this step). Any of these might be calibrated, i.e., adjusted so that a model output meets a specific target. Whatever modelling platform or programming language you choose, this step is where ideas go into the computer. Type *Run*. Hit *Enter*. Click *Go*.

Model evaluation

The evaluation of models is divided into two components, verification and validation, but both of these have internal parts and can be defined in multiple ways. Be aware: the definitions of the same term, especially validation, can mean quite different things! The verification stage is an evaluation of whether the model represents the system as we think it should and whether it behaves as expected. These steps are a review of the quantitative model as built with the conceptual model and a qualitative evaluation of the output of the model relative to how we think (thought) the system worked. For example, in creating a model of land use and population change in Ecuador, we needed to know the rate at which women married by age group. Not knowing, we decided to get the information later and, knowing the rate was between 0 and 1, entered 0.5 as a place-holder. But we forgot to look up the real number and plug it in. At the verification stage we ran the model longer than we otherwise might – and found that the area was depopulated and returning to forest. That led us to re-examine the model and find our error, although it would not have mattered over a short period (Mena et al., 2011).

Validation is the more controversial step. Hermann (1967) described five criteria of validity: 1) internal validity, 2) face validity (often subjective estimates), 3) variable-parameter validity, 4) event validity, and 5) hypothesis validity. The first two correspond to what we call verification; the third to sensitivity analyses, discussed below; the fourth is the gold standard discussed next; and fifth is a heuristic validity that asks if anything was learned in the modelling process.

The gold standard of validation is comparison of model output with observations. If the output is similar to observations in the field, then the model is validated – it works! No one disagrees that this is the preferred standard of validation, but difference comes in the degree to which this standard of validation can be achieved for a given system and model. For example, models that represent as yet unrealized systems would not have observations for comparison. Rastetter (1996) discussed alternative procedures in such cases and advocated a 'model-based evaluation' that amounts to a combination of the other four approaches. In such cases validations might be done for subsets of the model of for limited conditions. A critical step for event validity is that the observations used in the test must not have been used in the steps to parameterize or calibrate the model.

A further step of evaluation, or a step if validation is limited, is sensitivity analysis. Given the difficulties of event validation, sensitivity analysis is perhaps the primary means of model evaluation. How much does model output change for a given often +/– 10%) change in any parameter or combination? This step could examine every possible combination of parameters, but seldom does. In this way we may learn something useful about the system of interest or at least about the model. This is hypothesis or heuristic validity. If we think we have learned something, we are successful. Herein lies the controversy of validation. It is not definite when a modelling exercise is a success under this criterion; however, a model meeting gold standard event validation for a trivial problem may not be worth the effort either. Sensitivity analysis can extend into scenario analysis, in which the conditions or context of the model – as opposed to the parameters within the model – are varied. Scenario analysis can become prediction if the model is well-validated (confidence required!) or, more wishy-washy, 'projection,' i.e., answers to scenarios without certainty – and can combine scenario and sensitivity analysis.

Model communication

Grimm et al. (2006) recommended a common outline for the description of agent-based models (now widely known as ODD), which has been updated (Grimm et al., 2010). It has seven components in three domains:

- Overview
  - o Purpose
  - o Entities, state variables and scales
  - o Process overview and scheduling
- Design concepts (as domain and component)
  - o emergence
  - o stochasticity
  - o (+ 9 further concepts)
- Details
  - o Initialization
  - o Input data
  - o Submodels

ODD could be adapted to most simulations and should be reviewed by modellers in landscape ecology and biogeography. The pitfalls identified by Grant and Swannack (2008) at this stage are a failure to specifically link numerical results to ecological meaning and the tendency to overestimate the utility of a model or its correspondence with the system it simplifies, which are problems in scientific reporting in general.

28.5 Data

The analysis and investigation of earth and ecological systems including coupled human-natural systems often require using data from multiple disciplines. Whether assessing the impact potential of climate change projections on plant productivity or the effects of land management practices on nutrient cycling, collecting the data required is impractical and often impossible. First, research requiring data over large areas often relies on collections from multiple organizations or space-based remote sensing. Second, if historical data are needed, then the data cannot be collected but need to be obtained from an archive. Third, it is often impractical and unreasonable for an individual or a single research team to acquire the level of expertise in multiple fields to collect and process the data. Instead, it is usually more practical and productive to integrate existing datasets.

The interdisciplinary and multidisciplinary nature of this type of research makes it both exciting and challenging. Some of the challenges include identifying, locating, acquiring, combining and analysing data from multiple sources and disciplines in an accurate and rigorous manner. Furthermore, each dataset has its own spatial and temporal characteristics, as well as data quality issues that may need to be addressed. Thus combining the data is often more difficult and time consuming than it may appear.

Here we highlight major or unique datasets and databases to aid in the search for the appropriate data for this type of research. The types of data have been organized into the following groups: (1) organismal data including occurrence data and traits, (2) characterization of the landscape including land cover, soils, terrain, and hydrology, (3) biogeochemistry and energy exchange between the land surface and atmosphere, and (4) climate. While these datasets focus primarily on terrestrial environments, aquatic datasets will also be discussed.

Organismal data

The gathering of organisms to form collections in museums and universities to document life on earth dates back to the 18th century naturalists. While these collections have always been useful archives of the diversity, characteristics, and locations of species, the rapid advancements in bioinformatics and geographic information science have led to a new demand for large geo-located datasets of organismal data. While new collections are digitally archived with high resolution scans and GPS coordinates, collections from even as recently as 10 years ago may only exist on paper with a general description of location. The demand for digital collections has been recognized and there are several programs around the world to fund and promote the digitalization of collections. While many universities and museums have their own digital collections, searching through dozens of individual collections can still be a tedious and costly obstacle to research. There has been a trend over the last decade to create meta-databases or collections of collections.

The goal of these large meta-databases is to provide a single location to access biological data from around the globe. These meta-databases can be an invaluable tool by making large amounts of data freely and easily available to researchers around the world. Here, we focus on four large meta-collections of organismal data – GBIF (Global Biodiversity Information Facility), iDigBio (Integrated Digitized Biocollections), TRY (a global plant traits database), and DRYAD.

Selected organismal databases

GBIF (www.gbif.org) is the largest biodiversity database available via the Internet with over 640 million records of over 1.6 million species from over 15,000 datasets as of November, 2015, although only around 86% of those records are georeferenced. Currently, 54 countries participate in GBIF with another 40 international or non-governmental organizations. GBIF is often considered the primary source for occurrence data.

iDigBio (www.idigbio.org) is a United States-based meta-database that focuses primarily on the digitization of existing museum and natural history collections to enhance accessibility to these datasets that contain nearly 46.5 million specimen records in 664 record sets. iDigBio includes not only the occurrence data, but also digitized images and scans of the specimens, thus providing researchers access to the original specimens, albeit in a digital form. The iDigBio database is limited to collections that are based in the United States, although the records themselves may occur anywhere.

TRY (www.try-db.org) is a European-based meta-database with participants from around the world. The TRY database is focused on plant traits such as morphology, biochemistry, physiology, and phenology. It currently (November 2015) contains 5.6 million trait records for 2.2 million individual plants from more than 100,000 species. However, only about half of the data are georeferenced and the data are available only upon request.

DRYAD (datadryad.org) is a repository that allows anyone to store ecologically relevant data. It is meant to complement published work and has 74 affiliated journals. It allows data to be deposited in many formats and with varying kinds and amounts of information and metadata. It strengths are that it provides accessible storage, it makes the data citable by providing DOIs, and links to specific publications. Thus it is well-suited to the needs of researchers having completed a project and less so to those looking for data to start one.

Organismal database issues

It should be noted that while these meta-databases can provide large amounts of data using a single interface, they are not comprehensive. There are many collections, usually small, local collections that may not be integrated with the meta-databases. These smaller collections may have essential data especially if larger datasets omit occurrence data from large or unique geographical areas. Researchers should investigate local universities and biological stations for smaller collections that may be rich in local collections data. These data are often available via the Internet, but may require a written request for the data.

Another issue with these data is the accuracy and validity. While the records submitted to these collections are usually checked for missing metadata, the accuracy of the metadata may not meet the needs of individual research objectives. For example, while most of the data in GBIF are geo-located, historical occurrence data lack original GPS coordinates. Instead coordinates have been assigned to the specimens based on the location description. In some cases, that description can be very detailed. In other cases, it may simply list the state or province. In the latter case, the coordinates assigned to the specimen are the centroid of the state or province. For most geospatial analysis, this level of location uncertainty is unacceptable. While there will always be errors and some level of uncertainty in data, it is recommended that researchers check their data for obvious geospatial errors.

Landscape characterization

Characterization of the landscape has long been a research area of physical geographers and landscape ecologists. Both physical and biological processes are carried out temporally and spatially across the landscape. The landscape can be thought to contain the following characteristics: land cover/land use, soils, terrain, and hydrology. It should be noted that these characteristics are not independent of each other and often interact in complicated and interesting ways.

Land cover/Land use

Land cover and land use data are often produced using a combination of remote sensing and ground data. While land cover and land use are sometimes used interchangeably, these two terms have similar, yet distinct definitions. Land cover refers to the biophysical characteristics of the land surface, while land use includes anthropogenic activities and perception of the land. For example, an orchard and a forest may be classified as the same land cover, trees, but would have different land use classifications. Depending on the research question, the choice between land cover and land use may have little or significant impact. For example, the disturbance regime for a forest that is logged compared to a forest that is protected will affect both the biology and the physical processes. Land cover datasets are often easier to produce and more readily available since the physical land surface can often be identified using remote sensing, where the identification of land use is more nuanced and complex.

There are several datasets available that describe attributes of the landscape at global scales. Land cover is the most common type of global dataset and several different datasets have been developed. All of these datasets are based on remote sensing, although the sensor, resolution, methods, types of land cover classes and time period vary.

Selected land cover databases

The Global Land Cover Facility (http://glcf.umd.edu) based at the University of Maryland hosts several global land cover products including Advanced Very High

Resolution Radiometer (AVHRR) at 1 km, 8 km, and 1 degree data using imagery from 1981 to 1994 as well as annual MODIS-based land cover products at 0.5-degree and 5-degree resolution from 2001–2012. There are also several European-based alternative land cover products. The Land Resource Management Unit of the European Commission's Joint Research Centre produced Global Land Cover 2000 (http://www.eea.europa.eu/data-and-maps/data/global-land-cover-2000-europe) using the VEGETATION sensor on the SPOT 4 satellite with a spatial resolution of 1 km for the year 2000. The European Space Agency has two global land cover products with 300 m resolution based on the MERIS sensor – GlobCover which spans 2005 to 2006 and GlobCover 2009 which covers 2009 (http://due.esrin.esa.int/page_globcover.php). Recently, the United Nation's Food and Agriculture Organization launched 'Global Land Cover SHARE', a new global land cover product designed to consistently integrate national data and global satellite data sources with a resolution of ~1 km (http://www.glcn.org/databases/lc_glcshare_en.jsp).

While satellite remote sensing has provided several contemporary global land cover products, historical data are more difficult to find. One of the few gridded historical datasets was produced by Ramankutty and Foley (1999). They fused national and subnational records of land cover with contemporary satellite data, then hindcast the distribution of land cover based on historical inventory data. An updated version of this product spans from the year 1700 to 2000 with a resolution of five minutes (http://www.ramankuttylab.com).

There are several global datasets for specific land cover types such as urban development, croplands, and forests. The Global Rural-Urban Mapping Project (GRUMP) provides urban extents for 1990, 1995, and 2000 based on population counts, settlement points and night-time lights (http://sedac.ciesin.columbia.edu/data/collection/grump-v1). An alternative dataset from Schneider et al. (2009) at the Center for Sustainability and Global Environment (SAGE) is a MODIS-based urban extent product (https://nelson.wisc.edu/sage/data-and-models/schneider.php). Schneider et al. (2009) report over 90% accuracy for their product and that the GRUMP data have a similarly high producer's accuracy, but lower user's accuracy indicating an over-estimation of urban extent.

Conversion of land to agriculture is the largest type of land cover change globally. Pittman et al. (2010) have estimated global cropland extent using MODIS data available at 250 m and 1 km resolutions. This product is based on a per-pixel cropland probability allowing users to set their own threshold of uncertainty. Another dataset that looks at specific types of crops is described by Monfreda et al. (2008). This product includes global maps of the harvested area and yields for 175 crops. Another interesting global land use product is the Anthropogenic Biomes (or Anthromes) developed by Ellis and Ramankutty (2008) that combine natural vegetation and human use patterns to redefine the Earth's surface based more closely on land use than land cover (http://ecotope.org/products/datasets/).

At the course resolution of most global datasets, mixed pixels often occur. There are several datasets that provide a fuzzy classification such as percent vegetation cover. The Global Land Cover Facility hosts three of these products – AVHRR-based tree cover at 1 km using data from the 1990s, MODIS-based vegetation cover (and cover change) at 250 m, annually from 2000 to 2010, and a Landsat-based tree cover at 30 meters from 2000 and 2005.

At national and local scales, more specialized and place-specific products can be developed that better describe land cover and land use than more generalize global classifications. For example, in the United States, the USGS produced the GAP national land cover dataset using Landsat imagery that is based on the NatureServe Ecological Systems Classification (http://gapanalysis.usgs.gov/gaplandcover/). This dataset goes well beyond basic land cover types and provides consistent and detailed maps of vegetation communities that can be used for ecological modelling and assessment. Many national and local governments have produced place-specific datasets to describe the landscape. These datasets are often available online via governmental websites, but in some cases these data are only available through physical media. Efforts are underway to improve access.

Soils

Another attribute of the landscape is soils. Soils are important as they directly interact with the vegetation, hydrology, and biogeochemical processes. Generalized soils data are commonly available, especially in areas with agriculture or potential agriculture. Many nations have different soil classification systems, so it is important to ensure consistency in these data across national boundaries. Many soils datasets include attribute data about the soils in addition to the soil classification. While this may appear to provide a very rich dataset, researchers should be careful to read any documentation as these attributes can be average or typical values for soil types rather than independently measured attributes. While these averaged attributes may be generally correct, it does present a statistical problem since the desired variability is missing. It should also be noted that many soils datasets are based on field sampling that is extrapolated using aerial imagery and users should read the metadata of these products to fully understand the uncertainty of these products. For example, some soils datasets use visual interpretation of air photos to assign soil class based on vegetation patterns, which may produce circular logic for an analysis of vegetation based on soils.

Selected soils databases

Over the last decade, the Food and Agriculture Organization has been working to produce the Harmonized World Soil Database (HWSD) (http://webarchive.iiasa. ac.at/Research/LUC/External-World-soil-database). This dataset is based on regional and national soil information as well as the older 1:5,000,000 scale FAO-UNESCO Soil Map of the World. The current version of the dataset has variable reliability as some regions have not been updated. For example, the United States, Canada, and Australia all have national soils data products but these have not been harmonized and integrated into the HWSD yet. This raster dataset, which has a spatial resolution of 30 arc-seconds, is linked to soil property attribute data providing user-based query and visualization.

The NASA-based Distributed Active Archive Center (DAAC) for Biogeochemical Dynamics (http://daac.ornl.gov/index.shtml) also has gridded global soils products as well as archived data from field campaigns such as BOREAS and LBA. The Global Soils Collections has 16 global datasets including global soil types, soil phosphorus, soil respiration, and plant-extractable water capacity. Some of these datasets date back to 1998, so in some cases, a more up-to-date data source may be available elsewhere.

Terrain

Many processes and characteristics of the land surface are directly or indirectly affected by the elevation, slope, and aspect of the terrain. Terrain is often used as a proxy variable for hydrological and climatological factors, but may have more direct influences on organisms (Ying et al., 2014). Terrain data are usually provided as either gridded raster layer of elevation, or as contour lines. From these grids or lines, other terrain products such as slope and aspect can be derived using Geographic Information Systems. More sophisticated derived products include the identification of concave and convex hillslope features, potential annual or seasonal solar input, and hydrological characteristics such as watersheds or wetness indices. Global datasets are based on satellite remote sensing products using either stereo optical or synthetic aperture radar (SAR) methods. National and local datasets are available from a variety of sources including ground surveys, airborne sensors including aerial photographs and lidar, and satellite-based remote sensing. The resolution of these datasets can range from 1 km to less than 1 m depending on the data source. The Open Topography facility (http://www.opentopography.org), while limited in current coverage, exemplifies the direction that publicly available data will likely take in this domain.

Selected terrain databases

Freely available global digital elevation products include NASA's Shuttle Radar Topographic Mission (SRTM), METI/NASA's ASTER Global Digital Elevation Map (GDEM), and the USGS's Global Multi-resolution Terrain Elevation Data, 2010 (GMTED2010). SRTM (http://www2.jpl.nasa.gov/srtm/) data are available at 90 m resolution globally and were produced using SAR interferometry (InSAR) based on c-band SAR, which partially penetrates vegetation canopies. The ASTER GDEM (http://asterweb.jpl.nasa.gov/gdem.asp) is available at a finer spatial resolution (30 m). However, this dataset has been shown to have greater uncertainty than SRTM in many regions. The GMTED2010 (http://topotools.cr.usgs.gov/GMTED_viewer/) combines available elevation data from public sources to produce the best local product. It should be noted that this product does have variable resolutions between 7.5 and 30 arc-seconds, depending on the data product used. An x-band based InSAR product called TanDEM-X World DEM is currently being developed. This dataset will have a spatial resolution of 12 m and a 2 m relative vertical accuracy. However, this product will probably not be freely available.

   The distinction between a digital elevation model (DEM), digital surface model (DSM), and digital terrain model (DTM) is often not clear. In general, DSM refers to the Earth's surface including vegetation and man-made structures. DTM refers to the bare-earth surface. Unfortunately, while DEM and DTM should be equivalent, DEM products are sometimes actually DSM products. Users should also determine if any masks have been applied for features such as water bodies that may affect hydrological analysis.

Surface hydrology

Water is an essential ingredient for life and it is not surprising that water availability and water features are an important part of the landscape. The water cycle crosses disciplinary boundaries from atmospheric sciences, to oceanic sciences, physical geography, and geology, depending on the location of the water. Most globally available hydrology data focus on the inputs and outputs of basin hydrology – precipitation, evapotranspiration, and discharge rather than on the hydrologic features themselves. This section focuses on surface hydrology including rivers, lakes, dams, and soil moisture. Precipitation data can be found in the section on climate and evapotranspiration in the section on biogeochemistry.

Selected surface hydrological databases

HydroSHEDS (HYDROlogical data and maps based on SHuttle Elevation Derivatives at multiple Scales) was developed by the World Wildlife Federation and provides a comprehensive dataset of global rivers and watersheds at multiple scales (http://www.worldwildlife.org/pages/hydrosheds). This dataset is based on SRTM topographic data. The World Wildlife Fund has also produced the Global Lakes and Wetlands Database (GLWD – http://www.worldwildlife.org/pages/global-lakes-and-wetlands-database). This dataset is available at three levels; Levels 1 and 2 are polygon data and are divided based on water body size, where level 1 consists of the largest water bodies, and level 2 smaller water bodies. The level 3 data are a combination of the level 1 and level 2 data with additional information formatted in a 30-second resolution raster map. A Global Reservoir and Dam dataset (GRanD) was compiled by Lehner et al. (2011) and is available via SEDAC, the Socioeconomic Data and Application Center (http://sedac.ciesin.columbia.edu/data/set/grand-v1-reservoirs-rev01). This dataset contains over 6,800 records with a focus on reservoirs larger than 0.1 km^3 storage capacity, although the authors included many smaller reservoirs where data were available.

Soil moisture just below the surface is also often an important hydrological factor. NASA's Soil Moisture Active Passive (SMAP) data products include a 9 km resolution level 4 soil moisture product that includes surface and root zone (http://smap.jpl.nasa.gov/data/; http://nsidc.org /daac/). These data are generally available after 6 months as a beta product and after 12 months for a validated product.

Biogeochemistry

The exchange of chemicals between the atmosphere, land surface, and water is essential to understanding global environmental systems. These data are also very useful when combined with land cover, soils, hydrological, or climate data to understand how these attributes interact over time and space. However, given the complexity of these processes, the data are less common both in quantity and extent than land cover or terrain data.

Selected biogeochemistry databases

As mentioned earlier, The Oak Ridge National Laboratory (ORNL) Distributed Active Archive Center (DAAC) maintains a large database of biogeochemistry data products including gridded global products and detailed field campaigns. Another source of data on the exchange of energy, water, and CO_2 between the Earth's surface and the atmosphere is Fluxnet – a network of eddy covariance towers or flux towers (http://fluxnet.fluxdata.org). There are currently over 650 towers across the globe, with the highest concentrations in North America and Europe. Generally measurements are available at a 30-minute frequency. Although most of the Fluxnet data are available via the ORNL DAAC, the data are produced and maintained by individual towers or networks.

An interesting global biogeochemistry dataset related to agriculture is the Global Fertilizer and Manure dataset (http://sedac.ciesin.columbia.edu/data/set/ferman-v1-nitrogen-fertilizer-application). This was created by fusing national statistics about fertilizer and manure application with global maps of harvested area for 175 crops.

Climate

Climate is one of the driving factors in biogeography and the interaction between climate and biogeography is of great interest, especially in the context of global environmental change. Climate can be difficult to define, especially in terms relevant to biogeographical studies as each organism may respond in a unique way to different aspects of climate. Climate is often described in terms of mean values, but it is often the extreme conditions that have the greatest impact. For example, mangroves, a tropical inter-tidal woody plant, are intolerant of temperatures below 4 Celsius. It is important to understand how climate affects the organism you study and to find or create climate data that well represents the conditions that the studied organism is most sensitive to.

Selected climate databases

While numerous climate datasets are available via the meteorological and climate science community, one of the more popular climate datasets in biogeography and ecology is WorldClim (http://www.worldclim.org) – free 1 km gridded data developed by and for ecological researchers. The dataset includes monthly minimum, maximum, and average temperature, monthly total precipitation, and 19 other derived products such as annual temperature range, or precipitation seasonality. WorldClim is based on several major climate databases from the Food and Agriculture Organization, the World Meteorological Organization, the International Center for Tropical Agriculture, the Global Historical Climatology, R-Hydronet (which covers the Caribbean and Central and South America), and several national databases. Other climate data of use in biogeography and landscape ecology involve interpolation among meteorological stations or other spatially explicit representations. Among these are the PRISM and DAYMET sources in the US (http://www.prism.oregonstate.edu/; https://daymet.ornl. gov/); similar data for Canada are available (Price et al., 2011). Global coverage for different types of climate data and resolutions includes the NCEP/NCAR Global Reanalysis project (http://rda.ucar.edu/datasets/ds090.0/), which produces variables for vertical layers in the atmosphere, and CliMod at the MVZ GIS Portal (http://mvzgis.wordpress.com/gis-data/climate-data/), which has bioclimatic variables for use by modellers.

Oceans

So far, this chapter has focused on terrestrial datasets, but biogeography also occurs in the water. Since most global data are based on remote sensing, most global products focus on surface conditions. Even just a few meters below the water's surface, widespread measurement becomes more difficult, limiting the availability of global ocean data.

Selected oceans databases

NASA's Ocean Biology Processing Group (http://oceancolor.gsfc.nasa.gov) maintains several global data products including ocean colour, ocean biogeochemistry, sea surface temperature, and ocean biology. Ocean colour refers to the measurement of pigments associated with plankton, an important base of the ocean food web. NASA has a separate physical oceanography DAAC (http://podaac.jpl.nasa.gov) with data on ocean winds, circulations and currents, topography, and gravity. It should be noted that many of these data products differ in format from land products and may require different software to access and analyse the data.

28.6 Conclusion

Modelling is essential to the advance of biogeography and landscape ecology, and data are essential to models. Malanson et al. (2014) exhorted:

> Physical geographers need to 'model-up' in the areas of mathematical and computer-programming skills that are essential for explicitly specifying and evaluating meaningful theory …

This challenge, echoing others, is not easily answered. In addition to the mathematics and programming skills, learning the concepts and procedures for modelling, which we have tried to introduce, is necessary. Moreover, working with disparate data may require specialized abilities in remote sensing, field or lab (how many of you reading this know next-gen sequencing?) methods, or information management (including GIS). All this is in addition to knowing the system of interest. But this is the cutting edge. Be bold.

SUMMARY

- We focused on models and provided a starting point for the exploration of existing data for studies in biogeography and landscape ecology.
- We examined the importance of resolution, scale, model hierarchy, and spatial and temporal choices for model building.
- An array of organismal, landscape, soil, terrain, surface hydrology, biogeochemistry, climate, and ocean data and databases can be used to build models.
- Modelling is essential to the advancement of geographic science and evaluating meaningful theory.

Further Reading

- Borcard, D., Gillet, F., and Legendre, P. (2011). *Numerical Ecology with R*. Munich: Springer; a hands-on approach to examining species-environment relations with a strong spatial component.

- MacArthur, R.H. (1972). *Geographical Ecology*. Princeton: Princeton University Press; provides a nice background into quantitative geography and ecology.

- Mangel, M. (2006). *The Theoretical Ecologist's Toolbox*. Cambridge: Cambridge University Press; provides a modern view on how to think about models.

Note: Full details of the above can be found in the references list below.

References

Allen, T.F.H. and Starr, T.B. (1982) *Hierarchy: Perspectives for Ecological Complexity*. Chicago: University of Chicago Press.

Austin, M. (2007) 'Species distribution models and ecological theory: A critical assessment and some possible new approaches', *Ecological Modelling* 200: 1–19.

Bak, P. (1999) *How Nature Works*. New York: Springer.

Band, L.E., Patterson, P., Nemani, R. and Running, S.W. (1993) 'Forest ecosystem processes at the watershed scale: Incorporating hillslope hydrology', *Agricultural and Forest Meteorology* 63: 93–26.

Band, L.E., Peterson, D.L., Running, S.W., Coughlan, J., Lammers, R., Dungan, J. and Nemani, R. (1991) 'Forest ecosystem processes at the watershed scale. Basis for distributed simulation', *Ecological Modelling* 56: 171–96.

Bennett, D.A. and Tang, W. (2006) 'Modelling adaptive, spatially aware, and mobile agents: Elk migration in Yellowstone', *International Journal of Geographical Information Science* 20: 1039–66.

Bolker, B.M. (2008) *Ecological Models and Data in R*. Princeton: Princeton University Press.

Boone, R.B. and Galvin, K.A. (2014) 'Simulation as an approach to social-ecological integration, with an emphasis on agent-based modeling', in M.J. Manfredo, J.J. Vaske, A. Rechkemmer and E.A. Duke (eds) *Understanding Society and Natural Resources*. New York: Springer. 179–202.

Botkin, D.B. (1993) *Forest Dynamics: An Ecological Model*. Oxford: Oxford University Press.

Box, G.E.P. (1976) 'Science and statistics', *Journal of the American Statistical Association* 71: 791–9.

Cairns, D.M. (2005) 'Simulating carbon balance at treeline for krummholz and dwarf tree growth forms', *Ecological Modelling* 187: 314–28.

Cairns, D.M. and Malanson, G.P. (1998) 'Environmental variables influencing the carbon balance at the alpine treeline: A modeling approach', *Journal of Vegetation Science* 9: 679–92.

Courant, R., Friedrichs, K. and Lewy, H. (1967) 'On the partial difference equations of mathematical physics', *IBM Journal of Research and Development* 11(2): 215–34.

Ellis, E.C. and Ramankutty, N. (2008) 'Putting people in the map: Anthropogenic biomes of the world', *Frontiers in Ecology and the Environment* 6: 439–47.

Franklin, J. (2010) *Mapping Species Distributions: Spatial Inference and Prediction*. Cambridge: Cambridge University Press.

Gardener, M. (2014) *Community Ecology: Analytical Methods Using R and Excel*. Exeter: Pelagic.

Grant, W.E. and Swannack, T.M. (2008) *Ecological Modeling: A Common-Sense Approach to Theory and Practice*. Oxford: Blackwell.

Grimm, V., Berger, U., Bastiansen, F., et al. (2006) 'A standard protocol for describing individual based and agent-based models', *Ecological Modelling* 198: 115–26.

Grimm, V., Berger, U., DeAngelis, D.L., Polhill, G., Giske, J. and Railsback, S. F. (2010) 'The ODD protocol: A review and first update', *Ecological Modelling* 221: 2760–8.

Hagerstrand, T. (1965) 'A Monte Carlo approach to diffusion', *European Journal of Sociology* 6: 43–67.

Hanson, J.S., Malanson, G.P. and Armstrong, M.P. (1990) 'Landscape fragmentation and dispersal in a model of riparian forest dynamics', *Ecological Modelling* 49: 277–96.

Hermann, C.F. (1967) 'Validation problems in games and simulations with special reference to models of international politics', *Behavioral Science* 12: 216–31.

Hogeweg, P. (1988) 'Cellular automata as a paradigm for ecological modeling', *Applied Mathematics and Computation* 27: 81–100.

Hunt, E.R., Piper, S.C., Nemani, R., Keeling, C.D., Otto, R.D. and Running, S.W. (1996) 'Global net carbon exchange and intra-annual atmospheric CO_2 concentrations predicted by an ecosystem process model and three-dimensional atmospheric transport model', *Global Biogeochemical Cycles* 10: 431–56.

Hutchinson, G.E. (1957) 'Concluding remarks', *Cold Spring Harbor Symposia on Quantitative Biology* 22: 415–27.

Keane, R.E., Ryan, K.C. and Running, S.W. (1996) 'Simulating effects of fire on northern Rocky Mountain landscapes with the ecological process model FIRE-BGC', *Tree Physiology* 16: 319–31.

Kearney, M. and Porter, W. (2009) 'Mechanistic niche modelling: Combining physiological and spatial data to predict species' ranges', *Ecology Letters* 12: 334–50.

Kendall, B.E., Briggs, C.J., Murdoch, W.W., Turchin, P., Ellner, S.P., McCauley, E., Nisbet, R.M. and Wood, S.N. (1999) 'Why do populations cycle? A synthesis of statistical and mechanistic modeling approaches', *Ecology* 80: 1789–805.

Kupfer, J.A. and Runkle, J.R. (2003) 'Edge-mediated effects on stand dynamic processes in forest interiors: A coupled field and simulation approach', *Oikos* 101: 135–46.

Larsen, L., Thomas, C., Eppinga, M. and Coulthard, T. (2014) 'Exploratory modeling: Extracting causality from complexity', *EOS, Transactions of the American Geophysical Union* 95: 285–6.

Lehner, B., Liermann, C.R., Revenga, C., Vörösmarty, C., Fekete, B., Crouzet, P., Döll, P.,Endejan, M., Frenken, K., Magome, J., Nilsson, C., Robertson, J.C., Rödel1, R., Sindorf, N. and Wisser, D. (2011) 'High-resolution mapping of the world's reservoirs and dams for sustainable river-flow management', *Frontiers in Ecology and the Environment* 9: 494–502.

Malanson, G.P. (1996) 'Modelling forest response to climatic change: Issues of time and space', in S.K. Majumdar, E.W. Miller and F.J. Brenner (eds) *Forests – A Global Perspective*. Easton, PA: Pennsylvania Academy of Sciences. pp. 200–11.

Malanson, G.P. and Cramer, B.E. (1999) 'Landscape heterogeneity, connectivity, and critical landscapes for conservation', *Diversity and Distributions* 5: 27–40.

Malanson, G.P., Scuderi, L., Moser, K., Willmott, C., Resler, L., Warner, T. and Mearns, L.O. (2014) 'The composite nature of physical geography', *Progress in Physical Geography* 38: 3–18.

Malanson, G.P., Westman, W.E. and Yan, Y.L. (1992) 'Realized versus fundamental niche functions in a model of chaparral response to climatic change', *Ecological Modelling* 64: 261–77.

Mena, C.F., Walsh, S.J., Frizzelle, B.G., Yao, X. and Malanson, G.P. (2011) 'Land use change on household farms in the Ecuadorian Amazon: Design and implementation of an agent-based model', *Applied Geography* 31: 210–22.

Monfreda, C., Ramankutty, N. and Foley, J.A. (2008) 'Farming the planet: 2. Geographic distribution of crop areas, yields, physiological types, and net primary production in the year 2000', *Global Biogeochemical Cycles* 22. DOI: 10.1029/2007GB002947.

O'Sullivan, D. and Perry, G.L.W. (2013) *Spatial Simulation*. Oxford: Wiley-Blackwell.

Parker, D.C., Manson, S.M., et al. (2003) 'Multi-agent systems for the simulation of land-use and land-cover change: A review', *Annals of the Association of American Geographers* 93: 314–37.

Pittman, K., Hansen, M.C., Becker-Reshef, I., Potapov, P.V. and Justice, C.O. (2010) 'Estimating global cropland extent with multi-year MODIS data', *Remote Sensing* 2: 1844–63.

Price, D.T., McKenney, D.W., Joyce, L.A., Siltanen, R.M., Papadopol, P. and Lawrence, K. (2011) 'High-resolution interpolation of climate scenarios for Canada derived from general circulation model simulations', *Natural Resources Canada*, Canadian Forest Service, Northern Forestry Centre, Edmonton, Alberta. Information Report NOR-X-421.

Railsback, S.F. and Grimm, V. (2012) *Agent-Based and Individual-Based Modeling: A Practical Introduction*. Princeton: Princeton University Press.

Ramankutty, N. and Foley, J.A. (1999) 'Estimating historical changes in global land cover: Croplands from 1700 to 1992', *Global Biogeochemical Cycles* 13: 997–1027.

Rastetter, E.B. (1996) 'Validating models of ecosystem response to global change', *BioScience* 46: 190–8.

Reynolds, J.F., Hilbert, D.W. and Kemp, P.R. (1993) 'Scaling ecophysiology from the plant to the ecosystem: A conceptual framework', in: J.R. Ehlringer and C.B. Field (eds) *Scaling Physiological Processes Leaf to Globe*. San Diego: Academic Press. pp. 127–41.

Running, S.W. and Coughlan, J.C. (1988) 'A general model of forest ecosystem processes for regional applications. I. Hydrologic balance, canopy gas exchange and primary production processes', *Ecological Modelling* 42: 125–154.

Running, S.W. and Gower, S.T. (1991) 'FOREST-BGC, a general model of forest ecosystem processes for regional applications. II. Dynamic carbon allocation and nitrogen budgets', *Tree Physiology* 9: 147–60.

Running, S.W. and Hunt, E.R. (1993) 'Generalization of a forest ecosystem process model for other biomes, BIOME-BGC, and an application for global-scale models' in J.R. Ehlringer, and C.B. Field (eds) *Scaling Physiological Processes Leaf to Globe*. San Diego: Academic Press. pp. 141–58.

Schneider, A., Friedl, M.A. and Potere, D. (2009) 'A new map of global urban extent from MODIS satellite data', *Environmental Research Letters* 4 (4): 044003.

Shugart, H.H. (1998) *Terrestrial Ecosystems in Changing Environments*. Cambridge: Cambridge University Press.

Smith-McKenna, E., Malanson, G.P., Resler, L.M., Carstensen, L.W., Prisley, S.P. and Tomback, D.F. (2014) 'Cascading effects of feedbacks, disease, and climate change on alpine treeline dynamics', *Environmental Modelling and Software* 62: 85–96.

Soberon, J. (2007) 'Grinnellian and Eltonian niches and geographic distributions of species', *Ecology Letters* 10: 1115–23.

Stevens, M.H.H. (2009) *A Primer of Ecology with R*. New York: Springer.

Urban, D.L., Bonan, G.B., Smith, T.M. and Shugart, H.H. (1991) 'Spatial applications of gap models', *Forest Ecology and Management* 42: 95–110.

Yadav, V., Del Grosso, S.J., Parton, W.J. and Malanson, G.P. (2008) 'Adding ecosystem function to agent-based land use models', *Journal of Land Use Science* 3: 27–40.

Ying, L.X., Shen, Z.H., Piao, S.L., and Malanson, G.P. (2014) 'Terrestrial surface area increment: The effects of topography, DEM resolution, and algorithm', *Physical Geography* 35, 297–312.

Zeng,Y. and Malanson, G.P. (2006) 'Endogenous fractal dynamics at alpine treeline ecotones', *Geographical Analysis* 38: 271–87.

ON THE COMPANION WEBSITE…

Visit **https://study.sagepub.com/keymethods3e** for author videos, chapter exercises, resources and links, plus **free** access to the following recommended articles:

1. **Kupfer, J.A. (2012) 'Landscape ecology and biogeography: Rethinking landscape metrics in a post-FRAGSTATS landscape', *Progress in Physical Geography*, 36 (3): 400–20.**

This article links the general topic of landscape ecology to possible analyses of models. In this way it connects this chapter to other methods, and metrics are something readers should know about.

2. **Miller, J.A. (2012) 'Species distribution models: Spatial autocorrelation and non-stationarity', *Progress in Physical Geography*, 36 (5): 681–92.**

This is an updated view on the most widely used type of model in biogeography.

3. **Song, C. (2013) 'Optical remote sensing of forest leaf area index and biomass', *Progress in Physical Geography*, 37 (1): 98–113.**

This is a core report on the most widely-used remote sensing approach for biogeography.

29 Environmental Audit, Appraisal and Valuation

Peter Glaves

SYNOPSIS

A wide range of tools and methods has been developed over the last 40 years to assess how individuals, organizations and societies impact on their environment. Such methods are core to **sustainability** and can be classified into three broad types, firstly methods for assessing the environmental impact of new developments (**Environmental Impact Assessments/ EIA**), secondly methods for assessing the environmental impacts of existing organizations (**Environmental Audits/EA**), and thirdly methods for valuing the benefits society gains from the environment (**Ecosystem Services**). This chapter provides an overview of the range of methods that can be used and provides advice on their application.

This chapter is organized into the following sections:

- Introduction
- Environmental Impact Assessments
- Environmental Auditing
- Environmental Goods and Ecosystem Services
- Conclusion

29.1 Introduction

Rapid population and economic growth has lead to increasing demands for natural resources and new development (Glasson et al., 2005). Over the last century environmental impacts have occurred at a rate larger and more consequential than at any point during man's history (Millennium Ecosystem Assessment, 2005). Concerns that humans had reached a cross-road in their relationship with nature (WCED, 1987: 4) have led to the development of new concepts, approaches and methods. A key concept that emerged in the 1980s was sustainability, which has since been adopted as a global ethic and became ubiquitous in its use in development policies, legislation etc. (IEMA, 2011; Smith et al., 2010).

The idea that conflicts between development and the environment are not inevitable was first proposed at the United Nations Conference of the Human Environment in

1972. At the World Commission of Environment and Development in 1987 and the Rio Earth Summit (1992) the idea was formalized into the classic definition of sustainable development, known as the Brundtland definition: *development that meets the needs of present generations without compromising the opportunity for future generations to meet their needs.*

The planet operates as a largely enclosed system, human societies and our economic activities take place within a finite and non-growing Earth, and therefore we must adopt patterns of resource use and development that recognise such limits (UNCED, 1992). Sustainable development needs to take account of social and ecological factors, as well as economics, to consider the resource base on which society depends, and consider long-term as well as short-term impacts of the decisions we make (IUCN/UNEP/WWD, 1980). Such approaches have been referred to as the 'triple bottom line approach' (Elkington, 1997) where the three bottom lines represent society, the economy and the environment (Figure 29.1). Society depends on the economy for its current and future prosperity, and the economy depends on the global ecosystem, whose health represents the ultimate bottom line.

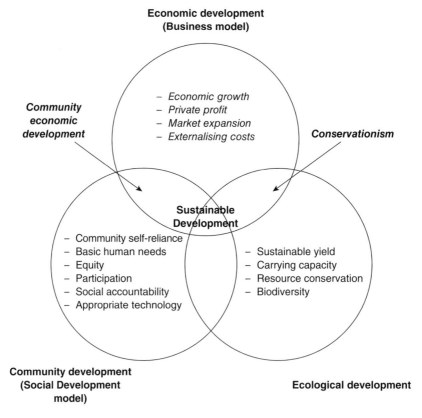

Figure 29.1   Triple bottom line approach to sustainability

Source: Adapted from Bell and Morse (2003)

Traditionally, economic factors alone were considered when measuring development. Such economy-based decisions tend to be disinterested in the environment, to dismiss anything that is not readily quantifiable and ignore finite environmental limits (Millennium Ecosystem Assessment, 2005). Sustainability-based decision-making considers economic, environmental and social factors and has been applied in three key areas:

1. Reviewing the sustainability of new developments as part of local and strategic planning i.e. Environmental Impact Assessments;

2. Measuring the impacts and sustainability of existing organizations i.e. Environmental Auditing; and

3. Recognising the wider range of benefits, goods and services obtained from the environment i.e. Ecosystem Services.

This chapter provides an overview of the three approaches above and outlines some of the standard methods used.

29.2 Environmental Impact Assessments

In the USA a series of environmental disasters led to the formulation of the National Environmental Policy Act (NEPA) in 1969. NEPA required that developmental projects should identify and predict the adverse impacts caused by the development and produce measures to prevent these. Such approaches have been adopted in many countries and are known as Environmental Impact Assessment (EIA) or in the UK as Environmental Assessment (EA). EIA have been described as being 'one of the most successful policy innovations of the 20th Century' (Sadler, 1998).

Since the implementation of the EIA Directive 85/337 in 1985, member states of the European Union are now required to systematically consider environmental impacts as part of planning. In UK planning law, 'development' means:

1. An act of construction (building); or

2. Extraction of minerals (mining); or

3. Significant change in land use (when existing uses of land change from one major land type to another e.g. converting a factory into flats).

The broadest definition of EIA is 'an assessment of a planned activity on the environment' (UN, 1990). The term EIA is most frequently used when assessing the impacts resulting from a local development. When an assessment is made of the impacts of strategic policies and programmes, these are normally referred to as Strategic Environmental Assessments (SEA). An EIA is

> a technique and a process by which information about the environmental effects of a project is collected ... and taken into account by the planning authority in forming their judgements on whether the development should go ahead. (DoE, 1989)

EIA is a systematic and predictive process for exploring the potential environmental effects of a development prior to development (Wood and Jones, 1997). The final output of an EIA is a report known as an Environmental Statement (ES) and should contain sufficient environmental information to make an informed decision of whether or not the planning application should be granted (Glasson et al., 2005).

An EIA is not a single process or method; EIAs involve several stages and encompass evaluations of impact on different aspects of the wider environment (Therivel and Morris, 2009). The stages involved in an Environmental Impact Assessment vary between countries; the approach used in the UK follows the eight stages set out in EIA Directive 2011/92/EU (Table 29.1). The time taken to complete a full EIA may be 12–18 months for major projects (DoE, 1995). Different methods are used in different stages of an EIA (Table 29.2), and the methods chosen also depend on the characteristics of the proposed development site, the type of development proposed and the resulting impacts.

Table 29.1   The EIA process used in the UK

1	Screening	The competent authority makes a decision – most often the local planning authority (LPA) – as to whether an EIA is required for the proposed development.
2	Scoping	Consultation with relevant statutory agencies in order to identify what environmental issues/factors should be covered by the EIA. Includes identification of the likely potential impacts, mitigation measures, monitoring activities and methods for obtaining information.
3	Production and Submission of the Environmental Statement (ES)	This is the largest stage in an EIA, which includes: • Collation and evaluation of existing site information • A series of surveys of the proposed development site and surroundings • Assessment of the significance of the environmental impacts • Proposed methods of mitigation and minimisation of negative impacts • Consultation with key agencies on the findings (see also Stage 5 below) • Summarising the above in a report (ES).
4	Review of ES	There may be a requirement for the ES to be reviewed before consent for the development is granted. This is not required under EU or UK legislation.
5	Consultation	The ES is made public for review. Public, interested organisations and authorities with environmental responsibilities are allowed to comment.
6	Consideration by Authority	The full EIA report is considered by the competent authority (LPA).
7	Decision	A decision is made by the LPA as to whether the proposed development can go ahead; the reasons for this decision are required to be made public. Local people and organisations can appeal against planning decisions and in such cases a Planning Enquiry may take place.
8	Monitoring	Monitoring of potentially serious environmental effects may be required.

Source: Adapted from EC (2001a)

Table 29.2   Example methods that can be used in key stages of an EIA

Stage	Examples of Methods
Baseline surveys of site	• Analysis of existing environmental reports • Environmental stock/setting • Points of reference • Baseline surveys e.g. Phase One Habitat Surveys
Screening/scoping	• Formal/informal checklists • Survey, case comparison • Effects networks • Public or expert consultation
Impact prediction	• Scenario development • Risk assessment • Environmental indicators and criteria • Policy impact matrix • GIS of capacity/habitat analysis • Cost/benefit analysis and other economic valuation techniques • Multi-criteria analysis
Documentation for decision making	• Cross-impact matrices • Sensitivity analysis • Decision trees

Source: Adapted from Sadler (1998)

Under UK planning law the environment consists of five broad areas:

- **Air**   (noise, light, chemical pollution etc.)

- **Water**   (water quality and quantity, surface and underground water, pollution, flood risk etc.)

- **Land**   (geology, soils, land stability, contaminated land etc.)

- **Biota**   (including species, habitats, biodiversity etc.)

- **People**   (socio-economics, demographics, quality of life, archaeology/history etc.)

An EIA will normally need to cover aspects of all/most of these areas. An EIA also has to consider how impacts on one aspect of the environment can result in impacts on other aspects e.g. how changes to river flow impact on species and existing businesses e.g. fish farms.

    An EIA should be undertaken by qualified, experienced and professionally-recognised practitioners and follow systematic standard methods; a full EIA will require a team of subject experts each looking at different types of impacts. Guidance on standard methods of survey and impact assessment has been produced by the relevant subject professional bodies e.g. the Chartered Institute for Ecology and Environmental Management has produced guidance on ecological impact assessment (IEEM, 2006), and the Landscape Institute on landscape impact assessment (Landscape Institute and IEMA, 2013). The following sections outline the standard methods used for the main stages in an EIA. For further information on the methods used in EIA see Morris and Therivel (2009) and Brady et al. (2011).

Stage 1: Screening

Stage 1 of an EIA is known as Screening, where development proposals are reviewed and a decision is made as to whether an EIA needs to be undertaken. If a development will have no likely significant negative environmental impacts then an EIA will not be required. Various factors determine if an EIA is required:

- The size of the development – larger developments are more likely to cause significant impacts.

- The type of development and the associated environmental impacts – a factory producing toxic chemicals has more associated environmental risks than a new home.

- The sensitivity of the proposed development site – a development next to a nature reserve will have more ecological impacts than one in an already build up area.

The EIA Directive 2011/92/EU provides guidance on whether an EIA is likely to be required. There are two types of development covered in the Directive:

- Annex I of the Directive detailing projects for which EIA is obligatory because of its size and associated risks e.g. nuclear power stations, new motorways.

- Annex II details types of development for which an EIA may be required, e.g. new housing and shop developments will require an EIA if they are above a certain size.

The EU have produced a checklist for screening of projects (European Commission, 2001a) which provides clear guidelines for the factors to consider and methods to follow: the guidance along with other key EIA method documents can be accessed via the www.europa.eu website.

Stage 2: Scoping

Stage 2 of an EIA is scoping, normally a desk-based review of existing information about the site and proposed development plan (Bond and Stewart 2002): 'Scoping is the process of determining the content and extent of matters that should be covered in the environmental information to be submitted to ... the planning authority' (European Commission, 2001b: n.p.). Scoping should:

- Identify the key issues and concerns of the affected and interested parties.

- Identify who is concerned.

- Identify what the concerns are.

- Identify why they are concerns.

- Identify when development becomes unacceptable (European Commission 2001b).

A useful tool to use in scoping is a scoping checklist; the EU has produced a checklist with detailed guidance (European Commission, 2001b). In scoping and later stages of an EIA the key factors to consider are the probability and significance of any impacts which may result from development. It is important to note that impacts can occur at any stage in the life of a new development, i.e. during construction, operation or decommissioning, and that impacts during construction may differ from those during operation.

Stage 3: Environmental Statement

Stage 3 of an EIA builds up to the production and submission of the Environmental Statement (ES). This is the longest stage in an EIA and involves a number of interlinked methods. The key questions to be covered in an ES are:

- What effects would this project have on the environment?

- Which of these effects are likely to be significant?

- Which alternatives and mitigating measures can be used to minimise the impacts of the project?

- What is the overall impact of the project after mitigation?

The first step in Stage 3 is to find out what information is already available about the site and its surroundings. Such data are held in various locations. Biodiversity data are usually held at Local Biodiversity Records Centres; archaeological records are held by local councils; historic information can be found in local study libraries; and information on previous planning applications and much other relevant information can be found via the local planning department. It is important to consider the current spatial plan and policies that apply to your proposed development. These plans can be accessed via the Planning Portal on your Local Authority website. Remember to check the date of any surveys as some environmental factors can change quite quickly, e.g. water quality and species. If a survey is several months or years old it may no longer reflect current site conditions.

When existing local data are absent or out of date, site surveys will need to be undertaken using standard professional body methods. When legally protected species or sites are involved, the work will need to be undertaken by professionally qualified and licenced individuals. Some surveys can only be undertaken at certain times of the year, e.g. plants and animals. Other impacts are also time dependent, e.g. traffic, recreation and tourism numbers vary during the day, between days and between months. In all surveys standardisation of both the method used and effort (number of surveyors, duration of survey etc.) is required. Note any limitations that could affect the survey results.

The next step of an EIA is assessing the significance of each of the resulting environmental impacts. In determining the scale or significance of any change the following parameters should be considered:

- What is the current condition of the site?

- How sensitive is the site to the proposed change?

- What is the likely magnitude of change?

- What is the duration of the impact?

- When will impacts occur and how frequently will they occur?

- Are impacts below critical thresholds or irreversible?

- Can any impacts be reduced, mitigated or compensated for?

- Are legal and planning issues involved?

- What is the level of confidence in the predicted impacts? (adapted from IEEM, 2006)

The type of impact and the criteria used to determine the significance of impacts vary depending on the environmental characteristic being considered. For most environmental impacts, these can be expressed using a seven-point ordinal significance scale:

Negligible	0
Minor (adverse or beneficial)	+1 / −1
Moderate (adverse or beneficial)	+2 / −2
Major (adverse or beneficial)	+3 / −3

Examples of the criteria used to determining significance of impacts are illustrated in Table 29.3. Once the significance of individual impacts has been determined, the next step is to consider how significant negative impacts can be overcome. In forward planning there is a hierarchy of options: avoid, mitigate, compensate, enhance. The ES then collates together the impacts for all environmental factors to assess the overall impact. There are two main approaches used to collate data, GIS and matrices (Table 29.4). An ES should contain a standard series of sections (Table 29.5 – for more detailed guidance on content see IEEM (2006); IEMA (2011)). For criteria for reviewing the content and quality of ES see Lee et al. (1999).

29.3 Environmental Auditing

Businesses have a central role to play in sustainability, both a cause of environmental problems and part of the solution. Businesses drive economic growth, are major consumers of environmental resources and causes of pollution. Sustainable development requires changes in business and industry as producers of goods and services (Morris and Therivel, 2009).

According to the Tomorrow's Company Global Business Think Tank, the role of a modern business is 'to provide ever better goods and services in a way that is profitable, ethical and respects the environment, individuals and the communities in which it operates' (Tomorrow's Company, 2007: 4). Businesses that consider the environment will benefit economically by reductions in expenditure on energy, transport, waste disposal etc.; organizations able to respond to changing social and environmental conditions are more likely to survive and succeed (Post et al., 2002).

Table 29.3 Example criteria for determining the significance of environmental impacts

Habitat Impacts

Impact	Criteria
+ **Beneficial Impact**	Increase in the diversity and extent of typical habitats
0 **No Impact**	No change in the diversity and extent of existing habitats
−1 **Low Adverse Impact**	Disturbance to and/or loss of wide-spread habitat types
−2 **Moderate Adverse Impact**	Temporary reversible loss or disturbance to scarce habitats
−3 **High Adverse Impact**	Permanent and irreversible loss of scarce habitats

Species Impacts

Impact Significance	Criteria
+ **Beneficial**	Increase in the diversity/abundance of native and/or protected species
0 **No Impact**	No change in existing diversity and abundance of native species
−1 **Low Adverse impact**	Reversible loss of common species – the magnitude of this loss below a level that would permanently reduce the existing population size
−2 **Moderate Adverse impact**	Irreversible loss of common species of fauna – the magnitude of this loss at a level that would permanently reduce the existing population size
−3 **High Adverse impact**	The loss of legally protected or rare species

Source: Adapted from IEEM (2006)

Table 29.4 Predicted impacts scale, mitigation and residual impacts

Environmental Impact Area	Baseline Survey Findings	Importance of Feature	Description of Unmitigated Impact	Significance of Impact (without mitigation)	Proposed Mitigation Measures	Residual Impact
Amphibians	Great Crested Newt	Nationally important Legally protected (Wildlife and countryside Act 1981)	Development is 1km from the ponds and will not directly affect ponds, but suitable grassland will be affected	Probable low to moderate negative impact	Create new ponds Newt friendly management of grassland and existing ponds Timing of works Pre-works check of area	No impact to low negative impact
Archaeology	Remains of medieval buildings recorded on development area	Local, possible county importance	Complete loss of any building remains during site construction	Certain low to moderate negative impact	Archaeological site survey prior to any construction, record any remains	Low impact

This enlightened self-interest has led to the development of a range of sustainable business approaches including environmental management systems (EMS) (Brady et al., 2011).

Table 29.5 Typical ES format in the UK

Non-technical summary

Part 1: Methods and Key Issues

   1. Methods statement
   2. Summary of key issues; monitoring programme statement

Part 2: Background to the Proposed Development

   3. Preliminary studies: need, planning, alternatives and site selection
   4. Site description, baseline conditions
   5. Description of proposed development
   6. Construction activities and programme

Part 3: EIA Topic Areas

   7. Land use, landscape and visual quality
   8. Geology, topography and soils
   9. Hydrology and water quality
 10. Air quality and climate
 11. Ecology: terrestrial and aquatic
 12. Noise
 13. Transport
 14. Socio-economic impact
 15. Interrelationships between effects

Source: Adapted from Glasson et al. (2005)

EMS provide a structured and comprehensive mechanism for ensuring that the activities and products of an enterprise do not cause unacceptable effects in the environment (International Chamber of Commerce, 1991) and for managing an organization's significant environmental impacts (Brady et al., 2011). EMS operate as a cyclical process involving identifying, improving, checking (Figure 29.2). An EMS assesses a company's environmental performance against internal or external environmental standards. A variety of environmental 'standards' have been developed, these include:

- BS 7750 (specification for an environmental management system) – British Standard Institute 1992, also BS 8555 (standard for the phased implementation of an environmental management system)

- EMAS (Community eco-management and audit scheme), Regulation 761/01 – the European Commission, 2001

- ISO 14001 (Environmental Management standard) – International Organization for Standardisation

Business and industry have found it difficult to accept that their environmental actions are meaningful. Tools such as environmental audits were developed to assess the worth of a company's environmental actions (Mirovitskaya and Asher, 2001). Environmental audits involve the 'systematic analysis of the environmental performance of organisations' (Post et al., 2002). Environmental auditing forms a key element of any EMS. It allows a company to verify its EMS and confirm that it meets

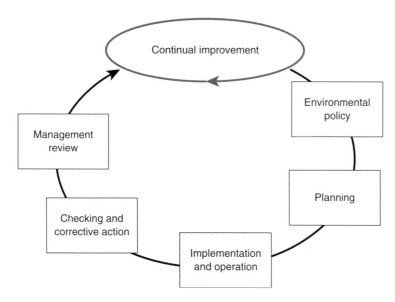

Figure 29.2 Typical cyclical process of Environmental Management Systems (EMS)

the requirements of external environmental certification. The term 'environmental audit' covers a range of methods for evaluating an organization's environmental compliance, identifying gaps and making recommendations.

Auditing tools (ICC, 1991) help safeguard the environment by facilitating management control of environmental practices and assessing compliance with company policies including whether regulatory requirements are met. Audits assess the efficiency and effectiveness of current environmental performance. Questions covered include: what are we doing in relation to the environment? Can we do it better? How can we do it better? Environmental audits are undertaken for a variety of reasons including: assessing the effectiveness of the current EMS; checking conformance with company policy and programmes; assessing compliance with environmental legislation; auditing environmental risk management; and checking the environmental performance of products.

An environmental audit is normally split into three phases, preparation, and execution and reporting (Figure 29.3). The preparation/planning stage of the audit involves setting objectives, selecting the audit team, an initial document review and preparation of the audit programme. In defining the scope of the audit it is important to identify the activities covered, the key issues considered, the criteria for assessment, and the timeframe. The audit team should have appropriate knowledge, skills and training and be independent of the activities being reviewed.

The execution of an on-site audit involves three elements: documentary analysis, site surveys, and the evaluation of the findings from these (Figure 29.4). The analysis of documents should be thematic and check that the content of the company's documents meets the requirements of the audit, e.g. all company documents meet the external environmental management system standards. Documents reviewed may

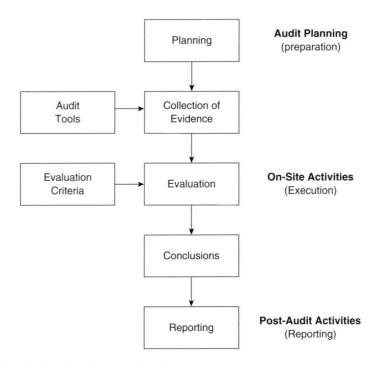

Figure 29.3 Outline of environmental audit process

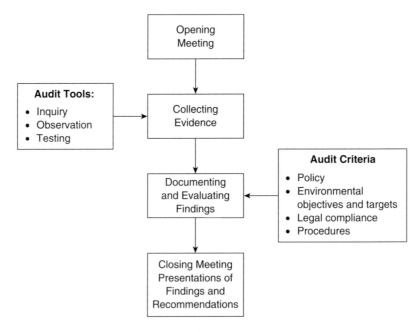

Figure 29.4 Outline of on-site environmental audit activities

include: site plans, organizational structures, organizational procedures, permits and consents, monitoring records and report, previous audit reports and all management system documentation.

When collecting evidence on-site a variety of tools can be used; normally a triangulation of methods is adopted to verify the robustness of the findings involving inquiry, observation and testing. Inquiry uses one-to-one meetings, interviews and questionnaires and seeks to confirm that the procedures followed are consistent with the documentation and appropriate for the EMS, and that staff understand their environmental responsibilities, have appropriate skills and are aware of relevant environmental policies and programmes. Inquiry also checks that regulations are being complied with. On-site observations of staff behaviour are required and these observations are used to verify the information gathered during the interviews. Several observation sessions should be undertaken. Testing forms an important element of site surveys and should cover both equipment and behaviour. Safety, control and monitoring equipment should be tested to ensure they meet the appropriate standards. Testing of employers and managers can use scenarios and role-play to check staff awareness of and compliance with company environmental policy and legislation.

Data gathered are collated into an environmental audit report which covers: details of (i) the organization and audit team; (ii) scope and objectives of the audit; (iii) a summary of audit methods used; and (iv) conclusions and recommendations all with supporting evidence. Common failings found in audits include: breaches of legislation, lack of evidence of training, and failure to meet company environmental objectives. Environmental auditing should be a cyclical process and regularly repeated to monitor a company's progress. Subsequent environmental audits should seek to test how the findings and recommendations from the previous audit have been enacted, to review the changes that have taken place and how these have affected the organization's environmental performance. For more information on auditing processes and methods see Brady et al. (2011).

29.4 Environmental Goods and Ecosystem Services

The ecologist Arthur Tansley first used the term 'ecosystem' in the 1930s to recognize the way in which physical and biological components of the environment operate as a single functioning ecological system or 'ecosystem'; a system which includes people (Tansley, 1935: 299). The ecosystem concept has been used in many fields, from urban planning to greening of business. Humans rely on ecosystems for a wide range of valuable and indeed essential benefits, services and goods. These include the fundamental resources we need to live, i.e. clean air, water, safe food, as well as those things that improve the quality of our lives, e.g. picturesque landscapes, birds in our gardens etc. Such benefits are known as 'environmental goods' or 'ecosystem services'. Some of the benefits that ecosystems provide may not be directly obvious. For example, the way wetlands absorb water and prevent our homes from being flooded; the way trees in urban areas cool the climate and reduce chemical and noise pollution. These benefits are, however, valuable to us in preventing harm, in reducing the amount of money spent on flood prevention and in maintaining human health and quality of life. Many of these benefits are not measured by traditional economic methods.

In the year 2000 the United Nations developed an approach which brought together the variety of ways in which nature provides benefits and support to humans in a single approach which could be used by decision makers, planners, policy makers etc. This was the Millennium Ecosystems Approach (MA). Its objective was to: 'assess the consequences of ecosystem change for human well-being and the scientific basis for action needed to enhance the conservation and sustainable use of those systems and their contribution to human well-being' (Alcamo et al., 2002).

The Millennium Ecosystem Assessment set out a typology of ecosystem services under four broad headings:

- **Provisioning services** (for example food, fuel, ornaments)

- **Regulating services** (for example air-quality maintenance, natural flood protection)

- **Cultural services** (for example spiritual value, recreation and tourism) and

- **Supporting services** (for example habitat provision, water cycling).

A full breakdown of the categories in the Millennium Ecosystem Assessment is shown in Table 29.6 and an example of the range of services provided by a wetland is shown in Figure 29.5.

The European Union has set itself the target of halting the loss of biodiversity and the degradation of ecosystem services in the EU by 2020. Achieving such targets means collecting scientific data on the state of ecosystem services and developing methods for measuring these services. Such methods are together known as Ecosystem Services Valuation (ESV), which is the process of assessing the contribution of ecosystem 'services to meeting a particular goal' (Costanza et al., 2006). The valuation is often given in economic terms, normally as benefits in pounds per hectare per year ($£\ ha^{-1}\ yr^{-1}$), using the following equation:

Total Value (V) of Ecosystem Services ES in £/ha/year for ecosystem type k is $V(ES)_k$

$$V(ES_k) = \sum_{i=1}^{x} A(LU_i) \times V(ES_{ki})$$

[29.1]

Where $A(LUi)$ = Area of i (Land Use in hectares)

$V(ES_{ki})$ = Annual value of k ES (Ecosystem Services) for each i LUi ($£\ ha^{-1}\ yr^{-1}$).

Environmental and resource economists have developed a range of methods (Table 29.7) for measuring the value of the different types of ecosystem services (e.g. deGroot et al., 2002; Freeman, 2003). The UK approach to valuing ecosystem services was set out in *An introductory guide to valuing ecosystem services*, published by Defra in 2007.

When measuring the ecosystem service value of a site, there is a need to understand the characteristics of the ecosystem(s), the uses, and the types of services/benefits

Table 29.6 Categories of services in the Millennium Ecosystem Assessment

Ecosystem Service	Service Type
Provisioning Services *(products obtained from ecosystems)*	Food
	Fibre and fuel
	Genetic resources/Biodiversity
	Biochemicals, natural medicines, pharmaceuticals
	Ornamental resources
	Fresh water
	Saline water
	New environmental products
Regulating Services *(benefits obtained from the regulation of ecosystem processes)*	Air quality regulation
	Climate regulation
	Water regulation
	Buffer
	Natural hazard regulation
	Pest regulation
	Disease regulation
	Erosion regulation
	Water quality regulation
	Pollination
	Fire regulation
Cultural services *(non-material benefits that people obtain through spiritual enrichment, cognitive development, recreation etc.)*	Cultural heritage
	Recreation and tourism
	Aesthetic value
	Employment
	Scientific
	Spiritual
	Educational
Supporting services *necessary for the production of all other ecosystem services*	Soil formation
	Primary production
	Nutrient cycling
	Water cycling

obtained (Figure 29.6). Traditional economics provides a range of methods for valuing marketed goods and such an approach can be adapted to value other ecosystem services/benefits (Figure 29.7). In ecosystem service valuation some of the benefits being valued are goods for which market data exist, e.g. food, timber. Other socially valuable benefits, such as aesthetic value of a beauty spot and the value of urban trees in providing shade and absorbing pollution, do not have market values, and in such cases their value can be 'revealed using alternative methods, e.g. willingness to pay (Table 29.7). Calculation of the individual ecosystem service values can be achieved in four ways:

- **Actual/market value** – local surveyed site specific data
- Transfer of **proxy values** – use of existing data from other comparable recent studies/areas

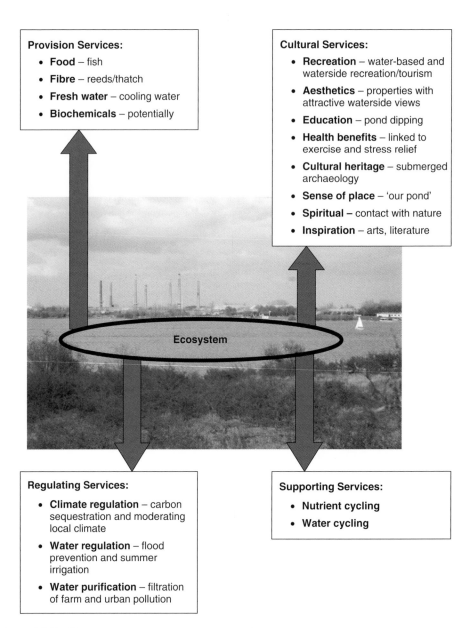

Provision Services:

- **Food** – fish
- **Fibre** – reeds/thatch
- **Fresh water** – cooling water
- **Biochemicals** – potentially

Cultural Services:

- **Recreation** – water-based and waterside recreation/tourism
- **Aesthetics** – properties with attractive waterside views
- **Education** – pond dipping
- **Health benefits** – linked to exercise and stress relief
- **Cultural heritage** – submerged archaeology
- **Sense of place** – 'our pond'
- **Spiritual** – contact with nature
- **Inspiration** – arts, literature

Ecosystem

Regulating Services:

- **Climate regulation** – carbon sequestration and moderating local climate
- **Water regulation** – flood prevention and summer irrigation
- **Water purification** – filtration of farm and urban pollution

Supporting Services:

- **Nutrient cycling**
- **Water cycling**

Figure 29.5 Illustrative ecosystem services provided by a wetland ecosystem

Source: Adapted from Glaves et al. (2010)

- Derivation of an **ordinal or indicative value** – scaled values indicating the importance of a service, e.g. locally important, regionally important, nationally important, or very important, important, unimportant etc.

- **Narrative / descriptive values** – word pictures which illustrate the local value of an ecosystem.

Table 29.7   Possible methods to use to value ecosystem services

Ecosystem Service Type	Method of Valuation
Provisioning Services	Market prices
	Replacement costs
	Gross value added
	Market-related estimates of opportunity
Regulatory Services	Losses avoided
	Welfare values
	Willingness to pay
	Hedonic pricing
	Values derived from meta-analyses
Cultural Services	Stated preferences
	Willingness to pay
	Recreational and tourism methods
	NOTE: there is some debate as to whether economic values should be applied to some cultural services.
Regulating Services	NOTE: there is debate regarding valuing supporting services, as such services support other service types and there is therefore a risk of double counting.

Valuation of ecosystem services can be undertaken using the following stages (Glaves et al., 2009):

1. Identify the ecosystems present in the case study.

2. Map and determine the extent and characteristics of each ecosystem.

3. For each ecosystem type identify the ecosystem services it provides, then confirm this via consultation with locals and experts.

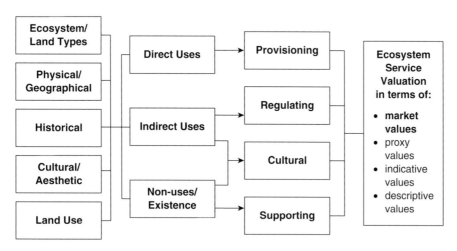

Figure 29.6   Framework approach for calculating Ecosystem Service Value (ESV)

Source: Adapted from Glaves et al. (2009)

Traditional Economic Valuation

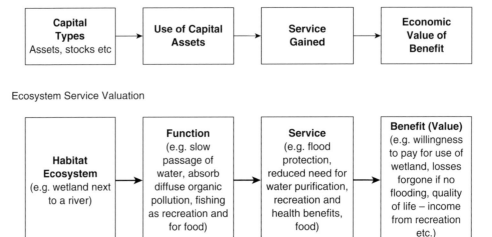

Ecosystem Service Valuation

Figure 29.7 Valuation of market goods and ecosystem services

Source: Adapted from Glaves et al. (2009)

4. Identifying the significant ecosystem services for each ecosystem type, then confirm this via consultation.

5. For the significant ecosystem services identifying if there are local data and/or values available.

6. Where local data are not available, identification of potential transferable values from other sites.

7. Establish the benefits values for each service per hectare of ecosystem and total benefits for each case study.

In some cases a full economic valuation of services is required, but in many cases a semi-qualitative ordinal scale is sufficient and only stages 1 to 4 need be completed. It must be noted that ecosystem services valuation is a relatively recent approach and methods of valuation are still being developed. At present some types of ecosystem services cannot easily be valued. It is however important to recognise the full range of services provided by local ecosystems and their significance in decision-making; failure to do so can lead to loss of key local ecosystem services and harm to local communities and economics, e.g. loss of flood storage habitats, loss of habitats important to tourism, poor air and water quality.

29.5 Conclusion

The International Union for Conservation of Nature (IUCN) stated that a sustainable society enables its members to achieve a high quality of life in ways that are ecologically sustainable (IUCN/UNEP/WWF, 1991). To measure progress towards a sustainable society, we need indicators of quality of life and of ecological sustainability. In addition we need methods and tools to enable us to measure whether we are meeting such indicators and standards. This chapter has introduced and outlined some of the key methods that can be used to measure our progress towards a more sustainable society.

Environmental appraisal tools such as EIA, environmental auditing and ecosystem service valuations are increasingly being used globally to achieve more sustainable land use planning, business development and financial decision-making. Such methods have real benefits to the environment, people and economic development:

1. As tools for improving the sustainability of businesses, governments and societies.

2. Incorporation of sustainable and environmental issues into decision-making.

3. Recognition of environment constraints, thresholds and limits in decision-making.

4. Identification of the best practicable environment options.

5. By minimising negative impacts and optimising the positive.

6. Identification of significant negative impacts and prevention of irreversible damage (adapted from Therivel, 2010).

SUMMARY

- A range of methods has been developed for assessing the sustainability of organizations and developments.
- Such tools assess environmental performance against set internal or external environmental standards.
- Effective environmental appraisals require the use of standard methods and often specialist subject expertise. Advice is available from relevant professional bodies.
- Environmental assessments require judgements to be made on the value, impact and significance of action.
- Environmental appraisals methods are used as standard tools by governments and business in predicting future environmental impacts and measuring the sustainability of past actions.

Further Reading

- Environmental Impact Assessments (EIA)

  Institute for Environmental Management & Assessment (2004) Guidelines for Environmental Impact Assessment, IEMA, Lincoln

  Morris P. and Therivel, R. (2009) *Methods in Environmental Impact Assessment.*

- Environmental Auditing (EA)

  Brady, J. et al. (2011) *Environmental Management in Organisations.*

- Ecosystem Services

  Department for Environment, Food and Rural Affairs (2007) *An Introductory Guide to Valuing Ecosystem Services.*

Note: Full details of the above can be found in the references list below.

References

Alcamo J. et al. (2002) *Ecosystems and Human Well-Being: A Framework for Assessment*. Washington, DC: Island Press. http://pdf.wri.org/ecosystems_human_wellbeing.pdf (accessed 29 November 2015).

Bell and Morse (2003) *Measuring Sustainability*. London: Earthscan Publications.

Bond, A. and Stewart, G. (2002) 'Environment Agency scoping guidance on the environmental impact assessment of projects', *Impact Assessment and Project Appraisal* 20 (2): 135–42.

Brady, J., Ebbage, A. and Lunn R. (2011) *Environmental Management in Organisations: The IEMA Handbook*, (2nd edition). Lincoln: IEMA.

Costanza, R., Wilson, M., Troy, A., Voinov, A., Liu, S. and D'Agostino, J. (2006) *The Value of New Jersey's Ecosystem Services and Natural Capital*. Trenton, NJ: Gund Institute for Ecological Economics, University of Vermont.

de Groot, R. S., Wilson, M.A. and Boumans. R.M.J. (2002) 'A typology for the classification, description and valuation of ecosystem functions, goods and service', *Ecological Economics* 41: 393–408.

Department for Environment, Food and Rural Affairs (2007) *An Introductory Guide to Valuing Ecosystem Services*, London: Defra Publications. https://www.gov.uk/government/uploads/system/uploads/attachment_data/file/191502/Introductory_guide_to_valuing_ecosystem_services.pdf (accessed 29 November 2015).

DoE (1989) *Environmental Assessment: A Guide to the Procedures*, Department of the Environment. London: HMSO.

DoE (1995) *Preparation Of Environmental Statements for Planning Projects that Require Environmental Assessment – A Good Practice Guide*. Department of the Environment. London: HMSO.

Elkington, J. (1997) *Cannibals With Forks: The Triple Bottom Line of Twenty-First Century Business*. London: Thompson.

European Commission (2001a) *Guidance on EIA: Screening*. Luxemburg: Office for Official Publications of the European Communities. http://ec.europa.eu/environment/archives/eia/eia-guidelines/g-screening-full-text.pdf (accessed November 29, 2015).

European Commission (2001b) *Guidance on EIA: Scoping*, European Union, Luxemburg: Office for Official Publications of the European Communities. http://ec.europa.eu/environment/archives/eia/eia-guidelines/g-scoping-full-text.pdf (accessed November 29, 2015).

Freeman III, A. K. (2003) *The Measurement of Environmental and Resources Values*. Washington, DC: Resource for the Future.

Glasson, J., Therivel, R. and Chadwick, A. (2005) *Introduction to Environmental Impact Assessment* (3rd edition). Abingdon: Routledge.

Glaves, P., Egan, D., Harrison, K. and Robinson, R. (2009) *Valuing Ecosystem Services in the East of England*, East of England Environment Forum, East of England Regional Assembly and Government Office East of England.

Glaves, P., Egan, D., Smith, S., Heaphy, D. Rowcroft, P. and Fessey, M. (2010) *Valuing Ecosystem Services in the East of England Phase Two Regional Pilot Technical Report*. Sustainability East.

IEEM (2006) *Guidelines for Ecological Impact Assessment in the UK*. Winchester: Chartered Institute for Ecology and Environmental Management. http://www.cieem.net/data/files/Resource_Library/Technical_Guidance_Series/EcIA_Guidelines/TGSEcIA-EcIA_Guidelines-Terestrial_Freshwater_Coastal.pdf (accessed 29 November 2015).

IEMA (2006) *Guidelines for Environmental Impact Assessment*, IEMA. Available from: http://www.iema.net/reading room/eia-guideline-updates/guidelines-environmental-impact-assessment.

IEMA (2011) *The State of Environmental Impact Assessment Practice in the UK*. Institute of Environmental Management and Assessment. Available from http://www.iema.net/iema-special-reports (accessed 29 November 2015).

International Chamber of Commerce (1991) *ICC Guide to Effective Environmental Auditing*. Paris: ICC Publishing,

IUCN/UNEP/WWF (1980) World Conservation Strategy, Gland, Switzerland. https://portals.iucn.org/library/efiles/documents/CFE-004.pdf (accessed 29 November 2015).

Landscape Institute and Institute of Environmental Management and Assessment (2013) *Guidelines for Landscape and Visual Impact Assessment*, (3rd edition). London: Routledge.

IUCN/UNEP/WWF (1991) *Caring for the Earth*. Gland, Switzerland. https://portals.iucn.org/library/efiles/documents/CFE-003.pdf (accessed 29 November 2015).

Lee, N., Colley, R., Bonde, J. and Simpson, J. (1999) *Reviewing the quality of environmental statements and environmental appraisals*. Occasional Paper 55. Manchester: EIA Centre, Manchester University.

Millennium Ecosystem Assessment (2005) *Ecosystems & Human Well-being: Synthesis*. Washington, DC: Island. http://www.millenniumassessment.org/documents/document.356.aspx.pdf (accessed 29 November 2015).

Mirovitskaya, N. and Asher, W. (2001) *Guide to Sustainable Development and Environmental Policy.* Durham: Duke University Press.

Morris, P. and Therivel, R. (eds) (2009) *Methods of Environmental Impact Assessment* (3rd edition). Abingdon: Routledge.

Post, J.E., Lawrence, A.T. and Weber, J. (2002) *Business and Society*, (10th edition). Boston: McGraw-Hill.

Sadler, B. (1998) 'Ex-post evaluation of the effectiveness of environmental assessment', in A.L. Porter and J.J. Fittipaldi (eds) *Environmental Methods Review: Retooling Impact Assessment for the New Century.* Fargo: AEPI/ The Press Club. pp. 30–40. http://www.iaia.org/publicdocuments/special-publications/Green%20Book_Environmental%20 Methods%20Review.pdf?AspxAutoDetectCookieSupport=1 (accessed 29 November 2015).

Smith, S., Richardson, R. and McNab, A. (2010) *Towards a more efficient and effective use of Strategic Environmental Assessment and Sustainability Appraisal in spatial planning.* Department of Communities and Local Government, London. http://webarchive.nationalarchives.gov.uk/20120919132719/http:/www.communities.gov.uk/documents/ planningandbuilding/pdf/1513010.pdf (accessed 29 November 2015).

Tansley, A.G. (1935) 'The use and abuse of vegetational terms and concepts', *Ecology* 16 (3): 284–307.

Therivel, R. (2010) *Strategic Environmental Assessment in Action.* Abingdon: Routledge.

Therivel, R. and Morris, P. (2009) 'Introduction', in P. Morris and R. Therivel (eds) *Methods of Environmental Impact Assessment* (3rd edition). Abingdon: Routledge.

Tomorrow's Company Global Business Think Tank. (2007) '*Tomorrow's Global Company: Challenges and Choices*'. Available from http://tomorrowscompany.com/tomorrows-global-company (accessed 29 November 2015).

UN (1990) *Post project analysis in environmental impact assessment* (Environmental Series 3). Report prepared for the Economic Commission for Europe, Geneva. New York: United Nations.

UNCED (1992) *Rio Declaration on Environment and Development.* The United Nations Conference on Environment and Development. Rio de Janeiro 3-14th June 1992. http://www.unep.org/Documents.Multilingual/Default.asp?docume ntid=78&articleid=1163 (accessed 29 November 2015).

Wood, C. and Jones, C.E. (1997) 'The effect of environmental assessment on UK local planning authority decisions', *Urban Studies*, 34 (8):1237–57.

World Commission on Environment and Development (1987) *Our Common Future, the Bruntland Report*, WCED. Available online at http://www.un-documents.net/our-common-future.pdf.

ON THE COMPANION WEBSITE…

Visit **https://study.sagepub.com/keymethods3e** for author videos, chapter exercises, resources and links, plus **free** access to the following recommended articles:

1. **Hudson, R. (2007) 'Region and place: Rethinking regional development in the context of global environmental change', *Progress in Human Geography*, 31 (6): 827–36.**

Hudson explores the environmental issues which underpin development and the need to assess the environment in development.

2. **Silvey, R. and Rankin, K. (2011) 'Development geography: Critical development studies and political geographic imaginaries', *Progress in Human Geography*, 35 (5): 696–704.**

Silvey and Rankin explore the link between developmental studies and other aspects of geography.

3. **Christophers, B. (2015) 'Geographies of finance II: Crisis, space and political-economic transformation', *Progress in Human Geography*, 39 (2): 205–13.**

Christophers explores the roots of the current financial crisis and makes links between economics and environment.

SECTION FOUR

Geographical Analysis: Representing, Visualising and Interpreting Geographical Data

30 Making Use of Secondary Data

Naomi Tyrrell

SYNOPSIS

Geographers make the distinction between primary data and secondary data. Primary data are information that is collected by you, the researcher; secondary data are information that already exists and was collected by someone else. Sometimes secondary data are collected for purposes not directly related to research but still can be used in a research project. Secondary data are particularly useful for *generating* research questions or research hypotheses because they often contain a lot of detailed information. You can also *analyse* secondary data to produce new results related to your research questions or hypotheses. Secondary data often exist on a vast scale as governments, companies, institutions, communities and individuals may produce secondary data related to a whole range of issues, and much of these are spatially referenced. Often these data can be accessed via the Internet quite cheaply or for free. This makes secondary data quite appealing for student research projects. Secondary data can be quantitative or qualitative, historical or contemporary.

This chapter is organized into the following sections:

- Introduction: The nature of secondary data
- Sources of secondary data
- Using secondary data in a research project
- Issues to consider when using secondary data
- Conclusion

30.1 Introduction: The Nature of Secondary Data

We are living in what has been called the 'Information Age' in which data are collected (and often made available) at a rapid pace. A simple journey such as leaving the house and getting a bus to university results in data being generated, collected, and stored. Ask yourself the questions in Box 30.1. This quick exercise shows that data, which is really just another word for information, are being generated constantly. Have you ever stopped to consider why? Or where they are stored, what happens to them and who uses them? Many, but not all, of these data are in numerical form (such as how much you paid for your coffee) and are therefore *quantitative*. Some of them are in text or visual form (such as the words and photos used in a Facebook or Instagram post) and are therefore *qualitative*. With the rise of integrated GPS technology in phones and other sensors, many of these secondary data sources are also spatially referenced, which generates additional opportunities for geographers (see Chapter 16). In research terms, all of these data are secondary data – they are collected by someone else. Some of them may be freely available to use in research; some of them are owned by interested parties and would cost a lot of money to purchase (if they were available at all).

Box 30.1   A simple journey with secondary data

- What happens to surveillance camera images that are collected as you pass through public and commercial areas on your walk to the bus stop?
- What happens to the information from your bus pass that you scan into the bus ticket machine?
- What happens to the information you share on social media during your bus journey?
- What happens to the information from the loyalty card you pass to the barista in your favourite coffee shop as you pop in for your essential dose of pre-lecture caffeine?
- What happens to the data from the debit card you use to pay for your coffee?
- What happens to the information collected from your student card as you swipe it to enter the library where you are meeting your friends?

Secondary data mean someone else has gone to the trouble of collecting information for their own purposes, but it may be relevant to your research project and it may be freely available. Secondary data can be valuable for *generating* your research hypotheses or research questions. You can use them on their own or they can help you to identify areas of interest that can be investigated by collecting primary data. You may use raw secondary data which you access in their original form (such as diaries or blogs[1]) or interpreted secondary data (such as survey data). However, Box 30.2 shows that there are several cautions you need to consider before considering making use of secondary data (and seeing them as a 'quick route' to research findings!).

Box 30.2   Questions to ask when considering whether to use secondary data

1 Is the secondary dataset you have found relevant to your study?
2 Is the secondary dataset important for your research questions or hypotheses?
3 Is the coverage of the secondary data (e.g. geographical area or population sub-group) relevant to your study?
4 Are the categories or variables in the secondary dataset relevant to your study?
5 Are the secondary data reliable? (Who has collected these? Who is making these available?)
6 Are there enough secondary data for your study or will you need to collect primary data as well?

When considering whether to use secondary data, it is also important to consider your analytical framework and epistemology (theory of knowledge) (see Mason, 2002; Bushin, 2008; DeLyser et al., 2010). Researchers take different views of the facts they are researching based on their ideas about what constitutes knowledge and what is 'worth' knowing (see Chapter 1). For example, positivists see facts as existing independent of interpretation (that is, 'objectively') and might consider a secondary data document to be an objective (true) reflection of reality. By contrast, interpretivists, social constructivists, and realists see 'reality' as influenced significantly by the social, political, and economic environments and open to manipulation and varying interpretations by those who are part of it. Indeed, they argue that there is no objective truth, but rather multiple 'realities' that can be experienced and analysed from different standpoints (see Chapter 4). They might consider that a secondary data document must be seen in its social context, and they would then attempt to make sense of that context. As a researcher, you need to consider how you view the nature of the secondary data you are thinking of using and how this makes a difference to how you will make use of them in your research project. First though, you need to know where secondary data are and how to find them.

Sources of Secondary Data

The *definition* of secondary data has remained fairly static in recent decades. However, *sources* of secondary data (where they come from and where you can get them) have increased and changed in recent times. Largely, this is due to the accessibility of the Internet and the ease with which many researchers can get their hands on a range of data produced by governments, companies, institutions, communities and individuals.

The starting point for deciding whether secondary data are going to be an important resource in your research project is finding out what secondary data relevant to your topic exist. The Internet is a very useful in this regard as a quick keyword search using a search engine may indicate several, or a plethora, of potential avenues for you to explore. For example, if you were interested in knowing what secondary data you could use in a project focussing on links between obesity and certain locations, ('obesogenic environments') and spelling out successful searches ('obesity data UK' looks promising, for instance), *obesogenic environments*, doing a keyword search on the Internet, might provide links to datasets on obesity collected by Public Health England and the Office for National Statistics (ONS) for the UK, or the National Institutes of Health (NIH) and Centers for Disease Control (CDC) in the US. Additionally, newspaper articles on the 'obesity epidemic', academic studies of increasing levels of obesity, and information about non-profit organizations' strategies to combat obesity may prove helpful. All of these are sources of secondary data and potentially useful for your research project.

Secondary data often are more vast and varied than the primary data a lone student researcher (with limited time and resources) could collect. For example, governments often carry out surveys of populations at a national scale (e.g. a population census) or representative sub-sections of a population which have particular characteristics they are interested in (e.g. people in employment). Governments also often carry out longitudinal surveys (surveys carried out over time), such as the US

Census Bureau's American Community Survey (ACS). The people who are involved in data collection and collation for governments are highly trained so this often makes secondary data published by governments a good source of reliable data for a student research project. However, whoever produces the data (including governments) will have their own reasons for doing so and you must consider these before using the data. For an important and sceptical discussion on the collection and use of statistics in society see Dorling and Simpson (1999).

Large amounts of secondary data are produced by governments who have the capacity and funds to gather a lot of information on a large scale. Population censuses are one example (see Table 30.1), and the UK population census is the most reliable and complete source of socio-economic data for the UK (Rees et al. 2002). It provides multifaceted data for every small area by targeting the whole population without sampling (Simpson and Brown, 2008). Government datasets like censuses are often strongly spatially referenced which makes them ideal for use in geographical research projects. However, they are also often under-utilised in student research projects despite being straightforward to access. In order to access them you need to find out the name of the statistical agency for the country relevant to your research project. For example, if you would like some internal migration data for England you would look up the website of the statistical agency for England and Wales, which is the Office for National Statistics (ONS) (www.ons.gov.uk). If you were interested in the number of traffic accidents in Germany you would look up the website of the statistical agency for Germany ('Statistisches Bundesamt') (https://www.destatis.de/EN/Homepage.html). Alternatively, passenger lists of immigrants to the US can be found at the US National Archives site (http://www.archives.gov/research/). Governments' statistics offices and agencies produce data on many topics that are often relevant to student research projects such as: population, health, employment, housing, crime and the environment. The data often are available at different scales – neighbourhoods, local authorities, provinces and regions.

Some datasets have their own geographical hierarchies so it is important that you understand the geographical references contained within the datasets before you use them. You will need to decide which layer of geography – which geographical unit(s) – in the dataset is relevant for your research. Datasets usually have separate and very detailed information files ('metadata') which give important information for the user so make sure you read these carefully (rather than ignoring them!) to avoid making false claims about the data you use.

Large-scale datasets often contain *aggregate* data rather than individual-level data. This is the case with many countries' population censuses. For example, it is possible for researchers to gain access to individual population census data in the UK but permission is restricted. These data are called microdata. The ONS produced a Sample of Anonymised Records (SARs) from 1991, 2001 and 2011 censuses which include one percent of all households in the UK and two percent of all individuals. There is also a census longitudinal survey which links one percent of individual records back to 1971. However, these datasets are not available to the general public and researchers have to apply to the ONS for permission to use them because of the sensitive individual information they may contain.

Quantitative datasets often reflect social ideas, situations, and values. For example, the questions asked in large-scale population surveys such as a census change over time, exemplified by the inclusion of questions on religion in recent UK censuses (Southworth, 2005),

Table 30.1 Some sources of official quantitative secondary data (national scale)

Country	(web address begins http://www unless stated)
Afghanistan	cso.gov.af/en
Argentina	indec.mecon.ar/
Australia	abs.gov.au/
Brazil	ibge.gov.br/english/default.php
Canada	statcan.gc.ca/start-debut-eng.html
China	stats.gov.cn/english/
Colombia	dane.gov.co/
Egypt	http://www.egypt.gov.eg/english/general/Open_Gov_Data_Initiative.aspx
France	insee.fr/en/
Germany	destatis.de/EN/Homepage.html
India	http://mospi.nic.in/Mospi_New/site/home.aspx
Ireland	cso.ie/en/index.html
Italy	istat.it/en/
Japan	stat.go.jp/english/index.htm
Malaysia	https://www.statistics.gov.my
Mexico	inegi.org.mx/default.aspx?
Netherlands (The)	cbs.nl/en-GB/menu/home/default.htm
New Zealand	stats.govt.nz/
Portugal	ine.pt/xportal/xmain?xpid=INE&xpgid=ine_main
South Africa	http://www.statssa.gov.za/
Spain	ine.es/en/
Sweden	scb.se/en_/
UK	ons.gov.uk/ons/index.html
USA	usa.gov/statistics

ethnicity in recent Republic of Ireland censuses (King-O'Riain, 2007), and the shift in the US census to allow respondents to claim multiple race identities (Omi and Winant, 2015). Datasets that are comprised of attitudinal data (e.g. the British Social Attitudes Survey) also often have new questions to reflect society at a given point in time in addition to standard questions which allow for longitudinal analyses.

Some governments and international agencies provide quantitative data specific to their role on their websites. For example, the website <www.police.uk> allows users to search for crime statistics for England and Wales at postcode level. Some private companies also make secondary data available to the public; sometimes these are just a small sample of the dataset they own as they may also sell data for profit. For example, the company Experian collects information about UK neighbourhoods using a range of data sources and allows members of the public limited use of its neighbourhood databases for free (Mosaic). Some companies also allow you to search for particular statistical information on their websites for free, such as UK house price data on Zoopla's website (http://www.zoopla.co.uk; the US equivalent is Zillow– http://www.zillow.com). Examples of some of these popular agencies, centres and companies and their websites are given in Table 30.2 but there are many more.

Table 30.2 A selection of agencies, centres and companies providing quantitative secondary data (some for free)

Agency/Centre/Company	Summary	Website (web address begins http://www unless stated)
National Scale (UK)		
CACI (Acorn)	A consumer classification system.	http://acorn.caci.co.uk/
Centre on Dynamics of Ethnicity (CoDE)	Downloadable statistics, census briefings and area profiles detailing ethnic inequalities and identities.	ethnicity.ac.uk
Experian (Mosaic)	A consumer classification system.	experian.co.uk/marketing-services/products/mosaic-uk.html
Neighbourhood Statistics (England and Wales)	Government website providing a range of local area information.	neighbourhood.statistics.gov.uk/dissemination/
Northern Ireland Neighbourhood Information Service	Statistical and locational information relating to small areas across Northern Ireland.	ninis2.nisra.gov.uk/
Police (UK)	Crime and policing data for England, Wales and Northern Ireland.	police.uk/
Scottish Neighbourhood Statistics	Statistics on health, education, poverty, unemployment, housing, population, crime and social/community issues in Scotland.	sns.gov.uk/
The Citizenship Survey (UK)	Bi-annual attitudinal and behavioural survey from 2001 to 2011.	http://discover.ukdataservice.ac.uk/series/?sn=200007
The British Social Attitudes Survey	Results of annual survey of 3,000 people who live in the UK. Survey has been conducted annually since 1983.	natcen.ac.uk/our-research/research/british-social-attitudes/
The General Household Survey (UK)	Multi-purpose annual survey on topical issues. (2001–2007).	http://www.statistics.gov.uk/ssd/surveys/general_household_survey.asp
The UK Labour Force Survey	A survey of the employment circumstances of the UK population. It is the largest household survey in the UK.	ons.gov.uk/ons/about-ons/get-involved/taking-part-in-a-survey/information-for-households/a-to-z-of-household-and-individual-surveys/labour-force-survey/index.html
UKCrimeStats	Crime data platform publishing monthly crime data for England and Wales.	ukcrimestats.com/
Zoopla	Property website for the UK; property listings with market data and local information.	zoopla.co.uk/
International Scale		
The CIA World Factbook	Provides information on the history, people, government, economy, geography, communications, transportation, military and transnational issues for 267 world entities.	https://www.cia.gov/library/publications/the-world-factbook/

Agency/Centre/Company	Summary	Website (web address begins http://www unless stated)
Eurobarometer	Surveys and studies address major topics concerning European citizenship.	ec.europa.eu/public_opinion
Europa	Statistics and opinion polls for EU countries.	http://europa.eu/publications/statistics/index_en.htm
Eurostats	Statistics for EU countries on a range of themes.	http://ec.europa.eu/eurostat
Food and Agriculture Organisation	Collates and disseminates food and agricultural statistics globally; produces publications, working papers and statistical yearbooks that cover food security, prices, production and trade and agri-environmental statistics.	fao.org/economic/ess/en/#.VFezC2ByZMs
International Labour Organization	Provides data on employed populations and unemployed populations.	ilo.org/global/statistics-and-databases/lang–en/index.htm
Social Science Information System	Information and resources relevant for social scientists.	sociosite.net/
UNESCO	Cross-nationally comparable statistics on education, science and technology, culture, and communication.	uis.unesco.org
UNICEF	Provides data on the situation of children and women around the world.	unicef.org/statistics/
United Nations	Provides demographic, economic, social and environmental data and reports.	http://unstats.un.org/unsd/default.htm
World Bank	Provides data on countries' development.	http://data.worldbank.org/
World Health Organisation	Provides data and analyses on global health priorities.	who.int/gho/en/
World Values Survey	A global network of social scientists studying changing values and their impact on social and political life.	http://www.worldvaluessurvey.org/wvs.jsp

Recently, historical census data have become more accessible in some countries as the records have become digitized and made available online. In the UK, the census has a 100 year rule which means that data at the level of an individual cannot be made public for 100 years. This privacy rule is important to remember, especially in an era when some people are very concerned about why and in what ways governments are collecting and using data about the general public; also it is the reason why you need special permission to access anonymised individual-level data. Historical census data for the UK are available on the Internet from 1841 to 1911 from various companies. The datasets are often free to search but there are charges for viewing and downloading data. This is because the cost of the digitization of the census records has been borne by private companies rather than the government. In the US, privacy rules extend to a slightly shorter 70 years; for example, the digital images for the 1940 census were released online in 2012 through a public-private partnership between the National Archives and a private company (see http://1940census.archives.gov/). If you are interested in genealogy you may have used some of these websites, such as ancestry.com. In some places, government records are available through specially designated libraries, such as in the US where many state universities act as official repositories for government documents, and thus may be a good portal for free or low-cost access to records. As a student researcher, the cost of accessing historical census datasets may need to be considered.

Another source of secondary data which potentially can be useful for student research projects is attitudinal surveys. There has been a large increase in the number of surveys of people's opinions and attitudes over the past decade, with many private companies conducting 'polls'. Have you ever been stopped by a market-researcher in the street and asked a series of questions about what toothpaste you like? Have you ever received a telephone call asking about your voting preferences? Have you ever seen '8 out of 10 cats' (a UK television programme on Channel 4) or even clicked 'Like' on Facebook? If so, you will have some understanding of what an attitudinal survey is and what they involve (to varying extents!).

There are costs associated with accessing many of these attitudinal surveys because many of them are conducted by private companies. Summaries of their results are often publicised quite widely, particularly if they relate to a popular or contentious contemporary issue. These attitudinal surveys use sampling strategies rather than surveying very large populations every time, so they are very different to population censuses that survey the whole population at a particular point in time. Those surveys conducted by governments and reputable companies usually use sophisticated methodologies to try to ensure representativeness and robustness. Researchers need to gather as much information as possible about the design and framework of the survey, as well as the sampling strategy. However, sometimes this information can be difficult to get hold of because it can be market-sensitive. The UK Data Archive (hosted by the University of Essex) provides access to a wide range of survey data, and also is a repository for interview transcripts from previous publically-funded qualitative research projects. The Social Science Information System (hosted by the University of Amsterdam) also contains information about databases which exist in many national contexts (see Table 30.2).

Table 30.3 shows that although secondary data often are thought of as large-scale and quantitative, there are many other types of secondary data that may be useful for your research project. Secondary data can be essential for projects of a historical nature, the focus of which may make collecting primary data impossible. For example, a project might focus on twentieth century colonial settlement in India; it is likely that the researcher would need to access archived secondary data which might be in the form of diaries, letters, and official documents (e.g. maps). Box 30.3 gives some examples of qualitative secondary data sources that you may find useful for your own research project.

Some of the types of qualitative secondary data given in Box 30.3 might not be what you immediately think of when someone says 'secondary data' to you. However, there are a variety of secondary data sources and an increasing diversity of geographical research topics. For example, if you are interested in researching patterns of nineteenth century migration from Ireland to Britain during the Potato Famine (Great Hunger) you might immediately think of using population census data. However, you may also be able to access letters, diaries and/or newspaper articles and analyse them using specific analytical techniques. If you are interested in researching the development of

Table 30.3 Some useful archives containing secondary data (UK)

Archive	Summary	Web address (http://www.)
British Library Sound Archive	Large collection of recorded sound and video: music, wildlife, drama, literature, oral history and BBC broadcasts.	bl.uk/soundarchive
Digimap	The most comprehensive maps and geospatial data available in UK Higher and Further Education.	http://digimap.edina.ac.uk/digimap/home
The National Archives (UK)	The UK government's official archive (one billion archived pages).	nationalarchives.gov.uk/
The National River Flow Archive	Collates quality controls, and archives hydrometric data from gauging station networks across the UK.	ceh.ac.uk/data/nrfa/index.html
The Thomas H. Manning Polar Archives	Large collection of manuscripts and other unpublished material relating to the Arctic and Antarctic regions, and to many persons who have worked there.	spri.cam.ac.uk/library/archives/
The Victoria and Albert Museum Archives	The Archive of Art and Design, the Beatrix Potter Collections, the V&A Archive and the V&A Theatre and Performance Archive.	vam.ac.uk/page/a/archives/
UK Data Archive	Largest collection of digital data in the social sciences and humanities in the United Kingdom.	data-archive.ac.uk/
UK Data Service	A single point of access to a wide range of secondary data including large-scale government surveys, international macrodata, business microdata, qualitative studies and census data from 1971 to 2011.	http://ukdataservice.ac.uk/about-us.aspx

hip-hop music in the USA, for example, you may decide to analyse sound recordings, memoirs and archived documents (Shabazz, 2014). Both of these topics are geographical and both may incorporate different types of secondary data.

Box 30.3 Examples of qualitative secondary data

- Data and/or findings produced from previous research
- Letters
- Official documents (e.g. reports, minutes of meetings, maps)
- Web-based documents
- Historical or archived documents
- Diaries
- Memoirs
- Newspaper articles
- Web-logs ('blogs')
- Online reader comments

(Continued)

(Continued)

- Photographs
- Film recordings
- Sound recordings
- Transcripts (of oral histories, interviews, etc.)

It is very important not to be tempted to think that it is easier to use secondary data than collecting primary data yourself; using secondary data is not a quick route to research success (see further discussion of this issue in Chapter 36). However, it may be that secondary data are the most appropriate data for you to use and it is perfectly acceptable to develop a research project which only uses secondary data (although you need to check that your institution allows this for a student research project).

At the beginning of your research project it is important to consider whether there are any potential sources of secondary data that are likely to be relevant to your study; reference librarians, archivists and experienced researchers are all worth consulting on this question to avoid missing a really obvious source of data of high relevance to your topic that you would be expected to have used. If you find there are relevant secondary data, make sure that they are *appropriate* for your study (i.e. this will enable you to test your hypotheses, address your research questions, justify your research topic or provide context for your study) and *available* for you to use. Once you have decided to use secondary data, there are several stages you must go through and several different ways in which you can make use of them. These are outlined in the next section.

30.1 Using Secondary Data in a Research Project

It is important to remember that the secondary data that you have access to are not likely to have been created with your research hypotheses or questions in mind, i.e. they were not collected and collated for your specific research project. They may be *useful* for your research project but they were *produced* for a different purpose. It is important to consider what that purpose was, who produced the data, and what methods were used. It is also important to consider the scale, scope, sampling framework and precise content (e.g. population or area covered and variables used) of the secondary data. Therefore when considering whether to use a specific dataset that is available to you, ask yourself the five 'Rs' outlined in Box 30.4.

Box 30.4   Questions to ask about a secondary dataset

- Is it *relevant* to my research?
- Is it *reliable*?
- Is it *robust*?
- Are the data *raw*?
- Are the data *representative*?

In your research project you may use quantitative secondary data, qualitative secondary data, or a combination of both. Consider how you will use these secondary data as there many different valid options (see Box 30.5). The most straightforward way of using secondary data is *to inform your research project.* The data can provide you with up-to-date information about the topic, issue, or location you are considering studying. There may even be secondary data which have been collected over time (longitudinal data) which highlight important changes you can explore in your research. Therefore, analysis of secondary data, even if only in descriptive terms, may help you to narrow down your research topic and/or focus on interesting issues or notable changes. Using secondary data in this way also enables you *to justify your choice of topic* because you are able to draw on data which suggest or confirm an important area of research. You also may decide to use secondary data *to provide a context for your study.* For example, if you were researching the impacts of second home ownership in a particular location it would strengthen the rationale for your study, and provide appropriate context, if you were able to present some data on the key features of second home ownership generally and then focus in on the location of your research in particular. Research projects that rely mainly on primary qualitative data often benefit from having some contextual secondary data analysis.

Box 30.5   Five ways of using secondary data

You can use secondary data to:

1 Inform your research project
2 Justify your choice of research topic
3 Provide a context for your study
4 Enable comparison between places or phenomena
5 Analyse phenomena

Secondary data also can be used in *comparative studies.* This can be particularly useful in student research projects because you can compare data from your own primary research with data from previous research (i.e. secondary data) that has investigated a similar issue or tested the same hypothesis. For example, you may want to explore public perceptions of the environmental impacts of a new waste incinerator in a particular town. You search the relevant website (see Table 30.2) and find out about similar projects that have been carried out. You decide to use some of the questions used in questionnaires in similar projects in different geographical areas, so that you can compare the findings of your questionnaire with the findings of the previous studies. In this way you are comparing the findings of your own primary research data with secondary data. This allows you to discuss the findings of your own research in a wider context – you can highlight any similarities and/or differences between your research findings and those of previous projects and then suggest possible reasons for these. You may also decide to repeat the methodology of a previous study in the same geographical location sometime after the previous study. This would allow you to explore an issue longitudinally. You may decide to use statistical tests to evaluate the significance of your findings (see Chapters 31, 32 and 35).

Another common way of using secondary data in a research project is to *analyse phenomena*. This means that you use data that have been collected by someone else (such as a government agency or a team of researchers) that you have been able to access (free or purchased) and analyse these in a similar way to how you would analyse primary data. An advantage to analysing secondary data is that these are often vaster in scale and range than a student researcher would be able to collect with the resources and time available to them. For example, you may be interested in comparing levels of car ownership across different geographical areas, testing the hypothesis that families living in rural areas are more likely to own a second car than families living in urban areas. You would need to find a dataset that included raw data on car ownership (think for yourself about what dataset might provide that information and who might produce it) that were disaggregated by geographical area (and decide the scale you were interested in). You could then analyse the data using appropriate statistical techniques (see Chapters 31 and 32). Another example would be carrying out a longitudinal analysis of ethnic group residential patterns in a particular city using population census data and GIS (see Chapter 37). Your research project can be based on analysis of secondary data on its own or you can include primary research as well. You may decide to do some statistical analyses using your quantitative secondary data and then use qualitative primary research techniques to investigate a phenomenon further; this is often referred to as using 'mixed methods' (see Chapter 16). Box 30.6 outlines some of the advantages and disadvantages of these ways of using secondary data in a student research project that are important to consider.

Box 30.6   Some advantages and disadvantages of using secondary data

Advantages:

- A good collection of data exists
- You are doing an historical study (primary research would be difficult)
- You are covering an extended period (a longitudinal study)
- The phenomenon/issue/unit may be difficult or too large to study directly
- Existing datasets are far larger than you could collect yourself
- Data may be of good quality

Disadvantages:

- It may be costly
- Datasets may be very large and complex
- Data may not match your precise research question
- There may be gaps in the data
- The measures between the units of analysis (e.g. populations or counties) may not be directly comparable
- You have no control over the quality of the data

To help you consider carefully whether secondary data are likely to be important in your own research project, Boxes 30.7 and 30.8 provide some case studies for you to consider. Use the information in the previous paragraphs to help you answer the reflective questions. In addition, you may find it helpful to ask the same questions when reading journal articles that you think might be relevant to your research project. You can build up a good idea of how to go about carrying out your own research project by examining the ways in which other researchers have investigated similar topics.

Box 30.7   A case study of secondary data use (I)

Smith, D. P. and Sage, J. (2014) 'The regional migration of young adults in England and Wales (2002–2008): A 'conveyor-belt' of population redistribution?', *Children's Geographies* 12(1): 102–17.

The authors explore the regional patterns of young adult migration. Using secondary data from the National Health Service Central Register for England and Wales between 2002 and 2008 they analyse migrations of young people between regions. They show that young adults are increasing as a proportion of regional migrants (i.e. taking the migrations of all ages into account) and that the migration flow of 16–24 year olds increased between 2002 and 2008, with there being major regional differences in these flows. They use their analysis of secondary data to investigate the idea of 'escalator regions' in Britain.

Reflective Exercise:

Using the information above, answer the following questions:

- What secondary dataset did they use in their research?
- What type of secondary data was it?
- In what ways did they use the secondary data?
- In what other ways might they have considered using these?

Box 30.8   A case study of secondary data use (II)

Baschieri, A. and Falkingham, J. (2009) 'Staying in school: Assessing the role of access, availability and economic opportunities – the case of Tajikistan', *Population, Space and Place* 15: 205–24.

A range of secondary data was used to investigate the roles of individual, household, and contextual factors in determining whether or not a child attends basic education in Tajikistan. The authors wanted to investigate what factors influenced children's attendance at school and they thought that the individual characteristics of children and their households, and the characteristics of the communities in which they lived, were likely to be important. They identified that there was not one single data source which contained all of the information they needed on the factors that were likely to influence school attendance so they used data from a range of sources to create a unique dataset of linked and simulated data.

(Continued)

(Continued)

They used data from the 2003 Tajikistan Living Standards Survey (TLSS). This was then supplemented by a questionnaire that was conducted in each of the primary sampling units of the TLSS. Key stakeholders within the communities such as village leaders, teachers and doctors were asked to complete the questionnaire. In addition to these data, the authors also matched the enumeration areas of the 1999 Census of Tajikistan to the primary sampling units of the 2003 TLSS. This enabled them to link additional data on socioeconomic characteristics of the population of each village. Lastly, they obtained a set of spatial data on land cover – derived from LandSat imagery – and linked these to the 2003 TLSS.

Through the use of multilevel modelling and geographical information system (GIS) techniques, the results of their research show that contextual factors have a strong effect on school attendance. In particular, the accessibility and availability of school services and the quality of education have a *positive* effect. However, the level of economic development of the community in which the child lives exerts a *negative* effect on school attendance, reflecting the influence of higher opportunity costs of education in terms of the opportunities for income-generating activities forgone.

Reflective Exercise:

Using the information above, answer the following questions:

- What secondary datasets did they use in their research?
- What type of secondary data was it?
- In what ways did they use the secondary data?
- What other types of data did they use?

30.3 Issues to Consider When Using Secondary Data

Reliable research must show consistency. For quantitative approaches this usually means that results are repeatable and consistent (Kitchin and Tate, 2000). For qualitative research this usually means that researchers must be consistent with the categorisation and coding of data and employ rigorous, transparent analysis techniques (see Ch. 36, also Baxter and Eyles, 1997). Reliability also can be increased by using multiple researchers but this is not often possible in student research projects. All data, primary and secondary, need to be analysed and interpreted systematically, using appropriate techniques. A common issue that can arise when secondary data are used in a student research project is that of the 'ecological fallacy'. This is when aggregate results for a geographical area are taken to suggest that they apply to all people within that area. In other words, an ecological fallacy takes place when you draw a claim about an individual based on your observation of grouped or combined data. The ecological fallacy is an example of the effect of spurious correlation. For example, a researcher might examine the aggregate data on income for a neighbourhood and discover that the average household income for the residents of that area is £30,000. To state that the average income for residents of that area is £30,000 is true and accurate. The ecological fallacy can occur when the researcher then states, based on this data, that all people living in the area earn about £30,000.

If you are using several different datasets in a research project, particularly if they include quantitative and qualitative data, you must consider how you will use them.

A traditional approach to multiple or mixed methods geographical research is triangulation of data (see the chapters in Section II of this volume). This allows you to consider different perspectives on the same issue. It may allow you to investigate any inconsistencies within and/or between the datasets, helping you to explain important issues arising in the data and complement or confirm your conclusions. However, fully achieving triangulation is difficult as data that are collected using different methods ultimately will not converge. This is because different methods are based on different ontological assumptions (see Chapter 1), which can make integration and comparability of methods and findings problematic (see Mason, 2002; Moran-Ellis et al., 2006). You may decide that it is appropriate to use each research method in sequence during your research project, except where their ontological assumptions are complementary, rather than seeking to triangulate your data. Multiple and mixed research methods may be used to increase the validity of your research findings, to increase understanding of a phenomenon and/or to uncover new interpretations. For further discussion on ways of analysing data different types of data in research projects see Davies and Heaphy (2011) and Elwood (2010).

If we reconsider our example of the student who takes a bus journey from home to university in the introduction to this chapter (see Box 30.1), we can see that ethical issues can arise from the generation, storage, and use of secondary data. Who has access to your personal data – the timing and length of the bus journey you took, the post you made on social media, the price of your coffee, your coffee-buying habits and the time you entered the library? Sometimes the ethics of using secondary data are overlooked but it is important to remember that ethical issues do not only pertain to using primary data or qualitative data. It is perhaps tempting to think that so much data are being generated in the 'Information Age' that questions over their collection, storage and usage are unimportant, i.e. it would take too long and cost too much money for the majority of them to be used in sinister way. However, contemporary global scandals such as 'Wikileaks' or 'ClimateGate', or recent debates about the availability of NHS patient data in the UK, demonstrate the power that can arise from owning or being able to access secondary data. Data collection is not benign and it is not necessarily problematic, but it is important that you consider the ethics of using a secondary dataset in your research (for more on research ethics see Chapter 3). At the very least you should make sure you have permission to use the data, that you acknowledge the data sources and permissions when you write up your research, and that the basic privacy of individuals is maintained.

30.4 Conclusion

There is a vast array of secondary data available for use in student research projects. Whatever your project is focussing on, you are likely to find some secondary data which are relevant; you just have to know where to look for them. When planning your research project you need to be aware of what secondary data there are as it may be that there is a dataset that is so relevant to your research questions or hypotheses that you will be expected to use it – research librarians are very helpful regarding this task. Using secondary data in a student research project can be advantageous if done so in the correct ways. You will need to consider whether there are any costs associated with accessing the data, the coverage of the data (e.g. the geographical units), the variables in the data (e.g. age, gender, ethnicity), the reliability of the data (i.e. who has produced them and why) and the ways in which the data have been collected (e.g. longitudinally, representative sample or not, etc.). You will also need to consider the ways in which you will use the data, ranging from providing the context of your research to forming the basis of your analysis, and what conclusions you can safely draw from your analysis and interpretations.

SUMMARY

This chapter has outlined:

- The nature of secondary data.
- The multiple sources of secondary data.
- The ways in which secondary data can be used.
- Some of the important issues to consider when using secondary data.

Secondary data can be very useful for student researchers but they must be utilised and analysed properly rather than being regarded as a quick route to research success.

Note

1   Alternatively, see Ch. 10 in which the discussion revolves around 'respondent diaries' that are directed by a researcher and thus constitute primary data.

Further Reading

- Cidell (2010) explores the use of content clouds as exploratory qualitative data analysis. Content clouds have become popular online tools as they provide visual summaries of the content of a document. In this article, the author uses two examples to demonstrate the possibilities of this method in research which utilises qualitative secondary data.

- Fotheringham, Brunsdon and Charlton (2000) discuss the application of quantitative methods with practical examples which are of particular relevance to student research projects. The authors also explain the philosophy of the new quantitative methodologies.

- Koteyko, Jaspal and Nerlich (2013) explore a framework for analysing a recent source of qualitative secondary data – online reader comments. They analyse comments published on a UK tabloid newspaper website before and after 'ClimateGate' to reveal the ways in which stereotypes of science and politics are appropriated in this type of discourse.

- Mason (2002) addresses some of the key issues involved in qualitative research, including ways of using qualitative secondary data sources. The philosophies underpinning qualitative research are discussed.

- Singleton (2012) uses geo-demographic analysis (using some of the sources discussed in this chapter) to reveal that pupils living in more affluent and less ethnically diverse areas record the highest rates of participation and attainment in GCSE Geography. The author uses several secondary datasets and provides a good starting point for student researchers interested in geo-demographics.

Note: Full details of the above can be found in the references list below.

References

Baschieri, A. and Falkingham, J. (2009) 'Staying in school: Assessing the role of access, availability and economic opportunities – the case of Tajikistan', *Population, Space and Place,* 15: 205–24.

Baxter, J. and Eyles, J. (1997) 'Evaluating qualitative research in social geography: Establishing rigour in interview analysis', *Transactions, Institute of British Geographers*, 22: 505–25.

Bushin, N. (2008) 'Quantitative datasets and children's geographies: Examples and reflections from migration research', *Children's Geographies*, 6 (4): 451–7.

Cidell, J. (2010) 'Content clouds as exploratory qualitative data analysis', *Area* 42(4): 514–23.

Davies, K. and Heaphy, B. (2011) 'Interactions that matter: Researching critical associations', *Methodological Innovations Online*, 6(3): 5–16. http://www.pbs.plym.ac.uk/mi/pdf/8-02-12/MIO63Paper11.pdf (accessed 30 November 2015).

DeLyser, D., Herbert, S., Aitken, S., Crang, M. and McDowell, L. (eds) (2010) *The Sage Handbook of Qualitative Geography*. London: Sage.

Dorling, D. and Simpson, S. (1999) 'Introduction to statistics in society', in D. Dorling and S Simpson (eds) *Statistics in Society: The Arithmetic of Politics*, London: Arnold. pp. 1–5.

Elwood, S. (2010) 'Mixed methods: Thinking, doing, and asking in multiple ways', in D. DeLyser, S. Herbert, S. Aitken, M. Crang and L. McDowell (eds) *The Sage Handbook of Qualitative Geography*. London: Sage. pp. 94–114.

Fotheringham, A. S., Brunsdon, C. and Charlton, M. (2000) *Quantitative Geography: Perspectives on Spatial Data Analysis*. London: Sage.

King-O'Riain, R. C. (2007) 'Counting on the Celtic Tiger: Adding ethnic census categories in the Republic of Ireland', *Ethnicities* 7: 516–42.

Kitchin, R. and Tate, N.J. (2000) *Conducting Research into Human Geography*. London: Prentice Hall.

Koteyko, N., Jaspal, R. and Nerlich, B. (2013) 'Climate change and "climategate" in online reader comments: A mixed methods study', *The Geographical Journal,* 179 (1): 74–86.

Mason, J. (2002) *Qualitative Researching* (2nd edition) London: Sage.

Mason, J. (2011) 'Facet methodology: The case for an Inventive Research Orientation', *Methodological Innovations Online* 6(3): 75–92. http://www.pbs.plym.ac.uk/mi/pdf/8-02-12/MIO63Paper31.pdf (accessed 30 November 2015).

Moran-Ellis, J., Alexander, V.D., Cronin, A., Dickenson, M., Fielding, J., Sleney, J. and Thomas, H. (2006) 'Triangulation and integration: Processes, claims and implications', *Qualitative Research* 6(1): 45–59.

Omi, M. and Winant, H. (2015) *Racial Formation in the United States*. New York: Routledge.

Rees, P., Martin, D. and Williamson, P. (2002) *The Census Data System*. Chichester: Wiley.

Shabazz, R. (2014) 'Masculinity and the mic: Confronting the uneven geography of hip-hop', *Gender, Place, and Culture* 21(3): 370–86.

Simpson, L. and Brown, M. (2008) 'Census fieldwork in the UK: The bedrock for a decade of social analysis', *Environment and Planning A* 40(9): 2132–48.

Singleton, A. D. (2012) 'The geodemographics of access and participation in Geography', *The Geographical Journal* 178 (3): 216–29.

Smith, D. P. and Sage, J. (2014) 'The regional migration of young adults in England and Wales (2002–2008): A "conveyor-belt" of population redistribution?', *Children's Geographies*, 12 (1): 102–17.

Southworth, J. (2005) '"Religion" in the 2001 census for England and Wales', *Population, Space and Place,* 11: 75–88.

ON THE COMPANION WEBSITE...

Visit **https://study.sagepub.com/keymethods3e** for author videos, chapter exercises, resources and links, plus **free** access to the following recommended articles:

1. **Sui, D. and DeLyser, D. (2012) 'Crossing the qualitative-quantitative chasm I: Hybrid geographies, the spatial turn, and volunteered geographic information (VGI)', *Progress in Human Geography*, 36 (1): 111–24.**

This report reviews broad trends that cross the traditional quantitative-qualitative divide in Human Geography and highlights some contemporary issues that require, transform and contest uses of secondary data.

2. **Smyth, F. (2008) 'Medical geography: Understanding health inequalities', *Progress in Human Geography*, 39 (2): 1–9.**

This report examines the ways in which quantitative data have been used by geographers to develop understandings of health inequalities. It highlights ways in which datasets can be utilised and the potential for contextualising research findings using mixed methods.

3. **Hulme, M. (2014) 'Attributing weather extremes to "climate change": A review', *Progress in Physical Geography*, 38 (4): 499–511.**

This report reviews previous research on extreme weather events and climate change, and discusses different approaches to the topic within the wider political context.

31 Using Statistics to Describe and Explore Spatial Data

Eric Delmelle

SYNOPSIS

The past two decades have witnessed two critical changes in the area of spatial data: (1) An increasing availability of spatial and temporal-explicit data and (2) the democratization of Geographic Information Systems (GIS). Georeferenced (or spatial) data are unique and characterized by a set of latitude and longitude coordinates, and come in different formats (point/lines/area – 'vector' – or raster). These datasets can be massive, and there is an increasing need for robust statistical and visualization methods that can integrate different dimensions, such as space and time. While traditional statistical methods can help explore trends, spatial statistical approaches have the potential to identify locally varying patterns. Particularly, the chapter focuses on spatial (and space-time) point pattern analysis, spatial autocorrelation and spatially-based regressions. The methods are illustrated on a set of epidemiological data and patterns of voter preference.

This chapter is organized into the following sections:

- Introduction
- Exploratory spatial data analysis
- Exploratory space-time data analysis
- Confirmatory analysis
- Conclusion

31.1 Introduction

The first decade of the twenty-first century has witnessed dramatic growth in the volume of available spatiotemporal data, coupled with increasing democratization of geospatial technologies (Delmelle et al., 2013b; Anselin, 2011). Data of different formats are widely available, and are generated from several key technological and social developments: (1) web-based geocoders (e.g. GoogleMaps™) can convert addresses to coordinates in a limited timeframe (Karimi et al., 2011); (2) remote sensing devices collect massive amounts of geospatial data from space every day (Goodchild, 2007); (3) the availability of Global Positioning System (GPS) devices embedded in mobile phones has enabled the tracking of millions of individuals in space and time; and

(4) citizens have increasingly participated in this data collection effort, knowingly or not (Sui, 2008). Social networks such as Twitter have experienced an unprecedented rise in popularity for decentralized information sharing and communication and can provide realistic representations of quickly changing phenomena, such as earthquakes (Crooks et al., 2013) or infectious diseases (Padmanabhan et al., 2014).

Given this rapid increase in the volume of spatial data, geographers and spatial scientists have recognized the need to develop and refine tools and techniques that can deal with georeferenced data (Fischer and Getis, 2009). There exists a palette of statistical and data mining methods specifically designed to handle geospatial data; these methods have found applications in diverse fields such as geography, public health, ecology, crime analysis, facility location modelling, and environmental sciences, to name a few. These methods have facilitated the extraction and detection of spatio-temporal *patterns*, eventually leading to a better understanding of complex spatial *relationships*. Developments in computational science and mapping technologies have enabled effective and efficient visualization of large geospatial data sets such as social media data on the Web (see Chapter 16 and examples at http://www.floatingsheep.org/). To that end, Geographical Information Systems (GIS) provide a robust platform to integrate these methods and visualization capabilities (Longley et al., 2005).

The nature of spatial data

Spatially referenced data are characterized by some form of locational information (e.g., longitude and latitude or x and y coordinates), a time stamp, and attribute information (Peuquet, 2002). For instance, a crime event is defined by the type of crime that occurred (non-spatial attribute), the location where the event took place (spatial attribute) and when it happened (time stamp). The question of accuracy often arises, and involves two primary components: *positional* accuracy refers to the deviation between the geographic positions on the map from its true locations, while *attribute* accuracy is the precision of the attribute database. Related to the question of accuracy, spatial data can either be *explicit* or *implicit*. Information gathered from GPS devices is readily observable, and is therefore explicit. An example of implicit spatial data is coordinates inferred from a geocoding procedure. These are of significant concern in all fields of practice; in epidemiology, for example, inaccurate spatial data may lead to false detection alarms and misguided responses. Although improvement in technology has facilitated our ability to collect massive amounts of data in (near) real time, many situations exist where gathering fine resolution information remains extremely expensive and careful sampling schemes must be designed accordingly. Relevant examples in spatial sciences include collecting observations on radioactive material (Melles et al., 2011), soil contamination (Van Groenigen et al., 2000) and census population (Spielman et al., 2014).

GIS facilitates the linking of temporal and non-spatial attributes to geospatial locations by means of a unique identifier (ID). Using structured query language (SQL) syntax, events occurring within a certain distance from one another or within a predefined time interval can be identified (see also Chapter 37). GIS allows users to detect spatial relationships among different layers of information; for instance whether patients exhibiting symptoms of cholera are located near a contaminated water pump

(Snow, 1855) or whether disease epidemics are more likely to appear in non-endemic regions due to increased air travel.

Structure of the chapter

The purpose of this chapter is to provide an overview of methods for the purpose of space-time analysis and modelling. The chapter attempts to answer the fundamental question of how spatially referenced datasets can be analysed both in space and time to reveal critical patterns. Specifically, in Section 31.2 (Exploratory Spatial Data Analysis), I review methodologies to explore and investigate patterns in both point and polygon (areal) datasets. In Section 31.3 (Exploratory Space-Time Data Analysis), I discuss methodologies to detect temporal trends and visualize patterns in spatiotemporal explicit datasets. Finally, in Section 31.4 (Confirmatory Analysis), I illustrate the benefits of geographically weighted regression to model the role of explanatory variables, spatially.

Datasets

I illustrate the concepts set forth in this chapter with two very distinct datasets. The dengue fever dataset is built around geocoded cases of dengue fever in the city of Cali, Colombia, during an outbreak in 2010. During that time period, a total of n = 11,760 cases were extracted from the Public Health Surveillance System (SIVIGILA, see Delmelle et al., 2013a). Individuals reported dengue fever symptoms at local hospitals on a daily basis (unit = Julian date). I use patients' data for the first six months of 2010 (see Figure 31.1a), which correspond to an outbreak with n = 7,111 cases, successfully geocoded and then geomasked to the street intersection level for confidentiality purposes (Delmelle et al., 2013a). Particularly, this dataset illustrates the spatial distribution of cases, point pattern, and space-time point patterns. The second dataset uses voting preferences in Virginia and Maryland during the February 12, 2008 *Potomac Primary* election aimed at determining the nominee of the Democratic Party for the US presidential campaign of 2008, between Barack Obama and Hillary Clinton[1] (Virginia Department of Elections, 2008). Voting scores are reported at the county polygon level, using a voting percentage (see Figure 31.2). I used this dataset to illustrate concepts of spatial clustering for areal data and confirmatory analysis.

31.2 Exploratory Spatial Data Analysis

The uniqueness of space-time data requires: (1) the development and applications of techniques for the identification of associations in space and time or in the attribute space, and (2) the development of visualization techniques, including scatter plots, graduated symbol, choropleth mapping and kernel density. *Exploratory Spatial Data Analysis* (ESDA) techniques facilitate the discovery of spatial patterns and identification of clusters (Anselin, 1999). ESDA is the first step in developing hypotheses of causal relationships, while confirmatory analysis forms the second step, which is used to statistically test whether the pattern of the phenomenon under study is not a product of a random process.

Spatial point pattern

A spatial point pattern refers to data in the form of points, where a point denotes the location of an event. It is usually desirable to analyse whether these particular events, such as, for instance, crimes, car accidents, emergency distress calls, or diseases, exhibit a specific pattern to better understand the underlying process that (may) have generated the events. Such events are considered *discrete*, because they occur at specific locations. A visual inspection of a map showing the locations of those events (e.g., scatter plot) may not always bring a correct interpretation of the true pattern, especially when events occur repeatedly at the same location. For instance, in Figure 31.1a, several dengue cases were reported in very close proximity to other cases; if several events occurred in the same apartment complex, these would be represented as points on top of one another. As such, true clusters may go unnoticed.

Although the magnitude of clustering can easily be estimated using a nearest neighbour or a *K*-function statistic (see Delmelle (2009) for a review), the location of those clusters is best identified using a Kernel Density Estimation (KDE) technique, which is supported in most GIS software. The KDE algorithm is a geospatial analytical technique that summarizes the intensity of the underlying point process into a surface. As a result, the KDE algorithm generates a so-called *heat map*, in which each grid cell reflects the intensity of the process at that location (Bailey and Gatrell, 1995; Delmelle, 2009). KDE is computed at each grid cell, which receives a higher weight if it has a larger number of observations in its surrounding neighbourhood. Let s ($s=x$, y) be a grid location where the kernel density estimation needs to be estimated, and $s_1...s_n$ the locations of n observed events. Then, the density $\hat{f}(x,y)$ at s is estimated by

$$\hat{f}(x,y) = \frac{1}{nh_s^2}\Sigma_i I(d_i < h_s)k_s\left(\frac{x-x_i}{h_s}, \frac{y-y_i}{h_s}\right) \qquad [31.1]$$

where $I(d_i < h_s)$ is an indicator function taking value 1 if $d_i < h_s$ and 0 otherwise. Here, h_s is the search radius (or bandwidth), governing the strength of smoothing. The bandwidth can either be calibrated with a *K*-function or cross-validation while its size can be modified to reflect the underlying heterogeneous population distributions (Carlos et al., 2010). A smaller bandwidth will result in more distinct events to be highlighted, while a larger radius can identify broad zones where a high number of incidents exist. A bandwidth that is too large will stretch the kernel and the surface will appear flat. The choice of the bandwidth may depend on the purpose of the study. The term k_s is a standardized kernel weighting function that determines the shape of the weighting function, and d_i is the distance between location s and event i, constrained by $d_i < h_s$, such that only points falling within the chosen bandwidth contribute to the estimation of the kernel density at s.

An example of a kernel density is given in Figure 31.1b, which uses a kernel bandwidth h_s = 1000 meters, roughly approximating the size of neighbourhood, where the peaks of the surface denote regions where there is a strong clustering of dengue fever cases.

Due to privacy concerns, sensitive epidemiological data – such as is the case with dengue fever information – can be *geomasked*, or aggregated to a certain level of

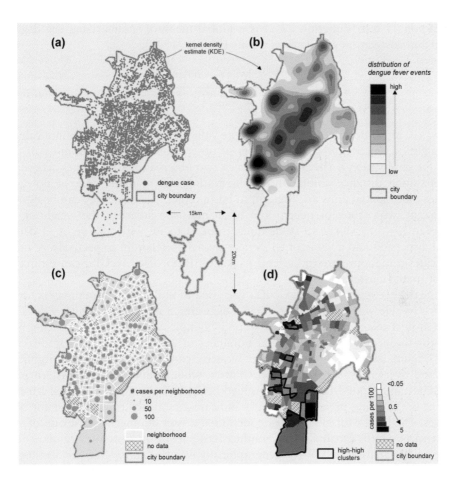

Figure 31.1 Spatial distribution of dengue fever rates (a) for the months of January through June 2010 included; (b) Kernel Density Estimation; (c) aggregated counts per neighbourhood; (d) divided by the population. Significant and positive clusters of dengue fever by population are highlighted in black.

Source: Sistema Nacional de Vigilancia en Salud Pública, Colombia (SIVIGILA)

census geography, for instance at the county or postal/zip code level. Figure 31.1c uses a proportional symbology to map the variation of dengue cases for each neighbourhood of the city of Cali, suggesting an uneven spatial distribution. Other techniques, such as choropleth mapping, are widely used to display disease rates across an area. Although Figure 31.1a and b suggest that dengue fever rates are clustered, the spatial variation in the *rates* is very different from Figure 31.1b after controlling for population (Figure 31.1d).

Spatial autocorrelation

Spatial autocorrelation measures the dependency of nearby data observations, assuming that measurements closer to one another should exhibit similar attribute values.

The *Moran's I* test (Equation 31.2) is a global measure summarizing the degree of spatial autocorrelation among data points (Moran, 1948):

$$I = \frac{n\sum_i^n\sum_j^m w_{ij}\left(u_i - \bar{u}\right)\left(u_j - \bar{u}\right)}{\sum_i^n\sum_j^m w_{ij}\sum_i^n\left(u_i - \bar{u}\right)}$$

[31.2]

with u_i the attribute value at i, u_j = attribute value at j, $\bar{u}$ = average attribute value, w_{ij} the weighting function, n = total number of observations, m = total number of neighbours for observation i. The weight between two observations w_{ij} varies according to the separating distance among those observations, or – in the case of polygons – whether those spatial units are adjacent to one another. The adjacency can be estimated in different ways (rook or queen) and when units are adjacent, the weighting score reduces to $w_{ij} = 1$. Moran's *I* value range from –1 to +1, with a value of –1 denoting a total dispersion and a value of +1 perfect spatial correlation, respectively. A value of 0 indicates a random pattern. Several packages (R, GeoDa, ArcGIS) can estimate the level of spatial clustering.

In the dengue fever example, GeoDa was used to build a weighting matrix, and the Moran's *I* value was estimated at 0.335, suggesting a tendency for high values to be clustered next to one another. In the second example, which highlights voting preferences for the democratic presidential nominees in 2008, there is a clear directional trend (Figure 31.2a); counties in the Western part of the states of Virginia and Maryland overwhelmingly supported Clinton, while counties in the Eastern part of both states and around DC strongly rallied behind Obama. With a global Moran's *I* value of 0.75, voting preferences clearly indicate a tendency to be similar for nearby counties. The local Moran's *I* statistic identified a very homogeneous group of voting preferences towards Obama in the Southeastern part of the state of Virginia.

For an explanation of 'GWR' (geographically weighed regression) see Section 31.4.

Local spatial autocorrelation

The *Local Moran's I* statistic (equation not included) detects local clusters of similar values; by evaluating the difference in data value of one unit i from its surrounding values j. To illustrate the concept of local spatial autocorrelation, I use the percentage of dengue cases per population at the neighbourhood level (Figure 31.2d). The *local Moran's I* statistic – computed in GeoDa and visualized in ArcGIS – identifies groups of high values of dengue rates next to each other (in black). The observed local sum of the attribute values is greater than what would be expected. I also illustrate the same concept on the voting example, where we observe a tight local cluster of voting preferences for candidate Obama (Figure 31.2a).

31.3 Exploratory Space-Time Data Analysis

The methods presented in the previous section do not account for time, and do not reflect inherent temporal patterns. Some phenomena, such as weather, migration, and the spread of infectious diseases, for instance, are known to vary significantly along the temporal dimension.

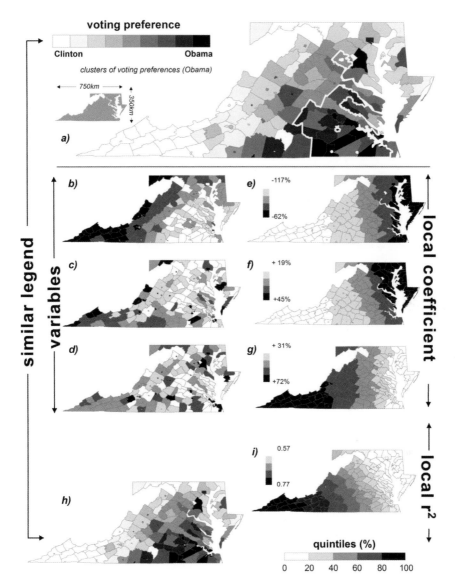

Figure 31.2   (a) Spatial distribution of voting preferences for the election of the democratic candidate (Obama and Clinton) during the primary of 2008. (b) Proportion of white individuals; (c) female voters; and (d) individuals of age 19–29. Figures (e)–(g) are the corresponding beta coefficients, following a GWR. Figures (h) and (i) reflect the predicted voting preferences using a GWR, and the associated local R^2 values, respectively (*Source:* Virginia Department of Elections 2008)

Temporal patterns

The cumulative distribution function (CDF) summarizes the probability that an event occurs before a certain date. Its derivative – the probability density function (PDF) – is the likelihood that an event occurs at a particular time. Figure 31.3a illustrates the cumulative

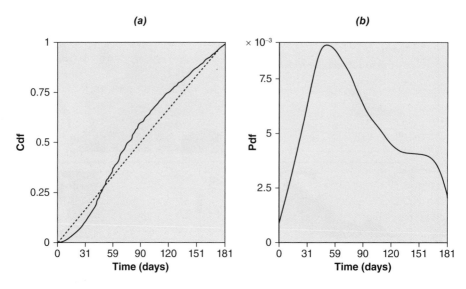

Figure 31.3 Cumulative distribution function for (a) dengue cases; and (b) probability density function

distribution function (CDF) of the dengue cases from January to June (X-axis units: Julian date), while Figure 31.3b reflects PDF values. The CDF indicates that the number of dengue cases were relatively low, but increased sharply in February. The PDF values indicate a rapid increase of cases in February with a peak at the end of the month, followed by a slow decline. Figure 31.3b is particularly important to better understand the distribution of cases and suggest hypotheses on the temporal variation of the disease.

Spatiotemporal trends

Exploratory spatial data analysis and temporal trend analysis provide insight on the spatial distribution of a phenomenon and its temporal variation, respectively. However, the methods fall short at understanding space-time *interaction* and process dynamics. There is an increasing need for novel statistical and visualization techniques that can explicitly integrate space and time.

To model temporal variation, it is possible to partition the data into different time intervals, and repeat the same ESDA technique. For instance, KDE can be repeated for several time periods, revealing whether the strength and location of clusters change over time (Delmelle et al., 2011). Alternatively, the methods presented earlier can be extended to explicitly incorporate the temporal dimension. An example is the Space-Time Kernel Density Estimation (STKDE) technique which extends the KDE methodology in time, and facilitates the mapping of space-time patterns (Demšar and Virrantaus, 2010; Nakaya and Yano, 2010). STKDE produces a raster volume where each volumetric pixel (voxel at x, y, t) is a density estimate based on the surrounding point data:

$$\hat{f}(x,y,t) = \frac{1}{nh_s^2 h_t} \Sigma_i I\,(d_i < h_s, t_i < h_t)\, k_s\left(\frac{x - x_i}{h_s}, \frac{y - y_i}{h_s}\right) k_t\left(\frac{t - t_i}{h_t}\right), \qquad [31.3]$$

where the density value $\hat{f}(x,y,t)$ of each voxel s is estimated based on point/event data (x_i, y_i, t_i) in the vicinity of the voxel. Each point in the neighbourhood of the voxel is weighted using a kernel function, k_t and k_s. Each data point is weighed based on its proximity in time and space to the voxel s, whose density is estimated. The distance in space and time between the voxel and an event data point is defined by d_i and t_i respectively.

The kernel density volume is visualized by colour-coding each voxel according to its kernel density value. The transparency level is also adjusted to concentrate the focus on those regions with higher density values (Figure 31.4). The KDE values range from 0 to 1 (on the map). The highest space-time kernel density values are found in a couple of cores within the voxel cloud (not at the edges). Different transparency levels are used depending on the KDE value of each voxel to reinforce the presence of space-time clusters. These clusters are further delineated using so-called 'egg shells', which are created by generating an iso-volume around voxels of a similar value, capturing those areas with higher kernel density values (Delmelle et al., 2015). Figure 31.4 should be compared with Figure 31.11b; we are now able to better understand the spatial and temporal signature of the disease, and identify regions where measures that were implemented to curb the expansion of the disease may have worked.

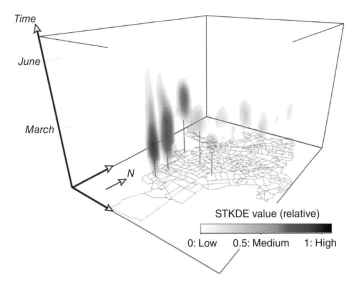

Figure 31.4   Space-time distribution of dengue fever cases from January to June 2010 in Cali, Colombia, using a cell size of 125 meters with shaded isovolumes

Source: Sistema Nacional de Vigilancia en Salud Pública, Colombia (SIVIGILA)

31.4 Confirmatory Analysis

The methods presented in Section 31.2 and 31.3 are of an exploratory nature; that is, we are using visualization techniques and statistics to identify patterns that merit further investigation. We anticipate that the spatial variation of the phenomena (dengue fever outbreaks, voting preferences) may be explained by several factors, and that this relationship may vary spatially. There are several tools we can use to begin identifying the causal factors that influence the patterns most significantly, which are discussed here.

Ordinary Least Squares

The ordinary least squares (OLS) regression (Equation 31.4) attempts to find an optimal set of variables which best explains the variation of the variable under study.

$$\hat{u}_i = b_0 + \sum_{k=1}^{N} b_k X_k + \varepsilon_i \qquad [31.4]$$

with $\hat{u}_i$ the predicted value at i (e.g. disease rate, voting percentage), b_0 the intercept and b_k the coefficient of the variables selected in the model ($k=1...N$). The term ε_i is the error term. The optimal set of variables is determined by minimizing the sum of the residuals r (difference between the observed and predicted values):

$$r_i = u_i - \hat{u}_i \qquad [31.5]$$

Positive [negative] residuals denote under-prediction [over-prediction]. To illustrate the variation in the voting patterns, we identify three key variables – as suggested by national newspapers – that could influence voting scores during the primary, specifically percentage of white voters, percentage of female voters, and percentage of young voters. One hypothesis set forth was that young voters would massively rally behind Barack Obama, while whites would support Hillary Clinton. The spatial variation of these variables is depicted in Figure 31.2b–d. Together, the variables explained 69% ($r^2 =.69$) of the variation in voting patterns.

It is often desirable to minimize the spatial pattern of the residuals (an assumption of the OLS is that errors are not correlated to one another); in this regard spatially explicit regressions (e.g. spatial regression, geographically weighted regression) may offer an alternative by capitalizing on spatial information provided by nearby samples.

Geographically weighted regression

A major shortcoming of the OLS is that predicted values of the variable, estimated at location i, are derived only from explanatory variables at i, although information provided by nearby samples j could explain the variation. An alternative is geographically weighted regression, which provides locally linear regression coefficient at every point i, using distance weighted samples:

$$\hat{u}_i = b_{0i} + \sum_{k=1}^{N} b_{ki} X_{ki} + \varepsilon_i \qquad [31.6]$$

For each location i, we have a set of neighbouring values j that we use to estimate by $\hat{u}_i$ while omitting observation i. The term b_{ki} represents the value of the kth parameter at location i. The parameter $k=1...N$ defines the number of explanatory variables. GWR estimates b_{ki} on observations taken at sample points close to i. An undeniable benefit of the GWR is that it allows us to produce maps of these coefficients for the study area, allowing estimation of the influence of each explanatory variable, locally.

Geographically Weighted Regression is applied on the set of voting preferences in Virginia, leading to an improvement of the average r^2 =.77. The coefficients of those variables used in the GWR technique are displayed in Figure 31.2e–g. In Figure 31.2e, the coefficient of the variable *proportion white* was negative throughout, confirming that white individuals did not vote for Obama, but that this pattern was stronger in the rural and western areas of Virginia. A similar pattern was observed for proportion of female voters, but the association was positive (weak association in the western part of the study region and stronger association in the eastern part). Finally, young voters overwhelmingly voted for Barack Obama, but this pattern was more pronounced in the western part of Virginia and Maryland. GWR also generates a map of the local r^2 values; the model predicted reality much better in the western part of Virginia, than in the eastern and northeastern part of the state. Predicted values following the GWR (in Figure 31.2h) compared relatively well with observed values (in Figure 31.2a).

Time series

When data vary temporally, it may be desirable to predict their outcome over time. In epidemiology for instance, predicting outbreaks of infectious diseases is particularly important when attempting to predict (and prevent) the re-occurrence of the disease. An autoregressive model allows us to predict future values based on previously observed values (an example of an autoregressive model – ARIMA for AutoRegressive Integrated Moving Average – on dengue fever outbreaks is given in Eastin et al., 2014). The extension of the ARIMA model in space is particularly suited for modelling space-time processes that exhibit stationarity in both space and time. Examples are given in Cheng et al. (2012) and Rey and Janikas (2006).

31.5 Conclusion

The development of methods for space and time analysis is a process that is continuously evolving. The increasing availability of spatiotemporal data has led to a pressing need to develop and integrate methods that can be used to reveal useful patterns, and this has been facilitated by the democratization of GIS. GIS can easily integrate geospatial data of different spatial scales and temporal granularity for a wide range of applications. This chapter has reviewed the importance of exploratory spatial data analysis to deepen our ability to detect spatial and temporal trends inherent to the data. Finally, the chapter has discussed the importance of confirmatory statistical approaches to model the role of explanatory variables.

SUMMARY

The past two decades have witnessed two critical changes in the area of spatial data: (1) An increasing availability of spatial and temporal-explicit data and (2) the democratization of Geographic Information Systems (GIS). Georeferenced (or spatial) data are unique and characterized by a set of latitude and longitude coordinates, and come in different formats (point/lines/area or raster). These datasets can be massive, and there is an increasing need for robust statistical and visualization methods that can integrate different dimensions, such as space and time. Traditional statistical methods help explore trends. Spatial point pattern techniques (*K*-function, Kernel Density Estimation) are well suited to identify the locations of clusters, particularly informative in an epidemiological context. Spatial statistical approaches have the potential to identify locally varying patterns when data are aggregated at the areal level, as is the case for several census units. It is particularly important to test for spatial autocorrelation before implementing a regression technique; indeed traditional non-spatial regression approaches do not capture the spatial variation of the phenomenon under study. Caution is recommended when the data exhibit temporal trends (e.g. epidemiological datasets). Suggestions are presented to visualize this information and incorporate it into regression techniques.

Note

1   John Edwards had officially suspended his campaign on 30 January 2008, but his name was still on the ballot.

Further Reading

Anselin (2011) 'From SpaceStat to CyberGIS: Twenty years of spatial data analysis software'.

Delmelle et al. (2013) 'Methods for space-time analysis and modeling: An overview'.

Fischer and Getis (2009) *Handbook of Applied Spatial Analysis: Software Tools, Methods and Applications.*

Note: Full details of the above can be found in the references list below.

References

Anselin, L. (1999) 'Interactive techniques and exploratory spatial data analysis', in P. Longley, M. Goodchild, D. Maguire and D. Rhind (eds), *Geographical Information Systems*. New York: Wiley. pp. 253–266.

Anselin, L. (2011) 'From SpaceStat to CyberGIS: Twenty years of spatial data analysis software', *International Regional Science Review* 35: 131–57.

Bailey, T. and Gatrell, Q. (1995) *Interactive Spatial Data Analysis*. Edinburgh Gate: Pearson Education Limited.

Carlos, H., Shi, X., Sargent, J., Tanski, S. and Berke, E. (2010) 'Density estimation and adaptive bandwidths: A primer for public health practitioners', *International Journal of Health Geographics* 9 (1): 39.

Cheng, T., Haworth, J. and Wang, J. (2012) 'Spatio-temporal autocorrelation of road network data', *Journal of Geographical Systems* 14 (4): 389–413.

Crooks, A., Croitoru, A., Stefanidis, A. and Radzikowski, J. (2013) '# Earthquake: Twitter as a distributed sensor system', *Transactions in GIS* 17 (1): 124–47.

Delmelle, E. (2009). 'Point Pattern Analysis', in R. Kitchin and N. Thrift (eds), *International Encyclopedia of Human Geography*. Kidlington: Elsevier. pp. 204–11.

Delmelle, E.M., Delmelle, E.C., Casas, I. and Barto, T. (2011) 'H.E.L.P: A GIS-based health exploratory analysis tool for practitioners', *Applied Spatial Analysis and Policy* 4 (2): 113–37.

Delmelle, E.M., Casas, I., Rojas, J.H. and Varela, A. (2013a) 'Spatio-temporal patterns of Dengue Fever in Cali, Colombia', *International Journal of Applied Geospatial Research* 4 (4): 58–75.

Delmelle, E.M., Kim, C., Xiao, N. and Chen, W. (2013b) 'Methods for space-time analysis and modeling: An overview', *International Journal of Applied Geospatial Research*.

Delmelle, E.M., Jia, M., Dony, C., Casas, I. and Tang, W. (2015) 'Space-time visualization of dengue fever outbreaks'. *Spatial Analysis in Health Geography*. Kent: Ashgate.

Demšar, U., and Virrantaus, K. (2010) 'Space-time density of trajectories: Exploring spatio-temporal patterns in movement data', *International Journal of Geographical Information Science* 24 (10): 1527–42.

Eastin, M.D., Delmelle, E.M., Casas, I., Wexler, J. and Self, C. (2014) 'Intra-and interseasonal autoregressive prediction of dengue outbreaks using local weather and regional climate for a tropical environment in Colombia', *The American Journal of Tropical Medicine and Hygiene* 91 (3): 598–610.

Fischer, M. and Getis, A. (eds) (2009) *Handbook of Applied Spatial Analysis: Software Tools, Methods and Applications.* Heidelberg: Springer.

Goodchild, M.F. (2007) 'Citizens as sensors: The world of volunteered geography', *GeoJournal* 69 (4): 211–21.

Karimi, H.A., Sharker, M. H. and Roongpiboonsopit, D. (2011) 'Geocoding recommender: An algorithm to recommend optimal online geocoding services for applications', *Transactions in GIS* 15 (6): 869–86.

Longley, P.A., Goodchild, M.F., Maguire, D.J. and Rhind, D.W. (2005) *Geographic Information Systems and Science.* Chichester: Wiley.

Melles, S., Heuvelink, G.B., Twenhöfel, C.J., Van Dijk, A., Hiemstra, P.H., Baume, O. and Stöhlker, U. (2011) 'Optimizing the spatial pattern of networks for monitoring radioactive releases', *Computers & Geosciences* 37 (3):280–288.

Moran, P.A. (1948) 'The interpretation of statistical maps', *Journal of the Royal Statistical Society. Series B (Methodological)* 10 (2): 243–51.

Nakaya, T. and Yano, K. (2010) 'Visualising crime clusters in a space-time cube: An exploratory data-analysis approach using space-time kernel density estimation and scan statistics', *Transactions in GIS* 14 (3): 223–39.

Padmanabhan, A., Wang, S., Cao, G.,Hwang, M., Zhang, Z., Gao, Y., Soltani, K., and Liu. Y. (2014) 'FluMapper: A cyberGIS application for interactive analysis of massive location-based social media', *Concurrency and Computation: Practice and Experience* 26 (13): 2253–65.

Peuquet, D. J. (2002) *Representations of Space and Time.* New York: Guilford Press.

Rey, S. J. and Janikas, M.V. (2006) 'STARS: Space–time analysis of regional systems', *Geographical Analysis* 38 (1): 67–86.

Snow, J. (1855) *On the Mode of Communication of Cholera.* London: John Churchill.

Spielman, S. E., Folch, D. and Nagle, N. (2014) 'Patterns and causes of uncertainty in the American Community Survey', *Applied Geography* 46: 147–57.

Sui, D. Z. (2008) 'The wikification of GIS and its consequences: Or Angelina Jolie's new tattoo and the future of GIS', *Computers, Environment and Urban Systems* 32 (1): 1–5.

Van Groenigen, J., Pieters, G. and Stein, A. (2000) 'Optimizing spatial sampling for multivariate contamination in urban areas', *Environmetrics* 11 (2): 227–44.

ON THE COMPANION WEBSITE...

Visit **https://study.sagepub.com/keymethods3e** for author videos, chapter exercises, resources and links, plus **free** access to the following recommended articles:

1. **Foody, G.M. (2006) 'GIS: Health applications', *Progress in Physical Geography*, 30 (5): 691–5.**

The example given in this chapter is health-related and so this should be of interest.

2. **Elwood, S. (2009) 'Geographic Information Science: New geovisualization technologies-emerging questions and linkages with GIScience research', *Progress in Human Geography*, 33 (2): 256–63.**

The importance of GIS and Geovisualization cannot be stressed enough and this is a useful summary.

32 Exploring and Presenting Quantitative Data

Richard Field

> What is to be sought in designs for the display of information is the clear portrayal of complexity. Not the complication of the simple; rather the task of the designer is to give visual access to the subtle and the difficult – that is, the revelation of the complex.
>
> Tufte (1983: 191)

SYNOPSIS

To achieve and communicate understanding of 'real-world' complexity, the ability to work with data is essential. A fundamental starting point is realizing that there are different levels of *abstraction* from the 'real world': data collection, graphical and numerical representation of data, and generalization from statistical or mathematical modelling. *Different types of data* are best explored, analysed and presented using different methods. In particular, the distinction between categorical and continuous variables is fundamental to the type of method chosen. Graphics (tables, graphs and maps) are essential for exploring data, refining analyses and presenting results. They can be primarily *descriptive* (e.g. tables with raw data or counts of cases, graphs displaying the distribution of data, maps showing geographical patterning) or more *interpretative* (e.g. graphs with regression lines, tables reporting results of statistical analyses, maps of errors from models). New methods are continually developed, including interactive online ones. Visualizing patterns of data and relationships between variables is a fundamental part of data analysis, and should be employed at all stages from the first look at the dataset (exploratory) through checking and refining models to presenting the results. The same sorts of *understanding* required for data analysis should be applied when deciding how to explore and present data. *How data are presented affects understanding, judgement and inference*. The presentation should be honest and as clear and uncluttered as possible, conveying the main messages efficiently. Good graphics make clear not only the main findings, but also the limits to their applicability (e.g. avoiding undue extrapolation).

This chapter is organized into the following sections:

- Introduction
- Abstraction
- The basics of your dataset
- Presenting data

- Examining the distribution of data
- Examining how variables relate to each other
- Explaining and interpreting
- Conclusion

32.1 Introduction

This chapter is about exploring and presenting data and results. A recurring theme is that doing these things well requires the same sorts of understanding and thought processes as analysing data does. Accordingly, the companion website contains material aimed at boosting these thought processes for those new to working with quantitative data. It is no surprise, then, that the history of development of analysis techniques goes hand-in-hand with the history of development of data exploration and presentation techniques.

Ever since people started thinking about the world, they have needed ways of coping with its inherent complexity. Early scientists relied primarily on logic: simplifying the world into a series of principles. Leonardo Da Vinci (1452–1519), along with other prominent figures of 'the Enlightenment', was instrumental in shifting the emphasis towards data collection and experiment. A period followed in which many leading figures, including some of the great 'natural philosophers' such as Lyell and Darwin, meticulously collected large amounts of information from observation and experiment. However, during much of this period, there was no generally accepted, objective way of either visualizing or drawing inferences from data. Instead, the subjective judgement of experts was relied upon, and this could vary considerably from expert to expert.

Tables were used quite routinely for much of this time, but were typically very lengthy: good for presenting detail but of little use as summaries. It seems that graphs did not occur to the Greeks or the Romans, nor to the likes of Newton and Leibnitz. Graphs were not invented until the great work of Descartes, who set up the Cartesian system in his book 'La Géométrie', published in 1637 (Spence and Lewandowsky, 1990). After that, the use of graphical means to explore or present data advanced little until the seminal work of Playfair (1786, 1801), who invented many of the graphs still popular today, including the histogram, the line graph and the pie chart (Spence and Lewandowsky, 1990). These visualization methods became more and more popular during the nineteenth century, and were linked to the rise of statistical thinking at the same time (Porter, 1986). Much important work was done in medicine. For instance, John Snow, a London anaesthetist and Queen Victoria's obstetrician, used inductive methods in conjunction with mapping cholera data across London to link its incidence to particular polluted water sources (most famously the Broad Street public water pump) and then determine the cause of cholera (see Snow, 1854). Florence Nightingale, who in 1859 became the first female member of the Royal Statistical Society, systematically gathered data on illness and fatalities, was meticulous in her attention to detail and careful to ensure that her premises were backed up statistically (Kopf, 1916) – that is, to grasp the essential, emergent trends from a mass of individual cases. To achieve this, she pioneered graphical representation of statistics, and her famous 'Diagram of the causes of mortality in the army in the East' revolutionized the care of wounded soldiers.

Probably the greatest period of development of classical statistical methods was towards the end of the nineteenth century and early in the twentieth century. Probability theory (developed to a considerable extent because gambling-loving French nobles had employed great mathematicians like Pascal to give them an edge) was taken and applied to data. The work of Jevons, Pearson and Fisher was instrumental during this period. During the twentieth century, the field of statistics was becoming increasingly theoretical and inaccessible to non-statisticians, until the considerable input of the great American statistician John Tukey. Tukey argued strongly that, as well as being subjected to formal statistical analysis, data should be explored graphically and visualized as much as possible. He pointed out the fuzzy nature of real-world data, which do not usually comply with theoretical ideals. He also passionately believed that statistical methods should be more accessible to ordinary people, and many of the techniques he developed go a long way towards achieving this goal.

Statistical methods are currently undergoing a new revolution, or series of revolutions. These include fundamentally different ways of inferring from data (e.g. the rise of Bayesian statistics, maximum likelihood and maximum entropy approaches). The recognition that real-world data are often not appropriate for classical significance testing (e.g. for geographical data, spatial autocorrelation often means that data are not truly independent) has led to new analysis methods (e.g. spatial regression). The recognition that most phenomena of interest are influenced by multiple causes, which cannot be adequately controlled in sampling or statistically, coupled with the increasing availability of very large datasets that contain variable-quality data, has led to often-complex computer simulations of null distributions against which to compare findings. The massive increases in computer power and data storage capacity have been associated with a rapid rise in methods of machine learning as alternatives to classical data analysis. In this age of 'Big Data' – and more analysis methods than you could ever hope to comprehend – the importance of clear, effective display of results has never been greater.

32.2 Abstraction

In trying to make sense of 'real-world' complexity, we go through several processes of abstraction (Figure 32.1). The first is the dataset itself. In most cases in geography, data are collected via a sampling process (see Chapters 8 and 20) and are therefore not the entire population of relevant units. These data consist of measurements that are subject to error and that cannot retain all possible information about the units sampled, no matter how careful and comprehensive the data collection. Thus the immense complexity of the 'real world' is reduced to a carefully chosen sample for which certain attributes are measured. Manipulative experimentation similarly simplifies the 'real world'. 'Big Data', often derived from crowd sourcing or other forms of citizen science, also sample from the 'real world', the resulting samples tending to be bigger but more prone to biases, data errors and inconsistencies.

In order to describe the information obtained in a more meaningful way than a large set of numbers, the raw data can be plotted on graphs or maps and summarized via descriptive statistics (e.g. averages and measures of variability), often arranged into tables. These methods represent further levels of abstraction because they summarize, but also lose, information. Graphs and maps are powerful tools in that they can simultaneously

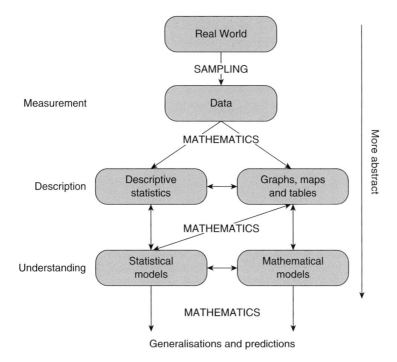

Figure 32.1 Abstraction of 'real-world' complexity in a quantitative study

The diagram is idealized and simplified, but illustrates the point that as you move down through the levels of abstraction, complexity is reduced. Detail (information) is lost, but understanding is enhanced – at least, that is the aim. Here, 'tables' refer to summary tables. Mathematics is the language that translates between the different levels of abstraction.

allow both an instant impression of the data and the display of detail – thus aiding understanding. However, for all except the simplest datasets, each graph or map depicts only a subset of the variables measured. In the case of maps, Geographical Information Systems (Chapters 18 and 27) provide powerful tools for displaying and analysing several variables at the same time, especially when the user can add or subtract layers interactively. Descriptive statistics allow an instant impression of the data but, again, do not convey all the detail present in the dataset.

The results of statistical analyses represent a higher level of abstraction still, for example the estimation of parameters within the 'real-world' population, based on the data collected, and the evaluation of hypotheses that aim to make generalizations about the 'real world'. Mathematical models occupy a similar level of abstraction because they incorporate estimated parameters in attempting to simulate real-world processes (see Chapters 19 and 23).

Graphical (including cartographic) and numerical means of exploring and presenting data can therefore sit in between the 'real world' (which is typically too complex to understand immediately) and the highly abstracted results of mathematical models and statistical analyses (which are typically of little use unless interpreted in the light of the data). Such techniques are of fundamental importance to most geographical enquiry, and crucial to understanding; it is essential that they are used carefully and competently. As seen in Figure 32.1, mathematics is the 'language' for translating between the levels of abstraction; thus numeracy and a basic grasp of mathematical principles are important skills for geographers. However, one can perform high-quality quantitative research without being a good mathematician, particularly with effective use of data visualization techniques.

32.3 The Basics of Your Dataset

Table 32.1 is a dataframe. Dataframes are essentially spreadsheets consisting of rows and columns and are the best way of organizing data in most cases. In fact, almost all statistical packages on computers require data in this format. Each row represents a 'case' – i.e. a unit of sampling. In this example, each case is an individual person, but it might also be a quadrat from a vegetation survey or a pebble on a shingle beach. Each entry in the row represents an attribute of that unit: an individual person's age, gender, income or occupation. The complete set of values for an attribute (one value for each case) makes up one column in the dataframe. This is called a 'variable', because the values vary between cases. Two of the variables in this dataframe are continuous (age and income) and the other two categorical (gender and occupation).

Table 32.1 An example of a dataframe

This dataframe comprises 18 units ('cases'; here they are individual people) and four variables (gender, age, occupation and income). 'Name' is not a true variable, but is a list of case identifiers. Notice how the cases are rows and the variables are columns. This is the standard way of organizing data, used by almost all statistical packages. Two of the variables are continuous and two categorical. The data in the table are fictional.

Name	Gender	Age (yrs)	Occupation	Income (£)
David	M	36.0	Doctor	55741
Justin	M	22.7	Social worker	19569
Lindsay	F	46.0	Doctor	42183
Vicki	F	60.3	Farmer	28293
Madeleine	F	59.6	Doctor	49658
Mark	M	63.0	Social worker	22485
Shelley	F	18.7	Lawyer	48627
Lizzie	F	37.1	Social worker	24630
Jessica	F	58.6	Lawyer	45268
Philip	M	24.5	Farmer	39228
Charles	M	29.5	Farmer	44165
Steve	M	20.1	Doctor	55182
Katherine	F	19.5	Lawyer	40677
Nicola	F	25.7	Farmer	40607
Charlotte	F	28.3	Lawyer	61191
Nicole	F	18.8	Doctor	50598
Nicholas	M	31.4	Farmer	44048
Daniel	M	34.1	Social worker	15878

See the companion website for more explanation about these types of data (https://study.sagepub.com/keymethods3e).

It is a good idea, whenever you are thinking about exploring, analysing or presenting data, to think first about what type of data they are. This helps you understand the dataset, and the context of any manipulation of the data that is performed. Knowing what type of data you are dealing with is an essential pre-requisite to deciding what type of graphic or statistical analysis is suitable. Always think about the numbers in a study, both when reading reports of other people's work and when conducting your own quantitative work. How were the numbers measured or arrived at? What do they mean? Are they what you expected? Do the results of an abstraction technique (e.g. a statistical analysis) make sense, or could there have been an error somewhere in the data manipulation process? Such thought processes also help develop a feeling of being comfortable with numbers, which itself is extremely useful both for spotting errors and for spotting points of interest in results.

32.4 Presenting Data

Most quantitative geographical studies involve a lot of data – too much to make much sense of simply by looking at the raw data. In almost all cases, it is a good idea to start by plotting the data graphically and/or cartographically, and by using simple descriptive statistics (see Chapter 31), often arranged in tables. These exploratory techniques help to identify underlying structure or pattern, and allow researchers to develop an understanding of their data. Later, once the data have been analysed and/or modelled, and we have an idea what they are actually telling us, we need ways of presenting the results in an effective and efficient way. Graphs, maps and tables are ideal. Thus they are used both to explore data and to present results.

A key point about data presentation is that another reader is likely to be much less familiar with the data than you are, and so needs your help in working out what everything means. For example, all axes of graphs should be labelled, including the units of measurement used. Maps should include a north arrow and have the scale clearly marked. Annotations to aid interpretation should be considered. Further explanation of graphs, maps and tables should be given in the legends (captions) of these illustrations. It is better to err on the side of giving too much information than to risk leaving the reader guessing. Note also that, when you are short of words, you can use these legends to convey useful information without adding to your word count, as long as the legends directly describe the illustrations they support. Generally, the supporting information (legends, captions, annotations, etc.) for tables, graphs, maps and other illustrations should be comprehensive enough that each illustration is understandable without reference to the main text.

Before examining particular types of graphic in more detail, it is worth illustrating how important our choice of presentation technique can be. At its most basic, the way that data are presented makes the difference between results being meaningful or not. Look at Figure 32.2. What does it show? It is very unclear, for several reasons, not least that the poor labelling makes it hard to work out what the point of the graph is in the first place. The most serious problem is that the chart type is completely inappropriate for the type of relationship being shown. The graph plots data

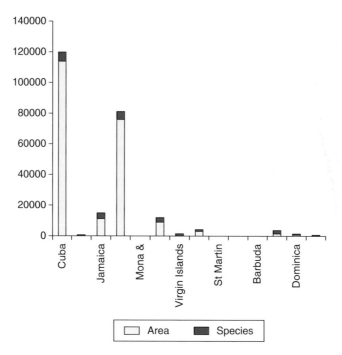

Figure 32.2   An example of poorly presented data

Many things are wrong with this chart. Some of the more important ones are: (1) Axis labelling – neither axis is labelled; not all of the islands are named on the x-axis and no units are given for the y-axis. (2) It is not clear, but the point of interest is the way in which the number of plant species increases as island area increases (a classic relationship in biogeography) for islands of the Caribbean. This type of graph is not appropriate for showing the relationship. (3) The scale of the y-axis is such that most of the data are lost in bars too small to discern. (4) The 3-D effect adds nothing to the plot and the legend is messily placed, so the graph looks unprofessional. Overall, we struggle to understand what message we are being given.

Source: Data taken from Frodin (2001)

on numbers of plant species and areas of islands. The relationship in question is a correlation – what we need to see is what happens to the number of plant species as island area increases. However, it is very hard to discern this relationship from the chart because (i) both variables are plotted on the same axis (the vertical axis), on top of each other, rather than having an axis each; (ii) the numeric scale is the same for both variables even though they are measured in very different units (number of species versus square kilometres) and span very different ranges of values; and (iii) most of the bars are too small to see. Other problems with the chart are noted in the legend of Figure 32.2.

What we need is a bivariate scatterplot, otherwise known as an x–y scatter graph. Figure 32.3 is such a plot, and displays the same relationship far more clearly. The axes are properly labelled, and the log–log scale allows us to see the variation in the data much more clearly. What emerges is a straight-line relationship (on these log–log axes), with a few notable exceptions: islands that deviate from the general trend. These islands are labelled (along with those at the extremes of the plot) because they are of

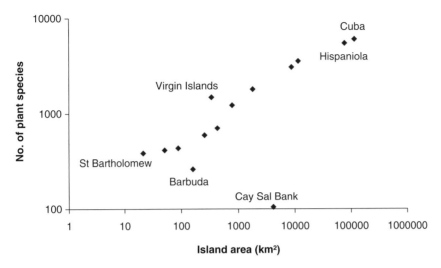

Figure 32.3 Better presentation of the data in Figure 32.2

Here, the message is much clearer than in Figure 32.2: The species richness tends to increase with island area in a predictable manner, though there are exceptions. These exceptions, which are labelled (along with the largest and smallest islands), can be examined in more detail to see whether an explanation is apparent. The type of graph is appropriate for the relationship being shown, and the log scale allows the trends within the data to be seen much more clearly than in Figure 32.2.

particular interest. The immediate question emerging is: why there are such obvious exceptions to an otherwise strong relationship? I return to this in the next section, but the important point here is that both the main 'species–area relationship' and the exceptions are immediately apparent in Figure 32.3, while being completely obscured in Figure 32.2. Extending this theme, most of the rest of this chapter comprises basic, guiding principles for presenting data.

Maps

Although a type of graphical display, maps deserve separate consideration in a geographical text! Most geographical data are spatial in some way or other. Typically, each datum is associated with two spatial coordinates, such as latitude and longitude, often obtained from global positioning systems. This spatial information can be very important for understanding the data, the relationships between the variables and many apparent outliers. For example, Figure 32.4a shows another species–area relationship, this time for reptiles and amphibians (herptiles) in some Caribbean islands. There is a remarkably strong relationship between species richness and island area for all islands except Trinidad. The obvious question is: why does Trinidad have so many more herptile species than the other Caribbean islands shown? Various answers are suggested by the map (Figure 32.4b), including the most likely one. Trinidad is much nearer to the South American mainland than the other islands are (and was in fact joined to the mainland when sea levels were lower). South America, being very large, has a very large number of herptile species, and acts as a source of immigrants to the Caribbean islands. However, it is difficult for terrestrial organisms to disperse long

(a)

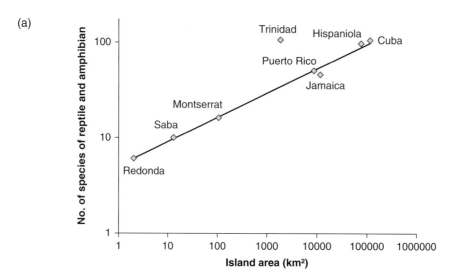

(b)

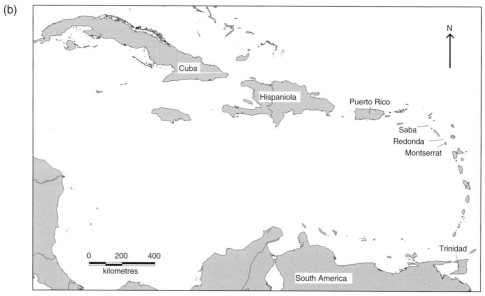

Figure 32.4   Species–area data for West Indian herptiles

(a) Scatterplot of the species–area relationship of herptiles (reptiles and amphibians) for some Caribbean islands. The fit-line shown excludes Trinidad, which appears to have far more herptile species than expected for its size. (b) Map showing the locations of the islands in (a). The map immediately suggests two possible explanations for the relatively high herptile species richness of Trinidad. First, it is further south than the other islands, and therefore nearer to the equator: it is a well-known trend for species richness to increase towards the equator. Second, Trinidad is very close to the South American mainland, while the other islands are not.

Sources: Data taken from MacArthur and Wilson (1967); base map taken from MapInfo Professional version 7.0 in-built map.

distances over salt water; amphibians in particular are poor at dispersing over the sea because they have semi-permeable skin and inundation in salt water leads to rapid dehydration. So it is very likely that Trinidad has more species than expected for its area because it has received and continues to receive more immigrants from South America than the other islands in the dataset. The geographical information, which is very obvious from looking at the map, is vital to this explanation.

Maps are also extremely useful for displaying results from statistical analysis or mathematical modelling of geographical data. For example, model predictions can be mapped. Residuals from statistical models (the errors: the difference between modelled values and actual values) should routinely be mapped as part of checking their assumptions and adequacy, and doing so often results in improved models.

Tables

There are two main types of table, as applied to quantitative data: data tables and summary tables. Summary tables can summarize large amounts of information, and should do so concisely, while being accurate, honest and clear. They are common in academic journal papers, books and lectures. There are also circumstances where very long, detailed data tables are required (e.g. tables of raw data in the appendices of dissertations, reports or journal papers). A guiding principle is to consider the needs of the reader as paramount. Think what the purpose of the table is and try to ensure that all attributes of the table help that purpose.

Data tables such as Table 32.1 are essentially spreadsheets whose purpose is usually to allow the reader to access the raw data, if required. This could be to allow checking of findings, looking for anomalies in the data, or to enable the raw data to be used in a meta-analysis or for some other analysis. The table should therefore be well organized, as a dataframe (see above), and as clear and uncluttered as possible. Rows and columns should be properly labelled and explained (good 'metadata'). When you have worked on something for some time, you become very familiar with it and it is extremely easy to take understanding of the data for granted. Good metadata will also help you if you ever need to revisit your dataset!

Summary tables are central to the presentation of results in most quantitative studies. The same principles of organization, explanation, clarity and precision apply as discussed above for detailed tables. However, in summary tables more thought is required to select what should be included and what excluded. Also, the number of decimal places (or significant figures) displayed for values in any one column should be (1) constant and (2) appropriate to the level of uncertainty of the data (see Taylor, 1982), especially Chapter 2, for a good explanation of uncertainty and how to report it).

Summary tables cannot be completed until some form of data processing and/or analysis has been done, which may simply be the calculation of descriptive statistics. For example, Table 32.2 gives the mean and standard deviation of pollutant concentrations measured for three streams on 10 different days. The key thing is to decide what information to include and what to omit. In Table 32.2, why give the standard deviations? Typically, pollutants are not a problem at very low concentrations but, above some threshold level, they may become toxic. Legislation often defines threshold levels, below which pollution levels are acceptable but above which they are not.

Let us assume that the legal threshold concentration for our pollutant is 10 units. The average level is clearly important, so it is reported, but this is not enough. In Table 32.2, if we were only to look at the means, we would conclude that the average state of each stream is within the acceptable limit. We might be concerned about stream B whose mean pollution level is only just below the threshold, but streams A and C appear to be well within acceptable limits. These would be incorrect conclusions! Look at the raw data (given in the legend of Table 32.2). While the average pollution level of stream B is relatively high, the 10 measurements are remarkably consistent, and based on these data a pollution concentration of 10 on any given day seems really quite unlikely. In contrast, the measured values are much more variable for streams A and C. The highest recorded value for stream A is 9.5, but based on the data it seems quite likely that values of 10 or more might occur reasonably often. Stream C, with the lowest mean concentration, actually has two of the 10 measured values above the critical concentration of 10, suggesting that toxic concentrations of the pollutant occur very often in this stream. If we work out the actual probabilities (assuming a normal distribution of the data), the chance of a concentration exceeding 10 on any given day is about 7% for Stream A, about 16% for Stream C, but only about 0.0003% for Stream B. Our conclusions obtained simply from looking at the means are therefore completely wrong.

The example just given concerns a general principle for presenting results: an associated measure of either variability or uncertainty should always be given when sample statistics (such as averages) are presented. Whether you present a measure of variability or one of uncertainty depends on the context. In the example in Table 32.2, standard deviation measures variability, which is appropriate here because the important thing is whether or not the threshold pollution level is likely to be exceeded. When inferential statistical analyses are performed, the focus is usually on the reliability of the parameter estimate (e.g. is a difference between two sample means 'significant'?), and so the appropriate measure is one of uncertainty, such as a confidence interval or the standard error of the mean. Further discussion of how to measure uncertainty is beyond the scope of this chapter, but it is important to note that all parameter estimates have associated measures of uncertainty. A good place to start learning about measuring and using uncertainty is Taylor (1982); also Chapter 33 of this volume.

It is very common to use summary tables to describe the results of statistical analyses, the outputs of mathematical modelling, or other quite complex procedures. *A key point*

Table 32.2   An example of a simple summary table

Here, sample means of pollutant concentrations (in parts per million) are presented, along with their associated standard deviations (std. dev.) and sample sizes (n), for a set of three streams, each sampled on 10 different days. A measure of variability or uncertainty should always accompany a sample statistic, as here. The fictional raw data behind these statistics are as follows. Stream A: 5.4, 3.4, 6.5, 8.6, 8.4, 9.5, 1.6, 5.5, 8.2, 3.8; Stream B: 9.1, 8.7, 9.2, 9.2, 9.2, 9.4, 8.8, 9.1, 9.1, 9.0; Stream C: 8.7, 3.1, 5.4, 3.3, 10.2, 6.0, 14.5, 4.4, 0.5, 0.0.

	Stream A	Stream B	Stream C
Mean	6.1	9.1	5.6
Std dev.	2.6	0.2	4.5
n	10	10	10

Table 32.3 An example of a summary table reporting the results of statistical analyses

This table reports the results of five simple, bivariate statistical analyses. Each analysis aims to account for variation in plant growth (g/year). The data come from a simulated experiment investigating effects of the mean temperature, rainfall, soil pH (all continuous variables), fertilizer addition and light level (both measured as categorical variables: fertilizer added or not; low, medium and high light levels). These data are used for many of the graphs in the following sections, and are available on the companion website. In the table, each statistical analysis (model) is summarized with: % of variation in the growth data accounted for by the variable (r^2); significance value for the model (P); number of data in the analysis (n); number of degrees of freedom used by the model (df); and the parameter estimates of the model ± 95% confidence interval (CI). Continuous variables were analysed using regression, which estimates the slope and the intercept of the best-fit line. Categorical variables were analysed using ANOVA, which estimates differences between means; the intercept for the fertilizer ANOVA is the mean growth of plants with fertilizer added and the difference is this value minus the mean growth of plants without fertilizer added.

Model	r^2	P	n	df	Parameter estimates	
					Slope ± 95% CI	*Intercept ± 95% CI*
Temperature (°C)	0.185	0.000	60	1	5.0 ± 2.7	3.5 ± 48.0
Rainfall (mm)	0.000	0.414	60	1	–	–
pH	0.000	0.550	60	1	–	–
					Difference ± 95% CI	*Intercept ± 95% CI*
Fertilizer	0.859	0.000	60	1	87.4 ± 9.2 g/yr	135.7 ± 6.5
Light	0.025	0.182	60	2	–	–

is that the same understanding of statistical procedures necessary for the analyses themselves should be used in deciding what to include in any summary table reporting their results. When designing summary tables, always ask yourself what information is necessary and what is not. Important information that is commonly left out includes sample sizes and/or degrees of freedom, significance values ('P-values') and measures of uncertainty. When you have decided what should go in the summary table, ask what is the most efficient way of presenting it. Table 32.3 is an example of a table summarizing statistical analyses, in this case for a dataset used in the remainder of this chapter. It efficiently summarizes the results of three regressions and two ANOVAs.

Graphical display

Graphical display is an excellent way of communicating results. When done properly, it is concise, memorable, persuasive and honest. For these reasons, graphics are almost always preferable to tables in oral presentations (Ellison, 2001). However, tables are generally more useful when exact values are important, and for certain types of information summary (e.g. reporting numerous statistical models simultaneously, as in Table 32.3). Graphical displays serve three main purposes: (1) preliminary exploration of patterns within data; (2) checking the quality and assumptions of statistical and mathematical models; and (3) communicating results to the reader or audience. The first two of these require accurate, revealing depiction of the data, but do not require great quality of presentation. They should be quick and easy to produce, and easy to interpret. Communicating results to the target audience, on the other hand, requires high-quality graphics. Again, clarity and ease of interpretation are important, though a relatively high level of complexity is often acceptable and necessary.

A wonderful example of a complex but very effective graphic is Minard's (1869) highly influential statistical depiction of Napoleon's disastrous Russian campaign of 1812–1813. This graphic presents six [or seven] continuous variables (longitude and latitude [and thus distance], time, temperature, number of men and direction of travel) as well as two categorical variables (army advancing vs. in retreat, and specific battles), yet is both visually striking and easy to interpret. Using modern enhancements, it becomes even more engaging, as in Landsteiner's (2013) interactive version.

It is important to understand the principles that underlie the graphical depiction of data. Tukey (1977), Cleveland (1985) and Tufte (1983, 1990) provide comprehensive discussions of the principles of graphing. A good summary can be found in Ellison (again note the link to understanding data structure and analysis):

> The question or hypothesis guiding the experimental design also should guide the decision as to which graphics are appropriate for exploring or illustrating the data set. Sketching a mock graph, without data points, *before* beginning the experiment usually will clarify experimental design and alternative outcomes ... Often, the simplest graph, without frills, is the best. However, graphs do not have to be simple-minded ... and they need not be assimilated in a single glance ... Besides the aesthetic and cognitive interest they provoke, complex graphs that are information-rich can save publication costs and time in presentations. (2001: 38)

Four principal guidelines for graphics may be recognized:

- Patterns of interest should be shown clearly.

- Graphics should be 'honest' – i.e. they should not distort, censor or exaggerate the data.

- It should be as easy as possible for the reader to read the data (as numbers) off the graphic.

- Figures should be efficient. That is, ink should be used only to show relevant information, and not unnecessary special effects ('chartjunk') like the 3-D effect in Figure 32.2. Try to attain a high 'data-to-ink ratio'.

Some of these principles are mutually reinforcing. In particular, efficient figures allow the patterns of interest to be shown most clearly. They also allow more space for annotations. Depending on the target audience, it may be desirable to annotate figures to point out particular features of interest (as in Figure 32.3); if this is done, it should again be efficient and avoid the use of gimmicks. All these principles apply to the mapping of data just as much as to other types of graphic. Tufte (1983: 51) summarizes the main principles of graphical data display: 'Graphical excellence is that which gives the viewer the greatest number of ideas in the shortest time with the least ink in the smallest space... And graphical excellence requires telling the truth about the data'.

Different types of graphic are, then, suitable for different types of data and should be chosen according to the type of relationship we are trying to show. Two common examples are: examining the distribution of data and exploring the relationships between variables. Before moving on to some of these graphics, some technical terms need explaining:

- The 'response variable' in a statistical analysis is the one under investigation – i.e. the one whose variation we are trying to explain. It is also known as the dependent variable, y-variable or ordinate.

- Each variable we use to try to account for variation in the response is called an 'explanatory variable', also known as an independent variable, x-variable or abscissa.

- The 'errors', otherwise known as the 'residuals', are the differences between the actual, observed values of the response variable and the modelled values; there is one such difference for each data point.

32.5 Examining the Distribution of Data

Graphs like those shown in Figure 32.5 depict the distribution of values of a *single variable* (usually a continuous one) within a sample. They are typically used for data exploration and model checking, but there are times when such graphs need to be presented formally. (The simulated dataset used for Figures 32.5–12 can be found in the online Appendix to this chapter: https://study.sagepub.com/keymethods3e)

The histogram

The most common graph type used to examine the distribution of data is the histogram (Figure 32.5a,b). Students often confuse histograms with bar charts (see below) because, to the beginner at least, they look similar. However, they are used to convey fundamentally different information. A histogram consists of categories of a variable along the horizontal axis and the number of data points of that same variable in each category on the vertical axis. The categories are often referred to as 'bins', using the analogy of physically sorting the (usually continuous) data into bins, each of which is for a different range of values. The number of data points in each category is known as the 'frequency'.

Histograms are often used to assess whether the distribution of a variable differs from the normal distribution. In Figure 32.5a, the data for plant growth rate are distributed more-or-less normally. To allow ease of comparison, the theoretical normal distribution curve is superimposed onto the histogram. Theoretical distributions can be represented as curves like this because an exact probability can be calculated for each possible value of the variable. Such curves are called probability density functions and can be thought of as histograms with infinitesimally small bin categories. In fact, the term 'density plot' is a more general one, which includes both histograms and curves representing theoretical distributions. Figure 32.5 panels b–e show a response variable whose measured values are not normally distributed (though the assumption of normality is not violated – see the companion website for why this is – https://study.sagepub.com/keymethods3e).

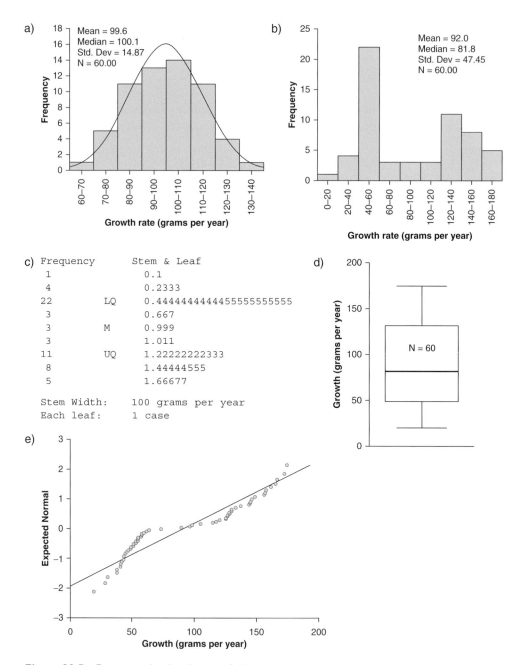

Figure 32.5 Examining the distribution of data in a single variable

(a) A histogram showing an approximately normal distribution for plant growth rate – a normal curve has been added to help compare the distribution of the data with the theoretical normal distribution. Histograms suffer several disadvantages. In particular, they hide the raw data, so data processing cannot be done from a histogram – such as calculating summary statistics (though the histogram can be annotated to include these, as here). Also, the division into categories ('bins') is arbitrary but can affect interpretation (Ellison, 2001). (b) A histogram showing a distinctly non-normal distribution, in this case very bimodal. (c) A stem-and-leaf plot, which is variant on the histogram; the main difference is that the raw data are displayed, not hidden. (d) A box-and-whisker plot, which is an efficient and informative way of displaying such data, but can tend to hide certain features of the data, particularly bimodality – as here. See text for further explanation. (e) A probability plot (this one is a Q–Q plot), displaying the actual data against the expectation from a given theoretical distribution (the straight line, here representing the normal distribution). Panels (c)–(e) show the same data as (b).

Stem-and-leaf plots

In Figure 32.5, panels c–e show different ways of depicting the distribution of the same data as Figure 32.5b. Figure 32.5c is a 'stem-and-leaf' plot, which is similar to a histogram (on its side), but has the advantage that the raw data are shown. Stem-and-leaf plots can be very useful, once you have got used to reading them. As with a histogram, the length of the 'bars' represents the number of data in the category or bin. Each of these data points is not only counted (as in a histogram) but also evaluated. The 'stems' are the first column of numbers, and represent the first figure(s) of the data values in the row. The 'leaves' are to the right of the stems, and comprise the next figure of the data value. In Figure 32.5c, the stems are in hundreds, and the leaves are tens. Thus, the lowest value in the dataset is somewhere between 10 and 20 (0×100 and 1×10). Its actual value is 19.2. The highest value reads off as somewhere between 170 and 180; its actual value is 174.2. As with a histogram, labels denoting the categories containing the mean, median and/or quartiles can be added to the plot, to provide further information, as in Figure 32.5c. However, most statistical packages do not offer the opportunity to superimpose normal distribution curves over stem-and-leaf plots, as they do for histograms.

Box-and-whisker plots

Figure 32.5d is a 'box-and-whisker' plot (or just 'boxplot'). The box in the middle spans the range between the 25th and 75th percentiles (respectively, the lower and upper quartiles or 'hinges') – in other words, the box shows the inter-quartile range. The line in the middle of that box shows the value of the median (the 50th percentile). In Figure 32.5d the SPSS default has been used, in which the whiskers extend at most a further 1.5 times the inter-quartile range, away from the box, stopping at the last data point within that range. Any points lying outside the range of the whiskers (beyond the 'inner fence') are 'outliers': points that are considerably different from the rest of the data. These outliers are sub-divided into two types. The less-extreme outliers lie no further from the end of the whisker than 1.5 times the inter-quartile range (the 'outer fence'). The more-extreme outliers are any points that are further still from the median. The two types of outlier, if present, are distinguished by different plotting symbols. Boxplots are extremely useful, information-rich graphics and are excellent ways of summarizing entire variables or categories within variables. However as ways of showing distribution of data, they suffer from the disadvantages of hiding data and tending to obscure bimodality (as demonstrated by Figure 32.5d).

Probability plots

Figure 32.5e is a probability plot. This is a very common way of examining how closely the distribution of a set of data conforms to a theoretical distribution. It is extremely useful for examining how well a statistical model conforms to the assumption of normality of errors. Probability plots are constructed so that the theoretical distribution plots as a straight, diagonal line, and the data are plotted as points. If the data were to conform perfectly to the theoretical distribution, all the points would lie on the line. In practice, of course, this does not happen. The key thing indicating difference from the theoretical distribution is systematic departure from the line, especially within the range in which most of the data lie (i.e. not so much at the two extremes). Some random scatter either side of the line is acceptable, but when the data clearly form a non-linear pattern (such as an S-shape, as in Figure 32.5e), it is a sure sign that the distribution is not normal – and that there is something wrong with the model, if the plot is being used to check errors for normality. Often (including in the case of Figure 32.5e), the pattern is caused by the influence of something not accounted for by the model, such as an influential but unmeasured variable.

32.6 Examining How Variables Relate to Each Other

Most quantitative studies in geography are concerned with trying to infer causation: we want to know what causes the patterns we see. Causation is determined by a combination of the research design and the interpretation of the results by the researcher or reader. It is not conferred by a statistical test; it is a common misconception among students that a significant regression indicates a causal relationship. Even so, good display of how variables relate to each other is essential to interpreting quantitative results.

Bivariate relationships: scatterplots, bar charts and boxplots again

The simplest case is the bivariate one: analysing the relationship between two variables. Often, there is a suggestion that the relationship might be causal, in which case usually the pattern being examined is within the response variable, and the possible cause of that pattern is measured via the explanatory variable. If both of these variables are continuous, by far the most useful type of graphic is the scatterplot (Figure 32.6). The explanatory variable goes on the x-axis and the response variable on the y-axis. Modelled relationships can be shown on scatterplots as lines, one of the simplest examples being a linear best-fit line. Scatterplots can also be used where no causation is being inferred – the focus instead being simply on whether there is any correlation between the two variables, and no regression line is added.

When one variable is continuous and the other categorical, scatterplots are not particularly helpful because they obscure many of the data (Figure 32.7a). If both variables are categorical, then there is usually little point in attempting to plot them against each other – devices such as tables of frequencies of each category combination tend to be more useful. Figure 32.7 shows perhaps the most common type of scenario: where the response variable is continuous and the explanatory variable categorical. Categorical explanatory variables are often called 'factors'.

There are many different ways of plotting relationships between factors and continuous response variables – where the focus is on whether the response variable differs significantly between categories. Scatterplots can be adapted to cope with factors by adding random variation along the x-axis (known as 'jitter'). This has been done in Figure 32.7b, in which the data are still clearly categorized, but the points are much less obscured than in Figure 32.7a. Another way of displaying factors, which tends to be more satisfactory, is the boxplot. In Figure 32.7c, a separate boxplot of the response variable (plant growth) is shown for each category within

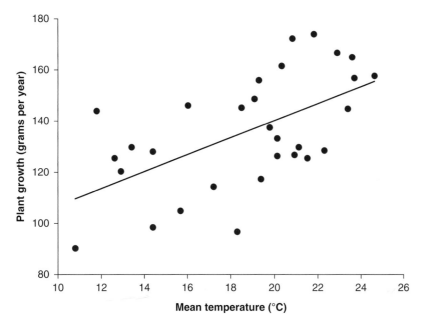

Figure 32.6 Scatterplot

This is one of the simplest and most useful of bivariate plots and one of the most commonly used. Values of one variable are spread along the x-axis, and those of another are spread along the y-axis. Both variables are usually continuous. By convention, the explanatory variable is the x-axis and the response variable is the y-axis (in cases where any sort of causality is inferred). A best-fit line from a simple linear regression is shown. This plot only shows data for plants to which fertilizer was added.

the factor. Alternatively, bar charts (or 'column charts') can be used to display the means, with error bars added to show the reliability of the estimates of the means; in Figure 32.7d, the 95% confidence intervals of the means are shown. Strictly speaking, the bars (columns) on the bar chart are unnecessary because the mean can be shown instead by a dot, but it is common practice to include them. Pie charts (Figure 32.7e) and stacked bar charts (Figure 7f) are more often used for percentage data, but are not recommended even for that: it is not easy to read the data off them, and it is hard to display any measures of uncertainty or variability on them. Indeed, Ellison (2001: 57) goes as far as to say 'I can think of no cases in which a pie chart should be used'.

Multivariate analyses: accounting for multiple possible causes

In most cases, a pattern will be caused by more than just a single influence. Analyses can be multivariate in the sense that they try to account for multiple influences on a given pattern: several explanatory variables are used to try to explain patterning in a response variable. An example could be the runoff rates of different streams in an area. The amount of rainfall, timing of rainfall, land-use types in the catchment and bedrock characteristics are just some of the things that are likely to affect runoff rates, and it

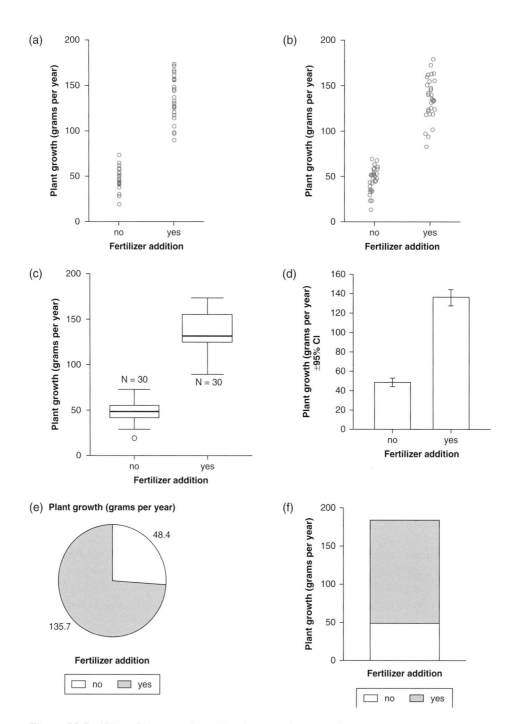

Figure 32.7 Ways of plotting relationships between factors and continuous response variables

A factor is a categorical explanatory variable. All the graphs show the same relationship, which is that between fertilizer addition and plant growth: (a) scatterplot; (b) scatterplot with jitter; (c) boxplots; (d) bar chart with error bars (representing the 95% confidence intervals); (e) pie chart; (f) stacked bar chart.

would be unrealistic to expect any one of these variables to explain all patterning in them. Our example of the growth of plants is also a case in point.

Sometimes, the pattern itself can be multivariate: too complex to measure as one single response variable. Instead, several attributes can be measured, to try to gain an overall picture. For example, we could measure a whole suite of attributes that could be construed as being related to the concept of quality of life, such as average income, equality of income, amount of free time, prevalence of disease and access to services. Numerous statistical techniques have been developed to allow analysis of multivariate patterns, including dimensionality-reducing procedures such as principal components analysis. These are beyond the scope of this chapter, but the reader is referred to statistical texts such as Crawley (2005), Pallant (2013) or the manuals of most standard statistical packages – I find the documentation for R particularly useful (http://www.r-project.org/).

Multiple influences on a single response variable can be analysed using extensions of the simple, bivariate techniques of ANOVA and regression. Again, it is not the purpose of this chapter to discuss these statistical methods, but some of the issues related to such analyses are relevant to this section. Two issues are especially important: (1) accounting for other influences when illustrating any given relationship; and (2) displaying interactions between explanatory variables.

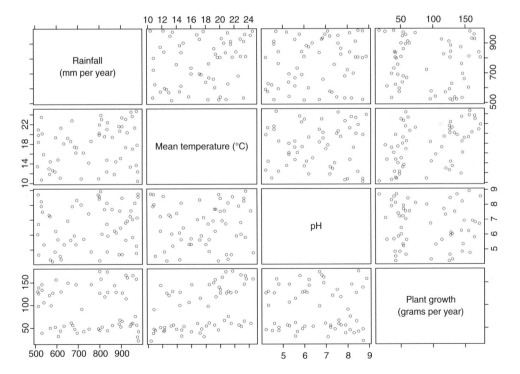

Figure 32.8 Scatterplot matrix

As its name suggests, this type of plot is a matrix of bivariate scatterplots. The variable names are listed in the diagonal, and apply to the vertical axes of the plots in the row in which the name is located, and the horizontal axes of the plots in the column containing the variable name.

The scatterplot matrix

When a pattern results from more than one cause, a simple bivariate plot of the response variable against one of the explanatory variables will not give the true picture of the relationship we are trying to show. We must try to remove variation accounted for by other variables, before we can display the 'true' modelled relationship. An extremely useful tool for the initial stages of exploring such higher-dimensional data is the scatterplot matrix (Figure 32.8). This comprises a bivariate scatterplot of each possible combination of two variables from the list of those under consideration. Usually, as in Figure 32.8, each of these plots is displayed twice: one way above the diagonal and the corresponding plot with the axes transposed below the diagonal.

Partial plots

Scatterplot matrices are very useful for exploring data and informing statistical analysis and mathematical modelling, but they do not achieve the goal of accounting for variation attributable to other putative causes when displaying a particular relationship. To achieve this, other graphical methods are commonly used. One is the partial plot (Figure 32.9), in which the response variable is re-calculated, 'correcting for' the modelled effects of other significant explanatory variables. This is very useful for interpreting the modelled effects of particular variables in relatively complex statistical

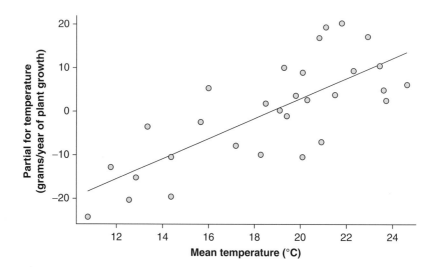

Figure 32.9   Partial plot

This shows the relationship between the two variables, after accounting for the modelled effects of other explanatory variables on the response variable. The response variable here, as before, is plant growth. This is a partial plot from a model in which the explanatory variables are rainfall, temperature and light, and in which only the plants given fertilizer are considered. The modelled effects of rainfall and light have been corrected for, and the y-axis therefore displays the change in plant growth directly related to temperature (according to the model). This plot is directly comparable to Figure 32.6; note how the regression line describes the data better than the one in Figure 32.6.

analyses, and as such one of its main uses is in model development. It can also be an effective way of presenting the meaning of a model to a viewer, but the scale of the response variable needs careful explanation.

3-D plots, clustered graphs and the separation of categories

It is possible to use 3-D plots to good effect, but they only allow the plotting of one extra variable, so their potential usefulness is limited. Also, it is typically hard to read off the data from such plots, because of the difficulty of projecting a 3-D plot onto the 2-D medium of a page or a screen. Online interactive plots can overcome these limitations, but in the absence of this facility I would rarely recommend a 3-D plot.

Another common method is the separation of categories in graphs. This can be achieved by plotting separate graphs for the different categories of a factor (e.g. separate graphs for men and women), or different symbols on the same graph with separate best-fit lines if appropriate (Figure 32.10). Using different symbols on the same graph is better for comparing the categories.

The use of different plotting symbols on the same graph for different categories of a factor is also a good way of illustrating an interaction between a continuous and a categorical explanatory variable – a form of statistical interaction. Statistical interactions are when the modelled effect of an explanatory variable on the response variable depends on the level of another explanatory variable. Figure 32.10 shows an example,

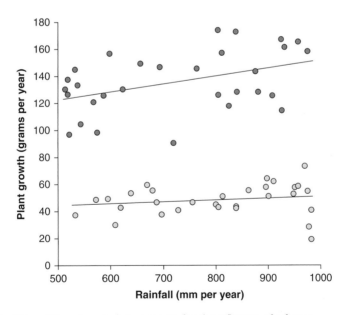

Figure 32.10 Using different symbols to account for the influence of a factor

Here, the relationship between rainfall and plant growth is shown separately for plants to which fertilizer has (closed circles) and has not (open circles) been given. If all the points were plotted with the same symbols and separate best-fit lines were not shown, it would appear that there is no relationship between rainfall and plant growth.

in which the relationship between rainfall and plant growth depends on whether or not fertilizer has been added. Where fertilizer has not been added, there is no significant relationship between rainfall and plant growth, but there is a positive relationship for plants given fertilizer. This interaction is statistically significant ($P = 0.002$). These results would be consistent with a situation in which nutrients, not water, represent the limiting factor for plant growth.

Where a statistical interaction involves two or more categorical explanatory variables, clustered graphs are often the best way of displaying the interaction. Figure 32.11 is an example of a clustered boxplot, which shows an interaction between the modelled effects of light and fertilizer on plant growth.

Conditioning plots and other types of graph

Statistical interactions may involve two or more continuous explanatory variables. By extension of the idea of the best-fit line on a 2-D scatterplot, a two-way interaction of this sort can be displayed as a best-fit surface within a 3-D scatterplot. Again, this suffers the problem of difficulty in reading off data. In addition, such a plot tends to imply, usually unrealistically, that all parts of the surface represent the valid range of

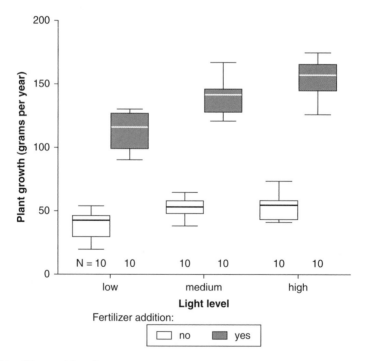

Figure 32.11   Clustered boxplot

This shows an interaction between the light and fertilizer treatments in their relationships with plant growth. In other words, the amount of increase in plant growth with increasing light levels depends on whether or not fertilizer has been added. The corollary is that the degree of difference in plant growth between the two fertilizer treatments depends on the light level. The interaction is statistically significant ($P = 0.0004$).

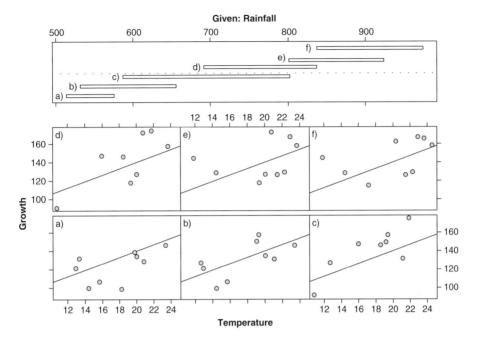

Figure 32.12   Conditioning plot (or coplot)

This examines the relationship between temperature and plant growth for the plants with fertilizer added, and how this relationship varies with rainfall. The plot comprises separate scatterplots of plant growth versus temperature, each of which corresponds to a different range of rainfall. The rainfall ranges are indicated by the top panel, which shows that the bottom-left graph in the bottom panel (labelled 'a) only includes sites that have annual rainfall between about 515 and 570 mm/year. The top-left scatterplot (d) only includes sites with rainfall in the range 690–835 mm/year, and so on. The dotted line in the top panel indicates the move from the graphs in the bottom row of the lower panel (a–c) to those at the top (d–f). There is no significant statistical interaction between temperature and rainfall in their modelled effects on plant growth for these data, so the fit lines drawn are for the overall regression – i.e. they are the same in all the graphs.

the model; in fact, some parts of graphs like this are typically data deficient. Instead, 'conditioning plots' (or 'coplots') are often better ways of showing these interactions. Figure 32.12 is an example of a coplot. The range of values for one of the explanatory variables is split into segments (rather like the bins of a histogram), and the data for each segment are plotted as separate 2-D scatterplots. This shows the way that the relationship between the x and y variables changes across the range of the other explanatory variable. In this case, there is no significant interaction; if there were, the regression lines would have different slopes from each other, indicating that the relationship between temperature and plant growth depends on the amount of rainfall. Advantages include the ability to add more explanatory variables, thereby being able to show higher-order interactions (e.g. three-way). Disadvantages include the fact that such plots focus on one of the explanatory variables, rather than being an equal depiction of all simultaneously.

Many other types of plot exist – triangular and time-series plots to name but two. The interested reader should refer to more specialist texts such as Tufte (1983),

Waltham (1994) and Heer et al. (2010) to learn more. Note also that some of the graphical techniques discussed above are not mutually exclusive. For instance, in trying to illustrate a very complex statistical model, we could draw separate graphs (to deal with the categories of one explanatory variable), each of which is a partial plot with different plotting symbols. All of these devices are aimed at illustrating most effectively the significant relationships found in the data analysis. These modelled effects are our best guess about the cause(s) of the pattern in which we are interested. In other words, when presenting results in graphs and tables, we usually try to adhere to the Shaker maxim: form follows function (Ellison, 2001).

32.7 Explaining and Interpreting

Figure 32.1 illustrates how our attempts to understand real-world complexity by quantitative study involve various levels of abstraction, all linked together by the 'language' of mathematics. Figure 32.1 also indicates that graphs, tables and maps lie right at the heart of quantitative treatment of data. When trying to explain patterns, and present and interpret results, we should use these illustrative tools to maximum effect. Graphs, in particular, can allow us to combine raw data, descriptive statistics and the output of statistical or mathematical models within a single illustration. (Note that the output of statistical modelling has been included in many of the graphs above – e.g. the regression lines, and also the legend of Figure 32.11 shows that the interaction was statistically significant, with a P-value of 0.0004.)

Each quantitative study is unique, and much thought is required to interpret the results and their implications, and to consider the generality of the findings. So far, our emphasis has been on presenting data and the results of analyses. The flip side of this is our reaction to illustrations presented by others. An understanding of data exploration and presentation processes (including the principles of good illustrative practice) is important for interpreting the meaning and context of work presented to us. We need to be able to judge quickly how well quantitative work has been done, what the value and implications of the results are and, indeed, whether we have been given enough information to make such judgements. When we are presented with a graph, we should seek to interpret it in as full and critical a way as possible. In doing this, we should consider the extent to which form follows function. Recall that illustrations are typically used to imply that the pattern we see (the form) results from the causes depicted (the function).

Returning to species–area relationships, in Figure 32.13a the implication is that greater area of islands tends to cause greater plant species richness. The regression line goes much further. It suggests that the increase of plant species richness with area is highly predictable: it follows a linear relationship on a log–log scale. What does this actually mean? Looking at the graph suggests that a tenfold change in island area is associated with much less than a tenfold change in species richness. The equation of the line quantifies this change in plant species number as 2.35-fold, which is consistent with the rule-of-thumb used by many conservationists that a 90% loss of habitat leads to a 50% loss of species. Figure 32.13b shows the same data on linear axes. The solid line is the same best fit line as in Figure 32.13a and represents a power relationship. The dashed line shows an alternative model for

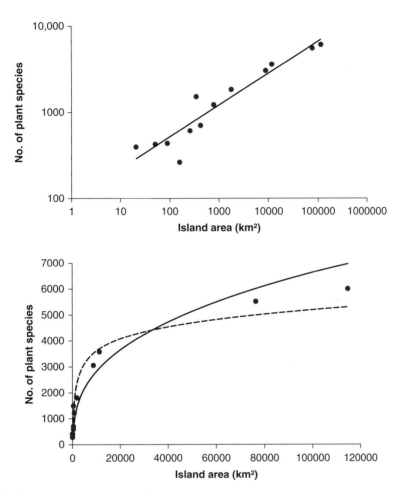

Figure 32.13 Species–area relationship

These two graphs show the same data as Figure 32.3 (plant species numbers plotted against island area for Caribbean islands), except that Cay Sal Bank has been removed as an outlier. Plot (a) is on log–log axes, while (b) is on untransformed axes. The solid line in both graphs shows the best-fit power model. The equation for this is y = 92.434$x^{0.371}$, and the model accounts for 90% of the variance in plant species richness (r^2 = 0.896). The dashed line in (b) is the best-fit logarithmic model: y = 693.3 ln(x) – 2810.5. This model accounts for 91% of the variance (r^2 = 0.913).

describing the data: a logarithmic function. There has been debate for decades in the biogeographical literature about which of these two models better represents the species–area relationship.

Further exploration of the two curves shown in Figure 32.13b gives greater insight into the relative merits of the two models on which they are based – for this dataset, at least. In statistical terms, the logarithmic model provides a very slightly better fit with the data, accounting for 91.3% of the variance in plant species numbers, compared with 89.6% for the power model (though we must bear in mind that the sample size is small). But how do the two models compare on theoretical grounds? It is clear from the graph that the two models predict vastly different numbers of species (in absolute terms) for large islands. It is also clear that, if we extrapolate beyond the range of the data to even larger islands, this difference will increase rapidly. Such differences could be very important if we wish to use these data to predict for other islands or if we wish to predict the effect of changes in habitat area on species numbers across continents. What is less clear from Figure 32.13b is what happens if we extrapolate beyond the range of the data in the other direction: to smaller islands. The equation for the logarithmic model yields negative species numbers for islands smaller than 57 km², spectacularly so for very small islands; for a 1-ha island, the predicted number of plant species is –6003. That is clearly nonsensical, and suggests that extrapolation in this direction is flawed in the case of the logarithmic model. The power model, in contrast, predicts 17 plant species on an island of 1 ha, and this model reaches 0 species at an area of 0 km² (undefined for the logarithmic model). At least such predictions are plausible! On the basis solely of these results, and despite the slightly poorer statistical fit, we might conclude that the power model is more theoretically sound and more likely to be useful as the basis for a general model of the species–area relationship than the logarithmic model.

This example raises a number of issues. For a start, it brings to the fore the differences between interpolation, extrapolation and prediction. Relationships that are qualitatively very different can look very similar in certain parts of their ranges. Figure 32.14 in the companion website shows fundamentally different relationships, all of which look very similar in the highlighted range (https://study.sagepub.com/keymethods3e).

The species–area example also raises the issue of what to do with outliers. Notice the difference in data between Figure 32.3 and Figure 32.13: Cay Sal Bank has been excluded from the data in Figure 32.13. It was clearly an outlier in Figure 32.3, but this alone is not sufficient reason to exclude it when we are interested in trying to explain the real world. In fact, there are good theoretical reasons to omit the Cay Sal Bank data. Unlike all the other data, this set of islands in the Bahamas is little more than coral reef and so is both qualitatively different from the other data points, and has a high degree of uncertainty attached to the measurement of its area (which changes greatly in every tidal cycle). The important point here is that the exclusion of this data point makes a big difference to the models' fit to the data.

Perhaps most importantly, the species–area example also demonstrates the important difference between a statistical fit (e.g. a regression line) and an underlying causal relationship. The fit of both regressions to the data is strong, but this does not prove that there is a causal relationship. Is the geographic pattern symptomatic of one or more geographic processes? In the above analysis, no mechanisms linking changes in area to changes in species numbers have been investigated, so the answer to this question is not clear. What is clear, though, is that we have to be very careful in deciding what we can, and cannot, conclude from any given study.

32.8 Conclusion

The enormous range of data exploration and presentation techniques available to geographers gives great scope for flexibility and power, but also for misunderstanding and deception. It is common knowledge that 'statistics' can be very misleading (the word 'statistics' here is usually used to mean facts and figures, rather than the parameter estimates or probability statements that emerge from statistical analysis or mathematical modelling). The same is true of models, graphs, maps and tables: the way that data are presented can seriously affect our inferences and conclusions. Models and illustrations can be used for deliberate deception or distortion of results, perhaps for a political purpose – Tufte (1983) gives many examples of this kind of deception – but in academic study it is vital that we strive to avoid such purposeful distortion. Unbiased exploration of data and honest graphical and numerical presentation of findings should be key aims of any quantitative study. Standard, repeatable techniques are used where possible, but there are always subjective decisions involved and it should be remembered that each graphic is both an abstraction of the 'real world' (a construct of the person producing it) and a simplification of reality. Further, misleading representation of data often occurs unintentionally through incomplete or incorrect analysis of data. The risk of this can be minimized by thoroughly exploring the data and carefully inspecting the outcomes (model fits, patterning in residuals, maps of predictions, etc.), both of which are best done graphically.

Carefully exploring and presenting data with graphical techniques are, therefore, fundamental activities in quantitative geography. Graphics provide not only our initial view of our own data (data exploration) but also the entry point for those reading and evaluating our results. Statistical analysis methods are arguably useful mainly to the extent that they can validate what we display in our graphics: as in Figure 32.11, we show the main results graphically and then use the results of the analyses to make it clear that what we see is significant. It is not exaggerating to say that, when we are working with quantitative data, our illustrative material forms the centrepiece around which our project reports, dissertations or research papers revolve.

We live in exciting times! New methods of visualising data are being developed all the time, facilitated by the extreme, sustained rises in computing power and data storage capacity. As research and teaching move increasingly online, we can expect both to increasingly harness the potential for interactivity. Some examples already exist, such as Field and Horton (2010) and Landsteiner (2013), but we are only just beginning to exploit the ability of the virtual environment to bring data to life.

SUMMARY

- Plotting the data in a variety of ways, in conjunction with calculating summary statistics, forms the basis of exploratory data analysis, a very powerful way of examining trends in data quickly and easily.
- Statistical analysis should typically include the use of graphs and maps to help examine the nature of the model fit and to check the validity of the underlying assumptions.
- Graphical display of the results is typically by far the best way of communicating the information in an elegant, concise and accurate way. This graphical display usually should incorporate elements of the statistical analysis, such as error bars (or other measures of uncertainty or variability), fit lines and indications of statistical significance on the graphs.
- Whatever we do, we need to be aware of the fact that all our data-presentation tools are constructs, and all our models are wrong to at least some extent. The important question is: how far from the truth are they?

Further Reading

- John Tukey was instrumental in developing and promoting useful, accessible ways of exploring and presenting data. His book (Tukey, 1977), although a bit idiosyncratic, is still well worth a read. The book edited by Fox and Long (1990) covers the topics I have discussed in much more detail. Tufte (1983) is a clearly written book containing much wisdom about the use and presentation of data; it encourages careful, honest and imaginative display of information and is a very interesting and informative read. The short chapter by Ellison (2001) is much more focused graphics in research (with particular emphasis on experimental work in ecology) than Tufte (1983). Although I do not agree with everything written in Ellison (2001), I strongly recommend it, along with much of the rest of the book containing it (Scheiner and Gurevitch, 2001) – even for geographers with little interest in ecology. Wainer (2005) and Wainer (2009) are written for a more popular (American) audience. The 2005 book covers the history of graphical display, as well as colourful examples of its use in the modern world, and is well worth reading. The 2009 book focuses more on the important topic of depicting uncertainty, while also extending into other areas of data presentation and interpretation. Heer et al. (2010) is a very up-to-date article on visualization techniques, and includes interactive graphics.

- Taylor (1982) provides a good introduction to error (uncertainty): how to estimate it, cope with it and report it. Even though it is written for physicists, it is a valuable read for undergraduate geographers. Taylor also has a useful section on dealing with outliers (Chapter 6, 'Rejection of data').

- Although an old book, Thornes and Brunsden (1977) contains much that is still important, including a good, short section on interpreting graphs with respect to the way that physical systems operate (in Chapter 7). Sayer (1992) provides a useful discussion of abstraction – from a social science viewpoint.

- Finally, it is very important to get at least some grasp of statistical methods. There are numerous books dealing with this, though not many specifically focus on geography. Crawley (2005) provides an excellent introduction to statistical techniques, starting from first principles and using the open-source statistical analysis package 'R', which is now the industry standard in most scientific disciplines. This is a general book, not just for geographers, but in my opinion it is hard to find a better statistics text. Zuur et al. (2009) provides a beginner's guide to R, which focuses heavily on learning the extremely powerful graphing capabilities of the software. Many social sciences students just starting statistics find Pallant (2013) and the previous editions very helpful. Rogerson (2014) also has useful material, especially on spatial statistics.

Note: Full details of the above can be found in the references list below.

References

Cleveland, W.S. (1985) *The Elements of Graphing Data*. Monterey, CA: Wadswork Advanced Books & Software.

Crawley, M.J. (2005) *Statistics: An Introduction Using R*. Chichester: Wiley.

Ellison, A.M. (2001) 'Exploratory data analysis and graphic display' in S.M. Scheiner and J. Gurevitch (eds) *Design and Analysis of Ecological Experiments* (2nd edition). Oxford: Oxford University Press. pp. 37–62.

Field, R. and Horton, J. (2010) *Statistics – an Intuitive Introduction*. Interactive online book, www.nottingham.ac.uk/toolkits/play_244 (accessed 07 December 2015 [in the process of being converted to Java-based]).

Fox, J. and Long, J.S. (eds) (1990) *Modern Methods of Data Analysis*. London: Sage.

Frodin, D.G. (2001) *Guide to Standard Floras of the World: An Annotated, Geographically Arranged Systematic Bibliography of the Principal Floras, Enumerations, Checklists, and Chorological Atlases of Different Areas* (2nd edition). Cambridge: Cambridge University Press.

Heer, J., Bostock, M. and Ogievetsky, V. (2010) 'A tour through the visualization zoo: A survey of powerful visualization techniques, from the obvious to the obscure', ACM Queue, 8(5): 20–30. (With interactive graphics using the Protovis javascript library.)

Kopf, E.W. (1916) 'Florence Nightingale as a statistician', *Journal of the American Statistical Association*, 15: 388–404.

Landsteiner, N. (2013) 'Charles Joseph Minard: Napoleon's retreat from Moscow (The Russian Campaign 1812–1813): an interactive chart', http://www.masswerk.at/minard/ (last accessed 1 December 2015).

MacArthur, R.H. and Wilson, E.O. (1967) *The Theory of Island Biogeography*. Princeton, NJ: Princeton University Press.

Minard, C.J. (1869) 'Carte Figurative des Pertes Successives en Hommes de l'Armée Française dans la Campagne de Russie 1812–1813'. Paris.

Pallant, J. (2013) *SPSS Survival Manual: a Step by Step Guide to Data Analysis Using IBM SPSS* (5th edition). Milton Keynes: Open University Press.

Playfair, W. (1786) *The Commercial and Political Atlas*. London: Corry.

Playfair, W. (1801) *Statistical Breviary*. London: Wallis.

Porter, T.M. (1986) *The Rise of Statistical Thinking, 1820–1900*. Princeton, NJ: Princeton University Press.

Rogerson, P.A. (2014) *Statistical Methods for Geography: a Student's Guide* (4th edn). London: Sage.

Sayer, A. (1992) *Method in Social Science: A Realist Approach* (2nd edition). London: Routledge

Scheiner, S.M. and Gurevitch, J. (eds) (2001) *Design and Analysis of Ecological Experiments* (2nd edn). Oxford: Oxford University Press.

Snow, J. (1854) *On the Mode of Communication of Cholera* (2nd edition). London: John Churchill.

Spence, I. and Lewandowsky, S (1990) 'Graphical perception' in J. Fox and J.S. Long (eds) *Modern Methods of Data Analysis*. London: Sage. pp. 13–57.

Taylor, J.R. (1982) *An Introduction to Error Analysis*. Mill Valley, CA: University Science Books.

Thornes, J.B. and Brunsden, D. (1977) *Geomorphology and Time*. London: Methuen.

Tufte, E.R. (1983) *The Visual Display of Quantitative Information*. Cheshire, CT: Graphics Press.

Tufte, E.R. (1990) *Envisioning Information*. Cheshire, CT: Graphics Press.

Tukey, J.W. (1977) *Exploratory Data Analysis*. Reading, MA: Addison–Wesley.

Wainer, H. (2005) *Graphic Discovery: a Trout in the Milk and Other Visual Adventures*. Princeton, NJ: Princeton University Press.

Wainer, H. (2009) *Picturing the Uncertain World: How to Understand, Communicate, and Control Uncertainty Through Graphical Display*. Princeton, NJ: Princeton University Press.

Waltham, D. (1994) *Mathematics: a Simple Tool for Geologists*. London: Chapman & Hall.

Zuur, A.F., Leno, E.N. and Meesters, E.H.W.G. (2009) *A Beginner's Guide to R*. New York: Springer.

ON THE COMPANION WEBSITE...

Visit **https://study.sagepub.com/keymethods3e** for author videos, chapter exercises, resources and links, plus **free** access to the following recommended articles:

1. **Gaston, K.J. (2006) 'Biodiversity and extinction: Macroecological patterns and people', *Progress in Physical Geography*, 30 (2): 258–69.**

This demonstrates that statistical methods are routine in that area of geography (biogeography), and that a basic understanding of data presentation is important for understanding the literature; it also includes discussion of species–area relationships, which are used as an example throughout this chapter.

2. **Song, C., Dannenberg, M.P. and Hwang, T. (2013) 'Optical remote sensing of terrestrial ecosystem primary productivity', *Progress in Physical Geography*, 37 (6): 834–54.**

This has a nice example (Figure 2 of the paper) of how choices made in presenting data are very important for interpretation – specifically, how to plot the temporal scale in a time series.

3. **Elwood, S. (2011) 'Geographic Information Science: Visualization, visual methods, and the Geoweb', *Progress in Human Geography*, 35 (3): 401–8.**

This focuses on the presentation of data, including new forms of information from the geoweb, for the purposes of visualisation.

33 Case Study Methodology

Liz Taylor

SYNOPSIS

Case study is a methodology which is ideal for small-scale, in-depth research. Bassey (1999: 47) sums up case study as 'a study of a singularity conducted in depth in natural settings'. It is a flexible research strategy that can fit with a range of theoretical perspectives. This chapter firstly introduces case study as a methodology, including different types as categorised by some key proponents. Next, the key issues of choosing, bounding, rigour and generalisation are considered.

Case study research is characterised by use of multiple data collection methods, employed to obtain a range of perspectives and insights into the case. There are various options in data analysis but this should be an iterative process balancing breadth and depth. Methods of data collection and analysis commonly used in case study will therefore be introduced, before finally turning to output from case study research. Case study reports can take a number of forms, but the intention is to communicate an in-depth understanding of the particular instance studied.

This chapter is organised into the following sections:

- Introduction
- Case study as a research strategy
- Key issues in case study research
- Data collection methods
- Data analysis methods
- Communicating case study research
- Conclusion

33.1 Introduction

Case study is a research methodology; an overall strategy which encompasses a range of empirical data collection and analysis methods. Most case study research includes data from observation, interviews and document review. It is a pragmatic and flexible methodology which can fit with a range of different aims. It is well suited to small-scale, in-depth studies, such as those that form the base of undergraduate dissertations.

However, at the other end of the scale, multiple case studies can also be used to complement large-scale surveys as part of mixed methods research.

Confusingly, the outcome of case study research also tends to be called a case study (Stake, 2005). This is, however, different from a case study written for a textbook, which is simplified for teaching purposes and may not be based directly on original research (Taylor, 2013a). Case study as an explicit methodology has been less prominent in geography research than in some other fields, but it has significant potential to be useful. It is well established in educational research, also in management studies.

33.2 Case study as a research strategy

It is not straightforward to distil the characteristics of case study research, because the key writers on case study (in particular I will refer to Yin, Bassey and Stake) come from a range of research backgrounds and each has their distinctive approaches and ways of constructing case study. All agree that case study is a form of naturalistic research, that is the case is being studied in its 'normal' context, rather than in a laboratory or other artificial environment (Bassey, 1999: 60). In addition, the timeframe of a case study is normally contemporary to the researcher (Yin, 2014: 12), so this approach is not suitable for historical research, although historical documents may, of course, provide context in the case study.

In any research project, is it important that there is congruence between the aims of the research, the questions being asked, the theoretical perspective, the methodology and the methods of data collection and analysis selected (Figure 33.1). For example, it is inappropriate to espouse the aims and approaches of ethnography, and then to conduct a controlled experiment as fieldwork. Case study as a methodology is not necessarily tied to a particular theoretical perspective or epistemological approach. However, different researchers are characteristically interested in different sorts of questions about the world, and often case study research is used by researchers who work within a constructivist/constructionist epistemology (that is they consider meaning to be constructed through interaction between subject and object, see Crotty, 2003) and are informed by an interpretivist theoretical perspective. This focuses their interest on understanding the complexities of a particular situation and the meaning that is constructed within it, without a pre-existing political commitment or key intention to uncover and challenge unequal power relations. Having said that, case study could also be conducted from a critical or positivist perspective. As case study methodology involves taking a holistic view of the case, as well as exploring many aspects of the case in depth, it is easy to see how it would be popular with researchers who want to learn about the meanings people construct within a complex socio-cultural context.

Different writers on case study have their own systems for classifying types of case study, based on the aims of the study, as summarised in Table 33.1.

There are some intersections between the different ways of classifying case study research shown in Table 33.1, for example between Stake's descriptive case study and Bassey's story-telling case study (Bassey, 1999: 27–30; 62–4). However, we also need

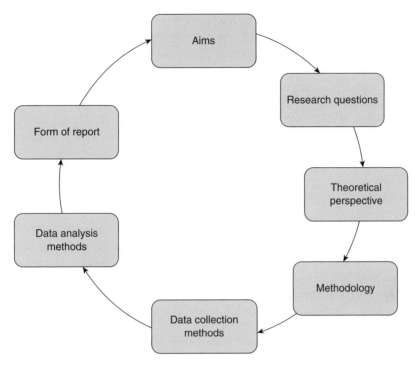

Figure 33.1 Congruence between different elements of the research

to exercise caution in making such links. As Bassey (1999: 35) comments, 'I cannot be certain that I have correctly elicited what these writers have meant by the terms they have used and, dare I say, neither can we be sure that these writers themselves had clear, unambiguous concepts in their minds and managed to express them coherently'. As always, classification systems provide useful labels to ease communication, but we must remember the variety and complexity of instances within each category. Any particular project may fall easily into a category, or sit uncomfortably between two, but it is helpful to spend time reflecting on the type of aim for a case study, as this will affect the questions asked and methods chosen.

Table 33.1 Types of case study as classified by three key proponents of the methodology

Yin (2014, p.238)	Bassey (1999, pp. 62–64)	Stake (1995 pp. 3–4; 2005 p. 445)
• *explanatory* (understanding how or why something came to be) • *descriptive* (describing a phenomenon within its context) • *exploratory* (considering what questions could be helpfully asked in a particular case or what procedures could be used)	• *theory-seeking/theory-testing* (case is expected to be somehow typical of something more general) • *story-telling/picture-drawing* (analytical accounts of cases designed to illuminate theory) • *evaluative*	• *intrinsic* (focus on case for its own interest) • *instrumental* (case as it illuminates a wider issue)

33.3 Key Issues in Case Study Research

Choosing and bounding the case

As the name suggests, case study revolves around the particular case, so a significant amount of thought has to go into choosing and distinguishing that case. There is much discussion in the literature about the choice of single versus multiple case studies. Yin (2014: 51–3) suggests that choice of a single case can be justified if that case is *critical* (meets all the conditions for testing a theory); *unusual*; or extreme (which may then shed light on the everyday); *common* or everyday (to understand more about normal circumstances and conditions); *revelatory* (in which the researcher has an unusual opportunity to access a phenomenon); or *longitudinal* (studied at two or more time points). There is a level at which all cases are unique, but it is important to think through the extent to which you are interested in your case study giving insight into a wider issue or phenomenon. This degree of focus on a wider issue is the difference between Stake's intrinsic and instrumental case study, as shown in Box 33.1. If your main interest is in a wider issue, then more than one case study across different sites is likely to give a richer insight. However, there will inevitably be a depth versus breadth trade-off, so a researcher with a smaller amount of time and resources available will often focus on a single, in-depth case study, and rely on cumulation to and from other people's work to build knowledge of the wider issue within the research community.

To find a suitable single case may well entail significant prior investigation, or it could be that the case is one to which the researcher already has privileged access (perhaps because they already live or work in that setting, or are involved with organising a particular event). It can take more time than might be initially imagined to locate and identify the case. For example, if a researcher's aim was to understand distribution practices in online bookstores, would it be best to focus on one or more firms? Would the case study consist of the company, the distribution section of that company, six distribution workers or one distribution worker? Much would depend on knowing more detail about the aims of the study and the envisaged methods. This type of initial planning involves an iterative process of refining the aims, research questions, theoretical perspective and methods of the research. Indeed, the process of refining is frequently continued into the empirical and writing stages of the work, as questions continue to be adjusted to reflect more accurately the research aims.

A case is usually seen as a concrete entity bounded within space and time, as Stake suggests: 'a specific, a complex, functioning thing' (1995: 2). So, for example, a case might be a person over one week, or an institution over three months, or an event over a weekend. Abstract ideas like 'power' are not normally seen as cases, though we might learn about a particular person's construction and deployment of power through a case study of that person in their workplace. Although it may sound simple, establishing the boundaries of a case is not always easy. As Goode and Hatt (cited in Stake, 2005: 444) recognise, a person is hard to bound: 'it is not always easy for the case researcher to recognise where the child ends and the environment begins'. This recognises the complex web of relational links which people have with their living and non-living environment.

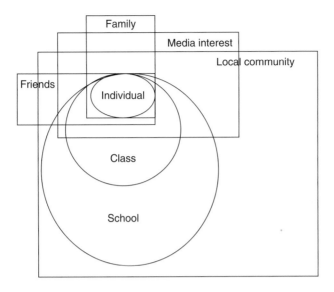

Figure 33.2   Nesting and intersecting groups within one case

---

Box 33.1   The Case as Space

Following a case study of one class of 13- to 14-year olds studying Japan in their geography lessons, I reflected on how the case could be constructed as space, and how that could inform methodology and methods (Taylor, 2013b). Using Doreen Massey's relational conceptualisation of place as a bundle of trajectories (Massey, 2005), I noted that:

- *the case may be bigger than we think* – a reminder to consider non-living, animal and plant elements of the case as well as people and for the researcher to reflect carefully on their own positioning in relation to the case.
- *the case is relationally complex* – if the case can be bigger than we think, then the level of inter-relationship between elements is potentially substantial. Each individual belongs to a series of nested and intersecting groups on a number of scales (Figure 33.2). The uniqueness of a case (as with a place) derives from its contemporary and historical links.
- *the case is radically dynamic* – each element of the 'bundle' of trajectories has come from somewhere and will go somewhere, over a range of timescales. To a greater or lesser degree, each element will also change and/or cause change. The case study researcher must ensure that their data collection methods are sufficiently sensitive to change, even over a short-term project.
- *the case involves radical heterogeneity* – whilst diversity within any group may not be immediately obvious, it will be there. Again, data collection methods need to be sufficiently sensitive to this.
- *the case is political* – all relationships have a power dimension. Even with an interpretivist project, the researcher must be alert to power relations within a case, particularly with regard to their own positioning. The practices of ethics are a response to inequalities of power.

Quality and rigour in case study research

All researchers want their work to have value. However, there is a considerable variety of ways in which value is defined. Each methodology has one, or a series, of established concepts and practices which are agreed (to whatever extent) by researchers to signal quality and rigour. The criteria for evaluating quality in a controlled experiment will necessarily be different from those in an ethnographic study. As case studies may be conducted from more than one theoretical perspective, different researchers will look to different ways of conceptualizing quality and different practices for achieving it, as they feel are consistent with ways of constructing knowledge in their work. A summary of some key approaches to quality and rigour in case study research is show in Table 33.2.

From Table 33.2, the difference between Yin's more positivist approach and Bassey's and, particularly, Stake's more interpretivist approach to case study can clearly be seen. Yin draws on ways of constructing quality common in social science research, though he does carefully consider the degree or way in which each will work in the context of case study methodology. Bassey, on the other hand, rejects reliability and validity as appropriate for case study research and instead suggests Guba and Lincoln's concept of trustworthiness (Bassey, 1999: 74–6). He poses eight questions, which when answered positively by the researcher will indicate quality and rigour in case study work. These centre around the key methods identified in Table 33.2.

All the researchers mentioned in Table 33.2 agree that checking data back with sources is important, though how feasible this is in practice will depend on the context (for example on age of respondents). They also stress the keeping of careful and methodical records of work undertaken. This is particularly important in case study research as a large amount of data is generated. It should be possible for another researcher, acting as a critical friend, to go back through your records and see how you arrived at your interpretations, even if they do not agree with them.

Another key feature of quality in case study work, mentioned by all three sources in Table 33.2, is the triangulation of data possible with multiple methods of data collection. Most case study research includes data from observation, interviews and document review. Data about the same issue but generated by different methods can be brought together and compared (triangulation). However, the aim of triangulation

Table 33.2 Ways of conceptualising and methods for attaining quality and rigour in case study research

	Yin	Bassey	Stake
key concepts	validity (3 types) reliability	trustworthiness	validation accuracy
key methods	Triangulation from multiple sources of evidence; chain of evidence; case study database; review by key informants; analytical strategy;	engagement with case; checking with sources; triangulation; systematic testing of interpretation; detail in account; audit trail	triangulation (defined as working to substantiate an interpretation *or* to clarify its different meanings); member checking
references	2014, pp. 45–49; 118–128	1999, pp. 60–2; 74–77	1995, p. 12, 45, 48, Ch. 7; 2005, p. 453

is viewed differently according to different theoretical perspectives. From a positivist perspective, the aim is to draw closer to a true and accurate understanding of a situation. From an interpretive perspective, it is to give a richer understanding of an issue. From the latter point of view, discrepancies in accounts from different people or sources are a point of interest and possibly further investigation to extend knowledge of the research issue (Flick, 2004). From the former, they suggest the existence of error or bias. It is important to use language consistent with the perspective in which you are working. For example, to say that a source is biased is not consistent with an interpretivist approach, in which all sources will be seen as positioned and all representations made in a particular context for particular reasons.

Generalisation

Case studies can be linked with issues or circumstances outside their own bounds, but the case study researcher has to be very careful to construct and to maintain consistent thinking on generalisation. As with issues of quality and rigour, approaches to generalisation vary according to the theoretical perspective within which the researcher is working. Yin (2014) makes a helpful distinction between statistical generalisation and analytic generalisation: 'In statistical generalisation, an inference is made about a population (or universe) on the basis of empirical data collected about a sample' (2014: 40). This is the type of generalisation commonly made in physical geography, or in some aspects of social science such as large-scale surveys. However, Yin explains that this is not the type of generalisation used in case study, as case studies are not chosen as samples in this way. Instead, he likens the case study researcher to a laboratory investigator testing a new theory. 'Analytic generalisation' is when the researcher compares results from a new case study to a theory to see if it can be supported (2014: 41). Not all case study researchers would see themselves working this way, but most would support Yin's assertion that as statistical generalisation is not being used, it is best to avoid terms such as 'sample' when discussing case study methodology.

In considering the complex and multivariate world of education, Bassey emphasises the need for any generalisations made from research to recognise 'built-in uncertainty' (1999: 52). He calls this 'fuzzy generalisation' and emphasises both the importance of reading a generalisation together with the report that supports it, as well as the value of cumulation from additional research in further contexts. Stake suggests that the process of reporting a case study inevitably involves some small level of generalisation within the case (1995: 7), but he sees understanding the particular rather than the general as the main aim of the methodology. However, he recognises that as the reader comes to know the case being reported, they will make links with other similar situations they have read about or experienced. In doing this, their own generalisations about the world may be supported or modified. He calls this process 'naturalistic generalisation' (1995: 85).

It is important to think through the issues of choosing and bounding a case, ensuring quality and rigour, as well as intentions for generalisation at the start of designing a case study project. These issues should be discussed in the methodology section of a formal report and appropriate language should be used throughout the report to reflect the choices made. In particular, it is important that any generalisations made to issues or contexts outside the study are appropriately tentative.

33.4 Data Collection Methods

As discussed, multiple methods of data collection are normally used in case study work. As the case study is contemporary to the researcher, observation is a key method. Immersion in first-hand experience of the case is important when the aim is to understand it in depth. The exact form the observation takes will depend on the type of case and the access you are able to negotiate. It may involve observing in a workplace, sitting in on meetings, joining in social events, just spending time in a particular location; methods that are similar to those in ethnography (see also Chapter 11). Recording observations in a clear way that enables you to access them later is very important. Longhand notes taken at the time of observation or very soon afterwards tend to be favoured, but you may also consider using audio or video recording or taking notes on an electronic device, depending on the context and access agreement. A research diary, whether hand-written or electronic, can be helpful for keeping an overview of your work and your emerging ideas (Figure 33.3). During the data-collection phase, particular care must be taken when working with children or vulnerable adults, and the ethical implications of all work must be carefully considered to ensure you are following the codes of your institution and agreed good practice (see Chapter 3).

When you enter the case study setting initially, you may wish to record your impressions quite broadly, as this will help you to make sense of the context. Fairly quickly, you should then start to focus on the key questions of your research, and your observations will themselves raise sub-questions which you may seek to answer. At this point, a process of bounding is important, especially if you only have a limited time in the setting. Your initial aims and questions for the research may need adjusting in the light of your deeper understanding, but it is important to have a plan for your focus. What is important to spend time on, who is it important to talk to, and what is less relevant for the moment?

As you spend time in a particular case study setting, some conversations are likely to arise naturally, but it is also likely that you will want to set up strategic conversations with individuals or groups as a part of your research methods. For case study work, semi-structured interviews or focus groups are commonly used. If you are not already experienced with these methods, it is important to read about suitable techniques and practices beforehand (see Chapter 9) and you may wish to try out your

Figure 33.3   Examples of pages from a research diary

intended framework of questions and prompts on a friend or colleague first. You will also need to think carefully about methods of recording.

A third common method in case study data collection is documentary analysis. Often, this stage can be started before you are in the case study location as materials may be available through publicly available sources, such as websites. Documents are often particularly helpful in understanding historical context, relevant policy, or in considering the 'public face' being presented by an institution. It is important to practise effective document management through recording the documents collected and read, and also storing copies electronically or in paper form in a clear and orderly way.

Indeed, the issue of data management generally is particularly important for case study work, as a significant volume of data can be collected. It is helpful to keep a separate record of all data collected, including time, date and location. Any recordings should be carefully logged and labelled, with backup copies being made and stored separately from the main archive.

Above all, do not be so busy collecting material that you miss out on time to think about it! As Stake (2005: 449) says: 'Perhaps the simplest rule for method in qualitative casework is this: Place your best intellect into the thick of what is going on'. This emphasises the role of careful reflection on observation, on the process of making sense of the particular issue being studied in the light of the researcher's disciplinary understanding. The initial stages of interpretation and reflection happen concurrently with data collection, even if you seek to separate them by using a different notebook or coloured pen, so the process of analysis becomes embedded and recursive.

Box 33.2   Young people's understandings of Brazil (Picton, 2008)

The case: One class of 13- to 14-year-olds at a mixed comprehensive school in England over a period of four weeks in their geography lessons. The class was predominantly white British and comprised 15 boys and 8 girls.

Aims: To develop understanding of how the young people learned about Brazil during a unit of study; to understand how teaching and resources influence their learning.

Choice of the case: A class taught by the researcher – pragmatic choice as relationships were already established and access possible. In Yin's terms this could be counted as a typical case (for the school and local area). As processes of learning were the focus, this design would have worked with any selected class.

Generalisation: to initial theory-building regarding the processes of learning about the distant place. Such theory then needs testing in alternative contexts (e.g. with a more ethnically diverse class).

Data collection methods: Observation and reflection on lessons, collecting classwork, questionnaire incorporating young people's drawings before and after the unit of work, semi-structured interviews with a sub-set of pupils, concept mapping.

Data analysis methods: Close reading, open coding to identify commonalities, themes and recurring patterns.

(Continued)

(Continued)

Findings: Football and TV reported as key sources of students' ideas; broad shift from essentialist stereotypes to binary oppositions/binary contrasts with reference to ideas of poverty, wealth and lifestyle observed; four-stage model of learning about distant places proposed.

Recommendations: Wealth, poverty and development would be better taught in less oppositional terms; need for further research, for example exploring the socio-economic and socio-cultural differences between children, and their impact on learning.

Report format: 8,000 word dissertation then written up as an article for *International Research in Geography and Environmental Education*

33.5 Data Analysis Methods

Stake (1995) emphasises how the process of analysis is a refinement of the everyday process of making sense of the situations and events we encounter. It is a matter of careful reflection, 'taking something apart' (Stake, 1995: 71) in order to understand it and give it meaning within the framework of our wider understandings of the world. He suggests that we reach new meanings about cases in two ways: 'direct interpretation of the individual instance and through aggregation of instances until something can be said about them as a class' (Stake, 1995: 74). Stake sees this process as drawing on intuitive processing, but this does not mean that it lacks systematicity and rigour. In some ways, the researcher acts as a detective piecing together evidence of a crime – with meticulous attention to detail, but also being open to innovative ideas and ways of thinking 'outside the box' about the problem. New knowledge is often created by spotting patterns and making links.

The initial stages of data preparation involve rigour and care, but also decisions about interpretation. For example, if you are transcribing audio data, what will you include? The process of interpretation begins as you decide whether or not to measure the lengths of pauses, to note inflections or include parts of words. Similarly if you are transcribing written work, will you include or correct spelling errors? These types of decisions tend to be a trade-off between closeness to the original and time available. Much depends on the purpose of your work and intended method of analysis. If you are dealing with a sensitive topic and you are interested in your respondents' reactions and emotions, or you are intending to perform detailed conversation analysis, then a very full transcription is necessary. Otherwise, you may go for a more basic approach. Decisions made to simplify expression or correction of spellings will make digital searching and collation easier, but they may obscure the 'voice' of your original respondent.

Once any necessary preparatory stages are completed, the way forward in analysis very much depends on the research questions asked and the type of data under consideration. You should read widely about issues of analysis and consider how studies similar to your own have proceeded (see Chapter 36). There is a balance to be struck in case study work between considering data as a whole and looking in detail at parts. This part to whole is usually an iterative process. For example, when analysing a set of young people's drawings of what they might see 'out of a window' in Japan, I used a four-part process of analysis, adapted from a method of analysing photographs by the anthropologists Collier and Collier (see Box 33.3). Stages 1 and 4 involve a holistic approach to the dataset, whilst Stages 2 and 3 involve detailed 'drilling down' to consider specific questions and issues. This approach complemented a more general process of categorisation, including coding, used across the full dataset, which comprised a range of oral, visual and written sources (Taylor, 2009).

Box 33.3   Collier and Colliers' (1986, pp. 178–9) model for analysis of photographs (abbreviated version)

1   Observe data as a whole, 'listen' to overtones and subtleties, discover connecting and contrasting patterns. Note feelings and impressions and what portions of the data they are in response to. Write down questions. Build context for research.
2   Inventory or log the evidence so that you know completely its general content. Design inventory around categories that reflect and assist your research goals.
3   Structured analysis responding to specific questions. Information gathered is often statistical in character. Detailed descriptions may also be made which abstract one situation for ready comparison to another.
4   Search for the overtones and significance of the details by returning to the complete field record. Try again to respond to the data in an open manner, so that details are placed in a more complete context that defines the significance of their patterns. Write your conclusions as influenced by this final exposure to the full context.

33.6 Communicating Case Study Research

In case study methodology, the act of writing the report receives as much attention from commentators as methods of data collection or analysis. There are different approaches to structure and genre, consistent with the different types of case study, but communicating the 'flavour' of the case is always seen as important. Yin stresses that 'the reporting phase makes great demands on a case study researcher' (2014: 177). Stake also emphasises the need for excellence in communicating the case, and provides a helpful checklist for evaluating the quality of a report (1995: 131).

Yin (2014: 179) recommends starting by considering the audience for the report – whether this is the general public, academic colleagues, research funders, policy-makers or other professionals. In many cases, you will write separate accounts for different audiences, for example a dissertation aimed at an academic review board, then a later summary paper for dissemination to academic colleagues, or professionals working in a particular field.

At an early stage in your work, it is worth thinking about the format of your final report. For example, it may be that this can employ a wider number of forms than just the written word. At a basic level, perhaps you can include photographs, but you may also be able to include video, audio clips or other multi-media formats. It is common for funded research to involve production of a website or other digital outcome. It is useful to be planning for this from the start so that you can collect suitable material to use, also so that your access and ethics agreements take this into account. For example, if you are working with children or vulnerable adults, particular permission will be needed for dissemination of any visual materials, and this is much easier to arrange at the time of initially negotiating permissions than later on.

Yin (2014) suggests six different structures for case study reports, each fitting with particular types of case study. For example, the researcher may choose a standard linear-analytic form (statement of the issue; literature review; methodology; findings; conclusions). An alternative is a chronological structure, which 'might follow the early, middle and late stages of a case' and can be particularly useful for showing causal sequences over time (2014: 189). Bassey (1999: 84 ff) talks about structured, narrative, descriptive and fictional reporting. If you are submitting a dissertation, it will be important to discuss any alternatives to the linear-analytic form with your supervisor. It may be necessary to make an argument for using a structure which is less standard within your institution's practice, but this should be possible if you can show why it is particularly appropriate for your methodology and data.

One way of engaging the reader and helping them gain a vicarious experience of the case is the vignette. Stake (1995: 128) defines these as: 'briefly described episodes to illustrate an aspect of the case, perhaps one of the issues'. They aim to convey a sense of place, time and character, making a selected incident or situation 'real' to the reader. As such, the writer has to pay careful attention to descriptive wording and relevant contextual detail (see the example in Box 33.4). Carefully chosen extended quotations from participants can also help readers understand the case more fully. Stake (1995: 86–7) emphasises the need to communicate carefully and fully about the case in order to aid naturalistic generalisation (the reader making wider links to their own experience).

An interesting issue in case study work is the decision whether to name or anonymise the case. This applies to individuals and institutions (Yin, 2014: 196 ff). Whereas in many research projects, the automatic and uncontentious procedure is to anonymise, in case study research, this may not be a straightforward decision. Firstly, particularly when the research is based on one case, you may need to give so much detail in the report that it would be hard to hide its identity. To avoid identification you would need to change further details, but would that obscure the case you are trying to illuminate? Secondly, the institution or people in the case may actually wish to be identified. This can raise issues of conflict with ethical codes in some research institutions, so this is an important issue to discuss with your supervisor and gatekeeper in the case at an early stage.

If you do decide to use pseudonyms, who should choose them? An argument for assigning them yourself is that you can roughly 'match' names popular within a similar age group and/or socio-economic group to still give a 'feel' of that person's context. However, this is clearly not an exact science, and can be particularly difficult to do when working in a cross-cultural situation. An alternative is to ask participants to choose their pseudonyms, which gives greater ownership to them, but risks the choice of names that might be distracting to the reader in the final case reports (for example names of cartoon characters or film stars). If you do use pseudonyms, it is important to ensure that you keep a careful record of their allocation and move into using them consistently at the most appropriate point in your work.

**Box 33.4   Vignette-style material in an academic article
(Excerpt from Taylor, 2009, pp. 183–4)**

Representations of Japan (2) – Lawrence

To develop some points from the previous section further, this section will focus on one student – Lawrence. Lawrence presented himself in a 'laid back' way and in class he sometimes played for laughs, but he kept a low profile if reprimanded. His attainment in Geography was about average for the mixed ability group, he liked Art and was also planning to take PE, History and Drama for his GCSE options. His interest in sport continued out of school, playing rugby, and also golf with friends and family. His mum was 'a qualified drama person' and he took part in pantomimes in his local town. Lawrence had some experience of travel to southern Europe on family holidays, and also to Ireland for a kick-boxing championship, a sport in which his father was also involved.

Lawrence's case gives strong exemplification of the culturally situated and relational nature of his learning about Japan, as mediated from stories of his family history. One of Lawrence's ambitions was to go to Burma. Indeed, his ideas about Burma were a strong influence on his thinking about Japan. The connection was that Lawrence's grandfather's family had been based in Burma, only returning to the UK during World War II. Thus one of Lawrence's narrative strands regarding Japan was that the family came to live in England

> 'cos of the Japanese ... my Grandad was a prisoner of them, and most of his family ... and he had to work for them, and do their engine ... and my Grandad was put in a cell – I don't know what for and I dunno how – he was about my age ...'.

These family stories had piqued his interest, and he also remembered a TV and a radio programme on the war in Asia. Some of the extended family still lived in Burma, and Lawrence's mum had visited a few years previously, bringing 'stuff' back with her. Pictures and other objects in his house were also a strong influence on Lawrence's ideas:

> I've seen pictures of Japan, and stuff like that ... and I've got things in my house ... so most of my house is, like quite continental ... I've seen ... painted umbrellas and ... houses like that on them ... and flowers and stuff like that

[...]

Comment: In the section from which this excerpt is taken, I chose one student and 'introduced' him to the reader, with the aim that they might feel they had started to know him. I was then able to give detailed illustration of some more general points that I wished to make about how students represent distant places within a socio-cultural context. The decision to use a series of quotations from interviews with the student, including some indication of pauses and retaining colloquialisms, was also with the aim of providing a sense of 'character'. Writing in this way uses quite a few words (always an issue when working within a word length prescription), but I felt this was necessary as I was aiming to balance a previous overview section with an insight into one student in depth.

33.7 Conclusion

Case study is a pragmatic and flexible methodology, but also one that is rigorous and demanding to undertake. A well-conducted and communicated case study can be illuminating, but this methodology must be selected for the right reasons and with careful consideration of issues such as bounding, generalisation and rigour. Assuming they are sufficiently rigorous, well-executed and reported, case studies are good for generating in-depth interpretations of complex systems of meanings positioned within their unique socio-cultural context. This is something that a breadth study (such as a large-scale survey) cannot do, though the two approaches can be combined to considerable effect in major research studies.

SUMMARY

- Case study is a flexible, practical research methodology.
- It focuses on in-depth understanding of an instance or a set of instances.
- It can fit with a wide range of projects and a number of theoretical perspectives.
- Issues of bounding the case, generalisation and rigour need to be addressed.
- Case study normally employs multiple data collection methods, which may be a mixture of quantitative and qualitative techniques.
- Analysis balances breadth and depth, considering the data holistically as well as focusing on particular parts in detail.
- The report may take a range of formats, including vignettes, to communicate in-depth understanding of the case to an audience.

Further Reading

- Yin (2014) – an excellent overview of case study from a more positivistic perspective. Particularly helpful material on generalisation, multiple case studies and case study design.

- Bassey (1999) – a more interpretivist approach to case study. Particularly helpful on ideas of trustworthiness, 'fuzzy generalisation' and the history of case study.

- Stake (1995) – an interpretivist approach to case study. Particularly useful material on ways of constructing generalisation and on communicating in a way that makes the case come to life for the reader.

Note: Full details of the above can be found in the references list below.

References

Bassey, M. (1999) *Case Study Research in Educational Settings*. Buckingham: Open University Press.
Collier, J., and Collier, M. (1986). *Visual Anthropology: Photography as a Research Method*. Albuquerque: University of New Mexico Press.
Crotty, M. (2003) *The Foundations of Social Research: Meaning and Perspective in the Research Process*. London: Sage Publications.
Flick, U. (2004) 'Triangulation in qualitative research', in U. Flick, E. von Kardorff and I. Steinke (eds), *A Companion to Qualitative Research* (Qualitative Forschung – Ein Handbuch [2000]). London: Sage. pp. 178–83.
Massey, D. (2005) *For Space*. London: Sage.
Picton, O. (2008) 'Teaching and learning about distant places: Conceptualising diversity', *International Research in Geographical and Environmental Education*, 17(3): 227–49.
Stake, R (1995) *The Art of Case Study Research*. London: Sage.
Stake, R (2005) 'Qualitative case studies', in N. Denzin and Y. Lincoln (eds), *The Sage Handbook of Qualitative Research* (3rd edition). Thousand Oaks, CA: Sage. pp. 443–66.

Taylor, L. (2009) 'Children constructing Japan: Material practices and relational learning', *Children's Geographies,* 7(2): 173–89.

Taylor, L. (2013a) 'Spotlight on ... case studies', *Geography,* 98(2): 100–4.

Taylor, L. (2013b) 'The case as space: Implications of relational thinking for methodology and method', *Qualitative Inquiry,* 19 (10): 807–17.

Yin, R. (2014) *Case Study Research: Design and Methods* (5th edition) Thousand Oaks, CA: Sage.

Acknowledgement

This chapter draws on material prepared for the MEd Researching Practice course at the University of Cambridge Faculty of Education. http://www.educ.cam.ac.uk

ON THE COMPANION WEBSITE...

Visit **https://study.sagepub.com/keymethods3e** for author videos, chapter exercises, resources and links, plus **free** access to the following recommended articles:

1. **Kirsch, S. (2014) 'Cultural geography 11: Cultures of nature (and technology)',** *Progress in Human Geography***, 38: 5.**

2. **Elwood, S. (2011) 'Geographic Information Science: Visualization, visual methods, and the geoweb',** *Progress in Human Geography***, 35: 3.**

34 Mapping and Graphicacy

Chris Perkins

SYNOPSIS

Maps are very powerful tools for representing our ideas and knowledge about places. As such, the skills of producing and reading maps are important within the discipline of Geography. Historically cartography has dealt with the development, production, dissemination and study of maps in a wide variety of forms, whereas graphicacy is the skills of reading and constructing graphic modes of communication, such as maps, diagrams and pictures. This chapter introduces the differing roles played by mapping and the changing social significance of the map, including the emergence of crowd-sourced collaborative mapping and the widespread democratization of mapping technologies. It discusses the availability of mapped information, explains how maps work and offers advice about how to design maps.

This chapter is organized into the following sections:

- Introduction
- The contested terrain of mapping
- Finding the map
- How maps work
- Practical suggestions for design
- Conclusion

34.1 Introduction

The map is a powerful medium for the representation of ideas and the communication of knowledge about places. It has been used by geographers to store spatial information, to analyse and generate ideas and to present results in a visual form. Maps are not just artefacts; mapping is a *process* reflecting a way of thinking. The quality of a printed map or map display on a screen is a reflection of the 'graphicacy' of its authors and readers. Fifty years or so ago, Balchin and Coleman (1966) argued that graphicacy should be placed alongside numeracy, literacy and articulacy as educational prerequisites – this chapter echoes that call and argues that technological, social and intellectual changes have made graphicacy increasingly important. It explains the changing social significance of the map, exploring roles maps might play and discussing how these may be relevant for the geographer. It introduces sources of mapping by discussing the availability of mapped information and introducing practical ways of finding out what maps exist.

With desktop mapping packages and web-served mapping, we can increasingly create our own maps, and contribute to the development of digital maps on mobile devices or on the desktop, through Volunteered Geographic Information (Chapter 16), but if we are to realize the creative power of the medium it is important to understand how maps work. Over half a century of cartographic research has led to some consensus, but there is still debate about issues of graphical quality. Most researchers accept the continuing need for a holistic and artistic approach to design.

The chapter concludes with some practical suggestions of how you might design better maps that use graphicacy in a creative and useful way. It reasserts the importance of mapping for anyone studying or researching in Geography.

34.2 The Contested Terrain of Mapping

The ability to construct and read maps is one of the most important means of human communication, as old as the invention of language and as significant as the discovery of mathematics. The mapping impulse seems to be universal across cultures, time and environments (Blaut, 1991). Histories of cartography have charted changing production technologies, from the earliest surviving map artefacts dated to 3500 BC. They have described changing world views, reflected on accuracy and design and increasingly examined the social context of these images (Harley and Woodward, 1987).

Contemporary official definitions encompass a wide diversity of maps but also extend the scope of cartography well beyond earlier narrow concerns with the technology of map making. In 1995, for example, the International Cartographic Association defined a map as 'a symbolized image of geographic reality resulting from the creative efforts of cartographers and designed for use when spatial relationships are of special relevance'. The ICA also defined cartography as 'the discipline dealing with the conception, production, dissemination and study of maps in all forms' (International Cartographic Association, 1995: 1). This chapter also adopts a very catholic view of mapping (see Figure 34.1).

Following the influential work of Arthur Robinson, academics sought to classify maps according to content and scale (Robinson et al., 1995). Thematic maps focus on one particular kind of information, such as solid geology or voting patterns. In contrast, general-purpose maps include diverse information ranging from large-scale planimetric coverage of a building, through official medium-scale topographic mapping, to smaller-scale maps of the world in reference atlases.

Others sought to classify by format of publication. Maps were formerly only issued as printed-paper publications. But most mapping is now available as digital data delivered on a wide variety of media and interfaces, from desktop machine to laptop and mobile device and held locally on DVD, hard drive, or network server, but increasingly also distributed over the Internet or stored in the Cloud (Peterson, 2014). Producers no longer control content or design; users can play a much more active role, and maps are increasingly also constructed to communicate by sound, to be read by touch, or even relate to smell.

Figure 34.1   The world of maps

Distinguishing maps by content becomes less significant when users can create their own maps. Emphasizing content also completely misses an essential aspect: it has been argued that all maps have a theme and every one represents an 'interest' (Wood, 1992). Understanding this interest may be much more useful than listing what the map appears to show. This view implies that maps might best be understood as propositions, instead of being seen as representations (Wood and Fels, 2008). The complexity of the medium is clarified if you appreciate the different roles played by maps (see Box 34.1).

Box 34.1   Some of the possible roles played by maps and mapping

- Practical tools
- Models
- Language and representation
- Inventories and databases

- Visualizations
- Cultural artefacts and practices
- Imaginings
- Political devices
- Metaphors for scientific knowledge
- Persuasive icons
- Contested texts
- Geography

Mapping is above all else a practical form of knowledge creation and representation, and is usually carried out with an end in mind. Cartography is a useful art, and the map is a tool to be used for informing, navigating, describing places, analysing spatial relationships, or many other purposes. Maps work as tools by simplifying and serving as guides to the much greater complexity they represent. This recognition was used by geographers in the 1970s to develop scientific approaches that focused upon how maps work as a form of communication and in the search for optimal map designs.

Maps communicate spatial information in a series of codes and these codes not only work within the map, but also allow us to use maps at a social level (Pickles, 2004). By the 1990s, there was an increasing concern that a narrow scientific view of the map as part of a cartographic communication system failed to reflect the diversity of roles it played. Instead, more recent scientific research has treated the map as a sign system functioning at different levels (MacEachren, 1995).

The map is also an efficient way of storing large amounts of spatial information (Tufte, 1983: 166). This 'visual inventorying' role has come to be supplanted by digital databases that no longer need to have a visual expression; we can follow the instructions given by the satnav, and navigate from a to b without seeing the bigger picture of the map. But even the simplest visual display still stores and communicates information very effectively: a picture is still 'worth a thousand words'.

In the last two decades of the twentieth century, the development of the digital computer and increasingly complex software allowed graphical images to be manipulated, and more interactive map use became possible (see Figure 34.2). Mapping also increasingly moved towards the private end of the use axis and became increasingly collaborative in design and use. In the 1990s, scientific visualization began to reintegrate map design with science. An ongoing raft of technological changes led to the development of mobile-based ubiquitous mapping, that placed the user on the screen and map, instead of separate from the map (Farman, 2014).

The development and uses of mapping reflect changing social and economic contexts. All maps are cultural, crafted works of the human mind. They are artefacts imbued with the cultural values of the society producing them. An Australian aboriginal bark painting map about aboriginal environmental relations might be incomprehensible to a western European (Turnbull, 1989). But the London Tube map familiar to Western users might also be difficult to understand for someone from another culture. Recent research has increasingly shown how mapping practices are

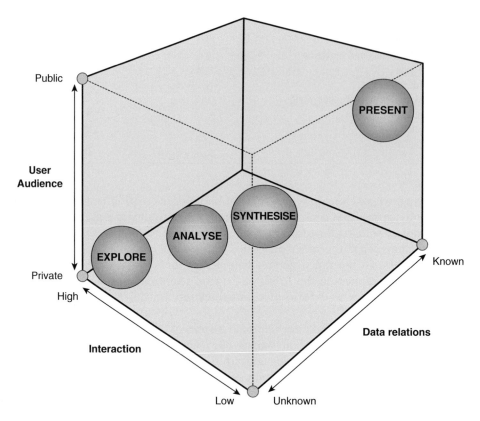

Figure 34.2　MacEachren's cubic map space

fundamental in understanding the significance of the medium (Dodge et al., 2009). Web-served applications now make collaborative mapping much easier than was the case, and crowd-sourced, shared resources such as OpenStreetMap.org disseminate community-sourced and free mapping (see for example Perkins, 2014). Meanwhile, everyday use of maps is increasing and modern artists frequently deploy mapping in their work (Harmon, 2010).

The way maps work in different cultures reflects individual imagined geographies. The power of the imagination is obvious in properly designed maps that reflect intuitive and individual artistic judgment: maps can decorate as well as inform.

It has also been argued that maps are all imbued with power and that the work they carry out is always political (Harley, 1989a). The map often stands for control and acts as a synonym for order. As a form of power knowledge, the map has often represented the interests of elite groups in society and served to reinforce social norms (Black, 1997). The power of many western maps resides in their apparent objectivity. The map appears to show everything, to offer a neutral 'view from above' (Cosgrove, 2001) that indexes the world and allows the unknown to be known. But while they may stand for science and factual knowledge, maps may also have an inherent ability to persuade – all the evidence suggests they are more likely to be believed than words. Mapping is widely used in political propaganda, in advertising, in cartoons and in the mass media where maps are associated with news stories and reinforce the narrative of the story line (Monmonier, 1996). Maps can also add authority to the people associated with them: military leaders and politicians often give their press briefings in front of maps. Others have argued that all maps are inherently persuasive, that maps are best read as rhetoric and that the map is a text that needs to be deconstructed and interpreted (Harley, 1989a). The problem is how to interpret such persuasive icons. Maps interact with other discourses, may be read in conflicting ways by different social groups and say different things to different people.

Perhaps above all else, mapping is still seen as something that distinguishes Geography from other disciplines. The map in the journal article or the geography dissertation sends a signal to the reader that the work is geographical (Harley, 1989b) and there is strong evidence that the discipline is returning to mapping (Dodge and Perkins, 2008), so deploying maps with your dissertation sends powerful messages.

34.3 Finding the Map

In the UK in 1996, there were 250 publishers releasing printed mapping (Perkins and Parry, 1996) and over 2,500 were listed worldwide in Parry and Perkins (2000). In September 2009 over 25,000 mapping sites were indexed on the now defunct but then definitive list of cartography websites (Oddens, 2009). Since 2005, there has been a widespread adoption of virtual globes such as Google Maps, or Google Earth, together with a profusion of mashups combining these backdrops with additional datasets. Quite simply, there are more maps available now than at any time in human history. These quantities make it difficult to find the right map – a problem exacerbated by the increasing diversity of types and by the infinite design possibilities offered by digital mapping. How do you decide which to use and how do you find that map?

It is difficult to find out what printed fixed-format maps have been published. Publishers provide descriptive information about their mapping, and increasingly these data are available over the web. So one option is to search the home pages of mapping organizations. A few detailed guides to mapping have been published for some nations, notably for the UK (Perkins and Parry, 1996). A basic but comprehensive introduction is provided in World Mapping Today, including lists of URLs and publisher details, pen portraits of the state of the art in different countries, simple graphic indexes and bibliographic information (Parry and Perkins, 2000). More current links to national mapping agencies are available at http://www.charlesclosesociety.org/organisations.

When searching through these sources, you might be interested in such factors as the spatial and temporal coverage, resolution, currency, consistency and reliability. But a more complex appreciation also helps. For example, who produced the map for which market? What does it include or omit? Are there legal restrictions on use? How is copyright enforced? Is the map affordable? Perhaps above all else, is it available? Despite globalization the nation-state and its legal framework continue to influence public civilian availability. Figure 34.3 shows that, although

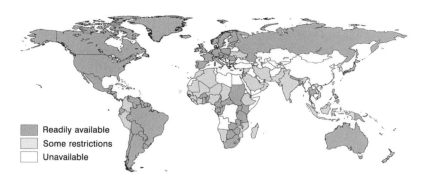

Readily available
Some restrictions
Unavailable

Figure 34.3   Official map availability (sourced from data provided in Parry and Perkins, 2000)

most terrestrial areas have been well mapped, it is very difficult to get access to larger scales of official mapping. Maps may still be reserved for military use (e.g. in India, much of the Islamic world and even in Greece) or priced at a market cost recovery rate that maximizes revenue and makes individual access expensive (e.g. in the UK until 2010).

Having identified the best-printed source, tracking down a copy involves consultation in a map library or purchase from the publisher or map seller (Parry, 1999). The quality of map libraries varies enormously according to funding, the nature of their parent organization and how they are staffed. In the UK, many schools of Geography or Earth Science departments hold collections of printed mapping. The currency and scope of many university collections have declined but a more comprehensive range of cartographic resources is to be found in major and national library collections, such as the six copyright libraries in the UK (see the British Cartographic Society, 2014, for a comprehensive list of over 400 collections in the UK). In North America, most significant map collections are to be found in central libraries on university campuses, where they serve as Federal depositories and where digital datasets are increasingly also being archived. National libraries such as the Library of Congress in Washington, DC, the National Library of Australia in Canberra, or the British Library in London offer the most significant map holdings and may be the best sources for the most difficult items. Information about the scope of major map collections is to be found via the History of Cartography Gateway site at http://www.maphistory.info/collections.html. Descriptive information may also increasingly be found in web-based online public access catalogues and may be used to discover whether the library holds a copy of the map you need. It may be necessary to visit the collection to seek advice from professional curatorial staff.

Parry (1999) suggests that buying a printed map may well be the last resort for students if resources are not available in library systems. In Great Britain and North America, mapping is relatively easy to acquire from many retailers. Wider ranges are available from specialist map shops (such as Stanfords), and almost all maintain mail-order operations and e-businesses with online ordering facilities (see Box 34.2 for a list of the most significant of these dealers). For more information about the relative merits of these competing sources, see Parry and Perkins (2000).

Box 34.2   Map dealers

International Map Industry Association

http://imiamaps.org/

Listings of map and data sellers worldwide

The Map Shop

http://www.themapshop.co.uk/

Elstead Maps

http://www.elstead.co.uk/

EastView Map Link Inc.

http://www.maplink.com/

East View Geospatial Inc.

http://www.geospatial.com/

ILH Internationales Landkartenhaus

http://www.ilh-stuttgart.de/

Maps Worldwide Ltd

http://www.mapsworldwide.com

OMNI Resources

http://www.omnimap.com/

Edward Stanford Ltd

http://www.stanfords.co.uk/

World of Maps

http://worldofmaps.com/

Increasingly of course, hard-copy printed mapping is being replaced by digital alternatives available over the Internet. Accessing digital-map data presents different challenges. It may require GIS software, storage media, hardware and output devices. Some data are distributed with viewing and interactive mapping software, but increasingly data are being served on the Internet. Even greater disparities of availability exist between the economically developed and less developed worlds, in part because of the economics of digital map production. Digital mapping is usually much more expensive to buy than printed maps and may only be available to use under strictly regulated licence conditions. In the UK, the Joint Information Services Committee has negotiated a number of deals with data providers to release digital data to universities and colleges. Students can register to use an increasing range of digital data in teaching, learning and research. Since 2000 selected Ordnance Survey data have been available via the DIGIMAP service from EDINA, and the scope of data available from this source has been expanded to encompass historical Ordnance Survey mapping, geological survey data, environmental mapping and marine charting. Other notable online sources of data relating to Great Britain are available through map libraries such as the National Library of Scotland (see http://maps.nls.uk/), and from portals such as the Vision of Britain Through Time (see http://www.visionofbritain.org.uk/). In the USA, federally produced digital map data are in the public domain and may be readily accessed over the Internet. Map libraries in North America, western Europe and Australasia increasingly offer access points to the more useful sources of digital mapping. The Internet also offers a plethora of copyright-free maps. Many mapping websites still serve static images, but notably rich online libraries include the University of Texas for contemporary mapping, and the David Rumsey collection for scanned historical mapping. Increasingly, the web is also a valuable source of more interactive mapping, such as the now ubiquitous virtual globes such as Google Maps. It also serves as a delivery gateway to data warehouses storing digital map data that may be imported in GIS (see Chapter 18). Box 34.3 lists a number of the more important online sources of digital mapping.

Box 34.3   Online sources of mapping

Gateway sites

History of Cartography Gateway

   http://www.maphistory.info/index.html

Over 6500 annotated links to historical sources relating to the history of cartography

Sources of online map data

University of Texas

   http://www.lib.utexas.edu/maps/

A rich source of scanned conventionally published mapping. Huge collection of maps, many produced by the CIA, available as gifs, jpegs or pdf files plus links to other sites (including historical maps, city plans and cartographic reference sources).

United Nations Cartographic Section

   http://www.un.org/Depts/Cartographic/english/htmain.htm

David Rumsey Historical Map Collection

http://www.davidrumsey.com/

The richest global collection of scans of old maps.

Ordnance Survey

http://www.ordnancesurvey.co.uk

The national mapping agency of the UK, with a wealth of information available about mapping products and map downloads.

Digimap

http://digimap.edina.ac.uk/digimap/home#

Digimap is a service offering free access to maps and geospatial data from a number of national data providers, including OS digital datasets and maps, historic topographic maps as well as geological, environmental and marine coverage.

Trails.com

http://www.trails.com/maps.aspx

Digital USGS mapping at 1:100 000, 1:25 000 and 1:24 000 scales and aerial coverage.

OpenStreetMap

http://www.openstreetmap.org

The most comprehensive global collaborative free map.

Virtual globes and mapping portals

Bing Maps

http://www.bing.com/maps/

Includes Ordnance Survey coverage from Explorer and Landranger series.

Google Earth

http://earth.google.com

Industry leading virtual globe.

Google Maps

http://maps.google.com

(Continued)

(Continued)

Global coverage and an ever-expanding range of mapping, applications, mashup capability and interface design *Streetview* photographic coverage.

Here

   http://here.com/

Desktop and mobile mapping from Nokia.

Mapquest

   http://www.mapquest.co.uk

AOL owned virtual globe

Placename Finding Aids

Geonames

   http://www.geonames.org/

10 million geographical names, global coverage, links to Google Maps, Creative Commons Licensed.

GEOnet Names Server

   http://geonames.nga.mil/gns/html/

Access to the National Geospatial Intelligence Agency database of foreign place and geographic feature names. 9 million feature names.

Getty Thesaurus of Geographic Names Online

   http://www.getty.edu/research/tools/vocabularies/tgn/index.html

Hierarchical database of nearly 1.5 million global place-names.

Census and Administrative Geographies

UK Data Service Census Support

   http://census.ukdataservice.ac.uk/about-us

Includes boundary and data selection and interactive thematic mapping tools for the last five UK censuses.

Neighbourhood Statistics

   http://www.neighbourhood.statistics.gov.uk/dissemination/

Interactive mapping tools for UK census data.

Data Archives

UK Data Archive

http://www.data-archive.ac.uk/

A specialist national resource containing the largest collection of accessible computer-readable data in the social sciences and humanities in the UK. Plus catalogue searching of other national archives for computer-readable data.

National Geophysical Data Center (NGDC)

http://www.ngdc.noaa.gov/

A wide range of science data services and information. Well documented databases from many sources, and value-added data services. NGDC acquires and exchanges global data through the World Data Center system and other international programs.

ESRI

http://www.esri.com

The software house responsible for industry-standard GIS data sources. Wide range of GIS data and digital map samples.

Air photographic data

National Collection of Aerial Photography

http://ncap.org.uk/

One of the largest collections of aerial photographs in the world with finding aids

Getmapping

http://www1.getmapping.com

Global and UK aerial photography, height data and mapping.

UK Aerial Photographs

http://www.ukaerialphotos.com/

All major UK suppliers on a single site.

Land cover and Thematic Data

Multi-Agency Geographic Information System (MAGIC)

http://www.magic.gov.uk/

Web-based interactive environmental mapping and data downloads for the UK.

(Continued)

(Continued)

Cranfield University National Soils Research Institute

 http://www.landis.org.uk/services/soilscapes.cfm

Interactive soils mapping of the UK.

British Geological Survey

 http://www.bgs.ac.uk/

The national earth science agency of the UK – rich source of geoscientific data.

National Atlas of the United States

 http://www.nationalatlas.gov/

The official interactive atlas of the USA, multi-thematic layered mapping and animations.

34.4 How Maps Work

Once you appreciate the complex roles maps play, it makes sense to try to understand how they work. This involves understanding basic spatial properties, and appreciating how maps simplify and represent the constraints of symbolizing data, as well as appreciating the wider social contexts in which mapping operates.

Spatial properties

All maps are about places and represent distance, direction and location in a graphical medium. For the tool to work, these spatial properties have to be mapped out in a consistent way. The surface of the Earth is not flat, so a mechanism is needed to translate the relative positions of places to the flat sheet of paper in a way that minimizes distortion of scale, area, direction and shape. The science of map projection regulates this process and for many years the mathematics and technology of projection dominated cartography. More recently, the politics of projection have been emphasized – for example, in the controversy over the use of the Peters projection (Monmonier, 1996). If you need to use a small-scale map of the world then take advice on the 'best' projection to use (American Cartographic Association, 1991).

 Projection maps out the position of the lines of latitude and longitude that comprise the graticule. Absolute locations of places on the Earth's surface may be defined in spherical co-ordinates, but the graticule does not intersect at right angles. Grids are a rectilinear net of lines that allow eastings and northings to be defined. They allow space to be structured and were first employed by ancient Greek cartographers. Grids may be arbitrary, like the alphanumeric references used by A–Z-style town atlases, or may be mathematically related to the projection, like the British National Grid.

Consistent representation of distance implies the use of a mathematical linear scale linking the map to the world, that may be expressed on the map as a representative fraction (1:50,000), as a bar scale or a verbal-scale statement (one and a quarter inches to one mile). Very few maps use scale consistently for every object; most will exaggerate the size of some features so they can be read as a symbol (for example, a road). Many small-scale maps of the world will also have to distort linear scale because of the projection used.

Generalization and classification

Scale regulates how much detail the map can show. Larger-scale maps depict more detail and with greater accuracy. As scales get smaller, so features must be more generalized (see Figure 34.4). There are a number of different ways of generalizing: simplifying, enlarging, displacing, merging and selecting, and so on. Classification can be another effective way to map out complexity. Alternatively, you may have to leave features off the map altogether. Matching the level of detail shown to the scale is an

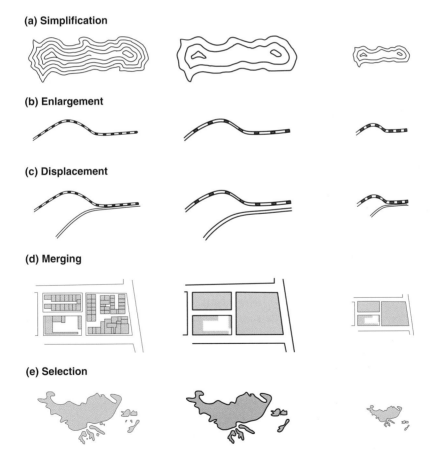

Figure 34.4 Generalization

important aspect of design: a cluttered map will not work as well as one with the right balance between content and space. The process of selection might be seen as a technical issue, but in many maps is also a political outcome, with omissions or 'silences' reflecting cultural values (Harley, 2001: 84–107).

Symbolization

The cartographer also has to symbolize the world in a regulated graphic language in which text and the visual properties of symbols are combined. This combination often takes place in quite standardized ways – common elements of a map on a screen or sheet of paper may be identified (see Figure 34.5; Dent et al., 2008: 242). Objects in the map itself may be thought of as having different numerical qualities. These *measurement levels* are important for design. Nominal data show the presence or absence of information; ordinal data imply that a feature is larger or smaller (but do not indicate how much larger). Interval data involve ordering with known distances between observations – e.g. Fahrenheit measurement – whereas ratio measurement is an interval scale with a known starting point. Symbols and objects also have *dimensions*: points, lines, areas, volumes and duration. They may be distributed in discrete, sequential or continuous patterns.

On the map itself, the geometry and measurement level of symbols have *attributes* that allow information to be communicated. For example, a road may be red and it may have a label indicating that it is the A57 Snake Pass. The effective use of lettering on maps, and the rules governing how it should be used, are one of the most difficult areas of cartographic design. Name placement is a complex and often intuitive process.

The map designer can use only a limited number of graphic variables. Figure 34.6 illustrates how they might be used for point symbols and suggests there are rules governing inappropriate use. For example, shape should not be used to suggest variation in quantitative data. Symbols constructed with these variables may be iconic, geometric or abstract (MacEachren, 1995).

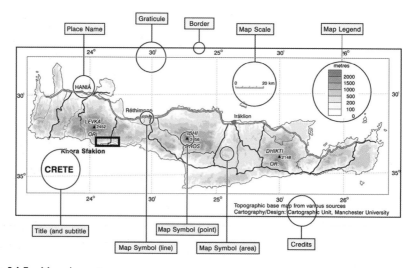

Figure 34.5 Map elements

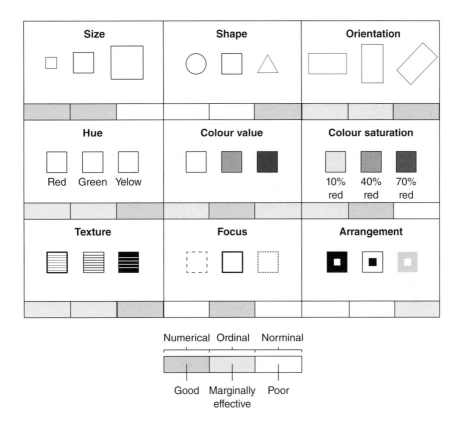

Figure 34.6 The graphic variables

34.5 Practical Suggestions for Design

Having obtained source material, how do you know whether it will make a 'good' map? Putting together all the elements in Figure 34.5 does not automatically result in a map that works well: these elements need to be combined in a meaningful aesthetically pleasing design. Can there be universal rules defining aesthetic quality? How might these be influenced by production technologies and are they transferable to different contexts?

Universal rules?

In 1983, Edward Tufte came up with a list of qualities that might define excellence in the design of statistical graphics (see Box 32.4). A survey carried out by the British Cartographic Society Map Design Group in 1991 revealed that professional cartographers also felt that quality resided in the overall perceptual qualities of the design rather than with the individual components of the map (see Box 34.5). Maps and other graphics seem to operate as wholes, greater than the sum of their parts. Artistic qualities are important. These qualities have been exemplified in a recent article highlighting the importance of design and drawing attention to some of the best-designed maps (see Demaj and Field, 2012).

Box 34.4   Tufte's principles of graphical excellence

- Show the data.
- Induce the reader to think about the substance rather than the methodology, graphic design, the technology of graphic production or something else.
- Avoid distorting what the data have to say.
- Present many numbers in a small space.
- Make large datasets coherent.
- Encourage the eye to compare different pieces of data.
- Reveal the data at several levels of detail, from a broad overview to the fine structure.
- Serve a reasonably clear purpose: description, exploration, tabulation or decoration.
- Be closely integrated with the statistical and verbal descriptions of a dataset.

Source: Tufte (1983: 13)

Box 34.5   Maps as communication graphics

- Contrast between symbol and background and between symbols is vital.
- The symbols themselves should be clearly legible and unambiguous.
- The amount and nature of the data depicted should be appropriate to the main purpose of the map.
- The overall appearance should be clear, simple and uncluttered.
- The metrical attributes of the map should be both appropriate and clear.
- The ordering of the data should be made clear by the hierarchical organization of the map image into recognizable visual levels.

Source: British Cartographic Society (1991)

There are two key areas involved. Our perceptual systems are programmed to respond to the *visual organization of images* (Dent et al., 2008). A good map should be balanced, with spatial layout allocated according to the Golden ratio. (The Golden section refers to rectangles with sides at a ratio of 1:1:6 that offer the most pleasing appearance to the eye.) It should be organized so that the eye's area of maximum attention (just above the geometric centre) corresponds to the central focus of the map. The individual elements in the design ought to work together as an integrated unit. The second key aim should be to maintain a *clear hierarchy between different visual levels* in the map (Dent et al., 2008). The 'figure' needs to stand out from the 'ground'. The most important objects should contrast most with their surroundings.

Production technology

By 2015, almost all maps produced by students in the UK were created using computer-based technologies. Choosing an appropriate type of software, and using it to best effect, is now probably the single most important impact on design. Be realistic and aware of some of the factors listed in Box 34.6.

Box 34.6   What Software Should I Use?

- How much time do you have?
- How computer literate are you?
- What software packages are you already aware of?
- How much support would you get for learning a new software?
- How does the software link to others?
- What file formats does it support, both for importing material and for creating completed graphics?
- What output devices does the software talk to?
- What kind of operating system do you intend to use: Mac or PC?
- Do you intend to use someone else's base material and edit it or design from scratch?
- Do you have access to a fast scanner?
- What role do you want the map to play?
- What kind of use is the map intended for: presentation, analysis or exploration?
- Is the map static or dynamic?
- How complex is the information you want to show?
- In which medium is the map going to be published: printed or electronic, black-and-white or colour?

Five examples illustrate this process:

1. Serving mapping on the web requires a number of different packages, in addition to the vehicle you use to design the map. The medium delivers maps to many users independent of platform, and maps are updatable. Production depends on the configuration of the site (whether processing is client- or server-based), the site format (whether maps are delivered in single pages, multiple pages, or a frames environment), the nature of the web interface (which browser and plug-ins are being used in web-applications), and the data type and the content interface. Be aware of the implications of using different file formats and the differences between serving maps in raster formats such as .jpeg, .tiff, or .png, as against vector formats like .sfw, .svg, or pdf. Seek advice from standard texts such as Muehlenhaus (2014) or Dent et al. (2008). Remember that the web environment offers different constraints to the designer. In particular, remember that screen resolution, the aspect ratio of the monitor viewing the site and the size of the monitor reduce the available space to view the map, and that monitors vary in the ways they display colours (Dent et al., 2008).

2. GIS software, such as MapInfo, ArcGIS or Idrisi, is designed for analysis but also offers a wide range of visualization options for map design. It is complex to learn, but if you know the software, have the data and want to display the results of your analysis then it offers a realistic choice. If you need to produce many statistical maps for a case study, designed to a common template and from a wide range of variables, this automated kind of production will save time. Advice and practical examples about this kind of software are included in Brewer (2005).

3. Some mapping software will also allow you to create maps automatically but will not include the range of analytical tools you would expect in a full GIS. Mapviewer from Golden Software is a good example of this kind of program. It offers a compromise between analytical and design capability, and is relatively straightforward to use. It, too, supports a number of predefined thematic map types that automatically create maps from data held in spreadsheets. The application also automatically creates associated support information such as legends. The designer simply has to work out which map type is appropriate for the data (see Figure 34.7).

4. If you want real design quality, have complex ideas to map and the time to spend learning a sophisticated piece of software, then a professional drawing package such as CorelDRAW or Illustrator is the best option. Open equivalents such as InkScape offer free but sophisticated alternatives.

5. The majority of static maps displayed in undergraduate dissertations do not need this sophistication. A simple drawing package will suffice. It needs to have drawing tools, layering capabilities, and to allow you to import and edit graphics created elsewhere by adding symbols, text and marginal information. Services such as Google Maps ('Virtual globes') almost all now also include simple annotation tools and Digimap allows users to produce publication quality output using OS base maps with additional annotations.

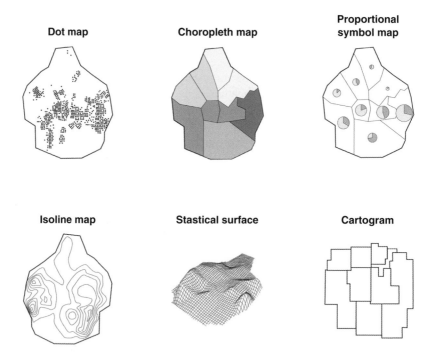

Figure 34.7 Different thematic map displays of the same data set

The context

Even if design rules are followed and the appropriate technology is used, maps still need to be matched to the medium in which they are to be published. Maps are not read in a cultural vacuum. They are interpreted in relation to what Denis Wood and John Fels have termed the *paramap*, comprising other printed elements associated with the image (the *perimap*), and wider cultural referents that allow us to understand the medium (the *epimap*; Wood and Fels, 2008). Most maps produced by students are still designed for presentation in printed reports: descriptors for these might specify how a map should be designed. But increasingly, they may be used in PowerPoint displays, slides or overhead transparencies, on posters, displayed on computer monitors, delivered via the web or as animated, dynamic visualizations. The medium may support colour, or perhaps designs will be limited to grey scale. The mapping may be static or animated. Practical decisions that need to be taken are listed in Box 34.7 and many of these flow from wider cultural concerns.

Box 34.7 Decision in design

- Should the base map be sourced from someone else's published map?
- How much should be included?
- How many graphic variables should be used?
- What type of map should be used?
- What symbols are needed? How many, what kinds?
- What sort of hierarchical structure will be used, how will the figure–ground relationship be used, how will the layout work?

With a coloured map, resolution may be inferior, but you can use two additional visual variables to improve visual structure. It makes no sense, though, to design in colour and print on a monochrome laser printer. Relating the map to other associated textual elements is also important. The more complex the medium, the more important the links. So, include mapping close to the words that relate to the graphic, rather than hidden in appendices.

Figure 34.8 shows a typical location map designed for incorporation into a report or dissertation and 'placing' the research by mapping a simple spatial context in which it is situated. The annotations draw attention to key aspects of the design. This book does not use colour, so the map has been designed in CorelDRAW and saved as a .eps file for incorporation into book design software employed by the publishers. Most maps created by students can be exported as .png or .pdf files that preserve the visual qualities of the image, such as line weight, area tones and lettering, and can be embedded as pictures in word-processed documents. The same map in a PowerPoint presentation would have to be much more generalized, and use larger lettering. Maps designed for display on paper also often translate badly to screen-based displays. On-screen colours are lit from behind and usually displayed using a red–green–blue colour model, in contrast to maps printed on paper. Screen resolution is still inferior to process colour-printed mapping, and the viewing area is much smaller than in printed versions of the same map. More complex multimedia versions of the same map will also have to reflect the design of the graphical user interface and incorporate marginalia concerned with screen navigation and manipulation. Display is no longer the only goal for design: functionality and navigation also become important.

Above all else when designing maps, seek out examples of graphical quality such as those referred to in the standard works by Cindy Brewer, Borden Dent, John Krygier and Denis Wood, Alan MacEachren and Edward Tufte. Use their advice, but be very aware of the context for which you are creating the map and what you want the map to say. Remember, someone will be reading the map, and reading it in the light of a very particular set of circumstances.

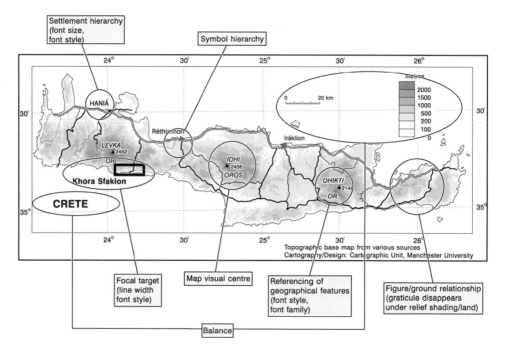

Figure 34.8 A location map placing a field course to Crete

34.6 Conclusion

The following statement concluded a text about British mapping published just as the Internet started to have a significant effect on the distribution of mapping, but well after digital map technologies had changed the world of visualization: 'Maps are important. Technology and society are rapidly changing the ways in which they are constructed, used and regarded, but as visual metaphors they will continue to provide statements which both reflect and shape our perception of the world' (Perkins and

Parry, 1996: 380). The nature of the representation and its role may have altered, but graphicacy is still central for geographers. We all have a responsibility to make better maps and use them in a more critical way.

SUMMARY

- Mapping works as a social process.
- Maps are available from publishers/map sellers, libraries and online.
- Accessing digital maps may require GIS software, storage media, hardware and output devices.
- Technological and social change has facilitated access to, and the making of, mapping.
- Maps work as representations, they highlight spatial properties and provide a visual means of creating generalizations, classifications and symbolization.
- Good maps should be organized so that the eye's area of maximum attention corresponds to the central focus of the map, and there should be a hierarchy between different visual levels of the map.
- There is a need for more critical use and design of mapping.

Further Reading

- The book by Dorling and Fairbairn (1997) is now dated but is still probably the single most accessible overview aimed at the undergraduate student, integrating critical and scientific approaches to how maps work as images.

- More challenging chapters brought together in Harley (2001) or in Dodge et al. (2011) explore the many roles played by the medium and establish the changing social context of the history of cartography.

- Parry and Perkins (2000) is still the best printed source about published map availability, providing vital publication details and contacts, but is increasingly being superceded by web sources.

- The most useful introduction to the practical issues in map design and production are Brewer (2005), Dent et al. (2008), Krygier and Wood (2011) and Muehlenhaus (2013).

- John Krygier's blog Making Maps (http://makingmaps.net/) and Ken Field's blog http://cartonerd.blogspot.co.uk/ are rich sources of recent design ideas.

- The Cartographic Communication section of the Geographers Craft home page (http://www.colorado.edu/geography/gcraft/notes/cartocom/cartocom_f.html) also offers useful practical advice on a wide range of map-design issues.

Note: Full details of the above can be found in the references list below.

References

American Cartographic Association (1991) *Matching the Map Projection to the Need.* Falls Church, VA: American Congress of Surveying and Mapping.
Balchin, W.G.V. and Coleman, A.M. (1966) 'Graphicacy should be the fourth ace in the pack', *The Cartographer*, 3: 23–8.
Black, J. (1997) *Maps and Politics.* London: Reaktion.
Blaut, J.M. (1991) 'Natural mapping', *Transactions, Institute of British Geographers*, 16: 55–74.
Brewer, C. (2005) *Designing Better Maps: A Guide for GIS Users.* Redlands: ESRI Press.
British Cartographic Society (2014) 'Directory of UK Map Collections'. London: British Cartographic Society. http://www.cartography.org.uk/downloads/UK_Directory/ukdirindex.html (accessed 2 December 2015).

Cosgrove, D. (2001) *Apollo's Eye*. Baltimore, MD: Johns Hopkins University Press.

Demaj, D. and Field, K. (2012) 'Reasserting design relevance in cartography: Part 1: Concepts; and part 2: Examples', *The Cartographic Journal,* 49: 70–76 and 77–93.

Dent, B.D., Torguson, J. and Hodler, T. (2008) *Cartography: Thematic Map Design* (6th edition). New York: McGraw Hill.

Dodge, M. and Perkins, C. (2008) 'Reclaiming the map: British Geography and ambivalent cartographic practice', *Environment and Planning A*, 40: 1271–6.

Dodge, M., Kitchin, R. and Perkins, C. (eds) (2009) *Rethinking Maps*. London: Routledge.

Dodge, M., Kitchin, R. and Perkins, C. (eds) (2011) *The Map Reader*. Chichester: Wiley.

Dorling, D. and Fairbairn, D. (1997) *Mapping: Ways of Representing the World*. Harlow: Longman.

Farman, J. (ed.) (2014) *The Mobile Story: Narrative Practices with Locative Technologies*. London: Routledge.

Harley, J.B. (1989a) 'Deconstructing the map', *Cartographica*, 26: 1–20.

Harley, J.B. (1989b) 'Historical geography and the cartographic illusion', *Journal of Historical Geography*, 15: 80–91.

Harley, J.B. (2001) *The New Nature of Mapping*. Baltimore, MD: Johns Hopkins University Press.

Harley, J.B. and Woodward, D. (1987) *The History of Cartography*. Chicago, IL: University of Chicago Press.

Harmon, K. (2010) *The Map as Art*. New York: Princeton Architectural Press.

International Cartographic Association (1995) *Achievements of the ICA 1991–1995*. Paris: Institute Géographique National.

Krygier, J. and Wood, D. (2011) *Making Maps: A Visual Guide to Map Design for GIS* (2nd edition). New York: Guilford Press.

MacEachren, A.M. (1995) *How Maps Work*. New York: Guilford Press.

Monmonier, M.S. (1996) *How to Lie with Maps* (2nd edition). Chicago, IL: University of Chicago Press.

Muehlenhaus, I. (2013) *Web Cartography: Map Design for Interactive and Mobile Devices*. CRC Press.

Oddens, R. (2009) 'Oddens Bookmarks' (main site at http://oddens.geog.uu.nl/index.html is defunct; archived version accessible at https://web.archive.org/web/20100105150644/http://oddens.geog.uu.nl/index.php [accessed 2 December 2015]).

Parry, R.B. (1999) 'Finding out about maps', *Journal of Geography in Higher Education*, 23: 265–272.

Parry, R.B. and Perkins, C.R. (2000) *World Mapping Today* (2nd edition). London: Bowker Saur.

Perkins, C. (2014) 'Plotting practices and politics: (im)mutable narratives in OpenStreetMap'. *Transactions Institute British Geographers* 39(2): 304–17.

Perkins, C.R. and Parry, R.B. (1996) *Mapping the UK*. London: Bowker Saur.

Peterson, M.P. (2014) *Mapping in the Cloud*. New York: Guilford Press.

Pickles, J. (2004) *History of Spaces*. London: Routledge.

Robinson, A., Morrison, J.L., Muehrke, P.C., Kimerling, A.J. and Guptill, S.C. (1995) *Elements of Cartography* (6th edition). Chichester: Wiley.

Tufte, E.R. (1983) *The Visual Display of Quantitative Information*. Cheshire, CT: Graphics Press.

Turnbull, D. (1989) *Maps are Territories, Science is an Atlas*. Geelong: Deakin University Press.

Wood, D. (1992) *The Power of Maps*. London: Routledge.

Wood, D. and Fels, J. (2008) *The Natures of Maps*. Chicago: University of Chicago Press.

ON THE COMPANION WEBSITE…

Visit **https://study.sagepub.com/keymethods3e** for author videos, chapter exercises, resources and links, plus **free** access to the following recommended articles:

1. **Foody, G.M. (2007) 'Map comparison in GIS', *Progress in Physical Geography*, 31 (4): 439–45.**

2. Sheridan, S.C. and Lee, C.C. (2011) 'The self-organizing map in synoptic climatological research', *Progress in Physical Geography*, 35 (1): 109–19.

3. Caquard, S. (2015) 'Cartography III: A post-representational perspective on cognitive cartography', *Progress in Human Geography*, 39 (2): 225–35.

35 Statistical Analysis Using MINITAB and SPSS

Stewart Barr

SYNOPSIS

Undertaking statistical analysis is one of the key skills geographers use to understand the natural and social world. In recent years, the MS Excel programme has been the default choice for simple data analysis and exploration, but beyond this, two powerful software packages are available and are periodically re-developed for personal computers which make analysing large numerical datasets simple and effective. MINITAB (http://www.minitab.com, Windows) and SPSS (Statistics Package for the Social Sciences; http://www-01.ibm.com/software/analytics/spss/; Windows, Mac) enable researchers to store, describe, present and analyse large datasets using commonly applied statistical techniques. In this way, they utilise the familiar spreadsheet format (used in programmes like MS Excel) and have the ability to produce a range of outputs that can be used in reports, dissertations and technical summaries.

This chapter outlines the ways in which geographers use spreadsheet data analysis packages, focusing initially on MS Excel and then describing the main features of MINITAB and SPSS. The chapter then explores the essential characteristics of each programme and explores some of the functions, before examining how different types of data analysis can be undertaken using these software packages.

This chapter is organized into the following sections:

- Using spreadsheets for data management and interpretation
- Why use Statistics vs. Excel software?
- What SPSS and MINITAB offer
- Importing and formatting data in SPSS and MINITAB
- Representing data in SPSS and MINITAB
- Doing analysis in SPSS and MINITAB
- Conclusion

35.1 Using Spreadsheets for Data Management and Interpretation

Within the social and natural sciences, the research process is often dedicated to the collection of large amounts of data that require storage, testing, manipulation, analysis and interpretation. Indeed, one of the key roles of researchers is to translate large and complex datasets into easily digestible summaries and conclusions that can contribute to knowledge about the world around us. In quantitative research, data are collected and coded in ways that enable them to be stored in numerical form and the way in which such data are now managed is almost exclusively by the use of spreadsheet programs, which organize data into rows and columns. In fact, to most readers of this book, this is now a familiar way of dealing with quantitative data – from the scheduling of a class or analysis of financial data, to the storage of large datasets on anything from household income to glacier movement.

Yet the use of spreadsheet programs on personal computers is a relatively recent development and earlier conventions for storing and organizing numerical data used different approaches, such as delimiting between data units by commas. Accordingly, it is worth considering how a common spreadsheet program works and what benefits it can offer to data management and presentation.

Probably the most widely used spreadsheet program is Microsoft's Excel, which has come to dominate the market since the 1990s. It is a highly powerful and versatile program, which is capable of handling both text and numerical data and can be used for simple functions like data storage and organization, as well as interrogating data using formulae and statistical techniques (see Figure 35.1). For many users, the horizontal and vertical organization of Excel is an ideal way to present data in an easily manageable way. For example, companies often use the program to display meeting schedules or timetables; these are functions that could easily be undertaken in a word-processing program, but can easily be edited and presented in the grid format offered in a spreadsheet environment.

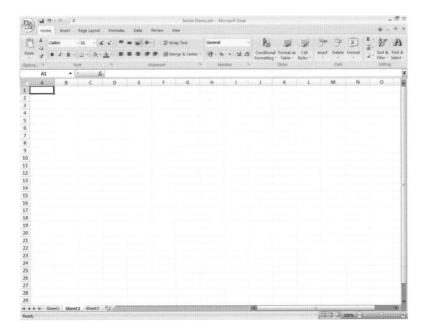

Figure 35.1   The Excel spreadsheet environment

However, spreadsheets are most often utilised as devices for storing and manipulating numerical data. As with most spreadsheet programs, the logic underlying Excel is the integration of variables and cases. Variables could be anything from questions in a survey to variables measured in the field, such as rock type, glacier extent or stream discharge. Cases are individual measurements, such as responses of participants in a questionnaire survey or measurements taken at different times or places. In Excel, variables are normally stored vertically, with cases being stored in rows.

Although programs like Excel are immensely powerful, there are some essential functions that make data manipulation simple. First, Excel affords the ability to move data within the spreadsheet through simple processes like highlighting the relevant cell(s) and clicking and dragging these to a desired location. Indeed, through using the Sort function on the Data menu, cases can be sorted according to numerical or alphabetical criteria. Second, Excel offers the capacity to store large amounts of data but for such datasets to still be easily searchable through either the simple Find function on the Home menu, or by using the Filter option on the Data menu, enabling key attributes of data to be used to display data. Both of these sets of functions enable simple data management and interrogation.

However, Excel is also useful for representing data through the graphics that it can produce. In several ways, Excel graphics are both more powerful and visually appealing that those produced in statistics packages; the flexibility of Microsoft programs means that they can be edited intuitively. The Chart function of the Insert menu is used to select different chart types (bar graphs, pie charts, line graphs and so on). Each chart type has a range of display options, including trend lines, and can be manipulated once drawn through double clicking on the area which requires editing. Because colours, line widths and text can all be edited, Excel charts are an effective way of presenting data and can ensure that reports and projects have consistency in approach and format.

Finally, Excel is excellent for data analysis at a relatively simple level. Calculations using numerical data can be undertaken using the equation function in the cell in which a result is required. For example, in calculating a final module mark for a university course, there may be several different components each student has taken for assessment. Each one will have a mark recorded out of 100, but each one will also have a different weighting (a presentation might be worth 20%, whilst a poster counts for 30% and so on). To derive the final module mark for each student, a function can be set up using an equation in the desired output cell, which also appears in the function script box above the main spreadsheet. In Excel, all equations start with an equals (=) sign and then use standard algebra to express the function. Components of the function are expressed by the cells in which different components lie. In the example provided in Figure 35.2, module marks for students are based on a calculation where a research proposal counts for 10%, a presentation for 20%, a poster 35% and a field notebook for 35%. In the expression, each component cell is multiplied by a fraction (0.1 for the 10% component and so on). When the equation is finished, the enter button is pressed and the final module mark appears in the cell, with the expression still visible in the function box above. The expression can then be copied and pasted into the cells below, so that all module marks for students can be calculated.

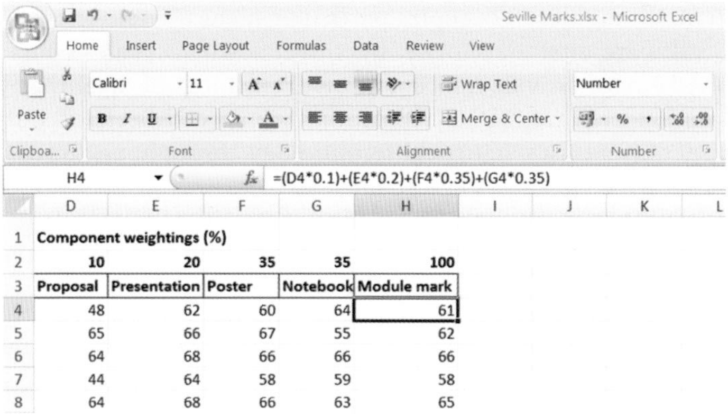

Figure 35.2 Numerical calculations in Excel

The ability to use Excel in this way makes generating descriptive statistics simple and if we wanted to know the mean (average) module mark for this course, then we could simply highlight the column of marks and press the Σ (sum) symbol and then create an expression that derived the arithmetic mean (i.e. the sum divided by the number of students).

Accordingly, Excel has a great deal of potential to provide an organizational framework for managing and simply interpreting data. For geographers, using a spreadsheet and using these skills for more specialist statistics software is essential, skills that are key for both future research and most careers. However, there are very good reasons why this chapter focuses not on Excel, but on specialized software that is dedicated to the manipulation of data using statistical methods.

35.2   Why Use Statistics vs. Excel Software?

For many of the requirements of studying geography, the standard software offered by Microsoft and Apple are more than adequate, but when it comes to statistical analysis, there are some important considerations to explore before deciding which piece of software adequately meets your needs. The software of choice for dealing with numerical data for most students is likely to be MS Excel, which enables users to conduct a variety of descriptive and analytical procedures, many of them statistical in nature. However, as regular users of Excel will attest, there is considerable prior knowledge required of the syntax (commands) and layout of the spreadsheet environment required for undertaking statistical analysis in Excel, not to mention the challenges of appropriately interpreting results and translating these into verifiable findings. In short, Excel's power is somewhat over-shadowed by the level of knowledge required by its users when it comes to statistical analysis.

By contrast, dedicated statistical software packages have progressively improved the user experience by attuning their products to the needs of particular kinds of users. Of interest to the geographer are two packages that are market competitors within industry, although each occupies particular niches. First, MINITAB is a piece of statistical software first developed by three academics at Pennsylvania State University in 1972 to provide more effective teaching of quantitative methods. Currently on its 17th release, MINITAB uses a dual spreadsheet and output window format that generally utilises statistical terminology throughout, drawing as it does from its tradition as a package for teaching statistics within the core disciplines of mathematics and statistics. As such, knowledge of statistical terminology is an advantage, although easily mastered with a good handbook and supporting textbook. Indeed, MINITAB offers a range of analysis tools that enable the effective and simple manipulation of data for undertaking some non-standard procedures.

Second, IBM's SPSS (Statistics Package for the Social Sciences) started as the SPSS Company in 1968 and incorporated in 1975. In 2009 IBM acquired SPSS and now uses the SPSS brand to market a range of software, of which IBM SPSS Statistics is the main product, offering the full range of spreadsheet-based statistical techniques. Readers should note that for a short period IBM re-named the software PASW (McKendrick, 2010), but since late 2010 it has reverted to using its original name. As its name suggests, SPSS was originally founded to provide dedicated computer support for analysing social science research data and this is reflected in the terminology and language used in both the programme and its manuals. There is more of an emphasis in SPSS on the outputs and presentation of results, which are anticipated to be used on presentations, reports and student dissertations. Indeed, the command syntax is often simpler to understand at first glance, although this does mean that an understanding of the statistical basis of procedures being undertaken is important to master.

There is no easy choice when deciding which product to use and in large part this is likely to be driven by the availability of the application at a host institution and the application of choice for teaching. This chapter aims to explore some of the basic elements of using these two applications for statistical analysis, starting with an overview of each application's interface and basic commands, then moving on to examining how data are imported and formatted, before describing how data can be presented and analysed. It should be noted that although subsequent releases of both pieces of software are likely within the timeframe of this book, these changes are likely to be cosmetic in nature and the syntax commands and screen shots of dialog boxes and outputs used in this chapter (from MINITAB Release 17 and IMB SPSS Statistics 19) are likely to remain similar to what you see when using the programs.

35.3   What SPSS and MINITAB offer

Both MINITAB and SPSS are spreadsheet-based applications and the basic interface is a familiar one, with data stored in columns and rows. In this section, the essential characteristics of each application will be examined and it is worth considering, as you read through this section, how each application's characteristics might suit your specific needs. In doing so, you should bear in mind that the applications differ in their assumptions about both the type of data analysis researchers will be undertaking and their basic knowledge of statistics. As a general rule, MINITAB is a application that has been written by and partly for physical scientists and assumes a much closer engagement with raw data from the outset (for example, MINITAB does not offer the

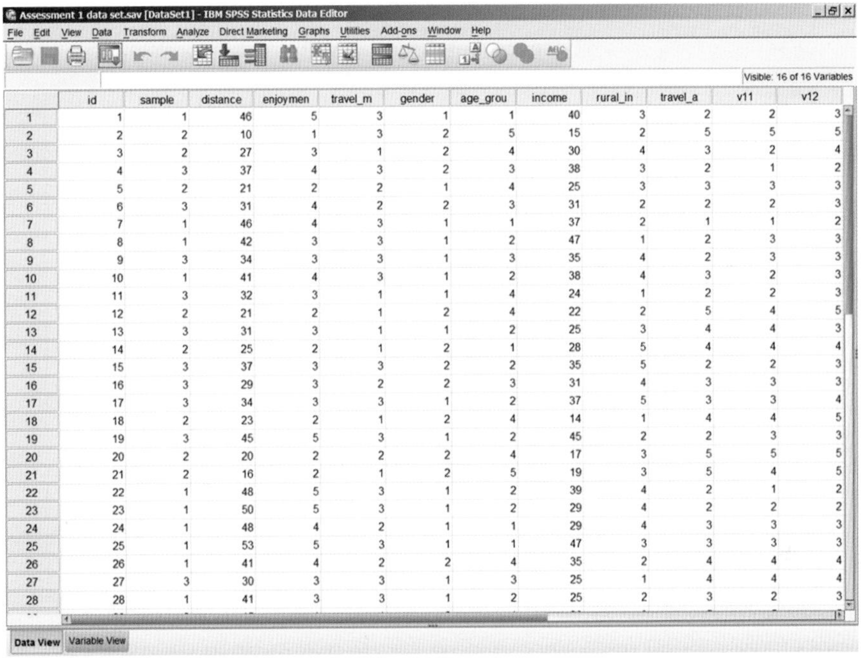

	id	sample	distance	enjoymen	travel_m	gender	age_grou	income	rural_in	travel_a	v11	v12
1	1	1	46	5	3	1	1	40	3	2	2	3
2	2	2	10	1	3	2	5	15	2	5	5	5
3	3	2	27	3	1	2	4	30	4	3	2	4
4	4	3	37	4	3	2	3	38	3	2	1	2
5	5	2	21	2	2	1	4	25	3	3	3	3
6	6	3	31	4	2	2	3	31	2	2	2	3
7	7	1	46	4	3	1	1	37	2	1	1	2
8	8	1	42	3	3	1	2	47	1	2	3	3
9	9	3	34	3	3	1	3	35	4	2	3	3
10	10	1	41	4	3	1	2	38	4	3	2	3
11	11	3	32	3	1	1	4	24	1	2	2	3
12	12	2	21	2	1	2	4	22	2	5	4	5
13	13	3	31	3	1	1	2	25	3	4	4	3
14	14	2	25	2	1	2	1	28	5	4	4	4
15	15	3	37	3	3	2	2	35	5	2	2	3
16	16	3	29	3	2	2	3	31	4	3	3	3
17	17	3	34	3	3	1	2	37	5	3	3	4
18	18	2	23	2	1	2	4	14	1	4	4	5
19	19	3	45	5	3	1	2	45	2	2	3	3
20	20	2	20	2	2	2	4	17	3	5	5	5
21	21	2	16	2	1	2	5	19	3	5	4	5
22	22	1	48	5	3	1	2	39	4	2	1	2
23	23	1	50	5	3	1	2	29	4	2	2	2
24	24	1	48	4	2	1	1	29	4	3	3	3
25	25	1	53	5	3	1	1	47	3	3	3	3
26	26	1	41	4	2	2	4	35	2	4	4	4
27	27	3	30	3	3	1	3	25	1	4	4	4
28	28	1	41	3	3	1	2	25	2	3	2	3

Figure 35.3a The Data Editor in SPSS

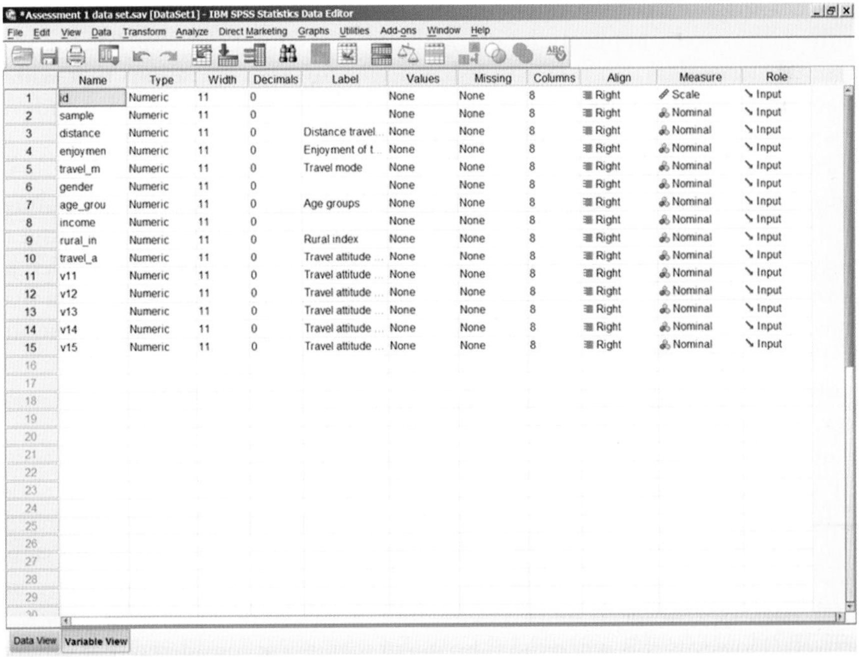

	Name	Type	Width	Decimals	Label	Values	Missing	Columns	Align	Measure	Role
1	id	Numeric	11	0		None	None	8	Right	Scale	Input
2	sample	Numeric	11	0		None	None	8	Right	Nominal	Input
3	distance	Numeric	11	0	Distance travel...	None	None	8	Right	Nominal	Input
4	enjoymen	Numeric	11	0	Enjoyment of t...	None	None	8	Right	Nominal	Input
5	travel_m	Numeric	11	0	Travel mode	None	None	8	Right	Nominal	Input
6	gender	Numeric	11	0		None	None	8	Right	Nominal	Input
7	age_grou	Numeric	11	0	Age groups	None	None	8	Right	Nominal	Input
8	income	Numeric	11	0		None	None	8	Right	Nominal	Input
9	rural_in	Numeric	11	0	Rural index	None	None	8	Right	Nominal	Input
10	travel_a	Numeric	11	0	Travel attitude ...	None	None	8	Right	Nominal	Input
11	v11	Numeric	11	0	Travel attitude ...	None	None	8	Right	Nominal	Input
12	v12	Numeric	11	0	Travel attitude ...	None	None	8	Right	Nominal	Input
13	v13	Numeric	11	0	Travel attitude ...	None	None	8	Right	Nominal	Input
14	v14	Numeric	11	0	Travel attitude ...	None	None	8	Right	Nominal	Input
15	v15	Numeric	11	0	Travel attitude ...	None	None	8	Right	Nominal	Input

Figure 35.3b Variable View

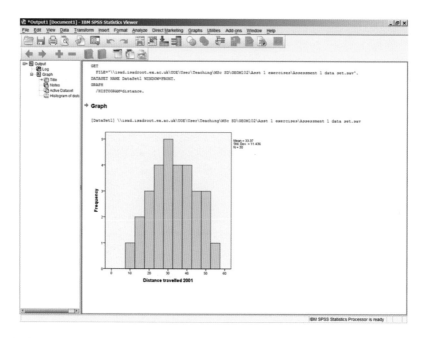

Figure 35.3c   Statistics Viewer

facility to code data using textual descriptors). Indeed, MINITAB's language assumes that users have a basic knowledge of statistics and tends to refer to operations by test name rather than the desired outcome. In contrast, SPSS has been developed as an explicitly social science application (although it is also entirely appropriate for analysis of physical science data). In this way, it has a bias towards operations that will be of use to social scientists. Indeed, it tends to assume less technical knowledge and overall is a less intimidating piece of software to start using.

In SPSS, the **Data Editor** is the spreadsheet (Figure 35.3a), but note that there are two tabs, located at the bottom left of the spreadsheet. In the **Data View** (as shown in Figure 35.3a), the raw data are stored, with columns representing variables (questions in a survey or units of measurement) and rows representing cases (respondents in a survey or measurements). The **Variable View** provides a description of the data that are presented in raw form in the Data View (Figure 35.3b). We will examine the importance of the Variable View in the next section. Finally, SPSS generates all of the outputs from procedures undertaken in the Data Editor through the use of an entirely separate window, known as the **Statistics Viewer** (Figure 35.3c).

To undertake operations in SPSS, the Data Editor and Statistics Viewer have slightly different menus, located at the top of each window. In the Output Viewer, these also vary when formatting tables and graphs, and provide more advanced formatting options. In the Data Editor, file management is undertaken in the standard way for Microsoft programs (using the File menu), with the Edit and View menus enabling basic spreadsheet functionality and a range of viewing sophistication, respectively. However, it is when the subsequent menus are considered that differences emerge between standard Microsoft Office software and SPSS. The Transform menu enables users to undertake a range of operations to manipulate data, in particular the re-coding of raw data (particularly useful for social researchers using survey data) and the creation of time series data (which is used frequently by physical geographers). The Analyze menu is the main tool for undertaking nearly all analysis in SPSS and provides descriptive as well as analytical tools for exploring measures of central tendency, dispersion and a host of inferential and advanced statistics for both parametric and non-parametric data. We will return to this menu when we examine analysis in SPSS in section 35.5. The other main menu of interest is entitled Graphs and as the name suggests, provides a way of presenting quantitative data graphically using both standard outputs (such as histograms and box plots) as well as in bespoke form (using the Chart Builder). Finally, the Help menu provides detailed and step-by-step guidance on menu-specific components of the program.

In MINITAB, the main interface is provided in a series of windows that are presented alongside each other, rather than using tabs, as in the case for SPSS. The two key windows are the worksheet and the session window (Figure 35.4). In terms of the worksheet, this is the spreadsheet-based format where the data are stored and has a very similar format to the SPSS Data View. Data are stored in columns of variables and rows of cases, with variable labels at the top of each column. Critically, it is possible to store data in a range of worksheets, which are either numbered or named, and which can be toggled between using the Window menu at the top of the interface.

Changing between the worksheet and session window can be undertaken by either clicking in the relevant area or by using the Window menu. The session window is somewhat of a historical element of MINITAB, because before the widespread using

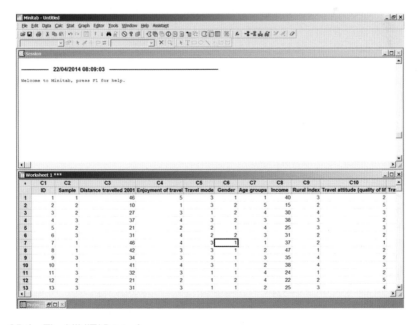

Figure 35.4 The MINITAB interface

of mice in computing, commands could be typed into this window to execute statistical procedures. This can still be activated through using the <u>E</u>nable commands function in the <u>E</u>ditor menu. However, the session window is now largely used as a means for presenting data outputs. In so doing, MINITAB will provide the command syntax, which can often be useful as a record of undertaking the procedures undertaken. Indeed, as the figures in this chapter demonstrate, the outputs provided by MINITAB are normally provided using a standard font and aesthetic that are quite different from SPSS.

The menu commands at the top of the MINITAB interface have some similarity to standard Windows programs. The <u>F</u>ile, <u>E</u>dit and <u>D</u>ata menus offer standard data management and file acquisition functions, whilst the <u>C</u>alc menu offers various ways of manipulating data in advance of statistical procedures (for example, transforming data onto different scales). However, the main menu used for the procedures likely to be used by geographers is the <u>S</u>tat menu, which provides a full range of descriptive statistics and analytical tests, arranged around a parametric/non-parametric interface. A critical distinction from SPSS is that MINITAB provides the test names rather than descriptions of the procedure (e.g. it uses ANOVA instead of 'Compare means'). The <u>G</u>raph menu provides a full range of graphics to support analysis (which will appear in new windows and which can be saved separately). Finally, the last five menus offer capability in editing worksheets and managing data files within the program, alongside both <u>H</u>elp and <u>A</u>ssistant functions, the latter of which is an advanced support service for particular types of test.

35.4 Importing and Formatting Data in SPSS and MINITAB

To start using SPSS and MINITAB, the process usually commences with the importation of existing raw data from a basic spreadsheet program like MS Excel. It is always prudent to have a raw data file available in a well-known spreadsheet program to enable simple graphics to be produced and to ensure that if changes made in either SPSS or MINITAB are not saved (or reversed, if necessary), a simple and clean data file still exists. This means that the first step for using either statistics program is importing the data from another source. This section will firstly outline the process of importing data and will then examine how these data are treated and can be formatted.

Importing data in SPSS and MINITAB

In SPSS, open the application by clicking on the Desktop icon or use **Start** > **Programs** > **IBM SPSS Statistics**. A sub-window will open that provides a range of options, for which **Open <u>e</u>xisting data set** should be selected (once there are saved SPSS data files, the option above can be used when opening the programme). Click on the **More files** tab and the standard Windows file management window will appear (Figure 35.5). Navigate to the folder where the raw data are stored and find the file, noting that it is important to ensure that you select the **<u>F</u>ile type** as Excel (or whichever file type you wish to import). Press <u>O</u>K and the data will be imported, although you may be asked to confirm that you only wish to import the data from one worksheet if the file you are using has multiple worksheets. The data will now be available in the Data Editor, albeit in a raw form that

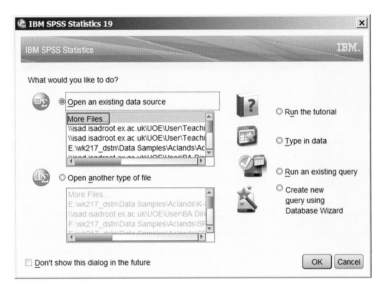

Figure 35.5 Importing data into SPSS

will require formatting and coding, which will be examined later in this section. To save your work, it is important to note that SPSS uses different file names for data (.sav files) and outputs (from the Output Viewer: .spv files). To save data, you will need to be using the Data Editor and should use the Save As function. Likewise, to save outputs, you will need to be using the Output Viewer and use the Save As function.

For MINITAB, open the application by clicking on the Desktop icon or use **Start** > **Programs** > **MINITAB 17**. Data can be accessed via the **File** menu using the **Open worksheet** function. This opens a file retrieval window (Figure 35.6) in which the appropriate file type is selected (for example, a MS Excel file) from the directory in which it is stored. By clicking the **Open** button, the data are added into a new worksheet window. At this point, it is important to save the data and any operations already undertaken in the session window. To save your work, click **File** and then **Save Current Project As**. This will enable you to save your work in the normal way and you should know that unlike SPSS, MINITAB will save both your data and the outputs in the session window (and it will also save associated graphs in they exist). In this way, this and some earlier editions of MINITAB differ from SPSS because it saves whole projects rather than separate pieces of data (although it should be noted that the historical legacy of saving separate worksheets, session windows and graphs is still possible by using the **File** menu).

Formatting data in SPSS and MINITAB

One of the major differences between the two computer packages is the way in which raw data are handled and formatted. In SPSS, numerical data are placed into columns and rows and the column headers (such as those used in Excel) are imported into the Data Editor. However, certain characters, long names and repetitions are not permitted and so often these headers require some adjustment. The editing of these

Figure 35.6 Opening data in MINITAB

headers and the formatting of the Data Editor must be undertaken in the Variable View (Figure 35.3b). This screen provides a set of columns that enable the data depicted in the Data View to be formatted and edited and its most essential function is to provide a coding scheme for numerical data that are stored as nominal or ordinal categories. Once assigned, these codes will then appear and give meaning to data outputs in the Statistics Viewer, obviating the need to remember the codes used for specific data. This is a critically important function provided in SPSS, because a considerable amount of natural and social science data are in the form of codes (for example, anything from rock types and tree classifications to political party affiliations and land uses). Importantly, it should be noted that MINITAB does not offer a mechanism for storing codes.

Read from left to right, data can be edited as follows. First variable **N**ames are those at the top of each column in the Data View and need to be whole words, short and ultimately descriptive of the variable concerned. The **T**ype column should be set to **N**umeric, to ensure that all data in the Data View can be analysed using statistics (if you are using SPSS to store textual data, set the selection to **S**tring). **W**idth and **D**ecimals specify the width of the columns in the Data View and the required number of decimal places for your data, respectively. The **L**abel column is used to provide an elongated and appropriate descriptor for each variable that will appear in all data outputs and therefore you should consider what the most appropriate terminology is to describe each variable at this stage. The **V**alues column is perhaps the most critical within the Variable View, because it provides meaning for the numerical codes that have been used to describe non-continuous data: anything from specifying gender composition to rock type. For each variable where this is the case, click on the right-hand part of the cell to open the sub-window, which will enable each numerical code to be sequentially assigned a text label, as shown in Figure 35.7. Missing data in the file can be managed through the **Missing** column, which enables the allocation of specific numerical codes to identify data which are missing, for example because a respondent did not answer a question or an observation was missed. Critically, a numerical code must be assigned to identify such missing values – simply having a blank space in the Data View is not permitted. When assigning missing values (which can be undertaken by clicking in the right-hand portion of each cell), there is the option to assign up to three missing value codes, which can be useful if there are several reasons why data are missing. Finally, the **Measure** column enables the allocation of each variable to a given measurement scale. This is somewhat of an academic exercise because assigning the wrong measurement scale to a variable will not prevent you from undertaken improper analyses, but it can be used as an aide memoire.

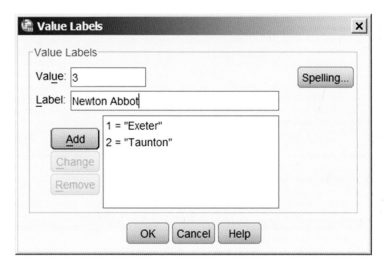

Figure 35.7   Coding variables using the **Values** column in the Variable View

In MINITAB, data are not formatted to the same degree and whilst some essential error and troubleshooting functions are available, the data remain raw and uncoded. This means that it is vitally important that you have a clear overview of the codes that are used for relevant data. For example, if you have collected questionnaire survey data and have coded questions (e.g. male and female respondents or agreement categories for attitude questions), you must be aware of these codes before proceeding to analysis. The lack of coding provided in MINITAB is largely a historical legacy from when the programme was invented, when its original intention was use by statisticians and physical scientists, where interpretative coding was not required.

Accordingly, the most effective means of checking data for consistency and dealing with missing values is to use the **Tally individual variables** function, available by using **Stat > Tables**. This opens the dialogue box featured in Figure 35.8a. By double clicking on the variables of interest in the list on the left-hand side, items appear in the Variables list. Press **OK** and the output will appear in the session window (Figure 35.8b). It is immediately apparent that in this example, there is a problem with the data as one case is identified as *. This denotes a missing value and is used in a similar way to the discrete missing values function in SPSS.

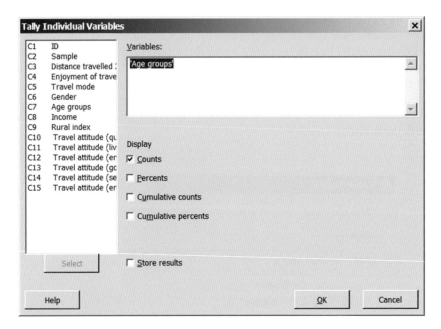

Figure 35.8a The **Tally** function in MINITAB: **Tally** dialogue box

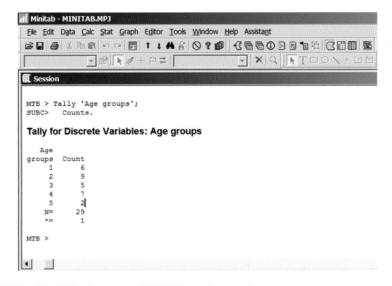

Figure 35.8b The **Tally** function in MINITAB: Session window output

35.5 Representing data in SPSS and MINITAB

Having imported data into SPSS and MINITAB, a useful way to start exploring the material is to do two things; first, to produce some simple frequency tables that enable the identification of errors in coding (often the simple result of mistyping data when initially importing results into a spreadsheet) and second to generate some basic graphics to highlight distributions (for example, histograms to illustrate the dispersion of data).

In SPSS, the simplest way to present data is by using the **Descriptive Statistics** menu of the **Analyze** menu. A range of options are available, but by far and away the best choice for initial descriptive exploration is the **Frequencies** function (Figure 35.9a). It is possible to add as many variables from your data set as you wish into the **Variable(s)** section, by using the right-facing arrow. On the right-hand side of the dialog box, a range of additional dialog boxes can be opened, each of which will provide a range of options to generate descriptive statistics about your data. However, it is often sufficient at this stage to simply press **OK**, which will prompt the Statistics Viewer to open and provide an overview of the data (Figure 35.9b). The output demonstrates

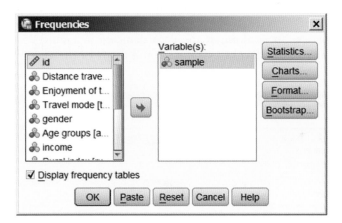

Figure 35.9a   The **Frequencies** dialog box in SPSS

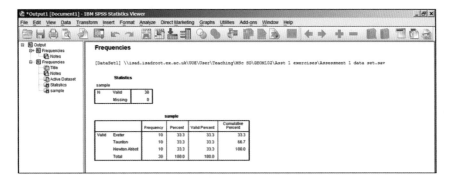

Figure 35.9b   SPSS Statistics Viewer showing **Frequencies** output

that there are no missing data and you should note how the labels assigned previously in the Variable View now appear in the table, rather than the numerical codes that are the basis for the labels.

Graphically, SPSS provides two ways of generating figures. By clicking on the **Graphs** menu in the Data Editor, you have the option of either using the **Chart Builder** or utilising what are termed **Legacy Dialogs**. The **Chart Builder** function provides an interface similar to that used in MS Excel and may be more familiar to some readers because of this. Essentially, chart types and associated variables can be dragged into the space at the top-right of the dialog box, to create the required output. However, building such charts in this way can often be problematic if you are not familiar with the program at first. Bu contrast then, the **Legacy Dialogs** function provides a simple and effective means of drawing graphs that represent the data you are interested in. The normal selection of outputs is presented and you should select the graphic most appropriate for your needs. For example, if we wanted to understand the dispersion of data, a Histogram would be appropriate. Using the menu to select a **Histogram,** the dialog box in Figure 35.10a is opened and an appropriate variable (measured on an interval or ratio scale) from the data set can be moved into the **Variable** section of the dialog box. By pressing **OK**, the output in Figure 35.10b is generated. Such histograms are a vital means of interpreting the measures of central tendency and dispersion of data, alongside appropriate descriptive statistics. So, looking at the data depicted in Figure 35.10b, it is clear that the data have a good level of symmetry and many of the visual properties of a

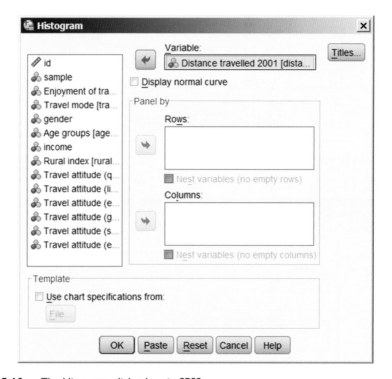

Figure 35.10a   The Histogram dialog box in SPSS

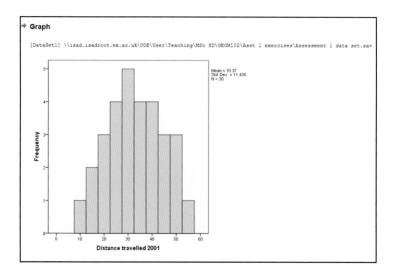

Figure 35.10b Histogram output in the SPSS Statistics Viewer

Normal distribution, although the data are clearly skewed to the right (a negative skew), which would require us to test the data for normality using an appropriate statistical test. By re-running the **Frequencies** procedure and selecting the **Statistics** option, it is possible to generate a range of numerical data that would enable examination of the data through the use of measures of central tendency (mean, median and mode) and dispersion (skewness and kurtosis), and alongside standard measures lie the standard deviation, range and variance.

The graphics SPSS provides can also be significantly edited by simply double clicking on the chart area. This opens the **Chart Editor** (Figure 35.10c), which is an intuitive tool enabling the simple formatting of the figure that is normally enabled by double-clicking on different portions of the graphics. Formatting such charts (the same process can also be undertaken by double-clicking on tables) means that data scale labels, graph colours and other features can be effectively edited into a style that suits the purpose of your work, for example an academic dissertation or consultancy report.

MINITAB provides similar descriptive functions to SPSS, accessed through the Tally individual variables function described in section 35.3 (see also Figure 35.8). Other ways of representing data in tabular form can also be accessed via the same Stat > Tables menu in MINITAB and one of the most useful is the cross-tabulation function, which enables data to be compared using both counts and percentages. To undertake this procedure, use **Stat > Tables > Cross tabulation and Chi-Square**. A dialogue box (Figure 35.11a) will be provided, which enables variables from the list on the left-hand side to be imported into 'columns' and 'rows' boxes on the right-hand side. By pressing **OK**, the output in the session window presents the cross-tabulation (Figure 35.11b). It should be noted that the explanatory information for interpreting the table is provided both above and below the data and are essential reading in order to accurately appreciate the differences between count data, percentages and totals. Indeed, this table illustrates the importance of understanding the codes that have been applied to data.

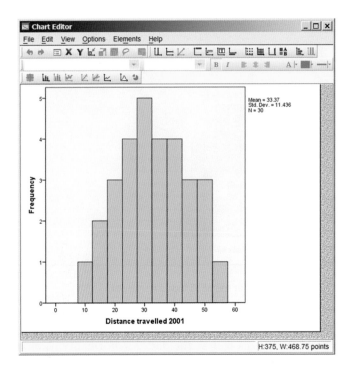

Figure 35.10c   The **Chart** Editor in SPSS

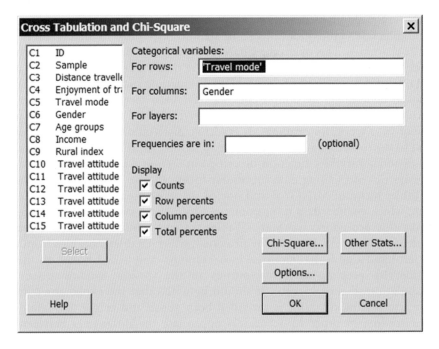

Figure 35.11a   Cross-tabulation in MINITAB: Dialogue box

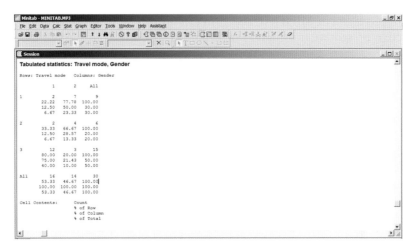

Figure 35.11b   Cross-tabulation in MINITAB: Dialogue Session window output

Graphically, MINITAB has several helpful functions for exploring and presenting data. In the <u>S</u>tat menu, it is possible to use **Basic Statistics** > **Graphical Summary** to provide an overview of data measured on interval and ratio scales. By placing a variable of interest into the <u>V</u>ariables box and pressing **OK**, the summary provided in Figure 35.12a is provided. This feature combines several graphics with key descriptive statistics and is an excellent means of rapidly appreciating the composition of data and the appropriateness of such data for parametric analysis (there is the provision of a normality test in the top-right corner of the output box).

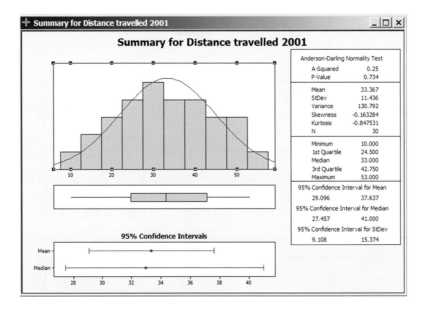

Figure 35.12a   <u>G</u>raphical Summary in MINITAB

The Graphs menu of MINITAB also provides a range of features that enable simple and effective presentation of data. For example, to draw a histogram, use **Graph** > **Histogram** and then select one of the four options in the resulting dialogue box (e.g. **With fit**). The output (Figure 35.12b) presents a standard histogram with a

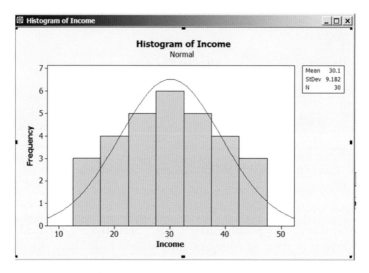

Figure 35.12b   Graphical output in MINITAB: Histogram

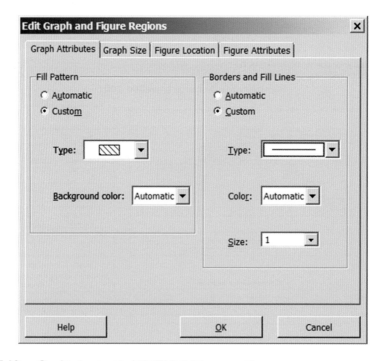

Figure 35.12c   Graphical output in MINITAB: Editing a graphic

normal curve fitted for ease of interpretation. As with SPSS, such graphics can be formatted using a range of design options and in MINITAB, these are activated by double clicking in the relevant area of the graph, which will permit the editing of data scale text, format and font, as well as the colour and lines used for the graphic (Figure 35.12c). As with most MS Windows programs, it is simplest just to experiment with the features available until the desired outcome is achieved.

35.6 Doing Analysis in SPSS and MINITAB

Having described data and formatted some of the outputs that can be generated through the basic functionality of the respective packages, many students wish to proceed directly with analysis. There are a couple of notes of caution that should be taken into account before pressing ahead with analysis. First, analysing data with packages like SPSS and MINITAB will only produce results as good as the background knowledge of the user. In other words, such packages cannot detect the poor reasoning or inappropriate choice of statistical procedures by users and as such it is all too easy to perform tests that are simply not valid or appropriate. Second, some basic understanding of the likely outputs and their statistical interpretation is required by users when utilising statistics packages; such applications rarely give an indication of the implications of results nor the caution required when interpreting findings. Third, packages like SPSS and MINITAB will process the data inputted into them; they will not automatically detect errors in coding that occur during data input and so comprehensive checking of data is necessary before proceeding with analysis.

Once these considerations have been taken into account, each package offers a range of analytical procedures that can handle the range of statistical procedures likely to be part of a project. It is helpful to break these down into the different types of procedures that could address particular research problems (Box 35.1). These can be relatively simple in nature (for example, describing data by means of measures of dispersion) or complex (utilising a statistical procedure to explain the variability in one variable by using another as a predictor). In most cases, the analysis undertaken needs to be driven by the research question being posed and should take into account, as the basis of test selection, the nature of the data being examined (parametric or non-parametric).

**Box 35.1   Types of statistical procedure for
specific research questions**

Choosing the correct statistical test depends on understanding what you want to find out (the analysis type) and the statistical procedure that will address the specific issue. Reading from left to right in the following table demonstrates different levels of analysis, the procedures that are used in such cases and examples of these.

(Continued)

Analysis type	Statistical procedure	Example
(Continued)		
Descriptive	Measures of central tendency, measures of dispersion, frequency analysis	Exploring the mean, median, mode, standard deviation, kurtosis[1] and skewness[2] of a distribution
Inferential	Tests of difference and association between samples and populations, and two or more samples	Examining the difference between a sample mean and the population mean
Relational	Tests of correlation between two variables	Establishing the strength of a relationship between two variables
Reductionist	Aggregating variables and understanding commonalities between variables	Grouping variables with similar characteristics and therefore being able to reduce the number of variables
Classificatory	Classifying cases into segments according to specific variables	Defining particular segments in a sample that have similar characteristics
Explanatory	Explaining and predicting a dependent variable using independent variable(s)	Understanding the explanatory power of an independent variable in explaining a dependent variable

1   Kurtosis is the extent to which a distribution of data on a histogram is peaked (the data are very concentrated) or flat (the data are spread out along the distribution).
2   Skewness is the extent to which data on a histogram are predominantly gathered towards one end of the distribution.

In SPSS, there is no simple demarcation between parametric and non-parametric tests that is initially obvious. For example, under the **Analyze** menu, statistical tests are to be found under the **Compare Means**, **Regression**, **Dimension Reduction**, **Classify** and **Non-parametrics** menus, notwithstanding the **Descriptives** menu which enables detailed exploration of some of the measures of central tendency and dispersion. Similarly, MINITAB offers a range of test menus that are not necessarily attuned to the language that might be found in standard statistics textbooks. Accordingly, Boxes 35.2 and 35.3 provide a guide to the tests related to specific research questions in each computer package, according to whether the data you are working with are parametric or non-parametric.

Box 35.2   Test selection in SPSS under the Analyze menu

Procedure	Parametric[1]	Non-parametric[2]
Descriptive	Utilise the **Descriptives** menu and **Frequencies** > **Statistics** options	
Inferential	**Compare Means** option for a range of tests	**Non-parametrics** > **Legacy dialogs** option for a range of tests

Procedure	Parametric[1]	Non-parametric[2]
Relational	**Correlate** option, selecting Pearson's correlation coefficient	**Correlate** option, selecting Spearman's correlation coefficient
Reductionist	**Dimension Reduction** > **Factor Analysis** and then select appropriate procedure according to data type	
Classificatory	**Classify** > **Hierarchical Cluster** and then select appropriate **Method** according to data type	
Explanatory	**Regression** > **Linear** and then select appropriate dependent and independent variable(s)	**Regression** and then select appropriate procedure from **Binary Logistic, Ordinal, Multinomial Logistic and Probit**

Note: the procedures indicated in this box are the most commonly used to address the research questions stated, but this is not a comprehensive list.

1   Parametric tests are those performed on data measured on continuous data scales where the difference between each data point is equal (e.g. rainfall in mm or age in years). To be parametric data, the data must also conform to a Normal distribution on a histogram (a symmetrical, bell-shaped distribution). Data can be tested for Normality in SPSS using a Kolmogorov-Smirnov test and in MINITAB using an Anderson-Darling test.
2   Non-parametric tests are those performed either on data that are measured on continuous data scales, but which are not normally distributed (see Parametric tests above), or on data that are measured on non-continuous data scales. There are two main types of non-continuous data scale. Ordinal scales are those where data are placed in an order but where the difference between data points is subjective or not equal (e.g. levels of agreement to question items in a survey). Nominal scales are those which contain data that cannot be placed in any logical order (e.g. rock types or family groupings).

Box 35.3   Test selection using MINITAB under the Stat menu

Procedure	Parametric	Non-parametric
Descriptive	Utilise the **Basic Statistics** or **Tables** functions under the **Stat menu**	
Inferential	**Basic Statistics** menu for a range of options	**Nonparametric tests** for a range of options
Relational	**Basic Statistics** > **Correlation**	**Basic Statistics** > **Correlation**
Reductionist	**Multivariate** > **Principal Components**	
Classificatory	**Multivariate** > **Cluster Observations**	
Explanatory	**Regression** > **Regression** and then select appropriate dependent and independent variable(s)	**Regression** and then select appropriate procedure from **Binary, Ordinal or Nominal**

Note: the procedures indicated in this box are the most commonly used to address the research questions stated, but this is not a comprehensive list.

Using these two boxes, it is evident that there are a wealth of procedures available to address key research questions and one of the most important things to consider before commencing an analysis is whether you have correctly selected the test you require according to the assumptions of the test (for example, the measurement scale of the data you are using and the sample size) and the analytical question you are seeking to address.

As an example of how to undertake a test in each computer package from test selection to interpretation, we can use an example of a hypothetical data set that has featured in the screen shots used throughout this chapter. These data relate to a Travel Attitudes survey, in which 30 people were asked to answer a series of questions about their travel habits for work and their attitudes about work-based travel. One of the most common analytical questions that geographers pose is whether there are differences between samples that we can infer from data that have been collected (Inferential procedures in Boxes 35.1 to 35.3). So, for this example, we might want to examine if there is a statistically significant difference between the mean distance travelled to work in an average week between males and females amongst our 30 individuals. On the basis that we know the two samples to conform to a Normal distribution, we would elect to undertake a two-sample t test.

In SPSS, this involves using the following menu syntax: **Analyze** > **Compare Means** > **Independent Samples T-test** (Figure 35.13a). As for most SPSS operations, variables are moved from the list on the left-hand side into the **Test Variable(s)** box (note that as many as you like can be moved across into this box). In this case, we move the Distance travelled variable into this box, which contains the mean distance travelled in an average week for all 30 respondents to our survey. Because the two sample t-test examines the difference between two sample means, SPSS needs to be told which variables in the Data Editor contain the sample groups. In this case, we wish to examine the difference between male and female respondents and so the Gender variable is moved into the **Grouping Variables** box. Because it is possible that a variable containing sample identifiers could have more than two components, the **Define Groups** option should be selected, into which the codes containing the samples we wish to analyse are inserted (1 for male, 2 for female, as per the Variable View coding). The output (Figure 35.13b) has two tables, the first being a set of descriptive statistics, which provide the respective sample sizes and mean distance travelled for the two samples. In this example, it is clear that the means are different from each other, with men more likely to travel further to work on an average week than women. The second table provides the test statistic (t), which is read on the top line of the table (4.437) and compared to the significance value (reported at Sig. (2-tailed) as 0.000). Critically, it should be noted that SPSS reports the actual probability (i.e. less than 0.001) and the figure generated should never be used when reporting results in reports or student work. Rather, it should be compared to your chosen significance level (for example, above or below your chosen P-value of 0.05, 0.01, or 0.001 and so on).

The procedures provided in MINITAB are focused on the **Stat** menu, and as noted previously, are divided between parametric tests (which are accessed via the **Basic Statistics** menu, as well as most of the other **Stat** menus) and non-parametric tests, accessed via the **Nonparametric tests** menu and also, for the **Chi-Square** test, via the **Tables** menu. As for SPSS, detailed instructions for undertaking a particular

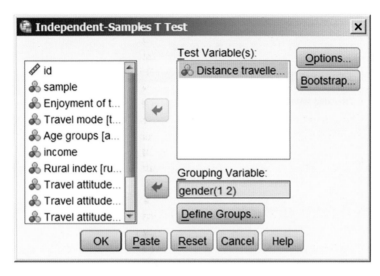

Figure 35.13a Independent Samples T-test dialog box in SPSS

Group Statistics

	gender	N	Mean	Std. Deviation	Std. Error Mean
Distance travelled 2001	Male	16	40.13	8.861	2.215
	Female	14	25.64	8.984	2.401

Independent Samples Test

		Levene's Test for Equality of Variances		t-test for Equality of Means						95% Confidence Interval of the Difference	
		F	Sig.	t	df	Sig. (2-tailed)	Mean Difference	Std. Error Difference		Lower	Upper
Distance travelled 2001	Equal variances assumed	.023	.880	4.437	28	.000	14.482	3.264		7.797	21.168
	Equal variances not assumed			4.433	27.366	.000	14.482	3.267		7.783	21.181

Figure 35.13b Output for Independent Samples T-test in SPSS

test relevant to your research question and data can be found in statistics textbooks, such as Wheeler et al. (2004). However, as an example of how another commonly applied test can be undertaken in MINITAB, we can use the same travel survey data applied in SPSS to explore if there is a statistically significant difference between gender in our sample and the travel mode used for commuting to work. In this case, because we are exploring the difference between two samples (males and females) and data measured in nominal categories (travel mode: car, bus and train), we will use the Chi-Square test, which is a non-parametric procedure (Box 35.2). To do this in MINITAB, we can use **Stat > Tables > Cross tabulation and Chi-Square**. In the resultant dialogue box, 'gender' and 'travel mode' are placed once in the 'columns' and 'rows' sections. By clicking on the **Chi-Square** button, check the 'Chi-Square analysis' and 'Expected cell counts' boxes. Press **OK** and **OK** again. The resultant output in the session window (Figure 35.14) provides an analysis of the data via

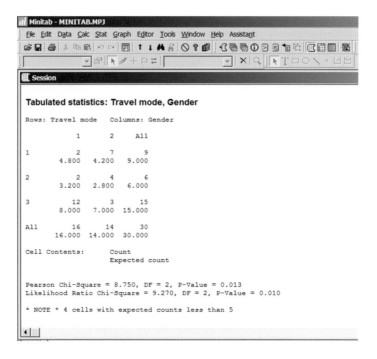

Figure 35.14   Chi-square output in the session window of MINITAB

counts (the top figure in each cell) and the expected value (the bottom figure in each cell). Underneath the table, the Chi-Square statistic and associated P-value are provided. Helpfully, MINITAB provides a warning at the end of the output, informing us that the number of expected values is over the maximum permitted for Chi-Square analysis (20% of expected values) and so the test result is not valid. If we wished to re-run the test to attain a valid result, we would need to re-code the data into a smaller number of travel mode groups (if possible) and this could be achieved through using the **Data** > **Code** > **Numeric to Numeric** function, which can store re-coded data in new columns.

35.7 Conclusion

In considering the role of computer software for analysing numerical data, programmes like MS Excel provide good, basic operations that are valuable to master. However, using SPSS and MINITAB for statistical analysis affords the user fast and reliable statistical analysis with small and large datasets alike. In fact, the procedures necessary to complete even a large research project do not take a great deal of time to complete once data have been checked for errors and manipulated to conform to the assumptions and requirements of specific statistical tests. Yet the ease and speed of using these programmes must be placed within the context of being knowledgeable and competent in understanding the data that are being examined and the suitability of particular tests for analytical questions. Put simply, it is vital to know why you are using statistical analysis for you research, what it can offer you, what characteristics the data have that you are working with and which tests will appropriately meet your requirements. These computer packages are therefore tools but not statistical solutions in their own right; they are only as good as the human user who is utilising them. The books cited in the following key readings section provide the necessary theoretical background to many of the statistical procedures used in SPSS and MINITAB and will support you in your further learning about both statistical analysis and these computer applications.

SUMMARY

- Spreadsheet programs like MS Excel are of great use for organizing and displaying numerical data, but to perform a wide variety of statistical tests, specialist statistics software like MINITAB and SPSS has a number of advantages.
- Statistics applications like MINITAB and SPSS have a great deal of power to analyse large and complex data sets, but it's important for the user to have a thorough understanding of why specific tests are being applied.
- Both MINITAB and SPSS offer very good functionality, but users should be aware of the different ways in which such programmes use and represent data.

Further Reading

There are three key texts that provide detailed descriptions of how to utilise MINITAB and SPSS for analysing statistical data. Although many more than this are available, the following three books contain examples and statistical tests that are relevant for the geographer and do so in an accessible manner. In particular, all three have a good balance between statistical theory and its application using software packages. Wheeler et al. (2004) utilise both MINITABL and SPSS as the basis for undertaking a range of both inferential and multivariate statistical texts, whereas Bryman and Cramer (2011) and Field (2013) utilise SPSS only. Bryman and Cramer's (2011) text is written exclusively for social scientists and largely uses non-technical language to understand even quite complex ideas. Field's (2013) approach is also accessible and humorous in places.

Note: Full details of the above can be found in the references list below.

Online resources

Websites: visit the MINITAB and SPSS websites for an indication of what the two programmes offer:
http://www.minitab.com/en-us/products/minitab/features/?WT.srch=1&WT.mc_id=SE004815
http://www-01.ibm.com/software/uk/analytics/spss/

References

Bryman, A. and Cramer, D. (2011) *Quantitative Data Analysis With IBM SPSS 17, 18 and 19.* London: Routledge.
Field, A. (2013) *Discovering Statistics Using IBM SPSS Statistics.* London: Sage.
McKendrick, J. H. (2010) 'Statistical Analysis Using PASW (formerly SPSS)', in N. Clifford, S. French and G. Valentine (eds) *Key Methods in Geography* (2nd edition). London: Sage. pp. 423–38.
Wheeler, D., Shaw, G. and Barr, S. (2004) *Statistical Techniques in Geographical Analysis.* London: Fulton.

ON THE COMPANION WEBSITE...

Visit **https://study.sagepub.com/keymethods3e** for author videos, chapter exercises, resources and links, plus **free** access to the following recommended articles:

1. **Longley, P. (2005) 'Geographical Information Systems: A renaissance of geodemographics for public service delivery', *Progress in Human Geography*, 29 (1): 57–63.**

This report makes a very strong case for why geographers can play a key role in influencing public policy through the analysis of spatial data, often based on large scale surveys. The techniques explored in this chapter form much of the foundation for these kinds of analyses.

2. **Poon, J.P.H. (2005) 'Quantitative methods: Not positively positivist', *Progress in Human Geography*, 29 (6): 766–72.**

This report enables us to consider what place the quantitative analyses we undertake have in our broader research framework. The report enables us to appreciate that we always have to be critical and to question why we are undertaking particular types of analysis.

36 Organizing, Coding, and Analyzing Qualitative Data

Meghan Cope and Hilda Kurtz

SYNOPSIS

Developing and maintaining a systematic process of interpreting and analysing data is an essential part of insuring the rigour of any qualitative research project. This chapter provides guidelines and references to further materials for three common analytical tools used in qualitative research in geography: coding, narrative analysis, and discourse analysis. Coding is the assigning of interpretive tags to text (or other material) based on categories or themes that are relevant to the research. The discussion here includes strategies surrounding the coding of qualitative text, such as how to evaluate your sources, identify topics and refine research questions, construct and fine-tune your coding structure, build themes, and maintain an accounting of the process of coding. Coding itself is broken down to include steps such as identifying patterns and forming categories as well as the procedures of defining first-level descriptive codes, developing second-level analytical codes and coding along a particular theme or concept. Narrative analysis and discourse analysis offer different but often congruent pathways for making conceptual sense of analytical codes.

This chapter is organized into the following sections:

- Introduction: Common interpretive and analytical practices in qualitative human geography research
- Beginning analysis: Evaluating your sources
- Practices of coding: When, how, why
- Building themes
- Cautions and considerations in coding
- Analysing qualitative data: Narrative and discourse analysis
- Conclusion

36.1 Introduction: Common Interpretive and Analytical Practices in Qualitative Human Geography Research

Qualitative research in geography uses primarily non-numerical materials, including texts, audio and video, artwork, and other forms of human expression as data; these present the researcher with unique challenges for performing rigorous and systematic interpretation, analysis and representation. In this chapter, we describe several approaches to organizing and analysing qualitative data. We begin with the process of *coding*, which is a method of data analysis that forms the building block of much of qualitative research in the social sciences and humanities. Coding refers to an iterative set of processes used to organize data, develop analytical structures, identify trends emerging from the text (or other) materials, and build themes that connect empirical findings with a broader conceptual literature. Although the process of interpreting and coding data does not follow a linear path, the strategies discussed here are presented in a roughly sequential order for clarity. We then review **narrative analysis** and **discourse analysis** as approaches that figure increasingly prominently in geographic research over the past decade, with attention given to how using one approach or the other can shape coding and analysis decisions. In order to demonstrate the analytical process, we take the reader through several stages of an actual research project – in this case, Hilda's work on controversies surrounding the sale of raw milk in the USA.

An important starting point is to consider how research questions are formed (see also Chapter 1 of this volume). To create robust and answerable research question(s), it is desirable to have some initial familiarity with both the *literature* (existing scholarship) that is relevant to your research topic and the potential sources and forms of *empirical* data to be used. New and innovative academic work often begins by bringing existing literature and theory together with an empirical context or dataset in a way that has not been done before. From that point insights start to flow as you become more familiar with the data, engage more deeply with existing scholarly literature, and rigorous analysis demonstrates that your project has something to contribute (Box 36.1).

Box 36.1   Hilda's process of forming research questions

As a critical food studies scholar, I became interested in controversies over raw milk. All milk is regulated at the state level; the purchase of fresh milk for human consumption is legal in 11 states and the District of Columbia, but market access is either banned or restricted in the other 39 states. The differential regulation of fresh milk among states in the US led me to wonder how and why such different regulations came into being, and what, if anything, raw milk drinkers do politically to open up access to this highly regulated foodstuff. Controversy about fresh milk centres on whether it is a healthful or harmful food, and my readings in the relevant scholarly literature suggested that state regulations that constrain access to fresh milk are an expression of *biopower*, or the management of population health **as a political object** (Foucault, 2007). Debates over fresh or raw milk and mandatory pasteurization had been identified in this literature as *biopolitics*, or **contestation** over the management of population health; my own reading of this body of work suggested that more attention was needed as to how and where such contestation happens in relation to food and food systems. I drew on a helpful schema for investigating biopolitics offered by Rabinow and Rose (2006) to form a series of related research questions: What are the truth and rights discourses shaping the biopolitics of raw milk in states with different raw milk regulatory regimes? How are these discourses authorized, and by what forms of authority? How do these discourses and forms of authority authorize and legitimate the interventions being made in the regulation and sale of raw milk? What modes of subjectification derive from these interventions, and how do raw milk activists contest these modes of subjectification? For a host of reasons, I decided to focus my work in Maine, where a local ordinance protecting direct sales of fresh milk and other farm foods that had recently been passed in six towns gave me a rich empirical focus for my project. Controversy around the ordinance gave me a case to incorporate into a case study research design, following Yin (2004). I was able to build a data set of published speeches and accounts, legislative materials, mainstream and blogosphere media accounts, videos, and in-depth semi-structured interviews that all say something about arguments for and against the local ordinance, and/or about fresh milk. Note that the close alignment of the data types and content to my research questions insured that I would be able to actually answer my questions in the course of the project.

Once you have some working research questions, and you have started to work on data collection (or 'data generation', depending on your approach – see next section), we encourage you to begin analysis immediately. Many researchers start organizing their data while they transcribe audio/video recordings; the process of transcription itself can be an important interpretive practice as one decides what to include or leave out, how to represent non-verbal expressions, and embarks on a first-level reflection on participants' representations (for specific guidance on transcription, see Cope, 2016, and Saldaña, 2013, which provide suggestions beyond the scope of this chapter). As you begin to look through your texts and other data, what themes do you see emerging? What commonalities or differences are easily identified, even in a cursory examination of the materials? What are the 'outliers' in your data (sometimes these are the most informative!)? While it is easy to think of research projects as progressing in an ideal-ized linear fashion from identification of question(s), collection of data, analysis, and finally to writing up results, few projects truly unfold in such tidy steps. Rather, you should assume that your research questions are somewhat tentative while you try them out and recognize that they may need to be tweaked as data collection, analysis, and further reflection proceed. An important step in that is understanding the potential and the limitations of your data sources.

36.2 Beginning Analysis: Evaluating Your Sources

Researchers are often confronted with two types of qualitative materials. One type of materials consists of pre-existing documents or visual records that were created without specific reference to the researcher's project (such as diaries, photos, maps, historical documents, secondary sources such as newspapers, oral histories and transcripts gener-ated from others' research, or digital sources such as blogs, social media, etc.). The other type is materials researchers themselves have constructed by carrying out interviews, focus groups, participatory research, or through on-line surveys or other digital interac-tions, which are then transcribed or catalogued. These two categories of materials (pre-existing and self-generated) require somewhat different approaches, particularly in the ways they connect to your research questions. Using pre-existing material tends to be an even more inductive process than using self-generated text – that is, the researcher's initial approach must be one of broad evaluation to see what trends are evident in the material. Often, particularly when using historical archives, the research questions must be somewhat flexible and open to change, depending on what emerges from the material.

On the other hand, having the opportunity to generate original data from interactive research you yourself are conducting means you can begin by linking questions for your respondents directly to your research interests. To consider how data collection is tailored, and taking an example from Meghan's interest in children's urban geographies, imagine you were interested in young people's daily negotiations of space as they pass through school, work, home and public space. You might ask them to keep diaries of their daily activities, draw maps of their patterns of movement and meaningful places, and then accompany them on a typical neighbourhood walk to listen to their own explanations of what challenges they face, how they experience mobility, and sources of friction in their negotiations of public space. In both these scenarios (using pre-existing materials or generating new ones), the development of strong research questions is essential. As stated above, the research questions will ideally reflect some element of what we already know (from related literature and theory) *and* incorporate initial findings or hypotheses of the empirical component of the research.

Whether using pre-existing sources or self-generated material, researchers doing qualitative work spend a lot of time reading and thinking about – *reflecting on* – their material. By approaching the data with an open mind, researchers allow the data to 'speak' to them to some degree. This is important because even in research with self-generated materials it may require several readings for the full diversity of topics and meanings to begin to reveal themselves. When allowing data to 'speak' researchers need to consider their own listening (reading) biases and decide whether to include an emerging theme or not. For example, in Meghan's work, if a young person's activity diary goes into great detail on family relationships but the researcher is primarily interested in daily mobility, a decision needs to be made on whether to stick to the original topic (mobility) or shift the research theme to family relationships, or perhaps combine elements of both themes and, say, examine mobility in the context of family relationships and household negotiations.

36.3 Practices of Coding: When, how, why

Coding is a way of evaluating and organizing data in an effort to identify and understand meanings, and thus is fundamentally an *analytical* practice. First, coding reveals categories and patterns, such as similarities and differences, if-then associations, and relationships between key factors or characteristics. Second, and at a more abstract level, coding helps to develop big-picture *themes* that are more closely tied to the conceptual frame of the project.

One way to begin systematically is to decide on a system of coding and taking notes. Many researchers operate entirely in the digital world for this (often using computer-aided qualitative data analysis software – CAQDAS), while others prefer traditional note-card and white-board methods, and still others blend these together. The medium of coding (digital, analog, or combined) does not determine the success or rigour of the analysis, the skill and commitment of the researcher do, so whatever you have ready access to and are comfortable with is best (on rigour, see Baxter and Eyles, 1997). The process of developing the coding structure for your project is one that is inevitably recursive, sporadic and, frankly, messy. Some scholars have tried to standardize the coding process (see, for example, Strauss, 1987) with some degree of success, but even they acknowledge that there is no clear, unidirectional process for which you can follow step-by-step instructions and at some point say that you are 'done'. Rather, coding involves reading and rereading, thinking and rethinking, and developing codes that are tentative and temporary along the way, even during an on-going research project. However, coding is also rewarding in that it enables the researcher to know his or her data intimately and see patterns and themes emerging in a way that would not be possible otherwise. For a lively account of some of the challenges and rewards, as well as two different strategies of coding (one digital and one analog), see Watson and Till (2010).

Typically, coding starts with the simpler 'descriptive' codes and then involves developing more complex 'analytic' codes, although in practice descriptive and analytical work is often overlapping and iterative. Descriptive codes are often categorical or refer to simple patterns; they contain many *in vivo* codes – that is, phrases that appear in the text, often as respondents' own words. Analytic codes are developed in relation to reflection on descriptive codes, and generally involve a return to the theoretical literature.

To begin, highlight interesting text segments and assign them relevant codes. An old-fashioned note-card example of codes from Meghan's dissertation (completed in 1995) appears in Figure 36.1, with codes along the left edge in pen and page numbers from oral history transcripts (as well as notes) in pencil. A digital version of coding, but still quite low-tech, from Hilda's milk project is demonstrated in Box 36.2; this could be done with a word processor or spreadsheet program. CAQDAS programs allow more elaborate visualization of coding, with highlighted text, color-coded categories and themes, and the ability to pull out multiple instances of the same codes. CAQDAS packages offer several different ways to schematize your coding structure, such as trees, hierarchies, or networks, though these visual arrangements can be produced in hardcopy materials too. Trial versions of CAQDAS programs are available on the companies' websites and many universities have site licenses for one or more of

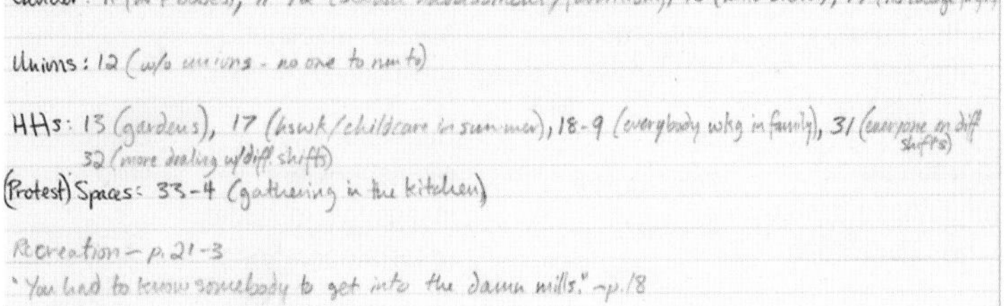

Figure 36.1 Example of low-tech coding of Meghan's dissertation project on textile mills of the 1920s and 1930s in Lawrence, Massachusetts

these, which is worth checking into. Some popular programs include Atlas.ti, NVivo, and HyperRESEARCH, each with their own strengths and weaknesses. Regardless of the level of technology employed in coding, the principles remain the same: to reveal and understand meanings from the material in rigorous and systematic ways.

Strauss (1987) pioneered the use of coding in sociology, and geographers borrow liberally from sociological methods such as his. He recommended combining three approaches to coding as needed: open coding, axial coding, and selective coding. Open coding 'is unrestricted coding of the data. This open coding is done by scrutinizing the text very closely: line by line or even word by word. The aim is to produce concepts that seem to 'fit the data' (1987: 28). The purpose of this stage is to 'open up' the data, fracturing them along the way if necessary, and breaking the data down so that conceptual implications can emerge in the later steps. Open coding mostly generates descriptive codes because it operates at a fairly superficial level. We recommend staging the open coding process by reading through your first text document, marking important sections, phrases or individual words and assigning those a code. As you read your second document, evaluate the relevance of these codes and use the text to develop more, and so on. After reading through all your materials with a critical eye, you should have a list of codes you think are important, along with your notes about them (keeping notes or 'memos' on your coding process is essential). You will find that some of your codes from 'open coding' qualify as descriptive but, as you go along, more of your codes will likely be *analytic* as you start to make new connections between empirical findings and the conceptual framework you developed (or are developing) from the relevant body of scholarly literature.

Box 36.2   Coding example

An example of Hilda's coding of an interview with a farmer-activist advocating for a local ordinance to insulate direct sales of farm food from costly regulations

Text	Descriptive codes	Analytic codes	Axial codes
Q: How do you think that the passage of the Food Safety Modernization Act relates to your work on the ordinance?			
A: it's the same thing, they're saying, how, you know, science is science,	FMSA-state regulator similarities	State-sponsored science insulated from society	
it doesn't matter if you're small or big, you can get sick from food raised from a small farm, just like from a big farm.	Scale Food-borne illness	Regulators and regulations are insensitive to scale of farm operations	Conditions – farm size

Text	Descriptive codes	Analytic codes	Axial codes
But they're totally not recognizing that we're talking about a totally different model of food production. You know? And the scale actually does have a big impact on food safety.	State regulators -farmer/activists differences. Scale Model of food production Food safety Risk	Respective knowledge contexts are worlds apart. Farmer-activists believe scale *of production* matters for food safety.	Interactions among actors (regulators-farmers) Consequences (large scale food production leads to risks)
And also, that the thing that makes the highest risk for food is when it's moving through multiple channels and chains of distribution. So each time it goes from one hand to another, from hands to the truck, they're increasing possibilities for problems to occur with that food. When it's going from this level to right here, there's no chain of distribution, and that's the safest food,	Risk Scale	Transactions as source of risk. Farmer-activists believe scale *of distribution* matters for food safety.	Conditions (descriptive) – food processing Tactics (analytical) – scale argument for food safety
so actually, you can calculate that, and you can quantify that that's actually a much safer transaction, because the risk is lower, you know?			Tactics (analytical) – discursive framing of risk/safety

Axial coding focuses the analyst's attention along an *axis*, a theme of particular interest. While there are many ways to identify an axis, we have found it most helpful to focus our work on four types of axial themes that are present in many forms of social data: *conditions, interaction among the actors, strategies and tactics,* and *consequences.* Many of these are indicated to the researcher directly by the subjects or participants. 'Conditions' can be indicated by such phrases as 'because' or 'on account of' (Strauss, 1987: 28) or passages like 'when I was in that situation ... ' In the example above, the respondent says 'you can get sick from food raised from a small farm, just like from a big farm', suggesting a set of conditions (scale of farm, potential foodborne illness). Similarly, 'interaction among the actors' means looking for how the informants engage with others, what they think of others, what others do to them. Thus, we might be interested in how farmers and activists interact with state regulatory agents. 'Strategies and tactics' refers to what people do in certain situations or how they handle particular events, or even how they frame their arguments. For example, the farmer/activist interviewed above uses tactics invoking 'science', the scale of farms, food safety, and risk to make the point. Again, there will usually be subcategories of strategies/tactics that become relevant, and these are very likely to be tied together with both 'conditions' and 'interaction with others'. Finally, 'consequences' are often easy to identify because the informant makes the connection for us: again using the example in Box 36.2, the respondent said 'So each time it goes from one hand to another, from hands to the truck, they're increasing possibilities for problems to occur with that food' and the flag here is the word 'so', indicating a cause and effect. Therefore, axial coding offers a means to identify *consequences* of certain *conditions*, but also indicate *interactions* and *strategies*. The idea, then, is to use these special types of categories to start analysing the data and pulling out new themes. Of course, you should not just code something as a 'strategy' but, rather, name that particular strategy or set of tactics that was used. In the above example, the tactic of framing the argument in terms of 'science' then becomes connected to other codes, such as the codes for scale, risk, and food safety. Codes do not stand alone but are part of a web of interconnected themes and categories, which can be visualized as hierarchies, networks, etc. to aid analysis. To round out the example in Box 36.2, imagine that as the open coding progresses, the axis of 'scale of farms related to food safety risk' catches our interest. We may follow that particular theme in a bout of axial coding by focusing on different ways the respondents framed their arguments, soon returning to the freer open coding.

Strauss's third type of coding is called 'selective coding'. This is a more systematic approach to coding that is done when a central or 'core' category has been identified and followed. For example, the putative connection between farm size and food safety may not be immediately apparent, but when it is considered in the context of the discursive arguments raised in the debates over raw milk, the relationship becomes more clear. The researcher is then sensitized to such connections and seeks to identify other similar relationships within the data, and may even decide to follow up with the individual respondents for more detail or ask about similar circumstances with other participants.

Alternatively, after some open coding and axial coding of the interviews, we might decide the core theme emerging is a more abstract one, such as the political struggle between protective legislation and people's freedom (to purchase raw milk), or the economic balance between regulation and the free market. We say it is 'core' because we have found that most of what the participants talk about is related to, or in various ways invokes, this theme. From that point onward other themes become secondary and the main lens through which the data are viewed is based on the core category (that is, we are being 'selective'). It is important to add that our identification of a core theme is related to our interest in and engagement with more academic literatures. In this example, we could take the project in the direction of political geography on one hand (as Hilda has done) or economic geography on the other, depending on which theme we found to be the core theme.

While there are many approaches to coding, Strauss's characterization of different types of codes and approaches to coding can be a helpful starting point in organizing what is a somewhat overwhelming process. Each of these approaches to coding can (and should) be done in tandem with data collection. The practices of data collection and analysis can be seen as blending together, affecting each other, and, through their mutual impact, contributing to more rigorous conclusions.

36.4 Building Themes

The coding process is fluid and dynamic, but it is not the end-product of analysis. As codes become more complex and more connected to the project's theoretical framework, they start building into *themes* that can then serve as the main topics for the final product (paper, report, dissertation, etc.). Connections between simple codes – for example, the belief that the scale of farms matters for food safety – can generate new paths of exploration, both into the data and back into the framing literature (see Box 36.3 for more detail on this). Themes can be thought of as small or emergent conceptual arguments that (hopefully!) grow in strength as the research progresses, as more corroborating evidence comes into play, and as coding demonstrates a level of robustness in the data to support the validity of theme.

The process of theme-building is central to qualitative, interpretive work because it allows for the organization of information into trends, categories, and common elements that are theoretically important. Themes may be based on similarities within the data or, conversely, on differences that appear and are interesting for some reason. The important thing to realize about theme building is that it is an ongoing process throughout the qualitative research project: themes may be identified before, during, or after the data collection and analysis stages. Indeed, many research projects are quite fluid in that they shift focus in response to the data and findings that emerge along the way.

One of the best approaches to theme building is to read across the materials being used rather than solely within them (Jackson, 2001). That is, after having spent time coding and interpreting individual documents, sit down with several or all of your materials, including your notes and memos on the coding process, third-party documents (such as reports, court proceedings, policy statements, etc.), and work with a particular topic or code while drawing from multiple texts. This process assists in seeing trends that manifest themselves in many different ways.

Box 36.3   An example of Hilda's iterative theme-building

The local ordinance that became the empirical focus of my research project challenges agricultural and food safety regulations, among other things, and so became itself an object of political debate. As time went on, I noticed that the emphasis in legislative debate and legal proceedings centred more and more emphatically on questions about food safety. The motivations for ordinance activists, to protect the viability of small-scale farmers in the state, had been all but eclipsed. In response to this insight, I conducted another round of interviews with stakeholders designed to bring their motivations back into view, by asking them what the ordinance was intended to protect. In addition, I combed back over the data I had already collected, looking for instances in which food safety was emphasized over other concerns, to see if I could recognize any pattern to the privileging of one set of concerns over another.

I found evidence of such privileging in many instances, including in the excerpt shown in Box 36.2. Comparison across subsequently collected data suggested that farmer-activists themselves could get caught up in a powerful pull away from questions about the viability of small-scale farms toward a relentless focus on food safety. This insight led me to begin to theorize that

(Continued)

(Continued)

food safety is an extraordinarily powerful meme that can be used to trump a host of other important concerns in public debate. Tracing back through my data, I used iterative coding to develop insights that range well beyond debate over raw milk *per se*, and speak to questions about the use of fear and fear-mongering in public policy-making. In addition, as Meghan observed about this provisional distinction, the 'protect small farms' message may resonate in the rural areas of northern New England as a place-specific/regional-scale argument, but 'food safety' has a more universal, larger-scale appeal. This suggested that the literature on the politics of scale may offer further insight into the project, and prompted me to read back through that literature as I moved forward with data analysis. It is also an example of how valuable it can be to share your preliminary work with others, either in writing or in conversation.

Once you have built some themes based on interesting coding insights, revisit your research questions to evaluate and perhaps refine them. It may be that what you had hoped to find really is not apparent in the data sources you are using but that another unexpected (and – one hopes – equally compelling) theme has emerged and you will adjust the direction of your project accordingly. This can happen both when using pre-existing documents and in cases in which the researcher structures the questions for the respondents. In the former, you may discover, for example, that a theme you saw emerging in the first document or two you looked at was not evident in any other materials – somewhat of a dead end. In the latter case, you might begin a set of interviews with one idea of what is happening but find that an initial idea is not borne out by respondents but a new (and even better?) theme emerges unexpectedly. In the case of Hilda's raw milk research, her review of pre-existing materials about the local ordinance suggested that it was a libertarian political strategy, but interviews with supporters of the ordinance did not consistently support that interpretation, leading her to look and listen more closely to her self-generated materials (interview transcripts, etc.) for signs of alternative political commitments on the part of ordinance supporters.

36.5 Cautions and Considerations in Coding

In the process of analysing qualitative data there are always additional issues and challenges to consider. First, it may be helpful (or even necessary, as in multi-investigator projects) to have multiple analysts. If you are working alone it may be constructive to give your code book and coded materials to a colleague or mentor to see if you are missing important themes and to test the strength of your interpretations. In group projects, a 'code book' is ideally developed with everyone who will be involved in the coding; if that isn't possible, at least have good notes attached to each code and discuss what different codes represent so there is minimal ambiguity and unevenness. Secondly, you may need to eliminate some codes as the project progresses. It is very easy to get caught up in designing codes for finer and finer resolutions of a theme, but having an unwieldy code book just means some codes won't be put to use very often and it may make sense to combine or consolidate them. This is another instance in which having a colleague or your tutor to serve as a sounding board is valuable.

Thirdly, if you are working on a project using self-constructed data materials (that is, if your respondents are available for further interaction), you may want to review your interpretations with some or all of your respondents as a way to check that your findings reflect what they intended. Note, for example, that Hilda discovered a gradual erosion of discussions around 'the viability of small farms' in favour of the theme of 'food safety' (Box 36.4) and then decided to go back to her participants to ask about this shift specifically. This practice, broadly called 'member checking,' is indicative of a larger shift in qualitative research in geography that has been growing in popularity over the past decade in which researchers assume a much deeper involvement with community members, such as through participatory action research (Kindon et al., 2007; see also Chapter 13). Indeed, some researchers have co-authored books and journal articles with the people they worked with in an effort to challenge typical power structures and issues of who 'speaks' for whom (see, for example, Pratt and the Philippine Women Centre, 1999; Sangtin Writers and Nagar, 2006). In the context of member checking, some questions to think about include: How do you deal with respondents who contradict their earlier statements? How much of your project do you want to reveal to respondents (especially that which may be critical or unflattering)? At what point (if at all) will you stop involving your subjects in the research process? What is your philosophy on the power relations set up between researcher and 'subjects'? How much time are you willing/able to devote to collaboration with your respondents? These questions all indicate a level of critical (self-) reflection that is increasingly expected to be part of rigorous qualitative research.

36.6 Analysing Qualitative Data: Narrative and Discourse Analysis

While coding qualitative data has become a popular and useful practice in human geography, particularly over the past two decades, questions about the ways that phenomena are framed in people's minds and through their communications in public arenas (such as in politics, economic exchanges, cultural practices, broadcast and social media, and legal proceedings), have also received a great deal of attention. We now turn to two methods that take a somewhat different approach from coding. This shift is perhaps best characterized as a move away from looking for social facts in the contents of participants' representations of their experiences toward thinking about the texts as objects of analysis. Thus, the driving question for this section is: *How do you understand the text itself as an object of analysis?* The answer to this question depends on your *epistemology* (your working theory of how knowledge is produced) and your interpretive strategy, both of which have tremendous consequences for your analysis. Feminist and social constructionist epistemologies have probably had the most influence on qualitative research in geography (see, for example, Moss, 2002; DeLyser et al., 2010). In this section, we discuss two interpretive strategies commonly used by feminists, social constructionists and other critical scholars: narrative analysis and discourse analysis.

Narrative Analysis

Narrative analysis as an interpretive strategy for making sense of textual data proceeds from the premise that humans are inherently story-tellers, and that stories have formal qualities that commonly relate the elements of a story. Generally speaking, stories have protagonists, antagonists, conflicts, character development/personal growth and resolutions. For research that concerns a series of events, a set of formal or informal processes, perhaps social movement struggle or protests, narrative analysis can be used to understand how the author of the text under analysis portrays these features of their narrative. In research interviews asking how people deal with ambiguity about the safety of consuming unpasteurized milk, Hilda found that they positioned themselves as protagonists wrestling with several sets of problems and tensions, and in doing so signalled the personal, social and other resources with which they address them.

Box 36.4 Narrative Analysis in the Raw Milk Project

What qualities does the speaker attribute to herself as a protagonist? What other narrative elements are identifiable in this text?

Interviewer:	So do you know many other people who drink raw milk? I mean, is this a feature of your social network?	Narrative Analytical Codes
Respondent:	(chuckling) No, most of the people here think I'm nuts. Although they're willing to try some things that I – I brought some cream cheese, and there were a few who were willing to try it.	Comfortable with being on the margin of a set of social norms in her workplace Is not totally marginalized; has social relationships such that some people try her unfamiliar foods
Interviewer:	Was it just because it was something outside their norm, or was it because they had some perception of a risky food?	Answers question with a portrayal of (warm) relationships Narrates pragmatic evaluation of risk
Respondent:	um, well, some of them, some of them that I'm close to that always laugh about me about what I eat, and how I eat. Every once in a while, they just want to give it a try, and see. And like, I told them that, you know, I had this 24 hours ago, so if anyone is going to keel over dead, I'll be the first one ... yeah, but some people are really scared of it. My sister and I go to the same doctor here...who's originally from India, and I remember when I was telling her I was drinking raw milk, she was horrified. But then [my sister] and I were saying if I were from India, I would be horrified at someone not drinking pasteurized milk, too. You know, it's kind of like, your frame of reference. I can understand it, so I just quit telling her I was doing it.	Shows sense of humour Willing to buck mainstream medical authority Doctor's position antagonistic to her own Understands risk from fresh milk as situational or contextual Resolves conflict by keeping worldviews intact and apart

In the coding stage, Hilda identified many of the same codes across a sample of fresh milk drinkers, in particular, a willingness to buck mainstream authority, a pragmatic and situational approach to evaluating risk, and a lack of social networks that included other raw milk drinkers. Many interview participants, like the one whose voice is shared in Box 36.4, described being scolded by friends and relatives for consuming a 'dangerous' food. By contrast, some participants offered a triumphal story of discovering raw milk's medicinal properties. Making sense of these different representations, sometimes by the same person, calls for iterative coding and analysis and an expanded conceptual framework which in Hilda's case centred on biopolitics (see Kurtz et al., 2012).

Discourse analysis

Discourse refers to 'a system of language which draws on a particular terminology and encodes specific forms of knowledge' (Tonkiss, 1998: 248). Discourse analysis is most appropriate in research in which discourse plays a recognizable and important role; thus, it is particularly well-suited to the study of policy change and public response, social/behavioural change in particular arenas, introduction of new programs, public debate over decisions affecting general welfare, among many other possibilities.

We can identify discourses that create conventional forms of knowledge in a variety of domains, ranging from the medical profession, professional sports, and party politics to kindergarten classrooms. Discourse analysis (DA) recognizes and analyses the ways in which discourses in such domains shape what is *sayable* and what is not, thus viewing language as a form of social practice that produces both discursive and material effects. That is, discourses not only spark further discourse, but also have tangible, 'real-world' impacts. Norman Fairclough's (1989) influential work in *critical* discourse analysis (that it, discourse analysis that identifies and critiques relations of power) draws attention to the ways in which discourses shape and are shaped by social institutions and structures. More operationally, discourses influence social identities, social relations, and social institutions, as well as systems of meaning. As a linguist, Fairclough demonstrates that these effects can be seen in the choices made about the use of language in a given unit of text.

John Paul Gee (2014) shares a linguistic focus with Fairclough, but offers a more comprehensive list of the functions of language in social interaction. He recommends focusing on building blocks or fundamental functions of language, examining how texts do the following: establish significance, delineate socially acceptable activities or practices, construct identities, effect relationships, affect politics, and foster connections. To take just three of these building blocks in a short example, if you were interested in how the text you are analysing relates to activities, identities and relationships, after developing a set of descriptive codes, you would revisit the contents of each code and ask questions such as:

- What activities are built or enacted in this text? **What social groups, institutions or cultures support and set norms** for these activities?

- What socially recognizable identity/identities does the speaker/writer **try to enact** or get others to recognize or take up?

- How do the linguistic characteristics (grammar, word choice, etc.) **build and sustain relationships** between the speaker/writer, other **people, social groups, cultures and institutions?**

To demonstrate the utility of discourse analysis, we return to the earlier example given in Box 36.2 but this time adding a column for discourse analysis to the descriptive and analytic codes. One thing to note in particular here is the above-mentioned shift from trying to identify and code the *contents* of the participant's account toward looking at how the farmer/activist uses various language strategies to effect understanding, claim legitimation, and even to 'bond' with the interviewer.

Box 36.5 Example of Hilda's coding with discourse analysis added

Text	Descriptive codes	Analytic codes	Discourse analysis *What is language doing?*
Q: How do you think that the passage of the Food Safety Modernization Act relates to your work on the ordinance?			A = Activities I = Identities R = Relationships
A: it's the same thing, they're saying, how, you know, science is science,	FMSA-state regulator similarities	State-sponsored science insulated from society	R: 'It's the same thing' and 'you know' assume common knowledge between speaker and listener A: Invokes norm of 'science' as objective and immutable
it doesn't matter if you're small or big, you can get sick from food raised from a small farm, just like from a big farm.	Scale Food-borne illness	Regulators and regulations are insensitive to scale of farm operations	R, I: Paired use of 'totally' emphasizes distance between regulators and farmer-activists, sketching a relationship of antagonism
But they're totally not recognizing that we're talking about a totally different model of food production. You know? And the scale actually does have a big impact on food safety.			R: Use of 'you know?' seeks alignment with listener I: Use of 'actually' positions speaker as an expert in her own right, challenging authority of scientists/regulators.

Text	Descriptive codes	Analytic codes	Discourse analysis *What is language doing?*
And also, that the thing that makes the highest risk for food is when it's moving through multiple channels and chains of distribution. So each time it goes from one hand to another, from hands to the truck, they're increasing possibilities for problems to occur with that food. When it's going from this level to right here, there's no chain of distribution, and that's the safest food,	State regulators-farmer/activists differences Scale Model of food production Food safety Risk	Respective knowledge contexts are worlds apart. Farmer-activists believe scale is important for food safety. Transactions as source of risk.	I: 'And also' evokes more expansive and detailed command of information, underscoring authority of speaker. R: Use of formal language like 'multiple channels and chains of distribution' signals a relationship to the institutions shaping debate (food safety policy) which is not totally dismissive or antagonistic. Signals pervasiveness of these models of thinking about food.
so actually, you can calculate that, and you can quantify that that's actually a much safer transaction, because the risk is lower, you know?	Food safety Risk		A, R: Use of 'calculate', 'quantify' and 'safer transaction' are activities demanded by speaker's antagonists. May speak to pervasiveness of quantification. May also signal an effort to find common ground in order to resolve differences.

It should be apparent that there is some conceptual overlap between narrative analysis and discourse analysis as outlined here. Each approach enables the researcher to think through how people (authors of texts) orient to an action, understand a context, and position themselves in relation to each. The focus of discourse analysis is generally broader than the focus of narrative analysis, however. Discourse analysis allows the researcher to examine how positions on a given issue (here, regulation of raw milk and other farm foods) are framed, justified, and contested. Discourses can be understood without necessarily focusing on the positioning of protagonists. At the same time, people speaking to a researcher about such issues do position themselves in relation to discourses, either explicitly or implicitly, and thus pairing discourse analysis with narrative analysis in the same project can be illuminating (and fun).

36.7 Conclusion

Coding, narrative analysis, and discourse analysis should be seen as active, thoughtful, interpretive processes that generate themes, elicit meanings, and identify participants' framings of their experiences; these in turn enable the researcher to produce representations of the data that are lively, robust and suggestive of some broader connections to the scholarly literature. By approaching qualitative analysis as both systematic and flexible, it becomes an enlightening, fruitful and revealing practice that allows for final products that are rich with meaning and as true as possible to the participants' experiences or intentions.

There are, of course, many other issues to address in the interpretation of qualitative data. For example, Jackson (2001) suggests that, at some point in the interpretation, researchers should consider what is *absent* from their respondents' accounts. In his work on masculinity, for example, he found that certain themes were rarely or never addressed by the men in his focus groups, including fatherhood, race, friendship and all things domestic. Whether working with coding, narrative analysis, or discourse analysis, the unsaid, and perhaps even the *unsayable*, are potentially important elements to explore. The way respondents frame the position of the researcher is also important. Participants in Jackson's focus groups displayed several different types of attitude towards the university researchers in his project. He also found evidence of respondents moderating what they said in his presence and in that of his co-investigators (particularly his female colleague). Indeed, feminist researchers have explored the issue of positionality (see also Chapter 9 in this volume) and the power relations between researcher and researched, but these issues are a matter of significance for all researchers and should be addressed seriously (see Valentine, 2002; Cope, 2002; Johnson, 2008).

All these issues are important to consider, depending on your types of sources, the scope of the project, the type of data collection that was used, who is involved in the research, who the participants are, and whether your materials were pre-existing or self-generated. Our final suggestion, then, is to keep a detailed account of which techniques you have used, what problems you have encountered and how you have dealt with issues such as those discussed in this section. Qualitative research has often been critiqued for its 'hidden' methodologies: researchers need to be transparent about their procedures of data collection, coding and analysis, and should always include an explicit discussion of their methods in any presentation, whether written or oral. The methods presented in this chapter attempt to balance the need for a systematic approach with the advantages of remaining flexible and open to emerging themes and multiple interpretations. Interpretation and analysis should not be seen as tedious and boring but rather as a type of detective work – we are trying to solve mysteries using varied clues and we are open to surprises in the data that generate those 'aha!' moments of investigation and inquiry.

SUMMARY

- Coding enables qualitative researchers to make sense of subjective data in a rigorous way. It is a way of evaluating and organizing data in order to understand meanings whereby researchers identify categories, patterns, themes and connections in the data.
- Narrative analysis and discourse analysis are techniques that take the text as the object of research.
- When interpreting data and texts, revisit your research questions, ask others to look at your codes, show your interpretations to informants, and consider what is absent.
- Researchers should be transparent about their procedures for interpretation and analysis.

Further Reading

There are many social science books and articles about how to code qualitative data, but we have found the following particularly useful:

- Jackson (2001) is a quick and easily comprehended chapter that raises many important issues for qualitative research, specifically the interpretation of results. His concluding 'checklist' is especially valuable.

- Silverman (2011) is a truly comprehensive treatment of how to go about coding and analysing all kinds of verbal and text-based data, with exhaustive step-by-step instructions and plenty of real-life research examples – this is the book!

- Flowerdew and Martin (2005) is the definitive guide for geography students with a wide range of coverage and suggestions for related sources.

- Hay (2010, 3rd edition; or forthcoming 4th edition in 2016) is an excellent collection of 'how-to' chapters on a range of methods.

Note: Full details of the above can be found in the references list below.

References

Baxter, J. and Eyles, J. (1997) 'Evaluating qualitative research in social geography: establishing rigour in interview analysis', *Transactions, Institute of British Geographers*, 22: 505–25.

Cope, M. (2002) 'Feminist epistemology in geography', in P. Moss (ed.) *Feminist Geography in Practice*. Oxford: Blackwell, pp. 43–56.

Cope, M. (2016/forthcoming) 'Transcripts (coding and analysis)', in M. Goodchild, A. Kobayashi, W. Liu, R. Marston, D. Richardson (eds) *International Encyclopedia of Geography: People, the Earth, Environment, and Technology*. Washington, DC: Wiley-AAG.

DeLyser, D., Herbert, S., Aitken, S., Crang, M. and McDowell, L. (eds) (2010) *The Sage Handbook of Qualitative Geography*. London: Sage.

Fairclough N. (1989) *Language and Power*. Essex: Longman.

Flowerdew, R. and Martin, D. (eds) (2005) *Methods in Human Geography: A Guide for Students Doing Research Projects* (2nd edition). Boston, MA: Addison-Wesley.

Foucault, M. (2007) 'Questions on geography' in J.W. Crampton and S. Elden (eds) *Space, Knowledge and Power: Foucault and Geography*. Aldershot: Ashgate, pp. 173-82.

Gee J. P. (2011) *An Introduction to Discourse Analysis: Theory and Method*. New York: Routledge.

Hay, I. (2010) *Qualitative Research Methods for Human Geographers* (3rd edition). South Melbourne: Oxford University Press.

Jackson, P. (2001) 'Making sense of qualitative data', in M. Limb and C. Dwyer (eds) *Qualitative Methodologies for Geographers*. Oxford: Oxford University Press, pp. 199–214.

Johnson, L. (2008) 'Re-placing gender? Reflections on fifteen years of gender, place and culture', *Gender, Place and Culture*, 15(6): 561–74.

Kindon, S., Pain, R. and Kesby, M. (eds) (2007) *Participatory Action Research Approaches and Methods: Connecting People, Participation and Place*. Abingdon: Routledge.

Kurtz, H., Trauger, A. and Passidomo, C. (2013) 'The contested terrain of biological citizenship in the seizure of raw milk in Athens, Georgia', *Geoforum*, 48: 136–44.

Moss, P. (ed.) (2002) *Feminist Geography in Practice*. Oxford: Blackwell.

Pratt, G. with the Philippine Women Centre (1999) 'Is this Canada? Domestic workers' experiences in Vancouver, BC', in J. Henshall Momsen (ed.) *Gender, Migration and Domestic Service*. Abingdon: Routledge. pp. 23–42.

Rabinow, P. and Rose, N. (2006) 'Biopower today', *BioSocieties* 1: 195–217.

Saldaña, J. (2013) *The Coding Manual for Qualitative Researchers* (2nd edition). London: Sage.

Sangtin Writers and Nagar, R. (2006) *Playing With Fire: Feminist Thought and Activism Through Seven Lives in India.* Minneapolis: University of Minnesota Press.

Silverman, D. (2011) *Interpreting Qualitative Data* (4th edition). Thousand Oaks, CA: Sage.

Strauss, A. (1987) *Qualitative Analysis for Social Scientists.* Cambridge: Cambridge University Press.

Tonkiss, F. (2004) 'Analysing text and speech: content and discourse analysis', in C. Seale (ed.) *Researching Society and Culture.* Thousand Oaks, CA: Sage. pp. 367–82.

Valentine, G. (2002) 'People like us: negotiating sameness and difference in the research process', in P. Moss (ed.) *Feminist Geography in Practice.* Oxford: Blackwell. pp. 116–26.

Watson, A. and Till, K. (2010) 'Ethnography and participant observation', in *The Sage Handbook of Qualitative Geography* (edited by D. DeLyser, S. Herbert, S. Aitken, M. Crang and L. McDowell). London: Sage. pp. 121–37.

Yin, C. (2004) *Case Study Research* (3rd edition). Thousand Oaks, CA: Sage.

ON THE COMPANION WEBSITE…

Visit **https://study.sagepub.com/keymethods3e** for author videos, chapter exercises, resources and links, plus **free** access to the following recommended articles:

1. **Dwyer, C. and Davies, G. (2010) 'Qualitative methods III: Animating archives, artful interventions and online environments', *Progress in Human Geography*, 34 (1): 88–97.**

As the last of their three insightful assessments of the state of qualitative geography, this article by Dwyer and Davies explores emerging qualitative methods that have roots in long-standing research practices in geography: archival work, art, and on-line 'virtual' geographies.

2. **Naughton, L. (2014) 'Geographical narratives of social capital: Telling different stories about the socio-economy with context, space, place, power and agency', *Progress in Human Geography*, 38:1 3–21.**

In this paper, Naughton identifies 'narrative' as the ways people use stories to make sense of the complexity of everyday life. Her work resonates well with our interest in qualitative methods being employed to clarify and make more transparent the connections between stories, agency and thick geographic context.

3. **Lees, L. (2004) 'Urban geography: Discourse analysis and urban research', *Progress in Human Geography*, 28 (1): 101–7.**

Lees helpfully distinguishes two strands of discourse analysis. The first emerges from the Marxist approach of political economy and the critique of ideology. The second strand pulls from Foucauldian poststructural theory. Discourse analysis is inherently a critical engagement with both the 'real' world and the ways that the world is framed, talked about, and made sense of in relation to multiple expressions of power, agency, and context.

37 Using Geographical Information Systems (GIS)

Nigel Walford

SYNOPSIS

Geographical Information Systems (GIS) have become the 'technology of choice' for capturing, storing, preserving, maintaining, analysing and visualising spatial and geographical information. The impact of transferring from a paper-based to a digital technology data over the last 40 years has produced many changes and delivered new ways of working with these types of data. Public and commercial organizations, and Geography and Planning departments in universities, no longer have extensive and expanding collections of paper maps catalogued on index cards and stored in large chests that can be viewed by one person at a time. The time when planning the route for a summer holiday involved leafing through the pages of a road atlas and drawing up a list of road numbers and distances has long since passed into history. GIS have enabled us to realize that geographical and spatial data are not just maps.

This chapter explores the emergence, development and diffusion of GIS during the closing decades of the twentieth and opening decades of the twenty-first centuries. The remarkable success and ubiquity of GIS, whose origins may be traced to academic research and the information needs of governmental bodies, owe much to its ability to accrete and adapt to developments in information and communications technologies: so GIS became an indispensable part of our daily lives, whether we realized it or not.

This chapter is organized into the following sections:

- Introduction
- What is GIS? Definition and characteristics
- Where did GIS come from? Brief history
- What is spatial science?
- What can you do with GIS? Applications explored
- Which GIS?
- Conclusion

37.1 Introduction

Using GIS you can follow directions dictated to you by a satellite navigation system while driving. Using GIS you can create a digital photographic tour of your holiday to share with family and friends. Using GIS you can find where the nearest restaurant serving French, Indian, Italian, Japanese or any other cuisine is located. Using GIS parents can track the movements of their children when they are away from home and out of sight. These scenarios show how GIS has infused our everyday lives, but use in the workplace and the working practices of different types of organization are where GIS started and these remain important applications. Using GIS meteorologists can track the course of weather systems to forecast the likely landfall location of hurricanes. Using GIS search and rescue services can reach accident and emergency sites efficiently avoiding delays. Using GIS planners can create visualisations of new developments and buildings for the public to imagine and 'experience' what they will be like. Using GIS companies selling mobility aids can spend their marketing budget more effectively by targeting residents in areas where potential customers for these products are likely to live. These examples illustrate how Geographical Information Systems, more commonly referred to as GIS, have become part of our daily lives both in the workplace and in our leisure time. But it was not always like this, or was it?

There was a view in the late 1980s that GIS simply seemed to allow us to do the things we had done with geographical and spatial data stored on paper in a more efficient, accurate, reliable and speedier fashion. However, continued advances in GIS and its incorporation into mainstream information and communications technologies mean that this view has expanded considerably. We are now able to use GIS to do things that were the 'stuff of dreams' 20 years ago. We will return to some of these 'dreams' later in this chapter, but for the time being we will attempt to define GIS, explore its origins and development, and examine what is meant by spatial science.

Once we realize that in one way or another GIS underpins many of the things we take for granted in our daily lives, there is a risk that we make GIS seem too easy. We can use a satellite navigation system, a spatial search engine or tracking device without understanding what lies behind them. Indeed, such applications are considered less successful if they leave exposed their underpinning technicalities. Desktop and web-served GIS software allow us to fashion our own maps and to undertake different forms of spatial analysis, but to appreciate and realize the full and growing potential of GIS we need to understand how these systems work. Without seeking to enable you to create an 'all singing, all dancing' GIS, this chapter concludes with some pointers to how you can make use of GIS when studying or researching in Geography.

37.2 What is GIS? Definition and Characteristics

It is time to clarify what is meant by the term Geographical Information System. The UK government set up a Commission of Enquiry into the *Handling of Geographic Information* in the mid-1980s, popularly known by the name of its Chair, Lord Chorley, as the Chorley Commission, which stated that a GIS is 'A system for capturing, storing, checking, integrating, manipulating, analysing and displaying data which are spatially referenced to the Earth' (Department of the Environment, 1987: 132). This became a *de facto* standard definition and still has some merit, although it arguably underemphasizes the role of spatial science in underpinning GIS as a practical tool or approach for carrying out the tasks listed in respect of geographical data. Any definition of GIS needs to reflect its existence as a computer-based system that includes hardware, software, spatial data and people (developers and users). These systems include two main types of spatial data: first, data providing a digital, computer readable definition of the spatial and geometrical properties of geographically distributed phenomena; and second data comprising descriptive attributes or measured variables that either quantitatively or qualitatively differentiate these phenomena or control how they are visualised (mapped) – see Box 37.1. A second use of the term GIS is as a specific application of such a system, which may be ascribed an epithet to denote its purpose (e.g. 'land information system') or simply be referred to as 'the GIS' (Rideout, 1992; Ireland, 1994).

Box 37.1   Definitions of a Geographical Information System

What is GIS?

- 'A powerful set of tools for collecting, storing, retrieving at will, transforming and displaying spatial data from the real world.' (Burrough, 1986: 6)
- 'GIS is designed to facilitate sorting, selective retrieval, calculation and spatial analysis and modelling.' (Mitchell, 1989: 54)
- 'Such integrated systems for the collection, storage, manipulation and presentation of geographical data, are referred to as Geographical Information Systems (GISs).' (Maguire, 1989: 171)
- 'A computer system that can hold and use data describing places on the Earth's surface.' (Rhind, 1989: 6)
- 'A computer-assisted information system to collect, store, manipulate and display spatial data within the context of an organization, with the purpose of functioning as a decision support system.' (Kraak and Ormeling, 1996: 9)
- 'The nature of the data used, and the attention given to the processing of these data, that should be at the centre of any definition of GIS.' (Heywood et al., 2006: 19)

But what is a GIS from a conceptual as distinct from an information systems perspective? We can start to answer this question by recalling that a common task for those creating a GIS during the early years of the technology's development was the capture and conversion of information shown on existing, published maps into a digital, computer readable format. Understanding what maps are may therefore help us to determine what GIS is. A map is commonly defined as a model that abstracts and represents information from the 'real world' in a visual, manageable, portable and understandable fashion, which Robinson and Petchenik summarize as 'a graphic representation of a milieu' (1976: 16–17). Topographic maps, for example, show where geographical phenomena are located, their proximity and size in relation to one another. Geographical phenomena include permanent and transient physical and man-made features, such as mountains, rivers, coastlines, forests, buildings, roads, shipping lanes, pipelines, land use zones, addresses, administrative and political boundaries. Historically different colours and symbols have been used on printed topographic maps to represent different themes: for example, blue is universally used to depict hydrological features and green to show areas of woodland. The typical conceptual model of a GIS sees the spatial data for these thematic categories of phenomena as separated out from their combined representation on traditional maps into a series of thematic layers. Figure 37.1 illustrates how this layered structure in a GIS holds data about the geometrical and spatial properties of geographical phenomena as well as any attributes that are of interest. Separation of different types of geographical phenomena into layers enables the information held in each layer to be recombined in novel and interesting ways: and not exclusively in the way they are assembled on a traditional map. The specific application or purpose of a GIS will determine which layers are included at the start, although additional layers can be added as use of the GIS proceeds: either due to the realisation that other geographical features and data may be necessary or after they have been created by geoprocessing and analytical operations.

If required most of us could produce a simple, if perhaps not very accurate, map showing the main geographical phenomena present in a defined an area using paper and coloured pens: indeed such sketch maps remain a useful tool in exploratory field work because they help to reveal through observation how geographical features are juxtaposed one with another. How would we set about creating such a map, or preferably a more accurate version, if the data are to be stored digitally on a computer rather than as markings on paper? There are two main ways in which the spatial representation of geographical phenomena can be converted from the analog form of representation shown on a paper map into a digital version. Figure 37.2, focusing on two types of surface feature (roads and woodland), shows these two methods in relation to converting data from a traditional printed map, although they also apply in situations where the data are obtained from other sources. Conceptually the raster method covers the entire study area with a regular grid of small units, known as 'cells', and through a process of tessellation a value code is recorded for each unit that denotes whether the type of geographical phenomenon being represented in the layer is present or absent. The simplest version of such a raster grid is therefore a series of square grid cells located 'in space' by their row and column number and containing a series of 1s and 0s signifying presence or absence. It is but simple step to imagine such

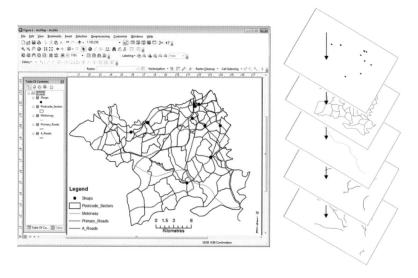

Figure 37.1   Conceptual model of a Geographical Information System

a grid where the cells hold a series of integer values (1, 2, 3, 4, etc.) denoting, for instance, different types of soil, or contain decimal values for variables such as mean annual rainfall or population density. The topographic form of a geographical feature (e.g. river or lake) is produced by (adjacent) cells with the same value or decimal values within a specified range forming linear or areal sets. This is sometimes called the field view of space and constructs features by means of a mathematical function that identifies areas with different densities or concentrations of an attribute, for example atmospheric pressure (high and low pressure systems), elevation (mountains and valleys) or land value (countryside and cities).

The vector view of the world starts from the idea that space is occupied by or filled with 'things': these things are the sorts of geographical phenomena listed above (mountains, rivers, addresses, administrative areas and so on) and are commonly known as entities. The location and geometry of these features 'in space' are recorded by means of a (usually) numerical grid referencing system that defines points, lines and areas. This enables us to classify features according to whether they exist at a single point (e.g. a post box), over a linear route (e.g. a river) or across an area (e.g. a lake). Each of these feature types (point, line and area) are defined by a minimum number of X,Y coordinate pairs that gives them a geometrical form and a location in space. This is sometimes referred to as the 'entity view of space' and is more efficient in terms of data storage than the raster system, since features are formed by telling the GIS how to join the points together to construct representations of real world features rather than requiring that each cell possess a value. The vector model may not explicitly store the grid coordinates of each pixel that make up the lake or road, but these are held ('known') implicitly by the GIS software, whereas the raster model requires the GIS to store the location and attribute value for each cell in each layer.

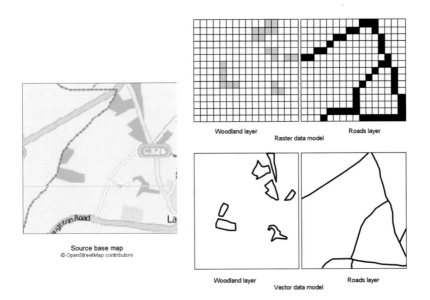

Figure 37.2 Raster and vector models for data storage in a Geographical Information System (roads and woodland feature)

The distinction between the raster and vector models has become less important to the extent that much GIS software is now capable of handling both types of data. Nevertheless these data models are still associated with certain types of data source. Remotely sensed data from satellite and airborne platforms, and digitally scanned or photographed images of other media (e.g. printed historical maps), are normally captured as raster data. Vector data may be obtained from devices capable of capturing grid coordinates of their position according to a global or local georeferencing system, or by digitizing from paper maps or raster images on screen ('heads up').

37.3 Where Did GIS Come From? Brief History

The earliest publication using the term GIS dates to 1968 (Tomlinson, 1968), which describes the Canada Land and Geographic Information System, which had started to collect different kinds for computer mapping in 1962 (Tomlinson, 1962; Switzer, 1975; Tomlinson, 1984; Goodchild, 1985). This system acted as an inventory of land-based resources and exemplified the role of GIS as a means of organizing and maintaining information relating to features above, on or under the Earth's surface. A GIS in this sense is rather like a filing cabinet of geographical and spatial data from which a user is able to find answers to geographical questions, such as through how many countries does the Mississippi flow or how many hectares of coniferous forests are there in British Columbia. GIS initially developed slowly from this starting point drawing on different aspects of information technology. Early histories of GIS published in the 1990s (e.g. Coppock and Rhind, 1991; and Foresman, 1997) regarded four aspects of information technology as central to the development of GIS: computer cartography, computer aided design, database management systems and remote sensing.

Early developers of GIS did not aim to replace the traditional skills of cartography, but they recognised that effective communication of information from geographical data should take account of spatial context. This desire coincided with the emergence of computer hardware and software capable of producing graphical representations of the geometry and attributes of geographical data. It resulted in the use of video monitors, large format printers and later pen plotters to visualise features and the patterns formed by thematic attributes. Allied with the development of computer graphics facilities was the emergence of Computer Aided Design (CAD) as a tool for designing and visualising 'new things'. CAD is used in a variety of applications areas with different levels of sophistication and the important overlap with GIS lies in the handling and display of the geometrical properties of features in space. Traditional paper-based data management involved storing records about things in a way that enabled retrieval, inspection and amendment by different people and organizations. The migration of such data management functions into the digital age resulted in the development of software (Database Management Systems) capable of allowing simultaneous, remote access to databases in a way that removed duplication, redundancy and conflict. These conditions are also important in a GIS and such systems need to be able to perform the crucial task of correctly and unambiguously linking and updating spatial and attribute data about geographical features. Remote sensing, the indirect capture of data relating to geographical phenomena, and image processing, the conversion of these data into a visual image, together comprise the fourth important contributor to GIS development. Burrough and McDonnell argued that 'a marriage between remote sensing, Earth-bound survey, and cartography [was enabled by] the class of spatial information handling tools known as GIS' (1998: 6).

Four important developments in information and computer technology have been crucial for the development, diffusion and commoditisation of GIS over the last 20 years: the Internet, mobile communications linked to global positioning systems, the creation of free, open source geoprocessing software and volunteered, crowd-sourced geo-data. In summary, the results or outputs (maps, text, images and so on) arising from carrying out various geoprocessing functions can now be delivered to devices other than a user's desktop PC. Developments in distributed computing and mobile communications mean that it is no longer necessary for the central processing power, the GIS software, geographical and attribute data and the user to be located in the same place. Changes in the relationship between GIS users and the 'system' have been reciprocal in the sense that not only can we receive the results of using geographical information processing tools remotely, but we can also contribute data to the database by devices capturing the coordinates of where we are using global positioning systems. We can either use this facility to carry out a specific task (e.g. to calculate the shortest or quickest route between our current location and that of a facility or service we want to use) or to provide data voluntarily, or, perhaps more worryingly, involuntarily, thus contributing to geographical databases. A notable example of such crowd-sourced geographical data is the OpenStreetMap organization (www.openstreetmap.org) that has built and continues to update a topographic map using the data provided by the public and mapping agencies under an Open Database Licence. People seeking to solve problems that involve some form of geoprocessing are no longer restricted to purchasing or accessing commercially produced GIS software. Box 37.2 shows that public domain GIS software was being developed in the USA during the 1970s and the MOSS system is still available. However, the creation of opensource tools and programming languages have prompted the development of freely available GIS software capable of being downloaded from the Internet and installed by users on their desktop, laptop and mobile computing devices. QGIS, formerly known as Quantum GIS, is a project of the Open Source Geospatial Foundation (OSGeo) and its creation and distribution illustrate how the Internet has not only allowed the sharing of geospatial data but also the collaborative development of software tools.

Box 37.2   Abbreviated history of Geographical Information Systems software

Selected milestones in the history of GIS

- 1960s: digital mapping programs (e.g. SYnagraphic MAPping (SYMAP) and CAMAP) developed to display maps on computer screens and to print maps on line printers respectively produced by the Harvard Lab for Computer Graphics and Spatial Analysis and the Department of Geography, University of Edinburgh. Environmental Systems Research Institute (ESRI) founded at Redlands, California.
- 1970s: (a) graphics subroutine libraries (e.g. Graphical INput Output (GINO) and National Algorithms Group (NAG) graphical supplement) written in programming language (typically

(Continued)

(Continued)

FORTRAN) are used as tools in specific applications; (b) specialist computer mapping and graphics software (GIMMS, MAPICS and ODYSSEY) included some geoprocessing functionality. Geographic Resources Analysis Support System (GRASS GIS) and Map Overlay and Statistical System (MOSS) developed separately as public domain software in the USA.
- 1980s: M&S Computing (later Intergraph), Bentley Systems, ESRI, MapInfo Corporation, Tydac Technologies Inc. and Earth Resource Data Analysis System (ERDAS)) emerge as commercial vendors of GIS software on mainframe computers (e.g. Arc-Info (vector) and SPANS (raster)). Mapping Display and Analysis System (MIDAS) is first desktop GIS in 1986, followed by IDRISI.
- 1990s: MIDAS transfers to Microsoft Windows operating system as MapInfo, ESRI produces ArcView and other vendors expand the range of desktop GIS software, but consolidation into small number of vendors and systems, and beginning of GIS data exploration and processing over the Internet, have occurred by 2000.
- 2000s: the Internet, globally positioned mobile computing, free open source software and volunteered, crowd-sourced data commoditise GIS applications and make geoprocessing available to everyone.

37.4 What is Spatial Science?

We have seen that different aspects of information and communications technologies over a period of 40 years have come together to create a geographical information industry and user community that defines and constitutes what we now recognise as GIS. However, the spatial science that underlies the processes and procedures comprising the functionality of GIS dates back much further. One of the first recorded examples of applying the principles of spatial science was in 1854 when Dr John Snow, a medical practitioner in the Soho district of central London, concerned about the high number of cholera cases he was dealing with, conceived the idea of there being a connection between these cases and the water people were drinking. Residents obtained their water from street pumps and he drew a map showing the locations of the cholera cases and the water pumps: he observed a clustering of cases around one pump. He tested his proposition that something in the water was the cause of cholera by removing the handle from this pump, thus stopping what he saw as the source of the disease. This resulted in a decrease in the number of new cases and a reduction in the cholera outbreak. Although Dr Snow was a doctor of medicine and not a geographer, we can gain some insights into what we mean by spatial science from this example (Box 37.3).

**Box 37.3   Features of Spatial Science learnt from Dr Snow's
Analysis of Cholera in Soho, London, in 1854**

What do we learn from the 1854 Soho cholera example?

- The location of the water pumps was pre-determined by the company supplying water to residents in the area.
- There were hundreds of addresses in the area where people lived who could have caught cholera, some did and some did not.

- Faced with a set of water pumps in fixed locations residents are likely to obtain their water from the nearest in terms of distance or travel time.
- Visualising the spatial distribution of the two sets of geographical phenomena (pumps and cholera cases) revealed a connection between them.
- This visualisation (map) indicates that people who caught cholera did not live at a random collection of addresses, but were clustered.
- The removal of the handle from the water pump around which the cholera cases were clustered led to a reduction in new cases.
- People caught cholera if they obtained their water from this pump which suggests the disease is caught from something in the water.

This identification of 'something in the water' as being the reason why there was a cluster of cholera cases around Dr Snow's suspect water pump relies on one important assumption about human behaviour: namely that people will try to minimise their travel time between two locations, in this case their residential address and street water pumps in mid-nineteenth century Soho. In other words, the process of obtaining water by the residents of mid-nineteenth century Soho was guided by a physical characteristic of space: the physical layout of the streets and the distance between two points. Space influences people's behaviour. Other processes concerned with human behaviour or with environmental-physical systems that produce geographical patterns in space are also guided by the physical characteristics of space. Individual and household decisions on whether to live near other people at the same stage in their lifecourse may be influenced by a desire to associated with like-minded people or to facilitate friendships amongst children, whereas the spread of molten lava flowing down the sides of a volcano or the date when winter snow melts from mountain slopes may respectively be connected with the gradient and aspect of the slopes in these areas. Underlying these ideas is what is now popularly known as the first law of geography, or Tobler's First Law, based on an assumption from Tobler used in simulation modelling of urban growth that 'Everything is related to everything else, but near things are more related than distant things'(1970: 234).

Most scientific disciplines attempt to establish universal laws concerning the behaviour of the phenomena with which they are concerned. During the 1960s and 1970s, the same time when GIS was developing, many academic geographers sought to establish the scientific credentials of their discipline by attempting to discover laws governing behaviour of human and physical phenomena in space and thus promote Geography as a nomothetic or 'law seeking' discipline. For an accessible history of these ideas see Cresswell (2013). This endeavour linked to the idea of specifying models that were able to explain human behaviour and physical processes: together these approaches resulted, for example, in various attempts to find settlements in developed and developing countries surrounded by the land use patterns of von Thünen's model and the settlement hierarchy described by Christaller's central place theory. Some of the ideas that emerged at this time, for example about determining optimal navigation routes through a transport network or the declining attraction of urban settlements with increased distance, otherwise known as the distance decay function, became part of the geoprocessing functionality of GIS, initially of mainframe and desktop systems, but now of distributed applications over the Internet.

A characteristic of the community of Geography scholars is a willingness to engage in debate about what they should be studying and how they should be studying it. This is not the place to offer definitive answers to these questions, if such answers exist. However, the arrival of GIS as a way of spatially defining things eligible for geographical study and providing a contemporary, even exciting means for doing so, as well as its permeation through undergraduate Geography degree programmes on a global scale, has re-kindled the quantitative versus qualitative debate in the discipline during the 1990s. During the last fifteen years there has been a reappraisal of this debate, which has led to the emergence of Critical Geographical Information Systems or Science (CGIS(c)) which has gone some way towards fusing quantitative and qualitative methodologies (see Chapter 18). Schuurman (2009: 140) comments that critics of GIS in the 1990s 'were concerned that GIS served large corporations, public agencies, and governments while eschewing the disenfranchised'. The four major changes in the nature of GIS during the early years of the twenty-first century outlined above (delivery of services over the Internet, ready access to global positioning systems, open source geoprocessing software and crowd-sourced geo-data) have all helped to make GIS something that is used by 'the people' rather than applied to 'the people'.

37.5 What Can You Do With GIS? Applications Explored

The introduction started with a number of scenarios in which people and organizations might be using the capabilities of GIS. Let us continue by deconstructing some of these examples to examine the basic questions underlying them and how answering them involves the application of spatial scientific concepts and techniques. Following directions when using a satellite navigation system requires that the routes and intersections along or through which you might travel between your origin and destination together with certain of their key features (e.g. length, flow direction, turn restrictions, speed limits, width and so on) are 'known' to the geoprocessing system. The GIS can then be instructed to determine a range of routes through this network, for example the shortest path between two locations or the optimal path between a series of scenic viewpoints. Finding the nearest restaurant serving a specific type of cuisine involves the GIS processing digital map data of your current position and the locations of all restaurants within a given search area together with the paths of different types of transportation route (walking, train and road vehicle) in combination with attributes relating to the type of food served and opening hours. Photographs and sound files captured with cameras and mobile phones can be tagged with the georeferenced location of the image and recording. Connecting these with a map database in a GIS can allow you and your friends, face-to-face or via the Internet, to view your pictures and hear your commentary in their geographical setting. Parents' ability to keep their children under surveillance when they are out of sight raises certain ethical issues, but a number of GIS-based products now on the market use network navigation and global positioning to provide this capability (e.g. http://lok8u.com/). Starting from the assumption that a higher percentage of the sales of mobility aids will be made to older people or those with a limiting long term illness and applying the principle underlying Tobler's First Law (see above), it seems logical for companies selling these items to target their marketing efforts in areas where above average numbers of such people live. Data collected by various means (e.g. population censuses, surveys, product registration and loyalty card schemes) about the demographic, economic and social characteristics of people and where they live can be analysed to generate descriptions of what areas are like. Creation of a three-dimensional computer generated model of a new building and the ability to alternately insert and remove this model from a similarly-dimensioned visualisation of the current cityscape allows planners and urban designers to present members of the public with existing and new versions of the built environment for assessment and feedback.

All this sounds quite complex and we should recall that the basic operations carried out in a GIS relate to measuring geometrical characteristics of digital map data and/or interrogating the attributes of the geographical features. Figure 37.3 encapsulates the main functional capabilities of a GIS starting from the relatively simple question 'what is at a location?', through to a more advanced question, 'what is the best route between A and B?' The illustration relates to a study area in south west London with Kingston upon Thames as the post town and shows selected roads, the postcode sectors and a set of shops selling cable TV services. Each section of Figure 37.3 is designed to a type of question and the GIS capability used to obtain an answer.

Clicking on a map at a location with the 'Info' tool cursor in a GIS is probably the simplest way of obtaining information. The example in Figure 37.3 shows the result of clicking on one of the shops selling cable TV facilities to discover, amongst other things, that its address is 35 Victoria Road in Surbiton. Using the 'Select by Attributes' tool in a GIS allows users to find the features in a geospatial database that possess a specific characteristic. Roads named as A245 have been selected using this tool in the second section of Figure 37.3(b) and are highlighted to make them stand out. Geographical features and their attributes are not necessarily stale and constant: their physical shape or location as well as the values and attributes captured can all change. The third section of Figure 37.3(c) shows two situations where such change has occurred: two of the original postcode sectors in the north-eastern part of the study area have each split, perhaps because the number of households and addresses increased; in the south-eastern corner four of the postcode sectors have merged into one. These changes are reflected not only by differences in the postcode sector map for each time period, but also respectively by an increase and decrease in the numbers of rows in the attribute table and the spatial database. The fourth section of Figure 37.3(d) illustrates searching for the best route between one residential location and, firstly the closest of four shops and, secondly the best route that links them, which exemplifies what is commonly known as the 'Travelling Salesperson Problem'. This illustration has been kept relatively simple with the 'best route' defined in each case as the shortest in length as opposed to journey time. It excludes travel on motorways or B roads and ignores any barriers or impedances, such as one-way streets, turn restrictions and speed limits. Geographical phenomena can display three basic patterns in their spatial location, namely dispersed, random or clustered, which may also be linked to a geographical pattern in their attribute values. Tobler's First Law (see above) says that near things are likely to be more closely related than distant things: this means that they are likely to possess similar values in respect of a given attribute. Section four of Figure 37.4 use the Getis-Ord spatial statistic G (Getis and Ord, 1992, 1996) to quantify the overall degree of clustering of postcode sector household total and to examine if there are clusters of high and low values close together.

Data entry, storage and interrogation   Location: What is at?

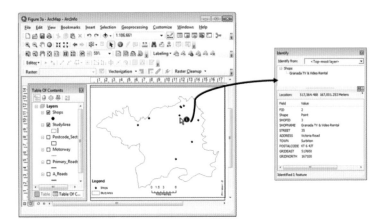

(a)

Map production   Condition: Where is it? (A245)

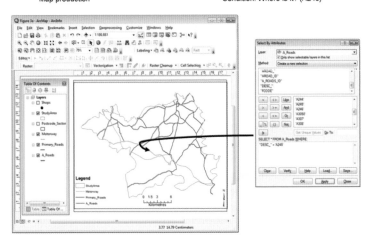

(b)

Database integration   Trend: What has changed?

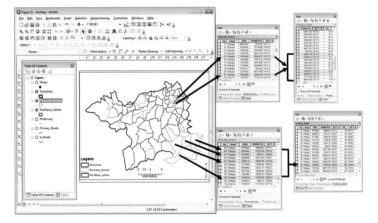

(c)

Data queries and searches   Routing: Which is the best route?

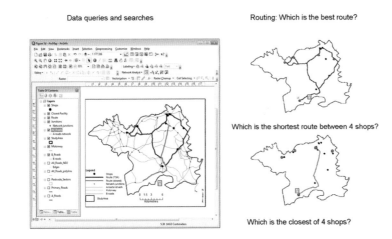

Which is the shortest route between 4 shops?

Which is the closest of 4 shops?

(d)

Figure 37.3a-d   Spatial questions and the capabilities of a GIS

Source: Adapted from Bahaire and Elliott-White (1999)

Spatial analysis (hot spot analysis)   Pattern: What is the pattern?
                                                Postcode Sector Households

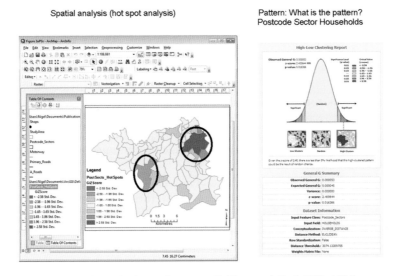

Figure 37.4   Use of the Getis-Ord spatial statistic G (Getis and Ord, 1992, 1996) to quantify the overall degree of clustering of postcode sector household total, and analysis of clusters of high and low values close together.

37.6 Which GIS?

Changes in the nature of GIS throughout its history but especially during the last fif-teen years have impacted in a number of different ways on who uses it, the way it is used and the range of uses. Early GIS software was developed in university Computing and Geography Departments on both sides of the Atlantic, with notable contributions coming from government laboratories, and was 'often made available first in the pub-lic domain' (Neteler et al., 2012: 124). Some of this software resulted in the development of proprietary, commercial products, some was available at minimal cost or for free, and others faded away. The recent developments in GIS outlined above have now led to a situation in which the range of proprietary GIS software has consolidated into a smaller number of products, when compared with a peak of over 200 in the 1990s, and open source GIS software that can be downloaded freely from the Internet. However, this division is not as straightforward as it might seem at first because several companies selling GIS software also provide freely available 'cut down' versions of their products on the Internet. There is not space to review all the GIS software available via these two routes, but some discussion of the relative merits of adopting proprietary versus free GIS software, or vice versa, is appropriate (see also Neteler et al., 2012; Steiniger and Hunter, 2013).

It is invidious to review the merits of these approaches by selecting examples of GIS software, but such a choice had to be made. Two of the leading proprietary products are ArcGIS from the Environmental Systems Research Institute (Esri) and MapInfo Professional from Pitney Bowes Software. The history of Esri and its GIS software dating back to 1969 can be found on the Institute's website (http://www.esri.com/about-esri/history), which indicates its engagement with 'GIS for everyone' in 2007 through ArcGIS Explorer. The ArcGIS suite of software with its extensions and products for specific applications (e.g. for developing Internet mapping or location analytics) together with a world-wide user community may be regarded as providing the 'complete solution' to GIS users' needs. Also in 2007, Pitney Bowes Software purchased the MapInfo Corporation, which in its original incorporation had launched MIDAS, the first desktop GIS on the marketplace in 1986 (see above). Pitney Bowes Software's MapInfo Professional is also not a single piece of software but a collection of products that link together to satisfy a comprehensive range of users' requirements. It is probably fair to say that ArcGIS through its toolbox extension includes a wider range of geostatistical and spatial analytic techniques than are present in MapInfo Professional, but some users not interested in such techniques may consider them an expensive encumbrance.

Two notable examples of free open source GIS software are GRASS (http://grass.osgeo.org/) and QGIS (formerly Quantum GIS) (http://qgis.org/en/site/index.html): each was developed as 'a volunteer project' (Steiniger and Hunter, 2013: 148) and demonstrates a degree of philanthropy in seeking to spread GIS capability to those parts of the world and user communities where resources are more scarce than in developed western economies and corporate business or governmental environments. The assertion from Neteler et al. (2012: 124) that GRASS 'can be used for geospatial data production, analysis, and mapping' applies equally to QGIS: their functionality allows querying/selection, storage, exploration, map creation, editing, transformation and analysis. Their open architecture means that users with sufficient skills and inclination are able to build on the core functionality to achieve customised solutions.

The decision on whether to use proprietary or free open source GIS software does not simply relate to matching what the software is capable of doing with what you want to achieve. It also concerns who you are, the context in which you are intending to use GIS and what resources you have at your disposal. An undergraduate student working on their final year project is a very different GIS user from an employee of a charity planning the delivery of emergency aid after a natural disaster, who is also unlike a GIS consultant contracted to identify land development sites to satisfy a supermarket's expansion plans. The support and backup available from commercial providers of GIS software undoubtedly make them more appealing to businesses and governments whose priorities may be to get answers to their 'geospatial questions' in a setting where 'time is money' or resources are limited. In contrast, users who can afford to spend more time using GIS in an exploratory fashion and perhaps who wish to develop solutions to 'geospatial questions' for themselves may find an open source GIS more appealing. Educational establishments, schools and colleges, as well as universities, are likely to adopt a mixed approach, especially in higher education where teaching programmes may aim to supply future generations of GIS software developers. Steiniger and Hunter (2013) suggest that four key sets of questions should guide the choice of GIS software for daily as distinct from developmental use, namely functionality, platform, support and other (e.g. cost). Overall the conclusion reached here is that the two types of GIS are complementary and, for example, learning to use GIS at school by means of Esri's Map Explorer or QGIS may stimulate an interest, perhaps even enthusiasm, that leads to engaging subsequently with other GIS software.

37.7 Conclusion

The history of GIS since its inception as a proposal in 1962 for computer-based mapping of land resources in Canada shows a degree of diversity and unpredictability. Who would confidently have predicted in 1962 that by 2014 many of the world's population would drive around in cars with a GIS on their dashboard that 'knew' where they were and could map out and dictate a route to their chosen destination? Such developments reinforce Kraak and Ormeling's (1996: 1) claim that 'maps are no longer the final products they used to be'. It therefore seems unwise to predict what GIS will be capable of achieving in 10 years' time, let alone another 40 years. The shift from using GIS exclusively in the professional realms of academia, public and commercial organizations into the everyday lives of the populace – what has been referred to as the commoditisation of GIS – is one of the most notable developments of recent times. Continuation of this trend towards personalised GIS applications will surely continue for the foreseeable future.

The consolidation of early developments using information technologies to capture, store, manipulate, maintain and map geospatial data into fully functional GIS programs in the 1970s precipitated debate about the difference between 'computer cartography' and 'automated cartography'. Rhind (1977) successfully defused the conflict over these and a series of similar terms – 'digital mapping', 'computer aided mapping' and 'computer mapping' – by using them synonymously in his review article. The focus of debate about GIS in the late 1980s and 1990s shifted towards its effect on geography as a discipline. The opening shots were fired in the editorial pages of *Political Geography Quarterly* with Taylor's (1990: 212) assertion that GIS represented 'a most naïve empiricism' that would diminish 'after the initial technological flush'. Goodchild's (1991: 336) riposte did not seek to laud the merits of GIS but asserted its greatest utility occurred when 'guided by people trained in the nature of geographical phenomena'. By the turn of the new millennium, open hostility had subsided and 'debates about GIS have had effects on the discipline of geography as well as the way we do GIS' (Schuurman, 2000: 664–5). The subsequent period has witnessed a certain measure of cooperation between critical and traditional proponents of

GIS. The growing number of applications involving the combination of qualitative and quantitative data, notably in the field of Historical GIS (e.g. Gregory et al., 2002; Great Britain Historical GIS Project, 2007; Knowles, 2008; Walford, 2013) together with the ubiquity of these systems outside academia testify to a future in which GIS has done more than simply survive the 'initial technological flush', but has touched the lives of most people in developed and developing economies.

SUMMARY

- GIS in the information and communications technology era dates from the early 1960s, but the underlying spatial scientific principles go back much further.
- GIS allows us to ask simple and complex questions of geospatial data and answers them by applying the principles of spatial science.
- GIS initially developed as a specialist field of information and communications technology, but the Internet, global positioning systems and crowd-sourced geospatial data now make it a central feature of the information age.
- GIS is not just about using information technology to produce maps.
- Commoditisation of GIS in the 2000s has made everyone with a mobile telecommunication or computing device into a potential user of GIS, although implementation and use of 'high end' systems in companies and public organizations still demands advanced GIS skills.

Further Reading

This list starts with the book widely regarded as the first GIS text book, which reflected the early emphasis on using these systems for land resources assessment. Since the 1980s many overarching, introductory GIS texts have been published as well as other books dealing with specific themes or applications. The texts listed here include some that have passed through several editions, others that provided an account of the subject at the date of publication and a sample of those dealing with a specific application area.

- Burrough (1986) *Principles of Geographical Information Systems for Land Resources Assessment.*
- Heywood et al. (2006) *An Introduction to Geographical Information Systems.*
- Longley et al. (2010) *Geographical Information Systems and Science.*
- Maguire et al. (1991) *Geographical Information Systems.*
- Star and Estes (1990) *Geographic Information Systems: An Introduction.*

GIS applications

- Brimicombe and Chao (2009) *Location-Based Services and Geo-engineering.*
- Chainey and Ratcliffe (2005) *GIS and Crime Mapping.*
- Harris et al. (2005) *Geodemographics, GIS and Neighbourhood Targeting.*

Note: Full details of the above can be found in the references list below.

Research articles are published in journals focusing on the applications, cartographic, spatial scientific and systems aspects of GIS. The main journals are the *International Journal of Geographical Information Science* (http://www.tandfonline.com/toc/tgis20/current#.U8ev1vldVv4), *Cartographica* (http://www.utpjournals.com/Cartographica.html),

Transactions in GIS (http://onlinelibrary.wiley.com/journal/10.1111/(ISSN)1467-9671). There are also a number of magazine publications aimed at GI professionals (e.g. http://www.geoconnexion.com/)

GIS is an international phenomenon and globally there are numerous websites where you can find out about GIS: some are run by educational or academic organizations, some by public sector bodies, and others by commercial companies. Some websites provide licensed or chargeable access to GIS services or data, others include freely available material. Examples include http://www.esri.com/training/main, http://www.mapinfo.com/, http://gisandscience.com/ and http://www.openstreetmap.org/about. UK Census and mapping data are available to students and researchers from UK Data Service (https://www.ukdataservice.ac.uk/get-data/key-data/census-data) and the University of Edinburgh (http://edina.ac.uk/).

References

Bahaire, T. and Elliot-White, M. (1999) 'The application of Geographical Information Systems (GIS) in sustainable tourism planning: A review', *Journal of Sustainable Tourism,* 7(2): 159–74.

Brimicombe, A. and Chao, Li. (2009) *Location-Based Services and Geo-engineering.* Chichester: John Wiley & Sons.

Burrough, P.A. (1986) *Principles of Geographical Information Systems for land resources assessment.* Oxford: Clarendon Press.

Burrough, P.A. and McDonnell, R.A. (1998) *Principles of Geographical Information Systems: Spatial Information Systems and Geostatistics.* Oxford: Oxford University Press.

Chainey, S. and Ratcliffe, J. (2005) *GIS and Crime Mapping.* Chichester: John Wiley & Sons.

Coppock, J.T. and Rhind, D.W. (1991) 'The history of GIS', in D.J. Maguire, M.F. Goodchild and D.W. Rhind (eds) *Geographical Information Systems: Principles and Applications.* Harlow: Longman. pp. 21–43.

Cresswell, T. (2013) *Geographic Thought: A Critical Introduction.* Chichester: Wiley-Blackwell.

Department of the Environment (1987) *Handling Geographic Information.* London: HMSO.

Foresman, T. (1997) *The History of GIS (Geographic Information Systems): Perspectives from the Pioneers* (1st edition). Upper Saddle River, NJ: Prentice Hall PTR.

Getis, A. and Ord, J.K. (1992) 'The analysis of spatial association by use of distance statistics', *Geographical Analysis* 24: 189–206.

Getis, A. and Ord, J.K. (1996) 'Local spatial statistics: An overview'. In P. Longley and M. Batty (eds) *Spatial Analysis: Modelling in a GIS Environment.* GeoInformation International, Cambridge (distributed by John Wiley and Sons: New York).

Goodchild, M.F. (1985) 'Geographic Information Systems in undergraduate geography: A contemporary dilemma', *Operational Geographer,* 8: 34–8.

Goodchild, M.F. (1991) 'Just the facts', *Political Geography Quarterly,* 10: 335–7.

Great Britain Historical GIS Project (2007) 'A Vision of Britain through Time' [Homepage of Great Britain Historic GIS Project]. http://www.visionofbritain.org.uk/ (accessed 4 December 2015).

Gregory, I.N., Bennett, C., Gilham, V.L. and Southall, H.R. (2002) 'The Great Britain historical GIS project: from maps to changing human geography', *Cartographic Journal* 39(1): 37–49.

Harris, R., Sleight, P. and Webber, R. (2005) *Geodemographics, GIS and Neighbourhood Targeting.* Chichester: John Wiley & Sons.

Heywood, I., Cornelius, S. and Carver, S. (2006) *An Introduction to Geographical Information Systems* (3rd edition). Harlow: Pearson International.

Ireland, P. (1994) 'GIS: another sword for St Michael', *Mapping Awareness* 8(3): 26–9.

Kraak, M. J. and Ormeling, F. J. (1996) *Cartography: Visualisation of Spatial Data.* Harlow: Longman.

Knowles, A.K. (ed.) (2008) *Placing History: How Maps, Spatial Data, and GIS are Changing Historical Scholarship.* Redlands, CA: ESRI Press.

Longley, P.A., Goodchild, M.F., Maguire, D.J. and Rhind, D.W. (2010) *Geographical Information Systems and Science* (3rd edition). New York: John Wiley and Sons.

Maguire, D.J. (1989) *Computers in Geography.* Longman: Harlow.

Maguire, D. J., Goodchild, M. F. and Rhind, D.W. (1991) *Geographical Information Systems*. Harlow: Longman.

Mitchell, B. (1989) *Geography and Resource Analysis*. Longman: Harlow.

Neteler, M., Bowman, M.H., Landa, M. and Metz, M. (2012) 'GRASS GIS: A multi-purpose open source GIS', *Environmental Modelling and Software*, 31: 124–30.

Rhind, D.W. (1977) 'Computer-assisted cartography', *Transactions of the Institute of British Geographers*, 2: 71–97.

Rhind, D.W. (1989) 'Why GIS?', *Arc News*, (Summer): 28–9.

Rideout, T.W. (ed.) (1992) *Geographical Information Systems and Urban and Rural Planning*, Planning and Environment Study Group of the Institute of British Geographers.

Robinson, A.H. and Petchenik, B. (1976) *The Nature of Maps: Essays towards Understanding Maps and Meaning*. Chicago: University of Chicago Press.

Schuurman, N. (2000) 'Trouble in the heartland: GIS and its critics in the 1990s', *Progress in Human Geography* 24(4): 569–90.

Schuurman, N. (2009) 'Critical GIScience in Canada in the new millennium', *The Canadian Geographer* 53 (2): 139–44.

Star, J. and Estes, J. (1990) *Geographic Information Systems: An Introduction*. Upper Saddle River, NJ: Prentice Hall.

Steiniger, S. and Hunter, A.J.S (2013) 'The 2012 free and open source GIS software map – a guide to facilitate research, development and adoption', *Computers, Environment and Urban Systems* 39: 136–50.

Switzer, W.A. (1975) 'The Canadian Geographic Information System', paper presented at the International Cartographic Association Conference, Enschede, The Netherlands, 21–25 April. Reprinted as CLDS Selected Papers II, #R001050 (Ottawa: Environment Canada).

Taylor, P.J. (1990) 'GKS', *Political Geography Quarterly*, 9: 211–2.

Tobler W.R. (1970) 'A computer movie simulating urban growth in the Detroit region', *Economic Geography*, 46(2): 234–40.

Tomlinson, R.F. (1962) 'Computer Mapping: An Introduction to the Use of Electronic Computers In the Storage, Compilation and Assessment of Natural and Economic Data for the Evaluation of Marginal Lands'. Report presented to the National Land Capability Inventory Seminar held under the direction of the Agricultural Rehabilitation and Development Administration of the Canada Department of Agriculture, Ottawa, 29-30 November.

Tomlinson, R.F. (1968) 'A Geographic Information System for regional planning', in G.A. Stewart (ed.) *Land Evaluation: Papers of a CSIRO symposium, organized in cooperation with UNESCO 26–31 August 1968*. South Melbourne: Macmillan. pp. 200–10.

Tomlinson, R.F. (1984) *Geographical Information Systems – a New Frontier*. Proceedings, International Symposium on Spatial Data Handling, Zürich, 1. pp. 2–3.

Walford, N.S. (2013) 'The extent and impact of the 1940 and 1941 "plough-up" campaigns on farming across the South Downs, England', *Journal of Rural Studies* 32: 38–49.

ON THE COMPANION WEBSITE...

Visit **https://study.sagepub.com/keymethods3e** for author videos, chapter exercises, resources and links, plus **free** access to the following recommended articles:

1. **Longley, P. (2005) 'Geographical Information Systems: A renaissance of geodemographics for public service delivery', *Progress in Human Geography*, 29 (1): 57–63.**

The leading exponent of GIS explains how geodemographics and GIS are improving people's lives.

2. Foody, G.M. (2006) 'GIS: Health applications', *Progress in Physical Geography*, 30 (5): 691–5.

GIS and health – well this report just had to mention Dr Snow's cholera map.

3. Foody, G.M. (2007) 'Map comparison in GIS', *Progress in Physical Geography*, 31 (4): 439–45.

The world does not stand still and so we need to know how to compare maps for one time with those of another.

38 Video, Audio and Technology-based Applications

Bradley L. Garrett

In this interconnection of embodied being and environing world, what happens in the interface is what is important.

Don Ihde (2001: 8)

SYNOPSIS

Video, audio and technology-based applications are components of what have recently been termed artful, creative and experimental geographies. These audio/visual approaches are conceived as both tools of engagement and methods of dissemination – they are also becoming not only more prevalent but almost impossible *not* to engage with in some way. In this chapter, I will demonstrate how the increasing ubiquity and utility of these technologies offer new ways to record and distribute research. However, as researchers, we have an obligation to engage with these technologies critically. The tension wrought between the social push behind audio/visual material and our urgent need to engage these methods judiciously is inspiring changes in the way we think about and do our research. These variations have dovetailed with emerging theoretical branches in geography that give consideration to sensory, more-than-textual and more-than-representational research praxes.

This chapter is organized into the following sections:

- Introduction
- Film and video as research method
- The audio in audio/visual
- From multi-sensory methods to more-than-textual methodologies

---

Box 38.1 Overview

- Video and audio recordings have long been a part of the work of geographers but are now coming more to the fore as part of emerging multimedia research methods
- Video opens out capacities for participatory work, documentary and ethnographic research, close analysis, and new ways to distribute these

- Audio in geography has two hurdles to jump – as an underused method and one seen as being subservient to visual methods
- These methods are more than new tools – they must be used critically
- Conceptualizing audio/visual work in the context of non-representational theory may be geography's most significant contribution to methodological reimagination

38.1 Introduction

Geographers have long experimented with forms of video, audio and technology in their work. The relationship between geography and these technologies has, however, also been riven by disputes, frustrations and concerns. These tensions primarily stem from the processual approach researchers have taken toward these methods as forms of 'data collection', rather than making the opportunity to generate intellectually robust aesthetic materials (Crang, 2003: 501). By using technology simply as tools, researchers missed a mark – for in working with video and audio in our research we undergo, whether we acknowledge it or not, an ontological reorientation that opens out new ways of being in, and engaging with, the world as well as new ways of creating and disseminating information. The process is inherently political. In re-framing these methods as methodologies, in acknowledgment of their foundational capacities to affect us, we find fertile ground that merges into modern (multi) media forms more innovatively. This type of work has been referred to as artful, experimental and creative geography at various points (Thompson et al., 2008; Paglen, 2009; Enigbokan and Patchett, 2012; Last, 2012; Hawkins, 2013). These labels might relay a sense that this type of work is happening at the margins – and that has been the case in the past. Increasingly however, we will find it impossible not to engage with these methodologies.

It is well recognized that people think and work effectively in different ways. Some people are, for instance, more attuned to listening to others speak. Others learn best sensorially, looking at, touching and feeling material objects. Still other people learn best by reading and writing. There are clearly many variations here. Geographers have been attentive to different ways of knowing and have made significant contributions to an understanding of 'visual geographies' in particular (Rose, 2016). However, the more we have come to understand the power of audio/visual products (maps, videos, photographs, audio recordings etc.), the more researchers feel the need to be critical of the hidden (or not-so-hidden!) agendas behind those productions. Consider the history of European exploration, where mapmakers acted, in a sense, as scouts for colonising forces. Often 'new' lands, such as those in North America during the nineteenth-century Westward expansion, were depicted on maps as being empty, when in fact they had been populated for thousands of years by Native Americans. We might consider this a kind of 'positioned spectatorship' where the position is wholly beneficial to those who commissioned the production of the map (Hearnshaw and Unwin, 1996; Rogoff, 2000: 11). Equally, photographs and videos of explorers 'conquering virgin territories', linked to organizations such as the Royal Geographical Society, also played a part in wielding visual methods as colonial tools (Ryan, 1997).

Though this history turned visual methods into a poisoned chalice for many geographers, this does not mean we should not engage with them. In fact, it may make it even more imperative that we do so.

Contemporary audio/visual researchers have reached for what Donna Haraway has termed 'situated knowledge', which undermines 'a conquering gaze from nowhere' (Haraway, 1991: 188) through a 'growing interest in multi-sensory methods' (Gallagher and Prior, 2014: 2). Some of this work used phenomenological frameworks to undermine the 'visual' by loading the balance more evenly among the senses and found that stretching the limits of, for instance, listening, opened up different potentialities for co-production (Ingham et al., 1999; also see Hoover, 2010, on scent). Researchers, in this vein, thus experimented with video diaries, video and photo elicitation, montage, performance, listening, sound recording and playback, and participatory video, amongst other techniques. In doing so, they realized that moving away from text-based methods text was not merely about undertaking a different type of 'field labour', producing records for analysis to prompt the production of yet more text, but that their relationship to these technologies, their co-producers and the world had been irrevocably altered. Thus audio/visual methods are

> ...practices that take on the production of space in a self-reflexive way, practices that recognize that cultural production and the production of space cannot be separated from each another, and that cultural and intellectual production is a spatial practice. (Paglen, 2009: np)

It follows then that when Gillian Rose tells us that visual materials should not be thought of as merely illustrative, or when Hayden Lorimer speaks of 'our self-evidently more-than-human, more-than-textual, multisensual worlds' (Lorimer, 2005: 83), or when Hester Parr writes that that making collaborative videos can 'reduce the "distance" between researcher and researched' (Parr, 2007: 115), we understand that we are working through something more suggestive than a simple discussion of how to use new tools. Gallagher and Prior suggest, further, that 'as these technologies become increasingly embedded into social worlds, it seems likely that the possibilities offered for geographers will continue to grow' (Gallagher and Prior, 2014: 6). In short, geographers are in the midst of a multi-sensory, multi-modal, multi-media, ontological reconfiguration (Garrett and Hawkins, 2014). In this chapter, I will begin by thinking about video methodologies, follow by a discussion of audio methodologies, and close by thinking about more-than-textual technologies more generally.

38.2 Film and Video as Research Method

> Neither words nor images but both of these and more besides
>
> Nigel Thrift (2011: 22)

As recently as 2010, Nicholas Bauch remarked that video as geographic method was a '... socially ubiquitous yet almost awkwardly absent medium in our discipline as a form of expression' (Bauch, 2010: 475). Bauch was particularly concerned about the absence of video as a form of research output, echoing Mike Crang where he wrote,

> [T]he spoken and written word constitute the primary form of 'data', whereas the world speaks in many voices through many different types of things that 'refuse to be reinvented as univocal witnesses'. Qualitative research, despite talking about the body and emotions, frames its enterprise in a particular way that tends to disallow other forms of knowledge. (Crang, 2005: 230; citing Whatmore, 2009, also see Garrett, 2010b)

Looking more closely, geographers have incorporated possibilities for film and video production into four main areas. First, participatory geographers (see Chapter 13) have used video as a way of levelling knowledge creation by co-producing filmic material (Kindon, 2003; Parr, 2007). Second, geographers have made documentary and ethnographic films (Gandy, 2007; SilentUK et al., 2010; Whatmore et al., 2011; Daniels and Veale, 2012). Third, researchers have created 'video articles', a documentary film and academic article hybrid form (Evans and Jones, 2008; Bauch, 2010; Garrett, 2010a). Finally, researchers have recorded video and then analysed distinct frames to tease out nuances of embodied, often auto-ethnographic workings (Laurier and Philo, 2006; Merchant, 2011; Simpson, 2011). These categories are obviously not exclusionary and much overlap exists.

Despite this body of work, it remains a small minority of research in the discipline. There is a reluctance to undertake research with video that may stem from technological anxieties. Matthew Gandy, who produced the beautifully shot and politically powerful film *Liquid City*, about water rights in Mumbai, outlined four 'warnings' for geographers using video on a future projects:

> First, a project of this kind involves a larger and more diverse team of people than is common for most academic projects. Second, the budgetary and time constraints pose a much greater risk of failure in comparison with most forms of academic research because filming schedules cannot easily be repeated or postponed. Third, the planning is complex and involves a greater range of responsibilities than is usually encountered in an academic project. And finally, there is a need for flexibility on the ground in order to respond to opportunities or difficulties that may arise (Gandy, 2009: 406–7).

These are important points. However, even in the past five years we have seen substantial changes in camera capability, portability, size, computing and storage capacities. More recently Daniels and Veale note, in the production of the film *Imagining Coastal Change*, that the cost of professional video production 'seems like a lot, but not compared to the costs of book and journal publishing' (Daniels and Veale, 2014: 502). Some of these films are also now being produced by researchers themselves, decreasing the price of production considerably. These changes are a by-product of unprecedented global technological literacy that has blossomed over the past decade. These evolving machineries, and our symbiotic relationships with machineries, are never only technological; the social ubiquity of a video camera in most mobile phones for instance has triggered ever-shifting codes of etiquette and how, when, where and why they are used

(see Fox, 2005). These newly-formed (and ever-morphing) relationships can complicate research projects greatly – but they can also assist them. Indeed, Gandy's concerns
above also work to highlight one of the major benefits of film and video making – that
'documentary filmmaking is by nature collaborative. Quite simply, it's impossible to
make a film about other people on your own' (Barbash and Taylor, 1997: 74). Which
leads to consideration of participatory video work.

Participatory video (PV) has been described by White (2003: 64) as a process-
oriented collaborative undertaking where 'interaction, sharing and cooperation' can
open up possibilities for 'personal, social, political and cultural change'. In this context, PV clearly fits into the broader remit of Participatory Action Research (Kindon
et al., 2008). The specificities here relate to the particularities of video as a medium:
the power of the tool to bear witness and its potential for wide distribution of
recorded material online. Development geographers, activists and participatory
geographers have thus turned to PV as a method of co-production which, contrary
to the comments of Gandy above, is seen as an accessible format where literacy is
not necessary, even if access to technology is. The simple fact is that many people in
economically disadvantaged areas are more comfortable co-creating images than
they are in co-creating text and that even in areas of relative economic affluence
children are more likely to own a mobile phone than a book (Paton, 2010). PV practitioners have recognized this and have brought many video-based projects to
communities around the world, giving them a chance to co-create material with
researchers. It is a frustrating process riven with power struggles – as all research
with human subjects should be (see Garrett and Brickell, 2014).

Figure 38.1   The author's video setup, with a 24mm lens, an external microphone and earbuds to
monitor sound

However, using video in this way, particularly because of the power and reach of the medium, has the capacity to outstrip the intentions of researchers or participants. As Haraway (2007: 263) explains, 'technologies are always compound. They are composed of diverse agents of interpretation, agents of recording, and agents for directing and multiplying relational action'. It is therefore vital that parameters everyone is comfortable with are built into these projects early on, with an understanding of what kind of (even unanticipated) work created materials might do. Data storage procedures, including encryption, deletion of unused footage, and anonymisation of imagery, are also important to discuss in the process of production. In a traditional PV model, the researcher would (gradually) give up control of the production. However, in some instances participatory goals are better met through co-authorship and collaborative work, as exemplified by the Understanding Environmental Knowledge Controversies project (see Box 38.2).

Box 38.2   The Understanding Environmental Knowledge Controversies Project

Perhaps best described as an academic documentary video production with participatory elements, Understanding Environmental Knowledge Controversies is an award-winning research project led by Sarah Whatmore, Stuart Lane and Neil Ward (see Whatmore, 2009). The project has been praised for its interdisciplinary methodology and scientific innovation, bridging human and physical geography in order to address flooding and pollution in two local areas of Sussex and Yorkshire. On the website of the project, the principal investigators describe their interests in the

> ...relationship between science and policy, and in particular how to engage the public with scientific research findings. We aim to develop a new approach to interdisciplinary environmental science, involving non-scientists throughout the process. (Whatmore et al., 2009: n.p.)

A six-minute documentary film that is, in many ways, the centrepiece of the project presentation, is interesting in that participatory methods are clearly in use yet it is not, by strict definition, a participatory film. The principal investigators have held consultation groups, had local people run the scientific modelling programmes and one participant in the film refers to Professor Whatmore as a 'facilitator' – classic PV language. However the film, rather than being made by the project participants, is clearly a documentary film about a participatory process. It has been showcased widely and is part of a suite of research outlets which included exhibitions, media appearances and numerous academic journal articles. It is the way that the film, though successful, cannot be strictly bounded or described that leads me to choose it as an interesting case study here. This project, from the involvement of local people in technical work to the well-executed (and succinct) presentation of the project on video to the construction of an interesting, up-to-date and easy to navigate website, without calling particular attention to any of these things, makes it a great model for video work.

Many participatory films are also ethnographic films. There is a small but robust body of work in geography of ethnographic filmmaking and a particularly interesting dovetail in geographers' interest in telling stories of people and places (see Chapter 6). Often, we find that as we try and tell these stories, we confront a difficulty in extending our own subjectivities to comprehend the way others apprehend the world. As Susan Hogan and Sarah Pink put it:

> Ethnographers are always faced with a problem: while we might aim to empathise with, understand, interpret and represent other people's experiences, imaginations and memories, their sensory and affective qualities are only accessible to us in limited ways. (Hogan and Pink, 2012: 236)

When we make an ethnographic film over a long period of time through a process of 'deep hanging out', we must makes sense of other people's realities (including how they relate to places) by telling stories that will communicate a feeling of being there, a feeling of time spent, a feeling of connecting across registers that is very often autobiographical. Inherently then, 'the film editor's job has always been metaphorically hypertextual' (Coover et al., 2012), a process of connecting people, time and places in a way that often defies linear logic. Ethnographic films are composed of webs of connections where distinctions between autobiography, empathetic observation and the subjective understandings of others cannot be clearly delineated. Because of the capacity for editing where moving images connect to myriad sounds, voices and fragments of events and where they relay a sense, a presence, film editing is inherently non-linear and multi-sensual and this changes the way we think about the process of research.

In the case of both video production and text: 'Ideas are formulated ahead of time, relevant evidence is sought and collected, then, both are converted by some form of technology and manipulated over and over until a compelling case can be strung together in a way that is intelligible to others' (Bauch, 2010: 481). And so, in working with audio/visual forms, we are, again, not making a move 'beyond' text but working in/alongside text to spark knowledge production across multiple registers. Importantly, the ethnographic film also involves the viewer as a player in the ethnography as soon as it is released, creating a sort of triangle of meaning between the researcher, the participant and the viewer (for more on the 'triangulation' of meaning construction through viewing images see Rose, 2016; Gold, 2002; and Rancière, 2011).

Constructing an ethnographic film about others (a film that may also be autobiographic) is a practice of 'contributing to an affective politics, a politics of relation' (Kanngieser, 2011: 337). It is in the doing, the making, the production, the distribution and the response, through all of these parts of the multi-sensual, multi-modal, multi-media research event, where ethnographic film finds its form. The 'layering and composting found in editing and effects programs have altered cinematic premises about the frame, continuity, and montage; they have expanded cinematic rhetoric and poetics, perhaps defining an increasingly hybrid language that fuses writing and images' (Coover et al., 2012).

Of course, variations in moving images are also a matter of scale. One key driver for geographical video-making specifically is the medium's recognised value for moving us beyond the textual and representational, in the context of 'facial expression, intonation, pauses, uncertainties, voice inflections, and body language', where these subtleties can be tended to with care (also see Laurier and Philo, 2006; Bauch, 2010: 482). Video can be shot at frame rates of 5,000 and 10,000 still images per second.

At this speed, the camera can see what the eye cannot, allowing for, in review, rendering the imperceptible *perceptible*. The benefits of frame-by-frame analysis for research on, for instance nuances of human interaction, are clear (Laurier and Philo, 2006; Simpson, 2011).

With these examples thus outlined, it is of course important to remind the reader that none of these notions of how film and video have, can, and should be used are bounded categories in any sense. Nor are they fixed in time – new technologies will undoubtedly break down and remake these categories, even between the time of writing this chapter and publishing it.

38.3 The Audio in Audio/Visual

> Sound, in my view, is neither mental nor material, but a phenomenon of experience – that is, of our immersion in, and comingling with, the world in which we find ourselves.
>
> Tim Ingold (2007: 11)

In the same way forms of film and video cannot be easily bounded, we also cannot draw a distinct line between the aural and visual (hence the ubiquity of the audio/visual abutment). Thinking about audio in geography has moved from Schafer's 'soundscapes' to Rodaway's early distinctions between the 'auditory' and the 'sonic' to Matless's more open-ended sonic formulations (Schafer, 1977; Rodaway, 1994; Matless, 2010). Recent work by Anderson has considered non-representational (see Chapter 12) potentialities for listening and remembering; Kanngieser has done work on the politics of speaking, tone, utterance and voice; and Gallagher and Prior have usefully summarized this disciplinary attentiveness in the context of experimental geographies of phonography (Anderson, 2004; Kanngieser, 2011; Gallagher and Prior, 2013). In our consideration of audio, the aural and the phonographic, we can revisit the question of audiencing and give ample room to both the act of recording (again as an act of sensual engagement with the world rather than as simply data collection) and the act of listening. In doing so, we run up against our lingering text barriers since as Gallagher and Prior write, 'the majority of geographically aligned research on sound has … been methodologically conventional, using techniques such as interviews, ethnography, archival research and discourse analysis, and has been disseminated via traditional written publications' (Gallagher and Prior, 2013: 1–2, citing Crang, 2003). I would therefore like, in this section, to look at some work which prioritises sound at a more sensorial level.

As with video, capturing sound is a process which can involve different forms of technology and contexts. From recording a spontaneous interview or conversation on one's phone to making sensually enticing binaural (two-channel, stereographic) recordings that mimic the way the ears and brain process sound to creating immersive soundscapes, audio recording can take many forms as a methodology. However, whatever the technology involved, Gallagher and Prior want to remind us that 'when phonographers go out into the field to "capture" sounds, they do not bring the sounds back with them' (Gallagher and Prior, 2013: 9). Sounds of course belong to the specific

Figure 38.2   An inexpensive audio recording set comprised of a recorder, a directional microphone and headphones

moments and contexts in which they exist. What we bring back are recordings – not false copies but new assemblages entirely, lush with layers of recall, anticipation and potential for replay and remix. Like video, replaying sounds will affect an audience as they bring their own experiences and expectation to bear on the process and so, with each listening, a sound is changed. This is why widely circulated sounds on the Internet become so saturated with meaning.

The Carrlands Project, led by ex-archaeologist and performance artist Mike Pearson, is an interesting aural intervention around notions of landscape and place harbouring similarities to the films of Patrick Keiller. The involvement of the Landscape and Environment program at the University of Nottingham Geography on the project (also involved in the *Imagining Coastal Change* film) is worth noting. Logging onto a website for the project, the listener is offered a selection of audio files. Each one is a rich mix of spoken word, landscape recording and dramatic music performance meshed into a tapestry of site-specific works that relay a profound a sense of place, especially if one were to listen to it near to where it was recorded. On the Carrland's website it describes the project as a

> small-scale case study that proposes site-specific performance as both inno-vative mode of enquiry and as tangible research output; performance is identified as a medium that can precipitate and encourage public visitation, inform presence, and illuminate the historically and culturally diverse ways in which a landscape is made, used, reused and interpreted. (Pearson et al. 2007: 1)

Carrlands is a reminder of course that sonic geographies are as much about listening as recording. Ben Anderson, in his research sitting with people as they listen to music in their homes, suggests that the act of listening and recalling, or indeed listening and encountering unbidden spontaneous memories, which may signify 'a not-yet conscious, visceral, logic of intensity… irreducible to wither a symbolic or emotional register', is linked to explicitly ontological questions (Anderson, 2004: 10). And so, like video, recording and listening to audio brings us once again to a point of repositioning – not at the point of distributing these productions, as is commonly assumed, but before even the moment of creation and in the anticipation of creation. For instance, Anderson writes that many people find it hard to explain why they choose to listen to certain songs when they do. Much of it has to do with a feeling embedded in that song specific to them that connects them to other times and places in their lives. Of course the anticipation of those connections is part of what guides the selection. This kind of memory might be considered *affective* (see Chapter 12) memory in that it rises not from cognition but from complex pre-cognitive feelings.

Sound is clearly very powerful. In thinking about how 'phonography is a form of writing – the inscription of sound – just as photography is the inscription of light', we must be careful not to position audio as subservient to, or indeed only visible in contrast to, video. As Gallagher and Prior tell us, 'audio is largely erased in human geography… except perhaps when subsumed within video, where it is stabilized by the referential qualities of the image' (Gallagher and Prior, 2013: 4). In response to just such prompting, I invite consideration of *Jute*, an audio/visual project led by sound.

Box 38.3   *Jute*: A Video that Foregrounds Audio

Jute is a film that was created in Dundee, Scotland, by myself, Brian Rosa and Jonathan Prior as part of an experimental geography workshop put together by Michael Gallagher and Jonathan Prior (Garrett et al., 2011). Dr Prior's research focuses on sound and listening. While co-creating the film, we made a conscious decision to preference the process of following sound over the search for 'visual' content in an attempt to subvert the normative priorities of filmmaking. In effect, we were working to pre-emptively prevent devaluing the sonic qualities we encountered as merely soundtracking. As a result, Dr Rosa and I ended up filming where Dr Prior led us with his microphone, often following sounds we could not hear with our ears or follow with our lenses, such as when he submerged contact microphones into the water to capture the claw clicks of crabs fighting at the bottom of the River Tay. We were forced then to find something to film that felt like the crabs fighting, a challenging (and fascinating!) process of visualizing unseen audio that was decontextualized from its context.

During the editing of the film, we also resolved to edit the video with the sound recordings as our starting point. We were curious to see how such an approach could lead to new possibilities in capturing the sensual flow and rhythm of the landscape similar to the way we had encountered it 'in the field'. We were interested in how this would translate as an audio/visual piece, where the viewer was not party to our direct experience of sound threading *in situ*. The final film, a six-minute drift through the city, is a film as much about sound and visuals and as much about process as place. It is, in a word, experimental.[1]

[1] The film can be viewed at http://liminalities.net/7-2/jute.mov, accessed 4 December 2015.

Figure 38.3 A still frame from *Jute*, immersed in reflection and layering

Sound-led walks have become an increasingly important way of experiencing places, especially in urban environments. Walks have been created of alternative museum narrations, street-level traversals of buried watercourses, hunting for the drone of air conditioners and ventilation shafts, and listening for urban wildlife. Many walks also just prompt the walkers to listen at particular points, waiting for a sound to come to them. Closing your eyes at these points, waiting for sounds to come at you, can be disconcerting at first but soon gives you a heightened sense of sound where vision is suppressed. Equally, recording sounds at a particular time and replaying them in the same place at a different time may also bring about a very distinct awareness of sound as the visual space empty of bodies (for instance) is populated with disembodied voices (see Gallagher, 2015). Techniques such as these can add a great deal to our understanding of the kinds of awareness cultivated by foregrounding sound.

As with video, we might assume that a certain aural literacy is required to 'properly' record and analyse sound. However, this is part of making our research more accessible to the public through participation in social practices. The public does not wait to be trained to begin recording sounds and we should not wait to be trained to begin critically recording. Further, most of us, as researchers, took the time to develop textual literacy to do our work. This likely included many fumbles, mistakes and a lot of hard work. With digital and web-based technologies pervasive nearly globally, it's important we take the same risks in learning to become aurally literate.

It is peculiar that although sound recording was developed long before the recording of moving images, moving images seem to have taken a preferential position in our interests as researchers. Even more strange of course is the fact that video recording technology costs more and generally has a steeper learning curve than audio. This would seem to reinforce suggestions of ocular-centrism in the discipline (Macpherson, 2005). However, this is perhaps to overlook the fact that audio has been used in our work to great extent in, for instance, the recording of interviews. The problem then is not one of under-recognition or utilization but, like video, one of not using these tools critically. And so we come full circle, where in the beginning of this article I suggested that the problem is not in access to or proficiency with audio/visual geographies but in our envisioning of them as tools or methods rather than as methodological ways of being-in-the-world. It is time that we take the audio recorder many of us have in our pockets at all times – our phone – and the ability we have to upload audio files to social media sharing sites like Soundcloud, and begin to experiment. And more – once we've become comfortable with that work-flow, we need to try our hand at more complicated forms of creating and editing, for it is in the tension of experimentation, which must include the hacking, subver-sion and reworking of these technologies, that discoveries are made. It is also vital of course that researchers push publishers, challenging the idea that these materials are not just 'data' to write from or mere supporting elements of publication but are in fact stand-alone forms of publication.

38.4 From Multi-Sensory Methods to More-Than-Textual Methodologies

> [A]udio-visual data [can] bear witness to phenomena that often escape talk and text based methods.
>
> Jamie Lorimer (2010: 251)

It is clear that audio/visual work in geography is becoming increasingly common, even if it is not yet ubiquitous. In concluding then, we might consider a few of the hurdles people still confront when working with audio and video. First, Clark has argued that 'visual material presents a significant challenge to conventional ethical practices surrounding the anonymity of participants' (Clark, 2012). This is less chal-lenging with audio and perhaps more challenging with video. But that process of negotiating methodological ethics is important. As I suggested above, part of the interesting thing about working with audio and video is that they are now in many ways more socially ubiquitous than text. Many project participants will likely have stronger feelings about how they are depicted in disseminated audio/visual material than in academic journal articles which are not widely read (Meho, 2007). This ten-sion requires a particular attentiveness not just to the desires of our project participants but also to the ways in which we deal with 'field' material. In other words, working with audio and video can, in some instances, prompt us to do more ethical work.

A second sticking point stems from a disjunction between desires to publish audio/visual material and reluctance by scholarly outputs to accommodate it. Expected modes of dissemination stem from university faculty being invested in a lifetime of reading and writing, from publishers using staid typesetting platforms and from constraining research frameworks that not only valorise text but also valorise the publication of it in niche outputs. These issues are undermining our capacity to do interesting, contemporary and ethical ethnographic, activist, documentary and artistic audio/visual research. In the case of journals, this is a problem easily reme-died since 'the technological capability for hyperlinked multimedia in e-journals is well established, but is at present under-utilized' (Gallagher and Prior, 2013: 13).

As researchers, we can catalyse the move to video, audio and technology-based outputs by requesting that journals accommodate those needs. If we act in an official capacity for any of these journals, we can also of course catalyse those changes from the inside, as it were.

More than ever before, we are surrounded by sounds and images. Work that once required access to special skills, tools and environments to undertake is now available to many of us in our pockets. I can, if I choose, walk away from this computer with nothing but my phone in my pocket, go shoot a film over the next few hours, edit it in my phone, upload it to the Internet and get feedback on the work from colleagues and the public, all before I even return to my computer. A decade from publication of this chapter, that process I've just described will no doubt sound clunky and ancient. This is the world we live in now – and so experimental geographies 'create important bridges between what were in the past the rather separate worlds of academic social-science scholarship on the one hand, and on the other, the domains of applied public research and of arts practice' (Pink, 2012). We are seeing an unprecedented upskilling of the general populace and, simultaneously, a need for researchers to meet those new skills sets to think through their terms of use at the cutting edge of critical praxis. It is at that edge that these methods bloom best (Dewsbury, 2009: 324). It is perhaps something about how we see these methods, how they cause us to change our perspective towards their potentialities, that means the sensorial, the phenomenological and even the collaborative inherently become part of their deployment, for

> works of art bring something new into the world rather than reflecting what's already there. This something new is constitutive rather than being merely representational, or reveals something already there but hidden. Works of art make culture. Each work makes different the culture it enters. (O'Neill, 2012: 158)

Human geographers are becoming increasingly brave in their experiments with, and critical reflection on, new methodologies (Shaw et al., 2015). Artistic approaches are 'part of a trend in social science and humanities research that focuses on the experiential, the sensory, and ways of knowing, being and remembering that cannot necessarily be articulated in words … these developments are also congruent with the concern with the senses in ethnography and a move beyond written text' (Hogan and Pink, 2012) and open out new potentials for the sensory, the affective, the atmospheric, the material and the relational. Grasseni suggests that 'sense-scapes require new reflections on the power of ethnographic representation' (2012: 98). However, as outlined in this chapter, it would be naïve to think that these sense-scapes are being 'captured' and accurately transmitted through these methods. What geographers uniquely have to offer to other humanities is our distinct ability to feel our way *through* the representational, which may very well be a by-product of our complicated relationship with visual geographies of the past relegated to the dustbin of colonialism. New work with video, audio and technology-based applications depends on our capacities to reimagine what is possible in the present without forgetting the past.

Figure 38.4   Researchers experimenting with a submersible camera and a boom pole

SUMMARY

- Geography has had a long, and sometimes fraught, relationship with audio/visual methodologies.
- Many early experiments of working with multimedia methods were problematic because of tendencies to treat audio/visual methods purely as observational tools.
- Audio/visual methodologies (as opposed to methods) are experiencing a resurgence in the discipline because of interest in their creative and more-than-representational capacities.
- These new methodologies require new platforms for publications.

Further Reading/Listening/Viewing

This further reading/listening/viewing is focused more on the why rather than the how of using audio and video as a geographic research methodology.

- There is no key book in geography about using video as method. However, *Doing Visual Ethnography: Images, Media and Representation in Research* by Sarah Pink (2007) is an excellent resource. I engage with Pink's work, and videographic methods more generally, in a 2010 article for *Progress in Human Geography* (Garrett, 2010b). I would also recommend Simpson (2011) for an interesting empirical case study paper on video analysis of busking and Lorimer (2010) for a more theoretical treatment of moving-image-methodologies.

- If you are interested in working through how one might go about constructing an academic article in audio/visual form, it's worth watching Evans and Jones (2008); Bauch (2010); and Garrett (2010b), all in *Geography Compass*.

- Three key readings on sonic or sonorous geographies include Ingham et al. (1999), Matless (2010) and Gallagher and Prior (2013).

- Those interested in the politics of listening and speaking should refer to Kanngieser (2011), winner of the 2013 *Progress in Human Geography* Essay Prize for its theoretical and creative innovativeness.

- Finally, if you are interested in artistic, creative and experimental geographies more generally, you can get a brief sketch from Thompson et al. (2008) or a book-length treatment from Hawkins (2013).

Note: Full details of the above can be found in the references list below.

References

Anderson, B. (2004) 'Recorded music and practices of remembering.' *Social and Cultural Geography* 5 (1): 3–20.

Barbash, I. and Taylor, L. (1997) *Cross Cultural Filmmaking: A Handbook for Making Documentaries and Ethnographic Films and Video.* Berkeley, CA: University of California Press.

Bauch, N. (2010) 'The academic geography video genre: A methodological examination.' *Geography Compass* 4 (5): 475–84.

Clark, A. (2012) 'Visual Ethics in a Contemporary Landscape', in S. Pink (ed.) *Advances in Visual Methodology.* London: Sage. pp. 17–36.

Coover, R., Badani, P., Caviezel, F., Marino, M., Sawhney, N. and Uricchio, W. (2012) 'Digital Technologies, Visual Research and the Non-Fiction Image', in S. Pink (ed.) *Advances in Visual Methodology.* London: Sage: 191–208.

Crang, M. (2003) 'Qualitative methods: Touchy, feely, look-see?.' *Progress in Human Geography* 27(4): 494–504.

Crang, M. (2005) 'Qualitative methods: There is nothing outside the text?.' *Progress in Human Geography,* 29 (2): 225–33.

Daniels, S. and Veale, L. (2012) Imagining Change: Coastal Conversations, Landscape and Environment Programme, University of Nottingham. http://www.landscape.ac.uk/landscape/impactfellowship/imagining-change/planetunder pressure.aspx (accessed 4 December 2015).

Daniels, S. and Veale, L. (2014) 'Imagining change: Coastal conversations.' *Cultural Geographies* (21): 3.

Dewsbury, J. D. (2009) 'Performative, Non-representational and Affect-based Research: Seven injunctions', in D. Delyser, S. Atkin, M. Crang, S. Herber and L. McDowell (eds) *The SAGE Handbook of Qualitative Research in Human Geography.* London: Sage. pp. 321–34.

Enigbokan, A. and M. Patchett (2012) 'Speaking with specters: Experimental geographies in practice.' *Cultural Geographies* 19(4): 535–46.

Evans, J. and P. Jones (2008) 'Towards Lefebvrian socio-nature? A film about rhythm, nature and science.' *Geography Compass* 2(3): 659–70. Film: https://www.youtube.com/watch?v=dQg86oSlVm4 (accessed 4 December 2015).

Fox, K. (2005) *Watching the English.* London: Hodder & Stoughton.

Gallagher, M. (2015) 'Working with Sound in Video: Producing an experimental documentary about school spaces', in C. Bates (ed.) *Video Methods: Social Science Research in Motion.* London: Routledge. pp. 165–86.

Gallagher, M. and Prior, J. (2013) 'Sonic geographies: Exploring phonographic methods.' *Progress in Human Geography* 38 (2): 267–84.

Gandy, M. (2007) *Liquid City* [Film]. United Kingdom, University College London.

Gandy, M. (2009) 'Liquid city: reflections on making a film.' *Cultural Geographies* 16: 403-08.

Garrett, B. L. (2010a) 'Urban explorers: Quests for myth, mystery and meaning.' *Geography Compass* 4(10): 1448–61. Film on https://vimeo.com/5366045 (accessed 4 December 2015).

Garrett, B.L. (2010b) 'Videographic geographies: Using digital video for geographic research.' *Progress in Human Geography* 35(4): 521–41.

Garrett, B. and Brickell, K. (2014) 'Participatory politics of partnership: Video workshops on domestic violence in Cambodia.' *Area* 47 (3): 230–6.

Garrett, B.L. and Hawkins, H. (2014) 'Creative video ethnographies: Video methodologies of urban exploration.' *Video Methods: Social Science Research in Motion.* Edited by C. Bates. London: Routledge. pp. 142–64.

Garrett, B. L., Rosa, B. and Prior, J. (2011) 'Jute: Excavating material and symbolic surfaces.' *Liminalities: A Journal of Performance Studies* 7 (2): 1–4.

Gold, J.R. (2002) 'The real thing? Contesting the myth of documentary realism through classroom analysis of films on planning and reconstruction'. *Engaging Film: Geographies of Mobility and Identity.* Lanham, MD: Rowman and Littlefield. pp. 209–25.

Grasseni, C. (2012) 'Community Mapping as Auto-Ethno-Cartography', in S. Pink (ed.) *Advances in Visual Methodology*. London: Sage. pp. 97–112.

Haraway, D. (1991) 'Situated Knowledges: The Science Question in Feminism and the Privilege of Partial Perspective', in D. Haraway (ed.) *Simians, Cyborgs, and Women*. New York: Routledge. pp. 183–202.

Haraway, D. (2007) *When Species Meet*. Minnesota: University Of Minnesota Press.

Hawkins, H. (2013) *For Creative Geographies: Geography, Visual Arts and the Making of Worlds*. London: Routledge.

Hearnshaw, H. M. and Unwin, D. J. (eds) (1996) *Visualization in Geographical Information Systems*. London: Wiley-Blackwell.

Hogan, S. and Pink, S. (2012) 'Visualising interior worlds: Interdisciplinary routes to knowing', in S. Pink (ed.) *Advances in Visual Methodology*. London: Sage. pp. 230–47.

Hoover, K. C. (2010) "The Geography of Smell: The international journal for Geographic Information and Geovisualization',' *Cartographica* 44(4): 237–9.

Ihde, D. (2001) *Bodies in Technology*. Minnesota: University of Minnesota Press.

Ingham, J., Purvis, M. and Clarke, D. (1999) 'Hearing place, making spaces: Sonorous geographies, ephemeral rythms, and the Blackburn warehouse parties'. *Environment and Planning D: Society and Space* 17: 283–305.

Ingold, T. (2007) 'Against Soundscape', in A. Carlyle (ed.) *Autumn Leaves*. Paris: Double Entendre. pp. 10–13.

Kanngieser, A. (2011) 'A sonic geography of voice: Towards an affective politics.' *Progress in Human Geography* 36(3): 336–53.

Kindon, S. (2003) 'Participatory video in geographic research: A feminist practice of looking?.' *Area* 35 (2): 142–53.

Kindon, S., Pain, R. and Kesby, M. (2008) 'Participatory action research'. *International Encyclopedia of Human Geography*. London: Elsevier: 90–95.

Last, A. (2012) 'Experimental geographies.' *Experimental Geographies* 6 (12): 706–24.

Laurier, E. and Philo, C. (2006) 'Natural problems of naturalistic video data', in H. Knoblauch, J. Raab, H. G. Soeffner and B. Schnettler (eds) *Video-analysis Methodology and Methods: Qualitative Audiovisual Data Analysis in Sociology*. Oxford: Peter Lang. pp. 183–92.

Laurier, E. and C. Philo (2006) 'Possible Geographies: A passing encounter in a café.' *Area* 38(4): 353–63.

Lorimer, H. (2005) 'Cultural geography: The busyness of being 'more-than-representational.' *Progress in Human Geography* 29 (1): 83–94.

Lorimer, J. (2010) 'Moving image methodologies for more-than-human geographies.' *Cultural Geographies* 17(2): 237–258.

Macpherson, H. (2005) 'Landscape's Ocular-centrism – and Beyond?', in B. Tress, G. Tress, G. Fry and P. Opdam (eds) *From Landscape Research to Landscape Planning: Aspects of Integration, Education and Application* (Wageningen UR Frontis Series*)*. New York: Springer: pp. 95–104.

Matless, D. (2010) 'Sonic geography in a nature region.' *Social and Cultural Geography* 6(5): 745–66.

Meho, L. I. (2007) 'The rise and rise of citation analysis.' *Physics World* 20 (1): 32–6.

Merchant, S. (2011) 'The body and the senses: Visual methods, videography and the submarine sensorium.' *Body & Society* 17: 53–72.

O'Neill, M. (2012) 'Ethno-Mimesis and Participatory Arts', in S. Pink (ed.) *Advances in Visual Methodology*. London, Sage. pp. 153–72.

Paglen, T. (2009) 'Experimental geography: From cultural production to the production of space', The Brooklyn Rail: critical perspectives on arts, politics, and culture. http://www.brooklynrail.org/2009/03/express/experimental-geography-from-cultural-production-to-the-production-of-space (accessed 4 December 2015).

Parr, H. (2007) "Collaborative film-making as process, method and text in mental health research.' *Cultural Geographies* 14: 114–38.

Paton, G. (2010) 'Children "more likely to own a mobile phone than a book"', *The Telegraph*. London: Telegraph Media Group Limited (26 May). http://www.telegraph.co.uk/education/educationnews/7763811/Children-more-likely-to-own-a-mobile-phone-than-a-book.html (accessed 4 December 2015).

Pearson, M., Hardy, J. and Fowler, H. (2007) 'Carrlands: Mediated manifestations of site-specific performance in the Ancholme valley, North Lincolnshire'. University of Aberystwyth, Department of Theatre, Film & Television Studies. Available from http://www.carrlands.org.uk/images/carrlands.pdf (accessed 4 December 2015).

Pink, S. (2007) *Doing Visual Ethnography: Images, Media and Representation in Research*. Manchester: Manchester University Press in association with the Granada Centre for Visual Anthropology.

Pink, S. (2012) 'Advances in Visual Methodology: An Introduction', in S. Pink (ed.) *Advances in Visual Methodology*. London: Sage. pp. 3–16.

Rancière, J. (2011) *The Emancipated Spectator*. London: Verso.

Rodaway, P. (1994) *Sensuous Geographies: Body, Sense and Place*. London: Routledge.

Rogoff, I. (2000) *Terra Infirma: Geography's Visual Culture*. London: Routledge.

Rose, G. (2016) *Visual Methodologies: An Introduction to Researching with Visual Materials* (4th edition). London: Sage.

Ryan, J. R. (1997) *Picturing Empire: Photography and the Visualization of the British Empire*. Chicago, IL: University of Chicago Press.

Schafer, R. M. (1977) *The Tuning of the World*. New York: Knopf.

Shaw W., DeLyser, D. and Crang, M. (2015) 'Limited by imagination alone: research methods in cultural geographies.' *Cultural Geographies* 22 (2): 211–15.

SilentUK, Sub-Urban and Place-Hacking (2010) 'Crack the Surface I', https://vimeo.com/26200018 (accessed 4 December).

Simpson, P. (2011) '"So, as you can see…": some reflections on the utility of video methodologies in the study of embodied practices.' *Area* 43 (3): 343–52.

Thompson, N., Kastner, J. and Paglen, T. (2008) *Experimental Geography: Radical Approaches to Landscape, Cartography and Urbanism*. New York: Melville House and Independent Curators International.

Thrift, N. (2011) 'Lifeworld Inc – and what to do about it.' *Environment and Planning D: Society and Space* 29 (1): 5–26.

Whatmore, S. (2009) 'Mapping knowledge controversies: Science, democracy and the redistribution of expertise.' *Progress in Human Geography* 33(5): 587–98.

Whatmore, S., Lane, S. and Ward, N. (2009) 'Understanding Environmental Knowledge Controversies', http://knowledge-controversies.ouce.ox.ac.uk/project (accessed 4 December 2015).

Whatmore, S., Lane, S. and Ward, N. (2011) 'Understanding Environmental Knowledge Controversies', http://knowledge-controversies.ouce.ox.ac.uk (accessed 4 December 2015).

White, S.A. (2003) *Participatory Video: Images That Transform and Empower*. London: Sage.

ON THE COMPANION WEBSITE…

Visit **https://study.sagepub.com/keymethods3e** for author videos, chapter exercises, resources and links, plus **free** access to the following recommended articles:

1. **Garrett, B. (2011) 'Videographic geographies: Using digital video for geographic research', *Progress in Human Geography*, 35 (4): 521–41.**

This article looks at how geographers have engaged with film and video over time and the ways in which videographic production is enriching geographical research, particularly in the context of non-representational theory.

2. **Gallagher, M. and Prior, J. (2013) 'Sonic geographies: Exploring phonographic methods', *Progress in Human Geography*, 38 (2): 267–84.**

This paper is a fantastic overview of the epistemological implications of phonographic methods as used by geographers writing about both recording and listening to sound.

3. **Dowling, R., Lloyd, K. and Suchet-Pearson, S. (2015) 'Qualitative methods 1: Enriching the interview',** *Progress in Human Geography*, **39. http://doi.org/10.1177/0309132515596880**

Here Dowling et al. give us an overview of the ways in which qualitative methods of interviewing are being transformed by technologies and mobilities. Their discussion of autoethnographies, more-than-textual methodologies and capturing social life is astute and useful.

Index